LESLIE O'BERGIN BRONER

TAROT COUNSELING
MASSAGE / BODYWORK

2166 FIFTEENTH STREET
SAN FRANCISCO, CA 94114   (415) 861-5412

PRINCIPLES OF HUMAN ANATOMY

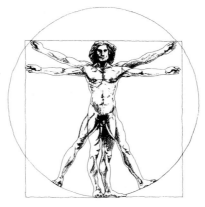

PRINCIPLES OF HUMAN ANATOMY

THIRD EDITION

GERARD J. TORTORA

Bergen Community College

1817

HARPER & ROW, PUBLISHERS, New York

Cambridge, Philadelphia, San Francisco,
London, Mexico City, São Paulo, Sydney

Sponsoring Editor: Claudia M. Wilson
Project Editor: Holly Detgen
Designer: T. R. Funderburk
Production Manager: Kewal K. Sharma
Photo Researcher: Mira Schachne
Compositor: Kingsport Press
Printer and Binder: Kingsport Press
Art Studio: J & R Art Services, Inc.
Medical Illustrators: Leonard Dank, George Weisbrod,
    Nelva B. Richardson, Marsha J. Dohrmann, Helen Gee Jeung
Cover Photomicrographs: © 1983 by Michael H. Ross. Used by permis-
    sion. Front cover: Renal tubules. Back cover (top to bottom): Cross
    section of muscle fibers, cross section of smooth muscle tissue,
    motor end plate.

PRINCIPLES OF HUMAN ANATOMY, Third Edition
Copyright © 1983 by Gerard J. Tortora

Library of Congress Cataloging in Publication Data

Tortora, Gerard J.
   Principles of human anatomy.

   Bibliography: p.
   Includes index.
   1. Human physiology.   2. Anatomy, Human.   I. Title.
[DNLM: 1. Anatomy.   2. Physiology.   QS 4 T712pa]
QP34.5.T68   1983   612   82–18714
ISBN 0–06–046634–0

In Memoriam
The third edition of *Principles of Human Anatomy* is dedicated to
Dr. Anthony Tortora
Father, Friend, and Mentor

CONTENTS
IN BRIEF

CONTENTS
IN DETAIL

PREFACE

AUDIENCE

Designed for the introductory course in human anatomy, the third edition of *Principles of Human Anatomy* assumes no previous study of the human body. The text is geared to students in biological, medical, and health-oriented programs. Among the students specifically served by this volume are those aiming for careers as nurses, medical assistants, physician's assistants, medical laboratory technologists, radiological technologists, respiratory therapists, dental hygienists, physical therapists, morticians, and medical record keepers. However, because of the scope of the text, *Principles of Human Anatomy,* Third Edition, is also useful for students in the biological sciences, premedical and predental programs, science technology, liberal arts, and physical education.

OBJECTIVES

The objectives of *Principles of Human Anatomy* remain unchanged in this third edition. Because human anatomy is such a large and complex body of knowledge to present in an introductory course, the first objective is to concentrate on unified concepts and data that contribute to a basic understanding of the structure of the human body. Data unessential to this objective have been minimized. The second objective is to present essential technical vocabulary and important, but difficult, concepts at a reading level that can be handled by the average student. Easy-to-comprehend explanation of terms and step-by-step development of concepts are presented to meet this objective.

THEMES

The third edition of this textbook departs from the approaches of most other anatomy texts in that somewhat more emphasis is given to physiology and applications to health. The basic content of the book is anatomy, but because structure and function are so inseparably related, it makes little sense to present students with anatomical detail without relating anatomy to function. The discussion of function gives the students a better understanding of anatomical concepts. This approach is especially valuable in describing certain systems of the body, such as the nervous system.

An understanding of structure and function may be enhanced by considering variations from the normal, or the defects and disorders observed in clinical situations. In turn, such conditions are better understood once the context of normal anatomy has been established. Disorders are treated under the special heading **Applications to Health** and throughout the text in boxed material labeled **Clinical Applications.**

ORGANIZATION

The book is organized by systems rather than by regions. The first chapter introduces the levels of structural organization, the general structural plan of the body, and methods of describing the body, including anatomical and directional terms, planes of the body, radiographical anatomy, and units of measurement. In this edition, the discussion of surface anatomy has been organized into a separate chapter and moved to the end of the book (Chapter 25), so that students will be familiar with the terms used by the time they approach the subject.

The more detailed discussion of levels of organization begins with a chapter on the cellular level. This chapter describes a generalized animal cell to demonstrate its basic structural features and activities. In this chapter an exhibit on the effects of aging on body cells has been added and the discussion of cells and cancer has been updated and expanded. Tissue organization is presented through descriptions of the structure, functions, and locations of the principal kinds of epithelium and connective tissue. The histology of bone, muscle, nervous tissue, and blood is considered, along with the relevant organ systems. The chapter on tissues includes a new section on cell junctions.

The discussion of the organ and system levels of organization begins with the chapter on the skin and the integumentary system. As in all the chapters on body systems, there is some reference to physiology, disorders, and relevant medical terminology. This chapter includes new sections on epidermal ridges and grooves, lines of cleavage, hair color, and hair replacement and revised sections on the structure of the dermis and skin color. Applications to health now consider acne, impetigo, systemic lupus erythematosus (SLE), psoriasis, decubitus, warts, cold sores, sunburn, skin cancer, and burns.

The first body system studied in detail is the skeletal system. This is accomplished by examining the principal features of osseous tissue, the axial skeleton, the appendicular skeleton, and articulations. Several changes have been made in these chapters. In the osseous tissue chapter, a new section has been added on the application of electricity to fracture repair. The chapter on the axial skeleton now contains sections dealing with microcephalus, hydrocephalus, slipped disc, curvatures, spina bifida, and fractures of the vertebral column. The chapter on the appendicular skeleton now contains an exhibit that compares the structural features of the female and male skeletons. The chapter on articulations has been shortened by the deletion of several exhibits dealing with less commonly studied joints of the body. A new disorder, septic arthritis, has been added.

The muscular system is analyzed through a study of muscle tissue and the locations and actions of the principal muscles of the body. The chapter on muscle tissue contains new sections on fast (white) and slow (red) muscle, muscular atrophy, muscular hypertrophy, and charleyhorse. The discussion of muscle tone has been revised. New to the chapter on the muscles of the body are exhibits dealing with the muscles of the pharynx, larynx, pelvic floor, and perineum.

The student is next introduced to the cardiovascular and lymphatic systems. In the previous edition, this was covered in a single chapter. In this edition, there are three separate chapters: blood, the heart and blood vessels, and the lymphatic system. The chapter on blood contains a new glossary of key medical terms. The chapter on the heart and blood vessels contains expanded discussions of arterioles, capillaries, venules, atherosclerosis, and hypertension and a new glossary of key medical terms. The chapter on the lymphatic system also has a new glossary of key medical terms.

The next major area of emphasis is the nervous system. Students are introduced to the structure and physiology of nervous tissue, the spinal cord and spinal nerves, the brain and cranial nerves, and the autonomic nervous system. The chapter on nervous tissue contains an expanded discussion of the synapse. The chapter on the spinal cord includes new sections on delayed nerve grafting and shingles. In the chapter on the brain, there are new discussions on brain lateralization, headache, and trigeminal neuralgia. An exhibit on clinical applications related to the cranial nerves has also been added. The sections dealing with the thalamus and hypothalamus have been revised. The chapter on the autonomic nervous system has been left basically intact.

The structure and function of visual, auditory, gustatory, and olfactory receptors are considered in the chapter on sensory structures. New to the chapter are discussions of acupuncture and pain, the muscles and ligaments of the auditory ossicles, the structure of the macula, labyrinthine disease, otitis media, and motion sickness.

Attention is then turned to the endocrine system. This chapter now contains an expanded and revised discussion of the histology of the thymus gland.

The chapter on the respiratory system now includes discussions of cardiopulmonary resuscitation (CPR), nasal polyps, sudden infant death syndrome (SIDS), and carbon monoxide poisoning.

New disorders added to the digestive system chapter are gallstones, appendicitis, diverticulitis, and anorexia nervosa.

The urinary system chapter has been expanded to include discussions of nephrosis and polycystic disease.

The chapter on the reproductive systems now precedes the new chapter on developmental anatomy. The discussions of spermatogenesis and oogenesis have been moved to the new chapter. New disorders added to the reproductive systems chapter are genital herpes, trichomoniasis, nonspecific urethritis (NSU), and pelvic inflammatory disease (PID).

A major addition to the third edition is a new chapter dealing with developmental anatomy. It has been placed near the end of the book because it should be easier for students to follow the process of embryonic and fetal development after they have learned about the final products of that development. The chapter considers gamete formation, fertilization and implantation, development of the primary germ layers, and embryonic membranes. Exhibits trace the development of body systems. Also considered are hormones of pregnancy, parturition and labor, adjustments of the infant at birth, potential hazards to the embryo and fetus, lactation, amniocentesis, and birth control. A glossary of key medical terms is given.

As noted earlier, surface anatomy is now studied in the last chapter of the text.

SPECIAL FEATURES

The text contains a number of special learning aids for students.

1. Student Objectives appear at the beginning of each chapter. Each objective describes a knowledge or skill the student should acquire while studying the chapter. (See **Note to the Student** for an explanation of how the objectives can be used.) At the end of each chapter a **Study Outline** provides a brief summary of the major topics and **Review Questions** provide a check to see if the stated objectives have been met.

2. The health science student is generally expected to learn the anatomical features of certain structures in great detail—bones, articulations, skeletal muscles, blood vessels, lymph nodes, and nerves. In these areas, many anatomical details have been removed from the narrative and placed in **Exhibits,** most of which are accompanied by illustrations. This method allows a clearer statement of general concepts in the narrative and organizes the specific details to be learned.

3. Common disorders associated with body systems are described in the section entitled **Applications to Health** in many chapters. The descriptions have been updated and new disorders have been added. The topics provide review of normal anatomy and physiology and show the student how fundamental the study of anatomy is to a career in any of the health fields.

4. Phonetic pronunciations are given for selected terms where the terms are introduced in the text. This feature is new to the third edition. (See **Note to the Student** for the pronunciation key.)

5. Glossaries of selected **Key Medical Terms** appear at the end of each chapter that deals with a major body system. Most of these glossaries have been revised for the third edition.

6. Throughout the text, **Clinical Applications** are now boxed off for greater emphasis. This feature is also new to the third edition. The number of applications has been significantly increased.

7. The **line drawings** are unusually large, so that details are clearly labeled and easily discernible. In the third edition we have greatly increased the use of full color to differentiate structures and regions. The tone-rendering technique for bones is also new to this edition.

8. The **photographs** amplify the narrative and the line drawings. Numerous photomicrographs (some in full color), newly added scanning electron micrographs, and transmission electron micrographs enhance the histological discussions. Many gross anatomy sections are made more meaningful by the addition of color photographs of specimens. Roentgenograms and CT scans reinforce clinical knowledge.

9. A comprehensive and expanded **Glossary** gives defi-

nitions and pronunciation of important terms used in the textbook.

10. A list of **Selected Readings** at the end of the book provides numerous references for instructors and students.

SUPPLEMENTARY MATERIALS

The following supplementary materials are available.

1. The *Learning Guide,* by Kathleen Schmidt Prezbindowski and Gerard J. Tortora, is designed to help students *learn* anatomy. Rather than asking for mere repetition of the text material, the guide requires students to perform activities, such as labeling a drawing or coloring parts of a diagram, as well as answering multiple-choice, matching, and fill-in questions. The book's emphasis is on *using* new information in order to learn it. Answers for key-concept exercises are provided so that the student has immediate feedback. A mastery test at the end of each chapter allows the student to evaluate his or her learning of the chapter material and to practice for classroom tests.

2. Each chapter in the complimentary *Instructor's Manual,* prepared by the author, contains a listing of key instructional concepts, selected audiovisual materials and distributors, and multiple-choice questions. The questions are carefully designed to evaluate students' understanding of data, concepts, clinical situations, and applications of this knowledge.

ACKNOWLEDGMENTS

I wish to thank the following people for the contributions to this edition of *Principles of Human Anatomy:* Franklyn F. Bolander, Jr., University of South Carolina; Stanley D. Boyer, St. Xavier College; Warren Burggren, University of Massachusetts, Amherst; Robert H. Catlett, University of Colorado, Colorado Springs; Charlotte M. Dienhart, Emory University; Larry Ganion, Ball State University; Henry N. McCutcheon, Rhode Island College; William Donald Newton, Arkansas State University; Gideon E. Nelson, University of South Florida; Anna Marie Parmley, California State University, Long Beach; and Ronald A. Schollenberger, California State University, Los Angeles.

In particular, I would like to thank Dr. Michael H. Ross of the University of Florida, Gainesville, for preparing the full-color photomicrographs that appear in this text, and Dr. Michael C. Kennedy of Hahnemann Medical College for his help in reviewing the illustration program.

All drafts of the manuscript were typed by Geraldine C. Tortora, my wife. She also handled all secretarial duties related to the project, for which I am deeply grateful.

I would like to invite readers of this book to send their reactions and suggestions concerning the book to me at the address given below. These responses will be helpful to me in formulating plans for subsequent editions.

GERARD J. TORTORA
Biology Department
Bergen Community College
400 Paramus Road
Paramus, NJ 07652

NOTE TO THE STUDENT

At the beginning of each chapter is a listing of **Student Objectives.** Before you read the chapter, please read the objectives carefully. Each objective is a statement of a skill or knowledge that you should acquire. To meet these objectives, you will have to perform several activities. Obviously, you must read the chapter carefully. If there are sections of the chapter that you do not understand after one reading, you should reread those sections before continuing. In conjunction with your reading, pay particular attention to the figures and exhibits; they have been carefully coordinated to the textual narrative.

At the end of each chapter are learning guides that you may find useful. The first, **Study Outline,** is a concise summary of important topics discussed in the chapter. This section is designed to consolidate the essential points covered in the chapter, so that you may recall and relate them to one another. The second guide, **Review Questions,** is a series of questions designed specifically to help you master the objectives. After you have answered the review questions, you should return to the beginning of the chapter and reread the objectives to determine whether you have achieved the goals.

A third aid, **Key Medical Terms,** appears at the end of many chapters. This is a listing of terms designed to build your medical vocabulary.

As a further aid, we have included pronunciations for many terms immediately following the first use of the term in the text and again in the glossary at the back of the book. Learning to pronounce a word correctly will help you to remember it and to make it a usable part of your vocabulary. The pronunciations are given according to the following key.

1. The strongest accented syllable appears in capital letters, for example, bilateral (bī-LAT-er-al) and diagnosis (dī-ag-NŌ-sis).

2. If there is a secondary accent, it is noted by a single quote mark ('), for example, constitution (kon'-sti-TOO-shun) and physiology (fiz'-ē-OL-ō-jē). Any additional secondary accents are also noted by a single quote mark, for example, decarboxylation (dē'-kar-bok'-si-Lā-shun).

3. Vowels marked with a line above the letter are pronounced with the long sound, as in the following common words.

ā as in *māke*
ē as in *bē*
ī as in *īvy*
ō as in *pōle*

4. Vowels not so marked are pronounced with the short sound, as in the following words.

e as in *bet*
i as in *sip*
o as in *not*
u as in *bud*

5. Other phonetic symbols are used to indicate the following sounds.

a as in *above*
oo as in *sue*
yoo as in *cute*
oy as in *oil*

PRINCIPLES OF HUMAN ANATOMY

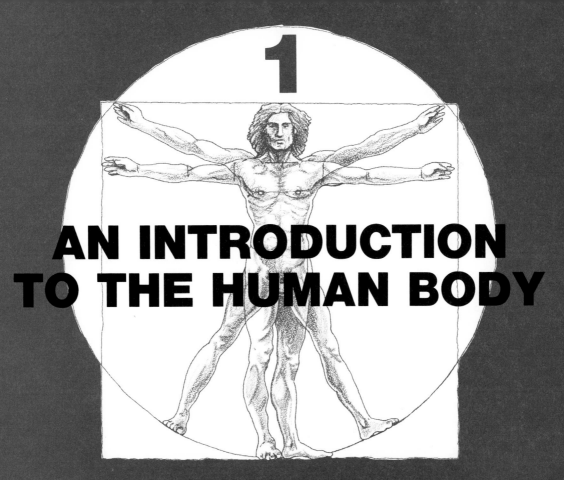

1

AN INTRODUCTION TO THE HUMAN BODY

STUDENT OBJECTIVES

■ Define anatomy, with its subdivisions, and physiology.

■ Compare the levels of structural organization that make up the human body.

■ Define a cell, a tissue, an organ, a system, and an organism.

■ Identify the principal systems of the human body, list the representative organs of each system, and describe the function of each system.

■ Describe the basic structural plan of the human body.

■ Define the anatomical position.

■ Compare common and anatomical terms used to describe the external features of the human body.

■ Define directional terms used in association with the human body.

■ Define the common anatomical planes of the human body.

■ List by name and location the principal body cavities and their major organs.

■ Describe how the abdominopelvic cavity is divided into nine regions and four quadrants.

■ Define the relationship of radiographic anatomy to the diagnosis of disease.

■ Point out the differences between a roentgenogram and a CT (computed tomography) scan.

■ Define the common metric units of length, mass, and volume, and their U.S. equivalents, that are used in measuring the human body.

You are about to begin a study of the human body in order to learn how your body is organized and how it functions. The study of the human body involves many branches of science. Each contributes to a comprehensive understanding of how your body normally works and what happens when it is injured, diseased, or placed under stress.

ANATOMY AND PHYSIOLOGY DEFINED

Two branches of science that will help you understand your body parts and functions are anatomy and physiology. **Anatomy** or **morphology** (*morpho* = form, shape) refers to the study of *structure* and the relationships among structures. Anatomy is a broad science, and the study of structure becomes more meaningful when specific aspects of the science are considered. **Surface anatomy** is the study of the form and markings of the surface of the body. **Gross anatomy** deals with structures that can be studied without a microscope. **Systemic** or **systematic anatomy** covers specific systems of the body, such as the system of nerves, spinal cord, and brain or the system of heart, blood vessels, and blood. **Regional anatomy** deals with a specific region of the body, such as the head, neck, chest, or abdomen. **Developmental anatomy** is the study of development from the fertilized egg to the adult form. **Embryology** is generally restricted to the study of development from the fertilized egg to the end of the eighth week in utero. Other branches of anatomy are **pathological** (*patho* = disease) **anatomy,** the study of structural changes caused by disease; **histology** (*hist, histio* = tissue), the microscopic study of the structure of tissues; and **cytology** (*cyt, cyto, cyte* = cell), the study of cells.

Whereas anatomy and its branches deal with structures of the body, **physiology** (fiz'-ē-OL-ō-jē) deals with *functions* of the body parts, that is, how the body parts work. However, the two sciences cannot be completely separated. Since each part of the body is custom-modeled to carry out a particular set of functions, structure and function are studied together. The structure of a part often determines the functions it will perform, and in turn, body functions often influence the size, shape, and health of the structures. For instance, bones function as rigid supports for the body because they are constructed of hard minerals. Glands perform the function of manufacturing chemicals, some of which stimulate bones to build up minerals so they become hard and strong. Other chemicals cause the bones to give up minerals to the blood for transportation to other parts of the body where they are used for activities such as muscle contraction and nervous activity.

LEVELS OF STRUCTURAL ORGANIZATION

The human body consists of several levels of structural organization that are associated with one another in several ways. The lowest level of organization, the **chemical level,** includes all chemical substances essential for maintaining life. All these chemicals are made up of atoms joined together in various ways (Figure 1-1).

The chemicals, in turn, are put together to form the next higher level of organization: the **cellular level. Cells** are the basic structural and functional units of the organism. Among the many kinds of cells in your body are muscle cells, nerve cells, and blood cells. Figure 1-1 shows several isolated cells from the lining of the stomach. Each has a different structure, and each performs a different job.

The next higher level of structural organization is the **tissue level. Tissues** are made up of groups of similarly specialized cells and their intercellular material that perform certain special functions. When the isolated cells shown in Figure 1-1 are joined together, they form a tissue called epithelium, which lines the stomach. Each cell in the tissue has a specific function. Mucous cells produce mucus, a secretion that lubricates food as it passes through the stomach. Parietal cells produce acid in the stomach. Chief cells produce enzymes needed to digest proteins. Other examples of tissues in your body are muscle tissue, connective tissue, and nervous tissue.

In many places in the body, different kinds of tissues are joined together to form an even higher level of organization: the **organ level. Organs** are structures of definite form and function composed of two or more different tissues. Organs usually have a recognizable shape. Examples of organs are the heart, liver, lungs, brain, and stomach. Figure 1-1 shows three of the tissues that make up the stomach. The serous layer (also called the serosa) around the outside protects the stomach and reduces friction when the stomach moves and rubs against other organs. The muscle tissue layers of the stomach contract to mix food and pass it on to the next digestive organ. The epithelial tissue layer lining the stomach produces mucus, acid, and enzymes.

The next higher level of structural organization in the body is the **system level. A system** consists of an association of organs that have a common function. The digestive system, which functions in the breakdown of food, is composed of the mouth, saliva-producing glands called salivary glands, pharynx (throat), esophagus (gullet), stomach, small intestine, large intestine, rectum, liver, gallbladder, and pancreas.

The highest level of organization is the **organismic level.** All the parts of the body functioning with one another constitute the total **organism**—one living individual.

In the chapters that follow, you will examine the anatomy and physiology of the major body systems. Exhibit 1-1

2

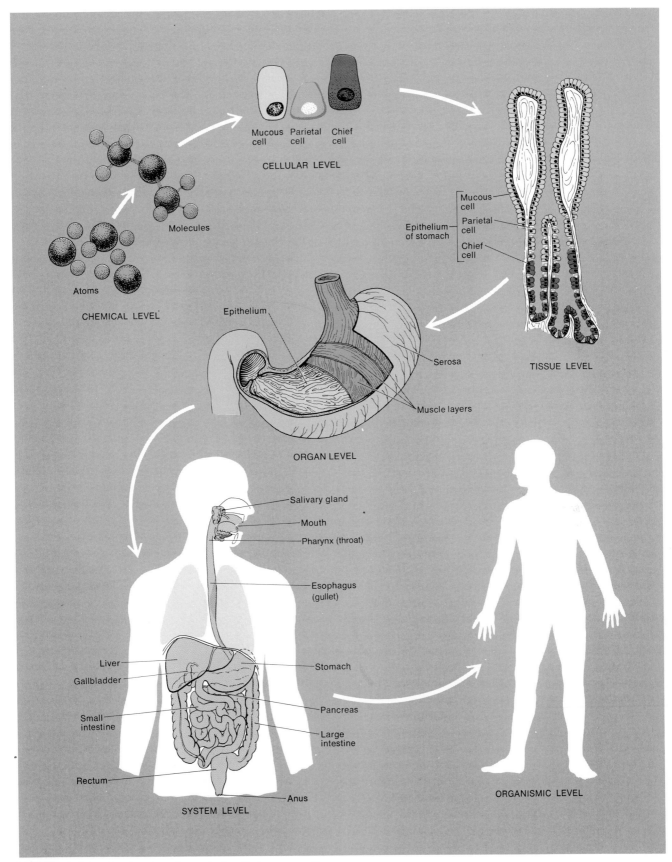

FIGURE 1-1 Levels of structural organization that compose the human body.

EXHIBIT 1-1 PRINCIPAL SYSTEMS OF HUMAN BODY: REPRESENTATIVE ORGANS AND FUNCTIONS

1. Integumentary

Definition: The skin and structures derived from it, such as hair, nails, and sweat and oil glands.

Function: Helps regulate body temperature, protects the body, eliminates wastes, and receives certain stimuli such as temperature, pressure, and pain.

2. Skeletal

Definition: All the bones of the body, their associated cartilages, and the joints of the body.

Function: Supports and protects the body, gives leverage, produces blood cells, and stores minerals.

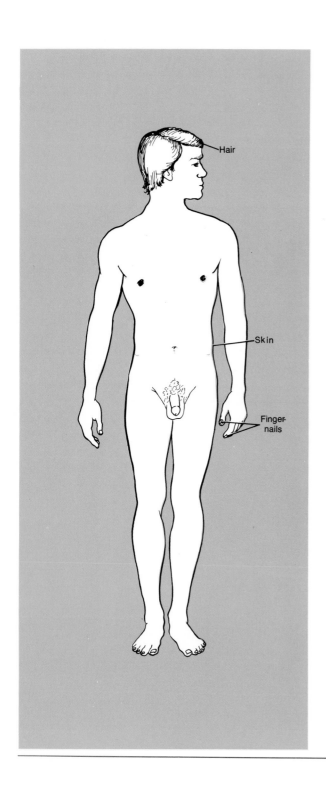

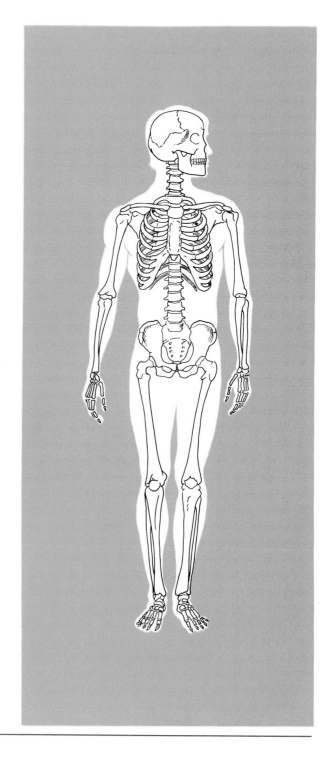

EXHIBIT 1-1 *(Continued)*

3. Muscular

Definition: All the muscle tissue of the body, including skeletal (shown in the illustration), visceral, and cardiac.

Function: Participates in bringing about movement, maintains posture, and produces heat.

4. Cardiovascular

Definition: Blood, heart, and blood vessels.

Function: Distributes oxygen and nutrients to cells, carries carbon dioxide and wastes from cells, maintains the acid–base balance of the body, protects against disease, prevents hemorrhage by forming blood clots, and helps regulate body temperature.

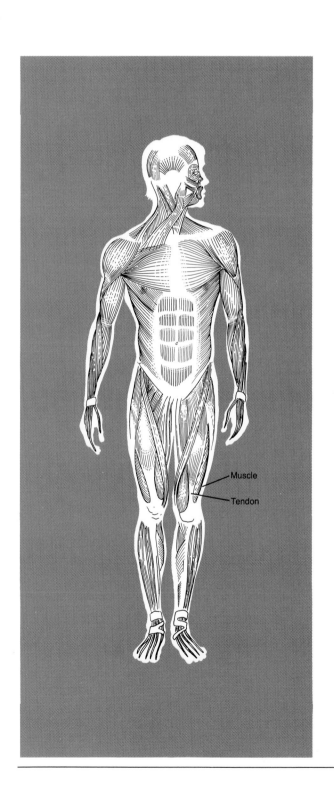

Muscle
Tendon

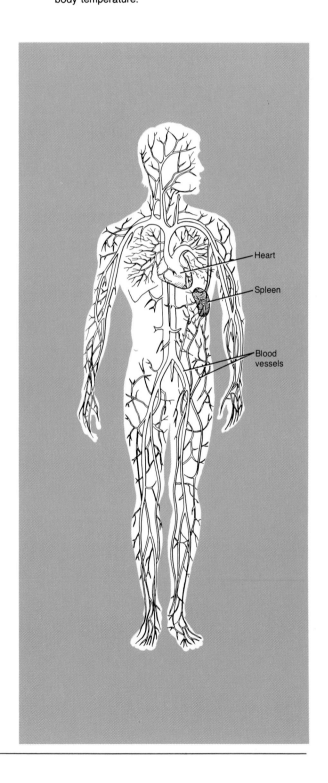

Heart
Spleen
Blood vessels

EXHIBIT 1-1 *(Continued)*

5. Lymphatic

Definition: Lymph, lymph nodes, lymph vessels and lymph glands, such as the spleen, thymus gland, and tonsils.

Function: Returns proteins and plasma to the cardiovascular system, transports fats from the digestive system to the cardiovascular system, filters the blood, produces white blood cells, and protects against disease.

6. Nervous

Definition: Brain, spinal cord, nerves, and sense organs, such as the eye and ear.

Function: Regulates body activities through nerve impulses.

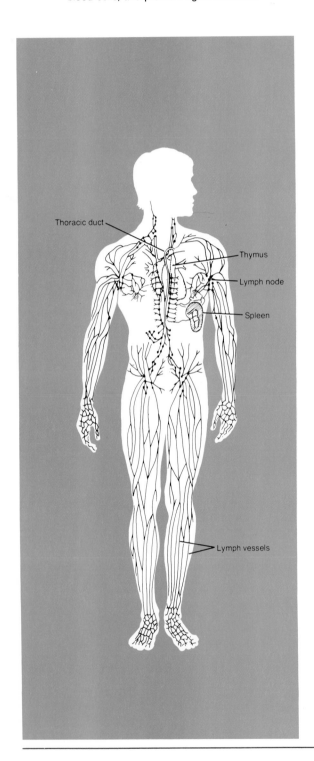

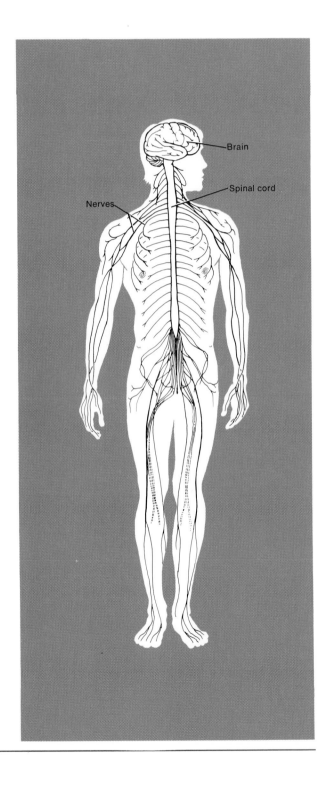

EXHIBIT 1-1 *(Continued)*

7. Endocrine

Definition: All glands that produce hormones.
Function: Regulates body activities through hormones transported by the cardiovascular system.

8. Respiratory

Definition: The lungs and a series of passageways leading into and out of them.
Function: Supplies oxygen, eliminates carbon dioxide, and helps regulate the acid–base balance of the body.

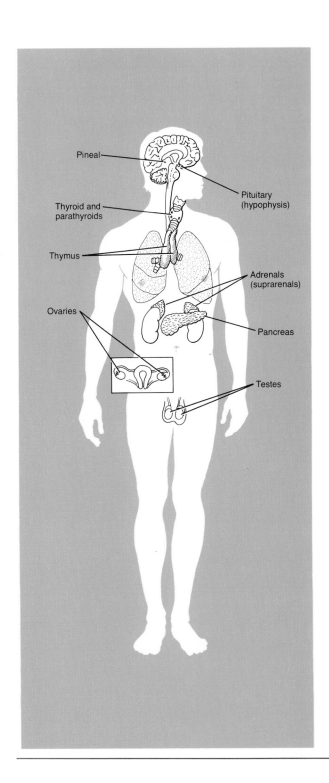

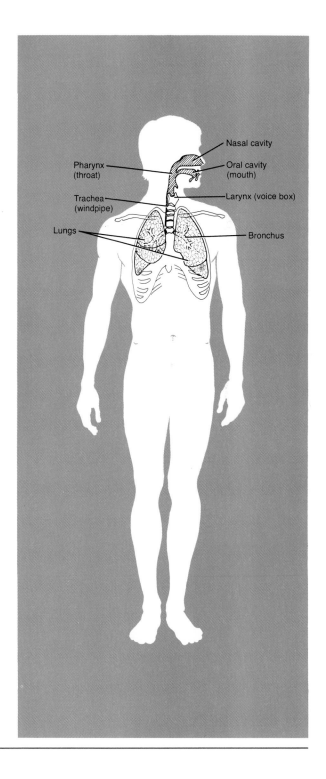

EXHIBIT 1-1 *(Continued)*

9. Digestive

Definition: A long tube and associated organs such as the salivary glands, liver, gallbladder, and pancreas.
Function: Performs the physical and chemical breakdown of food for use by cells and eliminates solid wastes.

10. Urinary

Definition: Organs that produce, collect, and eliminate urine.
Function: Regulates the chemical composition of blood, eliminates wastes, regulates fluid and electrolyte balance and volume, and helps maintain the acid–base balance of the body.

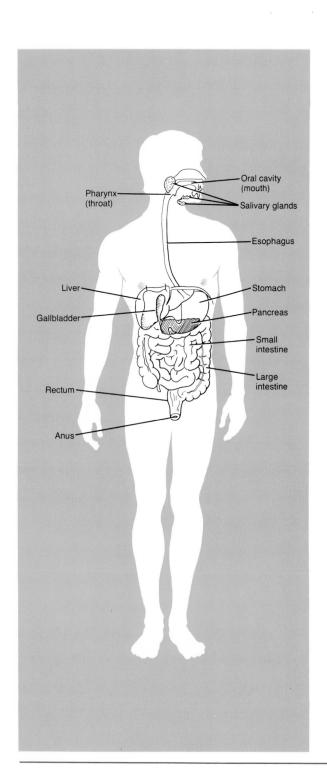

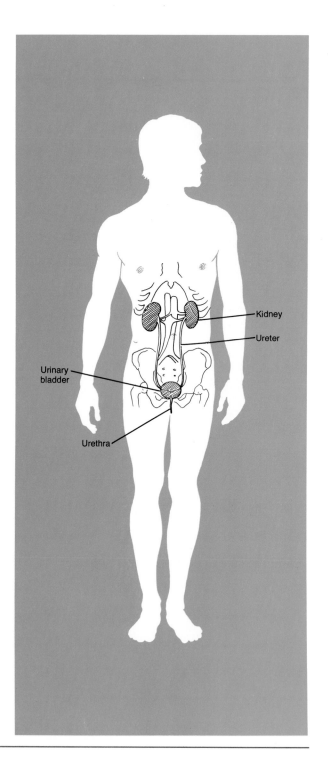

EXHIBIT 1-1 *(Continued)*

11. Reproductive

Definition: Organs (testes and ovaries) that produce reproductive cells (sperm and ova) and organs that transport and store reproductive cells.

Function: Reproduces the organism.

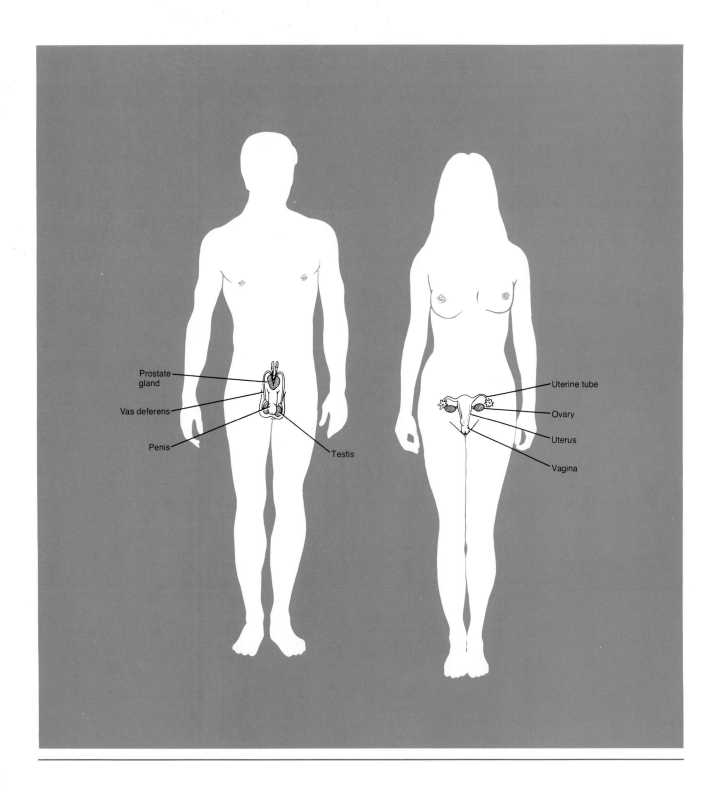

illustrates these systems, their representative organs, and their general functions. The systems are presented in the exhibit in the order in which they are discussed in later chapters.

STRUCTURAL PLAN

The human body has certain general **anatomical characteristics** that will help you to understand its overall structural plan. For example, humans have a *backbone,* or *vertebral column,* a characteristic that places them in a large group of organisms called vertebrates. Another characteristic of the body's organization is that it basically resembles a *tube within a tube* construction. The outer tube is formed by the body wall; the inner tube is the digestive tract. Moreover, humans are for the most part *bilaterally symmetrical*—externally the left and right sides of the body are mirror images.

ANATOMICAL POSITION AND REGIONAL NAMES

When a region of the body is described in an antomical text or chart, we assume that the body is in the **anatomical position**—erect and facing the observer. The arms are at the sides, and the palms of the hands are turned forward (Figure 1-2).

The common and anatomical terms, in parentheses, of certain body regions are also presented in Figure 1-2.

DIRECTIONAL TERMS

In order to explain exactly where various body structures are located relative to each other, anatomists use certain **directional terms.** If you want to point out the sternum (breastbone) to someone who knows where the clavicle (collarbone) is, you can say that the sternum is inferior (farther away from the head) and medial (toward the middle of the body) to the clavicle. As you can see, using the terms *inferior* and *medial* avoids a great deal of complicated description. Many directional terms are defined in Exhibit 1-2. The parts of the body used in the examples are labeled in Figure 1-3. Studying the exhibit and the figure together should make clear to you the directional relationships among various body parts.

PLANES

The structural plan of the human body may also be discussed with respect to **planes** (imaginary flat surfaces) that pass through it. Several of the commonly used planes are illustrated in Figure 1-4. A **midsagittal (or median) plane** through the midline of the body or an organ runs vertically and divides the body or organ into equal right

and left sides. A **sagittal** (SAJ-i-tal) **plane,** also called a **parasagittal** (*para* = near) **plane,** also runs vertically, but it divides the body or organ into unequal left and right portions. A **frontal** (or **coronal;** kō-RŌ-nal) **plane** runs vertically and divides the body or organ into anterior and posterior portions. Finally, a **horizontal** (or **transverse**) **plane** runs parallel to the ground and divides the body or organ into superior and inferior portions.

When you study a body structure, you will often view it in section, that is, look at the flat surface resulting from a cut made through the three-dimensional organ. It is important to know the plane of the section so that you can understand the anatomical relationship of one part to another. Figure 1-5 indicates how three different sections—a *cross section,* a *frontal section,* and a *midsagittal section*—are made through different parts of the brain.

BODY CAVITIES

Spaces within the body that contain internal organs are called **body cavities.** Specific cavities may be distinguished if the body is divided into right and left halves. Figure 1-6 shows the two principal body cavities. The **dorsal body cavity** is located near the posterior (dorsal) surface of the body. It is further subdivided into a **cranial cavity,** which is a bony cavity formed by the cranial (skull) bones and contains the brain, and a **vertebral** or **spinal canal,** which is a bony cavity formed by the vertebrae of the backbone and contains the spinal cord and the beginnings of spinal nerves.

The other principal body cavity is the **ventral body cavity.** This cavity, also known as the **coelom,** is located on the anterior (ventral) aspect of the body. The organs inside the ventral body cavity are called the **viscera** (VIS-er-a). Its walls are composed of skin, connective tissue, bone, muscles, and serous membranes. Like the dorsal body cavity, the ventral body cavity has two principal subdivisions—an upper portion, called the **thoracic** (thō-RAS-ik) **cavity** (or chest cavity), and a lower portion, called the **abdominopelvic cavity.** The anatomical landmark that divides the ventral body cavity into the thoracic and abdominopelvic cavities is the muscular diaphragm.

The thoracic cavity contains several divisions. There are two **pleural cavities** (Figure 1-7b), each containing a lung, and a **mediastinum** (mē'-dē-as-TĪ-num), a mass of tissue between the pleurae of the lungs that extends from the sternum to the vertebral column (Figure 1-7a, b). The mediastinum includes all of the contents of the thoracic cavity, except the lungs themselves. Within the mediastinum is the **pericardial** (per'-ē-KAR-dē-al; *peri* = around, *cardi* = heart) **cavity,** which encloses the heart (Figure 1-7b).

The abdominopelvic cavity is divided into two portions, although no wall separates them (see Figure 1-6).

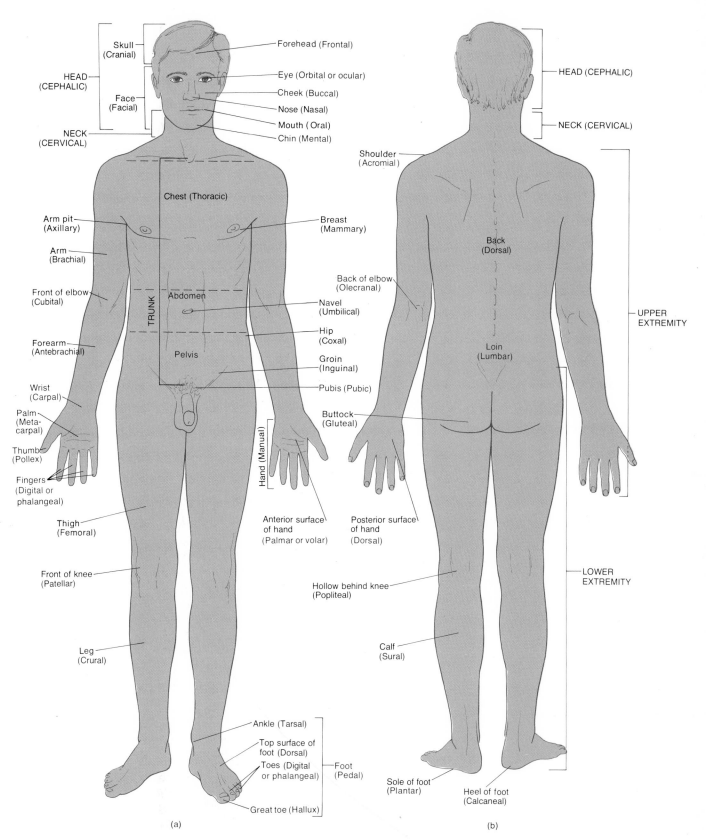

FIGURE 1-2 Anatomical position. The common names and anatomical terms for many of the regions of the body are indicated. (a) Anterior view. (b) Posterior view.

EXHIBIT 1-2 DIRECTIONAL TERMS

TERM	DEFINITION	EXAMPLE
Superior (cephalad or **craniad)**	Toward the head or the upper part of a structure; generally refers to structures in the trunk.	The heart is superior to the liver.
Inferior (caudad)	Away from the head or toward the lower part of a structure; generally refers to structures in the trunk.	The stomach is inferior to the lungs.
Anterior (ventral)	Nearer to or at the front of the body.	The sternum is anterior to the heart.
Posterior (dorsal)	Nearer to or at the back of the body.	The esophagus is posterior to the trachea.
Medial	Nearer the midline of the body or a structure.	The ulna is on the medial side of the forearm.
Lateral	Farther from the midline of the body or a structure.	The ascending colon of the large intestine is lateral to the urinary bladder.
Intermediate	Between two structures, one of which is medial and the other lateral.	The ring finger is intermediate between the little (medial) and middle (lateral) fingers.
Ipsilateral	On the same side of the body.	The gallbladder and ascending colon of the large intestine are ipsilateral.

FIGURE 1-3 Anatomical and directional terms. By studying Exhibit 1-2 with this figure, you should gain an understanding of the meanings of the terms *superior, inferior, anterior, posterior, medial, lateral, intermediate, ipsilateral, contralateral, proximal,* and *distal.*

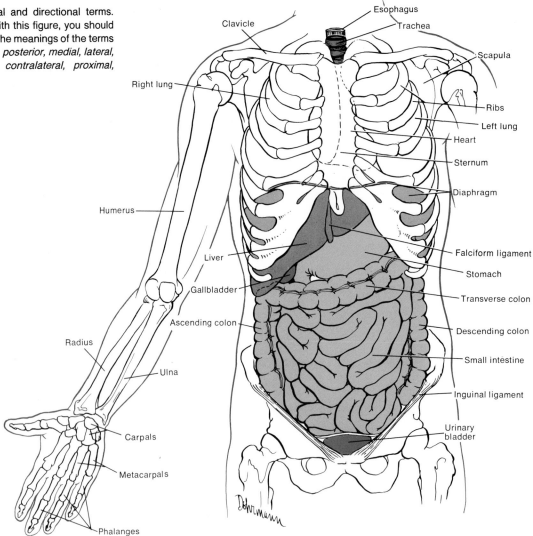

EXHIBIT 1-2 *(Continued)*

TERM	DEFINITION	EXAMPLE
Contralateral	On the opposite side of the body.	The ascending and descending colons of the large intestine are contralateral.
Proximal	Nearer the attachment of an extremity to the trunk or a structure.	The humerus is proximal to the radius.
Distal	Farther from the attachment of an extremity to the trunk or a structure.	The phalanges are distal to the carpals.
Superficial	Toward or on the surface of the body.	The muscles of the thoracic wall are superficial to the viscera in the thoracic cavity (see Figure 1-7b).
Deep	Away from the surface of the body.	The muscles of the arm are deep to the skin of the arm.
Parietal	Pertaining to the outer wall of a body cavity.	The parietal pleura forms the outer layer of the pleural sacs that surround the lungs (see Figure 1-7b).
Visceral	Pertaining to the covering of an organ (viscus).	The visceral pleura forms the inner layer of the pleural sacs and covers the external surface of the lungs (see Figure 1-7b).

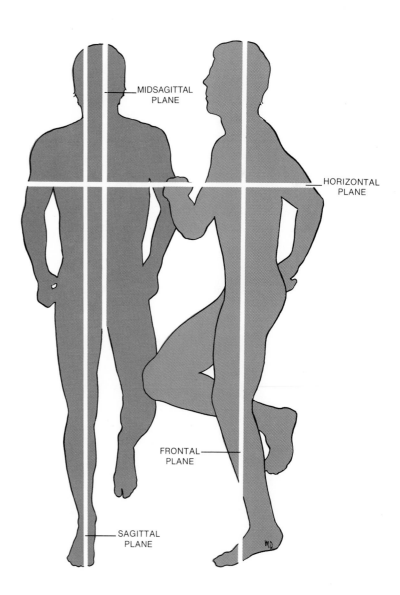

FIGURE 1-4 Planes of the human body.

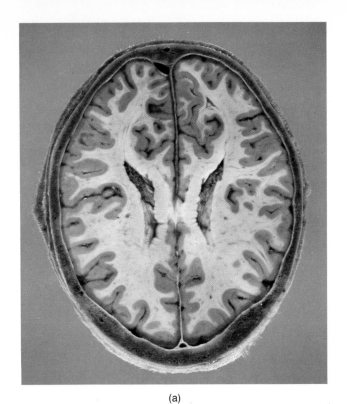

(a)

FIGURE 1-5 Sections through different parts of the brain. (a) Cross section through the brain. (Courtesy of Stephen A. Kieffer and E. Robert Heitzman, *An Atlas of Cross-Sectional Anatomy,* Harper & Row, Publishers, Inc., New York, 1979.) (b) Frontal section through the brain. (Courtesy of C. Yokochi and J. W. Rohen, *Photographic Anatomy of the Human Body,* 2nd ed., 1979, IGAKU-SHOIN, Ltd., Tokyo, New York.) (c) Midsagittal section through the brain. (Courtesy of C. Yokochi and J. W. Rohen, *Photographic Anatomy of the Human Body,* 2nd ed., 1979, IGAKU-SHOIN, Ltd., Tokyo, New York.)

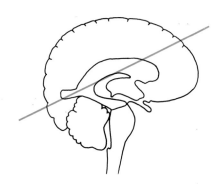

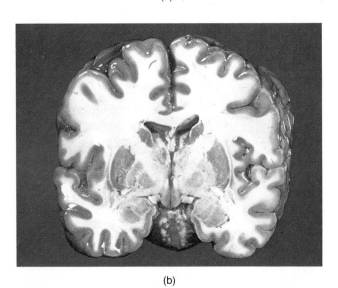

(b)

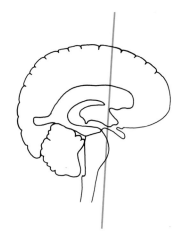

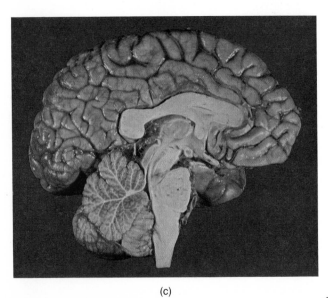

(c)

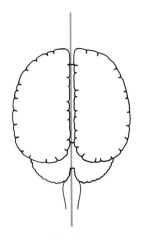

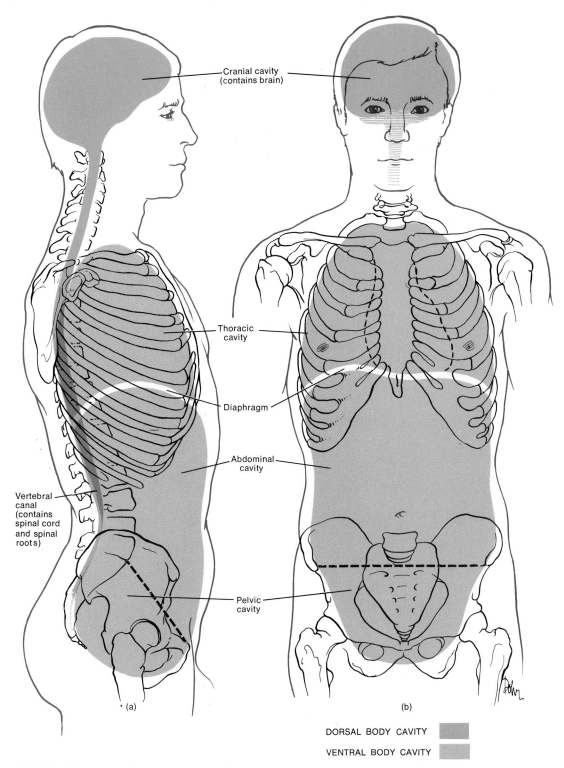

Cranial cavity
(contains brain)

Thoracic
cavity

Diaphragm

Abdominal
cavity

Vertebral
canal
(contains
spinal cord
and spinal
roots)

Pelvic
cavity

(a)

(b)

DORSAL BODY CAVITY

VENTRAL BODY CAVITY

FIGURE 1-6 Body cavities. (a) Location of the dorsal and ventral body cavities. (b) Subdivisions of the ventral body cavity.

The upper portion, the **abdominal cavity,** contains the stomach, spleen, liver, gallbladder, pancreas, small intestine, most of the large intestine, the kidneys, and the ureters. The lower portion, the **pelvic cavity,** contains the urinary bladder, sigmoid colon, rectum, and the internal male or female reproductive organs. One way to mark the division between the abdominal and pelvic cavities is to draw an imaginary line from the symphysis pubis (anterior joint between hipbones) to the superior border of the sacrum (sacral promontory).

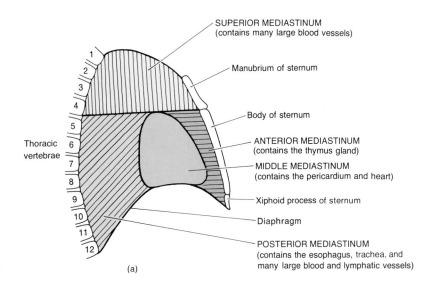

SUPERIOR MEDIASTINUM
(contains many large blood vessels)

Manubrium of sternum

Body of sternum

ANTERIOR MEDIASTINUM
(contains the thymus gland)

MIDDLE MEDIASTINUM
(contains the pericardium and heart)

Xiphoid process of sternum

Diaphragm

POSTERIOR MEDIASTINUM
(contains the esophagus, trachea, and
many large blood and lymphatic vessels)

Thoracic
vertebrae

(a)

FIGURE 1-7 Mediastinum. The mediastinum is the space between the pleurae of the lungs that extends from the sternum to the vertebral column. (a) Subdivisions of the mediastinum seen in right lateral view. (b) Mediastinum seen in a cross section of the thorax. Many of the structures shown and labeled are unfamiliar to you now. They are discussed in detail in later chapters.

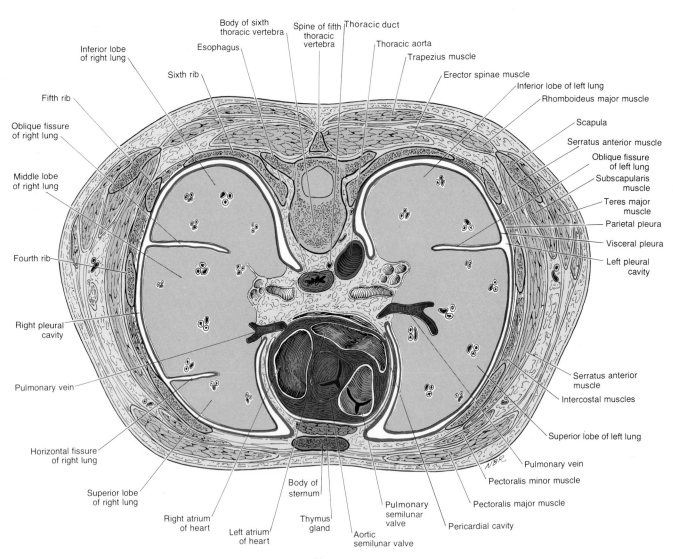

Body of sixth thoracic vertebra

Spine of fifth thoracic vertebra

Thoracic duct

Esophagus

Thoracic aorta

Inferior lobe of right lung

Trapezius muscle

Sixth rib

Erector spinae muscle

Fifth rib

Inferior lobe of left lung

Rhomboideus major muscle

Oblique fissure of right lung

Scapula

Serratus anterior muscle

Middle lobe of right lung

Oblique fissure of left lung

Subscapularis muscle

Teres major muscle

Parietal pleura

Fourth rib

Visceral pleura

Left pleural cavity

Right pleural cavity

Serratus anterior muscle

Intercostal muscles

Pulmonary vein

Superior lobe of left lung

Pulmonary vein

Horizontal fissure of right lung

Pectoralis minor muscle

Pericardial cavity

Pectoralis major muscle

Superior lobe of right lung

Body of sternum

Right atrium of heart

Thymus gland

Pulmonary semilunar valve

Left atrium of heart

Aortic semilunar valve

(b)

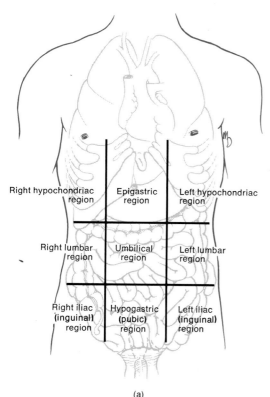

(a)

FIGURE 1-8 Abdominopelvic cavity. (a) The nine regions. The top horizontal line is drawn just below the bottom of the rib cage, and the bottom horizontal line is drawn just below the tops of the hipbones. The two vertical lines are drawn just medial to the nipples. The horizontal and vertical lines divide the area into a larger middle section and smaller left and right sections. (b) Anterior view of the abdomen and pelvis showing the most superficial organs. The greater omentum has been removed.

ABDOMINOPELVIC REGIONS

To describe the location of organs easily, the abdominopelvic cavity may be divided into the **nine regions** shown in Figure 1-8. Although some unfamiliar terms are used in describing the nine regions and their contents, follow the descriptions as well as you can. When the organs are studied in detail in later chapters, they will have more meaning. The **epigastric region** contains the left lobe and medial part of the right lobe of the liver, the pyloric part and lesser curvature of the stomach, the superior and descending portions of the duodenum, the body and upper part of the head of the pancreas, and the two adrenal (suprarenal) glands. The **right hypochondriac region** contains the right lobe of the liver, the gallbladder, and the upper third of the right kidney. The **left hypochondriac region** contains the body and fundus of the stomach, the spleen, the left colic (splenic) flexure, the upper two-thirds of the left kidney, and the tail of the pancreas. The **umbilical region** contains the middle of the transverse colon, the inferior part of the duodenum, the jejunum, the ileum, the hilar regions of the kidneys, and the bifurca-

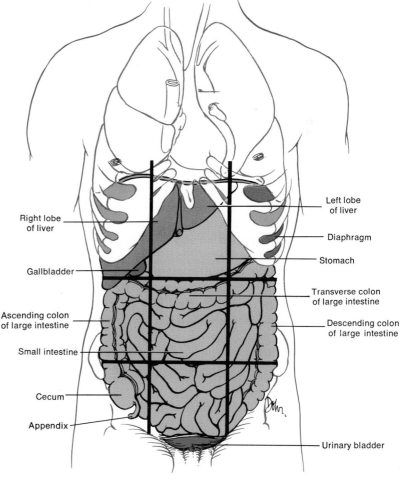

(b)

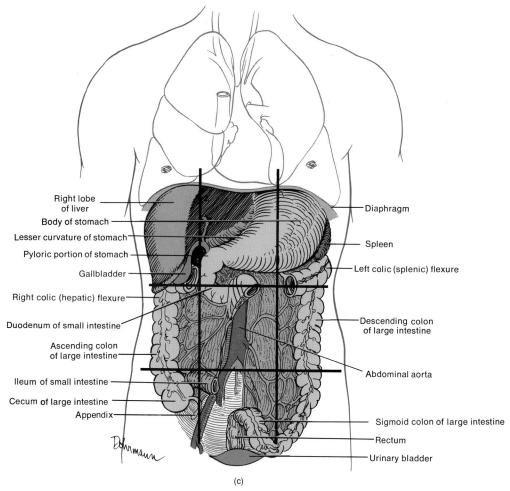

Right lobe of liver

Body of stomach

Lesser curvature of stomach

Pyloric portion of stomach

Gallbladder

Right colic (hepatic) flexure

Duodenum of small intestine

Ascending colon of large intestine

Ileum of small intestine

Cecum of large intestine

Appendix

Diaphragm

Spleen

Left colic (splenic) flexure

Descending colon of large intestine

Abdominal aorta

Sigmoid colon of large intestine

Rectum

Urinary bladder

(c)

FIGURE 1-8 *(Continued)* Abdominopelvic cavity. (c) Anterior view of the abdomen and pelvis in which most of the small intestine and transverse colon have been removed to expose deeper structures.

tions (branching) of the abdominal aorta and inferior vena cava. The **right lumbar region** contains the superior part of the cecum, the ascending colon, the right colic (hepatic) flexure, the lower lateral portion of the right kidney, and the small intestine. The **left lumbar region** contains the descending colon, the lower third of the left kidney, and the small intestine. The **hypogastric** (or **pubic**) **region** contains the urinary bladder when full, the small intestine, and part of the sigmoid colon. The **right iliac** (or **right inguinal**) **region** contains the lower end of the cecum, the appendix, and the small intestine. The **left iliac** (or **left inguinal**) **region** contains the junction of the descending and sigmoid parts of the colon and the small intestine.

ABDOMINOPELVIC QUADRANTS

The abdominopelvic cavity may be divided more simply into four **quadrants** (*quad* = four). These are shown in Figure 1-9. In this method, frequently used by clinicians, a horizontal line and a vertical line are passed through the umbilicus. These two lines divide the abdomen into

a *right upper quadrant* (*RUQ*), *left upper quadrant* (*LUQ*), *right lower quadrant* (*RLQ*), and *left lower quadrant* (*LLQ*). Whereas the nine-region designation is more widely used for anatomical studies, the four-quadrant designation is better suited for locating the site of an abdominopelvic pain, tumor, or other abnormality.

RADIOGRAPHIC ANATOMY

A specialized branch of anatomy that is essential for the diagnosis of many disorders is **radiographic anatomy,** which includes the use of x-rays. In the most common and familiar type of radiographic anatomy a single barrage of x-rays passes through the body and exposes an x-ray film, producing a photographic image called a **roentgenogram** (RENT-gen-ō-gram). A roentgenogram provides a two-dimensional shadow image of the interior of the body (Figure 1-10a). As valuable as they are in diagnosis, x-rays compress the body image onto a flat sheet of film, often resulting in an overlap of organs and tissues that could make diagnosis difficult. Moreover,

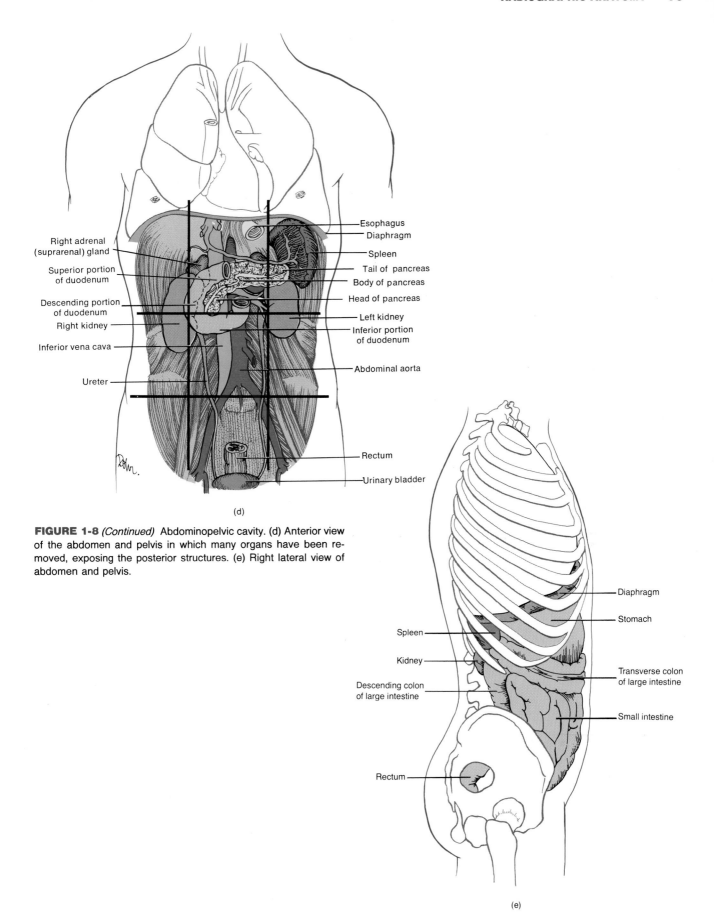

Right adrenal (suprarenal) gland
Superior portion of duodenum
Descending portion of duodenum
Right kidney
Inferior vena cava
Ureter

Esophagus
Diaphragm
Spleen
Tail of pancreas
Body of pancreas
Head of pancreas
Left kidney
Inferior portion of duodenum
Abdominal aorta
Rectum
Urinary bladder

(d)

FIGURE 1-8 *(Continued)* Abdominopelvic cavity. (d) Anterior view of the abdomen and pelvis in which many organs have been removed, exposing the posterior structures. (e) Right lateral view of abdomen and pelvis.

Diaphragm
Stomach
Spleen
Kidney
Transverse colon of large intestine
Descending colon of large intestine
Small intestine
Rectum

(e)

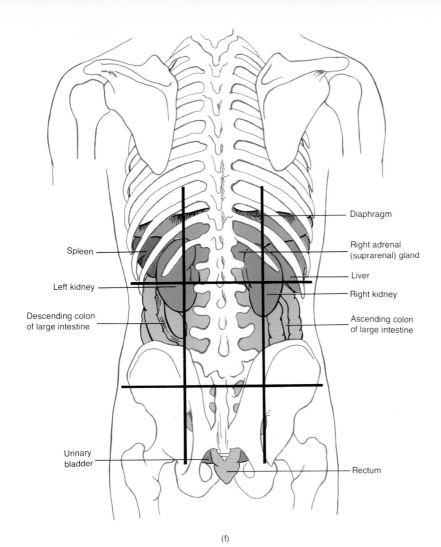

Spleen

Left kidney

Descending colon
of large intestine

Urinary
bladder

Diaphragm

Right adrenal
(suprarenal) gland

Liver

Right kidney

Ascending colon
of large intestine

Rectum

(f)

FIGURE 1-8 *(Continued)* Abdominopelvic cavity.
(f) Posterior view of abdomen and pelvis.

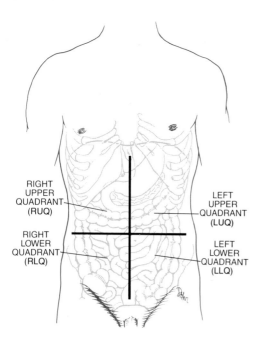

RIGHT
UPPER
QUADRANT
(RUQ)

RIGHT
LOWER
QUADRANT
(RLQ)

LEFT
UPPER
QUADRANT
(LUQ)

LEFT
LOWER
QUADRANT
(LLQ)

FIGURE 1-9 Four quadrants of the abdominopelvic cavity. The
two lines intersect at right angles at the umbilicus.

x-rays do not always differentiate between subtle differences in tissue density.

These diagnostic difficulties have been virtually eliminated by the use of an x-ray technique called **CT (computed tomography) scanning** or **CAT (computerized axial tomography) scanning.** First introduced in 1971, CT scanning combines the principles of x-ray and advanced computer technologies. An x-ray source moves in an arc around the part of the body being scanned and repeatedly sends out x-ray beams. The beams are then converted into electronic impulses that produce more than 46,000 readings of the density of the tissue in a one-centimeter slice of body tissue. Next, the computer produces a graphic, detailed sketch from these readings, called a **CT scan,** which can be viewed on a TV screen called the physician's console (Figure 1-11). The CT scan provides a very accurate cross-sectional picture of any area of the body (Figure 1-10b). The CT scan permits a significant differentiation of body parts that is not possible with conventional x-rays. The entire CT scanning process takes only seconds, it is completely painless, and the x-ray dose

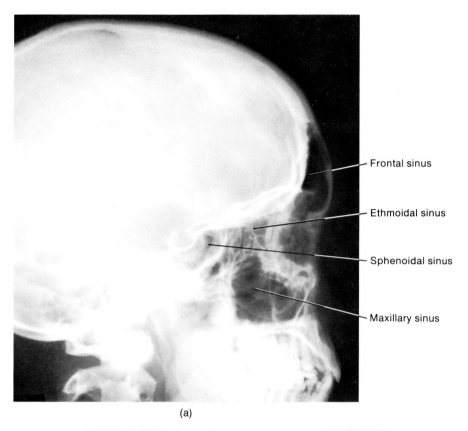

- Frontal sinus
- Ethmoidal sinus
- Sphenoidal sinus
- Maxillary sinus

(a)

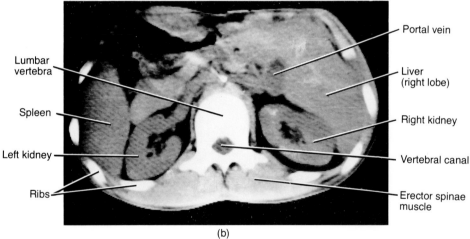

Lumbar vertebra

Spleen

Left kidney

Ribs

Portal vein

Liver (right lobe)

Right kidney

Vertebral canal

Erector spinae muscle

(b)

FIGURE 1-10 Radiologic anatomy. (a) Roentgenogram of the skull showing the paranasal sinuses. (Courtesy of Eastman Kodak Company.) (b) CT scan of the abdomen. (Courtesy of General Electric Company.)

is equal to or less than that of many other diagnostic procedures.

In subsequent chapters, references will be made to roentgenograms and CT scans. They will have more meaning to you once you have learned more about the anatomy of the body.

MEASURING THE HUMAN BODY

An important aspect of describing the body and understanding how it works is **measurement**—what are the dimensions of an organ, how much does it weigh, how long does it take for a physiological event to occur. Such questions also have clinical importance, for example, in determining how much of a given medication should be administered. As you will see, measurements involving time, weight, temperature, size, length, and volume are a routine part of your studies in a medical science program.

Whenever you come across a measurement in the text, the measurement will be given in metric units. The metric system is standardly used in sciences. To help you compare the metric unit to a familiar unit, the approximate U.S. equivalent will also be given in parentheses directly after the metric unit. For example, you might be told

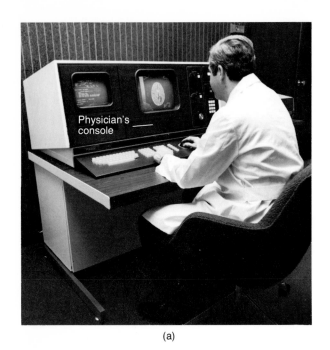

(a)

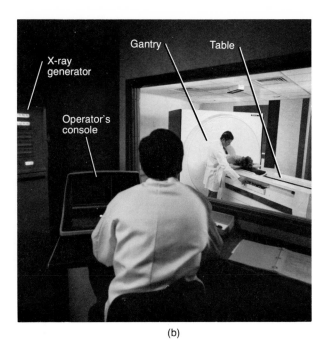

(b)

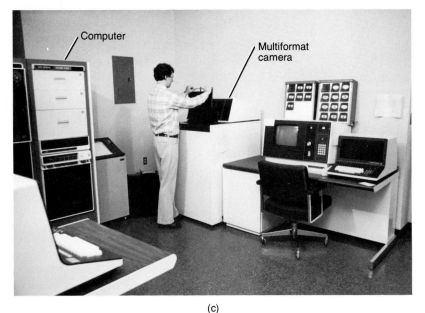

(c)

FIGURE 1-11 General Electric CT scanner system. (a) On the physician's console, images can be displayed, adjusted for contrast and density, magnified, and measured. In addition, previous scans stored in the computer may be instantly recalled and displayed. (b) The operator's console controls the scan procedure. The number of slices, slice thickness, scan speed, and slice location are determined here. The CT table and gantry perform the scans. After each slice, the table automatically moves to the next slice location. The gantry can angulate to scan from a variety of positions. The x-ray generator supplies power to the x-ray tube located in the gantry. (c) The computer processes thousands of bits of information to create each image. It also helps set up examinations, controls scan procedures, displays images, and stores and recalls previous studies. The multiformat camera can record CT scans on film for physician reference. (Courtesy of General Electric Company.)

that the length of a particular part of the body is 2.54 cm (1 inch).

As a first step in helping you understand the relationships of the metric to U.S. systems of measurement, three exhibits have been prepared. Exhibit 1-3 contains metric units of length and some U.S. equivalents. Exhibit 1-4 contains metric units of mass and some U.S. equivalents. Exhibit 1-5 contains metric units of volume and some U.S. equivalents. Carefully examine the exhibits. Even if you do not learn all the metric units and their equivalents at this point, you can refer back to the exhibits later.

EXHIBIT 1-3 METRIC UNITS OF LENGTH AND SOME U.S. EQUIVALENTS

METRIC UNIT (ABBREVIATION)	MEANING OF PREFIX	EQUIVALENT IN METERS	U.S. EQUIVALENT
1 kilometer (km)	kilo = 1,000	1,000 m	3,280.84 ft; 0.62 mi 1 mi = 1.61 km
1 hectometer (hm)	hecto = 100	100 m	328 ft
1 dekameter (dam)	deka = 10	10 m	3.28 ft
1 meter (m)	Standard unit of length	—	39.37 in.; 3.28 ft; 1.09 yd
1 decimeter (dm)	deci = 1/10	0.1 m	3.94 in.
1 centimeter (cm)	centi = 1/100	0.01 m	0.394 in. 1 in. = 2.54 cm
1 millimeter (mm)	milli = 1/1,000	0.001 m	0.0394 in.
1 micrometer (μm) [formerly **micron (μ)**]	micro = 1/1,000,000	0.000,0001 m	3.94×10^{-5} in.
1 nanometer (nm) [formerly **millimicron (mμ)**]	nano = 1/1,000,000,000	0.000,000,001 m	3.94×10^{-8} in.
1 angstrom (Å)		0.000,000,000,1 m	3.94×10^{-9} in.

EXHIBIT 1-4 METRIC UNITS OF MASS AND SOME U.S. EQUIVALENTS

METRIC UNIT (ABBREVIATION)	EQUIVALENT IN GRAMS	U.S. EQUIVALENT
1 kilogram (kg)	1,000 g	2.205 lb
1 hectogram (hg)	100 g	
1 dekagram (dag)	10 g	
1 gram (g)	—	1 lb = 453.6 1 oz = 28.35 g
1 decigram (dg)	0.1 g	
1 centigram (cg)	0.01 g	
1 milligram (mg)	0.001 g	
1 microgram (gmg)	0.000,001 g	

EXHIBIT 1-5 METRIC UNITS OF VOLUME AND SOME U.S. EQUIVALENTS

METRIC UNIT (ABBREVIATION)	METRIC EQUIVALENT	U.S. EQUIVALENT
1 liter (l)	1,000 ml	33.81 fl oz or 1.057 qt 1 qt = 946 ml
1 milliliter (ml)	0.001 liter	0.0338 fl oz; 1 fl oz = 30 ml 1 teaspoon = 5 ml
1 cubic centimeter (cm³)	0.999972 ml	0.0338 fl oz

STUDY OUTLINE

Anatomy and Physiology Defined
1. Anatomy is the study of biological structures and how these structures are related to one another.
2. Subdivisions of anatomy include surface anatomy (surface features), gross anatomy (macroscopic), regional anatomy (regions), systemic anatomy (systems), developmental anatomy (development from fertilization to adulthood), embryology (development of the embryo through the eighth week), pathological anatomy (disease), histology (tissues), and cytology (cells).
3. Physiology is the study of how body structures function.

Levels of Structural Organization
1. The human body consists of levels of structural organization, from the chemical level to the organismic level.
2. The chemical level is represented by all the atoms and molecules in the body. The cellular level consists of cells. The tissue level is represented by tissues. The organ level consists of body organs, and the system level is represented by organs that work together to perform a general function.
3. The human organism is a collection of structurally and functionally integrated systems.
4. The systems of the human body are the integumentary,

skeletal, muscular, cardiovascular, lymphatic, nervous, endocrine, respiratory, digestive, urinary, and reproductive.

Structural Plan
1. The human body has certain general characteristics.
2. Among the characteristics are a backbone, a tube within a tube organization, and bilateral symmetry.

Anatomical Position and Regional Names
1. The position from which directional terms and motions are studied and described is the anatomical position. The subject stands erect and faces the observer with arms at sides and palms turned forward.
2. Regional names are terms given to specific regions of the body for reference. Examples of regional names include cranial (skull), thoracic (chest), brachial (arm), patellar (knee), cephalic (head), and gluteal (buttock).

Directional Terms
1. Directional terms indicate the relationship of one part of the body to another.
2. Commonly used directional terms are superior (toward the head or upper part of a structure), inferior (away from the head or toward the lower part of a structure), anterior or ventral (near or at the front of the body), posterior or dorsal (near or at the back of the body), medial (nearer the midline of the body or a structure), lateral (farther from the midline of the body or a structure), proximal (nearer the attachment of an extremity to the trunk or a structure), distal (farther from the attachment of an extremity to the trunk or a structure), superficial (toward or on the surface of the body), deep (away from the surface of the body), parietal (pertaining to the outer wall of a body cavity), and visceral (pertaining to the covering of an organ).

Planes
1. Planes of the body are imaginary flat surfaces that are used to divide the body or organs into definite areas. The midsagittal (median) plane divides the body into equal right and left sides; the sagittal (parasagittal) plane, into unequal right and left sides; the frontal (coronal) plane, into anterior and posterior portions;

and the horizontal (transverse) plane, into superior and inferior portions.
2. Sections of body structures include cross sections, frontal sections, and midsagittal sections.

Body Cavities
1. Spaces in the body that contain internal organs are called cavities.
2. The dorsal and ventral cavities are the two principal body cavities. The dorsal cavity contains the brain and spinal cord. The organs of the ventral cavity are collectively called the viscera.
3. The dorsal cavity is subdivided into the cranial cavity and the vertebral canal.
4. The ventral body cavity is subdivided by the diaphragm into an upper, or thoracic, cavity and a lower, or abdominopelvic, cavity.
5. The thoracic cavity contains two pleural cavities and a mediastinum, which includes the pericardial cavity.

Abdominopelvic Regions and Quadrants
1. The abdominopelvic cavity, actually an interconnected upper abdominal cavity and a lower pelvic cavity, is superficially divided into nine imaginary anatomical regions.
2. It may also be divided into four anatomical quadrants.

Radiographic Anatomy
1. This branch of anatomy depends on the use of x-rays.
2. In conventional usage, x-rays produce a two-dimensional shadow image of the interior of the body called a roentgenogram.
3. CT (computed tomography) scanning employs the use of x-rays and advanced computer technology to produce an accurate cross-sectional image of the body called a CT scan.

Measuring the Human Body
1. Various kinds of measurements are important in understanding the human body.
2. Examples of such measurements include organ size, body weight, and amount of medication to be administered.
3. Measurements in this book are given in metric units followed by the approximate U.S. equivalents in parentheses.
4. Metric units of length may be reviewed in Exhibit 1-3, metric units of mass in Exhibit 1-4, and metric units of volume in Exhibit 1-5.

REVIEW QUESTIONS

1. Define anatomy. How does each subdivision of anatomy help you understand the structure of the human body? Define physiology.
2. Construct a diagram to illustrate the levels of structural organization that characterize the body. Be sure to define each level.
3. Outline the function of each system of the body, and list several organs that compose each system.

4. What does bilateral symmetry mean? Why is the body considered to be a tube within a tube? What is a vertebrate?

5. When is the body in the anatomical position? Why is the anatomical position used?

6. Review Figure 1-2. See if you can locate each region on your own body and give both its common and its anatomical name.

7. What is a directional term? Why are these terms important? Can you use each of the directional terms listed in Exhibit 1-2 in a complete sentence?

8. Name the various planes that may be passed through the body. Describe how each plane divides the body.

9. What is meant by the phrase "a part of the body has been sectioned"? Given a pear, can you make a cross section, frontal section, and midsagittal section with a knife?

10. Define a body cavity. List the body cavities discussed, and tell which major organs are located in each. What landmarks separate the various body cavities from one another?

11. Name and locate the nine regions of the abdominopelvic area. List the organs, or parts of organs, in each. Describe how the abdominopelvic cavity is divided into four quadrants and name each quadrant.

12. Explain the principle of CT scanning. Contrast it with simple x-ray procedures.

13. Convert the following lengths:
    a. A bacterial cell measures 100 μm in length. Give its length in nanometers.
    b. How many meters are in 1 mi?
    c. If a road sign reads 35 km/hour, what would your speedometer have to read to obey the sign?
    d. How many millimeters are in 1 km? In 1 inch?
    e. A person's arm measures 2 ft in length. How many centimeters is this?
    f. Convert 0.40 m to millimeters.
    g. How many millimeters are in 5 inches?
    h. If you ran 295.2 ft, how many meters would you have run?
    i. If the distance to the moon is 239,000 mi, what is it in meters?

14. Solve the following conversions of mass.
    a. Calculate the milligrams in 0.4 kg and in 1 lb.
    b. If a bottle contains 1.42 g, how many centigrams does it contain?
    c. The indicated dosage of a certain drug is 50 μg. How many milligrams is this?
    d. If you weigh 110 lb, how many kilograms do you weigh?
    e. How many centigrams are in 1 g?

15. Convert the following volumes.
    a. If you excrete 1,200 ml of urine in a day, how many liters is this?
    b. How many milliliters are in 2 liters?
    c. Convert 2 pt to milliliters.
    d. If you remove 15 cm³ of blood from a patient, how many milliliters have you removed?

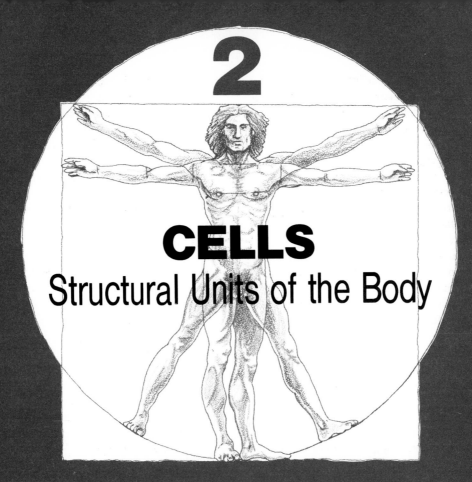

2

CELLS
Structural Units of the Body

STUDENT OBJECTIVES

- Define and list a cell's generalized parts.

- Describe the structure and molecular organization of the plasma membrane.

- Define diffusion, facilitated diffusion, osmosis, filtration, dialysis, active transport, phagocytosis, and pinocytosis.

- Describe the structure and function of several modified plasma membranes.

- Describe the chemical composition and list the functions of cytoplasm.

- Describe two general functions of a cell nucleus.

- Distinguish between agranular and granular endoplasmic reticulum.

- Define the function of ribosomes.

- Describe the role of the Golgi complex in synthesis, storage, and secretion.

- Describe the function of mitochondria as "powerhouses of the cell."

- Explain why a lysosome in a cell is called a "suicide packet," and describe microtubules.

- Describe the structure and function of centrioles in cellular reproduction.

- Differentiate between cilia and flagella.

- Define cell inclusion and give several examples.

- Define extracellular material and give several examples.

- Describe the stages and events involved in cell division.

- Discuss the significance of cell division.

- Describe the relationship of cancer and aging to cells.

- Define key medical terms associated with cells.

The study of the body at the cellular level of organization is important because many activities essential to life occur in cells and many disease processes originate there. A **cell** may be defined as the basic, living, structural and functional unit of the body and, in fact, of all organisms. **Cytology** is the branch of science concerned with the study of cells. This chapter concentrates on the structure, functions, and reproduction of cells.

A series of illustrations accompanies each cell structure that you study. A diagram of a generalized animal cell shows the location of the structure within the cell. An electron micrograph shows the actual appearance of the structure.* A diagram of the electron micrograph clarifies some of the small details by exaggerating their outlines. Finally, an enlarged diagram of the structure shows the details that make up the structure.

GENERALIZED ANIMAL CELL

A **generalized animal cell** is a composite of many different cells in the body. Examine the generalized cell illustrated in Figure 2-1, but keep in mind that no such single cell actually exists.

For convenience, we can divide the generalized cell into four principal parts:

1. Plasma (or **cell**) **membrane.** The outer, limiting membrane separating the cell's internal parts from the extracellular fluid and external environment.

2. Cytoplasm. The substance between the nucleus and the plasma membrane.

3. Organelles. The cellular components that are highly specialized for specific cellular activities.

4. Inclusions. The secretions and storage areas of cells.

Extracellular materials, which are substances external to the cell surface, will also be examined in connection with cells.

PLASMA (CELL) MEMBRANE
STRUCTURE

The exceedingly thin structure that separates one cell from other cells and from the external environment is called the **plasma membrane,** or **cell membrane.** Electron microscopy studies have shown that the plasma membrane ranges from 65 to 100 angstroms (Å) in thickness, a dimension below light-microscope limits.†

Scientists have known for a long time that plasma membranes are composed of phospholipid and protein

* An **electron micrograph** is a photograph taken with an electron microscope. The electron microscope can magnify an object more than 200,000 times. In comparison, most light microscopes magnify objects 100, 430, and 1,000 times their size.

† One **angstrom,** usually written as 1Å, = 0.000,000,000,1 m ($\frac{1}{250,000,000}$ inch). Another microscopic unit of measurement is the **micrometer** (μm), which is equal to 0.000,001 m ($\frac{1}{25,000}$ inch).

molecules. Recent studies of membrane structure suggest a new concept regarding the arrangement of these molecules. This new concept (unit membrane) of membrane structure is called the *fluid mosaic hypothesis*. These studies seem to indicate that the phospholipid molecules, which account for about half the mass of the membrane, are arranged in two parallel rows. This double row of phospholipid molecules is termed a *lipid bilayer* (Figure 2-2). The protein molecules are arranged somewhat differently. Some lie at or near the inner and outer surface of the membrane. Others penetrate the membrane partway, completely, singly, or in pairs. Such an arrangement suggests that membranes are not static but that the proteins and phospholipids have a considerable degree of movement, somewhat like ice cubes in water. It is quite possible that many key functions of membranes will be fully explained once the reasons for protein and phospholipid movements are understood. The fluid mosaic hypothesis may explain how specific receptor sites on membranes attach to hormones (for example, insulin), to transmitter substances produced by axons, and to antigens on the surfaces of red blood cells.

Other features of the membrane structure are areas that appear as breaks along the membrane surface. These breaks occur at intervals and range in size from 7 to 10 Å in diameter. Researchers suspect they may be pores.

FUNCTIONS

The basic functions of the plasma membrane are to enclose the components of the cell and to serve as a boundary through which substances must pass to enter or exit the cell. One important characteristic of the plasma membrane is that it permits certain ions and molecules to enter or exit the cell but restricts the passage of others. For this reason, plasma membranes are described as **semipermeable, differentially permeable,** or **selectively permeable.** Usually, plasma membranes are freely permeable to water. In other words, they let water into and out of the cell. However, they act as partial barriers to the movement of almost all other substances. The ease with which a substance passes through a membrane is called the membrane's *permeability* to that substance. The permeability of a plasma membrane appears to be a function of several factors.

1. Size of molecules. Large molecules cannot pass through the plasma membrane. Water and amino acids are small molecules and can enter and exit the cell easily. However, most proteins, which consist of many amino acids linked together, seem to be too large to pass through the membrane. Many scientists believe that the giant-sized molecules do not enter the cell because they are larger than the diameters of the suspected membrane pores.

2. Solubility in lipids. Substances that dissolve easily in lipids pass through the membrane more readily than

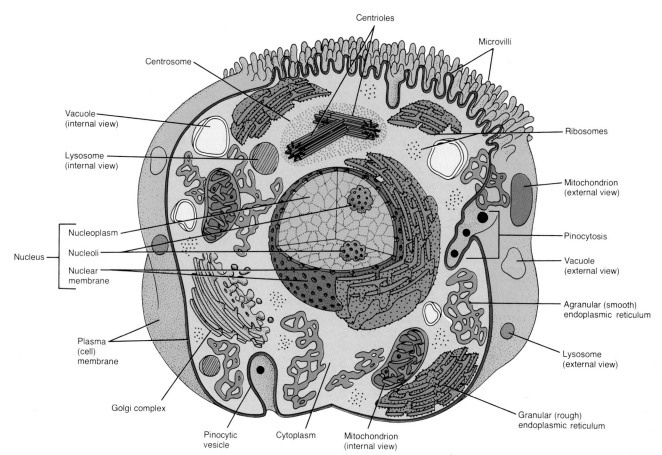

FIGURE 2-1 Generalized animal cell based on electron microscope studies.

other substances, since a major part of the plasma membrane consists of lipid molecules.

3. Charge on ions. The charge of an ion attempting to cross the plasma membrane can determine how easily the ion enters or leaves the cell. The protein portion of the membrane is capable of ionization. If an ion has a charge opposite that of the membrane, it is attracted to the membrane and passes through more readily. If the ion attempting to cross the membrane has the same charge as the membrane, it is repelled by the membrane and its passage is restricted. This phenomenon conforms to the rule of physics that opposite charges attract, whereas like charges repel each other.

4. Presence of carrier molecules. Plasma membranes contain special molecules called carriers that are capable of attracting and transporting substances across the membrane regardless of size, ability to dissolve in lipids, or membrane charge.

MOVEMENT OF MATERIALS ACROSS PLASMA MEMBRANES

The mechanisms whereby substances move across the plasma membrane are essential to the life of the cell. Certain substances, for example, must move into the cell

to support life, whereas waste materials or harmful substances must be moved out. The processes involved in these movements may be classed as either passive or active. In **passive processes,** substances move across plasma membranes without assistance from the cell. Their movement involves the kinetic energy of individual molecules. The substances move from an area where their concentration is greater to an area where their concentration is less. The substances can also be forced across the plasma membrane from an area where the pressure is greater to an area where it is less. In **active processes,** the cell contributes energy in moving the substance across the membrane.

Passive Processes

● *Diffusion* A passive process called **diffusion** occurs when there is a *net* or greater movement of molecules or ions from a region of high concentration to a region of low concentration. The movement from high to low concentration continues until the molecules are evenly distributed. At this point, they move in both directions at an equal rate. This point of even distribution is called *equilibrium.* The difference between high and low concentrations is called the *concentration gradient.* Molecules

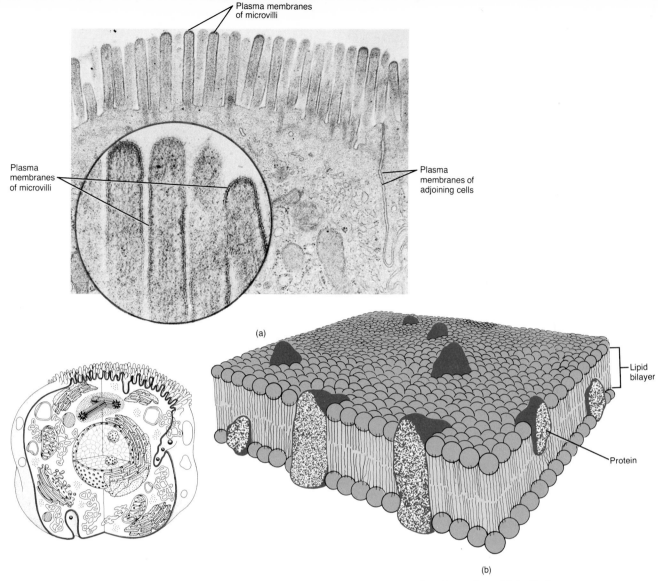

Plasma membranes of microvilli

Plasma membranes of microvilli

Plasma membranes of adjoining cells

(a)

Lipid bilayer

Protein

(b)

FIGURE 2-2 Plasma membrane. (a) Electron micrograph of the plasma membrane of portions of small intestinal cells at a magnification of 24,000×. Microvilli, as will be described later in the chapter, increase the surface area for absorption. The inset shows the plasma membrane of microvilli at a magnification of 96,000×. (© 1983 by Michael H. Ross. Used by permission.) (b) Enlargement of the plasma membrane showing the latest concept regarding the relationship of the lipid bilayer and protein molecules.

moving from the high-concentration area to the low-concentration area are said to move *down* or *with* the concentration gradient. If a dye pellet is placed in a beaker filled with water, the color of the dye is seen immediately around the pellet. At increasing distances from the pellet, the color becomes lighter (Figure 2-3). Later, however, the water solution will be a uniform color. The dye molecules possess kinetic energy (energy of motion) and move about at random. The dye molecules move down the concentration gradient from an area of high concentration to an area of low concentration. The water molecules also move from a high-concentration to a low-concentration area. When dye molecules and water molecules are evenly distributed among themselves, equilibrium is reached and diffusion ceases, even though molecular movements continue. As another example of diffusion, consider what would happen if you opened a bottle of

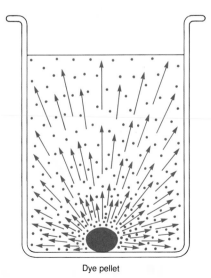

Dye pellet

FIGURE 2-3 Principle of diffusion.

29

perfume in a room. The perfume molecules would diffuse until an equilibrium is reached between the perfume molecules and the air molecules in the room.

In the examples cited, no membranes were involved. Diffusion may occur, however, through semipermeable membranes in the body. A good example of this kind of diffusion in the human body is the movement of oxygen from the blood into the cells and the movement of carbon dioxide from the cells back into the blood. This is essential in order for body cells to maintain homeostasis. It ensures that cells receive adequate amounts of oxygen and eliminate carbon dioxide as part of their normal metabolism.

• *Facilitated Diffusion*   Another example of diffusion through a semipermeable membrane occurs by a process called **facilitated diffusion.** Although some chemical substances are insoluble in lipids, they can still pass through the plasma membrane. Among these are different sugars, especially glucose. In the process of facilitated diffusion, glucose combines with a carrier substance. The combined glucose-carrier is soluble in the lipid layer of the membrane, and the carrier transports the glucose to the inside of the membrane. At this point, the glucose separates from the carrier and enters the cell. The carrier then returns to the outside of the membrane to pick up more glucose and transport it inside. The carrier makes the glucose soluble in the lipid portion of the membrane so it can pass through the membrane. By itself, glucose is insoluble and cannot penetrate the membrane. In the process of facilitated diffusion, the cell does not expend energy, and the movement of the substance is from a region of higher to lower concentration.

• *Osmosis*   Another passive process by which materials move across membranes is **osmosis.** It is the net movement of water molecules through a semipermeable membrane from an area of high water concentration to an area of lower water concentration. Once again, a simple apparatus may be used to demonstrate the process. The apparatus shown in Figure 2-4 consists of a tube constructed from cellophane, a semipermeable membrane. The cellophane tube is filled with a colored, 20 percent sugar (sucrose) solution. The upper portion of the cellophane tube is plugged with a rubber stopper through which a glass tubing is fitted. The cellophane tube is placed into a beaker containing pure water. Initially, the concentrations of water on either side of the semipermeable membrane are different. There is a lower concentration of water inside the cellophane tube than outside. Because of this difference, water moves from the beaker into the cellophane tube. The force with which the water moves is called osmotic pressure. Very simply, **osmotic pressure** is the force under which a solvent, usually water, moves from a solution of higher water concentration to a solution of lower water concentration when the solutions are separated by a semipermeable membrane. There is no movement of sugar from the cellophane tube inside the beaker, however, since the cellophane is impermeable to molecules of sugar—sugar molecules are too large to go through the pores of the membrane. As water moves into the cellophane tube, the sugar solution becomes increasingly diluted and the increased volume forces the mixture up the glass tubing. In time, the water that has accumulated in the cellophane tube and the glass tubing exerts a downward pressure that forces water molecules back out of the cellophane tube and into the beaker. When water molecules leave the cellophane tube and enter the tube at the same rate, equilibrium is reached. Osmotic pressure is an important force in the movement of water between various compartments of the body.

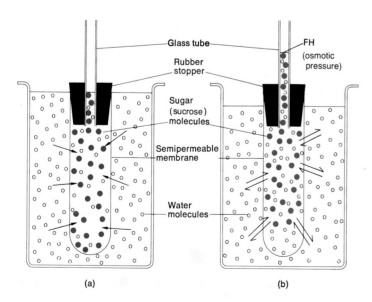

Glass tube — Rubber stopper — Sugar (sucrose) molecules — Semipermeable membrane — Water molecules — FH (osmotic pressure)

(a)   (b)

FIGURE 2-4 Principle of osmosis. (a) Apparatus at the start of the experiment. (b) Apparatus at equilibrium. In (a), the cellophane tube contains a 20 percent sugar solution and is immersed in a beaker of distilled water. The arrows indicate that water molecules can pass freely into the tube, but that sugar (sucrose) molecules are held back by the semipermeable membrane. As water moves into the tube by osmosis, the sugar solution is diluted and the volume of the solution in the cellophane tube increases. This increased volume is shown in (b), with the sugar solution moving up the glass tubing. The final height reached (FH) occurs at equilibrium and represents the osmotic pressure. At this point, the number of water molecules leaving the cellophane tube is equal to the number of water molecules entering the tube.

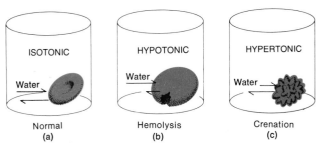

FIGURE 2-5 Principle of osmosis applied to red blood cells. Shown here are the effects on red blood cells when placed in (a) an isotonic solution, in which they maintain normal shape; (b) a hypotonic solution, in which they undergo hemolysis; and (c) a hypertonic solution, in which they undergo crenation.

● *Isotonic, Hypotonic, and Hypertonic Solutions* Osmosis may also be understood by considering the effects of different water concentrations on red blood cells. If the normal shape of a red blood cell is to be maintained, the cell must be placed in an **isotonic solution** (Figure 2-5a). This is a solution in which the total concentrations of water molecules and solute molecules are the same on both sides of the semipermeable cell membrane. The concentrations of water and solute in the extracellular fluid outside the red blood cell must be the same as the concentration of the intracellular fluid. Under ordinary circumstances, a 0.85 percent NaCl solution is isotonic for red blood cells. In this condition, water molecules enter and exit the cell at the same rate, allowing the cell to maintain its normal shape.

A different situation results if red blood cells are placed in a solution that has a lower concentration of solutes and, therefore, a higher concentration of water. This is called a **hypotonic solution.** In this condition, water molecules enter the cells faster than they can leave—causing the red blood cells to swell and eventually burst (Figure 2-5b). The rupture of red blood cells in this manner is called **hemolysis** (hē-MOL-i-sis) or **laking.** Distilled water is a strongly hypotonic solution.

A **hypertonic solution** has a higher concentration of solutes and a lower concentration of water than the red blood cells. One example of a hypertonic solution is a 10 percent NaCl solution. In such a solution, water molecules move out of the cells faster than they can enter. This situation causes the cells to shrink (Figure 2-5c). The shrinkage of red blood cells in this manner is called **crenation** (krē-NĀ-shun). Red blood cells may be greatly impaired or destroyed if placed in solutions that deviate significantly from the isotonic state.

● *Filtration* A third passive process involved in moving materials in and out of cells is **filtration.** This process involves the movement of solvents such as water and dissolved substances such as sugar across a semipermeable

membrane by mechanical pressure, usually hydrostatic (water) pressure. Such a movement is always from an area of higher pressure to an area of lower pressure and continues as long as a pressure difference exists. Most small to medium-sized molecules can be forced through a cell membrane.

An example of filtration occurs in the kidneys, where the blood pressure supplied by the heart forces water and small molecules like urea through thin cell membranes of tiny blood vessels and into the kidney tubules. In this basic process, protein molecules are retained by the body since they are too large to be forced through the cell membranes of the blood vessels. The molecules of harmful substances such as urea are small enough to be forced through and eliminated, however.

● *Dialysis* The final passive process to be considered is **dialysis.** Essentially, dialysis is diffusion and involves the separation of small molecules from large molecules by a semipermeable membrane. Suppose a solution containing molecules of various sizes is placed in a tube that is permeable only to the smaller molecules. The tube is then placed in a beaker of distilled water. Eventually, the smaller molecules will move from the tube into the water in the beaker and the larger molecules will be left behind.

The principle of dialysis is employed in artificial kidney machines. It does not occur within the human body. The patient's blood is passed across a dialysis membrane outside the body. The dialysis membrane takes the place of the kidneys. As the blood moves across the membrane, small-particle waste products pass from the blood into a solution surrounding the dialysis membrane. At the same time, certain nutrients can be passed from the solution into the blood. The blood is then returned to the body.

Active Processes

When cells actively participate in moving substances across membranes, they must expend energy. Cells can even move substances against a concentration gradient. The active processes considered here are active transport, phagocytosis, and pinocytosis.

● *Active Transport* The process by which substances, usually ions, are transported across plasma membranes typically from an area of lower concentration to an area of higher concentration is called **active transport** (Figure 2-6a). Although the exact mechanism is not known, the following sequence is believed to occur.

1. An ion in the extracellular fluid outside the plasma membrane is attached to an enzymelike carrier molecule located in or on the plasma membrane.

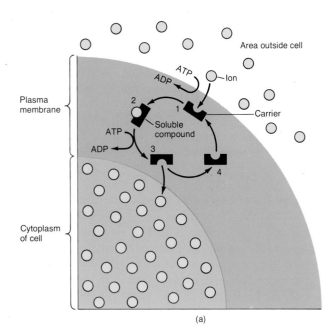

(a)

2. The ion-carrier complex is soluble in the lipid portion of the membrane.

3. The complex moves toward the interior of the membrane, where it is split by enzymes.

4. The ion is then transported into the intracellular fluid of the cell, and the carrier returns to the outer surface of the membrane to pick up another ion.

The energy for the attachment and release of the carrier molecule is supplied by ATP. Note the differences between active transport and facilitated diffusion.

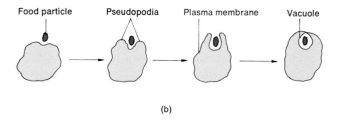

(b)

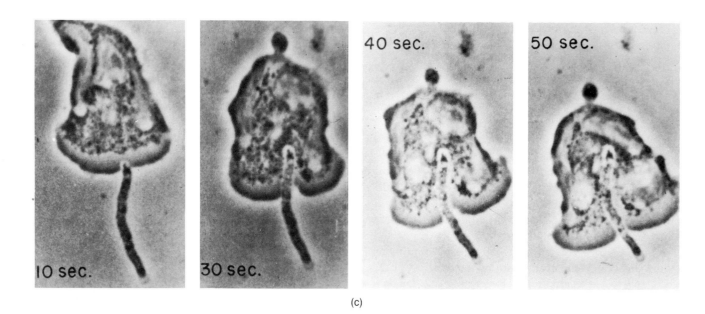

(c)

FIGURE 2-6 Active processes. (a) Mechanism of active transport. (b) Diagram of phagocytosis. (c) Photomicrographs of phagocytosis. In this time-lapse photography sequence, a bacterial cell (the rodlike structure) is ingested and destroyed by a human neutrophil. (Courtesy of Dr. James G. Hirsch.) (d) Two variations of pinocytosis. In the variation on the left, the ingested substance enters a channel formed by the plasma membrane and becomes enclosed in a vacuole at the base of the channel. In the variation on the right, the ingested substance becomes enclosed in a vacuole that forms and detaches at the surface of the cell.

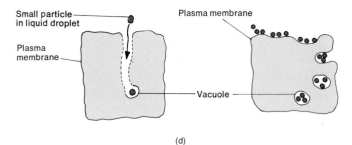

(d)

● *Phagocytosis* Another active process by which cells take in substances across the plasma membrane is called **phagocytosis** (fag'-ō-sī-TŌ-sis), or "cell eating" (Figure 2-6b). In this process, projections of cytoplasm, called *pseudopodia* (soo'-dō-PŌ-dē-a), engulf solid particles external to the cell. Once the particle is surrounded, the membrane folds inwardly, forming a membrane sac around the particle. This newly formed sac, called a *digestive vacuole,* breaks off from the outer cell membrane, and the solid material inside the vacuole is digested. Indigestible particles and cell products are removed from the cell by a reverse phagocytosis. This process is important because molecules and particles of material that would normally be restricted from crossing the plasma membrane can be brought into or removed from the cell. The phagocytic white blood cells of the body constitute a vital defense mechanism. Through phagocytosis, the white blood cells engulf and destroy bacteria and other foreign substances (Figure 2-6c).

● *Pinocytosis* In **pinocytosis** (pi'-nō-sī-TŌ-sis), or "cell drinking," the engulfed material consists of a liquid rather than a solid (Figure 2-6d). No cytoplasmic projections are formed. Instead, the liquid is attracted to the surface of the membrane. The membrane folds inwardly, surrounds the liquid, and detaches from the rest of the intact membrane. Few cells are capable of phagocytosis, but many cells may carry on pinocytosis. Examples include cells in the kidneys and urinary bladder.

The movement of materials into a cell by phagocytosis or pinocytosis is referred to as **endocytosis.** The reverse movement, that is, the export of materials from cells in vacuoles by either process, is referred to as **exocytosis.**

MODIFIED PLASMA MEMBRANES

Electron-microscope studies have revealed that plasma membranes of certain cells contain a number of modifications, that is, they contain different structures for specific purposes. For example, the membranes of some cells lining the small intestine have small, cylindrical projections called **microvilli** (Figure 2-7a, b). These fingerlike projections enormously increase the absorbing area of the cell surface. A single cell may have as many as 3,000 microvilli, and a 1 sq mm (0.04 sq inch) area of small intestine may contain as many as 200 million microvilli, which would increase the surface area for absorption by 20 times.

Another membrane modification is found in the rods and cones of the eye. They serve as photoreceptors, or light-receiving cells. The upper portion of each rod contains two-layered, disc-shaped membranes called **sacs** that contain the pigments involved in vision (Figure 2-7c, d).

Another example of a membrane modification is **stereocilia.** They are found only in cells lining a duct (ductus epididymis) of the male reproductive system. They appear by light microscopy as long, slender, branching processes at the free surfaces of the lining cells (Figure 2-7e, f). Electron micrographs show stereocilia to be microvilli.

A final example of a membrane modification is the **myelin sheath** that surrounds portions of certain nerve cells (Figure 2-7g, h). It is thought that the myelin sheath increases the velocity of impulse conduction, protects the portion of the nerve cell it surrounds, and is related to the nutrition of the nerve cell.

CYTOPLASM

The substance inside the cell's plasma membrane and external to the nucleus is called **cytoplasm** (Figure 2-8a, b). It is the matrix, or ground substance, in which various cellular components are found. Physically, cytoplasm may be described as a thick, semitransparent, elastic fluid containing suspended particles. Chemically, cytoplasm is 75 to 90 percent water plus solid components. Proteins, carbohydrates, lipids, and inorganic substances compose the bulk of the solid components. The inorganic substances and most carbohydrates are soluble in water and are present as a true solution. The majority of organic compounds, however, are found as colloids—particles that remain suspended in the surrounding ground substance. Since the particles of a colloid bear electrical charges that repel each other, they remain suspended and separated from each other.

Functionally, cytoplasm is the substance in which chemical reactions occur. The cytoplasm receives raw materials from the external environment and converts them into usable energy by decomposition reactions. Cytoplasm is also the site where new substances are synthesized for cellular use. It packages chemicals for transport to other parts of the cell or other cells of the body and facilitates the excretion of waste materials.

ORGANELLES

Despite the myriad chemical activities occurring simultaneously in the cell, there is little interference of one reaction with another. The cell has a system of compartmentalization provided by structures called **organelles.** These structures are specialized portions of the cell that assume specific roles in growth, maintenance, repair, and control.

NUCLEUS

The **nucleus** is generally a spherical or oval organelle (Figure 2-8a–c). In addition to being the largest structure in the cell, the nucleus contains hereditary factors of the cell, called genes, which control cellular structure and direct many cellular activities. Mature red blood cells do not have nuclei. These cells carry on only limited

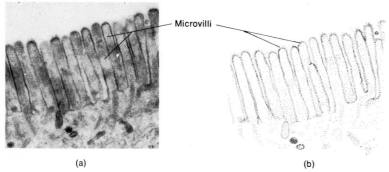

(a)

(b)

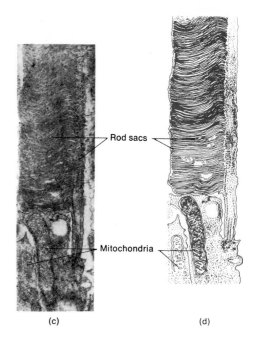

Rod sacs

Mitochondria

(c)   (d)

FIGURE 2-7 Modified plasma membranes. (a) Microvilli: electron micrograph of a portion of the small intestine at a magnification of 20,000×. (b) Microvilli: labeled diagram of the electron micrograph. (c) Rod sacs: electron micrograph of a portion of a rod of the eye at a magnification of 2,000×. (d) Rod sacs: labeled diagram of the electron micrograph. (e) Stereocilia: photomicrograph of stereocilia from the lining cells of the ductus epididymis at a magnification of 350×. (f) Stereocilia: labeled diagram of the photomicrograph. (g) Myelin sheath: electron micrograph of a myelin sheath in cross section at a magnification of 20,000×. (h) Myelin sheath: labeled diagram of the electron micrograph. (Electron micrographs courtesy of E. B. Sandburn, University of Montreal. Photomicrographs © 1983 by Michael H. Ross. Used by permission.)

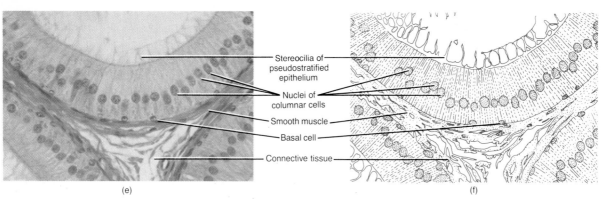

Stereocilia of pseudostratified epithelium

Nuclei of columnar cells

Smooth muscle

Basal cell

Connective tissue

(e)   (f)

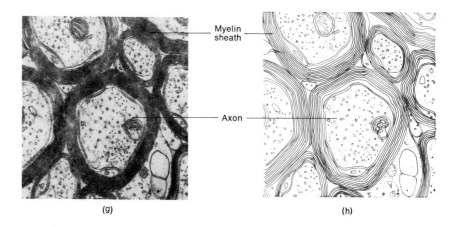

Myelin sheath

Axon

(g)   (h)

types of chemical activity and are not capable of growth or reproduction.

The nucleus is separated from the cytoplasm by a double membrane (two unit membranes) called the *nuclear membrane* or *envelope* (Figure 2-8c). Between the two layers of the nuclear membrane is a space called the *perinuclear cisterna.* This arrangement of each of the nuclear

membranes resembles the structure of the plasma membrane. Minute pores in the nuclear membrane allow the nucleus to communicate with a membranous network in the cytoplasm called the endoplasmic reticulum. Substances entering and exiting the nucleus are believed to pass through the tiny pores. Three prominent structures are visible internal to the nuclear membrane. The first

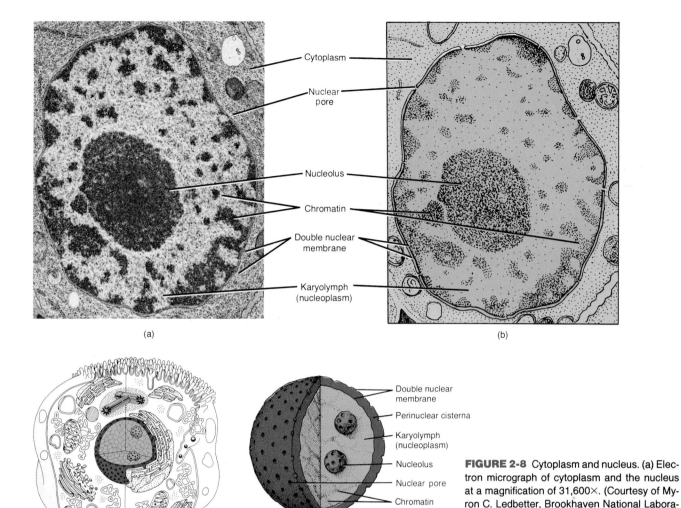

FIGURE 2-8 Cytoplasm and nucleus. (a) Electron micrograph of cytoplasm and the nucleus at a magnification of 31,600×. (Courtesy of Myron C. Ledbetter, Brookhaven National Laboratory.) (b) Diagram of the electron micrograph. (c) Diagram of a nucleus with two nucleoli.

of these is a gel-like fluid that fills the nucleus called *karyolymph* (*nucleoplasm*). One or more spherical bodies called the *nucleoli* are also present. These structures are composed primarily of RNA and protein and assume a function in protein synthesis. Finally, there is the *genetic material* consisting principally of DNA. When the cell is not reproducing, the genetic material appears as a threadlike mass called *chromatin.* Prior to cellular reproduction the chromatin shortens and coils into rod-shaped bodies called *chromosomes.*

ENDOPLASMIC RETICULUM AND RIBOSOMES

Within the cytoplasm, there is a system consisting of pairs of parallel membranes enclosing narrow cavities of varying shapes. This system is known as the **endoplasmic reticulum,** or **ER** (Figure 2-9). The ER, in other words, is a network of channels (*cisternae*) running through the entire cytoplasm. These channels are continuous with both the plasma membrane and the nuclear membrane. It is thought that the ER provides a surface area for chemical reactions, a pathway for transporting molecules

within the cell, and a storage area for synthesized molecules. And, together with the Golgi complex, the ER is thought to secrete certain chemicals.

Attached to the outer surfaces of some of the ER are exceedingly small, dense, spherical bodies called **ribosomes.** The ER in these areas is referred to as **granular,** or **rough, reticulum.** Ribosomes are thought to serve as the sites of protein synthesis in the cell. Portions of the ER that lack ribosomes are called **agranular,** or **smooth, reticulum.** One of the most important functions of the granular ER is the synthesis of protein for export. Agranular ER is associated with the synthesis of lipids and seems to have an important role in inactivating many toxic compounds.

GOLGI COMPLEX

Another structure found in the cytoplasm is the **Golgi** (GOL-jē) **complex.** This structure consists of four to eight flattened baglike channels, stacked upon each other with expanded areas at their ends. Like those of the ER, the stacked elements are called *cisternae,* and the expanded,

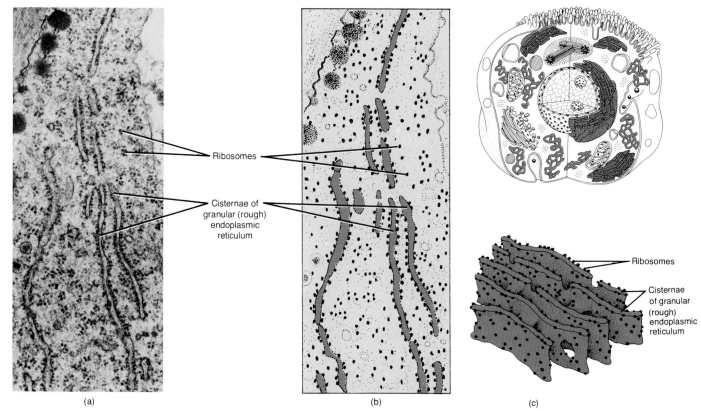

FIGURE 2-9 Endoplasmic reticulum and ribosomes. (a) Electron micrograph of the endoplasmic reticulum and ribosomes at a magnification of 76,000×. (Courtesy of Myron C. Ledbetter, Brookhaven National Laboratory.) (b) Diagram of the electron micrograph. (c) Diagram of the endoplasmic reticulum and ribosomes. See if you find the agranular (smooth) endoplasmic reticulum in Figure 2-10.

terminal areas are *vesicles* (Figure 2-10). Generally, the Golgi complex is located near the nucleus and is directly connected, in places, to the ER.

One function of the Golgi complex is the secretion and packaging of proteins. *Secretion* is the production and release from the cell of a fluid that usually contains a variety of substances. Proteins synthesized by the ribosomes associated with granular ER are transported into the ER cisternae. They migrate along the ER cisternae until they reach the Golgi complex. As proteins accumulate in the cisternae of the Golgi complex, the cisternae expand to form vesicles. After a certain size is reached, the vesicles pinch off from the cisternae. The protein and its associated vesicle is referred to as a *secretory granule.* The secretory granule then moves toward the surface of the cell where the protein is secreted. Cells of the digestive tract that secrete protein enzymes utilize this mechanism. The vacuole prevents "digestion" of the cytoplasm of the cell by the enzyme as it moves toward the cell surface.

Another function of the Golgi complex is associated with lipid secretion. It occurs in essentially the same way as protein secretion, except the lipids are synthesized by the agranular ER. The lipids pass through the ER into the Golgi complex. As in the mechanism just described,

the lipids migrate into the cisternae and vesicles and are discharged at the surface of the cell. In the course of moving through the cytoplasm, the vesicle may release lipids into the cytoplasm before being discharged from the cell. These appear in the cytoplasm as lipid droplets. Among the lipids secreted in this manner are the steroids.

The Golgi complex also functions in the synthesis of carbohydrates. Recent evidence indicates that carbohydrates synthesized by the Golgi complex are combined with proteins synthesized by ribosomes associated with granular ER to form carbohydrate-protein complexes. These complexes of carbohydrate and protein are called *glycoproteins.* As glycoproteins are assembled, they accumulate in the flattened channels of the Golgi complex. The channels expand and form vesicles. After a critical size is reached, the vesicles pinch off from the channel, migrate through the cytoplasm, and pass out of the cell through the plasma membrane. Outside the plasma membrane, the vesicles rupture and release their contents. Essentially, the Golgi complex synthesizes carbohydrates and combines them with proteins. It then packages the resulting glycoprotein and secretes it from the cell. The Golgi complex is well developed and highly active in secretory cells such as those found in the pancreas and the salivary glands.

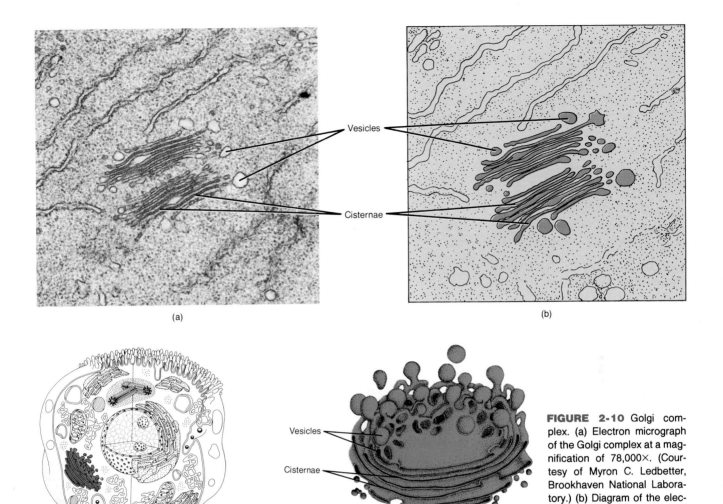

Vesicles

Cisternae

(a)

Vesicles

Cisternae

(b)

Vesicles

Cisternae

(c)

FIGURE 2-10 Golgi complex. (a) Electron micrograph of the Golgi complex at a magnification of 78,000×. (Courtesy of Myron C. Ledbetter, Brookhaven National Laboratory.) (b) Diagram of the electron micrograph. (c) Diagram of the Golgi complex.

MITOCHONDRIA

Small, spherical, rod-shaped, or filamentous structures called **mitochondria** (mī'-tō-KON-drē-a) appear throughout the cytoplasm. When sectioned and viewed under an electron microscope, each reveals an elaborate internal organization (Figure 2-11). A mitochondrion consists of two unit membranes, each unit of which is similar in structure to the plasma membrane. The outer mitochondrial membrane is smooth, but the inner membrane is arranged in a series of folds called *cristae*. The center of the mitochondrion is called the *matrix*.

Because of the nature and arrangement of the cristae, the inner membrane provides an enormous surface area for chemical reactions. Enzymes involved in energy-releasing reactions that form ATP are located on the cristae. Mitochondria are frequently called the "powerhouses of the cell" because they are the sites for the production of ATP. Active cells, such as muscle cells and liver cells, have a large number of mitochondria because of their high energy expenditure.

LYSOSOMES

When viewed under the electron microscope, **lysosomes** (*lysis* = dissolution; *soma* = body) appear as membrane-enclosed spheres. They are formed from Golgi complexes (Figure 2-12). Unlike mitochondria, lysosomes have only a single membrane and lack detailed structure. They contain powerful digestive enzymes capable of breaking down many kinds of molecules. These enzymes are also capable of digesting bacteria that enter the cell. White blood cells, which ingest bacteria by phagocytosis, contain large numbers of lysosomes. Scientists have wondered why these powerful enzymes do not also destroy their own cells. Perhaps the lysosome membrane in a healthy cell is impermeable to enzymes so they cannot move out into the cytoplasm. However, when a cell is injured, the lysosomes release their enzymes. The enzymes then promote reactions that break the cell down into its chemical constituents. The chemical remains are either reused by the body or excreted. Because of this function, lysosomes have been called "suicide packets."

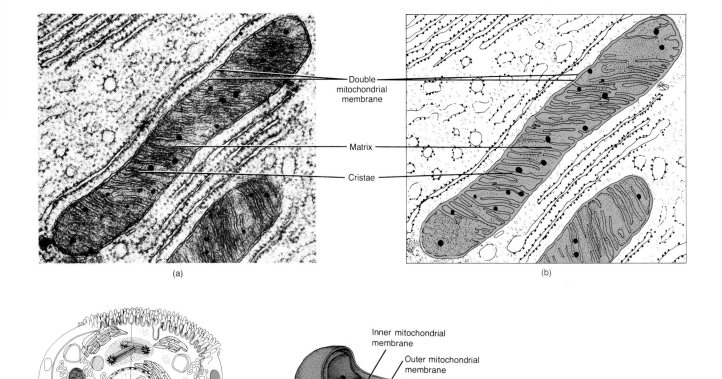

Double
mitochondrial
membrane

Matrix

Cristae

(a)

(b)

Inner mitochondrial
membrane

Outer mitochondrial
membrane

Cristae

Matrix

(c)

FIGURE 2-11 Mitochondria. (a) Electron micrograph of an entire mitochondrion and a portion of another. (Courtesy of Lester V. Bergman & Associates.) (b) Diagram of the electron micrograph. (c) Diagram of a mitochondrion.

Lysosomes are crucial in the removal of cell parts, whole cells, and even extracellular material, which may be the process underlying bone removal. In bone reshaping, especially during the growth process, special bone-destroying cells, called osteoclasts, secrete extracellular enzymes that dissolve bone. Bone tissue cultures given excess amounts of vitamin A remove bone apparently through an activation process involving lysosomes. Animals overfed on vitamin A interestingly develop spontaneous fractures, suggesting greatly increased lysosomal ac-

FIGURE 2-12 Lysosome. (a) Electron micrograph of a lysosome at a magnification of 55,000×. (Courtesy of F. Van Hoof, Université Catholique de Louvain.) (b) Diagram of the electron micrograph.

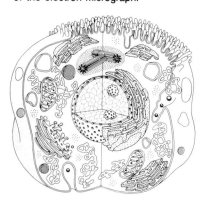

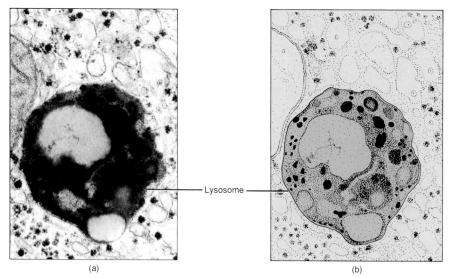

Lysosome

(a)

(b)

tivity. On the other hand, cortisone and hydrocortisone, steroid hormones produced by the adrenal gland, have a stabilizing effect on lysosomal membranes. The steroid hormones are well known for their anti-inflammatory properties, which suggests that they reduce destructive cellular activity by lysosomes.

In some metabolic diseases in which homeostasis is disrupted, the lack of certain lysosomal enzymes may lead to enormous accumulations of cellular products. Severe pathologic disturbances result when glycogen accumulates in lysosomes and is not metabolized for normal cell use.

MICROTUBULES

Small tubules, called **microtubules,** are found in most cells. They are made of protein and are remarkably uniform in size and shape. They range from 200 to 270 Å in diameter and do not branch.

Microtubules are thought to be internal "skeletons" that afford cell shape and stiffness to the regions they occupy. They are also believed to function as intracellular channels along which substances move from place to place. Microtubules are important in nerve cells; the internal structure of cilia, flagella, and centrioles; and the spindle fibers that appear in dividing cells (see Figures 2-13 and 2-14).

CENTROSOME AND CENTRIOLES

A dense area of cytoplasm, generally spherical and located near the nucleus, is called the **centrosome** or **centrosphere.** Within the centrosome is a pair of cylindrical structures: the **centrioles** (Figure 2-13). Each centriole is composed of a ring of nine evenly spaced bundles. Each bundle, in turn, consists of three microtubules. The two centrioles are situated so that the long axis of one is at right angles to the long axis of the other.

Centrioles assume a role in cell reproduction. Certain cells, such as most mature nerve cells, do not have a centrosome and so do not reproduce. This is why they cannot be replaced if destroyed.

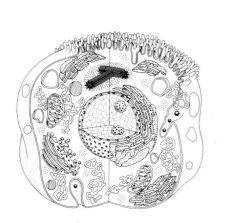
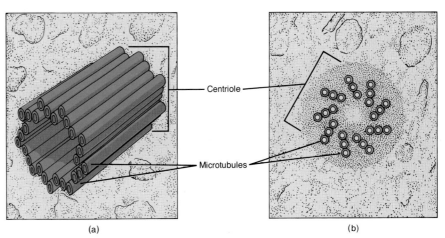

Centriole

Microtubules

(a)

(b)

FIGURE 2-13 Centrosome and centrioles. (a) Diagram of a centriole in longitudinal section. (b) Diagram of a centriole in cross section. (c) Electron micrograph of a centriole in cross section at a magnification of 67,000×. (Courtesy of Biophoto Associates.)

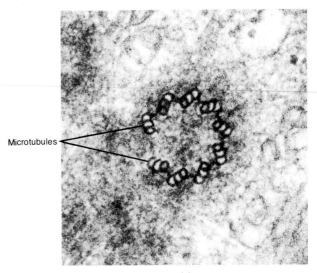

Microtubules

(c)

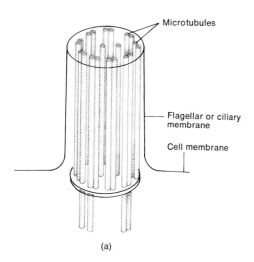

(a)

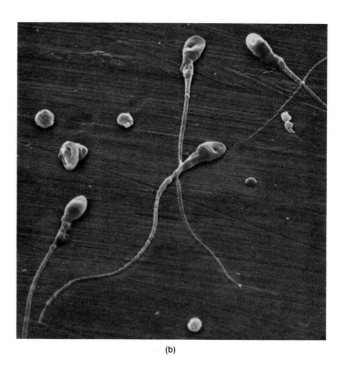

(b)

FIGURE 2-14 Flagella and cilia. (a) Structure of a flagellum or cilium. (b) Scanning electron micrograph of several sperm cells at a magnification of 2,000×. The tails of the sperm cells are flagella. (Courtesy of Fisher Scientific Company and S.T.E.M. Laboratories, Inc., Copyright 1975.) (c) Photomicrograph of the lining of the trachea at a magnification of 600×. The hairlike border at the top of the photo represents the cilia. (© 1983 by Michael H. Ross. Used by permission.)

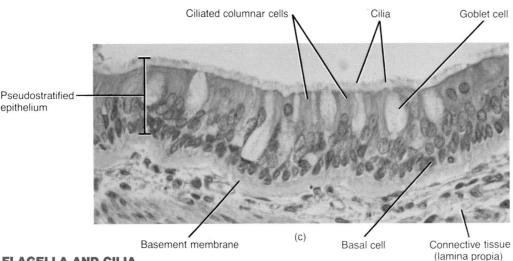

(c)

FLAGELLA AND CILIA

Some body cells possess projections for moving the entire cell or for moving substances along the surface of the cell. These projections contain cytoplasm and are bounded by the plasma membrane. If the projections are few and long in proportion to the size of the cell, they are called **flagella.** The only example of a flagellum in the human body is the tail of a sperm cell, used for locomotion. If the projections are numerous and short, resembling many hairs, they are called **cilia.** In humans, ciliated cells of the respiratory tract move lubricating fluids over the surface of the tissue and trap foreign particles. Electron microscopy has revealed no fundamental structural difference between cilia and flagella (Figure 2-14). Both consist of nine pairs of microtubules that form a ring around two microtubules in the center.

CELL INCLUSIONS

The **cell inclusions** are a large and diverse group of chemical substances, some of which have recognizable shapes. These products are principally organic and may appear or disappear at various times in the life of the cell. *Melanin* is a pigment stored in the cells of the skin, hair, and eyes. It protects the body by screening out harmful ultraviolet rays from the sun. *Glycogen* is a polysaccharide that is stored in liver and skeletal muscle cells. When the body requires quick energy, liver cells can break down the glycogen into glucose and release it. *Lipids,* which are stored in fat cells, may be decomposed when the body runs out of carbohydrates for producing energy. A final example of an inclusion is *mucus,* which is produced by cells that line organs. Mucus provides lubrication.

EXHIBIT 2-1 CELL PARTS AND THEIR FUNCTIONS

PART	FUNCTIONS
Plasma membrane	Protects and allows substances to enter or exit the cell through diffusion, facilitated diffusion, osmosis, filtration, dialysis, active transport, phagocytosis, and pinocytosis.
Cytoplasm	Serves as the ground substance in which chemical reactions occur.
Organelles **Nucleus**	Contains genes and controls cellular activities.
Endoplasmic reticulum	Provides a surface area for chemical reactions; provides a pathway for transporting chemicals; serves as a storage area.
Ribosomes	Act as sites of protein synthesis.
Golgi complex	Synthesizes carbohydrates, combines carbohydrates with proteins, packages materials for secretion, and secretes lipids and glycoproteins.
Mitochondria	Sites for production of ATP.
Lysosomes	Digest substances and foreign microbes.
Microtubules	Serve as intracellular "skeleton" and passageways; are components of cilia, flagella, centrioles, and spindle fibers.
Centrioles	Help organize spindle fibers during cell division.
Flagella and cilia	Allow movement of cell or movement of particles along surface of cell.
Inclusions	Principally organic molecules involved in overall body functions. Include materials such as melanin, glycogen, lipids, and mucus.

The major parts of the cell and their functions are summarized in Exhibit 2-1.

EXTRACELLULAR MATERIALS

The substances that lie outside cells are called **extracellular materials.** They include the body fluids, which provide a medium for dissolving, mixing, and transporting substances. Among the body fluids are interstitial fluid, the fluid that fills the microscopic spaces (interstitial spaces), and plasma, the fluid portion of blood. Extracellular materials also include secreted inclusions, such as mucus, and special substances that form the matrix in which some cells are embedded.

The matrix materials are produced by certain cells and deposited outside their plasma membranes. The matrix supports the cells, binds them together, and gives strength and elasticity to the tissue. Some matrix materials are *amorphous* (a-MOR-fus; *a* = without, *morpho* = shape); they have no specific shape. These include hyaluronic acid and chondroitin sulfate. **Hyaluronic** (hī-a-loo-RON-ik) **acid** is a viscous, fluidlike substance that binds cells together, lubricates joints, and maintains the shape of the eyeballs. **Chondroitin** (kon-DROY-tin) **sulfate** is a jellylike substance that provides support and adhesiveness in cartilage, bone, heart valves, the cornea of the eye, and the umbilical cord.

Other matrix materials are *fibrous,* or threadlike. Fibrous materials provide strength and support for tissues. Among these are **collagenous** (*kolla* = glue) **fibers** consisting of the protein *collagen.* These fibers are found in all types of connective tissue, especially in bones, cartilage, tendons, and ligaments. **Reticular** (*rete* = net) **fibers** represent a fibrous matrix consisting of the protein collagen and a coating of glycoprotein. Reticular fibers form a network around fat cells, nerve fibers, muscle cells, and blood vessels. They also form the framework or stroma for many soft organs of the body such as the spleen. **Elastic fibers,** consisting of the protein *elastin,* give elasticity to skin and to tissues forming the walls of blood vessels.

CELL DIVISION

Most of the cell activities mentioned thus far maintain the life of the cell on a day-to-day basis. However, cells become damaged, diseased, or worn out and then die. Thus new cells must be produced for growth.

Cell division is the process by which cells reproduce themselves. Cell division, or, more appropriately, nuclear division, may be of two kinds. In the first kind a single parent cell duplicates itself. This process consists of mitosis (nuclear division) and cytokinesis (cytoplasmic division). The process ensures that each new daughter cell has the same *number* and *kind* of chromosomes as the original parent cell. After the process is complete, the two daughter cells have the same hereditary material and genetic potential as the parent cell. This kind of cell division results in an increase in the number of body cells. Mitosis and cytokinesis are the means by which dead or injured cells are replaced and new cells are added for body growth.

The second kind of division is a mechanism by which sperm and egg cells are produced. This process, **meiosis,** is the mechanism that enables the reproduction of an entirely new organism. The process of meiosis is discussed in detail in Chapter 18.

MITOSIS

In a 24-hour period, the average adult loses trillions of cells from different parts of the body. Obviously, these cells must be replaced. Cells that have a short life span—

the cells of the outer layer of skin, the cornea of the eye, the digestive tract—are continually being replaced. The succession of events that takes place during mitosis and cytokinesis is plainly visible under a microscope after the cells have been stained in the laboratory.

When a cell reproduces, it must replicate its chromosomes so its hereditary traits may be passed on to succeeding generations of cells. A **chromosome** is a highly coiled DNA molecule that is partly covered by protein. The protein causes changes in the length and thickness of

the chromosome. Hereditary information is contained in the DNA portion of the chromosome in units called **genes.** Each human chromosome consists of about 20,000 genes.

Before taking a look at the relationship between chromosomes and cell division, it will first be necessary to briefly examine the structure of DNA, the basic component of chromosomes.

A molecule of DNA is a chain composed of repeating units called *nucleotides.* Each nucleotide of DNA consists of three basic parts (Figure 2-15a). (1) It contains one of four possible *nitrogen bases,* which are ring-shaped structures containing atoms of C, H, O, and N. The nitrogen bases found in DNA are named adenine, thymine, cytosine, and guanine. (2) It contains a sugar called *deoxyribose.* (3) It has a phosphoric acid called the *phosphate group.* The nucleotides are named according to the nitrogen base that is present. Thus, a nucleotide containing thymine is called a *thymine nucleotide,* one containing adenine is called an *adenine nucleotide,* and so on.

The chemical composition of the DNA molecule was known before 1900, but it was not until 1953 that a model of the organization of the chemicals was constructed. This model was proposed by J. D. Watson and F. H. C. Crick on the basis of data from many investigations. Figure 2-15b shows the following structural characteristics of the DNA molecule. (1) The molecule consists of two strands with crossbars. The strands twist about each other in the form of a *double helix* so that the shape resembles a twisted ladder. (2) The uprights of the DNA "ladder" consist of alternating phosphate groups and the deoxyribose portions of the nucleotides. (3) The rungs of the ladder contain paired nitrogen bases. As shown, adenine always pairs off with thymine, and cytosine always pairs off with guanine.

The process called **mitosis** is the distribution of the two sets of chromosomes into two separate and equal nuclei following the replication of the chromosomes of the parent nucleus. For convenience, biologists divide the process into four stages: prophase, metaphase, anaphase, and telophase. These are arbitrary classifications. Mitosis is actually a continuous process, one stage merging imperceptibly into the next. Interphase is the stage that occurs between consecutive cell divisions.

Interphase

When a cell is carrying on every life process except division, it is said to be in **interphase** or the **metabolic phase** (Figure 2-16a). One of the principal events of interphase is the replication of DNA. When DNA replicates, its helical structure partially uncoils (Figure 2-17). Those portions of DNA that remain coiled stain darker than the uncoiled portions. This unequal distribution of stain causes the DNA to appear as a granular mass called **chromatin** (see Figure 2-16a). During uncoiling, DNA

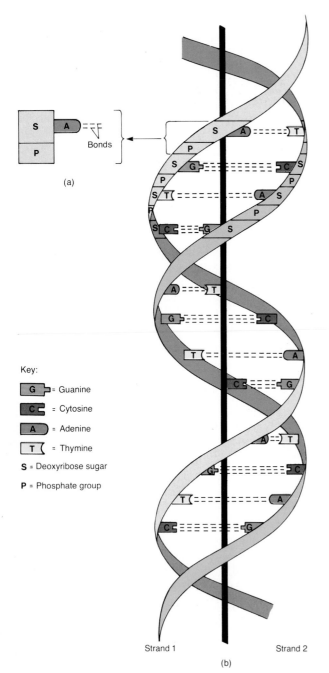

Key:

G = Guanine

C = Cytosine

A = Adenine

T = Thymine

S = Deoxyribose sugar

P = Phosphate group

Strand 1   Strand 2

(b)

FIGURE 2-15 DNA molecule. (a) Adenine nucleotide. (b) Portion of an assembled DNA molecule.

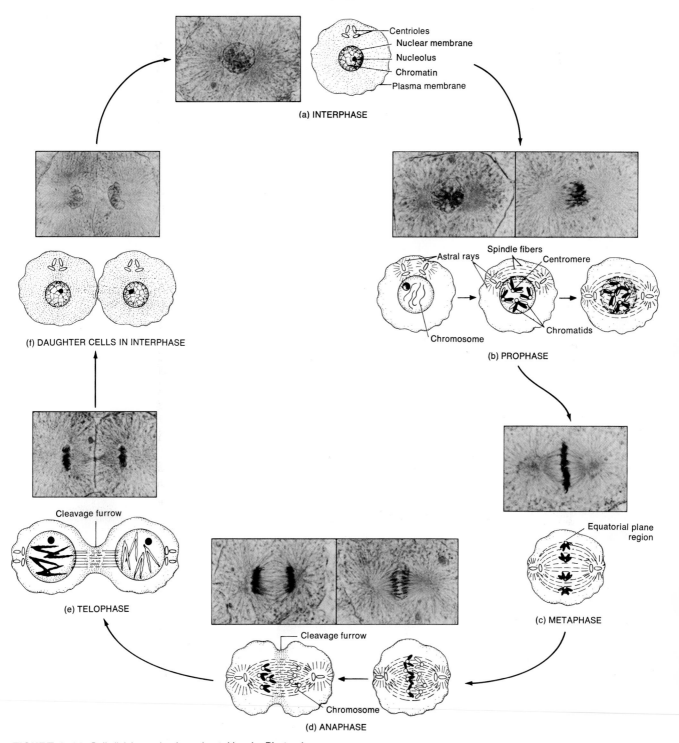

(a) INTERPHASE

Centrioles
Nuclear membrane
Nucleolus
Chromatin
Plasma membrane

Astral rays   Spindle fibers   Centromere

Chromosome   Chromatids

(b) PROPHASE

Equatorial plane region

(c) METAPHASE

Cleavage furrow

Chromosome

(d) ANAPHASE

Cleavage furrow

(e) TELOPHASE

(f) DAUGHTER CELLS IN INTERPHASE

FIGURE 2-16 Cell division: mitosis and cytokinesis. Photomicrographs and diagrammatic representations of the various stages of cell division in whitefish eggs. Read the sequence starting at (a), and move clockwise until you complete the cycle. (Courtesy of Carolina Biological Supply Company.)

separates at the points where the nitrogen bases are connected. Each exposed nitrogen base then picks up a complementary nitrogen base (with associated sugar and phosphate group) from the cytoplasm of the cell. This

uncoiling and complementary base pairing continues until each of the two original DNA strands is matched and joined with two newly formed DNA strands. The original DNA molecule has become two DNA molecules.

During interphase the cell is also synthesizing most of its RNA and proteins. It is producing chemicals so that all cellular components can be doubled during division. A microscopic view of a cell during interphase shows a clearly defined membrane, nucleoli, karyolymph, and chromatin. As interphase progresses, a pair of centrioles appears. Once a cell completes its replication of DNA and its synthesis of RNA and proteins during interphase, mitosis begins.

Prophase

During **prophase** (Figure 2-16b) the centrioles move apart and project a series of radiating fibers (microtubules) called *astral rays*. The centrioles move to opposite poles of the cell and become connected by another system of fibers (microtubules) called *spindle fibers*. Together, the centrioles, astral rays, and spindle fibers are referred to as the *mitotic apparatus*. Simultaneously, the chromatin has been shortening and coiling into chromosomes. The nucleoli have become less distinct, and the nuclear membrane has disappeared. Each prophase "chromosome" is actually composed of a pair of structures called *chromatids*. Each chromatid is a complete chromosome consisting of a double-stranded DNA molecule. Each chromatid is attached to its chromatid pair by a small spherical body called the *centromere*. During prophase, the chromatid pairs move toward the equatorial plane region or equator of the cell.

FIGURE 2-17 Replication of DNA. The two strands of the double helix separate by breaking the bonds between nucleotides. New nucleotides attach at the proper sites, and a new strand of DNA is paired off with each of the original strands. After replication, the two DNA molecules, each consisting of a new and an old strand, return to their helical structure.

Metaphase

During **metaphase** (Figure 2-16c), the second stage of mitosis, the chromatid pairs line up on the equatorial plane of the spindle fibers. The centromere of each chromatid pair attaches itself to a spindle fiber.

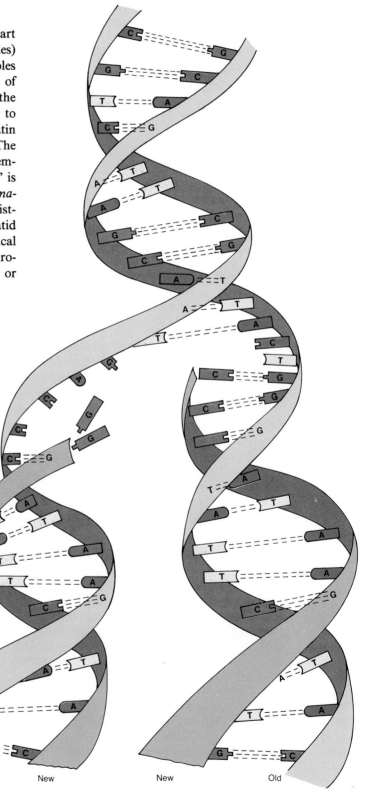

Old   New   New   Old

Anaphase

Anaphase (Figure 2-16d), the third stage of mitosis, is characterized by the division of the centromeres and the movement of complete identical sets of chromatids, now called chromosomes, to opposite poles of the cell. During this movement, the centromeres attached to the spindle fibers seem to drag the trailing parts of the chromosomes toward opposite poles.

Telophase

Telophase (Figure 2-16e), the final stage of mitosis, consists of a series of events nearly the reverse of prophase. By this time, two identical sets of chromosomes have reached opposite poles. New nuclear membranes begin to enclose them. The chromosomes start to assume their chromatin form. Nucleoli reappear, and the spindle fibers and astral rays disappear. The centrioles also replicate so that each cell has two centriole pairs. The formation of two nuclei identical to those of cells in interphase terminates telophase. A mitotic cycle has thus been completed (Figure 2-16f).

Time Required

The time required for a complete mitotic cycle varies with the kind of cell, its location, and the influence of factors such as temperature. Furthermore, the different stages of mitosis are not equal in duration. Prophase is usually the longest stage, lasting from one to several hours. Metaphase is considerably shorter, ranging from 5 to 15 minutes. Anaphase is the shortest stage, lasting from 2 to 10 minutes. Telophase lasts from 10 to 30 minutes. These lengths of time are only approximate, however.

CYTOKINESIS

Division of the cytoplasm, called **cytokinesis** (sī'-tō-ki-NĒ-sis), often begins during late anaphase and terminates at the same time as telophase. Cytokinesis begins with the formation of a *cleavage furrow* that extends around the cell's equator. The furrow progresses inward, resembling a constricting ring, and cuts completely through the cell to form two separate portions of cytoplasm (Figure 2-16d–f).

APPLICATIONS TO HEALTH

CELLS AND AGING

Aging is a progressive failure of the body's homeostatic adaptive responses. Although diseases such as cancer and atherosclerosis are frequently associated with it, it in itself is not a disease. However, it is true that disease and age seem to accelerate one another.

The obvious characteristics of aging are well known: graying and loss of hair, loss of teeth, wrinkling of skin, decreased muscle mass, and increased fat deposits. The physiological signs of aging are gradual deterioration in function and capacity to respond to environmental stress. Thus basic kidney and digestive metabolic rates decrease, as does the ability to respond effectively to changes in temperature, diet, and oxygen supply in order to maintain a constant internal environment. These manifestations of aging are related to a net decrease in the number of cells in the body (100,000 brain cells are lost each day) and to the disordered functioning of the cells that remain.

The extracellular components of tissues also change with age. Collagen fibers, responsible for the strength in tendons, increase in number and change in quality with aging. These changes in arterial walls are as much responsible for the loss of elasticity as those in atherosclerosis. Elastin, another extracellular substance, is responsible for the elasticity of blood vessels and skin. It thickens, fragments, and acquires a greater affinity for calcium with age—changes that may be associated with the development of arteriosclerosis.

Several kinds of cells in the body—heart cells, skeletal muscle cells, neurons—are incapable of replacement. Recent experiments have proved that certain other cell types are limited when it comes to cell division. Cells grown outside the body divided only a certain number of times and then stopped. The number of divisions correlated with the donor's age. The number of divisions also correlated with the normal life span of the different species from which the cells were obtained—strong evidence for the hypothesis that cessation of mitosis is a normal, genetically programed event.

Just as the factors that limit the life of an individual cell are unknown, so are those that restrict the growth or life of a tissue or organ. At menopause, the ovary ceases to function. Its cells die long before the rest of the female body. Perhaps similar mechanisms determine longevity.

Some investigators have studied aging from the standpoint of immunology. The ability to develop antibodies is said to diminish with age. Senescence, according to researchers, results in the older person's immunological system having a "shotgun," rather than a specific, response to foreign protein. This shotgun response may include an autoimmune reaction that attacks and gradually destroys the individual's normal tissue and organs.

The following generalizations on aging can be made.

1. Aging is a general process that produces observable changes in structure and function.

2. Aging produces increased vulnerability to environmental stress and disease.

3. Evidence suggests that the life span is about 110 years but maximum life expectancy is no greater than 85 years.

EXHIBIT 2-2 EFFECTS OF AGING ON VARIOUS BODY SYSTEMS

SYSTEM	AGING EFFECTS
Integumentary	Changes include wrinkling; loss of subcutaneous fat; atrophy (wasting away) of sweat glands; reduction of blood flow to skin; decrease in number of functioning pigment-producing cells, resulting in gray hair and atypical skin color; increase in size of some pigment cells, producing pigmented blotching; atrophy of sebaceous (oil) glands, producing dry and broken skin that is susceptible to infection. Aged skin also becomes susceptible to pathological conditions such as senile pruritus (itching), decubitus ulcers (bedsores), and herpes zoster (shingles).
Skeletomuscular	Changes and conditions include osteoarthritis, rheumatoid arthritis, gout, osteoporosis (decrease in bone mass), Paget's disease, loss of muscle mass, and diminishing of muscle reflexes.
Nervous	The number of nerve cells (neurons) decreases. Associated with this decline is a decreased capacity for sending nerve impulses to and from the brain. Conduction velocity decreases, voluntary motor movements slow down, and the reflex time for skeletal muscles is decreased. Degenerative changes and disease states involving the sense organs can alter vision, hearing, taste, smell, and touch. The disorders that represent the most common visual problems and which can be responsible for serious loss of vision are presbyopia (inability to focus on nearby objects), cataracts (cloudiness of the lens), and glaucoma (excessive fluid pressure in the eyeball). Impaired hearing associated with aging, known as presbycusis, is usually the result of changes in important structures of the inner ear. Parkinson's disease is the most common movement disorder involving the central nervous system.
Endocrine	The endocrine system exhibits a variety of changes, and many researchers look to this system with the hope of finding the key to the aging process. Disorders of the endocrine system are not frequent, and when they do occur, most often they are related to pathologic changes rather than age. Diabetes and thyroid disorders are the most important endocrine problems that have a significant effect on health and function.
Cardiovascular	Major physical changes include loss of elasticity of the aorta, reduction in cardiac muscle cell size, progressive loss of cardiac muscular strength, and a reduced output of blood by the heart. There is an increase in blood pressure. Coronary artery disease increases and represents the major cause of heart disease and death in older Americans. Congestive heart failure occurs and is viewed as a set of symptoms associated with the impaired pumping performance of the heart. Changes in blood vessels such as hardening of the arteries and cholesterol deposits in arteries that serve brain tissue can reduce its nourishment and result in the malfunction or death of brain cells, which is called cerebrovascular disease.

4. The mechanism behind aging is not known. Improvements in life expectancy may be due to an overall reduction in the number of life-threatening situations or to modification of the aging process.

5. Eventually the aging process may be modified and life span and life expectancy lengthened.

A detailed description of the aging process in humans, as it occurs in the various systems of the body, appears in Exhibit 2-2.

CELLS AND CANCER

Cancer is not a single disease but many. The human body contains more than a hundred different types of cells, each of which can malfunction in its own distinctive way to cause cancer. When cells in some area of the body duplicate unusually quickly, the excess of tissue that develops is called a growth, or **tumor.** Tumors may be cancerous and sometimes fatal or quite harmless. A cancerous growth is called a **malignant tumor,** or **malignancy.** A noncancerous growth is called a **benign growth.**

Cells of malignant growths duplicate continuously and very often quickly without control. The majority of cancer patients are not killed by the **primary tumor** that develops.

Rather, they are killed by **metastasis** (me-TAS-ta-sis), the spread of the disease to other parts of the body, since metastatic groups of cells are harder to detect and eliminate than primary tumors. One of the unique properties of a malignant tumor is its ability to metastasize.

In the process of metastasis, there is an initial invasion of the malignant cells into surrounding tissues. As the cancer grows, it expands and begins to compete with normal tissues for space and nutrients. Eventually, the normal tissue atrophies and dies. The invasiveness of the malignant cells may be related to mechanical pressure of the growing tumor, motility of the malignant cells, and enzymes produced by the malignant cells. Also, malignant cells lack what is called *contact inhibition.* When nonmalignant cells of the body divide and migrate (for example, skin cells that multiply to heal a superficial cut), their further migration is inhibited when they make contact on all sides with other skin cells. Unfortunately, malignant cells do not conform to the rules of contact inhibition; they have the ability to invade healthy body tissues with very few restrictions.

Following invasion, some of the malignant cells detach from the tumor and may enter a body cavity (abdominal or thoracic) or enter the blood or lymph. This latter condition can lead to widespread metastasis. In the next step

EXHIBIT 2-2 *(Continued)*

SYSTEM	AGING EFFECTS
Respiratory	The cardiovascular and respiratory systems operate as a unit, and damage or disease in one of these organ systems is often secondarily reflected in the other. In pulmonary heart disease the right side of the heart enlarges in response to certain lung diseases. This condition is known as right ventricular hypertrophy or cor pulmonale. The airways and tissues of the respiratory tract, including the air sacs, become less elastic and more rigid. The threat of serious respiratory infections such as pneumonia and tuberculosis increases, as does the threat of the obstructive conditions such as chronic bronchitis, emphysema, and lung cancer.
Digestive	Overall general changes include atrophy of the secretion mechanisms, decreasing motility (muscular movement) of the digestive organs, loss of strength and tone of the muscular tissue and its supporting structures, changes in neurosensory feedback on enzyme and hormone release, and the diminished response to pain and internal sensations. Specific changes include: reduced sensitivity to mouth irritations and sores, loss of taste, pyorrhea, difficulty in swallowing, hiatus hernia, cancer of the esophagus, gastritis, peptic ulcer, and gastric cancer. Changes in the small intestine include duodenal ulcers, appendicitis, malabsorption, and maldigestion. Other pathologies that increase in incidence are gallbladder problems, jaundice, cirrhosis, and acute pancreatitis. Large-intestinal changes such as constipation, cancer of the colon or rectum, hemorrhoids, and diverticular disease of the colon also occur.
Urinary	Urinary incontinence (lack of voluntary control over urination) and urinary tract infections are major problems. Other pathologies include polyuria (excessive urine production), nocturia (excessive urination at night), increased frequency of urination, dysuria (painful urination), retention of urine (failure to produce urine), and hematuria (blood in the urine). The prostate gland is often implicated in various disorders of the urinary tract and cancer of the prostate is the most frequent malignancy of old men. Changes and diseases in the kidney include acute and chronic kidney inflammations and renal calculi (kidney stones).
Reproductive	Although there are major age-specific physical changes in structure and function, few age-specific disorders are associated with the reproductive system. In the male the decreasing production of the hormone testosterone produces less muscle strength, fewer viable sperm, and decreased sexual desire. However, abundant spermatozoa may be found even in old age. Most of the pathologies produce uneventful recoveries, except for prostate problems that could become serious and fatal. The female reproductive system becomes less efficient, possibly as a result of less frequent ovulation and the declining ability of the uterine tubes and uterus to support the young embryo. There is a decrease in the production of the hormones progesterone and estrogen. Menopause is only one of a series of phases that leads to reduced fertility, irregular or absent menstruation, and a variety of physical changes. Uterine cancer peaks at about 65 years of age, but cervical cancer is more common in younger women, and breast cancer is the leading cause of death among women between the ages of 40 and 60. Prolapse (falling down or sinking) of the uterus is possibly the commonest complaint among female geriatric patients.

in metastasis, those malignant cells that survive in the blood or lymph invade adjacent body tissues and establish **secondary tumors.** It is believed that some of the invading cells involved in metastasis may have properties different from those of the primary tumor that enhance metastasis.

In the final stage of metastasis, the secondary tumors become vascularized, that is, they take on new networks of blood vessels that provide nutrients for their further growth. In all stages of metastasis, the malignant cells resist the antitumor defenses of the body. Usually death results from the atrophy of a vital organ. The pain associated with cancer develops when the growth puts pressure on nerves or blocks a passageway so that secretions build up pressure.

At present, cancers are classified by their microscopic appearance and the body site from which they arise. At least 100 different cancers have been identified in this way. If finer details of appearance are taken into consideration, the number can be increased to 200 or more. The name of the cancer is derived from the type of tissue in which it develops. **Sarcoma** is a general term for any cancer arising from connective tissue. **Osteogenic sarco-**

mas (*osteo* = bone; *genic* = origin), the most frequent type of childhood cancer, destroy normal bone tissue and eventually spread to other areas of the body. **Myelomas** (*myelos* = marrow) are malignant tumors, occurring in middle-aged and older people, that interfere with the blood-cell-producing function of bone marrow and cause anemia. **Chondrosarcoma** is a cancerous growth of the cartilage (*chondro* = cartilage).

Cancer has been observed in all species of vertebrates. In fact, cellular abnormalities that resemble cancer—such as crown gall of tomatoes—have been observed in plants as well. Cell masses that resemble cancers of higher animals have also been produced and studied in insects.

What triggers a perfectly normal cell to lose control and become abnormal? Scientists are uncertain. First there are environmental agents: substances in the air we breathe, the water we drink, the food we eat. The World Health Organization estimates that these agents—called **carcinogens**—may be associated with 60 to 90 percent of all human cancer. Examples of carcinogens are the hydrocarbons found in cigarette tar. Ninety percent of all lung cancer patients are smokers. Another environmental factor is radiation. Ultraviolet light from the sun, for example, may cause genetic mutations in exposed skin cells and lead to cancer, especially among light-skinned people.

Viruses are a second cause of cancer, at least in animals. These agents are tiny packages of nucleic acids that are capable of infecting cells and converting them to virus-producers. Virologists have linked tumor viruses with cancer in many species of birds and mammals, including primates. Since these experiments have not been performed on humans, there is no absolute proof that viruses cause human cancer. Nevertheless, with over 100 separate viruses identified as carcinogens in many species and tissues of animals, it is also probable that at least some cancers in humans are due to virus. In fact, the *Epstein-Barr (EB) virus,* the causative agent of infectious mononucleosis, has recently been linked as the causative agent of two human cancers—*Burkett's lymphoma* (a tumor of the jaw in African children and young adults) and *nasopharyngeal carcinoma* (common in Chinese males).

Currently, scientists are trying to establish a relationship between stress and cancer. Some believe that stress may play a role not only in the development but also in the metastasis of cancer.

Benign tumors, unlike malignant tumors, are composed of cells that do not metastasize, that is, the growth does not spread to other organs. Removing all or part of a tumor to determine whether it is benign or malignant is called a **biopsy.** A benign tumor may be removed if it impairs a normal body function or causes disfiguration.

A reminder: As mentioned in the preface, each chapter in this text that discusses a major system of the body is followed by a glossary of **key medical terms.** Both normal and pathological conditions of the system are included in these glossaries. You should familiarize yourself with the terms since they will play an essential role in your medical vocabulary.

Some of these disorders, as well as disorders discussed in the text, are referred to as local or systemic. A **local** disease, condition, or injury is one that involves one part or a limited area of the body. A **systemic** disease or systemic effects of a condition or injury involve several parts or the entire body.

KEY MEDICAL TERMS ASSOCIATED WITH CELLS

Atrophy (*a* = without; *tropho* = nourish) A decrease in the size of cells with subsequent decrease in the size of the affected tissue or organ; wasting away.

Biopsy (*bio* = life; *opsis* = vision) The removal and microscopic examination of tissue from the living body for diagnosis.

Carcinogen (*carc* = cancer) A chemical or other environmental agent that produces cancer.

Carcinoma (*oma* = tumor) A malignant tumor or cancer made up of epithelial cells.

Geriatrics (*geri* = aged) The branch of medicine devoted to the medical problems and care of elderly persons.

Hyperplasia (*hyper* = over; *plas* = grow) Increase in the number of cells due to an increase in the frequency of cell division.

Hypertrophy Increase in the size of cells without cell division.

Metaplasia (*meta* = change) The transformation of one cell into another.

Metastasis (*stasis* = standing still) The transfer of disease from one part of the body to another that is not directly connected with it.

Necrosis (*necros* = death; *osis* = condition) Death of a group of cells.

Neoplasm (*neo* = new) Any abnormal formation or growth, usually a malignant tumor.

Progeny (*progignere* = to bring forth) Offspring or descendants.

Senescence The process of growing old.

STUDY OUTLINE

Generalized Animal Cell

1. A cell is the basic, living, structural and functional unit of the body.
2. Cytology is the science concerned with the study of cells.
3. A generalized cell is a composite that represents various cells of the body.
4. The principal parts of a cell are the plasma membrane, cytoplasm, organelles, and inclusions. Extracellular

materials are manufactured by the cell and deposited outside the plasma membrane.

Plasma (Cell) Membrane

1. The plasma membrane, or cell membrane, surrounds the cell and separates it from other cells and the external environment.
2. The plasma membrane is composed of proteins and a bilayer of lipids. It is believed that the membrane contains pores.
3. The membrane's semipermeable nature restricts the passage of certain substances. Substances can pass through the membrane depending on their molecular size, lipid solubility, electrical charges, and the presence of carriers.

Movement of Materials Across Plasma Membranes

1. Passive processes involve the kinetic energy of individual molecules.
2. Diffusion is the net movement of molecules or ions from an area of higher concentration to an area of lower concentration until an equilibrium is reached.
3. In facilitated diffusion, certain molecules, such as glucose, combine with a carrier to become soluble in the lipid portion of the membrane, so that they can penetrate the membrane.
4. Osmosis is the movement of water through a semipermeable membrane from an area of higher water concentration to an area of lower water concentration.
5. Osmotic pressure is the force under which a solvent moves from a solution of lower solute concentration to a solution of higher solute concentration when the solutions are separated by a semipermeable membrane.
6. Filtration is the movement of water and dissolved substances across a semipermeable membrane by pressure.
7. Dialysis is the separation of small molecules from large molecules by diffusion through a semipermeable membrane.
8. Active processes involve the use of ATP by the cell.
9. Active transport is the movement of ions across a cell membrane from lower to higher concentration. This process relies on the participation of carriers.
10. Phagocytosis is the ingestion of solid particles by pseudopodia. It is an important process used by white blood cells to destroy bacteria that enter the body.
11. Pinocytosis is the ingestion of a liquid by the plasma membrane. In this process, the liquid becomes surrounded by a vacuole.

Modified Plasma Membranes

1. The membranes of certain cells are structured for specific functions.
2. Microvilli are microscopic, fingerlike projections of the plasma membrane that increase the surface area for absorption.
3. Rod and cone cells of the eye contain sacs of light-sensitive pigments.
4. Stereocilia are long, slender, branching cells which line the epididymis.
5. The myelin sheath of nerve cells protects, aids impulse conduction, and provides nutrition.

Cytoplasm

1. The cytoplasm is the substance inside the cell's membrane and external to the nucleus. It contains organelles and inclusions.
2. It is composed mostly of water plus proteins, carbohydrates, lipids, and inorganic substances. The chemicals in cytoplasm are either in solution or in a colloid form.
3. Functionally, cytoplasm is the medium in which chemical reactions occur.

Organelles

Organelles are specialized portions of the cell that carry on specific activities.

Nucleus

1. Usually the largest organelle, the nucleus controls cellular activities.
2. Cells without nuclei, such as mature red blood cells, do not grow or reproduce.
3. The parts of the nucleus include the nuclear membrane, nucleoplasm, nucleoli, and genetic material (DNA).

Endoplasmic Reticulum and Ribosomes

1. The ER is a network of parallel membranes continuous with the plasma membrane and nuclear membrane.
2. It functions in chemical reactions, transportation, and storage.
3. Granular or rough ER has ribosomes attached to it. Ribosomes are small spherical bodies that serve as sites of protein synthesis. Agranular or smooth ER does not contain ribosomes.

Golgi Complex

1. The Golgi complex consists of four to eight flattened channels vertically stacked on one another.
2. In conjunction with the ER, the Golgi complex synthesizes glycoproteins and secretes lipids.
3. It is particularly prominent in secretory cells such as those in the pancreas or salivary glands.

Mitochondria

1. A mitochondrion consists of a smooth outer membrane and a folded inner membrane. The inner folds are

called cristae. Two unit membranes surround the matrix.
2. The mitochondria are called "powerhouses of the cell" because ATP is produced in them.

Lysosomes
1. Lysosomes are membrane-enclosed spherical structures containing digestive enzymes.
2. They are found in large numbers in white blood cells, which carry on phagocytosis.
3. If the cell is injured, lysosomes release enzymes and digest the cell. Thus they are called "suicide packets."

Microtubules
1. Microtubules consist of small protein tubules.
2. They form intracellular "skeletons" and intracellular channels and compose cilia, flagella, centrioles, and spindle fibers.

Centrosome and Centrioles
1. The dense area of cytoplasm containing the centrioles is called a centrosome.
2. Centrioles are paired cylinders arranged at right angles to one another. They assume an important role in cell reproduction.

Flagella and Cilia
1. Flagella and cilia are cell projections used in movement. They have the same basic structure.
2. If projections are few and long, they are called flagella. If they are numerous and hairlike, they are called cilia.
3. The flagellum on a sperm cell moves the entire cell. The cilia on cells of the respiratory tract move foreign matter along the cell surfaces toward the throat for elimination.

Cell Inclusions
1. Inclusions are chemical substances produced by cells. They may be stored, may participate in chemical reactions, and may have recognizable shapes.
2. Examples of cell inclusions are glycogen, mucus, and melanin.

Extracellular Materials
1. Extracellular materials are all the substances that lie outside the cell membrane.

2. They provide support and a medium for the diffusion of nutrients and wastes.
3. Some, like hyaluronic acid, are amorphous, or have no shape. Others, like collagen, are fibrous, or thread-like.

Cell Division
1. Cell division that results in the formation of new cells involves mitosis (nuclear division) and cytokinesis (cytoplasmic division). The kind of nuclear division that results in the production of sperm and egg cells is termed meiosis.
2. Mitosis and cytokinesis replace and add body cells. Prior to mitosis and cytokinesis, the DNA molecules, or chromosomes, replicate themselves so the same chromosomal complement can be passed on to future generations of cells.

Mitosis
1. Mitosis consists of prophase, metaphase, anaphase, and telophase.
2. A cell carrying on every life process except division is said to be in interphase or metabolic phase.

Cytokinesis
Cytokinesis begins in late anaphase and terminates in telophase.

Applications to Health
Cells and Aging
1. Aging is a progressive failure of the body's homeostatic adaptive responses.
2. Many theories of aging have been proposed, including genetically programed cessation of cell division and excessive immune responses, but none successfully answers all the experimental objections.
3. All of the various body systems exhibit definitive and sometimes extensive changes with aging.

Cells and Cancer
1. Cancerous tumors are referred to as malignant; noncancerous tumors are called benign.
2. The spread of cancer from its primary site is called metastasis.
3. Carcinogens include environmental agents and viruses.

REVIEW QUESTIONS

1. Define a cell. What are the four principal portions of a cell? What is meant by a generalized cell?
2. Discuss the structure of the plasma membrane. What determines the permeability of the plasma membrane? How are plasma membranes modified for various functions?
3. What are the major differences between active processes and passive processes in moving substances across plasma membranes?
4. Define and give an example of each of the following: diffusion, facilitated diffusion, osmosis, filtration, active transport, phagocytosis, pinocytosis.

5. Compare the effect on red blood cells of an isotonic, hypertonic, and hypotonic solution. What is osmotic pressure?

6. Describe the structure and function of microvilli, rod sacs, myelin sheaths, and stereocilia as membrane modifications.

7. Discuss the chemical composition and physical nature of cytoplasm. What is its function?

8. What is an organelle? Indicate the structure by means of a diagram, and describe the function of the following organelles: nucleus, endoplasmic reticulum, ribosomes, Golgi complex, mitochondria, lysosomes, microtubules, centrosome, centrioles, cilia, flagella.

9. Define a cell inclusion. Provide examples and indicate their functions.

10. What is an extracellular material? Give examples and the functions of each.

11. How does DNA replicate itself?

12. Discuss mitosis and cytokinesis with regard to stages. What are the characteristics of each stage, the relative duration, and the importance?

13. List the five generalizations on aging that can be made based on current information.

14. Starting with the integumentary system, list the major effects of aging on all of the body systems.

15. What is a tumor? Distinguish between malignant and benign tumors.

16. Define metastasis. Why is it clinically important?

17. Discuss how certain carcinogens may be related to cancer.

18. Refer to the glossary of key medical terms at the end of the chapter and be sure that you can define each term.

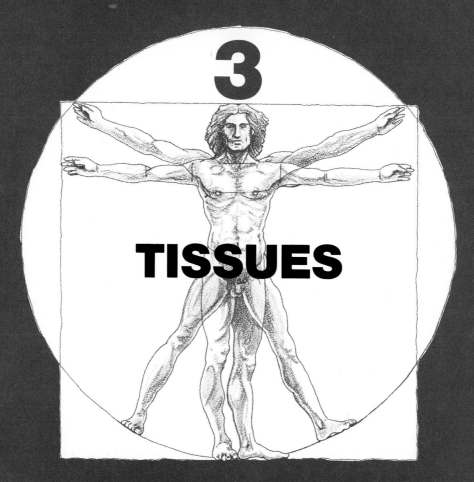

3

TISSUES

STUDENT OBJECTIVES

- Define a tissue.

- Classify the tissues of the body into four major types.

- Describe the distinguishing characteristics of epithelial tissue.

- Contrast the structural and functional differences of covering, lining, and glandular epithelium.

- Compare the shape of cells and the layering arrangements of covering and lining epithelium.

- Explain the various kinds of cell junctions that exist between epithelial cells.

- List the structure, function, and location of simple, stratified, and pseudostratified epithelium.

- Define a gland.

- Distinguish between exocrine and endocrine glands.

- Classify exocrine glands according to structural complexity and physiology.

- Describe the distinguishing characteristics of connective tissue.

- Contrast the structural and functional differences between embryonic and adult connective tissues.

- Describe the ground substance, fibers, and cells that constitute connective tissue.

- List the structure, function, and location of loose connective tissue, adipose tissue, and dense, elastic, and reticular connective tissue.

- List the structure, function, and location of the three types of cartilage.

- Distinguish between the interstitial and appositional growth of cartilage.

- Define an epithelial membrane.

- List the location and function of mucous, serous, synovial, and cutaneous membranes.

The cells discussed in the preceding chapter are highly organized units, but they do not function as isolated units. Instead, they work together as part of a group of similar cells, called a tissue. We shall now examine how the body operates at the tissue level of organization.

TYPES OF TISSUES

A **tissue** is a group of similar cells and their intercellular substance functioning together to perform a specialized activity. Certain tissues function to move body parts. Others move food through body organs. Some tissues protect and support the body. Others produce chemicals such as enzymes and hormones. The various tissues of the body are classified into four principal types according to their function and structure.

1. **Epithelial** (ep'-i-THĒ-lē-al) **tissue** covers body surfaces or tissues, lines body cavities, and forms glands.

2. **Connective tissue** protects and supports the body and its organs and binds organs together.

3. **Muscular tissue** is responsible for movement.

4. **Nervous tissue** initiates and transmits nerve impulses that coordinate body activities.

EPITHELIAL TISSUE

The tissues in this main category perform many activities in the body, including protection, absorption, and secretion. **Absorption** is the intake of fluids or other substances by cells of the skin or mucous membranes. **Secretion** is the production and release by cells of a fluid that may contain a variety of substances, such as mucus, perspiration, or enzymes.

Epithelial tissue, or more simply **epithelium,** may be divided into two subtypes: (1) *covering and lining epithelium* and (2) *glandular epithelium.* Covering and lining epithelium forms the outer covering of external body surfaces and the outer covering of some internal organs. It lines the body cavities and the interiors of the respiratory and digestive tracts, blood vessels, and ducts. It makes up, along with nervous tissue, the parts of the sense organs that are sensitive to the stimuli that produce smell and hearing sensations. Glandular epithelium constitutes the secreting portion of glands.

Both types of epithelium consist largely or entirely of closely packed cells with little or no intercellular material between adjacent cells. (Such intercellular material is also called the matrix.) Epithelial cells are arranged in continuous sheets that may be either single or multilayered. Nerves may extend through these sheets, but blood vessels do not. Thus they are *avascular* (*a* = without, *vascular* = blood vessels). The vessels that supply nutrients and remove wastes are located in underlying connective tissue. Epithelium overlies and adheres firmly to

the connective tissue, which holds the epithelium in position and prevents it from being torn. The surface of attachment between the epithelium and the connective tissue is a thin extracellular layer called the **basement membrane.** With few exceptions, epithelial cells secrete along their basal surfaces a material consisting of a special type of collagen and glycoproteins. This structure averages 500 to 800 Å in thickness and is referred to as the *basal lamina.* Frequently, the basal lamina is reinforced by an underlying *reticular lamina* consisting of reticular fibers and glycoproteins. This lamina is produced by cells in the underlying connective tissue. The combination of the basal lamina and reticular lamina constitutes the basement membrane. Since all epithelium is subjected to a certain degree of wear, tear, and injury, its cells must divide and produce new cells to replace those that are destroyed. These general characteristics are found in both types of epithelial tissue.

COVERING AND LINING EPITHELIUM

Arrangement of Layers

Covering and lining epithelium is arranged in several different ways related to location and function. If the epithelium is specialized for absorption or filtration and is in an area that has minimal wear and tear, the cells of the tissue are arranged in a single layer. Such an arrangement is called **simple epithelium.** If the epithelium is not specialized for absorption or filtration and is found in an area with a high degree of wear and tear, then the cells are stacked in several layers. This tissue is referred to as **stratified epithelium.** A third, less common, arrangement of epithelium is called **pseudostratified.** Like simple epithelium, pseudostratified epithelium has only one layer of cells. However, some of the cells do not reach the surface—an arrangement that gives the tissue a multilayered, or stratified, appearance. The pseudostratified cells that do reach the surface either secrete mucus or contain cilia that move mucus and foreign particles for eventual elimination from the body.

Cell Shapes

In addition to classifying covering epithelium according to the number of its layers, we may also categorize it by cell shape. The cells may be flat, cubelike, or columnar or may resemble a cross between shapes. **Squamous** (SKWĀ-mus) cells are flattened and scalelike. They are attached to each other like a mosaic. **Cuboidal** cells are usually cube-shaped in cross section. They sometimes appear as hexagons. **Columnar** cells are tall and cylindrical, appearing as rectangles set on end. **Transitional** cells often have a combination of shapes and are found where there is a great degree of distention or expansion in the body. Transitional cells on the bottom layer of an epithelial tissue may range in shape from cuboidal to columnar. In the intermediate layer, they may be cuboidal or polyhedral. In the superficial layer, they may range from cuboidal to squamous, depending on how much they are pulled out of shape during certain body functions.

Cell Junctions

Epithelial cells, unlike connective tissue cells, are tightly joined to form a close functional unit. The points of attachment between adjacent plasma membranes of epithelial cells (and a few other types of cells as well) are called **cell junctions** or **junctional complexes.** In addition to providing cell-to-cell attachments, cell junctions prevent the movement of materials into certain cells and provide channels for communication between other cells. Four categories of cell junctions are recognized: (1) desmosome (macula adherens), (2) tight junction (zonula occludens), (3) intermediate junction (zonula adherens), and (4) gap junction (nexus). These cell junctions are illustrated in Figure 3-1.

• *Desmosome (Macula adherens)* One of the most frequent types of cell junctions is called a **desmosome** (*desmos* = bond; *soma* = body) or **macula adherens** (*macula* = spot; *adhaereo* = to stick). Desmosomes are numerous in stratified epithelium, especially the epidermis of the skin. They are also found in the epithelial lining of the small intestine. Desmosomes are scattered over adjacent cell surfaces somewhat like spot welds and form firm

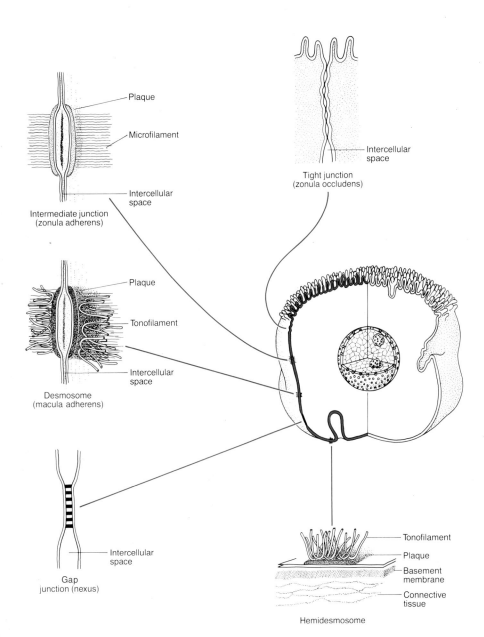

FIGURE 3-1 Cell junctions.

intercellular attachments between cells. The adjacent plasma membranes at a desmosome appear thickened because there is a dense proteinaceous *plaque* of cytoplasmic material on the inner surfaces of the neighboring membranes. Cytoplasmic filaments, called *tonofilaments,* converge in the region of the plaques and make a U-turn back into the cytoplasm. The plasma membranes of the adjacent cells at a desmosome are separated by an intercellular space measuring about 220 to 240 Å. In addition to serving as firm sites of adhesion between adjacent cells, desmosomes also help to support the entire epithelial tissue.

In the basal cells of epithelium, the plasma membranes are in contact with the underlying basement membrane rather than adjacent plasma membranes. Here, there are structures that look like half a desmosome, called **hemidesmosomes** (*hemi* = half). Hemidesmosomes are assumed to provide anchorage for the basal epithelial cell plasma membranes to the basement membrane.

● *Tight Junction* (*Zonula occludens*)   The **tight junction** or **zonula occludens** (*zonula* = small zone; *occludens* = occlude) is typically located along the lateral surfaces of adjacent plasma membranes, closest to the luminal (free) surface just below microvilli. In a tight junction, the outer layers of adjacent plasma membranes are fused together at various points. The junction completely encircles each cell that is joined and occludes the intercellular spaces. Tight junctions are common in epithelial cells lining the small intestine. Here, they prevent the movement of substances across epithelial cells via the intercellular route. Some substances, however, can selectively enter the small intestinal cells through microvilli by absorption.

● *Intermediate Junction* (*Zonula adherens*)   The **intermediate junction** or **zonula adherens,** like the tight junction, completely encircles each cell that is joined. The intermediate junction is located just below the tight junction between small intestinal epithelial cells. In an intermediate junction, the adjacent plasma membranes are separated by an intercellular space measuring about 200 Å. The inner surfaces of the neighboring plasma membranes contain a moderately dense area of cytoplasmic material (not as dense as the plaques in desmosomes) in which are embedded *microfilaments.* These filaments are smaller in diameter than the tonofilaments of desmosomes. Intermediate junctions are believed to form firm connections between adjacent cells.

● *Gap Junction* (*Nexus*)   In a **gap junction** or **nexus** (*nexus* = bond), the outer layers of adjacent plasma membranes approach each other, leaving a gap of about 20 Å between them. The gap is bridged by structures that link the cells together. These structures may be minute channels formed by membrane proteins that enable adja-

cent cells to communicate. It is postulated that ions and small molecules pass rapidly through gap junctions from one cell to the next, resulting in spontaneous impulse conduction. Gap junctions are found between smooth and cardiac muscle cells, as well as between epithelial cells. The presence of gap junctions between muscle cells presumably enables the cells to conduct impulses rapidly and contract in an orderly and coordinated manner.

Classification

Considering layers and cell type in combination, we may classify covering and lining epithelium as follows:

Simple
   1. Squamous
   2. Cuboidal
   3. Columnar
Stratified
   1. Squamous
   2. Cuboidal
   3. Columnar
   4. Transitional
Pseudostratified

Each of the epithelial tissues described in the following sections is illustrated in Exhibit 3-1.

Simple Epithelium

● *Simple Squamous Epithelium*   This type of simple epithelium consists of a single layer of flat, scalelike cells. Its surface resembles a tiled floor. The nucleus of each cell is centrally located and oval or spherical. Since simple squamous epithelium has only one layer of cells, it is highly adapted to diffusion, osmosis, and filtration. Thus it lines the air sacs of the lungs, where oxygen is exchanged with carbon dioxide. It is present in the part of the kidney that filters the blood. It is also found in delicate structures such as the crystalline lens of the eye and the lining of the eardrum. Simple squamous epithelium is found in body parts that have little wear and tear.

A tissue similar to simple squamous epithelium is endothelium. **Endothelium** lines the heart, the blood vessels, and the lymph vessels and forms the walls of capillaries. Another tissue similar to simple squamous epithelium that forms the epithelial layer of serous membranes is **mesothelium.** Mesothelium lines the thoracic and abdominopelvic cavities and covers the viscera within them.

● *Simple Cuboidal Epithelium*   Viewed from above, the cells of simple cuboidal epithelium appear as closely fitted polygons. The cuboidal nature of the cells is obvious only when the tissue is sectioned at right angles. Like simple squamous epithelium, these cells possess a central nucleus. Simple cuboidal epithelium covers the surface of the ova-

EXHIBIT 3-1 EPITHELIAL TISSUES

COVERING AND LINING EPITHELIUM
Simple squamous (250×)

Description: Single layer of flat, scalelike cells; large, centrally located nuclei.

Location: Lines air sacs of lungs, glomerular capsule of kidneys, crystalline lens of eyes, and eardrum. Called endothelium when it lines heart, blood and lymphatic vessels, and forms capillaries. Called mesothelium when it lines the ventral body cavity and covers viscera as part of a serous membrane.

Function: Filtration, absorption, and secretion in serous membranes.

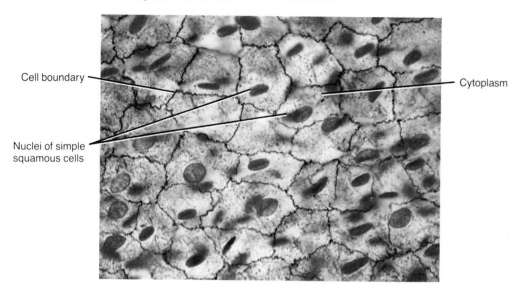

Cell boundary

Cytoplasm

Nuclei of simple squamous cells

Simple cuboidal (450×)

Description: Single layer of cube-shaped cells; centrally located nuclei.

Location: Covers surface of ovary, lines inner surface of capsule of the lens of eye, forms pigmented epithelium of retina, and lines kidney tubules and smaller ducts of many glands.

Function: Secretion and absorption

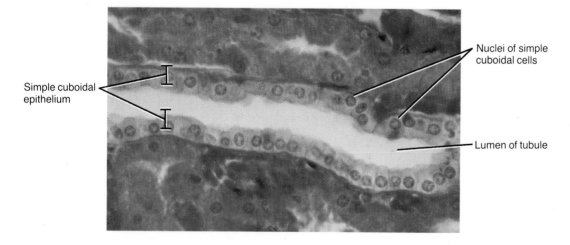

Simple cuboidal epithelium

Nuclei of simple cuboidal cells

Lumen of tubule

EXHIBIT 3-1 (*Continued*)

Simple columnar (nonciliated) (600×)

Description: Single layer of nonciliated, rectangular cells; contains goblet cells; nuclei at bases of cells.
Location: Lines the digestive tract from the cardia of the stomach to the anus, excretory ducts of many glands, and gallbladder.
Function: Secretion and absorption.

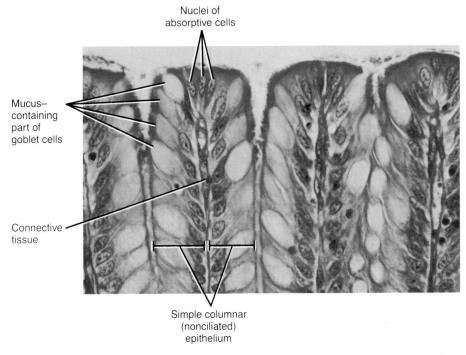

Simple columnar (ciliated) (400×)

Description: Single layer of ciliated, columnar cells; contains goblet cells; nuclei at bases of cells.
Location: Lines a few portions of upper respiratory tract, uterine (fallopian) tubes, uterus, some paranasal sinuses, and central canal of
                spinal cord.
Function: Moves mucus by ciliary action.

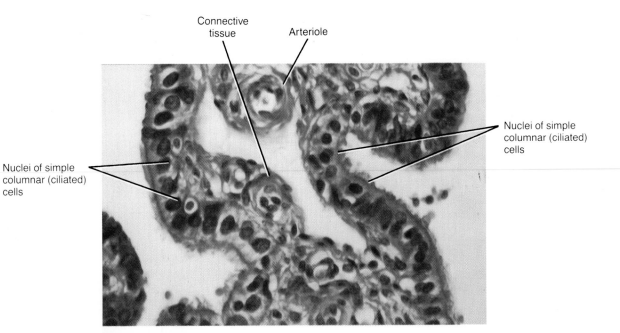

EXHIBIT 3-1 (*Continued*)

Stratified squamous (185×)

Description: Several layers of cells; cuboidal to columnar shape in deep layers; scalelike shape in superficial layers. Continual multiplication of basal cells replaces surface cells.
Location: Nonkeratinizing variety lines wet surfaces such as mouth, esophagus, part of epiglottis, and vagina. Keratinizing variety forms outer layer of skin.
Function: Protection.

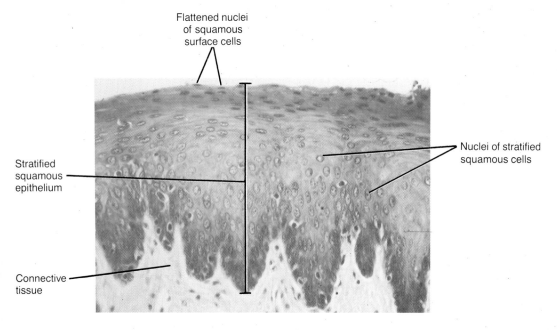

Stratified cuboidal (185×)

Description: Two or more layers of cells in which the surface cells are cube-shaped.
Location: Ducts of adult sweat glands, fornix of conjunctiva, cavernous urethra, and epiglottis.
Function: Protection.

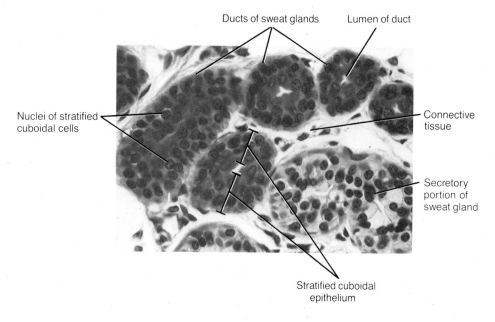

EXHIBIT 3-1 (*Continued*)

Stratified columnar (600×)

Description: Several layers of polyhedral cells; columnar cells only in superficial layer.
Location: Lines part of male urethra and some larger excretory ducts.
Function: Protection and secretion.

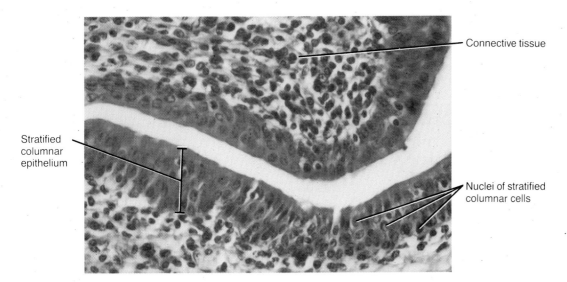

Connective tissue

Stratified
columnar
epithelium

Nuclei of stratified
columnar cells

Stratified transitional epithelium (185×)

Description: Resembles nonkeratinized stratified squamous tissue, except that superficial cells are larger and more rounded.
Location: Lines urinary bladder.
Function: Permits distention.

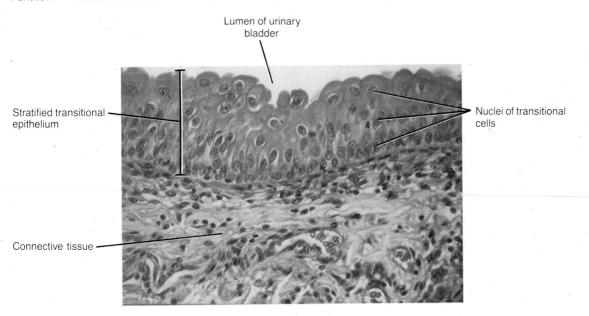

Lumen of urinary
bladder

Stratified transitional
epithelium

Nuclei of transitional
cells

Connective tissue

EXHIBIT 3-1 (*Continued*)

Pseudostratified (500×)

Description: Not a true stratified tissue; nuclei of cells at different levels; all cells attached to basement membrane, but not all reach
      surface.
Location: Lines larger excretory ducts of many glands, male urethra and eustachian tubes; ciliated variety with goblet cells lines most
      of the upper respiratory tract and some ducts of male reproductive system.
Function: Secretion and movement of mucus by ciliary action.

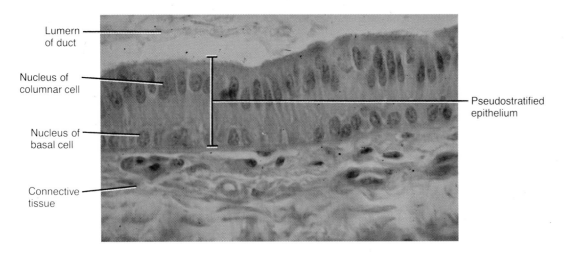

Lumern of duct

Nucleus of columnar cell

Nucleus of basal cell

Connective tissue

Pseudostratified epithelium

GLANDULAR EPITHELIUM
Exocrine portion of pancreas (300×)

Description: Secretes products into ducts.
Location: Sweat, oil, wax, and mammary glands of the skin; digestive glands such as salivary glands which secrete into mouth cavity.
Function: Produce perspiration, oil, wax, and digestive enzymes.

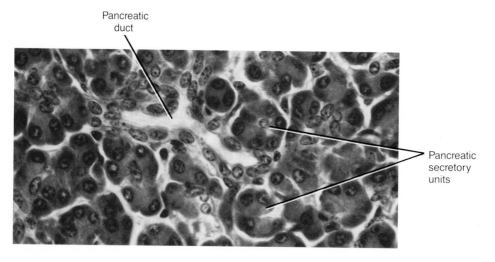

Pancreatic duct

Pancreatic secretory units

EXHIBIT 3-1 (*Continued*)

Endocrine gland (180×)

Description: Secretes hormones into blood.
Location: Pituitary gland at base of brain; thyroid gland near larynx; adrenal (suprarenal) glands above kidneys, among others.
Function: Produce hormones that regulate various body activities.

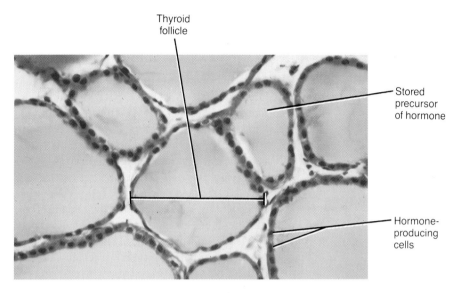

Thyroid follicle

Stored precursor of hormone

Hormone-producing cells

Photomicrographs © 1983 by Michael H. Ross. Used by permission.

ries, lines the inner surface of the capsule of the lens of the eye, and forms the pigmented epithelium of the retina. In the kidneys, where it forms the kidney tubules and contains microvilli, it functions in water reabsorption. It also lines the smaller ducts of some glands and the secreting units of glands, such as the thyroid. This tissue performs the functions of secretion and absorption.

● *Simple Columnar Epithelium* The surface view of simple columnar epithelium is similar to that of simple cuboidal tissue. When sectioned at right angles to the surface, however, the cells appear somewhat rectangular. The nuclei are located near the bases of the cells.

The luminal surfaces of simple columnar epithelial cells are modified in several ways, depending on location and function. Simple columnar epithelium lines the digestive tract from the cardia of the stomach to the anus, the gallbladder, and excretory ducts of many glands. In such sites, the cells protect the underlying tissues. Many of them are also modified to aid in food-related activities. In the small intestine especially, the plasma membranes of the cells are folded into **microvilli.** (See Figure 2-7a.) The microvilli arrangement increases the surface area of

the plasma membrane and thereby allows larger amounts of digested nutrients and fluids to diffuse into the body.

Interspersed among the typical columnar cells of the intestine are other modified columnar cells called **goblet cells.** These cells, which secrete mucus, are so named because the mucus accumulates in the upper half of the cell, causing the area to bulge out. The whole cell resembles a goblet or wine glass. The secreted mucus serves as a lubricant between the food and the walls of the digestive tract.

A third modification of columnar epithelium is found in cells with hairlike processes called **cilia.** In a few portions of the upper respiratory tract, ciliated columnar cells are interspersed with goblet cells. Mucus secreted by the goblet cells forms a film over the respiratory surface. This film traps foreign particles that are inhaled. The cilia wave in unison and move the mucus, with any foreign particles, toward the throat, where it can be swallowed or eliminated. Air is filtered by this process before entering the lungs. Ciliated columnar epithelium is also found in the uterus and uterine tubes of the female reproductive system, some paranasal sinuses, and the central canal of the spinal cord.

Stratified Epithelium

In contrast to simple epithelium, stratified epithelium consists of at least two layers of cells. Thus it is durable and can protect underlying tissues from the external environment and from wear and tear. Some stratified epithelium cells also produce secretions. The name of the specific kind of stratified epithelium depends on the shape of the surface cells.

● *Stratified Squamous Epithelium* In the more superficial layers of this type of epithelium, the cells are flat, whereas the cells of the deep layers vary in shape from cuboidal to columnar. The basal, or bottom, cells are continually multiplying by cell division. As new cells grow, they compress the cells on the surface and push them outward. According to this growth pattern, basal cells continually shift upward and outward. As they move farther from the deep layer and their blood supply, they become dehydrated, shrink, and grow harder. At the surface, the cells are rubbed off. New cells continually emerge, are sloughed off, and replaced.

One form of stratified squamous epithelium is called **nonkeratinized stratified squamous epithelium.** This tissue is found on wet surfaces that are subjected to considerable wear and tear and do not perform the function of absorption—the linings of the mouth, the esophagus, and the vagina. Another form of stratified squamous epithelium is called **keratinized stratified squamous epithelium.** The surface cells of this type are modified into a tough layer of material containing keratin. **Keratin** is a protein that is waterproof and resistant to friction and helps to resist bacterial invasion. The outer layer of skin consists of keratinized tissue.

● *Stratified Cuboidal Epithelium* This relatively rare type of epithelium is found in the ducts of the sweat glands of adults, fornix of the conjunctiva, cavernous urethra, pharynx, and epiglottis. It sometimes consists of more than two layers of cells. Its function is mainly protective.

● *Stratified Columnar Epithelium* Like stratified cuboidal epithelium, this type of tissue is also uncommon in the body. Usually the basal layer or layers consist of shortened, irregularly polyhedral cells. Only the superficial cells are columnar in form. This kind of epithelium lines part of the male urethra and some larger excretory ducts such as lactiferous (milk) ducts in the mammary glands. It functions in protection and secretion.

● *Transitional Epithelium* Transitional epithelium is very much like nonkeratinized stratified squamous epithelium. The distinction is that the outer layer of cells in transitional epithelium tend to be large and rounded rather than flat. This feature allows the tissue to be stretched (distended) without the outer cells breaking apart from one another. When stretched, they are drawn out into squamouslike cells. Because of this arrangement, transitional epithelium lines hollow structures that are subjected to expansion from within, such as the urinary bladder. Its function is to help prevent a rupture of the organ.

Pseudostratified Epithelium

The third category of covering and lining epithelium consists of columnar cells and is called pseudostratified epithelium. The nuclei of the cells in this kind of tissue are at varying depths. Even though all the cells are attached to the basement membrane in a single layer, some do not reach the surface. This feature gives the impression of a multilayered tissue, the reason for the designation *pseudo*stratified epithelium. It lines the larger excretory ducts of many glands, parts of the male urethra, and the eustachian tubes. Pseudostratified epithelium may be ciliated and may contain goblet cells. In this form, it lines most of the upper respiratory tract and certain ducts of the male reproductive system.

GLANDULAR EPITHELIUM

The function of glandular epithelium is secretion, accomplished by glandular cells that lie in clusters deep to the covering and lining epithelium. A **gland** may consist of one cell or a group of highly specialized epithelial cells that secrete substances into ducts or into the blood. The production of such substances always requires active work by the glandular cells and results in an expenditure of energy.

All glands of the body are classified as exocrine or endocrine according to whether they secrete substances into ducts or into the blood (Exhibit 3-1). **Exocrine glands** secrete their products into ducts or tubes that empty at the surface of the covering and lining epithelium. The product of an exocrine gland may be released at the skin surface or into the lumen (cavity) of a hollow organ. The secretions of exocrine glands include enzymes, oil, and sweat. Examples of exocrine glands are sweat glands, which eliminate perspiration to cool the skin, and salivary glands, which secrete a digestive enzyme. **Endocrine glands,** by contrast, are ductless and secrete their products into the blood. The secretions of endocrine glands are always hormones, chemicals that regulate various physiological activities. The pituitary, thyroid, and adrenal glands are endocrine glands.

Structural Classification of Exocrine Glands

Exocrine glands are classified into two structural types: unicellular and multicellular. **Unicellular glands** are single-celled. A good example of a unicellular gland is a goblet cell (Exhibit 3-1). Goblet cells are found in the epithelial lining of the digestive, respiratory, urinary, and reproductive systems. They produce mucus to lubricate the free surfaces of these membranes.

Multicellular glands (many-celled glands) occur in several different forms (Figure 3-2). If the secretory portions of a gland are tubular, it is referred to as a **tubular gland.** If flasklike, it is called an **acinar** (AS-i-nar) **gland.** If the gland contains both tubular and flasklike secretory portions, it is called a **tubuloacinar gland.** Further, if the duct of the gland does not branch, it is referred to as a **simple gland;** if the duct does branch, it is called a **compound gland.** By combining the shape of the secretory portion with the degree of branching of the duct, we arrive at the following structural classification for exocrine glands.

Unicellular. One-celled gland that secretes mucus. Example: goblet cell of the digestive system.

Multicellular. Many-celled gland.

  Simple. Single, nonbranched duct.

  1. **Tubular.** The secretory portion is straight and tubular. Example: crypts of Lieberkuhn of intestines.
  2. **Branched tubular.** The secretory portion is branched and tubular. Examples: gastric and uterine glands.
  3. **Coiled tubular.** The secretory portion is coiled and tubular. Example: sudoriferous (sweat) glands.

  4. **Acinar.** The secretory portion is flasklike. Example: seminal vesicle glands.
  5. **Branched acinar.** The secretory portion is branched and flasklike. Example: sebaceous (oil) glands.

  Compound. Branched duct.

  1. **Tubular.** The secretory portion is tubular. Examples: bulbourethral glands, testes, and liver.
  2. **Acinar.** The secretory portion is flasklike. Examples: salivary glands (sublingual and submandibular).
  3. **Tubuloacinar.** The secretory portion is both tubular and flasklike. Examples: salivary glands (parotid) and pancreas.

Functional Classification of Exocrine Glands

The functional classification of exocrine glands is based on how the gland releases its secretion. The three recognized categories are holocrine, merocrine, and apocrine glands. **Holocrine glands** accumulate a secretory product in the cytoplasm of each secreting cell. The cell then dies and is discharged with its contents as the glandular secretion (Figure 3-3a). The discharged cell is replaced by a new cell. One example of a holocrine gland is a sebaceous gland of the skin. **Merocrine glands** simply form and discharge the secretory product from the cell (Figure 3-3b). Examples of merocrine glands are the pancreas and the salivary glands. **Apocrine glands** accumulate their secretory product at the apical (outer) margin of

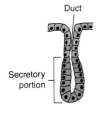
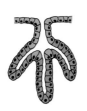

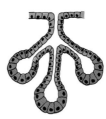

Duct

Secretory portion

Simple tubular   Simple branched tubular   Simple coiled tubular   Simple acinar   Simple branched acinar

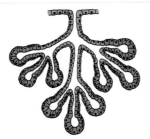

Compound tubular   Compound acinar   Compound tubuloacinar

FIGURE 3-2 Structural types of multicellular exocrine glands. The secretory portions of the glands are indicated in purple. The blue areas represent the ducts of the glands.

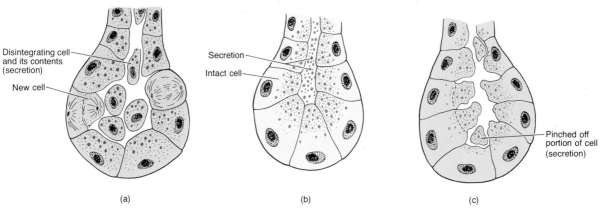

FIGURE 3-3 Functional classification of multicellular exocrine glands. (a) Holocrine gland. (b) Merocrine gland. (c) Apocrine gland.

the secreting cell. That portion of the cell pinches off from the rest of the cell to form the secretion (Figure 3-3c). The remaining part of the cell repairs itself and repeats the process. An example of an apocrine gland is the mammary gland.

CONNECTIVE TISSUE

The most abundant tissue in the body is **connective tissue.** This binding and supporting tissue usually has a rich blood supply. Thus it is highly *vascular.* An exception is cartilage, which is avascular. The cells are widely scattered, rather than closely packed, and there is considerable intercellular material, or matrix. In contrast to epithelium, connective tissues do not occur on free surfaces, such as the surfaces of a body cavity or the external surface of the body. The general functions of connective tissues are protection, support, and the binding together of various organs.

The intercellular substance in a connective tissue largely determines the tissue's qualities. These substances may consist of fluid, semifluid, or mucoid (mucuslike) material. In cartilage, the intercellular material is firm but pliable. In bone, it is quite hard and not pliable. The cells of connective tissue produce the intercellular substances. The cells may also store fat, ingest bacteria and cell debris, form anticoagulants, or give rise to antibodies that protect against disease.

CLASSIFICATION

Connective tissue may be classified in several ways. We will classify them as follows.

Embryonic connective tissue
    Mesenchyme
    Mucous connective tissue
Adult connective tissue
    Connective tissue proper
        1. Loose connective (areolar) tissue

        2. Adipose tissue
        3. Dense connective (collagenous) tissue
        4. Elastic connective tissue
        5. Reticular connective tissue
    Cartilage
        1. Hyaline cartilage
        2. Fibrocartilage
        3. Elastic cartilage
    Bone (osseous) tissue
    Vascular (blood) tissue

Each of the connective tissues described in the following sections is illustrated in Exhibit 3-2.

EMBRYONIC CONNECTIVE TISSUE

Connective tissue that is present primarily in the embryo or fetus is called **embryonic connective tissue.** The term *embryo* refers to a developing human from fertilization through the first two months of pregnancy; *fetus* refers to a developing human from the third month of pregnancy to birth.

One example of embryonic connective tissue found almost exclusively in the embryo is **mesenchyme** (MEZ-en-kīm)—the tissue from which all other connective tissues eventually arise. Mesenchyme may be observed beneath the skin and along the developing bones of the embryo. Some mesenchymal cells are scattered irregularly throughout adult connective tissue, most frequently around blood vessels. Here mesenchymal cells differentiate into fibroblasts that assist in wound healing.

Another kind of embryonic connective tissue is **mucous connective tissue,** found primarily in the fetus. This tissue, also called **Wharton's jelly,** is located in the umbilical cord of the fetus, where it supports the wall of the cord.

ADULT CONNECTIVE TISSUE

Adult connective tissue is connective tissue that exists in the newborn and that does not change after birth. It is subdivided into several kinds.

EXHIBIT 3-2

EMBRYONIC
Mesenchymal (180×)

Description: Consists of highly branched mesenchymal cells embedded in a fluid substance.
Location: Under skin and along developing bones of embryo; some mesenchymal cells found in adult connective tissue, especially along blood vessels.
Function: Forms all other kinds of connective tissue.

Blood vessels

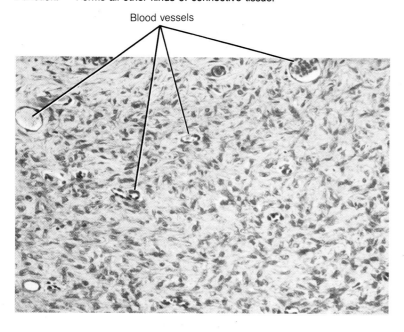

Mucous (320×)

Description: Consists of flattened or spindle-shaped cells embedded in a mucuslike substance containing fine collagenous fibers.
Location: Umbilical cord of fetus.
Function: Support.

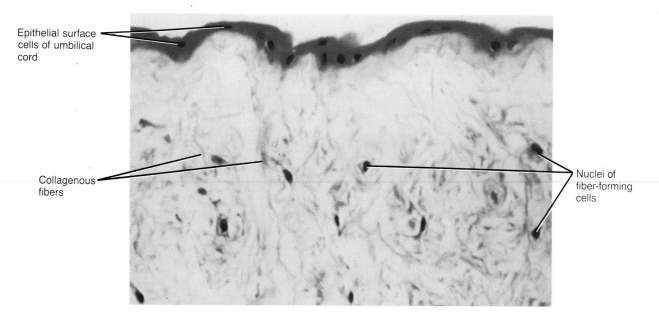

Epithelial surface cells of umbilical cord

Collagenous fibers

Nuclei of fiber-forming cells

EXHIBIT 3-2 (*Continued*)

ADULT
Loose (areolar) (180×)

Description: Consists of fibers (collagenous, elastic, and reticular) and several kinds of cells (fibroblasts, macrophages, plasma cells, and mast cells) embedded in a semifluid ground substance.
Location: Subcutaneous layer of skin, mucous membranes, blood vessels, nerves, and body organs.
Function: Strength, elasticity, and support.

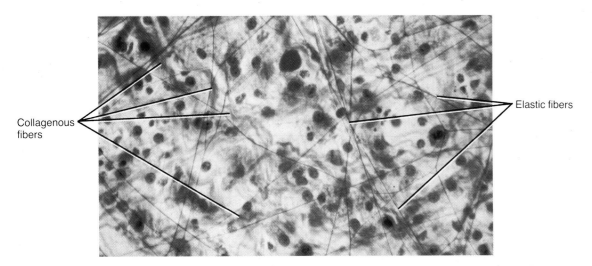

Adipose (250×)

Description: Contains adipocytes, "signet-ring" shaped cells which are specialized for fat storage.
Location: Subcutaneous layer of skin, around heart and kidneys, marrow of long bones, padding around joints.
Function: Reduces heat loss through skin, serves as an energy reserve, supports, and protects.

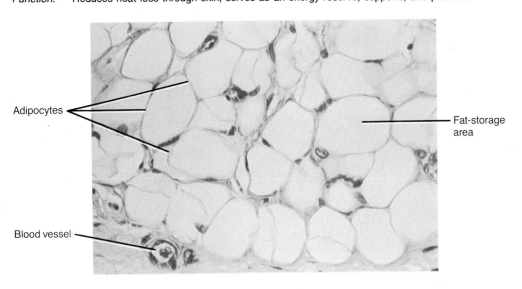

EXHIBIT 3-2 (*Continued*)

Dense (collagenous) (250×)

Description: Consists of predominately collagenous, or white, fibers arranged in bundles; fibroblasts present in rows between bundles.
Location: Forms tendons, ligaments, aponeuroses, membranes around various organs, and fasciae.
Function: Provides strong attachment between various structures.

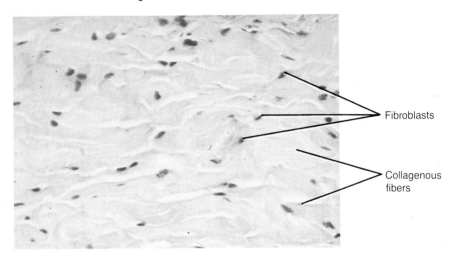

Fibroblasts

Collagenous fibers

Elastic (180×)

Description: Consists of predominately freely branching elastic, or yellow, fibers; fibroblasts present in spaces between fibers.
Location: Lung tissue, cartilage of larynx, walls of arteries, trachea, bronchial tubes, true vocal cords, and ligamenta flava of vertebrae.
Function: Allows stretching of various organs.

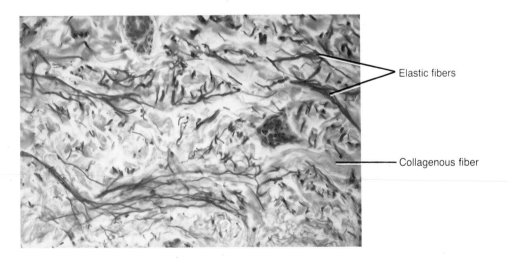

Elastic fibers

Collagenous fiber

EXHIBIT 3-2 *(Continued)*

Reticular (250×)

Description: Consists of network of interlacing reticular fibers with thin, flat cells wrapped around fibers.
Location: Liver, spleen, and lymph nodes.
Function: Forms stroma of organs; binds together smooth muscle tissue cells.

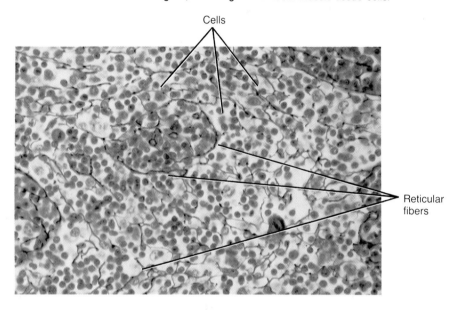

Hyaline cartilage (350×)

Description: Also called gristle; appears as a bluish-white, glossy mass; contains numerous chondrocytes; is the most abundant type of cartilage.
Location: Ends of long bones, ends of ribs, nose, parts of larynx, trachea, bronchi, bronchial tubes, and embryonic skeleton.
Function: Provides movement at joints, flexibility, and support.

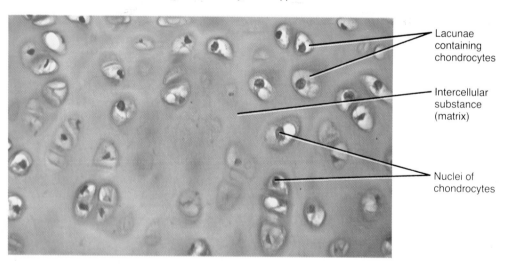

EXHIBIT 3-2 (*Continued*)

Fibrocartilage (180×)

Description: Consists of chondrocytes scattered among bundles of collagenous fibers.
Location: Symphysis pubis and intervertebral discs.
Function: Support and fusion.

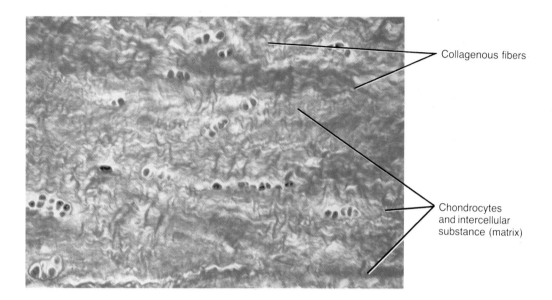

Collagenous fibers

Chondrocytes
and intercellular
substance (matrix)

Elastic cartilage (350×)

Description: Consists of chondrocytes located in a threadlike network of elastic fibers.
Location: Epiglottis, parts of larynx, external ear, and eustachian (auditory) tubes.
Function: Gives support and maintains shape.

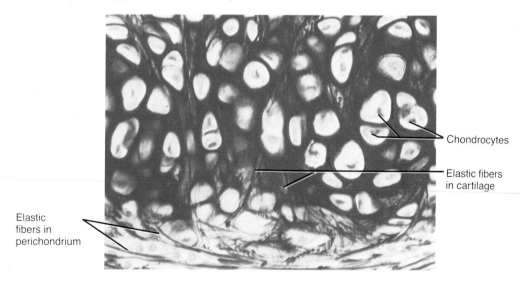

Chondrocytes

Elastic fibers
in cartilage

Elastic
fibers in
perichondrium

Connective Tissue Proper

Connective tissue that has a more or less fluid intercellular material and a fibroblast as the typical cell is termed **connective tissue proper.**

● *Loose Connective (Areolar) Tissue* Loose or areolar (a-RĒ-ō-lar) connective tissue is one of the most widely distributed connective tissues in the body. Structurally, it consists of fibers and several kinds of cells embedded in a semifluid intercellular substance. The term "loose" refers to the loosely woven arrangement of the fibers in the intercellular substance. The fibers are neither abundant nor arranged to prevent stretching. This intercellular substance consists of a viscous material called **hyaluronic acid.** Hyaluronic acid normally facilitates the passage of nutrients from the blood vessels of the connective tissue into adjacent cells and tissues, although the thick consistency of this acid may impede the movement of some drugs. But if an enzyme called **hyaluronidase** is injected into the tissue, hyaluronic acid changes to a watery consistency. This feature is of clinical importance because the reduced viscosity hastens the absorption and diffusion of injected drugs and fluids through the tissue and thus can lessen tension and pain. Some bacteria, white blood cells, and sperm cells produce hyaluronidase.

The three types of fibers embedded between the cells of loose connective tissue are collagenous, elastic, and reticular. **Collagenous,** or **white, fibers** are very tough and resistant to a pulling force, yet are somewhat flexible because they are usually wavy. These fibers often occur in bundles. They are composed of many minute fibers called fibrils lying parallel to one another. The bundle arrangement affords a great deal of strength. Chemically, collagenous fibers consist of the protein collagen. **Elastic,** or **yellow, fibers,** by contrast, are smaller than collagenous fibers and freely branch and rejoin one another. Elastic fibers consist of a protein called elastin. These fibers also provide strength and have great elasticity, up to 50 percent of their length. **Reticular fibers** also consist of collagen, plus some glycoprotein, but they are very thin fibers that branch extensively. Some authorities believe that reticular fibers are immature collagenous fibers. Like collagenous fibers, reticular fibers provide support and strength and form the *stroma* (framework) of many soft organs.

Cells in loose connective tissue are numerous and varied. Most are **fibroblasts**—large, flat cells with branching processes. If tissue is injured, the fibroblasts are believed to form collagenous fibers. Evidence suggests that fibroblasts also form elastic fibers and the viscous ground substance. Mature fibroblasts are referred to as **fibrocytes.** Fibroblasts (or *"blasts"* of any form of cell) are involved in the formation of immature tissue or repair of mature tissue, and fibrocytes (or *"cytes"* of any form of cell) are involved in maintaining health of mature tissue.

Other cells found in loose connective tissue are called **macrophages** (MAK-rō-fā-jēz), or **histiocytes** (HIS-tē-ō-sīts). They are irregular in form with short branching projections. These cells are derived from undifferentiated (unspecialized) mesenchymal cells called hemocytoblasts and are capable of engulfing bacteria and cellular debris by the process of phagocytosis. Thus they provide a vital defense for the body.

A third kind of cell in loose connective tissue is the **plasma cell.** These cells are small and either round or irregular. Plasma cells probably develop from a type of white blood cell called a lymphocyte. They give rise to antibodies and, accordingly, provide a defensive mechanism through immunity. Plasma cells are found in many places of the body, but most are found in connective tissue, especially that of the digestive tract and the mammary glands.

Another cell in loose connective tissue is the **mast cell.** It may develop from another type of white blood cell called a basophil. The mast cell is somewhat larger than a basophil and is found in abundance along blood vessels. It forms heparin, an anticoagulant that prevents blood from clotting in the vessels. Mast cells are also believed to produce histamine, a chemical that dilates, or enlarges, small blood vessels.

Other cells in loose connective tissue include **melanocytes** (or pigment cells), fat cells, and white blood cells.

Loose connective tissue is continuous throughout the body. It is present in all mucous membranes and around all blood vessels and nerves. And it occurs around body organs and in the papillary region of the dermis of the skin. Combined with adipose tissue, it forms the **subcutaneous** (sub'-kyoo-TĀ-nē-us; *sub* = under, *cut* = skin) **layer**—the layer of tissue that attaches the skin to underlying tissues and organs. The subcutaneous layer is also referred to as the **superficial fascia** (FASH-ē-a).

● *Adipose Tissue* Adipose tissue is basically a form of loose connective tissue in which the cells, called **adipocytes,** are specialized for fat storage. Adipocytes have the shape of a "signet ring" because the cytoplasm and nucleus are pushed to the edge of the cell by a large droplet of fat. Adipose tissue is found wherever loose connective tissue is located. Specifically, it is in the subcutaneous layer below the skin, around the kidneys, at the base and on the surface of the heart, in the marrow of long bones, as a padding around joints, and behind the eyeball in the orbit. Adipose tissue is a poor conductor of heat and therefore reduces heat loss through the skin. It is also a major energy reserve and generally supports and protects various organs.

● *Dense Connective (Collagenous) Tissue* Dense connective tissue is characterized by a close packing of fibers and less intercellular substance than in loose connective tissue. The fibers can be either irregularly arranged or regularly arranged. In areas of the body where tensions are exerted in various directions, the fiber bundles are interwoven and without regular orientation. Such a dense connective tissue is referred to as **irregularly arranged** and occurs in sheets. It forms most fasciae, the reticular region of the dermis of the skin, the periosteum of bone, the perichondrium of cartilage, and the *membrane (fibrous) capsules* around organs, such as the kidneys, liver, testes, and lymph nodes. In other areas of the body, dense connective tissue is adapted for tension in one direction, and the fibers have an orderly, parallel arrangement. Such a dense connective tissue is known as **regularly arranged.** The commonest variety of dense regularly arranged connective tissue has a predominance of collagenous (white) fibers arranged in bundles. Fibroblasts are placed in rows between the bundles. The tissue is silvery white, tough, yet somewhat pliable. Because of its great strength, it is the principal component of *tendons,* which attach muscles to bones; many *ligaments* (collagenous ligaments), which hold bones together at joints; and *aponeuroses* (ap'-ō-noo-RŌ-sēs), which are flat bands connecting one muscle with another or with bone.

● *Elastic Connective Tissue* Unlike collagenous connective tissue, elastic connective tissue has a predominance of freely branching elastic fibers. These fibers give the tissue a yellowish color. Fibroblasts are present only in the spaces between the fibers. Elastic connective tissue can be stretched and will snap back into shape. It is a component of the cartilages of the larynx, the walls of elastic arteries, the trachea, the bronchial tubes to the lungs, and the lungs themselves. Elastic connective tissue provides stretch and strength, allowing structures to perform their functions efficiently. Yellow elastic ligaments, as contrasted with collagenous ligaments, are composed mostly of elastic fibers, and form the ligamenta flava of the vertebrae (ligaments between successive vertebrae), the suspensory ligament of the penis, and the true vocal cords.

● *Reticular Connective Tissue* This kind of connective tissue consists of interlacing reticular fibers. It helps to form the framework, or stroma, of many organs, including the liver, spleen, and lymph nodes. Reticular connective tissue also helps to bind together the cells of smooth muscle tissue. It is especially adapted to providing strength and support.

Cartilage

One type of connective tissue, called **cartilage,** is capable of enduring considerably more stress than the tissues just discussed. Unlike other connective tissues, cartilage has no blood vessels or nerves of its own. Cartilage consists of a dense network of collagenous fibers and elastic fibers firmly embedded in a gel-like substance. The cells of mature cartilage, called **chondrocytes** (KON-drō-sīts), occur singly or in groups within spaces called *lacunae* (la-KOO-nē) in the intercellular substance. The surface of cartilage is surrounded by irregularly arranged dense connective tissue called the *perichondrium* (per'-i-KON-drē-um; *chondro* = cartilage, *peri* = around). Three kinds of cartilage are recognized: hyaline cartilage, fibrocartilage, and elastic cartilage (Exhibit 3-2).

● *Growth of Cartilage* The growth of cartilage follows two basic patterns. In **interstitial (endogenous) growth,** the cartilage increases rapidly in size through the division of existing chondrocytes and continuous deposition of increasing amounts of intercellular matrix by the chondrocytes. The formation of new chondrocytes and their production of new intercellular matrix causes the cartilage to expand from within, thus the term interstitial growth. This growth pattern occurs while the cartilage is young and pliable—during childhood and adolescence.

In **appositional (exogenous) growth,** the growth of cartilage occurs because of the activity of the inner chondrogenic layer of the perichondrium. The deeper cells of the perichondrium, the fibroblasts, divide. Some differentiate into chondroblasts (immature cells that develop into specialized cells) and then into chondrocytes. As differentiation occurs, the chondroblasts become surrounded with intercellular matrix and become chondrocytes. As a result, the matrix is deposited on the surface of the cartilage, increasing its size. The new layer of cartilage is added beneath the perichondrium on the surface of the cartilage, causing it to grow in width. Appositional growth starts later than interstitial growth and continues throughout life.

● *Hyaline Cartilage* This cartilage, also called gristle, appears as a bluish-white, glossy, homogeneous mass. The collagenous fibers, although present, are not visible with ordinary staining techniques, and the prominent chondrocytes are found in lacunae. Hyaline cartilage is the most abundant kind of cartilage in the body. It is found at joints over the ends of long bones (where it is called *articular cartilage*) and forms the *costal cartilages* at the ventral ends of the ribs. Hyaline cartilage also helps to form the nose, larynx, trachea, bronchi, and bronchial tubes leading to the lungs. Most of the embryonic skeleton consists of hyaline cartilage. It affords flexibility and support.

● *Fibrocartilage* Chondrocytes scattered through many bundles of visible collagenous fibers are found in this type of cartilage. Fibrocartilage is found at the symphysis pubis, the point where the coxal (hip) bones fuse anteriorly at the midline. It is also found in the discs between the vertebrae. This tissue combines strength and rigidity.

● *Elastic Cartilage* In this tissue, chondrocytes are located in a threadlike network of elastic fibers. Elastic cartilage provides strength and maintains the shape of certain organs—the larynx, the external part of the ear (the pinna), and the eustachian (auditory) tubes (the internal connection between the middle ear cavity and the upper throat).

Bone Tissue

The details of bone, or osseous, tissue, another kind of connective tissue, are discussed in Chapter 5 as part of the skeletal system.

Vascular Tissue

This kind of connective tissue, also known as blood, is treated in Chapter 11 as a component of the circulatory system.

MUSCLE TISSUE AND NERVOUS TISSUE

Epithelial and connective tissue can take a variety of forms to provide a variety of body functions. They are all-purpose tissues. By contrast, the third major type of tissue, **muscle tissue,** consists of highly modified cells that perform one basic function: contraction (Chapter 9). The fourth major type, **nervous tissue,** is specialized to conduct electrical impulses (Chapter 14).

MEMBRANES

The principal membranes of the body are mucous, serous, cutaneous, and synovial membranes. The first three consist of an epithelial layer and an underlying connective tissue layer and are called **epithelial membranes.** Another kind of membrane, a synovial membrane, does not contain epithelium.

MUCOUS MEMBRANES

Mucous membranes line the body cavities that open to the exterior. They line the entire digestive, respiratory, excretory, and reproductive tracts. The surface tissue of a mucous membrane may vary in type—it is stratified squamous epithelium in the esophagus and simple columnar epithelium in the intestine.

The epithelial layer of a mucous membrane secretes mucus, which prevents the cavities from drying out. It also traps dust in the respiratory passageways and lubricates food as it moves through the digestive tract. In addition, the epithelial layer is responsible for the secretion of digestive enzymes and the absorption of food.

The connective tissue layer of a mucous membrane binds the epithelium to the underlying structures and allows some flexibility of the membrane. It is referred to as the *lamina propria.* It holds the blood vessels in place, protects underlying muscles from abrasion or puncture, provides the epithelium covering it with oxygen and nutrients, and removes wastes.

SEROUS MEMBRANES

Serous membranes line the body cavities that do not open to the exterior, and they cover the organs that lie within those cavities. They consist of thin layers of loose connective tissue covered by a layer of mesothelium. Serous membranes are in the form of invaginated double-walled sacs. The part attached to the cavity wall is called the *parietal* (pa-RĪ-i-tal) *portion.* The part that covers the organs inside these cavities is the *visceral portion.* The serous membrane lining the thoracic cavity and covering the lungs is called the *pleura.* The membrane lining the heart cavity and covering the heart is the *pericardium* (*cardio* = heart). The serous membrane lining the abdominal cavity and covering the abdominal organs and some pelvic organs is called the *peritoneum.*

The epithelial layer of a serous membrane secretes a lubricating fluid that allows the organs to glide easily against one another or against the walls of the cavities. The connective tissue layer of the serous membrane consists of a relatively thin layer of loose connective tissue.

CUTANEOUS MEMBRANE

The **cutaneous membrane,** or skin, constitutes an organ of the integumentary system and is discussed in the next chapter.

SYNOVIAL MEMBRANES

Synovial membranes line the cavities of the joints. Like serous membranes, they line structures that do not open to the exterior. Unlike mucous, serous, and cutaneous membranes, they do not contain epithelium and are therefore not epithelial membranes. They are composed of loose connective tissue with elastic fibers and varying amounts of fat. Synovial membranes secrete *synovial fluid,* which lubricates the ends of bones as they move at joints and nourishes the articular cartilage covering the bones that form the joint.

STUDY OUTLINE

Types of Tissues

1. A tissue is a group of similar cells and their intercellular substance specialized for a particular function.
2. Specialized activities of different types of tissue include moving body parts, moving food through organs, protecting and supporting the body, and producing chemicals such as enzymes and hormones.
3. Depending on their function and structure, the various tissues of the body are classified into four principal types: epithelial, connective, muscular, and nervous.

Epithelial Tissue

1. Epithelium covers and lines body surfaces and forms glands.
2. Epithelium has many cells, little intercellular material, and no blood vessels. It is attached to connective tissue by a basement membrane. It can replace itself.
3. Cell junctions found between epithelial cells include desmosomes, tight junctions, intermediate junctions, and gap junctions.

Covering and Lining Epithelium

1. Types of epithelium based on arrangement of cell layers are simple epithelium, a single layer of cells adapted for absorption or filtration; stratified, several layers of cells adapted for protection; and pseudostratified, a single layer in which some cells do not reach the surface.
2. Epithelial cell shapes include squamous (flat), cuboidal (cubelike), columnar (rectangular), and transitional (variable).
3. Simple squamous epithelium is adapted for diffusion and filtration and is found in lungs and kidneys. Endothelium lines the heart and blood vessels. Mesothelium lines body cavities and covers internal organs.
4. Simple cuboidal epithelium is adapted for secretion and absorption in kidneys and glands.
5. Simple columnar epithelium lines the digestive tract. Specialized cells containing microvilli perform absorption. Goblet cells perform secretion. In some portions of the respiratory tract, the cells are ciliated to move foreign particles out of the body.
6. Stratified squamous epithelium is protective. It lines the upper digestive tract and forms the outer layer of skin.
7. Stratified cuboidal epithelium is found in adult sweat glands.
8. Stratified columnar epithelium protects and secretes. It is found in the male urethra and excretory ducts.
9. Transitional epithelium lines the urinary bladder and is capable of stretching.
10. Pseudostratified epithelium has only one layer of co-

lumnar cells but gives the appearance of many. It lines excretory and upper respiratory structures where it protects and secretes.

Glandular Epithelium

1. A gland is a single cell or a mass of epithelial cells adapted for secretion.
2. Exocrine glands (sweat, oil, and digestive glands) secrete into ducts.
3. Exocrine glands may be classified structurally as unicellular or multicellular.
4. Exocrine glands may be classified functionally as holocrine, merocrine, or apocrine.
5. Endocrine glands secrete hormones into the blood.

Connective Tissue

1. Connective tissue is the most abundant body tissue. It has few cells, an extensive matrix, and a rich blood supply.
2. It protects, supports, and binds organs together.
3. It can be classified into specific types of tissue: embryonic and adult. Adult connective tissue includes connective tissue proper, cartilage, bone, and vascular tissue.

Embryonic Connective Tissue

1. Mesenchyme forms all other connective tissue.
2. Mucous connective tissue is found only in the umbilical cord of the fetus, where it gives support.

Adult Connective Tissue

1. Adult connective tissue is connective tissue that exists in the newborn and that does not change after birth. It is subdivided into several kinds: connective tissue proper, cartilage, bone tissue, and vascular tissue.
2. Connective tissue proper has a more or less fluid intercellular material, and a typical cell is the fibroblast. Five examples of such tissues may be distinguished.
3. Loose connective (areolar) is one of the most widely distributed connective tissues in the body.
4. Adipose tissue is a form of loose connective tissue in which the adipocytes are specialized for fat storage.
5. Dense connective (collagenous) tissue has a predominance of collagenous or white fibers arranged in bundles.
6. Elastic connective tissue has a predominance of freely branching elastic fibers that give it a yellow color.
7. Reticular connective tissue consists of interlacing reticular fibers.
8. Cartilage has a gel-like matrix of collagenous and elastic fibers that contains chondrocytes. Cartilage

grows by interstitial growth (expansion from within) and appositional growth (surface growth).

9. Hyaline cartilage is found at the ends of bones, in the nose, and in respiratory structures. It is flexible, allows movement, and provides support.

10. Fibrocartilage connects the pelvic bones and the vertebrae. It provides strength.

11. Elastic cartilage maintains the shape of organs such as the external ear.

Muscle Tissue and Nervous Tissue

1. Muscle tissue performs one major function: contraction.

2. Nervous tissue is specialized to conduct electrical impulses.

Membranes

1. An epithelial membrane is an epithelial layer overlying a connective tissue layer. Examples are mucous, serous, and cutaneous.

2. Mucous membranes line cavities that open to the exterior, such as the digestive tract.

3. Serous membranes (pleura, pericardium, peritoneum) line closed cavities and cover the organs in the cavities. These membranes consist of parietal and visceral portions.

4. The cutaneous membrane is the skin.

5. Synovial membranes line joint cavities and do not contain epithelium.

REVIEW QUESTIONS

1. Define a tissue. What are the four basic kinds of human tissue?

2. What characteristics are common to all epithelium? Distinguish covering and lining epithelium from glandular epithelium.

3. Discuss the classification of epithelium based on layering and cell type.

4. Define a cell junction. List several functions of cell junctions.

5. Compare desmosomes, hemidesmosomes, tight junctions, intermediate junctions, and gap junctions on the basis of structure, location, and function.

6. For each of the following kinds of epithelium, briefly describe the microscopic appearance, location in the body, and functions: simple squamous, simple cuboidal, simple columnar, stratified squamous, stratified cuboidal, stratified columnar, transitional, and pseudostratified.

7. What is a gland? Distinguish between endocrine and exocrine glands. Describe the classification of exocrine glands according to structural complexity and function and give at least one example of each.

8. Enumerate the ways in which connective tissue differs from epithelium. How are connective tissues classified?

9. Compare embryonic connective tissue with adult connective tissue.

10. Describe the following connective tissues with regard to microscopic appearance, location in the body, and function: loose (areolar), adipose, dense (collagenous), elastic, reticular, hyaline cartilage, fibrocartilage, and elastic cartilage.

11. Distinguish between interstitial and appositional growth as it applies to cartilage.

12. Define the following kinds of membranes: mucous, serous, cutaneous, and synovial. Where is each located in the body? What are their functions?

13. Name the tissue described in each of the following statements.

    A stratified epithelium that permits distention.

    A single layer of flat cells concerned with filtration and absorption.

    Forms all other kinds of connective tissue.

    Specialized for fat storage.

    An epithelium with waterproofing qualities.

    Forms the framework of many organs.

    Produces perspiration, wax, oil, and digestive enzymes.

    Cartilage that shapes the external ear.

    Contains goblet cells and lines the intestine.

    Most widely distributed connective tissue.

    Forms tendons, ligaments, and aponeuroses.

    Specialized for the secretion of hormones.

    Provides support in the umbilical cord.

    Lines kidney tubules and is specialized for absorption and secretion.

    Permits extensibility of lung tissue.

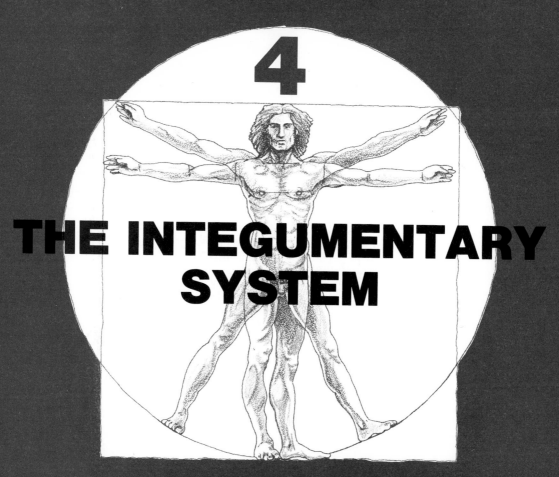

4

THE INTEGUMENTARY SYSTEM

STUDENT OBJECTIVES

■ Define the skin as the organ of the integumentary system.

■ Explain how the skin is structurally divided into epidermis and dermis.

■ List the various layers of the epidermis and describe their functions.

■ Explain the basis for skin color.

■ Describe the composition and function of the dermis.

■ Describe the development, distribution, structure, color, and replacement of hair.

■ Compare the structure, distribution, and functions of sebaceous and sudoriferous glands.

■ List the parts of a nail and describe their composition.

■ Describe the causes, effects, and treatment (where applicable) for the following skin disorders: acne, impetigo, systemic lupus erythematosus, psoriasis, decubitus, warts, cold sores, sunburn, and skin cancer.

■ Classify burns into first, second, and third degrees.

■ Define the "rule of nines" for estimating the extent of a burn.

■ Define key medical terms associated with the integumentary system.

An aggregation of tissues that performs a specific function is an **organ.** The next higher level of organization is a **system**—a group of organs operating together to perform specialized functions. The skin and its derivatives—such as hair, nails, glands, and several specialized receptors—constitute the **integumentary** (in'-teg-yoo-MEN-tar-ē) **system** of the body.

The developmental anatomy of the integumentary system is considered in Exhibit 24-1.

SKIN

The **skin** or **cutis** is an organ because it consists of tissues structurally joined together to perform specific activities. It is not just a simple thin covering that keeps the body together and gives it protection. The skin is quite complex in structure and performs several vital functions. In fact, this organ is essential for survival.

The skin is one of the larger organs of the body in terms of surface area. For the average adult, the skin occupies a surface area of approximately 7,620 sq cm (3,000 sq inches). It varies in thickness from 0.05 to 3 mm (0.02 to 0.12 inch). In general, skin is thicker on the dorsal surface of a part of the body than it is on the ventral surface, but this condition is reversed in the hand and the foot. The skin of the palmar surface of the hand and the plantar surface of the foot is thicker than on their dorsal aspects. Thick skin has a relatively thick epidermis whereas thin skin has a relatively thin epidermis. On most of the body, the epidermis is from 0.07 to 0.12 mm thick but can be 0.8 mm on the palms and 1.4 mm on the soles.

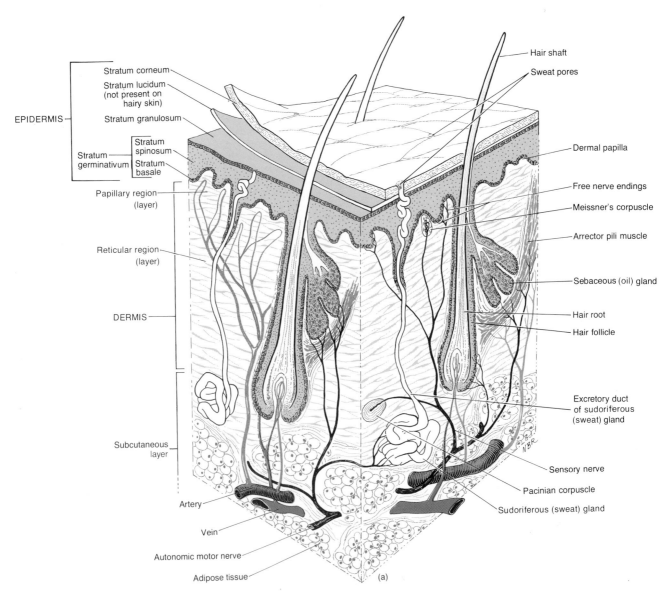

FIGURE 4-1 Skin. (a) Structure of the skin and underlying subcutaneous layer.

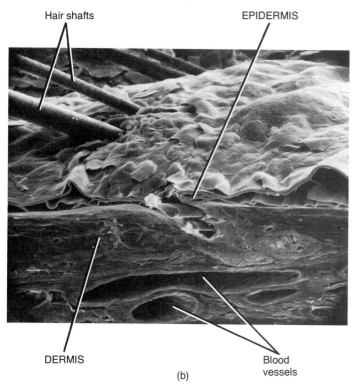

Hair shafts

EPIDERMIS

DERMIS

Blood vessels

(b)

FIGURE 4-1 *(Continued)* Skin. (b) Scanning electron micrograph of the skin and several hairs at a magnification of 260×. (Courtesy of Richard G. Kessel and Randy H. Kardon, from *Tissues and Organs: A Text-Atlas of Scanning Electron Microscopy,* W. H. Freeman and Company, San Francisco, 1979.)

FUNCTIONS

The skin covers the body and protects the underlying tissues, not only from bacterial invasion, but also from drying out and from harmful light rays. Moreover, the skin helps to control body temperature, prevents excessive loss of inorganic and organic materials, receives stimuli from the environment, stores chemical compounds, excretes water and salts, and synthesizes several important compounds, including vitamin D.

STRUCTURE

Structurally, the skin consists of two principal parts (Figure 4-1). The outer, thinner portion, which is composed of epithelium, is called the **epidermis.** The epidermis is cemented to the inner, thicker, connective tissue part called the **dermis.** Beneath the dermis is a **subcutaneous layer.** This layer, also called the **superficial fascia,** consists of areolar and adipose tissues. Fibers from the dermis extend down into the superficial fascia and anchor the skin to the subcutaneous layer. The superficial fascia, in turn, is firmly attached to underlying tissues and organs.

EPIDERMIS

The **epidermis** is composed of stratified squamous epithelium organized in four or five cell layers, depending on its location in the body (see Figure 4-1 and Figure 4-2). Where exposure to friction is greatest, such as the palms and soles, the epidermis has five layers. In all other parts it has four layers. The names of the five layers from the inside outward are as follows.

1. Stratum basale. This single layer of cuboidal to columnar cells is capable of continued cell division. As these cells multiply, they push up toward the surface and become part of the layers to be described next. Their nuclei degenerate, and the cells die. Eventually, the cells are shed from the top layer of the epidermis.

2. Stratum spinosum. This layer of the epidermis contains eight to ten rows of polyhedral (many-sided) cells

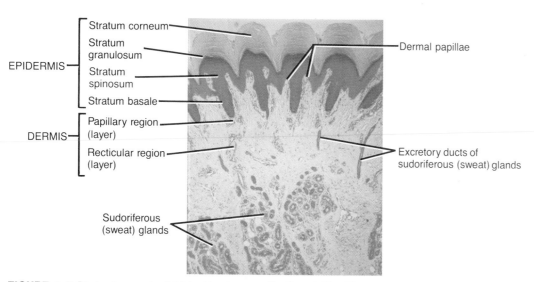

EPIDERMIS
- Stratum corneum
- Stratum granulosum
- Stratum spinosum
- Stratum basale

DERMIS
- Papillary region (layer)
- Recticular region (layer)

Sudoriferous (sweat) glands

Dermal papillae

Excretory ducts of sudoriferous (sweat) glands

FIGURE 4-2 Photomicrograph of thick skin at a magnification of 60×. The stratum lucidum is not shown here but is illustrated in Fig 4-1a. (© 1983 by Michael H. Ross. Used by permission.)

that fit closely together. The surfaces of these cells may assume a prickly appearance when prepared for microscope examination (*spinosum* = prickly). The stratum basale and stratum spinosum are sometimes collectively referred to as the **stratum germinativum** (jer'-mi'na-TĒ-vum) to indicate the layers where new cells are germinated. The deeper layers of the epidermis of hairless skin contain nerve endings sensitive to light touch called *Merkel's* (*tactile*) *discs* (see Figure 18-1).

3. Stratum granulosum. This third layer of the epidermis consists of three to five rows of flattened cells that contain darkly staining granules of a substance called *keratohyalin* (ker'-a-tō-HĪ-a-lin). This compound is involved in the first step of keratin formation. *Keratin* is a waterproofing protein found in the top layer of the epidermis. The nuclei of the cells in the stratum granulosum are in various stages of degeneration. As these nuclei break down, the cells are no longer capable of carrying out vital metabolic reactions and die.

4. Stratum lucidum. This layer is quite pronounced in the thick skin of the palms and soles. It consists of several rows of clear, flat, dead cells that contain droplets of a translucent substance called *eleidin* (el-Ē-i-din). Eleidin is formed from keratohyalin and is eventually transformed to keratin. This layer is so named because of the translucent property of eleidin (*lucidum* = clear).

5. Stratum corneum. This layer consists of 25 to 30 rows of flat, dead cells completely filled with keratin. These cells are continuously shed and replaced. The stratum corneum serves as an effective barrier against light and heat waves, bacteria, and many chemicals.

SKIN COLOR

The color of the skin is due to melanin, a pigment in the epidermis; carotene, a pigment mostly in the dermis; and blood in the capillaries in the dermis. The amount of **melanin** varies the skin color from pale yellow to black. This pigment is found primarily in the basale and spinosum layers. Since the number of melanocytes (melanin-producing cells) is about the same in all races, differences in skin color are due to the amount of pigment the melanocytes produce and disperse. An inherited inability of an individual in any race to produce melanin results in **albinism** (AL-bi-nizm). The pigment is absent in the hair and eyes as well as the skin. An individual affected with albinism is called an **albino** (al-BĪ-no). In some people, melanin tends to form in patches called **freckles.**

Melanin is synthesized in cells called **melanocytes** (me-LAN-ō-sīts), located either just beneath or between cells of the stratum basale. Melanocytes synthesize melanin from the amino acid *tyrosine* in the presence of an enzyme called *tyrosinase*. Exposure to ultraviolet radiation increases the enzymatic activity of melanocytes and leads to increased melanin production. The cell bodies of melanocytes send out long processes between epidermal cells (Figure 4-3). Upon contact with the processes, epidermal cells take up the melanin by phagocytosis. When the skin is again exposed to ultraviolet radiation, both the amount and the darkness of melanin increase, tanning and further protecting the body against radiation. Thus melanin serves a vital protective function. In mammals, the melanocyte-stimulating hormone (MSH) produced by the an-

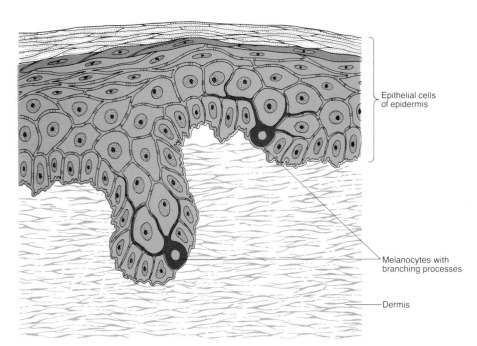

Epithelial cells of epidermis

Melanocytes with branching processes

Dermis

FIGURE 4-3 Melanocytes. Note the branching processes that project from the melanocytes and supply the epithelial cells of the epidermis with melanin.

terior pituitary gland causes increased melanin synthesis and dispersion through the epidermis. The exact role of MSH in humans is not clear.

CLINICAL APPLICATION

Overexposure of the skin to the ultraviolet light of the sun may lead to skin cancer. Among the most malignant and lethal skin cancers is **melanoma** (*melano* = dark-colored; *oma* = tumor), cancer of the melanocytes. Fortunately, most skin cancers are operable.

Another pigment called **carotene** (KAR-o-tēn) is found in the stratum corneum and fatty areas of the dermis in Oriental people. Together carotene and melanin account for the yellowish hue of their skin. The pink color of Caucasian skin is due to blood in capillaries in the dermis. The redness of the vessels is not heavily masked by pigment. The epidermis has no blood vessels, a characteristic of all epithelia.

DERMIS

The second principal part of the skin, the **dermis,** is composed of connective tissue containing collagenous and elastic fibers (see Figure 4-1). It is about 0.5 to 2.5 mm (0.02 to 0.11 inch) thick. The dermis is very thick in the palms and soles and very thin in the eyelids, penis, and scrotum. It also tends to be thicker on the dorsal aspects of the body than the ventral and thicker on the lateral aspects of extremities than medial aspects. Numerous blood vessels, nerves, glands, and hair follicles are embedded in the dermis.

The upper region of the dermis, about one-fifth of the thickness of the total layer, is named the **papillary region** or **layer.** Its surface area is greatly increased by small, fingerlike projections called **dermal papillae** (pa-PIL-ē). These structures project into the concavities between the ridges in the deep surface of the epidermis and many contain loops of capillaries. Some dermal papillae contain *Meissner's* (MĪS-nerz) *corpuscles,* nerve endings sensitive to light touch. The papillary region consists of loose connective tissue containing fine elastic fibers.

The remaining portion of the dermis is called the **reticular region** or **layer.** It consists of dense, irregularly arranged connective tissue containing interlacing bundles of collagenous and coarse elastic fibers. It is named the reticular layer because the bundles of collagenous fibers interlace in a netlike manner. Spaces between the fibers are occupied by a small quantity of adipose tissue, hair follicles, nerves, oil glands, and the ducts of sweat glands. Varying thicknesses of the reticular region are responsible for differences in the thickness of the skin.

The combination of collagenous and elastic fibers in the reticular region provides the skin with strength, exten-

sibility, and elasticity. The ability of the skin to stretch can readily be seen during conditions of pregnancy, obesity, and edema. The small tears that occur during extreme stretching are initially red and remain visible afterward as silvery white streaks called *striae* (STRĪ-ē).

The reticular region is attached to underlying organs, such as bone and muscle, by the subcutaneous layer. The subcutaneous layer also contains nerve endings called *Pacinian* (pa-SIN-ē-an) *corpuscles* that are sensitive to deep pressure (see Figure 18-1).

EPIDERMAL RIDGES AND GROOVES

The outer surface of the skin of the palms and fingers and soles and toes is marked by a series of ridges and grooves which appear either as fairly straight lines or as a pattern of loops and whorls, as on the tips of the digits.

The **epidermal ridges** develop during the third and fourth fetal months as the epidermis conforms to the contours of the underlying dermal papillae (Figure 4-4). Since the ducts of sweat glands open on the summits of the epidermal ridges, fingerprints (or footprints) are left when a smooth object is touched. The ridge pattern, which is genetically determined, is unique for each individual. It does not change throughout life, except to enlarge, and thus can serve as the basis for identification through fingerprints or footprints. The function of the ridges is to increase the grip of the hand or foot by increasing friction.

Epidermal grooves on other parts of the skin divide the surface into a number of diamond-shaped areas. Examine the dorsum of the hand as an example. Note that hairs typically emerge at the points of intersection of the grooves. Note also that the grooves increase in frequency and depth as regions of free joint movement are approached.

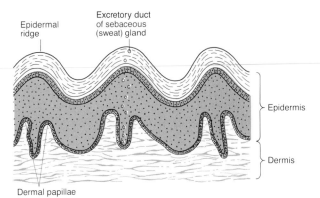

FIGURE 4-4 Relationship between epidermal ridges and dermal papillae.

BLOOD SUPPLY

The dermis is well supplied with blood; the epidermis is avascular. The arteries supplying the dermis are generally derived from branches of arteries supplying skeletal muscles in a particular region. Some arteries supply the skin directly. One plexus of arteries, the **cutaneous plexus,** is located at the junction of the dermis and subcutaneous layer and sends branches that supply the sebaceous and sudoriferous glands, the deep portions of hair follicles, and adipose tissue. The **papillary plexus,** formed at the level of the papillary layer, sends branches that supply the capillary loops in the dermal papillae, sebaceous glands, and superficial portions of hair follicles. The arterial plexuses are accompanied by venous plexuses that drain blood from the dermis into larger subcutaneous veins.

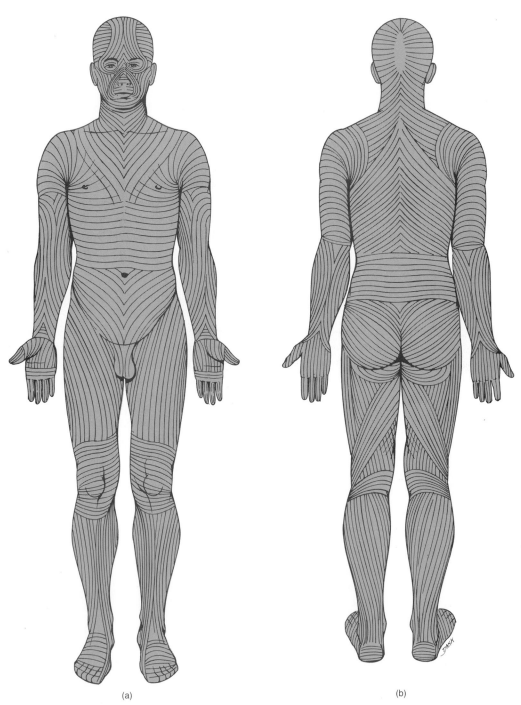

(a)   (b)

FIGURE 4-5 Lines of cleavage (Langer's lines). (a) Anterior view. (b) Posterior view.

SKIN AND TEMPERATURE REGULATION

Humans, like other mammals, are *homeotherms,* or warm-blooded organisms. This means that we are able to maintain a remarkably constant body temperature of 37° C (98.6° F) even though the environmental temperature may vary over a broad range.

Suppose you are in an environment where the temperature is 37.8° C (100° F). A sequence of events is set into operation to counteract this above-normal temperature. Sensing devices in the skin called receptors pick up the stimulus—in this case, heat—and activate nerves that send the message to the brain. A temperature-regulating area of the brain then sends nerve impulses to the sudoriferous glands, which produce more perspiration. As the perspiration evaporates from the surface of the skin, the skin is cooled and your body temperature is lowered.

Temperature regulation by perspiration represents only one mechanism by which we maintain normal body temperature. Other mechanisms include adjusting blood flow to the skin, regulating metabolic rate, and regulating skeletal muscle contractions.

CLINICAL APPLICATION

The collagenous fibers in the dermis run in all directions, but in particular regions of the body they tend to run more in one direction than another. The predominate direction of the underlying collagenous fibers is indicated in the skin by **lines of cleavage (Langer's lines)** or **tension lines.** The lines are especially evident on the palmar surfaces of the fingers where they run parallel to the long axis of the digit. The characteristic lines for each part of the body are shown in Figure 4-5.

Lines of cleavage are of particular interest to a surgeon because an incision running parallel to the collagen fibers will heal with only a fine scar. An incision made across the rows of fibers disrupts the collagen, and the wound tends to gape open and heal in a broad, thick scar.

EPIDERMAL DERIVATIVES

Structures developed from the embryonic epidermis—hair, glands, nails—perform functions that are necessary and sometimes vital. Hair and nails protect the body. The sweat glands help regulate body temperature.

HAIR

Growths of the epidermis variously distributed over the body are **hairs** or **pili.** The primary function of hair is protection. Though the protection is limited, hair guards the scalp from injury and the sun's rays. Eyebrows and eyelashes protect the eyes from foreign particles. Hair in the nostrils and external ear canal protects these structures from insects and dust.

Development and Distribution

During the third and fourth months of fetal life, the epidermis develops downgrowths into the dermis called *follicles.* Originating from these follicles are small hairs. By the fifth or sixth month, the follicles produce delicate hairs called **lanugo** (lan-YOO-gō; *lana = wool*)—hair that covers the fetus. The lanugo is usually shed prior to birth, except in the regions of the eyebrows, eyelids, and scalp. Here they persist and become stronger. Several months after birth, these hairs are shed and are replaced by still coarser ones, while the remainder of the body develops a new hair growth called the **vellus** (fleece). At puberty, coarse hairs develop in the axillary and pubic regions of both sexes and on the face and to a lesser extent on other parts of the body in the male. The coarse hairs that develop at puberty, plus the hairs of the scalp and eyebrows, are referred to as **terminal hairs.** The life span of scalp hair is from two to six years. If hairs are not replaced when they are shed, baldness results. Eyelashes are replaced every three to five months.

Hairs are distributed on nearly all parts of the body. They are absent from the palms of the hands, the soles of the feet, the dorsal surfaces of the distal phalanges, lips, nipples, clitoris, glans penis, inner surface of the prepuce, inner surfaces of the labia majora and minora, and outer surfaces of the labia minora. Straight hairs are oval or cylindrical in cross section, whereas curly hairs are flattened. Straight hairs are stronger than curly ones.

Structure

Each hair consists of a shaft and a root (Figure 4-6a). The **shaft** is the superficial portion, most of which projects above the surface of the skin. The shaft of coarse hairs consists of three principal parts. The inner **medulla** is composed of rows of polyhedral cells containing granules of eleidin and air spaces. The medulla is poorly developed or not present at all in fine hairs. The second principal part of the shaft is the middle **cortex.** It forms the major part of the shaft and consists of elongated cells that contain pigment granules in dark hair, but only air in white hair. The **cuticle of the hair,** the outermost layer, consists of a single layer of thin, flat, scalelike cells that are the most heavily keratinized and arranged like shingles on the side of a house, but the free edges of the cuticle cells point upward rather than downward like shingles (Figure 4-6f). The **root** is the portion below the surface that penetrates into the dermis and even into the subcutaneous layer and, like the shaft, also contains a medulla, cortex, and cuticle (Figure 4-6b, c).

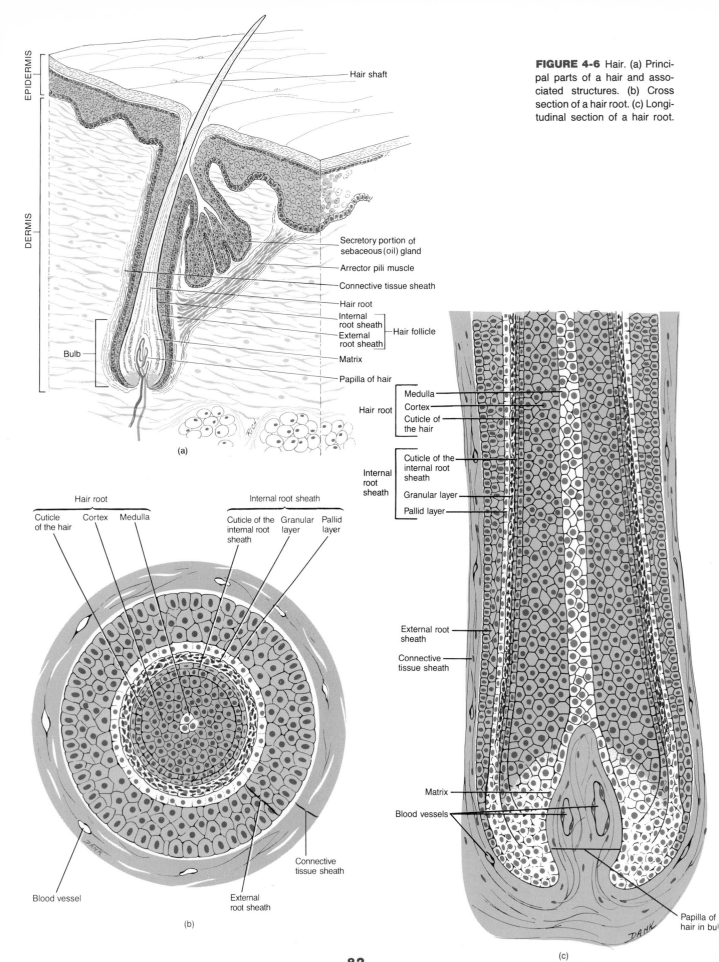

EPIDERMIS

DERMIS

Hair shaft

Secretory portion of sebaceous (oil) gland

Arrector pili muscle

Connective tissue sheath

Hair root

Internal root sheath

External root sheath

} Hair follicle

Bulb

Matrix

Papilla of hair

(a)

FIGURE 4-6 Hair. (a) Principal parts of a hair and associated structures. (b) Cross section of a hair root. (c) Longitudinal section of a hair root.

Hair root

Cuticle of the hair | Cortex | Medulla

Internal root sheath

Cuticle of the internal root sheath | Granular layer | Pallid layer

Blood vessel

External root sheath

Connective tissue sheath

(b)

Hair root

Medulla
Cortex
Cuticle of the hair

Internal root sheath

Cuticle of the internal root sheath

Granular layer

Pallid layer

External root sheath

Connective tissue sheath

Matrix

Blood vessels

Papilla of hair in bu

(c)

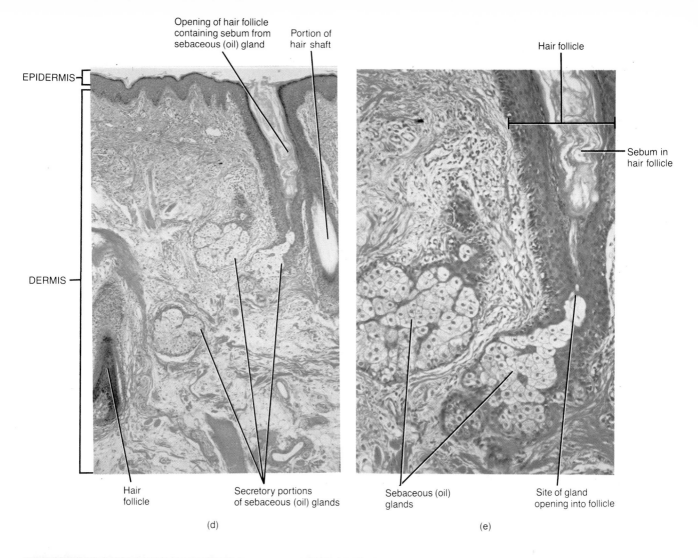

EPIDERMIS

DERMIS

Opening of hair follicle
containing sebum from
sebaceous (oil) gland

Portion of
hair shaft

Hair
follicle

Secretory portions
of sebaceous (oil) glands

(d)

Hair follicle

Sebum in
hair follicle

Sebaceous (oil)
glands

Site of gland
opening into follicle

(e)

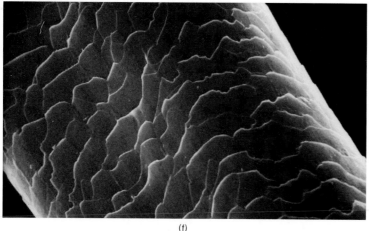

(f)

FIGURE 4-6 *(Continued)* Hair. (d) Photomicrograph of thin skin of the face at a magnification of 60×. (© 1983 by Michael H. Ross. Used by permission.) (e) Photomicrograph of a sebaceous gland opening into a hair follicle at a magnification of 150×. (© 1983 by Michael H. Ross. Used by permission.) (f) Scanning electron micrograph of the surface of a hair shaft showing the shinglelike cuticular scales at a magnification of 1,000×. (Courtesy of Fisher Scientific Company and S.T.E.M. Laboratories, Inc., Copyright, 1975.)

Surrounding the root is the **hair follicle,** which is made up of an external zone of epithelium (the external root sheath) and an internal zone of epithelium (the internal root sheath). The **external root sheath** is a downward continuation of the epidermis. Near the surface it contains all the epidermal layers. As it descends, it does not exhibit the superficial epidermal layers. At the bottom of the hair follicle, the external root sheath contains only the stratum basale. The **internal root sheath** is formed from proliferating cells of the matrix (described shortly) and takes the form of a cellular tubular sheath that separates the hair from the external root sheath. It extends only partway up the follicle. The internal root sheath consists of (1) an inner layer, the **cuticle of the internal root sheath,** which is a single layer of flattened cells with atrophied nuclei, (2) a middle **granular layer,** which is one to three layers of flattened nucleated cells, and (3) an outer **pallid layer,** which is a single layer of cuboidal cells with flattened nuclei.

The base of each follicle is enlarged into an onion-shaped structure, the **bulb.** This structure contains an indentation, the **papilla of the hair,** filled with loose connective tissue. The papilla of the hair contains many blood vessels and provides nourishment for the growing hair. The base of the bulb also contains a region of cells called the **matrix,** a germinal layer. The cells of the matrix produce new hairs by cell division when older hairs are shed. This replacement occurs within the same follicle. Hair grows about 1 mm (0.04 inch) every 3 days. Hair loss in an adult is about 70 to 100 hairs a day. But the rate of growth may be altered by illness and diet, among other factors.

Sebaceous glands and a bundle of smooth muscle are also associated with hair. The smooth muscle is called **arrector pili;** it extends from the dermis of the skin to the side of the hair follicle (Figure 4-6a, d). In its normal position hair is arranged at an angle to the surface of the skin. The arrectores pilorum muscles contract under stresses of fright and cold and pull the hairs into a vertical position. This contraction results in "goosebumps" or "gooseflesh" because the skin around the shaft forms slight elevations.

Around each hair follicle are nerve endings, called *root hair plexuses,* that are sensitive to light touch (see Figure 18-1). They respond if a hair shaft is moved.

Color

The color of hair is due primarily to melanin. It is formed by melanocytes distributed in the matrix of the bulb of the follicle. There are only three colors of hair pigment—black, brown, and yellow, due to the pigments black melanin, brown melanin, and pheomelanin, respectively. The many variations in hair color are combinations of different amounts of the three pigments. Graying of hair is the loss of pigment believed to be the result of a progressive inability of the melanocytes to make tyrosinase, the enzyme necessary for the synthesis of melanin.

Replacement

Hair replacement occurs according to a cyclic pattern, alternating between growing and resting periods. During the growing phase of the cycle, the cells of the matrix increase in number by cell division. This causes the hair to increase in length. During the resting phase, the matrix becomes inactive and undergoes atrophy. At this point the root of the hair detaches from the matrix and slowly moves up the follicle. It may remain there for some time

until pulled out, shed, or pushed up by a replacing hair. Meanwhile, the external root sheath retracts upward toward the surface. Either before or after the hair comes out of the follicle, proliferation of cells by the external root sheath forms a new matrix. The new matrix undergoes cell division, forming a new hair that grows up the follicle and replaces the old hair. You might be interested to know that shaving or cutting the hair has no effect on its growth.

GLANDS

Two kinds of glands associated with the skin are sebaceous glands and sweat glands.

Sebaceous (Oil) Glands

Sebaceous (se-BĀ-shus) or **oil glands,** with few exceptions, are connected to hair follicles (see Figure 4-6a, d, e). The secreting portions of the glands lie in the dermis and those glands associated with hairs open into the necks of hair follicles. Sebaceous glands not associated with hair follicles open directly onto the surface of the skin (lips, glans penis, labia minora, and meibomian glands of the eyelids). Sebaceous glands are simple branched acinar glands. Absent in the palms and soles, they vary in size and shape in other regions of the body. For example, they are small in most areas of the trunk and extremities but large in the skin of the breasts, face, neck, and upper chest. The sebaceous glands secrete an oily substance called **sebum** (SĒ-bum), a mixture of fats, cholesterol, proteins, and inorganic salts. These glands keep the hair from drying and becoming brittle and also form a protective film that prevents excessive evaporation of water from the skin. The sebum keeps the skin soft and pliable, too.

CLINICAL APPLICATION

When sebaceous glands of the face become enlarged because of accumulated sebum, **blackheads** develop. Since sebum is nutritive to certain bacteria, **pimples** or **boils** often result. The color of blackheads is due to melanin and oxidized oil, not dirt.

Sudoriferous (Sweat) Glands

Sudoriferous (soo'-dor-IF-er-us; *sudor* = sweat; *ferre* = to bear) or **sweat glands** are distributed throughout the skin except on the nail beds of the fingers and toes, margins of the lips, eardrums, and tips of the penis and clitoris. In contrast to sebaceous glands, sudoriferous glands are most numerous in the skin of the palms and the soles, where their density can be as high as 3,000/square inch. They are also found in abundance in the armpits and forehead. Each gland consists of a coiled end embedded primarily in the subcutaneous layer plus a single tube that projects upward through the dermis and epidermis. This tube, the excretory duct, terminates in a pore at the surface of the epidermis (see Figure 4-1). The base of each sudoriferous gland is surrounded by a network of small blood vessels. In the axillary region, sudoriferous glands are of the simple branched tubular type. Elsewhere they are simple coiled tubular glands.

Perspiration, or **sweat,** is the substance produced by sudoriferous glands. It is a mixture of water, salts (mostly NaCl), urea, uric acid, amino acids, ammonia, sugar, lactic acid, and ascorbic acid. Its principal function is to help regulate body temperature. It also helps to eliminate wastes.

Ceruminous Glands

In certain parts of the skin, sudoriferous glands are modified as **ceruminous** (se-ROO-mi-nus) **glands.** Such modified glands are simple, coiled tubular glands present in the external auditory meatus (canal). The secretory portions of ceruminous glands lie in the submucosa, deep to sebaceous glands, and the ceruminous excretory ducts open either directly onto the surface of the external auditory meatus or into sebaceous ducts. The combined secretion of the ceruminous and sebaceous glands is called **cerumen** (*cera* = wax). Cerumen may accumulate resulting in impacted cerumen (Chapter 18).

NAILS

Hard keratinized cells of the epidermis are referred to as **nails.** The cells form a clear, solid covering over the dorsal surfaces of the terminal portions of the fingers and toes. Each nail (Figure 4-7) consists of a **nail body, a free edge,** and a **nail root.** The nail body is the portion of the nail that is visible, the free edge is the part that projects beyond the distal end of the digit, and the nail root is the portion that is hidden in the nail groove. Most of the nail body is pink because of the underlying vascular tissue. The whitish semilunar area at the proximal end of the body is called the **lunula** (LOO-nyoo-la). It appears whitish because the vascular tissue underneath does not show through.

The fold of skin that extends around the proximal and lateral borders of the nail is known as the **nail fold,** and the epidermis beneath the nail constitutes the **nail bed.** The furrow between the two is the **nail groove.**

The **eponychium** (ep'-ō-NIK-ē-um) or **cuticle** is a narrow band of epidermis that extends from the margin of the nail wall (lateral border), adhering to it. It occupies the proximal border of the nail and consists of stratum corneum. The thickened area of stratum corneum below the free edge of the nail is referred to as the **hyponychium** (hī'-pō-NIK-ē-um).

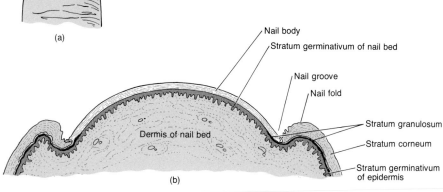

(a)

(b)

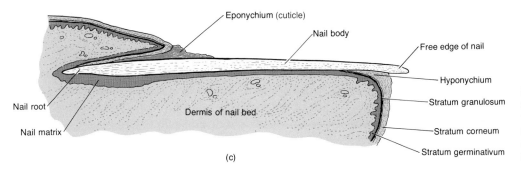

(c)

FIGURE 4-7 Structure of nails. (a) Fingernail viewed from above. (b) Cross section of a fingernail and the nail bed. (c) Sagittal section of a fingernail.

The epithelium of the proximal part of the nail bed is known as the **nail matrix.** Its function is to bring about the growth of nails. Essentially, growth occurs by the transformation of superficial cells of the matrix into nail cells. In the process, the outer, harder layer is pushed forward over the stratum germinativum. The average growth in the length of fingernails is about 1 mm (0.04 inch)/week. The growth rate is somewhat slower in toenails.

APPLICATIONS TO HEALTH

ACNE

Acne is an inflammation of sebaceous glands and usually begins at puberty. In fact, a few comedones (blackheads) on the face may be the first signs of approaching puberty.

Common acne is called **acne vulgaris.** It occurs mostly in individuals between the ages of 14 and 25 and affects more than 80 percent of teenagers. It appears to affect an increasing number of older people, however, and women in their 20s who had little or none as teenagers. This latter variety, **acne cosmetica** or adult acne, is most likely to occur around the chin area. Causes may include taking birth control pills, going off the "pill," oily cosmetics, stress, and iodides found in many shellfish and other salt-water fish.

At puberty the sebaceous glands, under the influence of androgens (male hormones) grow in size and increase production of their complex lipid product, sebum. Although testosterone, a male hormone, appears to be the most potent circulating androgen for sebaceous cell stimulation, adrenal and ovarian androgens can stimulate sebaceous secretions as well. Acne occurs predominantly in sebaceous follicles. The sebaceous follicles are rapidly colonized by microorganisms such as staphylococcal species and *Propionibacterium acnes* bacteria that thrive in the lipid-rich environment of the follicles. When this occurs, the cyst or sac of connective tissue cells can destroy and displace epidermal cells, resulting in permanent scarring. Care must be taken to avoid squeezing, pinching, or scratching the lesions. The four basic types of acne lesions in order of increasing severity are comedones (open blackheads, closed whiteheads), papules, pustules, and cysts.

IMPETIGO

Impetigo (im'-pe-TĪ-gō) is a superficial skin infection caused by staphylococci or streptococci. It is characterized by isolated pustules that become crusted and rupture. Occurring principally around mouth, nose, and hands, the inflammation is located in the papillary layer of the skin and involves the capillary network and stratum corneum. The disease is most common in children, and epidemics in nurseries (due to staphylococci) may be serious.

SYSTEMIC LUPUS ERYTHEMATOSUS (SLE)

Systemic lupus erythematosus (er-i'-them-a-TŌ-sus), **SLE,** or **lupus** is an autoimmune, inflammatory disease of connective tissue, occurring mostly in young women during their 20s. An *autoimmune disease* is one in which the body attacks its own tissues, failing to differentiate between what is foreign and what is not. Although the cause of SLE is unknown, it affects virtually every body system.

SLE is not contagious. The inflammatory process occurs in collagenous connective tissue throughout the body. Symptoms include low-grade fever, aches, fatigue, and sometimes an eruption across the bridge of the nose and cheeks called a "butterfly rash." Other skin lesions may occur with blistering and ulceration. The disease may be triggered by medication, such as penicillin, sulfa, or tetracycline, exposure to excessive sunlight, injury, emotional upset, infection, or other stress. These triggering factors, once recognized, are to be avoided by the patient in the future.

Although SLE is not thought to be hereditary, there seems to be a strong incidence of other connective tissue disorders—especially rheumatoid arthritis and rheumatic fever—in relatives of SLE victims. The two most serious complications of SLE involve inflammation of the kidneys and the central nervous system.

PSORIASIS

Psoriasis (sō-RĪ-a-sis) is a chronic, occasionally acute, relapsing skin disease borne by 6 to 8 million people in the United States. In its severe forms, psoriasis is a disabling and disfiguring affliction. It may begin at any age, although it is most severe between ages 10 and 50. The cause is unknown. It is thought, however, to be hereditary in at least one-third of the patients.

Psoriasis is characterized by distinct, reddish, slightly raised plaques or papules (small, round skin elevations) covered with scales. Itching is seldom severe, and the lesions heal without scarring. Psoriasis ordinarily involves the scalp, the elbows and knees, the back, and the buttocks. Occasionally the disease is generalized. Psoriasis involves a high rate of mitosis, either as the primary defect or as secondary to other phenomena. Any antimitotic drug is effective against this disease.

DECUBITUS

Decubitus (dē-KYOO-be-tus), **bedsore, pressure sore,** or **trophic ulcer** is caused by a constant deficiency of blood to tissues overlying a bony projection that has been subjected to prolonged pressure against an object such as a bed, cast, or splint. It causes tissue ulceration and tissue destruction and is seen most frequently in patients who are bedridden for long periods of time.

The most common areas involved are the skin over the sacrum, heels, ankles, buttocks, and other large bony projections. The chief causes are pressure from infrequent turning of the patient, trauma and maceration of the skin, and malnutrition. It begins with damage to the sensitive subcutaneous and deeper tissues; later the skin dies. Small breaks in the epidermis become infected. Maceration of the skin often follows soaking of bed and clothing by perspiration, urine, or feces.

WARTS

Certain viruses (papovaviruses) can cause epithelial skin cells to proliferate into uncontrolled growths called **warts.** Warts are generally benign (noncancerous), and most regress spontaneously. It is possible to spread warts by contact-transfer of the viruses, and sexual contact has been implicated in the transmission of warts on the genitals. After infection, there is an incubation period of several weeks before the appearance of the wart.

Warts may be removed by liquid nitrogen cryotherapy; by cauterization by chemicals such as cantharidin, salicylic acid, trichloroacetic acid, dinitrochlorobenzene, bleomycin sulfate; or by surgical excision.

COLD SORES (FEVER BLISTERS)

The herpes simplex virus has the ability to lay dormant for extended periods of time without expressing any signs or symptoms of disease. During its periods of dormancy, the virus probably infects only a few cells at a time. Although the initial infection usually occurs during infancy, it is for the most part subclinical. The few individuals who express the disease during infancy have characteristic lesions, usually in the oral mucous membrane. These infections subside, but recur as what are called **cold sores (fever blisters).** The appearance of cold sore lesions is associated with stimuli such as exposure to the ultraviolet (UV) radiation from the sun, hormonal changes related to the menstrual cycle, or even emotional upset. Cold sores are caused by type 1 herpes simplex virus, which is transmitted by oral or respiratory routes. Type 2 herpes simplex virus causes genital herpes infections and is transmitted primarily by sexual contact (Chapter 23). There is not yet any effective treatment for cold sores or genital herpes infections.

SUNBURN

Sunburn is injury to the skin as a result of acute, prolonged exposure to the ultraviolet (UV) rays of sunlight; it usually occurs 2 to 8 hours after exposure. Acute redness and pain are maximal in about 12 hours and dissipate approximately 72 to 96 hours later. Sunburn is without doubt one of the most common and painful skin afflictions. The damage to skin cells caused by sunburn is due to inhibition of DNA and RNA synthesis which leads to cell death. There can also be damage to blood vessels as well as other structures in the dermis. Overexposure over a period of years results in a leathery skin texture, wrinkles, skin folds, sagging skin, warty growths called keratoses, freckling, a yellow discoloration due to abnormal elastic tissue, and premature aging of the skin.

The tendency to sunburn is determined by type of skin. There are basically four skin groups. Those who burn and never tan, those who burn but keep some tan, those who burn slightly and develop a good tan, and those who never burn and always tan.

Covering the body remains the best protection against the sun's rays, while a topical sunscreen will help protect uncovered parts. Many such agents indicate their screening effectiveness with a numerical rating—the *sun protection factor* (SPF). United States Food and Drug Administration guidelines set 15 as the ceiling for SPF values on labeling. Agents with an SPF greater than 12 afford a high degree of protection against sunburn and allow only very slow tanning. These are the products of choice for persons who are fair-skinned, tan poorly, and burn easily. Screening agents with an SPF of 6 to 12 provide moderate protection and should be used by sports participants who wish to tan while minimizing sunburn.

Extreme and severe sunburn may produce *sunstroke—* a disruption of the body's heat-regulating mechanism. It is characterized by high fever and collapse and sometimes by convulsions, coma, and death.

SKIN CANCER

Excessive sun exposure can result in **skin cancer.** Some 600,000 people have diagnosed skin cancer in the United States and 300,000 more will have diagnosed skin cancer this year. About 9,300 of these new cases will be melanoma.

Everyone, regardless of skin pigmentation, is a potential victim of skin cancer if exposure to sunlight is sufficiently intense and continuous. Natural skin pigment can never give complete protection. If you must be in direct sunlight for long periods of time, use a suitable sunscreen. One of the best agents for protection against overexposure to ultraviolet rays of the sun is para-aminobenzoic acid (PABA). The alcohol preparations of PABA are best because the active ingredient binds to the stratum corneum of the skin.

A number of widely prescribed drugs (tetracyclines, sulfa drugs, and others) and constituents of foods (such as riboflavins) are potential photosensitizers. When activated in the body by light, they may produce substances that can damage the tissues in sensitive individuals. A typical response is the appearance of a rash on parts of the body exposed to the sun. Repeated and persistent exposure could produce permanent changes leading to skin cancer.

BURNS

Tissues may be damaged by thermal (heat), electrical, radioactive, or chemical agents. These agents can destroy the proteins in the exposed cells and cause cell injury or death. Such damage is a **burn.** The injury to tissues directly or indirectly in contact with the damaging agent, such as the skin or the linings of the respiratory and digestive tracts, is the local effect of a burn. Generally, however, the systemic effects of a burn are a greater threat to life than the local effects. The systemic effects of a burn may include:

1. A large loss of water, plasma, and plasma proteins, which causes shock.
2. Bacterial infection.
3. Reduced circulation of blood.
4. Decreased production of urine.

Classification

A burn may extend through the entire thickness of the skin or it may damage or destroy only part of the skin. Clinically, the depth of a burn is determined by its color, the presence or absence of sensation, blister formation, or loss of elasticity. A **first-degree burn** is characterized by mild pain and erythema (redness) and involves only the surface epithelium. Generally, a first-degree burn will heal in about two to three days and may be accompanied by flaking or peeling. A typical sunburn is an example of a first-degree burn.

A **second-degree burn** involves the deeper layers of the epidermis or the upper levels of the dermis. In a superficial second-degree burn, the deeper layers of the epidermis are injured and there is characteristic erythema, blister formation, edema, and pain. Blisters beneath or within the epidermis are called *bullae* (BYOOL-ē), meaning bubbles. Such an injury usually heals within seven to ten days with only mild scarring. In a deep second-degree burn, there is destruction of the epidermis as well as the upper levels of the dermis. Epidermal derivatives, such as hair follicles, sebaceous glands, and sweat glands are usually not injured. If there is no infection, deep second-degree burns heal without grafting in about three to four weeks. Scarring may result.

First- and second-degree burns are collectively referred to as **partial-thickness burns.** A **third-degree burn** or **full-**

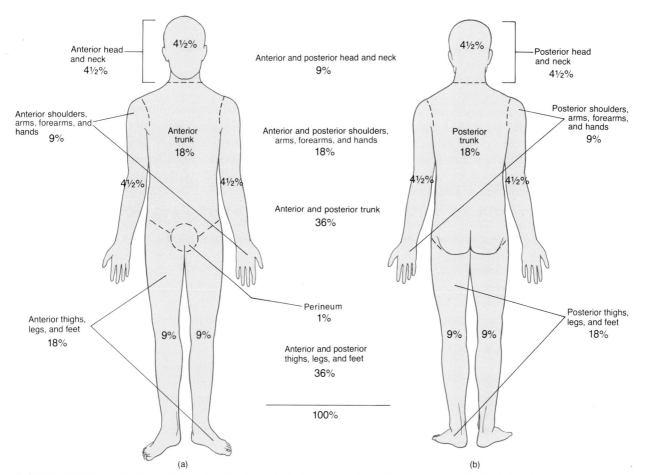

FIGURE 4-8 The "rule of nines" for estimating the extent of burns in adults. (a) Anterior view. (b) Posterior view.

thickness burn involves destruction of the epidermis, dermis, and the epidermal derivatives. Such burns vary in appearance from marble-white to mahogany colored to charred, dry wounds. There is little if any edema and such a burn is usually not painful to the touch due to destruction of nerve endings. Regeneration is slow and much granulation tissue forms before being covered by epithelium. Even if skin grafting is quickly begun, third-degree burns quickly contract and produce scarring.

"Rule of Nines"

A fairly accurate method for estimating the extent of a burn in adults is to apply the **"rule of nines"** (Figure 4-8).

1. If the anterior and posterior surfaces of the head and neck are affected, the burn covers 9 percent of the body surface.

2. The anterior and posterior surfaces of each shoulder, arm, forearm, and hand constitute 9 percent of the body surface.

3. The anterior and posterior surfaces of the trunk, including the buttocks, constitute 36 percent.

4. The anterior and posterior surfaces of each foot, leg, and thigh as far up as the buttocks total 18 percent.

5. The perineum (per-i-NĒ-um) consists of 1 percent. The perineum includes the anal and urogenital regions.

Treatment

A severely burned individual should be moved as quickly as possible to a hospital. Treatment may then include:

1. Cleansing the burn wounds thoroughly.

2. Removing all dead tissue (debridement) so antibacterial agents can directly contact the wound surface and thereby prevent infection.

3. Replacing lost body fluids and electrolytes (ions).

4. Covering wounds with grafts as soon as possible.

KEY MEDICAL TERMS ASSOCIATED WITH THE INTEGUMENTARY SYSTEM

Albinism (*alb* = white; *ism* = condition) Congenital (existing at birth) absence of pigment from the skin, hair, and parts of the eye.

Anhidrosis (*an* = without; *hidr* = sweating; *osis* = condition) A rare genetic condition characterized by inability to sweat.

Athlete's foot A superficial fungus infection of the skin of the foot.

Callus (callosity) An area of hardened and thickened skin that is usually seen in palms and soles and is due to pressure and friction.

Carbuncle (*carbunculus* = little coal) A hard, round, deep, painful inflammation of the subcutaneous tissue that causes necrosis (death) and pus formation (abscess).

Comedo (*comedo* = to eat up) A collection of sebaceous material and dead cells in the hair follicle and excretory duct of the sebaceous gland. Usually found over the face, chest, and back, and more commonly during adolescence. Also called blackhead or whitehead.

Corn A painful conical thickening of the skin found principally over toe joints and between the toes. It may be hard or soft, depending on the location. Hard corns are usually found over toe joints, and soft corns are usually found between the fourth and fifth toes.

Cyst (*cyst* = sac containing fluid) A sac with a distinct connective tissue wall, containing a fluid or other material.

Dermabrasion (*derm* = skin) Removal of acne scars, tattoos, or nevi (see below) by sandpaper or a high-speed brush.

Dermatome (*tomy* = cut) An instrument for excising areas of skin to be used for grafting.

Eczema (*ekzein* = to boil out) An acute or chronic superficial inflammation of the skin, characterized by redness, oozing, crusting, and scaling. Also called chronic dermatitis.

Epidermophytosis (*epi* = upon; *phyto* = pertaining to plants; *osis* = condition) Any fungus infection of the skin producing scaliness with itching.

Erythema (*erythema* = redness) Redness of the skin caused by engorgement of capillaries in lower layers of the skin. Erythema occurs with any skin injury, infection, or inflammation.

Fulminate (*fulminare* = to flare up) To occur suddenly with great intensity.

Furuncle A boil; an abscess resulting from infection of a hair follicle.

Hypodermic (*hypo* = under) Relating to the area beneath the skin.

Intradermal (*intra* = within) Within the skin. Also called intracutaneous.

Keratosis (*kera* = horn) Formation of a hardened growth of tissue.

Maceration Softening of a solid by soaking; wasting away.

Nevus A round, pigmented, flat, or raised skin area that may be present at birth or develop later. Varying in color from yellow-brown to black. Also called mole or birthmark.

Nodule (*nodulus* = little knot) A large cluster of cells raised above the skin but extending deep into the tissues.

Papilloma Benign tumor, such as a wart, derived from epithelium.

Papule A small, round skin elevation varying in size from a pinpoint to that of a split pea. One example is a pimple.

Pathogenesis (*patho* = disease; *genesis* = giving origin to) Origination and development of a disease.

Polyp A tumor on a stem found especially on mucous membranes.

Pustule A small, round elevation of the skin containing pus.

Subcutaneous (*sub* = under) Beneath the skin.

Topical Pertaining to a definite area; local.

STUDY OUTLINE

Skin

1. The skin and its derivatives (hair, glands, and nails) constitute the integumentary system.
2. The skin is one of the larger organs of the body. It performs the functions of protection, maintaining body temperature, receiving stimuli, synthesis or storage of chemical compounds, and excretion.
3. The principal parts of the skin are the outer epidermis and inner dermis. The dermis overlies the subcutaneous layer.
4. The epidermal layers, from the inside outward, are the stratum basale, spinosum, granulosum, lucidum, and corneum. The basale and spinosum undergo continuous cell division and produce all other layers.
5. The color of skin is due to melanin, carotene, and blood in capillaries in the dermis.
6. The dermis consists of a papillary region and a reticular region. The papillary region is loose connective tissue containing blood vessels, nerves, hair follicles, and dermal papillae. The reticular region is dense, irregularly arranged connective tissue containing adipose tissue, hair follicles, nerves, oil glands, and ducts of sweat glands.
7. Epidermal ridges reduce friction and provide the basis for fingerprints and footprints.
8. Lines of cleavage indicate the direction of collagenous fiber bundles in the dermis and are considered during surgery.

Epidermal Derivatives

1. The epidermal derivatives of the skin are hair, sebaceous glands, sudoriferous glands, ceruminous glands, and nails.
2. Hair consists of a shaft above the surface, a root anchored in the dermis, and a hair follicle.
3. Sebaceous glands are connected to hair follicles. They secrete sebum, which moistens hair and waterproofs the skin.
4. Sudoriferous glands produce perspiration, which car-

ries small amounts of wastes to the surface and assists temperature regulation.

5. Ceruminous glands are modified sudoriferous glands that secrete cerumen. They are found in the external auditory meatus.
6. Nails are modified epidermal cells. The principal parts of a nail are the body, free edge, root, eponychium, and matrix.

Applications to Health

1. Acne is an inflammation of sebaceous glands.
2. Impetigo is a superficial skin infection caused by bacteria and characterized by pustules that become crusted and rupture.
3. Systemic lupus erythematosus (SLE) is an autoimmune disease of connective tissue.
4. Psoriasis is a chronic skin disease characterized by reddish, raised plaques or papules.
5. Decubitus is tissue ulceration and destruction caused by a chronic deficiency of blood to tissues subjected to prolonged pressure.
6. Warts are uncontrolled growths of epithelial skin cells caused by a virus. Most warts are benign.
7. Cold sores are lesions caused by type 1 herpes simplex virus. The dormant infection is triggered by certain stimuli.
8. Sunburn is a skin injury as a result of prolonged exposure to the ultraviolet rays of sunlight.
9. Skin cancer is caused by excessive exposure to sunlight.
10. Tissue damage that destroys protein is called a burn. Depending on the extent of damage, skin burns are classified as first-degree and second-degree (partial-thickness) and third-degree (full-thickness). One method employed for determining the extent of a burn is to apply the "rule of nine." Burn treatment may include cleansing the wound, removing dead tissue, replacing lost body fluids, and covering wounds with grafts.

REVIEW QUESTIONS

1. Define an organ. In what respect is the skin an organ? What is the integumentary system?
2. List the principal functions of the skin.
3. Compare the structures of epidermis and dermis. What is the subcutaneous layer?
4. List and describe the epidermal layers from the inside outward. What is the importance of each layer?
5. Explain the factors that produce skin color. What is an albino?
6. Describe how melanin is synthesized and distributed to epidermal cells.
7. Contrast the structural differences between the papillary and reticular regions of the dermis.
8. How are epidermal ridges formed? Why are they important?
9. What are lines of cleavage? What is their importance during surgery?
10. Describe which receptors are located in the epidermis, dermis, and subcutaneous layer and indicate the role of each receptor.
11. Describe the development and distribution of hair.
12. Describe the structure of a hair. How are hairs moistened? What produces "goosebumps" or "gooseflesh"?
13. Explain the basis for hair color. Why does hair turn gray?
14. Describe how hair is replaced.
15. Contrast the locations and functions of sebaceous glands, sudoriferous glands, and ceruminous glands. What are the name and chemical components of the secretions of each?
16. From what layer of the skin do nails form? Describe the principal parts of a nail.
17. Define each of the following disorders of the integumentary system: acne, impetigo, systemic lupus erythematosus, psoriasis, decubitus, warts, cold sores, sunburn, and skin cancer.
18. Define a burn. Classify burns according to degree.
19. What is the "rule of nines"?
20. How are burns treated?
21. Refer to the glossary of key medical terms associated with the integumentary system. Be sure that you can define each term.

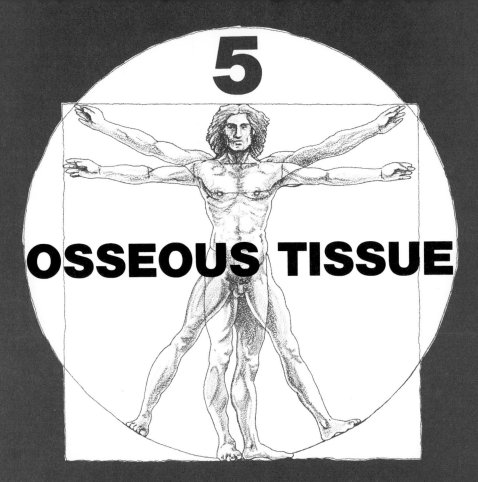

5

OSSEOUS TISSUE

STUDENT OBJECTIVES

- Identify the components of the skeletal system.
- Explain the gross features of a long bone.
- Describe the histological features of dense bone tissue.
- Compare the histological characteristics of spongy and dense bone.
- Contrast the steps involved in intramembranous and endochondral ossification.
- Identify the zones and growth pattern of the epiphyseal plate.
- Describe the processes of bone construction and destruction involved in bone replacement.
- Describe the blood and nerve supply of bone tissue.
- Describe the conditions necessary for normal bone growth.
- Define rickets and osteomalacia as vitamin deficiency disorders.
- Contrast the causes and clinical symptoms associated with osteoporosis, Paget's disease, and osteomyelitis.
- Define a fracture and describe several common kinds of fractures.
- Describe the sequence of events involved in fracture repair.
- Define key medical terms associated with the skeletal system.

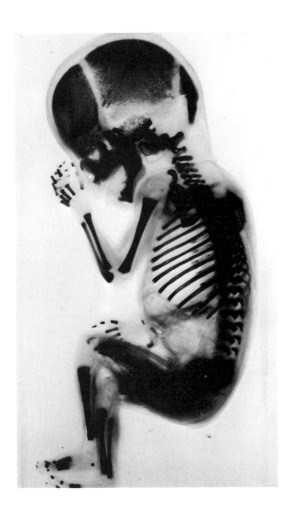

FIGURE 5-4 Diagrammatic representation (opposite page) and photograph of a 10-week-old human embryo. Colored bones are those that develop by intramembranous ossification. Gray bones are those that develop by endochondral ossification. (Photograph courtesy of Carolina Biological Supply Company.)

The epiphyseal (ep'-i-FIZ-ē-al) plate consists of four zones (Figure 5-6). The *zone of reserve cartilage* is adjacent to the epiphysis of the bone. It consists of small chondrocytes that are scattered irregularly throughout the intercellular matrix. The cells of this zone do not function in bone growth. The zone of reserve cartilage functions in anchoring the epiphyseal plate to the bone of the epiphysis. Its blood vessels also provide nutrients for the other zones of the epiphyseal plate.

The second zone, the *zone of proliferating cartilage,* consists of slightly larger chondrocytes that are arranged like stacks of coins. The function of this zone is to make new chondrocytes by cell division to replace those that die at the diaphyseal surface of the epiphyseal plate.

The third zone, the *zone of hypertrophic* (hī'-per-TRŌF-ik) *cartilage,* consists of even larger chondrocytes also arranged in columns. The cells are in various stages of maturation, with the more mature cells closer to the diaphysis. Lengthwise expansion of the epiphyseal plate is the result of cellular proliferation of the zone of proliferating cartilage and maturation of the cells in the zone of hypertrophic cartilage. Near the diaphyseal end of the bone, the intercellular matrix of the cells of the zone of hypertrophic cartilage becomes calcified and dies.

The fourth zone, the *zone of calcified matrix,* is only a few cells thick and consists mostly of dead cells because the intercellular matrix around them has calcified. The calcified matrix is taken up by osteoclasts, and the area is invaded by osteoblasts and capillaries from the bone in the diaphysis. These cells lay down bone on the calcified cartilage that persists. As a result, the diaphyseal border of the epiphyseal plate is firmly cemented to the bone of the diaphysis.

The region between the diaphysis and epiphysis of a bone where the calcified matrix is replaced by bone is called the *metaphysis* (me-TAF-i-sis). The activity of the epiphyseal plate is the only mechanism by which the diaphysis can increase in length. Unlike cartilage, which can grow by both interstitial and appositional growth, bone can grow only by appositional growth.

The epiphyseal plate allows the diaphysis of the bone to increase in length until early adulthood. As the child

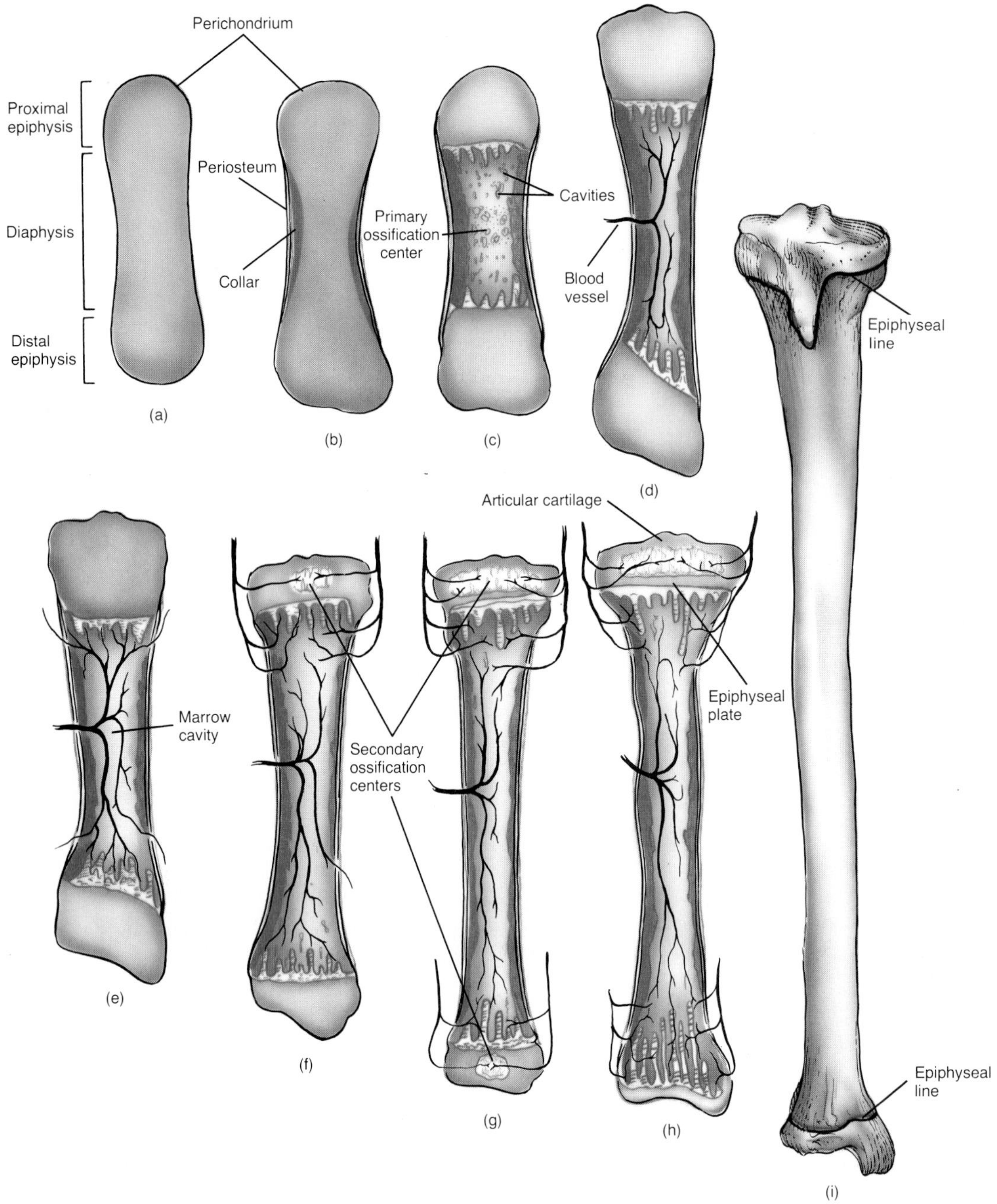

FIGURE 5-5 Endochondral ossification of the tibia. (a) Cartilage model. (b) Collar formation. (c) Development of primary ossification center. (d) Entrance of blood vessels. (e) Marrow-cavity formation. (f) Thickening and lengthening of collar. (g) Formation of secondary ossification centers. (h) Remains of cartilage as the articular cartilage and epiphyseal plate. (i) Formation of the epiphyseal lines.

grows, cartilage cells are produced by mitosis on the epiphyseal side of the plate. Cartilage cells are then destroyed and the cartilage is replaced by bone on the diaphyseal side of the plate. In this way, the thickness of the epiphyseal plate remains fairly constant but the bone on the diaphyseal side increases in length. Growth in diameter occurs along with growth in length. In this pro-

cess, the bone lining the marrow cavity is destroyed so that the cavity increases in diameter. At the same time, osteoblasts from the periosteum add new osseous tissue around the outer surface of the bone. Initially, diaphyseal and epiphyseal ossification produce only spongy bone. Later, by reconstruction, the outer region of spongy bone is reorganized into compact bone.

Ossification of all bones is usually completed by age 25. Figure 5-7 consists of a series of roentgenograms that show ossification progress in the epiphyses of two long bones at the knee.

Bones undergoing either intramembranous or endochondral ossification are continually remodeled from the time that initial calcification occurs until the final structure appears. Compact bone is formed by the transformation of spongy bone. The diameter of a long bone is increased by the destruction of the bone closest to the marrow cavity and the construction of new bone around the outside of the diaphysis. However, even after bones have reached their adult shapes and sizes, old bone is perpetually destroyed and new osseous tissue is formed in its place.

BONE REPLACEMENT

Bone shares with skin the feature of replacing itself throughout adult life. This **remodeling** takes place at different rates in various body regions. The distal portion of the femur (thighbone) is replaced about every 4 months. By contrast, bone in certain areas of the shaft will not be completely replaced during the individual's life. Remodeling allows worn or injured bone to be removed and replaced with new tissue. It also allows bone to serve as the body's storage area for calcium. Many other tissues in the body need calcium in order to perform their functions. For example, nerve cells need calcium for their activities and muscle needs calcium in order to contract. The nerve and muscle cells take their calcium from the blood. However, the blood itself needs calcium in order to clot. The blood continually trades off calcium with the bones, removing calcium when other tissues are not receiving enough of this element and resupplying the bones with dietary calcium when available to keep them from losing too much bone mass.

The cells believed to be responsible for the destruction of bone tissue are called **osteoclasts** (*clast* = break). In the healthy adult, a delicate homeostasis is maintained between the action of the osteoclasts and the removal of calcium, on one hand, and the action of the bone-making osteoblasts and the deposition of calcium on the other. Should too much new tissue be formed, the bones become abnormally thick and heavy. If too much calcium

is deposited in the bone, the surplus may form thick bumps, or spurs, on the bone that interfere with movement at joints. A loss of too much tissue or calcium weakens the bones and allows them to break easily or to become very flexible.

Normal bone growth in the young and bone replacement in the adult depend on several factors. First, sufficient quantities of calcium and phosphorus, components of the primary salt that makes bone hard, must be included in the diet.

Second, the individual must obtain sufficient amounts of vitamins A, C, and D, substances that are responsible for the proper utilization of calcium and phosphorus by the body. Vitamin D, for example, is required for the absorption of calcium from the digestive tract into the blood and has important effects on bone deposition and reabsorption.

Third, the body must manufacture the proper amounts of the hormones responsible for bone tissue activity. Pituitary growth hormones are responsible for the general

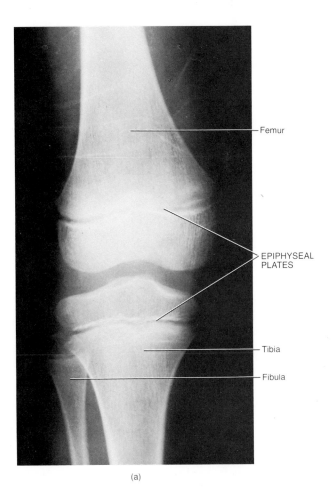

Femur

EPIPHYSEAL PLATES

Tibia

Fibula

(a)

FIGURE 5-6 Epiphyseal plate. (a) Anteroposterior projection showing epiphyseal plates in the femur and tibia of a 10-year-old child. (Courtesy of Arthur Provost, R.T.)

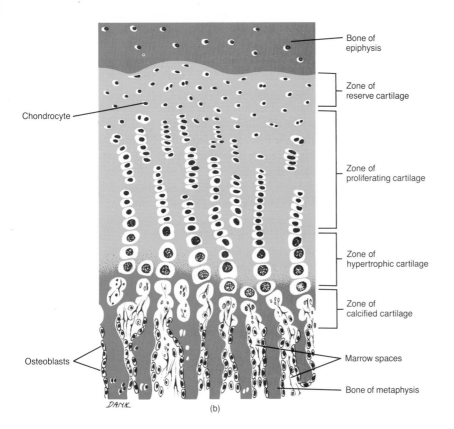

Bone of
epiphysis

Zone of
reserve cartilage

Chondrocyte

Zone of
proliferating cartilage

Zone of
hypertrophic cartilage

Zone of
calcified cartilage

Osteoblasts

Marrow spaces

Bone of metaphysis

DANK

(b)

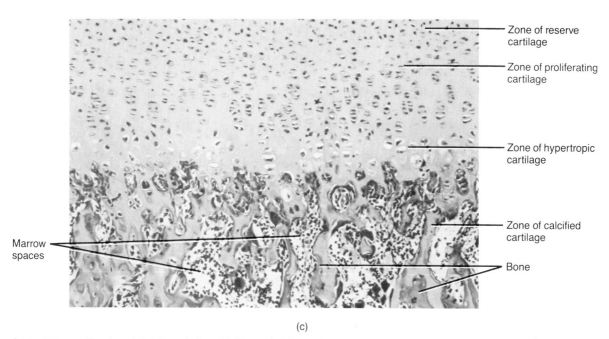

Zone of reserve
cartilage

Zone of proliferating
cartilage

Zone of hypertropic
cartilage

Zone of calcified
cartilage

Marrow
spaces

Bone

(c)

FIGURE 5-6 *(Continued)* Epiphyseal plate. (b) Diagram of the various zones of the epiphyseal plate. (c) Photomicrograph of the epiphyseal plate at a magnification of 160×. (© 1983 by Michael H. Ross. Used by permission.)

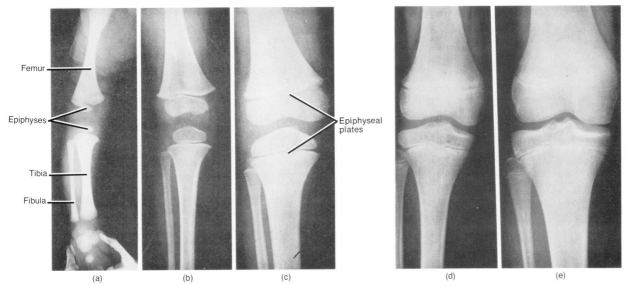

FIGURE 5-7 Anteroposterior projections of normal ossification at the knee. (a) One-month-old infant. The epiphyses at the knee are mostly cartilage. Ossification centers have formed for the femur and tibia. The space between the ends of the bones is occupied by epiphyseal cartilage. (b) Two-year-old child. Centers of ossification have grown. The white transverse zones marking the ends of the shafts are areas where mineral salts are deposited temporarily around the degenerating cartilage cells. (c) Five-year-old child. The epiphyses have assumed the shape of adult bones. The epiphyseal plates are clearly visible between the epiphyses and diaphyses of all three bones. (d) Eight-year-old child. The epiphyseal plates are still distinct as ossification continues. (e) Twelve-year-old child. The epiphyses have ossified almost completely. The epiphyseal plates are assuming the character of epiphyseal lines. (Courtesy of Lester W. Paul and John H. Juhl, *The Essentials of Roentgen Interpretation,* 3d ed., Harper & Row, Publishers, Inc., New York, 1972.)

growth of bones. Too much or too little of these hormones during childhood makes the adult abnormally tall or short. Other hormones specialize in regulating the osteoclasts. And still others, especially the sex hormones, aid osteoblastic activity and thus promote the growth of new bone. The sex hormones act as a double-edged sword. They aid in the growth of new bone, but they also bring about the degeneration of all the cartilage cells in the epiphyseal plates. Because of the sex hormones, the typical adolescent experiences a spurt of growth during puberty, when sex hormone levels start to increase. The individual then quickly completes the growth process as the epiphyseal cartilage disappears. Premature puberty can actually prevent one from reaching an average adult height because of the simultaneous premature degeneration of the plates. A hormone produced by the parathyroid glands controls the removal of calcium and phosphorus from the bones, while a hormone produced by the thyroid gland controls storage of calcium and phosphorus in bones.

BLOOD AND NERVE SUPPLY

Bone is richly supplied with blood, and blood vessels are especially abundant in portions of bone containing red bone marrow. Blood vessels pass into bones from the periosteum. Although the blood supply to bones varies according to their shape, here we will consider the blood supply to a long bone only.

Because the artery to the diaphysis of a long bone is usually the largest, it is referred to as the **nutrient artery.**

The artery first enters the bone during early development and passes through an opening in the diaphysis, the *nutrient foramen,* which leads into a *nutrient canal* (see Figure 5-1a). On entering the medullary cavity, the nutrient artery divides into a proximal and a distal branch, each of which supplies most of the marrow, inner portion of compact bone of the diaphysis, and metaphysis. As the nutrient artery passes through compact bone on its way to the medullary cavity, it sends branches into the Haversian canals. Branches of the **articular arteries** enter many vascular foramina in the epiphysis and supply the marrow, metaphysis, and bony tissue of the epiphysis. **Periosteal arteries,** accompanied by nerves, enter the diaphysis at numerous points through Volkmann's canals (see Figure 5-1c). These blood vessels run in the Haversian canals and supply the outer part of the compact bone of the diaphysis. Veins accompany the several types of arteries; the principal veins leave the bone by numerous vascular foramina at the epiphyses of the bone.

Although the nerve supply to bone is not extensive, some nerves (vasomotor) accompany blood vessels and some (sensory) occur in the periosteum. The nerves of the periosteum are primarily concerned with pain, as might be associated with a fracture or tumor.

APPLICATIONS TO HEALTH

You have already learned how important certain vitamins, minerals, and hormones are to bone growth and development. Many bone disorders result from deficiencies in vitamins or minerals or from too much or too little of the hormones that regulate bone homeostasis. Infection and tumors which disrupt homeostasis are also responsible for certain bone disorders.

VITAMIN DEFICIENCIES

Vitamin D is important to normal bone growth and maintenance. It is essential for the synthesis of a protein that transports the calcium obtained from foods across the lining of the intestine and into the extracellular fluid. When the body lacks this vitamin, it is unable to absorb calcium and phosphorus. A deficiency of vitamin D produces rickets in children and osteomalacia in adults.

Rickets

In the condition called **rickets,** there is a deficiency of vitamin D that results in an inability of the body to transport calcium and phosphorus from the digestive tract into the blood for utilization by bones. As a result, epiphyseal cartilage cells cease to degenerate and new cartilage continues to be produced. Epiphyseal cartilage thus becomes wider than normal. At the same time, the soft matrix laid down by the osteoblasts in the diaphysis fails to calcify. As a result, the bones stay soft. When the child walks, the weight of the body causes the bones in the legs to bow. Malformations of the head, chest, and pelvis also occur.

The cure and prevention of rickets consist of adding generous amounts of calcium, phosphorus, and vitamin D to the diet. Exposing the skin to the ultraviolet rays of sunlight also aids the body in manufacturing additional vitamin D.

Osteomalacia

A deficiency of vitamin D in the adult causes the bones to give up excessive amounts of calcium and phosphorus. This loss, called *demineralization,* is especially heavy in the bones of the pelvis, legs, and spine. Demineralization caused by vitamin D deficiency is called **osteomalacia** (os'-tē-ō-ma-LĀ-shē-a; *malacia* = softness). After the bones demineralize, the weight of the body produces a bowing of the leg bones, a shortening of the backbone, and a flattening of the pelvic bones. Osteomalacia mainly affects women who live on poor cereal diets devoid of milk, are seldom exposed to the sun, and have repeated pregnancies that deplete the body of calcium. The condition responds to the same treatment as rickets: if the disease is severe enough and threatens life, large doses of vitamin D should be given.

Osteomalacia may also result from failure to absorb fat (steatorrhea) because vitamin D is soluble in fats and calcium combines with fats. As a result, vitamin D and calcium remain with the unabsorbed fat and are lost in the feces.

OSTEOPOROSIS

Osteoporosis (os'-tē-ō-pō-RŌ-sis) is a bone disorder affecting the middle-aged and elderly—white women more than men of black or white ancestry. Between puberty and the middle years, the sex hormones maintain osseous tissue by stimulating the osteoblasts to form new bone. Women produce smaller amounts of sex hormones after menopause, and both men and women produce smaller amounts during old age. As a result, the osteoblasts become less active and there is a decrease in bone mass

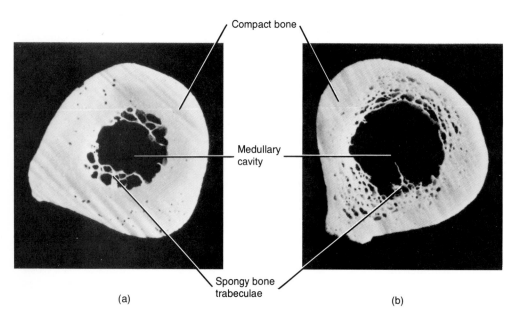

Compact bone

Medullary cavity

Spongy bone trabeculae

(a)   (b)

FIGURE 5-8 Osteoporosis. Photographs of a cross section of the femur from (a) a normal female aged 25 and (b) a female with osteoporosis aged 82. In (b) note the decreased thickness of the compact bone of the diaphysis, the increase in the diameter of the medullary cavity, and the thinning of the spongy bone trabeculae. (Photographs courtesy of D. P. Jenkins, from *Introduction to the Musculoskeletal System* by Cornelius Rosse and D. Kay Clawson, Harper & Row, Publishers, Inc., New York, 1970.)

(Figure 5-8). Osteoporosis affects the entire skeletal system, especially the spine, legs, and feet. As the spine collapses and curves, the thorax drops, and the ribs fall on the pelvic rim. This condition leads to gastrointestinal distension and an overall decrease in muscle tone.

Among the factors implicated in bone loss is a decrease in estrogen, calcium deficiency and malabsorption, vitamin D deficiency, loss of muscle mass, inactivity, and high-protein diets.

PAGET'S DISEASE

Paget's disease is characterized by irregular thickening and softening of the bones. It rarely occurs in individuals under 50. The cause, or **etiology** (ē'-tē-OL-ō-jē), of the disease is unknown. Bone-producing osteoblasts and bone-destroying osteoclasts apparently become uncoordinated. This alters the balance between bone formation and bone destruction. Paget's disease affects the skull, the pelvis, and the bones of the extremities.

OSTEOMYELITIS

The term **osteomyelitis** (os'-tē-ō-mī-i-LĪ-tis) includes all the infectious diseases of bone. These diseases may be localized or widespread and may also involve the periosteum, marrow, and cartilage. Various microorganisms may give rise to bone infection, but the most frequent are bacteria known as *Staphylococcus aureus,* commonly called "staph." These bacteria may reach the bone by various means: the bloodstream, an injury such as a fracture, or an infection such as a sinus infection or a tooth abscess. The infection may destroy extensive areas of bone, spread to nearby joints, and, in rare cases, lead to death by producing abscesses. Antibiotics have been effective in treating the disease and in preventing it from spreading through extensive areas of bone.

FRACTURES

In simplest terms, a **fracture** is any break in a bone. Usually, the fracture is restored to normal position by manipulation without surgery. This procedure of setting a fracture is called *closed reduction.* In other cases, the fracture must be exposed by surgery before the break is rejoined. This procedure is known as *open reduction.*

Types

Although fractures of bones of the extremities may be classified in several different ways, the following scheme is useful (Figure 5-9):

1. Partial. A fracture in which the break across the bone is incomplete.

2. Complete. A fracture in which the break across the bone is complete, so that the bone is broken into two pieces.

3. Closed or **simple.** A fracture in which the bone does not break through the skin.

4. Open or **compound.** A fracture in which the broken ends of the bone protrude through the skin.

5. Comminuted (KOM-i-nyoo'-ted). A fracture in which the bone is splintered at the site of impact and smaller fragments of bone are found between the two main fragments.

6. Greenstick. A partial fracture in which one side of the bone is broken and the other side bends; occurs only in children.

7. Spiral. A fracture in which the bone is usually twisted apart.

8. Transverse. A fracture at right angles to the long axis of the bone.

9. Impacted. A fracture in which one fragment is firmly driven into the other.

10. Pott's. A fracture of the distal end of the fibula, with serious injury of the distal tibial articulation.

11. Colles'. A fracture of the distal end of the radius in which the distal fragment is displaced posteriorly.

12. Displaced. A fracture in which the anatomical alignment of the bone fragments is not preserved.

13. Nondisplaced. A fracture in which the anatomical alignment of the bone fragments is preserved.

Fracture Repair

Unlike the skin, which may repair itself within days, or muscle, which may mend in weeks, a bone sometimes requires months to heal. A fractured femur, for example, may take six months to heal. Sufficient calcium to strengthen and harden new bone is deposited only gradually. Bone cells also grow and reproduce slowly. Moreover, the blood supply to bone is decreased, which helps to explain the difficulty in the healing of an infected bone.

The following steps occur in the repair of a fracture (Figure 5-10). The healing of a fractured femur is illustrated by sequential roentgenograms in Figure 5-11.

1. As a result of the fracture, blood vessels crossing the fracture line are broken. These vessels are found in the periosteum, Haversian systems, and marrow cavity. As blood pours from the torn ends of the vessels, it coagulates and forms a clot in and about the site of the fracture. This clot, called a **fracture hematoma** (hē'-ma-TŌ-ma), usually occurs 6 to 8 hours after the injury. Since the circulation of blood ceases when the fracture hematoma forms, bone cells and periosteal cells at the fracture line die.

2. A growth of new bone tissue—a **callus**—develops in and around the fractured area. It forms a bridge between separated areas of bone. The part of the callus

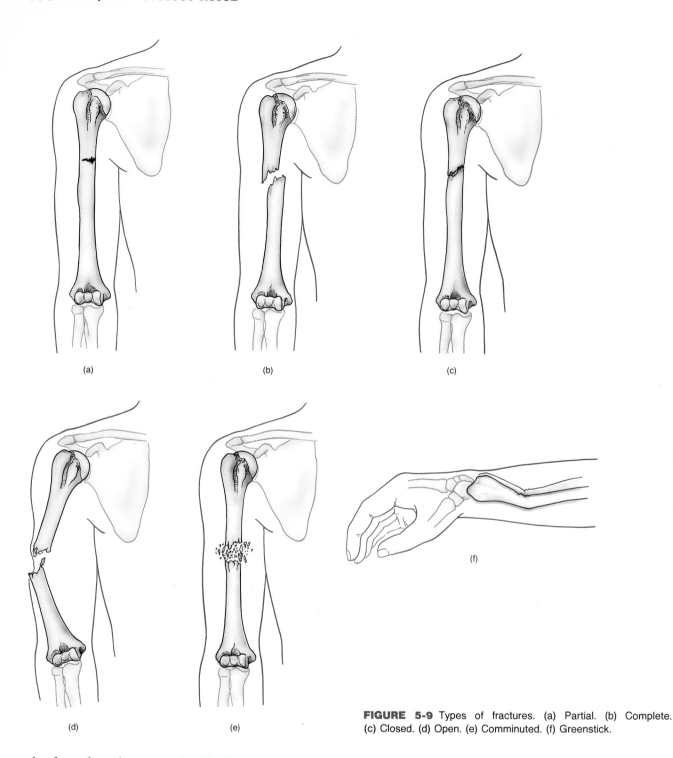

(a) (b) (c)

(d) (e) (f)

FIGURE 5-9 Types of fractures. (a) Partial. (b) Complete. (c) Closed. (d) Open. (e) Comminuted. (f) Greenstick.

that forms from the osteogenic cells of the torn periosteum and develops around the outside of the fracture is called an *external callus.* The part of the callus that forms from the osteogenic cells of the endosteum and develops between the two ends of bone fragments and between the two marrow cavities is called the *internal callus.*

Approximately 48 hours after a fracture occurs, the cells that ultimately repair the fracture become actively mitotic. These cells come from the osteogenic layer of

the periosteum, the endosteum of the marrow cavity, and the bone marrow. As a result of their accelerated mitotic activity, the cells of the three regions grow toward the fracture. During the first week following the fracture, the cells of the endosteum and bone marrow form new trabeculae in the marrow cavity near the line of fracture. This is the internal callus. During the next few days, osteogenic cells of the periosteum form a collar around each bone fragment. The collar, or external callus, is re-

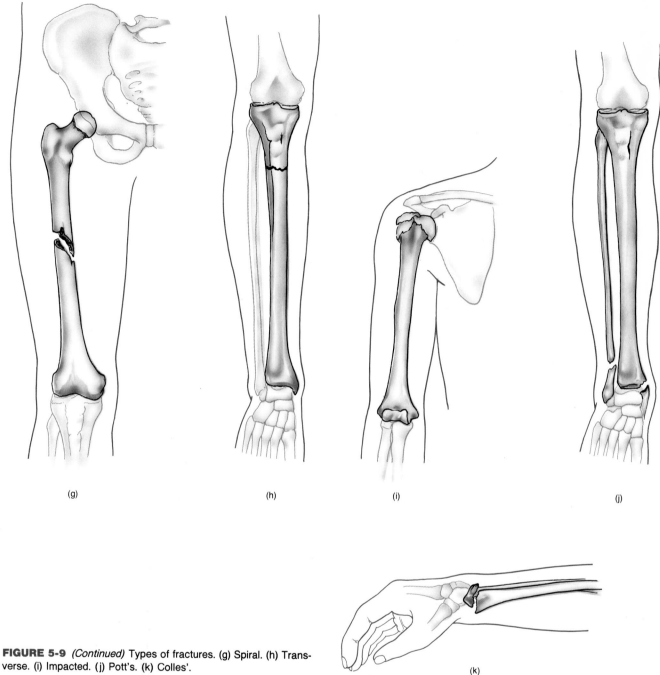

(g)   (h)   (i)   (j)

(k)

FIGURE 5-9 *(Continued)* Types of fractures. (g) Spiral. (h) Transverse. (i) Impacted. (j) Pott's. (k) Colles'.

placed by trabeculae. The trabeculae of the calli are joined to living and dead portions of the original bone fragments.

3. The final phase of fracture repair is the **remodeling** of the calli. Dead portions of the original fragments are gradually resorbed. Compact bone replaces spongy bone around the periphery of the fracture. In some cases, the healing is so complete that the fracture line is undetectable, even by x-ray. However, a thickened area on the surface of the bone usually remains as evidence of the fracture site.

A relatively new field of scientific research called **electrobiology** has provided some very dramatic results as a treatment for poorly healing fractures. Essentially, the fracture is exposed to an electrical current generated from a pair of coils that are fastened around the cast. In response to electrical stimulation, cells adjacent to the fracture site become more active metabolically as a result of an alteration of electrical charges on the surfaces of their membranes. This apparently causes an acceleration

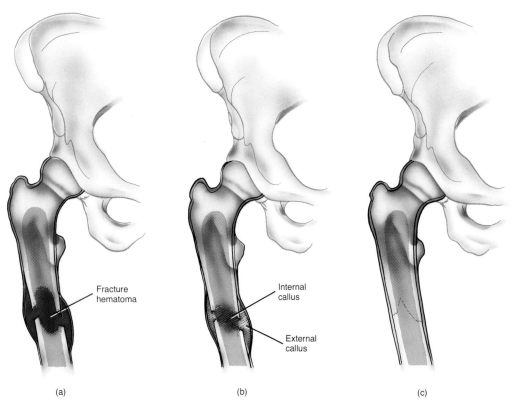

(a)   (b)   (c)

FIGURE 5-10 Fracture repair. (a) Formation of fracture hematoma. (b) Formation of external and internal calli. (c) Completely healed fracture.

of calcification and a subsequent acceleration of fracture repair.

Although the original application of electrobiology was to treat improperly healing fractures, it is now being ap-plied to limb regeneration and killing tumor cells. Re-search is now underway to determine if electrical stimula-tion can be used to grow new limbs and to stop the growth of tumor cells in humans.

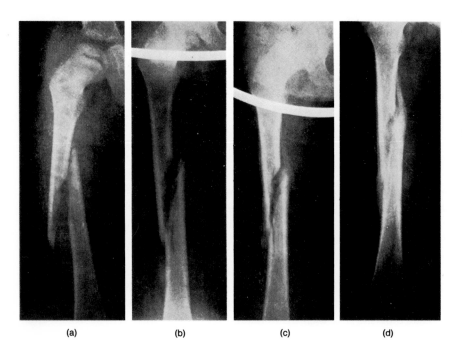

(a)   (b)   (c)   (d)

FIGURE 5-11 Anteroposterior projections of the repair of a fractured femur. (a) Immedi-ately after fracture. (b) Two weeks later, a hazy external callus is visible around the margins of the fracture. (c) About 3½ weeks after the fracture, the internal and external calli begin to bridge the separated frag-ments. (d) Almost 9 weeks after the fracture, the bridge between fragments is fairly well developed. In time, remodeling will occur and the fracture will be repaired. (Courtesy of Ralph C. Frank, Eau Claire, Wisconsin, from Lester W. Paul and John H. Juhl, *The Essentials of Roentgen Interpretation*, 3d ed., Harper & Row, Publishers, Inc., New York, 1972.)

KEY MEDICAL TERMS ASSOCIATED WITH OSSEOUS TISSUE

Achondroplasia (*a* = without; *chondro* = cartilage; *plasia* = growth) Imperfect ossification within cartilage of long bones during fetal life; also called *fetal rickets.*

Brodie's abscess Infection in the spongy tissue of a long bone, with a small inflammatory area.

Craniotomy (*cranium* = skull; *tome* = a cutting) Any surgery that requires cutting through the bones surrounding the brain.

Necrosis (*necros* = death; *osis* = condition) Death of tissues or organs; in the case of bone, necrosis results from deprivation of blood supply; could result from fracture, extensive removal of periosteum in surgery, exposure to radioactive substances, and other causes.

Osteitis (*osteo* = bone) Inflammation or infection of bone.

Osteoarthritis (*arthro* = joint) A degenerative condition of bone and also the joint.

Osteoblastoma (*oma* = tumor) A benign tumor of the osteoblasts.

Osteochondroma (*chondro* = cartilage) A benign tumor of the bone and cartilage.

Osteoma A benign bone tumor.

Osteomyelitis (*myel* = marrow) Infection that involves bone marrow.

Osteosarcoma (*sarcoma* = connective tissue tumor) A malignant tumor composed of osseous tissue.

Pott's disease Inflammation of the backbone, caused by the microorganism that produces tuberculosis.

STUDY OUTLINE

Functions

1. The skeletal system consists of all bones attached at joints, cartilage between joints, and cartilage found elsewhere (nose, larynx, and outer ear).
2. The functions of the skeletal system include support, protection, leverage, mineral storage, and blood-cell formation.

Histology

1. Osseous tissue consists of widely separated cells surrounded by large amounts of intercellular substance. The intercellular substance contains collagenous fibers and abundant hydroxyapatites (mineral salts).
2. Parts of a typical long bone are the diaphysis (shaft), epiphyses (ends), metaphysis, articular cartilage, periosteum, medullary or marrow cavity, and endosteum.
3. Cancellous, or spongy, bone has many marrow-filled pores and does not contain Haversian systems. It consists of trabeculae containing osteocytes and lacunae.
4. Dense, or compact, bone has fewer pores and fewer networks of passageways called Haversian systems. Dense bone lies over spongy bone and composes most of the bone tissue of the diaphyses.

Ossification

1. Bone forms by a process called ossification or osteogenesis, which begins when mesenchymal cells become transformed into osteoblasts.
2. Intramembranous ossification occurs within fibrous membranes of the embryo and the adult.
3. Endochondral ossification occurs within a cartilage model.

4. The primary ossification center of a long bone is in the diaphysis. Cartilage degenerates, leaving cavities that merge to form the marrow cavity. Osteoblasts lay down bone. Next, ossification occurs in the epiphyses, where bone replaces cartilage, except for the epiphyseal plate.
5. The anatomical zones of the epiphyseal plate are the zones of reserve cartilage, proliferating cartilage, hypertrophic cartilage, and calcified matrix.
6. Because of the activity of the epiphyseal plate, the diaphysis of a bone increases in length by appositional growth.
7. In both types of ossification, spongy bone is laid down first. Dense bone is later reconstructed from spongy bone.

Bone Replacement

1. The growth and development of bone depends on a balance between bone formation and destruction.
2. Old bone is constantly destroyed by osteoclasts, while new bone is constructed by osteoblasts. This process is called remodeling.
3. Normal growth depends on calcium, phosphorus, and vitamins (A, C, and D) and is controlled by hormones that are responsible for bone mineralization and reabsorption.

Blood and Nerve Supply

1. Long bones are supplied by nutrient, articular, and periosteal arteries; veins accompany the arteries.
2. The nerve supply to bones consists of vasomotor and sensory nerves.

Applications to Health

1. Rickets is a vitamin D deficiency in children in which the body does not absorb calcium and phosphorus. The bones soften and bend under the body's weight.
2. Osteomalacia is a vitamin D deficiency in adults that leads to demineralization.
3. Osteoporosis is a decrease in the amount and strength of bone tissue due to decreases in hormone output.
4. Paget's disease is the irregular thickening and softening of bones, apparently related to an imbalance between osteoclast and osteoblast activities.
5. Osteomyelitis is a term for the infectious diseases of bones, marrow, and periosteum. It is frequently caused by "staph" bacteria.
6. A fracture is any break in a bone.
7. The types of fractures include: partial, complete, simple, compound, comminuted, greenstick, spiral, transverse, impacted, Pott's, Colles', displaced, and nondisplaced.
8. Fracture repair consists of forming a fracture hematoma, forming a callus, and remodeling.
9. Electrobiology has provided dramatic results in healing fractures that would otherwise not have mended properly. It is also being researched in regeneration and stopping the growth of tumor cells.

REVIEW QUESTIONS

1. Define the skeletal system. What are its five principal functions?
2. Diagram the parts of a long bone, and list the functions of each part. What is the difference between compact and spongy bone tissue? Diagram the microscopic structure of bone.
3. What is meant by ossification? When does the process begin and end?
4. Distinguish between the two principal kinds of ossification.
5. Outline the major events involved in intramembranous and endochondral ossification.
6. Describe the various zones of the epiphyseal plate. How does the plate grow?
7. What is a roentgenogram?
8. List the primary factors involved in bone growth.
9. How does osteoblast activity in balance with osteoclast activity control the replacement of bone?
10. Describe the blood and nerve supply of a long bone.
11. Define rickets and osteomalacia in terms of symptoms, cause, and treatment. What do these two diseases have in common?
12. What are the principal symptoms of osteoporosis, osteomyelitis, and Paget's disease? What is the etiology of each?
13. What is a fracture? Distinguish 13 principal kinds.
14. Outline the steps involved in fracture repair.
15. Refer to the glossary of key medical terms associated with the skeletal system. Be sure that you can define each term.

THE SKELETAL SYSTEM
The Axial Skeleton

STUDENT OBJECTIVES

■ Define the four principal types of bones in the skeleton.

■ Describe the various markings on the surfaces of bones.

■ Relate the structure of the marking to its function.

■ List the components of the axial and appendicular skeletons.

■ Identify the bones of the skull and the major markings associated with each.

■ Identify the sutures and fontanels of the skull.

■ Identify the paranasal sinuses of the skull in projection diagrams and roentgenograms.

■ Identify the principal foramina of the skull.

■ Identify the bones of the vertebral column and their principal markings.

■ List the defining characteristics and curves of each region of the vertebral column.

■ Identify the bones of the thorax and their principal markings.

■ Contrast microcephalus, hydrocephalus, slipped disc, curvatures, spina bifida, and fractures of the vertebral column as disorders associated with the skeletal system.

The skeletal system forms the framework of the body. For this reason, a familiarity with the names, shapes, and positions of individual bones will help you to understand some of the other organ systems. For example, movements such as throwing a ball, typing, and walking require the coordinated use of bones and muscles. To understand how muscles produce different movements, you need to learn the parts of the bones to which the muscles attach. The respiratory system is also highly dependent on bone structure. The bones in the nasal cavity form a series of passageways that help clean, moisten, and warm inhaled air. Furthermore, the bones of the thorax are specially shaped and positioned so the chest can expand during inhalation. Many bones also serve as landmarks to students of anatomy as well as to surgeons. Blood vessels and nerves often run parallel to bones. These structures can be located more easily if the bone is identified first.

We shall study bones by examining the various regions of the body. For instance, we shall look at the skull first and see how the bones of the skull relate to each other. Then we shall move on to the chest. This regional approach will allow you to see how all the many bones of the body relate to each other.

EXHIBIT 6-1 BONE MARKINGS

MARKING	DESCRIPTION	EXAMPLE
DEPRESSIONS AND OPENINGS		
Fissure (FISH-ūr)	A narrow, cleftlike opening between adjacent parts of bones through which blood vessels or nerves pass.	Superior orbital fissure of the sphenoid bone (Figure 6-2).
Foramen (fō-RĀ-men; *foramen* = hole)	An opening through which blood vessels, nerves, or ligaments pass.	Infraorbital foramen of the maxilla (Figure 6-2).
Meatus (mē-Ā-tus; *meatus* = canal)	A tubelike passageway running within a bone.	External auditory meatus of the temporal bone (Figure 6-2).
Paranasal sinus (*sin* = cavity)	An air-filled cavity within a bone connected to the nasal cavity.	Frontal sinus of the frontal bone (Figure 6-8).
Groove or sulcus (*sulcus* = ditchlike groove)	A furrow or groove that accommodates a soft structure such as a blood vessel, nerve, or tendon.	Intertubercular sulcus of the humerus (Figure 7-4).
Fossa (*fossa* = basinlike depression)	A depression in or on a bone.	Mandibular fossa of the temporal bone (Figure 6-4).
PROCESSES	Any prominent projection.	Mastoid process of the temporal bone (Figure 6-2).
Processes that form joints		
Condyle (KON-dil; *condylus* = knucklelike process)	A large articular prominence.	Medial condyle of the femur (Figure 7-10).
Head	A rounded articular projection supported on the constricted portion (neck) of a bone.	Head of the femur (Figure 7-10).
Facet	A smooth, flat surface.	Articular facet for the tubercle of rib on a vertebra (Figure 6-18).
Processes to which tendons, ligaments, and other connective tissues attach		
Tubercle (TOO-ber-kul; *tuber* = knob)	A small, rounded process.	Greater tubercle of the humerus (Figure 7-4).
Tuberosity	A large, rounded, usually roughened process.	Ischial tuberosity of the hipbone (Figure 7-8).
Trochanter (trō-KAN-ter)	A large, blunt projection found only on the femur.	Greater trochanter of the femur (Figure 7-10).
Crest	A prominent border or ridge on a bone.	Iliac crest of the hipbone (Figure 7-7).
Line	A less prominent ridge.	Linea aspera of the femur (Figure 7-10).
Spinous process or spine	A sharp, slender process.	Spinous process of a vertebra (Figure 6-12).
Epicondyle (*epi* = above)	A prominence above a condyle.	Medial epicondyle of the femur (Figure 7-10).

TYPES OF BONES

The bones of the body may be classified into four principal types: long, short, flat, and irregular. **Long bones** have greater length than width and consist of a diaphysis and two epiphyses. They are slightly curved for strength. A curved bone is structurally designed to absorb the stress of the body weight at several different points so the stress is evenly distributed. If such bones were straight, the weight of the body would be unevenly distributed and the bone would easily fracture. Examples of long bones include bones of the thighs, legs, toes, arms, forearms, and fingers. Figure 5-1a shows the parts of a long bone.

Short bones are somewhat cube-shaped and nearly equal in length and width. Their texture is spongy except at the surface, where there is a thin layer of compact bone. Examples of short bones are the wrist and ankle bones.

EXHIBIT 6-2 DIVISIONS OF THE SKELETAL SYSTEM

REGIONS OF THE SKELETON	NUMBER OF BONES
AXIAL SKELETON	
Skull	
Cranium	8
Face	14
Hyoid	1
Auditory ossicles	6
(3 in each ear)*	
Vertebral column	26
Thorax	
Sternum	1
Ribs	24
	80
APPENDICULAR SKELETON	
Shoulder girdles	
Clavicle	2
Scapula	2
Upper extremities	
Humerus	2
Ulna	2
Radius	2
Carpals	16
Metacarpals	10
Phalanges	28
Pelvic girdle	
Coxal, hip, or pelvic bone	2
Lower extremities	
Femur	2
Fibula	2
Tibia	2
Patella	2
Tarsals	14
Metatarsals	10
Phalanges	28
	126

* Although the auditory ossicles are not considered part of the axial or appendicular skeleton, but rather as a separate group of bones, they are placed with the axial skeleton for convenience.

Flat bones are generally thin and composed of two more or less parallel plates of compact bone enclosing a layer of spongy bone. The term *diploe* is applied to the spongy bone of the cranial bones. Flat bones afford considerable protection and provide extensive areas for muscle attachment. Examples of flat bones include the cranial bones, which protect the brain, the sternum, and ribs, which protect organs in the thorax, and the scapulas.

Irregular bones have complex shapes and cannot be grouped into any of the three categories just described. They also vary in the amount of spongy and compact bone present. Such bones are the vertebrae and certain facial bones.

Besides these four principal types of bones, two other kinds are recognized. **Wormian**, or **sutural** (SOO-chur-al), **bones** are small clusters of bones between the joints of certain cranial bones (see Figure 6-2e). Their number varies greatly from person to person. **Sesamoid bones** are small bones in tendons where considerable pressure develops, for instance, in the wrist. These bones, like the Wormian bones, are also variable in number. Two sesamoid bones, the patellas, or kneecaps, are present in all individuals.

SURFACE MARKINGS

The surfaces of bones reveal various **markings.** The structure of many of these markings indicates their functions. Long bones that bear a great deal of weight have large, rounded ends that can form sturdy joints. Other bones have depressions that receive the rounded ends. Rough areas serve for the attachment of muscles, tendons, and ligaments. Grooves in the surfaces of bones provide for the passage of blood vessels, and openings occur where blood vessels and nerves pass through the bone. Exhibit 6-1 describes the different markings and their functions.

DIVISIONS OF THE SKELETAL SYSTEM

The adult human skeleton usually consists of 206 bones grouped in two principal divisions: the **axial** and the **appendicular.** The longitudinal *axis,* or center, of the human body is a straight line that runs vertically along the body's center of gravity. This imaginary line runs through the head and down to the space between the feet. The midsaggital section is drawn through this line. The axial division of the skeleton consists of the bones that lie around the axis: ribs, breastbone, the bones of the skull, and backbone.

The appendicular division contains the bones of the free *appendages,* which are the upper and lower extremities, plus the bones called *girdles,* which connect the free appendages to the axial skeleton.

The 80 bones of the axial division and the 126 bones of the appendicular division are typically grouped as shown in Exhibit 6-2.

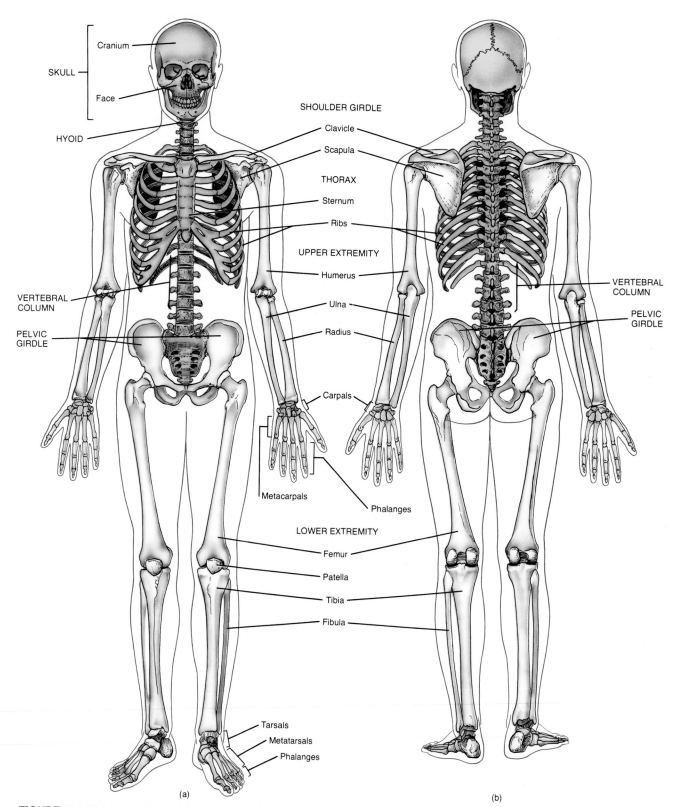

FIGURE 6-1 Divisions of the skeletal system. (a) Anterior view. (b) Posterior view.

Now that you understand how the skeleton is organized into axial and appendicular divisions, refer to Figure 6-1 to see how the two divisions are joined to form the skeleton. The bones of the axial skeleton are shown in gray. Be certain to locate the following regions of the skeleton: skull, cranium, face, hyoid bone, vertebral column, thorax, shoulder girdle, upper extremity, pelvic girdle, and lower extremity.

SKULL

The **skull,** which contains 22 bones, rests on the superior end of the vertebral column and is composed of two sets of bones: cranial bones and facial bones. The **cranial bones** enclose and protect the brain and the organs of sight, hearing, and balance. The eight cranial bones are the frontal bone, parietal bones (two), temporal bones (two), occipital bone, sphenoid bone, and ethmoid bone. There are 14 **facial bones:** nasal bones (two), maxillae (two), zygomatic bones (two), mandible, lacrimal bones (two), palatine bones (two), inferior nasal conchae (two), and vomer. Be sure you can locate all the skull bones in the anterior, lateral, median, and posterior views of the skull (Figure 6-2).

SUTURES

A **suture** (SOO-chur; = seam or stitch) is an immovable joint found only between skull bones. Very little connective tissue is found between the bones of the suture. Four prominent skull sutures are:

1. Coronal suture between the frontal bone and the two parietal bones.
2. Sagittal suture between the two parietal bones.
3. Lambdoidal (lam-BOY-dal) **suture** between the parietal bones and the occipital bone.
4. Squamosal (skwa-MŌ-sal) **suture** between the parietal bones and the temporal bones.

Refer to Figures 6-2 and 6-3 for the locations of these sutures. Several other sutures are also shown. Their names are descriptive of the bones they connect. For example, the frontonasal suture is between the frontal bone and the nasal bones. These sutures are indicated in Figures 6-2 to 6-5.

FONTANELS

The "skeleton" of a newly formed embryo consists of cartilage or fibrous membrane structures shaped like bones. Gradually the cartilage or fibrous membrane is replaced by bone. At birth, membrane-filled spaces called

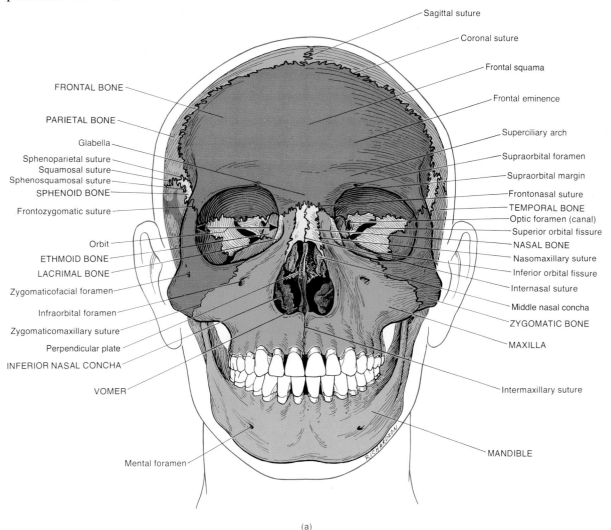

(a)

FIGURE 6-2 Skull. (a) Anterior view.

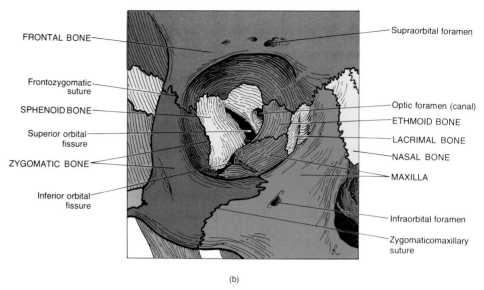

FRONTAL BONE

Frontozygomatic suture

SPHENOID BONE

Superior orbital fissure

ZYGOMATIC BONE

Inferior orbital fissure

Supraorbital foramen

Optic foramen (canal)

ETHMOID BONE

LACRIMAL BONE

NASAL BONE

MAXILLA

Infraorbital foramen

Zygomaticomaxillary suture

(b)

FIGURE 6-2 *(Continued)* Skull. (b) Detail of the right orbit in anterior view.

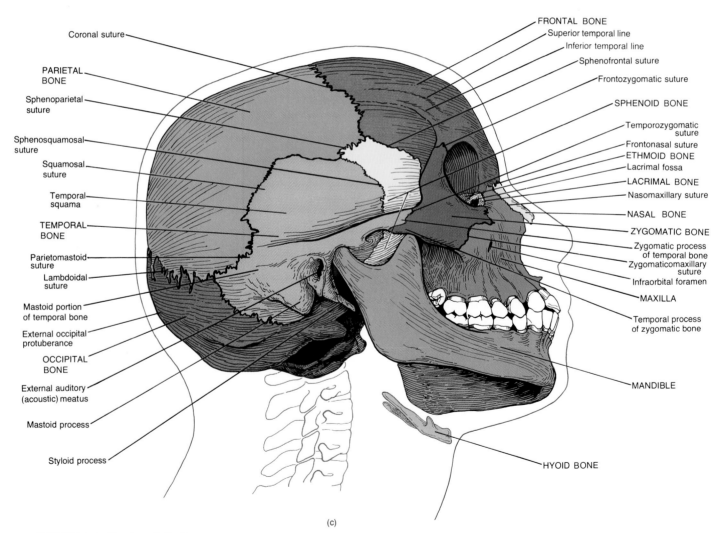

Coronal suture

PARIETAL BONE

Sphenoparietal suture

Sphenosquamosal suture

Squamosal suture

Temporal squama

TEMPORAL BONE

Parietomastoid suture

Lambdoidal suture

Mastoid portion of temporal bone

External occipital protuberance

OCCIPITAL BONE

External auditory (acoustic) meatus

Mastoid process

Styloid process

FRONTAL BONE

Superior temporal line

Inferior temporal line

Sphenofrontal suture

Frontozygomatic suture

SPHENOID BONE

Temporozygomatic suture

Frontonasal suture

ETHMOID BONE

Lacrimal fossa

LACRIMAL BONE

Nasomaxillary suture

NASAL BONE

ZYGOMATIC BONE

Zygomatic process of temporal bone

Zygomaticomaxillary suture

Infraorbital foramen

MAXILLA

Temporal process of zygomatic bone

MANDIBLE

HYOID BONE

(c)

FIGURE 6-2 *(Continued)* Skull. (c) Right lateral view.

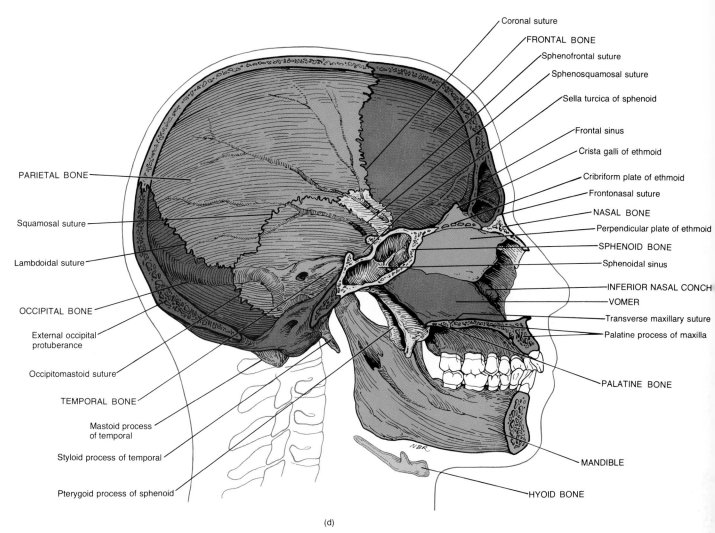

Coronal suture
FRONTAL BONE
Sphenofrontal suture
Sphenosquamosal suture
Sella turcica of sphenoid
Frontal sinus
Crista galli of ethmoid
Cribriform plate of ethmoid
Frontonasal suture
NASAL BONE
Perpendicular plate of ethmoid
SPHENOID BONE
Sphenoidal sinus
INFERIOR NASAL CONCH
VOMER
Transverse maxillary suture
Palatine process of maxilla
PALATINE BONE
MANDIBLE
HYOID BONE

PARIETAL BONE
Squamosal suture
Lambdoidal suture
OCCIPITAL BONE
External occipital protuberance
Occipitomastoid suture
TEMPORAL BONE
Mastoid process of temporal
Styloid process of temporal
Pterygoid process of sphenoid

(d)

FIGURE 6-2 *(Continued)* Skull. (d) Median view.

fontanels (fon'-ta-NELS; = little fountains), are found between cranial bones (Figure 6-3). These "soft spots" are areas where the bone-making process is not yet complete. They allow the skull to be compressed during birth. Physicians find the fontanels helpful in determining the position of the infant's head prior to delivery. Although an infant may have many fontanels at birth, the form and location of six are fairly constant.

The **anterior (frontal) fontanel** is located between the angles of the two parietal bones and the two segments of the frontal bone. This fontanel is roughly diamond-shaped, and it is the largest of the six fontanels. It usually closes 18 to 24 months after birth.

The **posterior (occipital;** ok-SIP-i-tal) **fontanel** is situated between the two parietal bones and the occipital bone. This diamond-shaped fontanel is considerably smaller than the anterior fontanel. It generally closes about two months after birth.

The **anterolateral (sphenoidal;** sfē-NOY-dal) **fontanels** are paired. One is located on each side of the skull at the junction of the frontal, parietal, temporal, and sphenoid bones. These fontanels are quite small and irregular in shape. They normally close three months after birth.

The **posterolateral (mastoid) fontanels** are also paired. One is situated on each side of the skull at the junction of the parietal, occipital, and temporal bones. These fontanels are irregularly shaped. They begin to close 1 or 2 months after birth, but closure is not generally complete until 12 months.

FRONTAL BONE

The **frontal bone** forms the forehead, the anterior part of the cranium; the roofs of the *orbits* (eye sockets); and most of the anterior part of the cranial floor. Soon after birth the left and right parts of the frontal bone are united

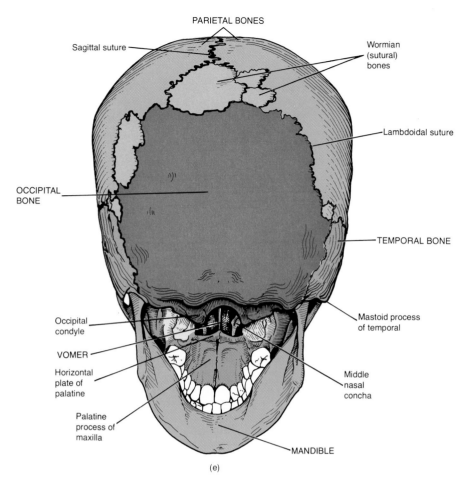

(e)

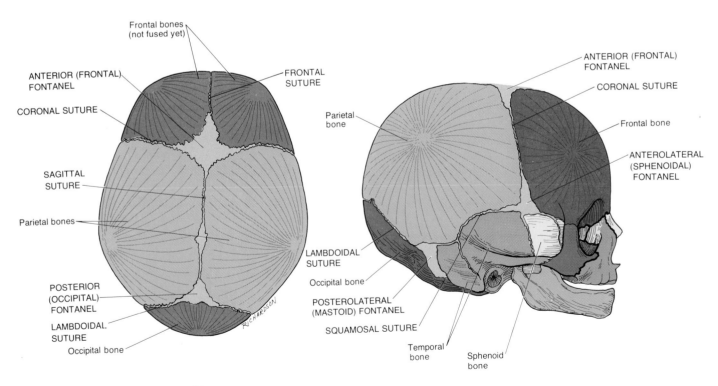

FIGURE 6-2 *(Continued)* Skull. (e) Posterior view.

FIGURE 6-3 Fontanels of the skull at birth. (a) Superior view. (b) Right lateral view.

(a)

(b)

by a suture. The suture usually disappears by age 6. If, however, the suture persists throughout life, it is referred to as the **metopic suture.**

If you examine the anterior and lateral views of the skull in Figure 6-2, you will note the **frontal squama** (SKWĀ-ma), or vertical plate (*squam* = scale). This scalelike plate, which corresponds to the forehead, gradually slopes down from the coronal suture, then turns abruptly downward. It projects slightly above its lower edge on either side of the midline to form the **frontal eminences.** Inferior to each eminence is a horizontal ridge, the **superciliary arch,** caused by the projection of the frontal sinuses posterior to the eyebrow.

Between the eminences and the arches just superior to the nose is a flattened area, the **glabella.** A thickening of the frontal bone inferior to the superciliary arches is called the **supraorbital margin.** From this margin the frontal bone extends posteriorly to form the roof of the orbit and part of the floor of the cranial cavity.

Within the supraorbital margin, slightly medial to its midpoint, is a hole called the **supraorbital foramen** (*foramen* = hole). The supraorbital nerve and artery pass through this foramen. The **frontal sinuses** lie deep to the superciliary arches. These mucus-lined cavities act as sound chambers which give the voice resonance.

PARIETAL BONES

The two **parietal bones** (pa-RĪ-i-tal; *paries* = wall) form the greater portion of the sides and roof of the cranial cavity.

The external surface contains two slight ridges that may be observed by looking at the lateral view of the skull in Figure 6-2. These are the **superior temporal line** and a less conspicuous **inferior temporal line** below it. The internal surface has many eminences and depressions that accommodate the blood vessels supplying the outer meninx (covering) of the brain called the *dura mater.*

TEMPORAL BONES

The two **temporal bones** form the inferior sides of the cranium and part of the cranial floor. The term *tempora* pertains to the temples.

In the lateral view of the skull in Figure 6-2, notice the **squama** or **squamous portion**—a thin, large, expanded area that forms the anterior and superior part of the temple. Projecting from the inferior portion of the squama is the **zygomatic process,** which articulates with the temporal process of the zygomatic bone. The zygomatic process of the temporal bone together with the temporal process of the zygomatic bone constitutes the **zygomatic arch.**

At the floor of the cranial cavity, shown in Figure 6-5, is the **petrous portion** of the temporal bone. This portion is triangular and located at the base of the skull

between the sphenoid and occipital bones. The petrous portion contains the internal ear, the essential part of the organ of hearing. It also contains the **carotid foramen (canal)** through which the internal carotid artery passes (see Figure 6-4). Posterior to the carotid foramen and anterior to the occipital bone is the **jugular foramen (fossa)** through which the internal jugular vein and the glossopharyngeal nerve (IX), vagus nerve (X), and accessory nerve (XI) pass. (As you will see later, the Roman numerals associated with cranial nerves indicate the order in which the nerves arise from the brain, from front to back.)

Between the squamous and petrous portions is a socket called the **mandibular fossa.** Anterior to the mandibular fossa is a rounded eminence, the **articular tubercle.** The mandibular fossa and articular tubercle articulate with the condylar process of the mandible (lower jawbone) to form the temporomandibular joint. The mandibular fossa and articular tubercle are seen best in Figure 6-4.

In the lateral view of the skull in Figure 6-2, you will see the **mastoid portion** of the temporal bone, located posterior and inferior to the external auditory meatus, or ear canal. In the adult, this portion of the bone contains a number of **mastoid air "cells."** These air spaces are separated from the brain only by thin bony partitions.

CLINICAL APPLICATION

If **mastoiditis,** the inflammation of these bony cells, occurs, the infection may spread to the brain or its outer covering. The mastoid air cells do not drain as do the paranasal sinuses.

The **mastoid process** is a rounded projection of the temporal bone posterior to the external auditory meatus. It serves as a point of attachment for several neck muscles. Near the posterior border of the mastoid process is the **mastoid foramen** through which a vein (emissary) to the transverse sinus and a small branch of the occipital artery to the dura mater pass. The **external auditory meatus** is the canal in the temporal bone that leads to the middle ear. The **internal acoustic meatus** is superior to the jugular foramen. It transmits the facial and acoustic nerves and the internal auditory artery. The **styloid process** projects downward from the undersurface of the temporal bone and serves as a point of attachment for muscles and ligaments of the tongue and neck. Between the styloid process and the mastoid process is the **stylomastoid foramen,** which transmits the facial nerve (VII) and stylomastoid artery (see Figure 6-4).

OCCIPITAL BONE

The **occipital** (ok-SIP-i-tal) **bone** forms the posterior and prominent portion of the base of the cranium (Figure 6-4).

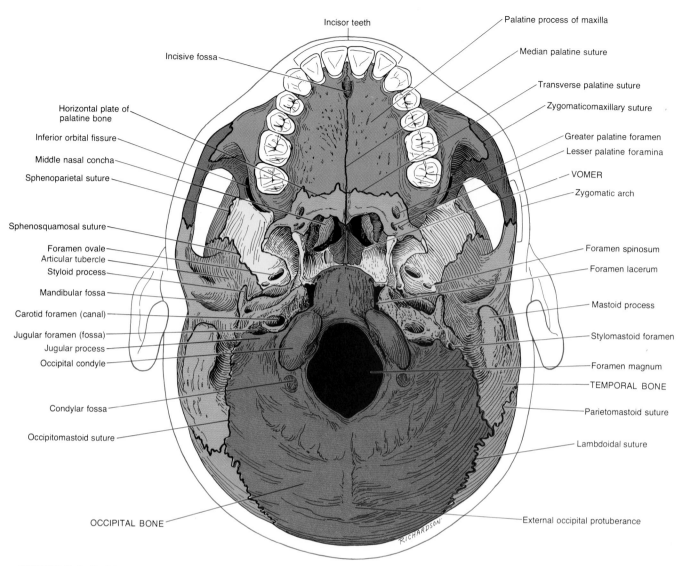

Incisor teeth

Incisive fossa

Horizontal plate of palatine bone

Inferior orbital fissure

Middle nasal concha

Sphenoparietal suture

Sphenosquamosal suture

Foramen ovale

Articular tubercle

Styloid process

Mandibular fossa

Carotid foramen (canal)

Jugular foramen (fossa)

Jugular process

Occipital condyle

Condylar fossa

Occipitomastoid suture

OCCIPITAL BONE

Palatine process of maxilla

Median palatine suture

Transverse palatine suture

Zygomaticomaxillary suture

Greater palatine foramen

Lesser palatine foramina

VOMER

Zygomatic arch

Foramen spinosum

Foramen lacerum

Mastoid process

Stylomastoid foramen

Foramen magnum

TEMPORAL BONE

Parietomastoid suture

Lambdoidal suture

External occipital protuberance

FIGURE 6-4 Skull in inferior view.

The **foramen magnum** is a large hole in the inferior part of the bone through which the medulla oblongata and its membranes, the spinal portion of the accessory nerve (XI), and the vertebral and spinal arteries pass.

The **occipital condyles** are oval processes with convex surfaces, one on either side of the foramen magnum, which articulate with depressions on the first cervical vertebra. Extending laterally from the posterior portion of the condyles are quadrilateral plates of bone called the **jugular processes.** At the base of the condyles is the **hypoglossal canal (fossa)** through which the hypoglossal nerve (XII) passes (see Figure 6-5).

The **external occipital protuberance** is a prominent projection on the posterior surface of the bone just superior to the foramen magnum. You can feel this structure as a definite bump on the back of your head, just above your neck. The protuberance is also visible in Figure 6-2d.

SPHENOID BONE

The **sphenoid** (SFĒ-noyd) **bone** is situated at the middle part of the base of the skull (Figure 6-5). The combining form *spheno* means wedge. This bone is referred to as the keystone of the cranial floor because it articulates with all the other cranial bones. If you view the floor of the cranium from above, you will note that the sphenoid articulates with the temporal bones anteriorly and the occipital bone posteriorly. It lies posterior and slightly superior to the nasal cavities and forms part of the floor and sidewalls of the eye socket. The shape of the sphenoid is frequently described as a bat with out-stretched wings.

The **body** of the sphenoid is the cubelike central portion between the ethmoid and occipital bones. It contains the **sphenoidal sinuses,** which drain into the nasal cavity (see Figure 6-8). On the superior surface of the sphenoid body is a depression called the **sella turcica** (SEL-a TUR-si-

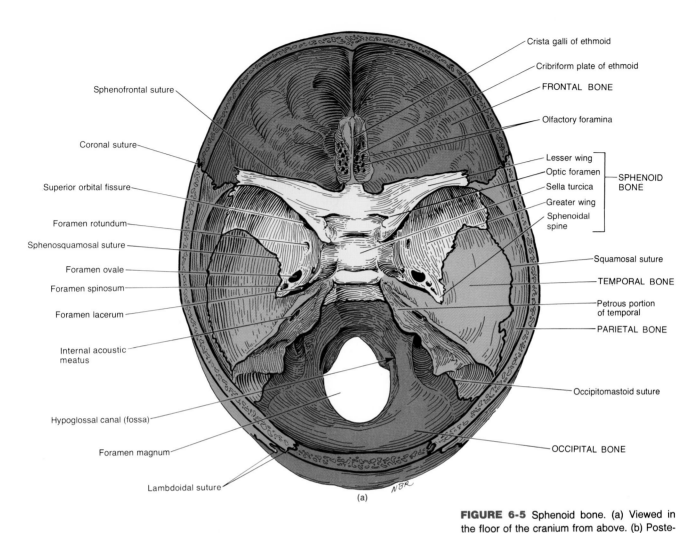

Sphenofrontal suture

Coronal suture

Superior orbital fissure

Foramen rotundum

Sphenosquamosal suture

Foramen ovale

Foramen spinosum

Foramen lacerum

Internal acoustic
meatus

Hypoglossal canal (fossa)

Foramen magnum

Lambdoidal suture

Crista galli of ethmoid

Cribriform plate of ethmoid

FRONTAL BONE

Olfactory foramina

Lesser wing
Optic foramen
Sella turcica   SPHENOID
Greater wing   BONE
Sphenoidal
spine

Squamosal suture

TEMPORAL BONE

Petrous portion
of temporal

PARIETAL BONE

Occipitomastoid suture

OCCIPITAL BONE

(a)

FIGURE 6-5 Sphenoid bone. (a) Viewed in the floor of the cranium from above. (b) Posterior view. (c) Anterior view.

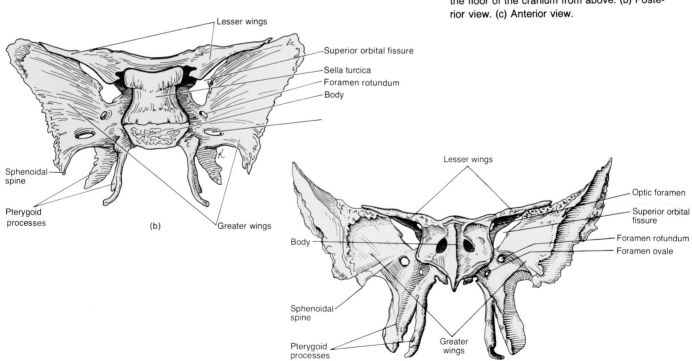

Lesser wings

Superior orbital fissure

Sella turcica
Foramen rotundum
Body

Sphenoidal
spine

Pterygoid
processes

Greater wings

(b)

Lesser wings

Optic foramen

Superior orbital
fissure

Foramen rotundum
Foramen ovale

Body

Sphenoidal
spine

Pterygoid
processes

Greater
wings

(c)

ka; = Turk's Saddle). This depression houses the pituitary gland.

The **greater wings** of the sphenoid are lateral projections from the body and form the anterolateral floor of the cranium. The greater wings also form part of the lateral wall of the skull just anterior to the temporal bone. The posterior portion of each greater wing contains a triangular projection, the **sphenoidal spine,** that fits into the angle between the squama and petrous portion of the temporal bone.

The **lesser wings** are anterior and superior to the greater wings. They form part of the floor of the cranium and the posterior part of the **orbit,** or eye socket. Between the body and lesser wing, you can locate the **optic foramen** through which the optic nerve (II) and ophthalmic artery pass. Lateral to the body between the greater and lesser wings is a somewhat triangular slit called the **superior orbital fissure.** It is an opening for the oculomotor nerve (III), trochlear nerve (IV), ophthalmic branch of the trigeminal nerve (V), and abducens nerve (VI). This fissure may also be seen in the anterior view of the skull in Figure 6-2.

On the inferior part of the sphenoid bone you can see the **pterygoid** (TER-i-goyd) **processes.** These structures project inferiorly from the points where the body and greater wings unite. The pterygoid processes form part of the lateral walls of the nasal cavities.

At the base of the lateral pterygoid process in the greater wing is the **foramen ovale** through which the mandibular branch of the trigeminal nerve (V) passes. Another foramen, the **foramen spinosum,** lies at the posterior angle of the sphenoid and transmits the middle meningeal vessels. The **foramen lacerum** is bounded anteriorly by the sphenoid bone, and medially by the sphenoid and occipital bones. Although the foramen is covered in part by a layer of fibrocartilage in living subjects, it transmits the internal carotid artery and the meningeal branch of the ascending pharyngeal artery. A final foramen associated with the sphenoid bone is the **foramen rotundum** through which the maxillary branch of the trigeminal nerve (V) passes. It is located at the junction of the anterior and medial parts of the sphenoid bone.

ETHMOID BONE

The **ethmoid bone** is a light, spongy bone located in the anterior part of the floor of the cranium between the orbits. It is anterior to the sphenoid and posterior to the nasal bones (Figure 6-6). This bone forms part of the anterior portion of the cranial floor, the medial wall of the orbits, the superior portions of the nasal septum, or partition, and most of the sidewalls of the nasal roof. The ethmoid is the principal supporting structure of the nasal cavity.

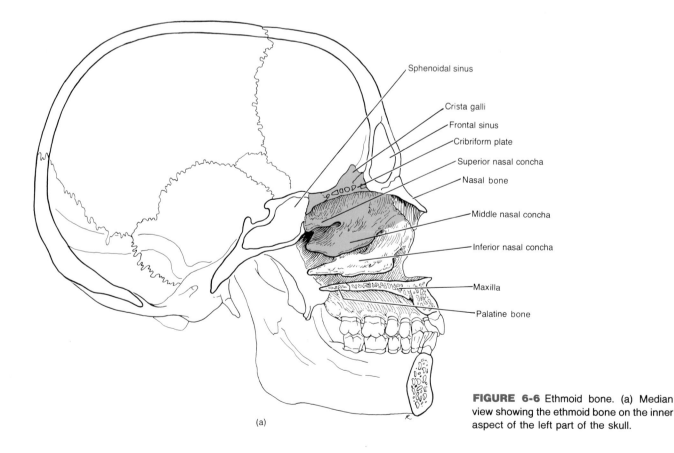

Sphenoidal sinus
Crista galli
Frontal sinus
Cribriform plate
Superior nasal concha
Nasal bone
Middle nasal concha
Inferior nasal concha
Maxilla
Palatine bone

(a)

FIGURE 6-6 Ethmoid bone. (a) Median view showing the ethmoid bone on the inner aspect of the left part of the skull.

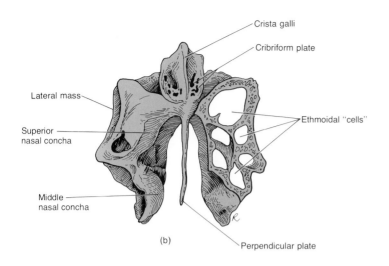

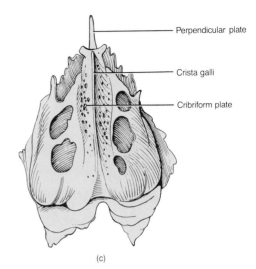

(c)

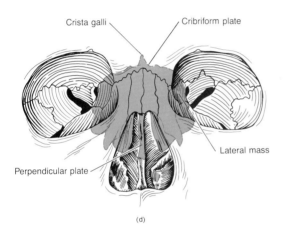

(d)

FIGURE 6-6 *(Continued)* Ethmoid bone. (b) Anterior view. A frontal section has been made through the left side to expose the ethmoidal "cells." (c) Superior view. (d) Highly diagrammatic representation showing the approximate position of the ethmoid bone in the skull in anterior view.

NASAL BONES

The paired **nasal bones** are small, oblong bones that meet at the middle and superior part of the face (see Figures 6-2 and 6-7). Their fusion forms the superior part of the bridge of the nose. The inferior portion of the nose, indeed the major portion, consists of cartilage.

MAXILLAE

The paired maxillary bones unite to form the upper jaw-bone (Figure 6-7). The **maxillae** (mak-SIL-ē) articulate with every bone of the face except the mandible, or lower jawbone. They form part of the floors of the orbits, part of the roof of the mouth (most of the hard palate), and part of the lateral walls and floor of the nasal cavities. The two portions of the maxillary bones unite, and the fusion is normally completed before birth.

> **CLINICAL APPLICATION**
> If the palatine processes of the maxillary bones do not unite before birth, a condition called **cleft palate** results. The condition may also involve incomplete fusion of the horizontal plates of the palatine bones (see Figure 6-4). Another form of this condition, called **cleft lip,** involves a split in the upper lip. Cleft lip is often associated with cleft palate. Depending on the extent and position of the cleft, speech and swallowing may be affected. A surgical procedure can sometimes improve the condition.

Its **lateral masses** or **labyrinths** compose most of the wall between the nasal cavity and the orbits. They contain several air spaces, or "cells," ranging in number from 3 to 18. It is from these "cells" that the bone derives its name (*ethmos* = sieve). The ethmoid "cells" together form the **ethmoidal sinuses.** The sinuses are shown in Figure 6-8. The **perpendicular plate** forms the superior portion of the nasal septum (see Figure 6-7). The **cribriform plate,** or **horizontal plate,** lies in the anterior floor of the cranium and forms the roof of the nasal cavity. The cribriform plate contains the **olfactory foramina** through which the olfactory nerves (I) pass. These nerves function in smell. Projecting upward from the horizontal plate is a triangular process called the **crista galli,** which means cock's comb. This structure serves as a point of attachment for the membranes that cover the brain.

The labyrinths contain two thin, scroll-shaped bones on either side of the nasal septum. These are called the **superior nasal concha** (KONG-ha; *concha* = shell) and the **middle nasal concha.** The conchae allow for the efficient circulation and filtration of inhaled air before it passes into the trachea, the bronchi, and the lungs.

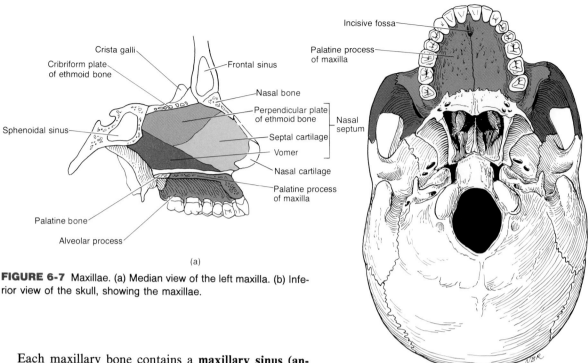

FIGURE 6-7 Maxillae. (a) Median view of the left maxilla. (b) Inferior view of the skull, showing the maxillae.

Each maxillary bone contains a **maxillary sinus (antrum of Highmore)** that empties into the nasal cavity (see Figure 6-8). The **alveolar** (al-VĒ-ō-lar) **process** (*alveolus* = hollow) contains the bony sockets into which the maxillary (upper) teeth are set. The **palatine process** is a horizontal projection of the maxilla that forms the anterior three-fourths of the hard palate, or anterior portion of the roof of the oral cavity.

The **infraorbital foramen,** which can be seen in the anterior view of the skull in Figure 6-2, is an opening in the maxilla inferior to the orbit. The infraorbital nerve and artery are transmitted through this opening. Another prominent fossa in the maxilla is the **incisive fossa** just posterior to the incisor teeth. Through it pass branches of the descending palatine vessels and the nasopalatine nerve. A final fossa associated with the maxilla and sphenoid bone is the **inferior orbital fissure.** It is located between the greater wing of the sphenoid and the maxilla (see Figure 6-4). It transmits the maxillary branch of the trigeminal nerve (V), the infraorbital vessels, and the zygomatic nerve.

PARANASAL SINUSES

Paired cavities, called **paranasal sinuses,** are located in certain bones near the nasal cavity (Figure 6-8). The paranasal sinuses are lined with mucous membranes that are continuous with the lining of the nasal cavity. Cranial bones containing paranasal sinuses are the frontal bone, the sphenoid, the ethmoid, and the maxillae. (The sinuses were described in the discussion of each of these bones.) Besides producing mucus, the paranasal sinuses lighten the skull bones and serve as resonant chambers for sound.

CLINICAL APPLICATION

Secretions produced by the mucous membranes of the paranasal sinuses drain into the nasal cavity. An inflammation of the membranes due to an allergic reaction or infection is called **sinusitis.** If the membranes swell enough to block drainage into the nasal cavity, fluid pressure builds up in the paranasal sinuses and a sinus headache results.

ZYGOMATIC BONES

The two **zygomatic bones (malars),** commonly referred to as the cheekbones, form the prominences of the cheeks and part of the outer wall and floor of the orbits (Figure 6-2b).

The **temporal process** of the zygomatic bone projects posteriorly and articulates with the zygomatic process of the temporal bone. These two processes form the **zygomatic arch.** A foramen associated with the zygomatic bone is the **zygomaticofacial foramen** near the center of the bone (see Figure 6-2). It transmits the zygomaticofacial nerve and vessels.

MANDIBLE

The **mandible** or lower jawbone is the largest, strongest facial bone (Figure 6-9). It is the only movable bone in the skull.

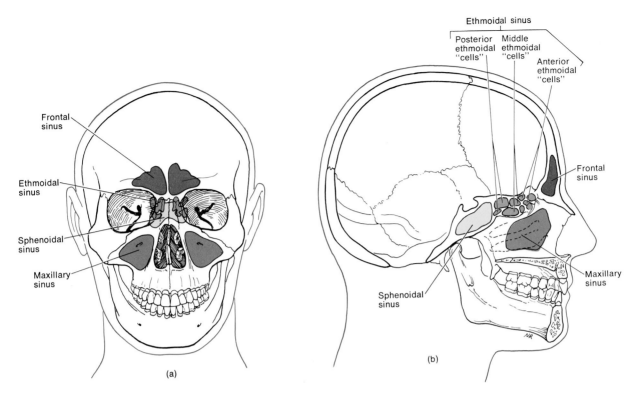

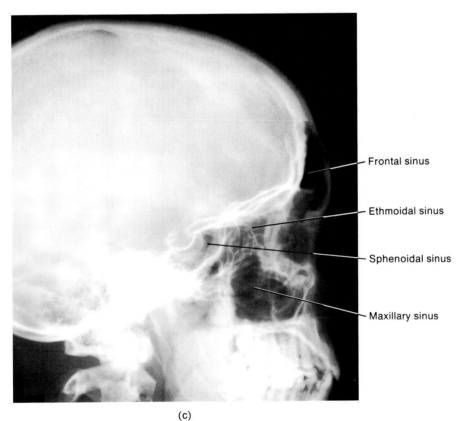

FIGURE 6-8 Paranasal sinuses. (a)
Anterior view. (b) Median view. (c) Roent-
genogram of a lateral view of the skull.
(Courtesy of Eastman Kodak Company.)

In the lateral view you can see that the mandible consists of a curved, horizontal portion called the **body** and two perpendicular portions called the **rami.** The **angle** of the mandible is the area where each ramus meets the body. Each ramus has a **condylar** (KON-di-lar) **process** that articulates with the mandibular fossa and articular tubercle of the temporal bone to form the temporomandibular joint. It also has a **coronoid** (KOR-ō-noyd) **pro-**

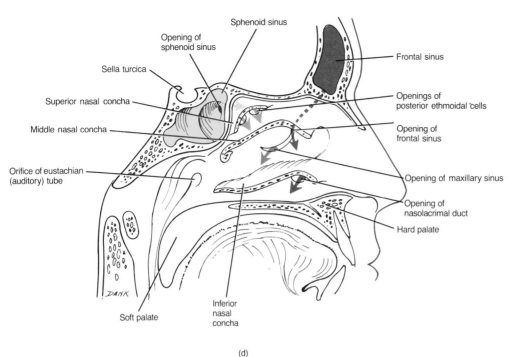

(d)

FIGURE **6-8** *(Continued)*
Paranasal sinuses. (d) Median
view showing the openings of
the paranasal sinuses into the
nasal cavity.

cess to which the temporalis muscle attaches. The depression between the coronoid and condylar processes is called the **mandibular notch.** The **alveolar process** is an arch containing the sockets for the mandibular (lower) teeth.

The **mental foramen** (*mentum* = chin) is approximately below the first molar tooth. The mental nerve and vessels pass through this opening. It is through this foramen that dentists sometimes reach the nerve when injecting anesthetics. Another foramen associated with the mandible is the **mandibular foramen** on the medial surface of the ramus, another site frequently used by den-

tists to inject anesthetics. It transmits the inferior alveolar nerve and vessels.

CLINICAL APPLICATION

Opening of the mouth very wide, as in a very large yawn, may cause displacement of the condylar process of the mandible from the mandibular fossa of temporal bone, thereby producing a **dislocation of the jaw.** In this condition the patient is unable to close the mouth.

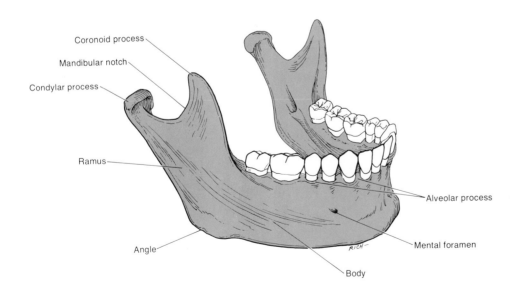

FIGURE 6-9 Mandible in right lateral view.

LACRIMAL BONES

The paired **lacrimal bones** (LAK-ri-mal; *lacrima* = tear) are thin bones roughly resembling a fingernail in size and shape. They are the smallest bones of the face. These bones are posterior and lateral to the nasal bones in the medial wall of the orbit. They can be seen in the anterior and lateral views of the skull in Figure 6-2.

The lacrimal bones form a part of the medial wall of the orbit. They also contain the **lacrimal fossae** through which the tear ducts pass into the nasal cavity (see Figure 6-2).

PALATINE BONES

The two **palatine** (PAL-a-tīn) **bones** are L-shaped and form the posterior portion of the hard palate, part of the floor and lateral walls of the nasal cavities, and a small portion of the floor of the orbit. The posterior portion of the hard palate, which separates the nasal cavity from the oral cavity, is formed by the **horizontal plates** of the palatine bones. These can be seen in Figure 6-4.

Two foramina associated with the palatine bones are the greater and lesser palatine foramina (Figure 6-4). The **greater palatine foramen,** at the posterior angle of the hard palate, transmits the greater palatine nerve and descending palatine vessels. The **lesser palatine foramina,** usually two or more on each side, are posterior to the greater palatine foramina. They transmit the lesser palatine nerve.

INFERIOR NASAL CONCHAE

Refer to the views of the skull in Figures 6-2a and 6-6a. The two **inferior nasal conchae** (KONG-kē) are scroll-like bones that form a part of the lateral wall of the nasal cavity and project into the nasal cavity inferior to the superior and middle nasal conchae of the ethmoid bone. They serve the same function as the superior and middle nasal conchae, that is, they allow for the circulation and filtration of air before it passes into the lungs. The inferior nasal conchae are separate bones and not part of the ethmoid.

VOMER

The **vomer** (= *plowshare*) is a roughly triangular bone that forms the inferior and posterior part of the nasal septum. It is clearly seen in the anterior view of the skull in Figure 6-2a and the inferior view in Figure 6-4.

The inferior border of the vomer articulates with the cartilage septum that divides the nose into a right and left nostril. Its superior border articulates with the perpendicular plate of the ethmoid bone. Thus the structures that form the **nasal septum,** or partition, are the perpendicular plate of the ethmoid, the septal cartilage, and the vomer (Figure 6-7a). If the vomer is deviated, that is, pushed to one side, the nasal chambers are of unequal size.

CLINICAL APPLICATION

A **deviated septum** is deflected laterally from the midline of the nose. The deviation usually occurs at the junction of bone with the septal cartilage. If the deviation is severe, it may entirely block the nasal passageway. Even though the blockage may not be complete, infection and inflammation develop and cause nasal congestion, blockage of the paranasal sinus openings, and chronic sinusitis.

FORAMINA

The various foramina of the skull were mentioned along with the descriptions of the cranial and facial bones with which they are associated. As preparation for studying other systems of the body, especially the nervous and circulatory systems, the foramina and the structures passing through them are listed in Exhibit 6-3. For your convenience and for future reference, the foramina are listed alphabetically.

HYOID BONE

The single **hyoid bone** (*hyoid* = U-shaped) is a unique component of the axial skeleton because it does not articulate with any other bone. Rather, it is suspended from the styloid process of the temporal bone by ligaments and muscles. The hyoid is located in the neck between the mandible and larynx. It supports the tongue and provides attachment for some of its muscles. Refer to the anterior and lateral views of the skull in Figure 6-2 to see the position of the hyoid bone.

The hyoid consists of a horizontal **body** and paired projections called the **lesser cornu** (*cornu* = horns) and the **greater cornu** (Figure 6-10). Muscles and ligaments attach to these paired projections.

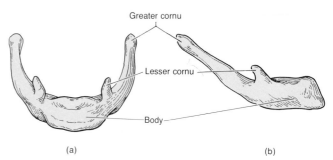

Greater cornu

Lesser cornu

Body

(a)

(b)

FIGURE 6-10 Hyoid bone. (a) Anterior view. (b) Right lateral view.

EXHIBIT 6-3 SUMMARY OF FORAMINA OF THE SKULL

FORAMEN	LOCATION	STRUCTURES PASSING THROUGH
Carotid (Figure 6-4)	Petrous portion of temporal.	Internal carotid artery.
Greater palatine (Figure 6-4)	Posterior angle of hard palate.	Greater palatine nerve and greater palatine vessels.
Hypoglossal (Figure 6-5)	Superior to base of occipital condyles.	Hypoglossal nerve (XII)* and branch of ascending pharyngeal artery.
Incisive (Figure 6-7)	Posterior to incisor teeth.	Branches of descending palatine vessels and nasopalatine nerve.
Inferior orbital (Figure 6-4)	Between greater wing of sphenoid and maxilla.	Maxillary branch of trigeminal nerve (V), zygomatic nerve, and infraorbital vessels.
Infraorbital (Figure 6-2a)	Inferior to orbit in maxilla.	Infraorbital nerve and artery.
Jugular (Figure 6-4)	Posterior to carotid canal between petrous portion of temporal and occipital.	Internal jugular vein, glossopharyngeal nerve (IX), vagus nerve (X), accessory nerve (XI), and sigmoid sinus.
Lacerum (Figure 6-5)	Bounded anteriorly by sphenoid, posteriorly by petrous portion of temporal, and medially by the sphenoid and occipital.	Internal carotid artery and branch of ascending pharyngeal artery.
Lacrimal (Figure 6-2c)	Lacrimal bone.	Lacrimal (tear) duct.
Lesser palatine (Figure 6-4)	Posterior to greater palatine foramen.	Lesser palatine nerves and artery.
Magnum (Figure 6-4)	Occipital bone.	Medulla oblongata and its membranes, the spinal accessory nerve (XI), the vertebral and spinal arteries and meninges.
Mandibular	Medial surface of ramus of mandible.	Inferior alveolar nerve and vessels.
Mastoid (Figure 6-4)	Posterior border of mastoid process of temporal bone.	Emissary vein to transverse sinus and branch of occipital artery to dura mater.
Mental (Figure 6-9)	Inferior to second premolar tooth in mandible.	Mental nerve and vessels.
Olfactory (Figure 6-5)	Cribriform plate of ethmoid.	Olfactory nerve (I).
Optic (Figure 6-5)	Between upper and lower portions of small wing of sphenoid.	Optic nerve (II) and ophthalmic artery.
Ovale (Figure 6-5)	Greater wing of sphenoid.	Mandibular branch of trigeminal nerve (V).
Rotundum (Figure 6-5)	Junction of anterior and medial parts of sphenoid.	Maxillary branch of trigeminal nerve (V).
Spinosum (Figure 6-5)	Posterior angle of sphenoid.	Middle meningeal vessels.
Stylomastoid (Figure 6-4)	Between styloid and mastoid processes of temporal.	Facial nerve (VII) and stylomastoid artery.
Superior orbital (Figure 6-5)	Between greater and lesser wings of sphenoid.	Oculomotor nerve (III), trochlear nerve (IV), ophthalmic branch of trigeminal nerve (V), and abducens nerve (VI).
Supraorbital (Figure 6-2a)	Supraorbital margin of orbit.	Supraorbital nerve and artery.
Zygomaticofacial (Figure 6-2a)	Zygomatic bone.	Zygomaticofacial nerve and vessels.

* As you will see later, the roman numerals associated with cranial nerves indicate the order in which the nerves arise from the brain, from front to back.

VERTEBRAL COLUMN

DIVISIONS

The **vertebral column,** or **spine,** together with the sternum and ribs, constitutes the skeleton of the **trunk** of the body. The vertebral column is composed of a series of bones called **vertebrae.** In the average adult, the column measures about 71 cm (28 inches) in length. In effect, the vertebral column is a strong, flexible rod that moves anteriorly, posteriorly, and laterally. It encloses and protects the spinal cord, supports the head, and serves as a point of attachment for the ribs and the muscles of the back. Between the vertebrae are openings called **intervertebral foramina.** The nerves that connect the spinal cord to various parts of the body pass through these openings.

The adult vertebral column typically contains 26 vertebrae (Figure 6-11). These are distributed as follows: 7 **cervical vertebrae** in the neck region; 12 **thoracic vertebrae** posterior to the thoracic cavity; 5 **lumbar vertebrae** supporting the lower back; 5 **sacral vertebrae** fused into

one bone called the **sacrum;** and usually four **coccygeal** (kok-SIJ-ē-al) **vertebrae** fused into one or two bones called the **coccyx** (KOK-six). Prior to the fusion of the sacral and coccygeal vertebrae, the total number of vertebrae is 33.

Between adjacent vertebrae from the axis to the sacrum are fibrocartilaginous **intervertebral discs.** Each disc is composed of an outer fibrous ring consisting of fibrocartilage called the *anulus fibrosus* and an inner soft, pulpy, highly elastic structure called the *nucleus pulposus* (see Figure 8-1b). The discs form strong joints, permit various movements of the vertebral column, and absorb vertical shock. Under compression, they flatten, broaden, and bulge from their intervertebral spaces.

CURVES

When viewed from the side, the vertebral column shows four **curves.** From the anterior view, these are alternately convex, meaning they curve toward the viewer, and con-

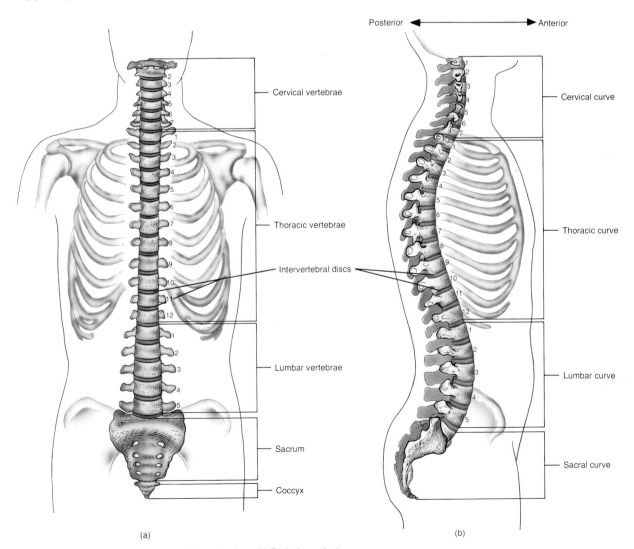

FIGURE 6-11 Vertebral column. (a) Anterior view. (b) Right lateral view.

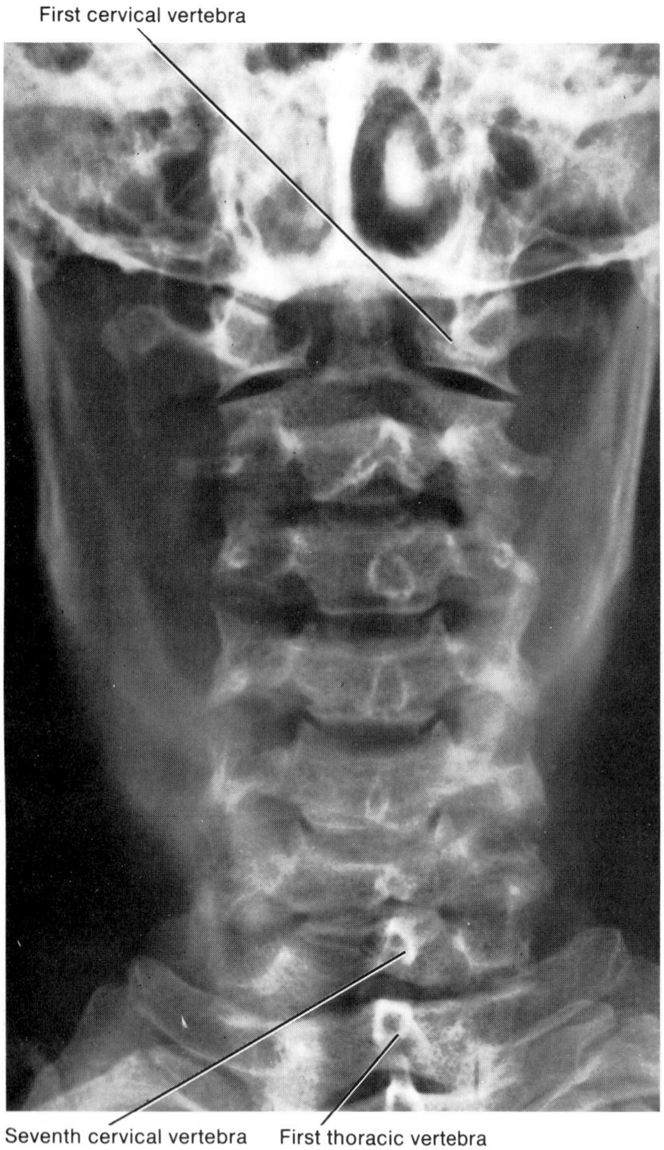

First cervical vertebra

Seventh cervical vertebra

First thoracic vertebra

Seventh cervical vertebra   First thoracic vertebra

(c)

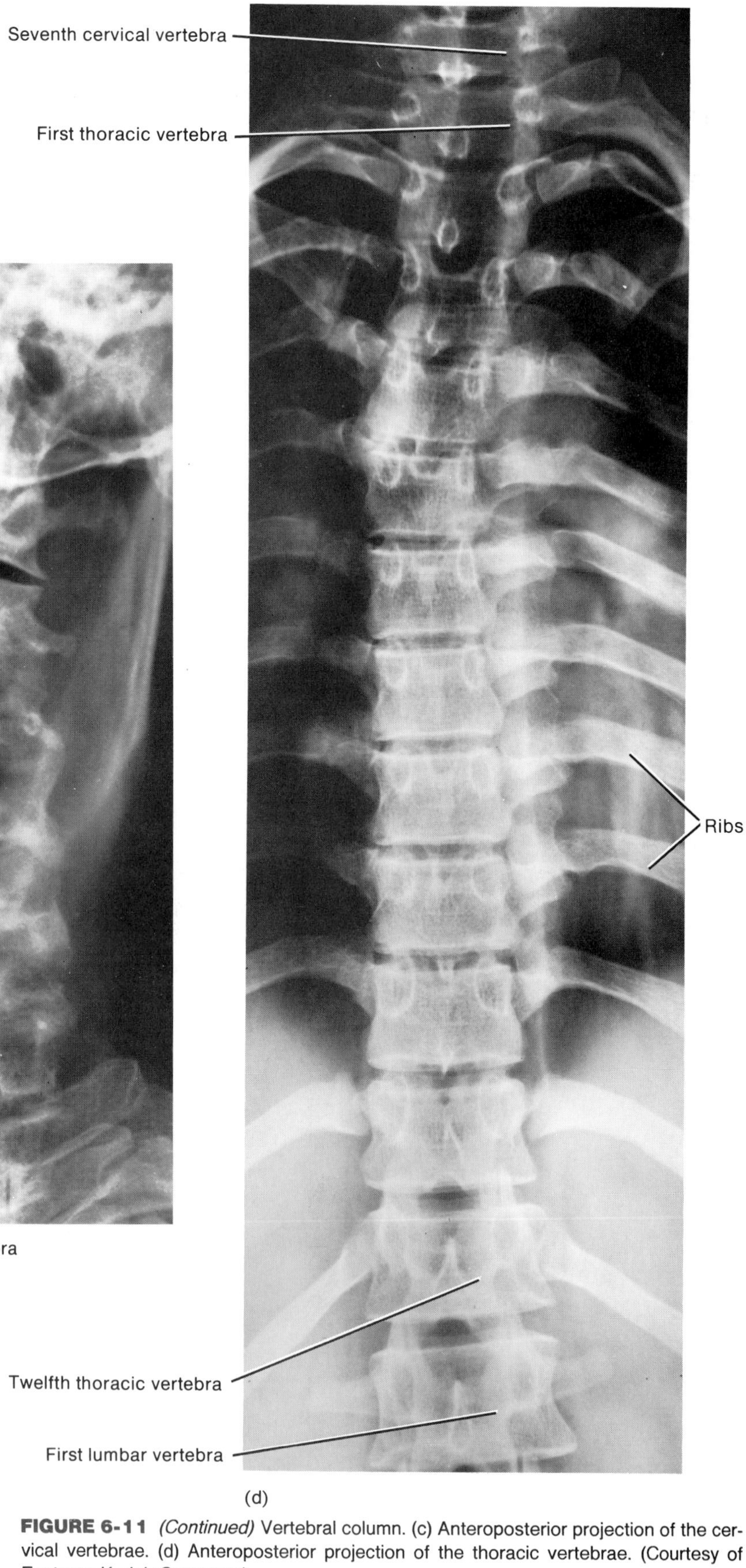

Ribs

Twelfth thoracic vertebra

First lumbar vertebra

(d)

FIGURE 6-11 *(Continued)* Vertebral column. (c) Anteroposterior projection of the cervical vertebrae. (d) Anteroposterior projection of the thoracic vertebrae. (Courtesy of Eastman Kodak Company.)

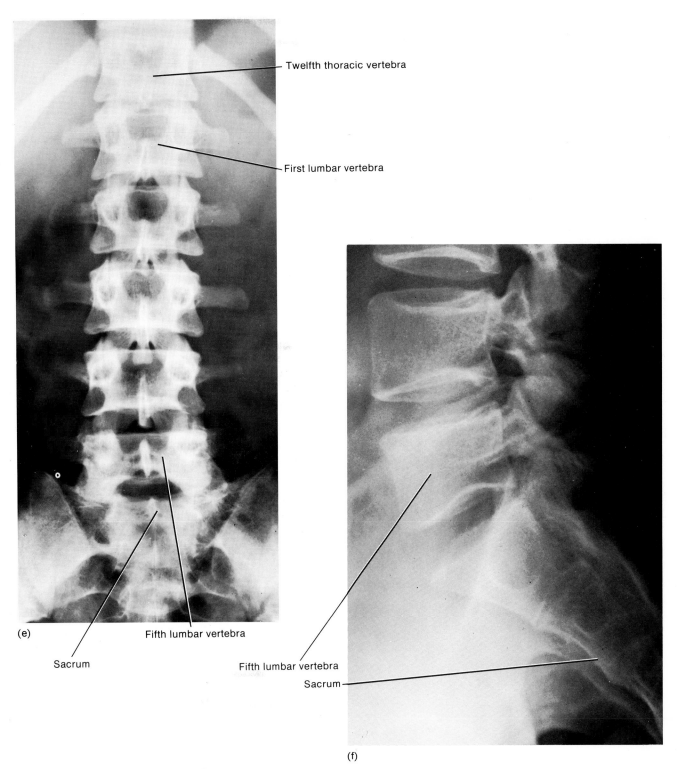

Twelfth thoracic vertebra

First lumbar vertebra

(e)

Fifth lumbar vertebra

Sacrum

Fifth lumbar vertebra

Sacrum

(f)

FIGURE 6-11 *(Continued)* Vertebral column. (e) Anteroposterior projection of the lumbar vertebrae. (f) Lateral view of the sacrum. (Courtesy of Eastman Kodak Company.)

cave, meaning they curve away from the viewer. The curves of the column, like the curves in a long bone, are important because they increase its strength, help maintain balance in the upright position, absorb shocks from walking, and help protect the column from fracture.

In the fetus, the four curves of the vertebrae are not present. There is only a single curve that is anteriorly concave. At approximately the third postnatal month, when an infant begins to hold its head erect, the **cervical curve** develops. Later, when the child stands and walks, the **lumbar curve** develops. The cervical and lumbar curves are convex anteriorly. Because they are modifications of the fetal positions, they are called **secondary curves.** The other two curves, the **thoracic curve** and the **sacral curve,** are anteriorly concave. Since they retain the anterior concavity of the fetus, they are referred to as **primary curves.**

TYPICAL VERTEBRA

Although there are variations in size, shape, and detail in the vertebrae in different regions of the column, all the vertebrae are basically similar in structure (Figure 6-12). A typical vertebra consists of the following components.

1. The **body** is the thick, disc-shaped anterior portion that is the weight-bearing part of a vertebra. Its superior and inferior surfaces are roughened for the attachment of intervertebral discs. The anterior and lateral surfaces contain nutrient foramina for blood vessels.

2. The **vertebral arch (neural arch)** extends posteriorly from the body of the vertebra. With the body of the vertebra, it surrounds the spinal cord. It is formed by two short, thick processes, the **pedicles** (PED-i-kuls), which project posteriorly from the body to unite with the laminae. The **laminae** (LAM-i-nē) are the flat parts that join to form the posterior portion of the vertebral arch. The space that lies between the vertebral arch and body contains the spinal cord. This space is known as the **vertebral foramen.** The vertebral foramina of all verte-

brae together form the **vertebral,** or **spinal, canal.** The pedicles are notched superiorly and inferiorly in such a way that, when they are arranged in the column, there is an opening between vertebrae on each side of the column. This opening, the **intervertebral foramen,** permits the passage of the spinal nerves.

3. Seven **processes** arise from the vertebral arch. At the point where a lamina and pedicle join, a **transverse process** extends laterally on each side. A single **spinous process** or **spine** projects posteriorly and inferiorly from the junction of the laminae. These three processes serve as points of muscular attachment. The remaining four processes form joints with other vertebrae. The two **superior articular processes** of a vertebra articulate with the vertebra immediately superior to them. The two **inferior articular processes** of a vertebra articulate with the vertebra inferior to them.

CERVICAL REGION

When viewed from above, it can be seen that the bodies of **cervical vertebrae** are smaller than those of the thoracic vertebrae (Figure 6-13). The arches, however, are larger. The spinous processes of the second through sixth cervical vertebrae are often *bifid,* that is, with a cleft. Each cervical transverse process contains an opening, the **transverse foramen.** The vertebral artery and its accompanying vein and nerve fibers pass through it.

The first two cervical vertebrae differ considerably from the others. The first cervical vertebra, the **atlas,** is named for its support of the head. Essentially, the atlas

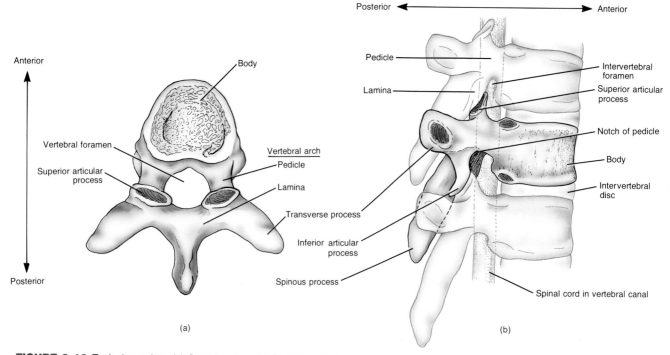

FIGURE 6-12 Typical vertebra. (a) Superior view. (b) Right lateral view.

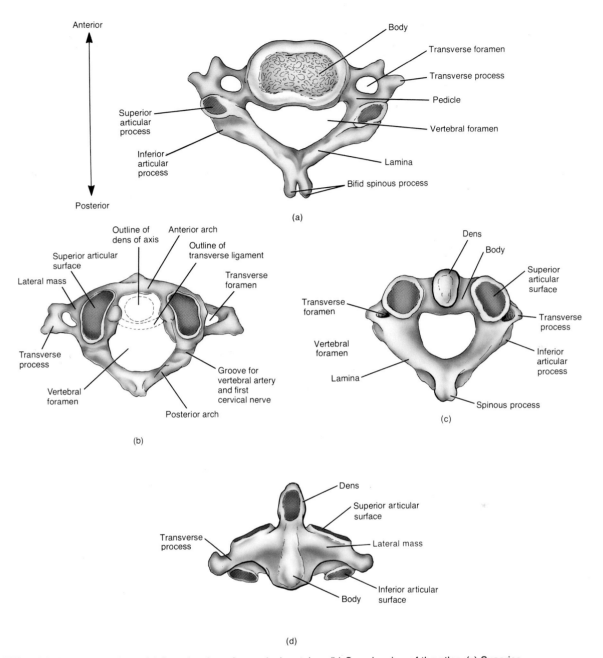

FIGURE 6-13 Cervical vertebrae. (a) Superior view of a cervical vertebra. (b) Superior view of the atlas. (c) Superior view of the axis. (d) Anterior view of the axis.

is a ring of bone with **anterior** and **posterior arches** and large **lateral masses.** It lacks a body and a spinous process. The superior surfaces of the lateral masses, called **superior articular surfaces,** are concave and articulate with the occipital condyles of the occipital bone. This articulation permits the movement seen when nodding the head. The inferior surfaces of the lateral masses, the **inferior articular surfaces,** articulate with the second cervical vertebra. The transverse processes and **transverse foramina** of the atlas are quite large.

The second cervical vertebra, the **axis,** does have a **body.** A peglike process called the **dens,** or **odontoid process,** projects up through the ring of the atlas. The dens makes a pivot on which the atlas and head rotate. This arrangement permits side-to-side rotation of the head.

CLINICAL APPLICATION

In various instances of trauma, the dens of the axis may be driven into the medulla oblongata of the brain, usually with instantly fatal results. This injury is the usual cause of fatality in **whiplash injuries.**

The third through sixth cervical vertebrae correspond to the structural pattern of the typical cervical vertebra shown.

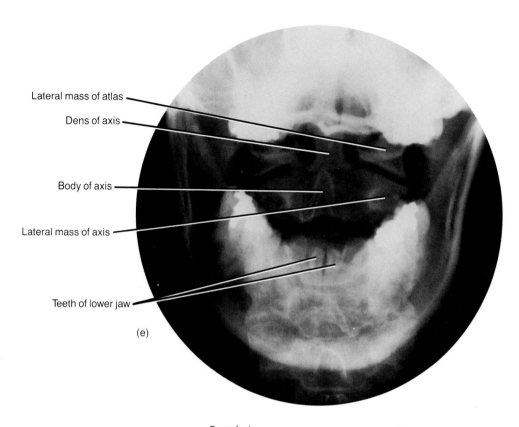

Lateral mass of atlas

Dens of axis

Body of axis

Lateral mass of axis

Teeth of lower jaw

(e)

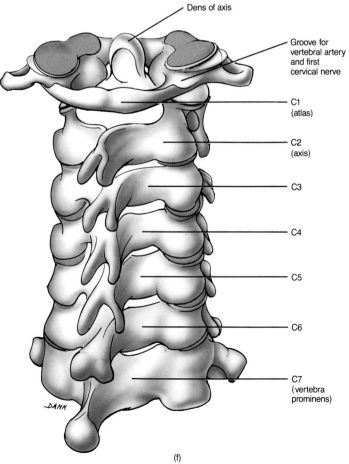

Dens of axis

Groove for vertebral artery and first cervical nerve

C1 (atlas)

C2 (axis)

C3

C4

C5

C6

C7 (vertebra prominens)

—DANK

(f)

FIGURE 6-13 *(Continued)* Cervical vertebrae. (e) Anteroposterior projection of the atlas and axis taken through the open mouth. (Courtesy of John C. Bennett, St. Mary's Hospital, San Francisco.) (f) Articulated in posterior view.

Posterior ◄──────────────────────────► Anterior

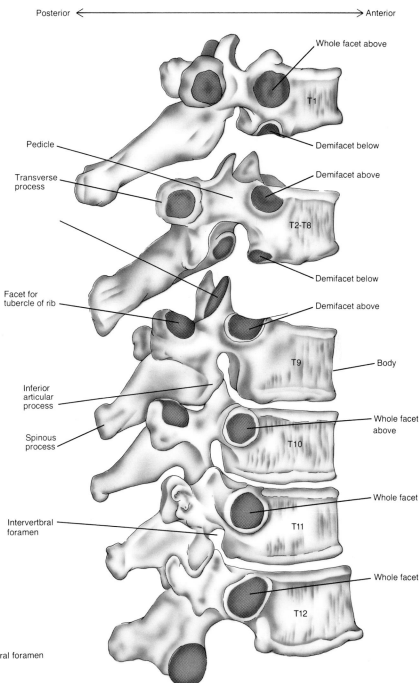

Whole facet above

T1

Demifacet below

Pedicle

Transverse process

Demifacet above

T2-T8

Demifacet below

Facet for tubercle of rib

Demifacet above

T9   Body

Inferior articular process

Whole facet above

T10

Spinous process

Whole facet

Intervertbral foramen

T11

Whole facet

T12

(b)

FIGURE 6-14 Thoracic vertebrae. (a) Superior view. (b) Articulated in right lateral view.

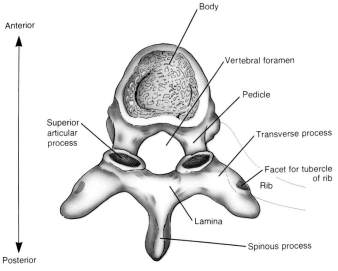

Anterior

Body

Vertebral foramen

Pedicle

Transverse process

Superior articular process

Facet for tubercle of rib

Rib

Lamina

Spinous process

Posterior

(a)

The seventh cervical vertebra, called the **vertebra prominens,** is somewhat different. It is marked by a large, nonbifid spinous process that may be seen and felt at the base of the neck (see Figure 6-11).

THORACIC REGION

Viewing a typical **thoracic vertebra** from above, you can see that it is considerably larger and stronger than a vertebra of the cervical region (Figure 6-14). In addition, the spinous process on each vertebra is long, pointed, and directed inferiorly. Thoracic vertebrae also have longer and heavier transverse processes than cervical vertebrae.

Except for the eleventh and twelfth thoracic vertebrae, the transverse processes have **facets** for articulating with the tubercles of the ribs. Thoracic vertebrae also have whole **facets** or half facets, called **demifacets,** on the sides of their bodies for articulation with the heads of the ribs. The first thoracic vertebra (T1) has, on either side of its body, a superior whole facet and an inferior demifacet. The superior facet articulates with the first rib, and the inferior demifacet, together with the superior demifacet of the second thoracic vertebra (T2), forms a facet for articulation with the second rib. The second through eighth thoracic vertebrae (T2–T8) have two demifacets on each side, a larger superior demifacet and a smaller inferior demifacet. When the vertebrae are articulated, they form whole facets for the heads of the ribs. The ninth thoracic vertebra (T9) has a single superior demifacet on either side of its body. Thoracic vertebrae ten through twelve (T10–T12) have whole facets on either side of their bodies (Figure 6-14b).

LUMBAR REGION

The **lumbar vertebrae** are the largest and strongest in the column (Figure 6-15). Their various projections are short and thick. The superior articular processes are di-rected medially instead of superiorly. The inferior articular processes are directed laterally instead of inferiorly. The spinous process is quadrilateral in shape, thick and broad, and projects nearly straight posteriorly. The spinous process is well adapted for the attachment of the large back muscles.

SACRUM AND COCCYX

The **sacrum** is a triangular bone formed by the union of five sacral vertebrae (Figure 6-16). These are indicated in the figure as S1 through S5. The sacrum serves as a strong foundation for the pelvic girdle. It is positioned at the posterior portion of the pelvic cavity between the two hipbones.

The concave anterior side of the sacrum faces the pelvic cavity. It is smooth and contains four **transverse lines** that mark the joining of the vertebral bodies. At the ends of these lines are four pairs of **anterior sacral (pelvic) foramina.**

The convex, posterior surface of the sacrum is irregular. It contains a **median sacral crest,** the fused spinous processes of the upper sacral vertebrae; a **lateral sacral crest,** the transverse processes of the sacral vertebrae; and four pairs of **posterior sacral (dorsal) foramina.** These foramina communicate with the anterior sacral foramina through which nerves and blood vessels pass. The **sacral canal** is a continuation of the vertebral canal. The laminae

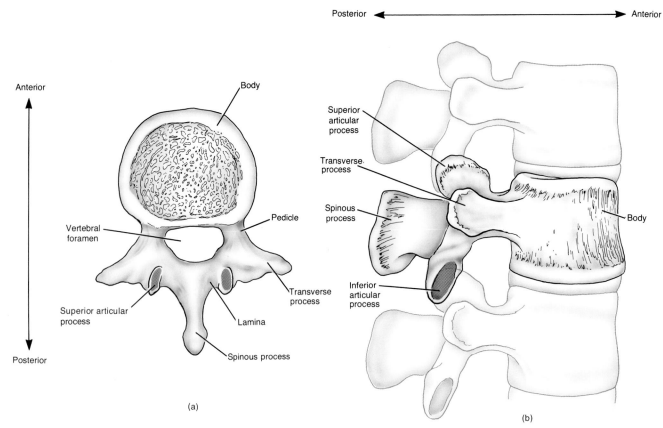

FIGURE 6-15 Lumbar vertebrae. (a) Superior view. (b) Articulated in right lateral view.

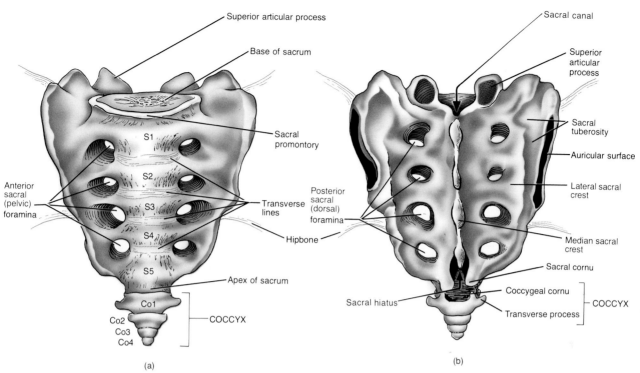

FIGURE 6-16 Sacrum and coccyx. (a) Anterior view. (b) Posterior view.

of the fifth sacral vertebra, and sometimes the fourth, fail to meet. This leaves an inferior entrance to the vertebral canal called the **sacral hiatus** (hi-Ā-tus). On either side of the sacral hiatus are the **sacral cornua,** the inferior articular processes of the fifth sacral vertebra. They are connected by ligaments to the coccygeal cornua of the coccyx.

The superior border of the sacrum exhibits an anteriorly projecting border, the **sacral promontory** (PROM-on-tō'-rē). It is an obstetrical landmark for measurements of the pelvis. An imaginary line running from the superior surface of the symphysis pubis to the sacral promontory separates the abdominal and pelvic cavities. Laterally, the sacrum has a large **auricular surface** for articulating with the ilium of the hipbone. Dorsal to the auricular surface is a roughened surface, the **sacral tuberosity,** which contains depressions for the attachment of ligaments. Its **superior articular processes** articulate with the fifth lumbar vertebra.

The **coccyx** is also triangular in shape and is formed by the fusion of the coccygeal vertebrae, usually the last four. These are indicated in Figure 6-16 as Co1 through Co4. The dorsal surface of the body of the coccyx contains two long **coccygeal cornua** that are connected by ligaments to the sacral cornua. The coccygeal cornua are the pedicles and superior articular process of the first coccygeal vertebra. On the lateral surfaces of the body of the coccyx are a series of **transverse processes,** the first pair being the largest. The coccyx articulates superiorly with the sacrum. The coccyx is the most rudimentary part of the column, representing the vestige of a tail.

THORAX

Anatomically, the term **thorax** refers to the chest. The skeletal portion of the thorax is a bony cage formed by the sternum, costal cartilage, ribs, and the bodies of the thoracic vertebrae (Figure 6-17a).

The thoracic cage is roughly cone-shaped, the narrow portion being superior and the broad portion inferior. It is flattened from front to back. The thoracic cage encloses and protects the organs in the thoracic cavity. It also provides support for the bones of the shoulder girdle and upper extremities.

STERNUM

The **sternum,** or breastbone, is a flat, narrow bone measuring about 15 cm (6 inches) in length. It is located in the median line of the anterior thoracic wall.

The sternum (see Figure 6-17a, b) consists of three basic portions: the **manubrium** (ma-NOO-brē-um), the triangular, superior portion; the **body,** the middle, largest portion; and the **xiphoid** (ZĪ-foyd) **process,** the inferior, smallest portion. The manubrium has a depression on its superior surface called the **jugular (suprasternal) notch.** On each side of the jugular notch are **clavicular notches** that articulate with the medial ends of the clavicles. The manubrium also articulates with the first and second ribs. The body of the sternum articulates directly or indirectly with the second through tenth ribs. The xiphoid process has no ribs attached to it but provides attachment for some abdominal muscles.

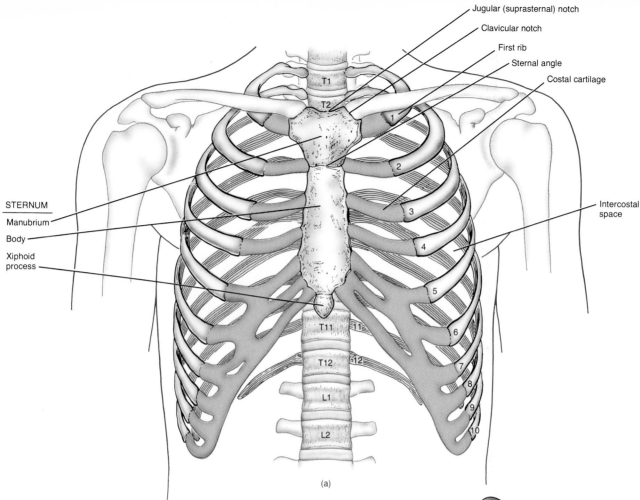

Labels on figure (a):
- Jugular (suprasternal) notch
- Clavicular notch
- First rib
- Sternal angle
- Costal cartilage
- T1
- T2
- STERNUM
  - Manubrium
  - Body
  - Xiphoid process
- Intercostal space
- T11
- T12
- L1
- L2

(a)

FIGURE 6-17 Skeleton of the thorax. (a) Anterior view. (b) Right lateral view of the sternum.

RIBS

Twelve pairs of **ribs** make up the sides of the thoracic cavity (see Figure 6-17a). The ribs increase in length from the first through seventh. Then they decrease in length to the twelfth rib. Each rib articulates posteriorly with its corresponding thoracic vertebra.

The first through seventh ribs have a direct anterior attachment to the sternum by a strip of hyaline cartilage, called **costal cartilage** (*costa* = rib). These ribs are called **true ribs** or **vertebrosternal ribs.** The remaining five pairs of ribs are referred to as **false ribs** because their costal cartilages do not attach directly to the sternum. The cartilages of the eighth, ninth, and tenth ribs attach to each other and then to the cartilage of the seventh rib. These are thus called **vertebrochondral ribs.** The eleventh and twelfth ribs are designated as **floating ribs** because their anterior ends do not attach even indirectly to the sternum. They attach only to the muscles of the body wall.

Although there is some variation in rib structure, we will examine the parts of a typical (third through ninth) rib when viewed from the right side and from behind (Figure 6-18). The **head** of a typical rib is a projection at the posterior end of the rib. It is wedge-shaped and consists of one or two **facets** that articulate with facets

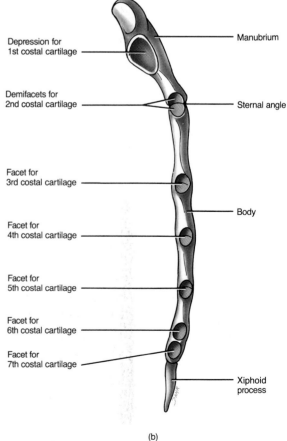

Labels on figure (b):
- Depression for 1st costal cartilage
- Manubrium
- Demifacets for 2nd costal cartilage
- Sternal angle
- Facet for 3rd costal cartilage
- Body
- Facet for 4th costal cartilage
- Facet for 5th costal cartilage
- Facet for 6th costal cartilage
- Facet for 7th costal cartilage
- Xiphoid process

(b)

on the bodies of adjacent thoracic vertebrae. The facets on the head of a rib are separated by a horizontal **interarticular crest.** The inferior facet on the head of a rib is larger than the superior facet. The **neck** is a constricted portion just lateral to the head. A knoblike structure on the posterior surface where the neck joins the body is called a **tubercle.** It consists of a **nonarticular part** which affords attachment to the ligament of the tubercle and an **articular part** which articulates with the facet of a transverse process of the inferior of the two vertebrae to which the head of the rib is connected. The **body** or **shaft** is the main part of the rib. A short distance beyond the tubercle, there is an abrupt change in the curvature of the shaft. This point is called the **angle** and represents the greatest change in curvature of the rib. The inner surface of the rib has a **costal groove** that protects blood vessels and a small nerve, artery, and vein.

The posterior portion of the rib is connected to a thoracic vertebra by its head and articular portion of a tubercle. The facet of the head fits into a facet on the body of a vertebra, and the articular portion of the tubercle articu-

lates with the facet of the transverse process of the vertebra. Each of the second through ninth ribs articulates with the bodies of two adjacent vertebrae. The first, tenth, eleventh, and twelfth ribs articulate with only one vertebra each. On the eleventh and twelfth ribs, there is no articulation between the tubercles and the transverse processes of their corresponding vertebrae.

Spaces between ribs, called **intercostal spaces,** are occupied by intercostal muscles, blood vessels, and nerves.

APPLICATIONS TO HEALTH
MICROCEPHALUS AND HYDROCEPHALUS

In the disorder called **microcephalus** (mī-krō-SEF-a-lus), meaning small head, the anterior fontanel closes earlier than normal. The brain consequently does not have adequate space for growth, and mental retardation results. In **hydrocephalus** (hī-drō-SEF-a-lus), meaning water on the brain, pressure from excess cerebrospinal fluid inside the skull may cause the anterior fontanel to remain open. Approximately 1 of every 500 babies is born with hydrocephalus, which can lead to mental retardation or even death.

One treatment for hydrocephalus involves the installation of a flexible silicone tube in the largest ventricle at the base of the brain. (A ventricle is a cavity in the brain where fluid is produced and circulated.) The other end is inserted into a body cavity, such as the chest or abdomen, or into a large salivary gland duct emptying into the floor of the mouth. The shunt enables the excess cerebrospinal fluid to drain off and be absorbed.

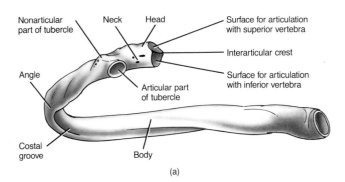

(a)

FIGURE 6-18 Typical rib. (a) A left rib viewed from below and behind. (b) Photograph of the inner aspect of a portion of the fifth right rib. (Courtesy of Vincent P. Destro, Mayo Foundation.)

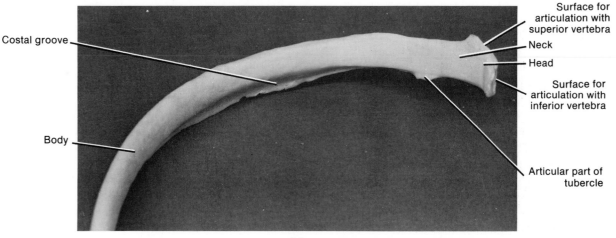

(b)

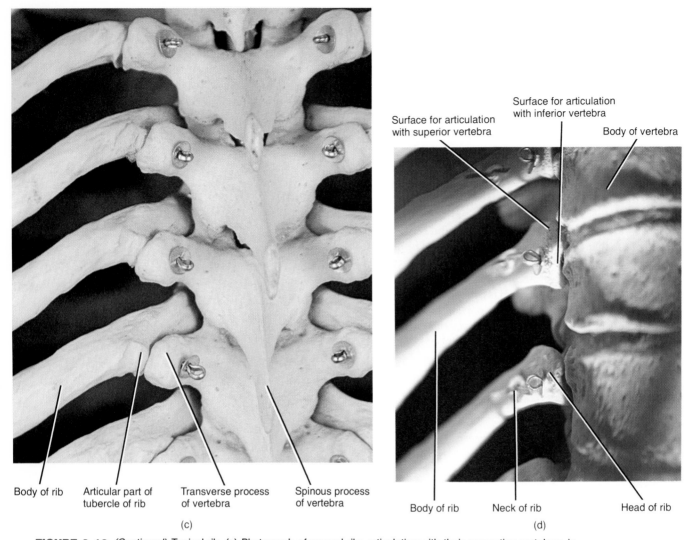

Body of rib

Articular part of
tubercle of rib

Transverse process
of vertebra

Spinous process
of vertebra

(c)

Surface for articulation
with superior vertebra

Surface for articulation
with inferior vertebra

Body of vertebra

Body of rib

Neck of rib

Head of rib

(d)

FIGURE 6-18 *(Continued)* Typical rib. (c) Photograph of several ribs articulating with their respective vertebrae in posterior view. (d) Photograph of several ribs articulating with their respective vertebrae in anterior view. (Courtesy of Matt Iacobino.)

SLIPPED DISC

In their function as shock-absorbers, intervertebral discs are subject to compressional forces. The discs between the fourth and fifth lumbar vertebrae and between the fifth lumbar vertebra and sacrum usually are subject to more forces than other discs. If the anterior and posterior ligaments of the discs become injured or weakened, the pressure developed in the nucleus pulposus may be great enough to rupture the surrounding fibrocartilage (see Figure 8-1b). If this occurs, the nucleus pulposus may protrude posteriorly or into one of the adjacent vertebral bodies (herniate). This condition is called a **slipped disc.**

Most often the nucleus pulposus slips posteriorly toward the spinal cord and spinal nerves. This movement exerts pressure on the spinal nerves, causing considerable, sometimes very acute, pain. If the roots of the sciatic nerve, which passes from the spinal cord to the foot, are pressured, the pain radiates down the back of the thigh, through the calf, and occasionally into the foot. If pressure is exerted on the spinal cord itself, nervous tissue may be destroyed. Traction, bed rest, and analgesia usually relieve the pain. If such treatment is ineffective, surgical decompression of the spinal nerves by laminectomy or removal of some of the nucleus pulposus may be necessary to relieve pain.

CURVATURES

As a result of various conditions, the normal curves of the vertebral column may become exaggerated or the column may acquire a lateral bend. Such conditions are called **curvatures** of the spine.

Scoliosis (skō'-lē-Ō-sis; *scolio* = bent) is a lateral bending of the vertebral column, usually in the thoracic region. This is the most common of the curvatures. It may be congenital, due to an absence of the lateral half of a vertebra (hemivertebra). It may also be acquired from a persistent severe sciatica. Poliomyelitis may cause scoliosis by the paralysis of muscles on one side of the body, which produces a lateral deviation of the trunk toward the unaffected side. Poor posture is also a contributing factor.

Kyphosis (kī-FŌ-sis; *kypho* = hunchback) is an exaggeration of the thoracic curve of the vertebral column. In tuberculosis of the spine, vertebral bodies may partially collapse, causing an acute angular bending of the vertebral column. In the elderly, degeneration of the intervertebral discs leads to kyphosis. Kyphosis may also be caused by rickets and poor posture. The term "round-shouldered" is an expression for mild kyphosis.

Lordosis (lor-DŌ-sis; *lordo* = swayback) is an exaggeration of the lumbar curve of the vertebral column. It may result from increased weight of abdominal contents as in pregnancy or extreme obesity. Other causes include poor posture, rickets, and tuberculosis of the spine.

SPINA BIFIDA

Spina bifida (SPĪ-na BIF-i-da) is a congenital defect of the vertebral column in which laminae fail to unite at the midline. The lumbar vertebrae are involved in about 50 percent of all cases. In less serious cases, the defect is small and the area is covered with skin. Only a dimple or tuft of hair may mark the site. Symptoms are mild, with perhaps intermittent urinary problems. An operation is not ordinarily required. Larger defects in the vertebral arches with protrusion of the membranes around the spinal cord or spinal cord tissue produce more serious problems, such as partial or complete paralysis, partial or complete loss of urinary bladder control, and the absence of reflexes.

FRACTURES OF THE VERTEBRAL COLUMN

Fractures of the vertebral column most commonly involve T12, L1, and L2. They usually result from a flexion-compression type of injury such as might be sustained in landing on the feet or buttocks after a fall from a height or having a heavy weight fall on the shoulders. The forceful compression wedges the involved vertebrae. If, in addition to compression, there is forceful forward movement, one vertebra may displace forward on its adjacent vertebra below with either dislocation or fracture of the articular facets between the two (fracture dislocation) and with rupture of the interspinous ligaments.

The cervical vertebrae may be fractured or, more commonly, dislodged by a fall on the head with acute flexion of the neck, as might happen on diving into shallow water. Dislocation may even result from the sudden forward jerk which may occur when an automobile or airplane crashes ("whiplash"). The relatively horizontal intervertebral facets of the cervical vertebrae allow dislocation to take place without their being fractured, whereas the relatively vertical thoracic and lumbar intervertebral facets nearly always fracture in forward dislocation of the thoracolumbar region. Spinal nerve damage may occur as a result of fractures of the vertebral column.

STUDY OUTLINE

Types of Bones
1. On the basis of shape, bones are classified as long, short, flat, or irregular.
2. Wormian or sutural bones are found between the sutures of certain cranial bones. Sesamoid bones develop in tendons or ligaments.

Surface Markings
1. Markings are definitive areas on the surfaces of bones.
2. Each marking is structured for a specific function—joint formation, muscle attachment, or passage of nerves and blood vessels.
3. Terms that describe markings include fissure, foramen, meatus, fossa, process, condyle, head, facet, tuberosity, crest, and spine.

Divisions of the Skeletal System
1. The axial skeleton consists of bones arranged along the longitudinal axis. The parts of the axial skeleton are the skull, hyoid bone, auditory ossicles, vertebral column, sternum, and ribs.
2. The appendicular skeleton consists of the bones of the girdles and the upper and lower extremities. The parts of the appendicular skeleton are the shoulder girdle, the bones of the upper extremities, the pelvic girdle, and the bones of the lower extremities.

Skull
1. The skull consists of the cranium and the face. It is composed of 22 bones.
2. Sutures are immovable joints between bones of the skull. Examples are coronal, sagittal, lambdoidal, and squamosal sutures.
3. Fontanels are membrane-filled spaces between the cranial bones of fetuses and infants. The major fontanels are the anterior, posterior, anterolaterals, and posterolaterals.
4. The 8 cranial bones include the frontal, parietal (2), temporal (2), occipital, sphenoid, and ethmoid.
5. The 14 facial bones are the nasal (2), maxillae (2), zygomatic (2), mandible, lacrimal (2), palatine (2), inferior nasal conchae (2), and vomer.

6. Paranasal sinuses are cavities in bones of the skull that communicate with the nasal cavity. They are lined by mucous membranes. The cranial bones containing the paranasal sinuses are the frontal, sphenoid, ethmoid, and maxilla.

7. The foramina of the skull bones provide passages for nerves and blood vessels.

Vertebral Column

1. The vertebral column, the sternum, and the ribs constitute the skeleton of the trunk.

2. The bones of the adult vertebral column are the cervical vertebrae (7), thoracic vertebrae (12), lumbar vertebrae (5), the sacrum (5, fused) and the coccyx (4, fused).

3. The vertebral column contains primary curves (thoracic and sacral) and secondary curves (cervical and lumbar). These curves give strength, support, and balance.

4. The vertebrae are similar in structure, each consisting of a body, vertebral arch, and seven processes. Vertebrae in the different regions of the column vary in size, shape, and detail.

Thorax

1. The thoracic skeleton consists of the sternum, the ribs and costal cartilages, and the thoracic vertebrae.

2. The thorax protects vital organs in the chest area.

Applications to Health

1. In microcephalus, the anterior fontanel closes earlier than normal; in hydrocephalus, the fontanel may remain open due to excess cerebrospinal fluid.

2. Protrusion of the nucleus pulposus of an intervertebral disc posteriorly or into an adjacent vertebral body is called a slipped disc.

3. Exaggeration of a normal curve of the vertebral column is called a curvature. Examples include scoliosis, kyphosis, and lordosis.

4. The imperfect union of the vertebral laminae at the midline, a congenital defect, is referred to as spina bifida.

5. Fractures of the vertebral column most often involve T12, L1, and L2.

REVIEW QUESTIONS

1. What are the four principal types of bones? Give an example of each.

2. What are surface markings? Describe and give an example of each.

3. Distinguish between the axial and appendicular skeletons. What subdivisions and bones are contained in each?

4. What are the bones that compose the skull? The cranium? The face?

5. Define a suture. What are the four principal sutures of the skull? Where are they located?

6. What is a fontanel? Describe the location of the six major fontanels.

7. What is a paranasal sinus? Give examples of cranial bones that contain paranasal sinuses.

8. Identify the following: sinusitis, cleft palate, cleft lip, and mastoiditis.

9. What is the hyoid bone? In what respect is it unique? What is its function?

10. What bones form the skeleton of the trunk? Distinguish between the number of nonfused vertebrae found in the adult vertebral column and that of a child.

11. What are the normal curves in the vertebral column? How are primary and secondary curves differentiated? What are the functions of the curves?

12. What bones form the skeleton of the thorax? What are the functions of the thoracic skeleton?

13. Distinguish between microcephalus and hydrocephalus. How is hydrocephalus treated?

14. What is a slipped disc? Why does it cause pain? How is it treated?

15. What is a curvature? Describe the symptoms and causes of scoliosis, kyphosis, and lordosis.

16. Define spina bifida.

17. What are some common causes of fractures of the vertebral column?

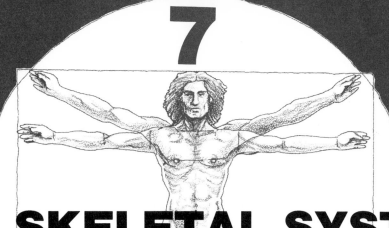

7

THE SKELETAL SYSTEM
The Appendicular Skeleton

STUDENT OBJECTIVES

■ Identify the bones of the shoulder girdle and their major markings.

■ Identify the upper extremity, its component bones, and their markings.

■ Identify the components of the pelvic girdle and their principal markings.

■ Identify the lower extremity, its component bones, and their markings.

■ Define the structural features and importance of the arches of the foot.

■ Compare the principal structural differences between male and female skeletons, especially those that pertain to the pelvis.

This chapter discusses the bones of the appendicular skeleton, that is, the bones of the shoulder and pelvic girdles and extremities. The differences between male and female skeletons are also compared.

SHOULDER GIRDLES

The **shoulder,** or **pectoral** (PEK-tō-ral), **girdles** attach the bones of the upper extremities to the axial skeleton (Figure 7-1). Each of the two shoulder girdles consists of two bones: a clavicle and a scapula. The clavicle is the anterior component of the shoulder girdle and articulates with the sternum at the sternoclavicular joint. The posterior component, the scapula, which is positioned freely by complex muscle attachments, articulates with the clavicle and humerus. The shoulder girdles have no articulation with the vertebral column. Although the shoulder joints are not very stable, they are freely movable and thus allow movement in many directions.

CLAVICLES

The **clavicles** (KLAV-i-kuls), or collarbones, are long, slender bones with a double curvature (Figure 7-2). The two bones lie horizontally in the superior and anterior part of the thorax superior to the first rib.

The medial end of the clavicle, the **sternal extremity,** is rounded and articulates with the sternum. The broad, flat, lateral end, the **acromial** (a-KRŌ-mē-al) **extremity,** articulates with the acromion process of the scapula. This joint is called the **acromioclavicular joint.** (Refer to Figure 7-1 for a view of these articulations.) The **conoid tubercle** on the inferior surface of the lateral end of the bone serves as a point of attachment for a ligament. The **costal tuberosity** on the interior surface of the medial end also serves as a point of attachment for a ligament.

CLINICAL APPLICATION

Because of its position, the clavicle transmits forces from the upper extremity to the trunk. If such forces are excessive, as in falling on one's outstretched arm, a **fractured clavicle** may result. In fact, it is the most frequently broken bone in the body.

Separation of the shoulder simply means dislocation of the acromioclavicular joint.

SCAPULAE

The **scapulae** (SCAP-yoo-lē), or shoulder blades, are large, triangular, flat bones situated in the dorsal part of the thorax between the levels of the second and seventh ribs (Figure 7-3). Their medial borders are located about 5 cm (2 inches) from the vertebral column.

A sharp ridge, the **spine,** runs diagonally across the dorsal surface of the flattened, triangular **body.** The end

of the spine projects as a flattened, expanded process called the **acromion** (a-KRŌ-mē-on). This process articulates with the clavicle. Inferior to the acromion is a depression called the **glenoid cavity.** This cavity articulates with the head of the humerus to form the shoulder joint.

The thin edge of the body near the vertebral column is the **medial** or **vertebral border.** The thick edge closer to the arm is the **lateral** or **axillary border.** The medial and lateral borders join at the **inferior angle.** The superior edge of the scapular body is called the **superior border.** It joins the vertebral border at the **superior angle.**

At the lateral end of the superior border is a projection of the anterior surface called the **coracoid** (KOR-a-koyd) **process** to which muscles attach. Above and below the spine are two fossae: the **supraspinatous** (soo'-pra-SPĪ-na-tus) **fossa** and the **infraspinatous fossa,** respectively. Both serve as surfaces of attachment for shoulder muscles. On the ventral (costal) surface is a lightly hollowed-out area called the **subscapular fossa,** also a surface of attachment for shoulder muscles.

UPPER EXTREMITIES

The **upper extremities** consist of 60 bones. The skeleton of the right upper extremity is shown in Figure 7-1. It includes a humerus in each arm, an ulna and radius in each forearm, carpals, or wrist bones, metacarpals, which are the palm bones, and phalanges in the fingers of each hand.

HUMERUS

The **humerus** (HYOO-mer-us), or arm bone, is the longest and largest bone of the upper extremity (Figure 7-4). It articulates proximally with the scapula and distally at the elbow with both the ulna and radius.

The proximal end of the humerus consists of a **head** that articulates with the glenoid cavity of the scapula. It also has an **anatomical neck,** which is an oblique groove just distal to the head. The **greater tubercle** is a lateral projection distal to the neck. The **lesser tubercle** is an anterior projection. Between these tubercles runs an **intertubercular sulcus (bicipital groove).** The **surgical neck** is a constricted portion just distal to the tubercles and is named because of its liability to fracture.

The **body** or shaft of the humerus is cylindrical at its proximal end. It gradually becomes triangular and is flattened and broad at its distal end. Along the middle portion of the shaft, there is a roughened, V-shaped area called the **deltoid tuberosity.** This area serves as a point of attachment for the deltoid muscle.

The following parts are found at the distal end of the humerus. The **capitulum** (ka-PIT-yoo-lum) is a rounded knob that articulates with the head of the radius. The **radial fossa** is a depression that receives the head of the radius when the forearm is flexed. The **trochlea** (TRŌK-

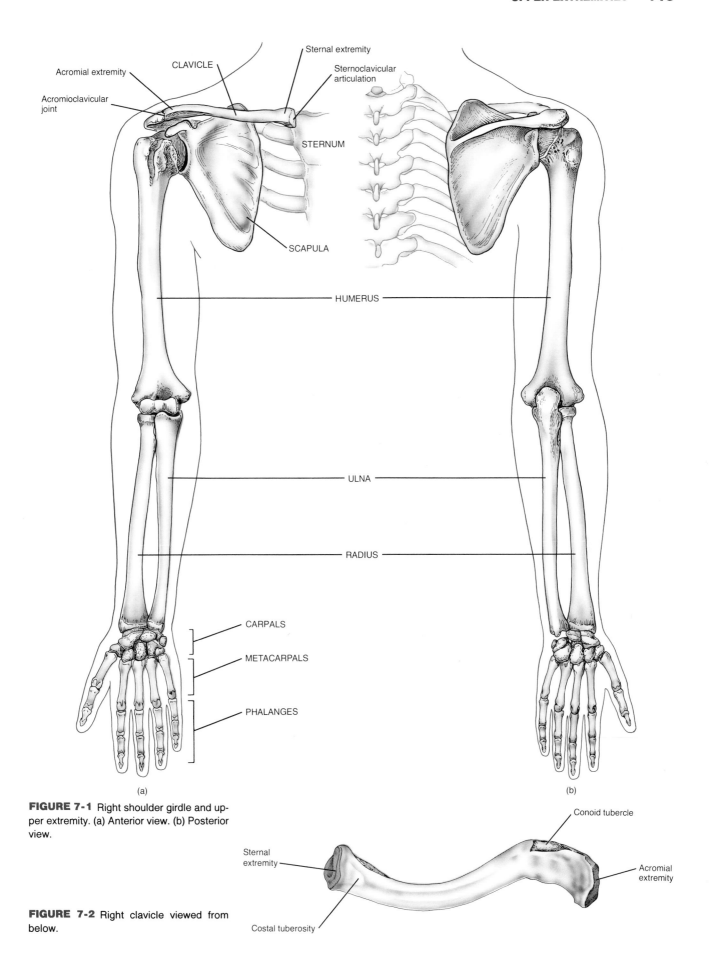

FIGURE 7-1 Right shoulder girdle and upper extremity. (a) Anterior view. (b) Posterior view.

FIGURE 7-2 Right clavicle viewed from below.

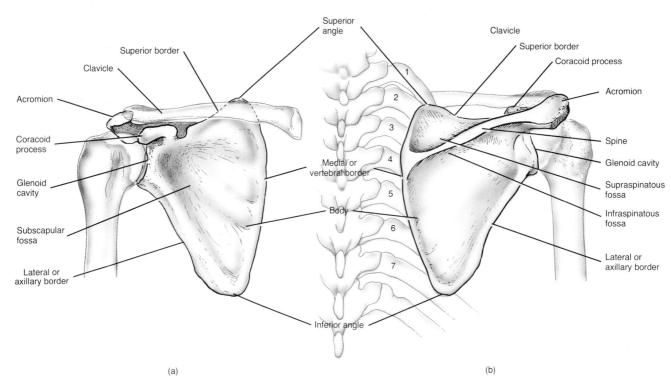

(a)

(b)

FIGURE 7-3 Right scapula. (a) Anterior view. (b) Posterior view. (c) Lateral border view. (d) Anteroposterior projection. (Courtesy of Garret J. Pinke, Sr., R.T.)

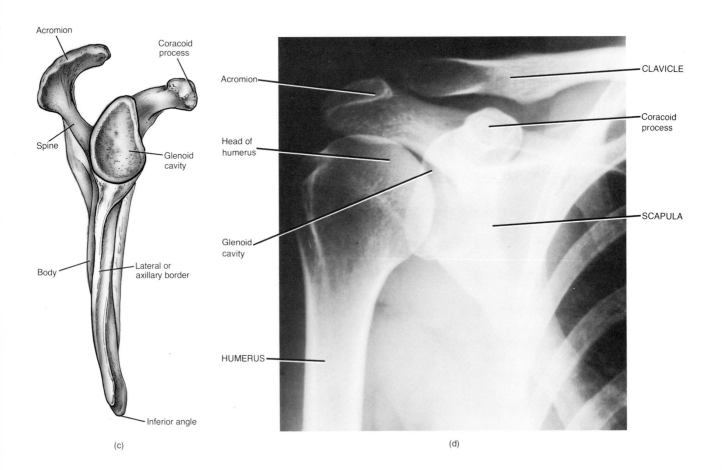

(c)

(d)

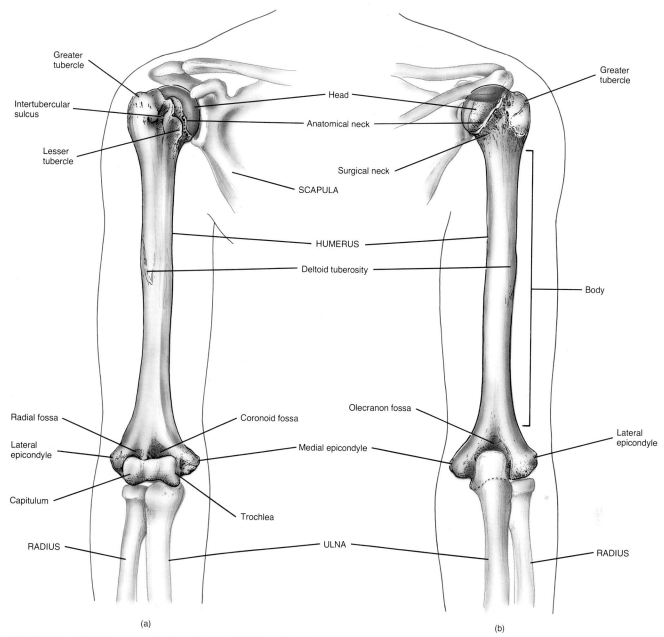

Greater tubercle

Intertubercular sulcus

Lesser tubercle

Head

Anatomical neck

Surgical neck

SCAPULA

Greater tubercle

HUMERUS

Deltoid tuberosity

Body

Radial fossa

Lateral epicondyle

Capitulum

RADIUS

Coronoid fossa

Medial epicondyle

Trochlea

ULNA

Olecranon fossa

Lateral epicondyle

RADIUS

(a)

(b)

FIGURE 7-4 Right humerus. (a) Anterior view. (b) Posterior view.

lē-a) is a pulleylike surface that articulates with the ulna. The **coronoid fossa** is an anterior depression that receives part of the ulna when the forearm is flexed. The **olecranon** (ō-LEK-ra-non) **fossa** is a posterior depression that receives the olecranon of the ulna when the forearm is extended. The **medial epicondyle** and **lateral epicondyle** are rough projections on either side of the distal end.

ULNA AND RADIUS

The **ulna** is the medial bone of the forearm (Figure 7-5). In other words, it is located on the small finger side. The proximal end of the ulna presents an **olecranon** (ole-

cranon process), which forms the prominence of the elbow. The **coronoid process** is an anterior projection that, together with the olecranon, receives the trochlea of the humerus. The **trochlear (semilunar) notch** is a curved area between the olecranon and the coronoid process. The trochlea of the humerus fits into this notch. The **radial notch** is a depression located laterally and inferiorly to the trochlear notch. It receives the head of the radius. The distal end of the ulna consists of a **head** that is separated from the wrist by a fibrocartilage disc. A **styloid process** is on the posterior side of the distal end.

The **radius** is the lateral bone of the forearm, that is, it is situated on the thumb side. The proximal end of

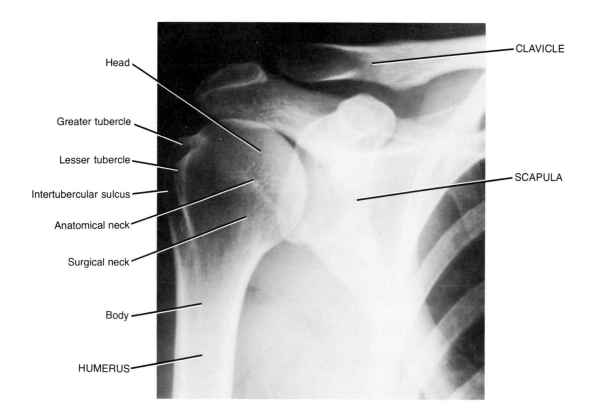

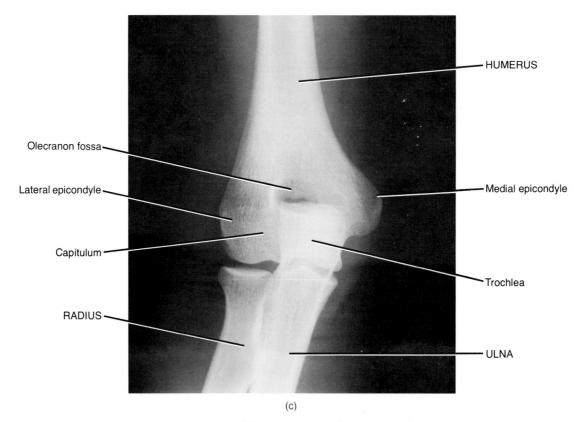

(c)

FIGURE 7-4 (*Continued*) Right humerus. (c) Anteroposterior projection of the proximal end of the humerus (top) and the distal end of the humerus and proximal ends of the ulna and radius (bottom). (Courtesy of Garret J. Pinke, Sr., R.T.)

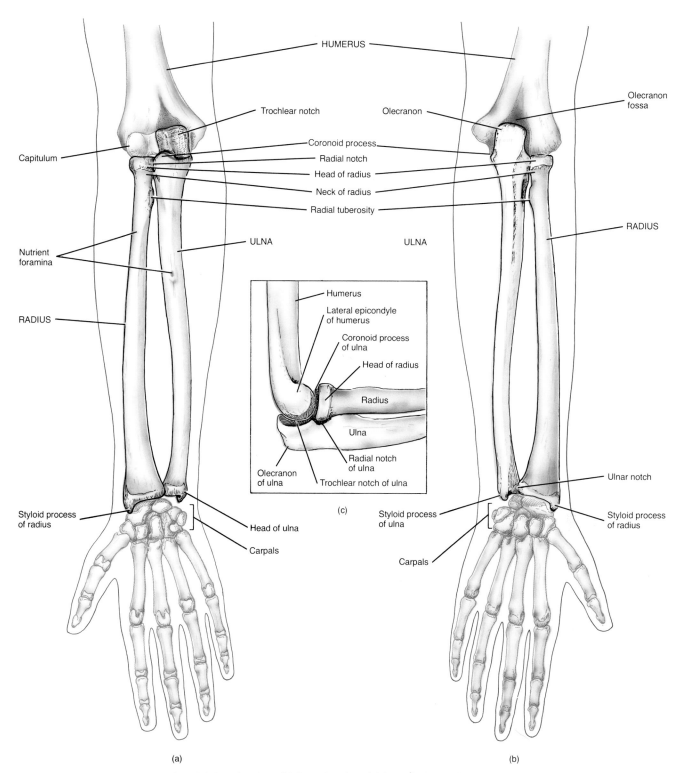

HUMERUS

Trochlear notch

Olecranon

Olecranon fossa

Capitulum

Coronoid process
Radial notch
Head of radius
Neck of radius
Radial tuberosity

RADIUS

Nutrient foramina

ULNA

ULNA

RADIUS

Humerus

Lateral epicondyle of humerus

Coronoid process of ulna

Head of radius

Radius

Ulna

Radial notch of ulna

Olecranon of ulna

Trochlear notch of ulna

(c)

Styloid process of radius

Styloid process of ulna

Ulnar notch

Styloid process of radius

Head of ulna

Carpals

Carpals

(a)

(b)

FIGURE 7-5 Right ulna and radius. (a) Anterior view. (b) Posterior view. (c) Lateral view of the right elbow.

the radius has a disc-shaped **head** that articulates with the capitulum of the humerus and radial notch of the ulna. It also has a raised, roughened area on the medial side called the **radial tuberosity.** This is a point of attachment for the biceps muscle. The shaft of the radius widens distally to form a concave inferior surface that articulates with two bones of the wrist called the lunate and scaphoid bones. Also at the distal end is a **styloid process** on the lateral side and a medial, concave **ulnar notch** for articulation with the distal end of the ulna.

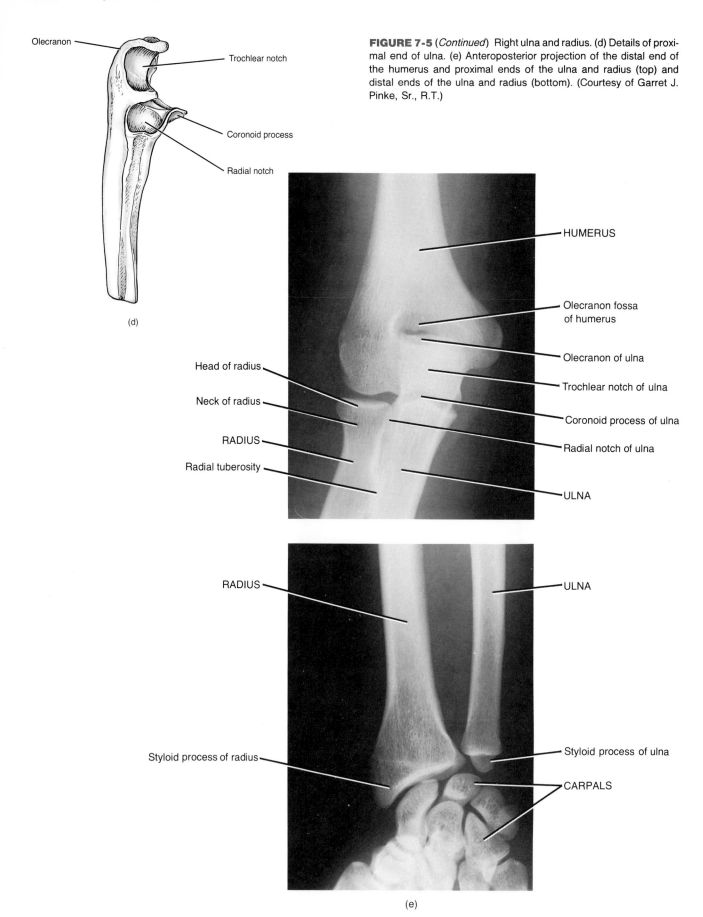

Olecranon

Trochlear notch

Coronoid process

Radial notch

(d)

FIGURE 7-5 (*Continued*) Right ulna and radius. (d) Details of proximal end of ulna. (e) Anteroposterior projection of the distal end of the humerus and proximal ends of the ulna and radius (top) and distal ends of the ulna and radius (bottom). (Courtesy of Garret J. Pinke, Sr., R.T.)

HUMERUS

Olecranon fossa of humerus

Olecranon of ulna

Trochlear notch of ulna

Coronoid process of ulna

Radial notch of ulna

ULNA

Head of radius

Neck of radius

RADIUS

Radial tuberosity

RADIUS

ULNA

Styloid process of radius

Styloid process of ulna

CARPALS

(e)

CARPUS, METACARPUS, AND PHALANGES

The **carpus,** or wrist, consists of eight small bones, the **carpals,** united to each other by ligaments (Figure 7-6). The bones are arranged in two transverse rows, with four bones in each row. The proximal row of carpals, from the lateral to medial position, consists of the following bones: **scaphoid, lunate, triquetral,** and **pisiform.** In about 70 percent of cases involving carpal fractures, only the scaphoid is involved. The distal row of carpals, from the lateral to medial position, consists of the following: **trapezium, trapezoid, capitate,** and **hamate.**

The five bones of the **metacarpus** constitute the palm of the hand. Each metacarpal bone consists of a proximal **base,** a **shaft,** and a distal **head.** The metacarpal bones are numbered I to V, starting with the lateral bone. The bases articulate with the distal row of carpal bones and with one another. The heads articulate with the proximal phalanges of the fingers. The heads of the metacarpals are commonly called the "knuckles" and are readily visible when the fist is clenched.

The **phalanges** (fa-LAN-jēz), or bones of the fingers, number 14 in each hand. Each consists of a proximal **base,** a **shaft,** and a distal **head.** There are two phalanges in the first digit, called the thumb or *pollex,* and three phalanges in each of the remaining four digits. The first row of phalanges, the **proximal row,** articulates with the metacarpal bones and second row of phalanges. The second row of phalanges, the **middle row,** articulates with the proximal row and the third row. The third row of phalanges, the **distal row,** articulates with the middle row. A single finger bone is referred to as a **phalanx** (FĀ-lanks). The thumb has no middle phalanx.

PELVIC GIRDLE

The **pelvic girdle** consists of the two **coxal** (KOK-sal) **bones,** commonly called the pelvic, innominate, or hipbones (Figure 7-7). The pelvic girdle provides a strong

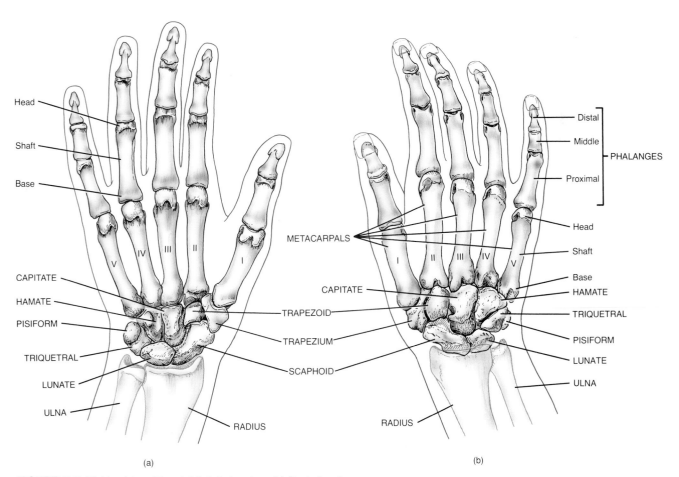

Head

Shaft

Base

CAPITATE

HAMATE

PISIFORM

TRIQUETRAL

LUNATE

ULNA

V IV III II I

TRAPEZOID

TRAPEZIUM

SCAPHOID

RADIUS

(a)

Distal

Middle

Proximal

PHALANGES

Head

Shaft

Base

HAMATE

TRIQUETRAL

PISIFORM

LUNATE

ULNA

METACARPALS

CAPITATE

I II III IV V

RADIUS

(b)

FIGURE 7-6 Right wrist and hand. (a) Anterior view. (b) Posterior view.

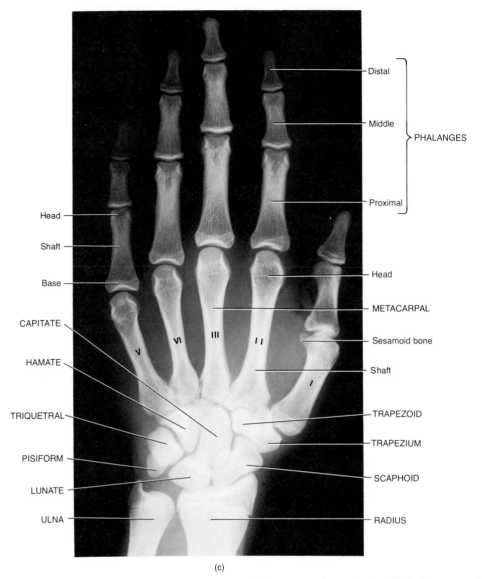

(c)

FIGURE 7-6 (*Continued*) Right wrist and hand. (c) Anteroposterior projection. Note the sesamoid bone. (Courtesy of Daniel Sorrentino, R.T.)

and stable support for the lower extremities on which the weight of the body is carried. The coxal bones are united to each other anteriorly at the symphysis (SIM-fi-sis) pubis. They unite posteriorly to the sacrum.

Together with the sacrum and coccyx, the pelvic girdles form the basinlike structure called the **pelvis.** The pelvis is divided into a greater pelvis and a lesser pelvis by an oblique plane that passes through sacral promontory (posterior), iliopectineal lines (laterally), and symphysis pubis (anteriorly). The circumference of this oblique plane is called the **brim of the pelvis.** The **greater** or **false pelvis** represents the expanded portion situated superior to the brim of the pelvis. The greater pelvis consists laterally of the superior portions of the ilia and posteriorly of the superior portion of the sacrum. There is no bony component in the anterior aspect of the greater

pelvis. Rather, the front is formed by the walls of the abdomen.

The **lesser** or **true pelvis** is inferior and posterior to the brim of the pelvis. It is formed by the inferior portions of the ilia, the sacrum, the coccyx, and the pubes. The lesser pelvis contains a superior opening called the **pelvic inlet** and an inferior opening called the **pelvic outlet.**

CLINICAL APPLICATION
Pelvimetry is the measurement of the size of the inlet and outlet of the birth canal. Measurement of the pelvic cavity is important to the physician, because the fetus must pass through the narrower opening of the lesser pelvis at birth.

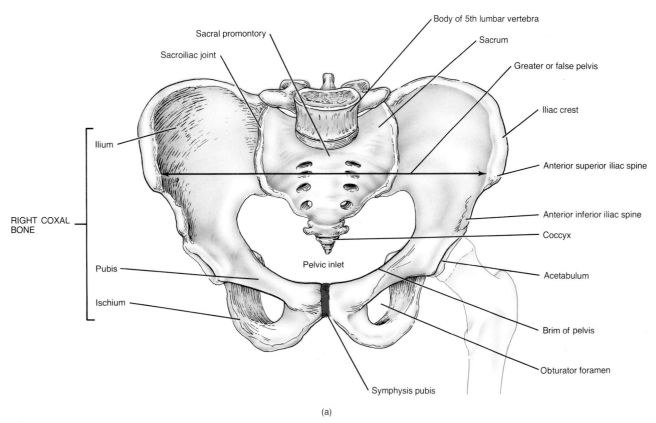

Sacral promontory

Sacroiliac joint

Ilium

RIGHT COXAL
BONE

Pubis

Ischium

Body of 5th lumbar vertebra

Sacrum

Greater or false pelvis

Iliac crest

Anterior superior iliac spine

Anterior inferior iliac spine

Coccyx

Acetabulum

Brim of pelvis

Obturator foramen

Pelvic inlet

Symphysis pubis

(a)

FIGURE 7-7 Pelvic girdle. (a) Anterior view.

COXAL BONES

Each of the two **coxal bones (os coxae)** of a newborn consists of three components: a superior **ilium,** an inferior and anterior **pubis,** and an inferior and posterior **ischium** (Figure 7-8). Eventually, the three separate bones fuse into one. The area of fusion is a deep, lateral fossa called the **acetabulum** (as'-e-TAB-yoo-lum). Although the adult coxae are both single bones, it is common to discuss the bones as if they still consisted of three portions.

The ilium is the largest of the three subdivisions of the coxal bone. Its superior border, the **iliac crest,** ends anteriorly in the **anterior superior iliac spine.** The **anterior inferior iliac spine** is located inferior to the anterior superior spine. Posteriorly, the iliac crest ends in the **posterior superior iliac spine.** The **posterior inferior iliac spine** is just inferior. The spines serve as points of attachment for muscles of the abdominal wall. Slightly inferior to the posterior inferior iliac spine is the **greater sciatic** (sī-AT-ik) **notch.** The internal surface of the ilium seen from the medial side is the **iliac fossa.** It is a concavity where the iliacus muscle attaches. Posterior to this fossa are the **iliac tuberosity,** a point of attachment for the sacroiliac ligament, and the **auricular surface,** which articulates with the sacrum. The other conspicuous markings of the ilium are three arched lines on its gluteal (buttock) surface called the **posterior gluteal line,** the **anterior gluteal line,**

and the **inferior gluteal line.** The gluteal muscles attach to the ilium between these lines.

The ischium is the inferior, posterior portion of the coxal bone. It contains a prominent **ischial spine,** a **lesser sciatic notch** below the spine, and an **ischial tuberosity.** The rest of the ischium, the **ramus,** joins with the pubis and together they surround the **obturator** (OB-tyoo-rā'-ter) **foramen.**

The pubis is the anterior and inferior part of the coxal bone. It consists of a **superior ramus,** an **inferior ramus,** and a **body** that contributes to the formation of the symphysis pubis.

The **symphysis pubis** is the joint between the two coxal bones (Figure 7-7). It consists of fibrocartilage. The **acetabulum** is the fossa formed by the ilium, ischium, and pubis. It is the socket for the head of the femur. Two-fifths of the acetabulum is formed by the ilium, two-fifths by the ischium, and one-fifth by the pubis. On the inferior portion of the acetabulum is the **acetabular notch.**

LOWER EXTREMITIES

The **lower extremities** are composed of 60 bones (Figure 7-9). These include the femur of each thigh, each kneecap, the fibula and tibia in each leg, and ankle bones in each ankle, and the metatarsals and phalanges of each foot.

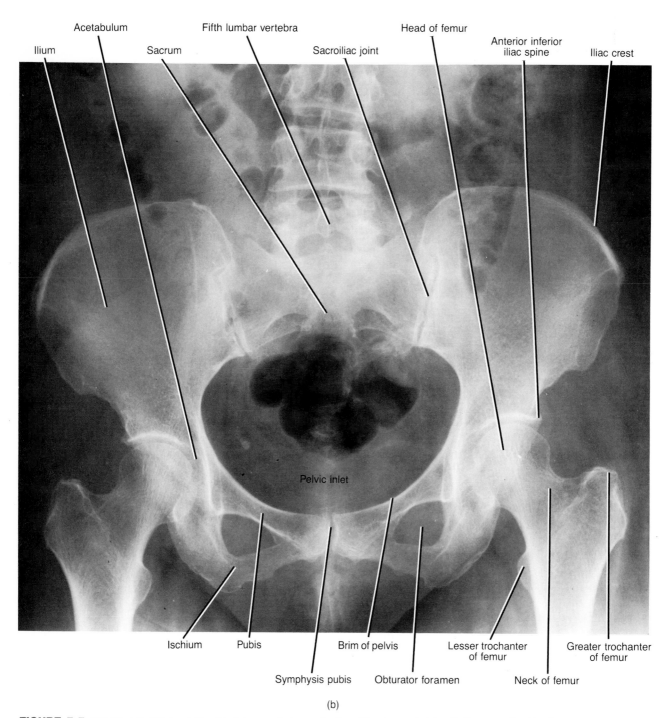

Acetabulum   Fifth lumbar vertebra   Head of femur   Anterior inferior iliac spine   Iliac crest

Ilium   Sacrum   Sacroiliac joint

Pelvic inlet

Ischium   Pubis   Brim of pelvis   Lesser trochanter of femur   Greater trochanter of femur

Symphysis pubis   Obturator foramen   Neck of femur

(b)

FIGURE 7-7 (*Continued*) Pelvic girdle. (b) Anteroposterior projection. (Courtesy of John C. Bennett, St. Mary's Hospital, San Francisco.)

FEMUR

The **femur,** or thigh bone, is the longest and heaviest bone in the body (Figure 7-10). Its proximal end articulates with the coxal bone. Its distal end articulates with the tibia. The shaft of the femur bows medially so that it approaches the femur of the opposite thigh. As a result of this convergence, the knee joints are brought nearer to the body's line of gravity. The degree of convergence

is greater in the female because the female pelvis is broader.

The proximal end of the femur consists of a rounded **head** that articulates with the acetabulum of the coxal bone. The **neck** of the femur is a constricted region distal to the head. A fairly common fracture in the elderly occurs at the neck of the femur. Apparently the neck becomes so weak that it fails to support the body. The **greater trochanter** (trō-KAN-ter) and **lesser trochanter**

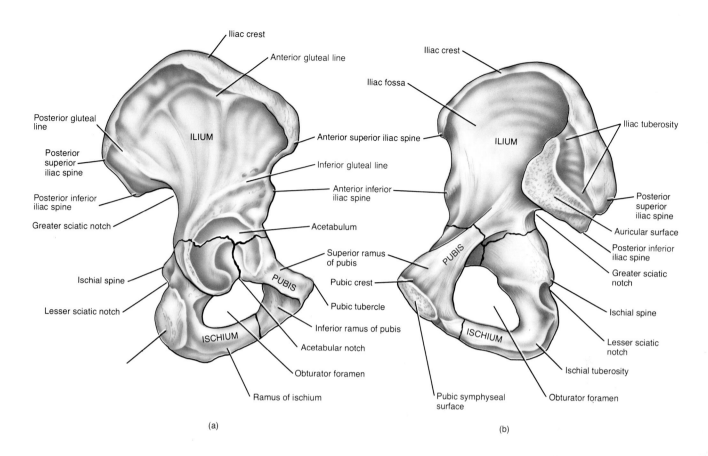

(a)

(b)

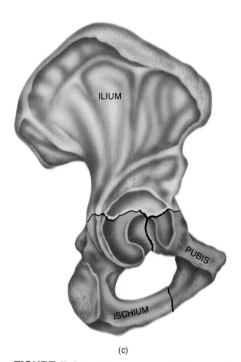

(c)

FIGURE 7-8 Right coxal bone. (a) Lateral view. (b) Medial view. The lines of fusion of the ilium, ischium, and pubis that are shown in color in (a) and (b) are not actually visible in an adult bone. (c) Three divisions of the coxal bone.

are projections that serve as points of attachment for some of the thigh and buttock muscles. Between the trochanters on the anterior surface is a narrow **intertrochanteric line.** Between the trochanters on the posterior surface is an **intertrochanteric crest.**

The shaft of the femur contains a rough vertical ridge on its posterior surface called the **linea aspera.** This ridge serves for the attachment of several thigh muscles.

The distal end of the femur is expanded and includes the **medial condyle** and **lateral condyle.** These articulate with the tibia. A depressed area between the condyles on the posterior surface is called the **intercondylar** (in'-ter-KON-di-lar) **fossa.** The **patellar surface** is located between the condyles on the anterior surface. Lying superior to the condyles are the **medial epicondyle** and **lateral epicondyle.**

PATELLA

The **patella,** or kneecap, is a small, triangular bone anterior to the knee joint (Figure 7-11). It develops in the tendon of the quadriceps femoris muscle. A bone that forms in a tendon, such as the patella, is called a *sesamoid bone.* The broad superior end of the patella is called the **base.** The pointed inferior end is the **apex.** The posterior surface contains two **articular facets,** one for the medial condyle and the other for the lateral condyle of the femur.

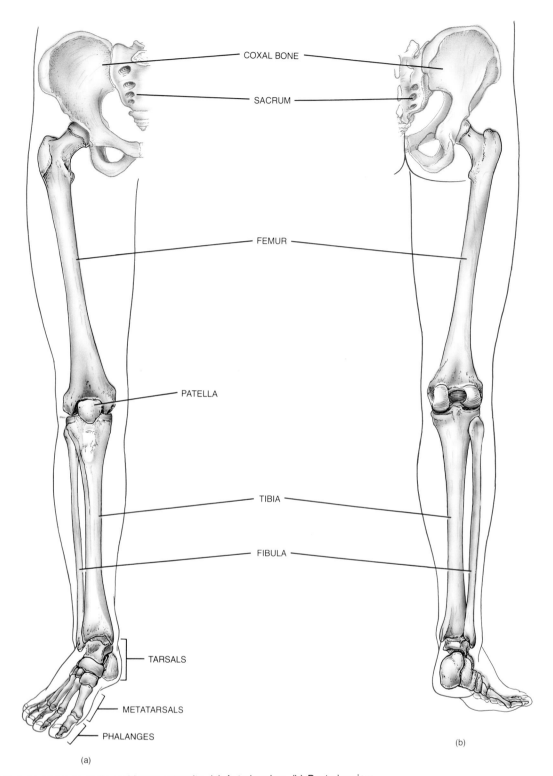

FIGURE 7-9 Right pelvic girdle and lower extremity. (a) Anterior view. (b) Posterior view.

TIBIA AND FIBULA

The **tibia,** or shinbone, is the larger, medial bone of the leg (Figure 7-12). It bears the major portion of the weight of the leg. The tibia articulates at its proximal end with the femur and tibia and at its distal end with the fibula of the leg and talus of the ankle.

The proximal end of the tibia is expanded into a **lateral condyle** and a **medial condyle.** These articulate with the condyles of the femur. The inferior surface of the lateral

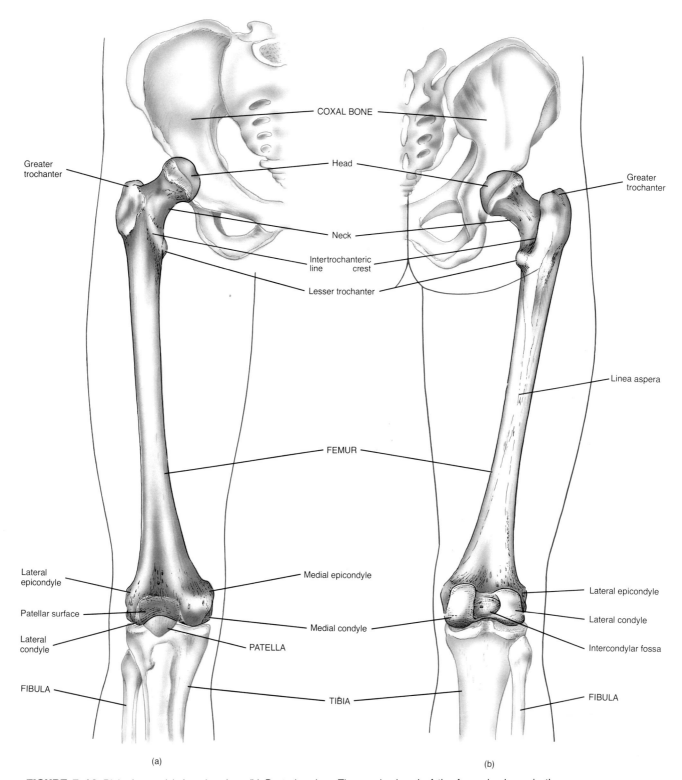

Greater trochanter

COXAL BONE

Head

Greater trochanter

Neck

Intertrochanteric line   crest

Lesser trochanter

Linea aspera

FEMUR

Lateral epicondyle

Medial epicondyle

Patellar surface

Lateral epicondyle

Lateral condyle

Medial condyle

Lateral condyle

PATELLA

Intercondylar fossa

FIBULA

TIBIA

FIBULA

(a)

(b)

FIGURE 7-10 Right femur. (a) Anterior view. (b) Posterior view. The proximal end of the femur is shown in the anteroposterior projection in Figure 7-7b.

condyle articulates with the head of the fibula. The slightly concave condyles are separated by an upward projection called the **intercondylar eminence.** The **tibial tuberosity** on the anterior surface is a point of attachment for the patellar ligament.

The medial surface of the distal end of the tibia forms the **medial malleolus** (mal-LĒ-ō-lus). This structure articulates with the talus bone of the ankle and forms the prominence that can be felt on the medial surface of your ankle. The **fibular notch** articulates with the fibula.

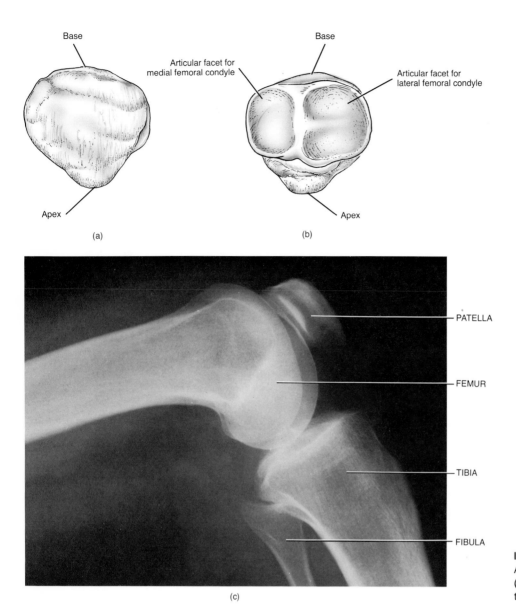

FIGURE 7-11 Right patella. (a) Anterior view. (b) Posterior view. (c) Lateral roentgenogram. (Courtesy of Daniel Sorrentino, R.T.)

The **fibula** is parallel and lateral to the tibia. It is considerably smaller than the tibia.

The **head** of the fibula, the proximal end, articulates with the inferior surface of the lateral condyle of the tibia below the level of the knee joint. The distal end has a projection called the **lateral malleolus,** which articulates with the talus bone of the ankle. This forms the prominence on the lateral surface of the ankle. The inferior portion of the fibula also articulates with the tibia at the fibular notch. A fracture of the lower end of the fibula with injury to the tibial articulation is called a **Pott's fracture.**

TARSUS, METATARSUS, AND PHALANGES

The **tarsus** is a collective designation for the seven bones of the ankle called **tarsals** (Figure 7-13). The term *tarsos* pertains to a broad, flat surface. The **talus** and **calcaneous**

(kal-KĀ-nē-us) are located on the posterior part of the foot. The anterior part contains the **cuboid, navicular,** and three **cuneiform bones** called the first (medial), second (intermediate), and third (lateral) cuneiform. The talus, the uppermost tarsal bone, is the only bone of the foot that articulates with the fibula and tibia. It is surrounded on one side by the medial malleolus of the tibia and on the other side by the lateral malleolus of the fibula. During walking, the talus initially bears the entire weight of the extremity. About half the weight is then transmitted to the calcaneus. The remainder is transmitted to the other tarsal bones. The calcaneus, or heel bone, is the largest and strongest tarsal bone.

The **metatarsus** consists of five metatarsal bones numbered I to V from the medial to lateral position. Like the metacarpals of the palm of the hand, each metatarsal consists of a proximal **base,** a **shaft,** and a distal **head.** The metatarsals articulate proximally with the first, sec-

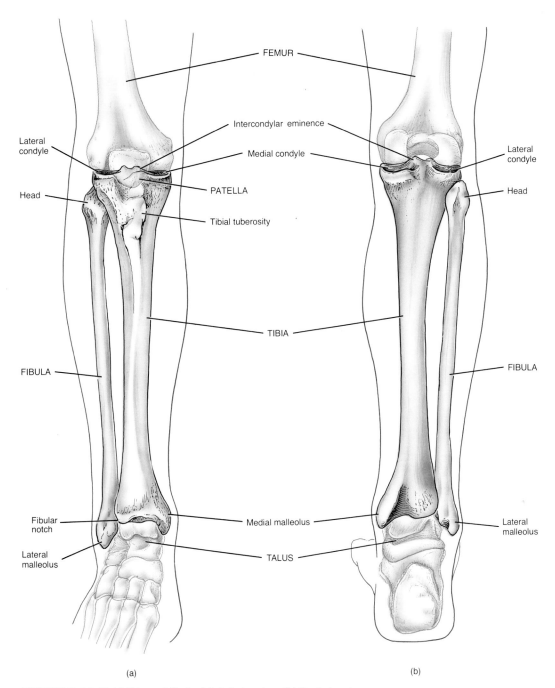

(a)   (b)

FIGURE 7-12 Right tibia and fibula. (a) Anterior view. (b) Posterior view.

ond, and third cuneiform bones and with the cuboid. Distally, they articulate with the proximal row of phalanges. The first metatarsal is thicker than the others because it bears more weight.

The **phalanges** of the foot resemble those of the hand both in number and arrangement. Each also consists of a proximal **base,** a **shaft,** and a distal **head.** The great (big) toe, or *hallux,* has two large, heavy phalanges called proximal and distal phalanges. The other four toes each have three phalanges—proximal, middle, and distal.

ARCHES OF THE FOOT

The bones of the foot are arranged in two arches (Figure 7-14). These arches enable the foot to support the weight of the body and provide leverage while walking. The arches are not rigid. They yield as weight is applied and spring back when the weight is lifted.

The **longitudinal arch** has two parts. Both consist of tarsal and metatarsal bones arranged to form an arch from the anterior to the posterior part of the foot. The

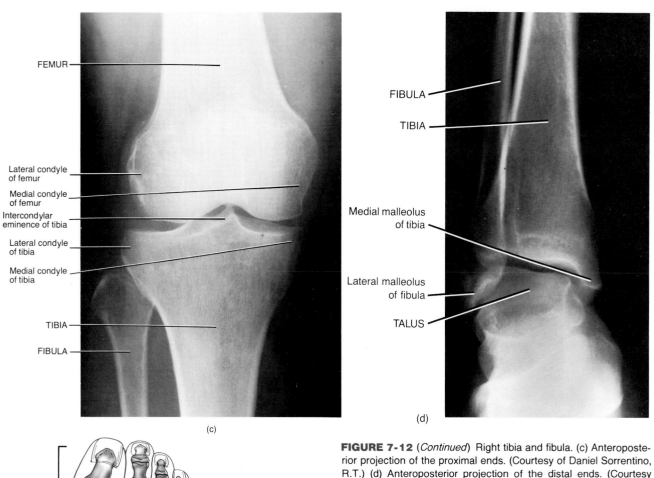

(c)

(d)

FIGURE 7-12 (*Continued*) Right tibia and fibula. (c) Anteroposterior projection of the proximal ends. (Courtesy of Daniel Sorrentino, R.T.) (d) Anteroposterior projection of the distal ends. (Courtesy of John C. Bennett, St. Mary's Hospital, San Francisco.)

FIGURE 7-13 Right foot. (a) Superior view. (b) Medial view.

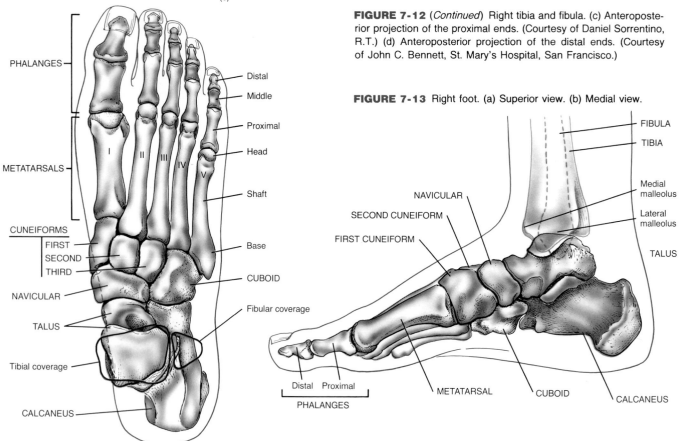

(a)

(b)

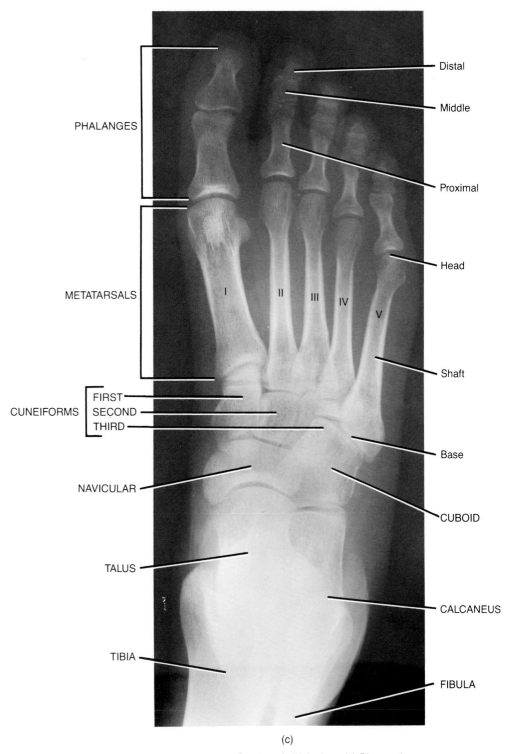

PHALANGES

Distal

Middle

Proximal

METATARSALS

Head

I II III IV V

Shaft

CUNEIFORMS
FIRST
SECOND
THIRD

Base

NAVICULAR

CUBOID

TALUS

CALCANEUS

TIBIA

FIBULA

(c)

FIGURE 7-13 (*Continued*) Right foot. (c) Plantar view. (Courtesy of John C. Bennett, St. Mary's Hospital, San Francisco.)

medial, or inner, part of the longitudinal arch originates at the calcaneus. It rises to the talus and descends anteriorly through the navicular, the three cuneiforms, and the three medial metatarsals. The talus is the keystone of this arch. The **lateral,** or outer, part of the longitudinal arch also begins at the calcaneus. It rises at the cuboid and descends to the two lateral metatarsals. The cuboid is the keystone of the arch.

The **transverse arch** is formed by the calcaneus, navicular, cuboid, and the posterior parts of the five metatarsals.

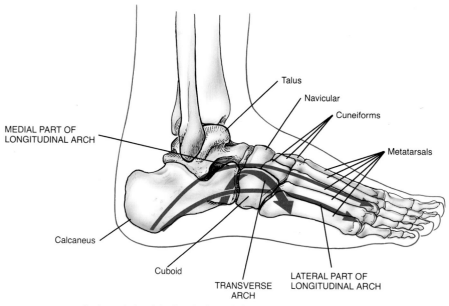

FIGURE 7-14 Arches of the right foot in lateral view.

EXHIBIT 7-1 COMPARISON OF TYPICAL MALE AND FEMALE PELVIS

POINT OF COMPARISON	MALE	FEMALE
General structure	Heavy and thick.	Light and thin.
Joint surfaces	Large.	Small.
Muscle attachments	Well marked.	Rather indistinct.
Greater pelvis	Deep.	Shallow.
Pelvic inlet	Heart-shaped.	Larger and more oval.
Pelvic outlet	Comparatively small.	Comparatively large.
First piece of sacrum	Superior surface of the body spans nearly half the width of sacrum.	Superior surface of the body spans about a third the width of sacrum.
Sacrum	Long, narrow, with smooth concavity.	Short, wide, flat, curving forward in lower part.
Auricular surface	Extends well down the third piece of the sacrum.	Extends only to upper border of third piece of the sacrum.
Pubic arch	Less than a right (90°) angle.	Greater than a right (90°) angle.
Inferior ramus of pubis	Presents strong everted surface for attachment of the crus of the penis.	Everted surface not present.
Pubic symphysis	More deep.	Less deep.
Ischial spine	Turned inward.	Turned inward less.
Ischial tuberosity	Turned inward.	Turned outward.
Ilium	More vertical.	Less vertical.
Iliac fossa	Deep.	Shallow.
Iliac crest	More curved.	Less curved.
Anterior superior iliac spine	Closer.	Wider apart.
Acetabulum	Large.	Small.
Obturator foramen	Round.	Oval.
Greater sciatic notch	Narrow.	Wide.

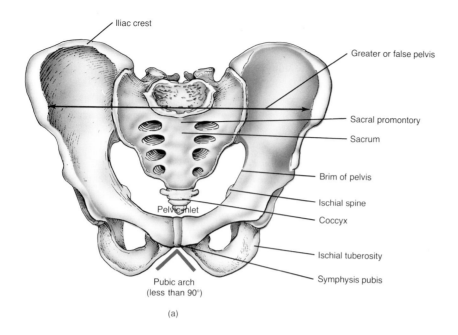

Iliac crest

Greater or false pelvis

Sacral promontory

Sacrum

Brim of pelvis

Ischial spine

Pelvic inlet

Coccyx

Ischial tuberosity

Symphysis pubis

Pubic arch
(less than 90°)

(a)

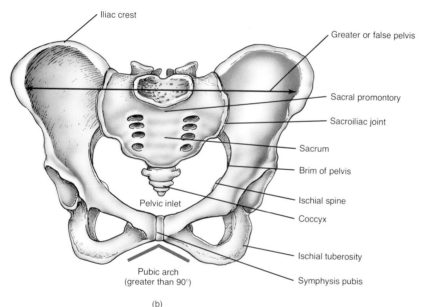

Iliac crest

Greater or false pelvis

Sacral promontory

Sacroiliac joint

Sacrum

Brim of pelvis

Ischial spine

Pelvic inlet

Coccyx

Ischial tuberosity

Symphysis pubis

Pubic arch
(greater than 90°)

(b)

FIGURE 7-15 Pelvis. (a) Male pelvis in anterior view. (b) Female pelvis in anterior view.

CLINICAL APPLICATION

The bones composing the arches are held in position by ligaments and tendons. If these ligaments and tendons are weakened, the height of the longitudinal arch may decrease or "fall." The result is **flatfoot.**

A **bunion** is an abnormal lateral displacement of the big toe from its natural position. This condition produces an inflammatory reaction of the bursae that results in the formation of abnormal tissue.

MALE AND FEMALE SKELETONS

The bones of the male are generally larger and heavier than those of the female. Also, the articular ends are thicker in relation to the shafts. In addition, since certain muscles of the male are larger than those of the female, the male skeleton has larger tuberosities, lines, and ridges for the attachment of larger muscles.

Many significant structural differences between male and female skeletons are noted in the pelvis; most are related to pregnancy and childbirth. The typical differences are listed in Exhibit 7-1 and illustrated in Figure 7-15.

STUDY OUTLINE

Shoulder Girdles

1. Each shoulder, or pectoral, girdle consists of a clavicle and scapula.
2. Each attaches the upper extremity to the trunk.

Upper Extremities

The bones of each upper extremity include the humerus, ulna, radius, carpals, metacarpals, and phalanges.

Pelvic Girdle

1. The pelvic girdle consists of two coxal bones or hip-bones.
2. It attaches the lower extremities to the trunk at the sacrum.
3. Each coxal bone consists of three fused components—ilium, pubis, and ischium.

Lower Extremities

1. The bones of each lower extremity include the femur, tibia, fibula, tarsals, metatarsals, and phalanges.
2. The bones of the foot are arranged in two arches, the longitudinal arch and the transverse arch, to provide support and leverage.

Male and Female Skeletons

1. Male bones are generally larger and heavier than female bones and have more prominent markings for muscle attachment.
2. The female pelvis is adapted for pregnancy and childbirth. Differences in pelvic structure are listed in Exhibit 7-1.

REVIEW QUESTIONS

1. What is the shoulder girdle? Why is it important?
2. What are the bones of the upper extremity? What is a Colles' fracture?
3. What is the pelvic girdle? Why is it important?
4. What are the bones of the lower extremity? What is a Pott's fracture?
5. What is pelvimetry? What is its clinical importance?
6. In what ways do the upper extremity and lower extremity differ structurally?
7. Define an arch of the foot. What is the function of an arch? Distinguish between a longitudinal arch and a transverse arch.
8. How do bunions and flatfeet arise?
9. What are the principal structural differences between typical male and female skeletons? Use Exhibit 7-1 as a guide in formulating your response.

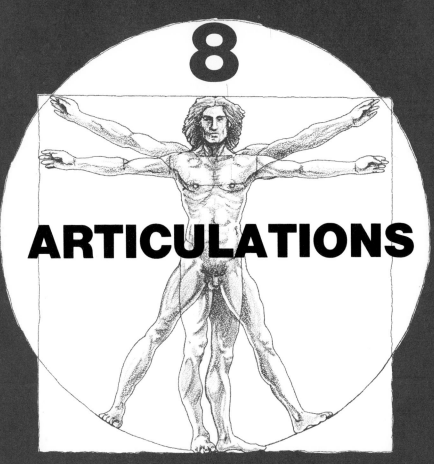

8

ARTICULATIONS

STUDENT OBJECTIVES

■ Define an articulation and identify the factors that determine the degree of movement at a joint.

■ Contrast the structure, kind of movement, and location of fibrous, cartilaginous, and synovial joints.

■ Discuss and compare the movements possible at various synovial joints.

■ Describe selected joints of the body with respect to the bones that enter into their formation, structural classification, and anatomical components.

■ Describe the causes and symptoms of common joint disorders, including arthritis, rheumatism, rheumatoid arthritis, osteoarthritis, gouty arthritis, septic arthritis, bursitis, and tendinitis.

■ Define dislocation and sprain.

■ Define key medical terms associated with joints.

Bones are too rigid to bend without damage. Fortunately, the skeletal system consists of many separate bones, which are held together at joints by flexible connective tissue. All movements that change the positions of the bony parts of the body occur at joints. You can understand the importance of joints if you imagine how a cast over the knee joint prevents flexing the leg or how a splint on a finger limits the ability to manipulate small objects.

The terms **articulation** and **joint** refer to a point of contact between bones or between cartilage and bones. The joint's structure determines its function. Some joints permit no movement, others permit slight movement, and still others afford considerable movement. In general, the closer the fit at the point of contact, the stronger the joint. At tightly fitted joints, however, movement is restricted. The looser the fit, the greater the movement. Unfortunately, loosely fitted joints are prone to dislocation. Movement at joints is also determined by the flexibility of the connective tissue that binds the bones together and by the position of ligaments, muscles, and tendons.

CLASSIFICATION

FUNCTIONAL

The functional classification of joints takes into account the degree of movement they permit. Functionally, joints are classified as **synarthroses** (sin'-ar-THRŌ-sēz), which are immovable joints; **amphiarthroses** (am'-fē-ar-THRŌ-sēz), which are slightly movable joints; and **diarthroses** (dī'-ar-THRŌ-sēz), which are freely movable joints.

STRUCTURAL

The structural classification of joints is based on the presence or absence of a joint cavity (a space between the articulating bones) and the kind of connective tissue that binds the bones together. Structurally, joints are classified as **fibrous,** in which there is no joint cavity and the bones are held together by fibrous connective tissue; **cartilaginous,** in which there is no joint cavity and the bones are held together by cartilage; and **synovial,** in which there is a joint cavity and the bones forming the joint are united by a surrounding articular capsule and frequently by accessory ligaments (described in detail later). We will now discuss the joints of the body based upon their structural classification, but with reference to their functional classification as well.

FIBROUS JOINTS

Fibrous joints lack a joint cavity, and the articulating bones are held very closely together by fibrous connective tissue. They permit little or no movement. The three types

of fibrous joints are (1) sutures, (2) syndesmoses, and (3) gomphoses.

Sutures are found between bones of the skull. In a suture, the bones are united by a thin layer of dense fibrous connective tissue. Based upon the form of the margins of the bones, several types of sutures can be distinguished. In a *serrate suture,* the margins of the bones are serrated like the teeth of a saw. An example is the sagittal suture between the two parietal bones (see Figure 6-2a, e). In a *squamous suture,* the margin of one bone overlaps that of the adjacent bone. The squamosal suture between the parietal and temporal bones is an example (see Figure 6-2c). In a *plane suture,* the even, fairly regular margins of the adjacent bones are brought together. An example is the intermaxillary suture between maxillary bones (see Figure 6-2a). Since sutures are immovable, they are functionally classified as synarthroses.

Some sutures, present during growth, are replaced by bone in the adult. In this case they are called **synostoses** (sin'-os-TŌ-sēz), or bony joints—joints in which there is a complete fusion of bone across the suture line. An example is the frontal suture between the left and right sides of the frontal bone (see Figure 6-3a). Synostoses are also functionally classified as synarthroses.

A **syndesmosis** (sin'-dez-MŌ-sis) is a fibrous joint in which the uniting fibrous connective tissue is present in a much greater amount than in a suture, but the fit between the bones is not quite as tight. The fibrous connective tissue forms an interosseous membrane or ligament. A syndesmosis is slightly movable because the bones are separated more than in a suture and some flexibility is permitted by the interosseous membrane or ligament. Thus, syndesmoses are functionally classified as amphiarthrotic. Examples of syndesmoses include the distal articulation of the tibia and fibula (see Figure 7-12) and the articulations between the shafts of the ulna and radius.

A **gomphosis** (gom-FŌ-sis) is a type of fibrous joint in which a cone-shaped peg fits into a socket. The intervening substance is the periodontal ligament. A gomphosis is functionally classified as synarthrotic. Examples are the articulations of the roots of the teeth with the alveolar processes of the maxillae and mandible.

CARTILAGINOUS JOINTS

Another joint that has no joint cavity is the **cartilaginous joint.** Here the articulating bones are tightly connected by cartilage. Like fibrous joints, they allow little or no movement (Figure 8-1).

A **synchondrosis** (sin'-kon-DRŌ-sis) is a cartilaginous joint in which the connecting material is hyaline cartilage. The most common type of synchondrosis is the epiphyseal plate. Such a joint is found between the epiphysis and diaphysis of a growing bone and is immovable. Thus it is synarthrotic. Since the hyaline cartilage is eventually

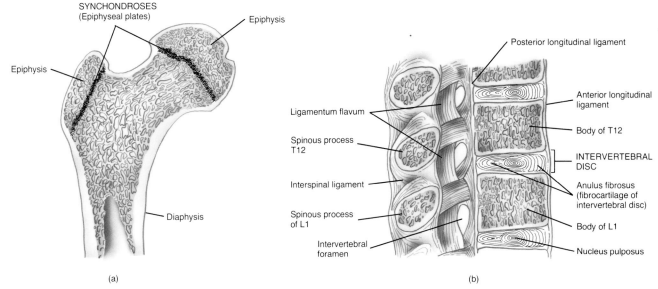

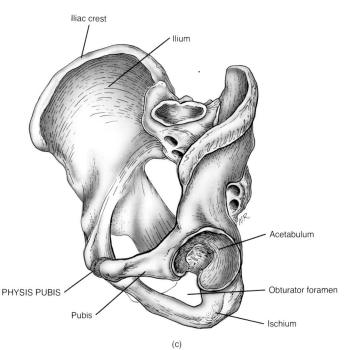

FIGURE 8-1 Cartilaginous joints. (a) Synchondrosis between the diaphysis and epiphysis of a growing femur seen in sagittal section. (b) Symphysis joint between the bodies of vertebrae seen in sagittal section. (c) Symphysis joint (symphysis pubis) between the coxal bones seen in an oblique view.

SYNOVIAL JOINTS

STRUCTURE

A joint in which there is a space between articulating bones is called a **synovial** (si-NŌ-vē-al) **joint.** The space, or **joint cavity,** is also called a **synovial cavity** (Figure 8-2). Because of this cavity and because of the arrangement of the articular capsule and accessory ligaments, synovial joints are freely movable. Thus, synovial joints are functionally classified as diarthrotic.

Synovial joints are also characterized by the presence of **articular cartilage.** Articular cartilage covers the surfaces of the articulating bones, but does not bind the bones together. The articular cartilage of synovial joints is hyaline cartilage.

Synovial joints are surrounded by a tubular **articular capsule** that encloses the joint cavity and unites the articulating bones. The articular capsule is composed of two layers. The outer layer, the *fibrous capsule,* consists of dense connective (collagenous) tissue. The fibrous capsule is attached to the periosteum of the articulating bones at a variable distance from the edge of the articular cartilage. The flexibility of the fibrous capsule permits movement at a joint, whereas its great tensile strength resists dislocation. The fibers of some fibrous capsules are highly adapted to resist recurrent strain. In these instances, they are arranged in parallel bundles. Here they are called *ligaments* and are given special names. The fibrous capsule is one of the principal structures that holds bone to bone.

replaced by bone when growth ceases, the joint is temporary. It is replaced by a synostosis. Another example of a synchondrosis is the joint between the first rib and the sternum. The cartilage in this joint undergoes ossification during adult life.

A **symphysis** (SIM-fi-sis) is a cartilaginous joint in which the connecting material is a broad, flat disc of fibrocartilage. This joint is found between bodies of vertebrae. A portion of the intervertebral disc is cartilaginous material. The symphysis pubis between the anterior surfaces of the coxal bones is another example of a symphysis-type joint. These joints are slightly movable, or amphiarthrotic.

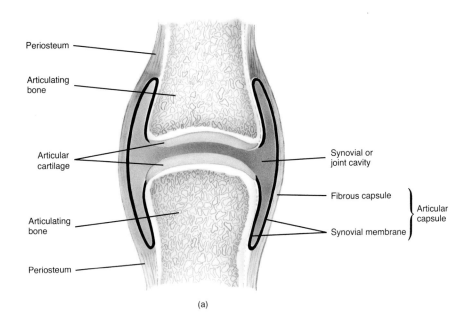

(a)

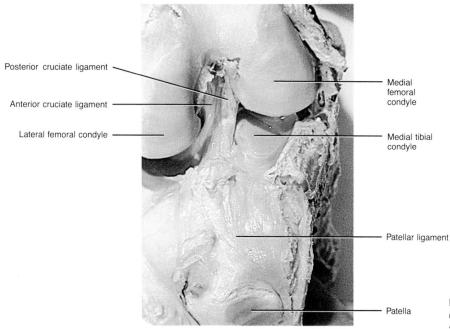

(b)

FIGURE 8-2 Synovial joint. (a) Diagram. (b) Photograph of the internal structure of the right knee joint. (Courtesy of John W. Eads.)

The inner layer of the articular capsule is formed by a *synovial membrane.* The synovial membrane is composed of loose connective tissue with elastic fibers and a variable amount of adipose tissue. It secretes *synovial fluid,* which lubricates the joint and provides nourishment for the articular cartilage. Synovial fluid also contains phagocytic cells that remove microbes and debris resulting from wear and tear in the joint. Synovial fluid consists of hyaluronic acid and interstitial fluid formed from blood plasma and is similar in appearance and consistency to egg white. When there is no joint movement, the fluid is viscous, but as movement increases, the fluid becomes less viscous. Although the amount of synovial fluid is variable in different joints of the body, even a large joint such as the knee contains only 0.5 ml. The amount present in each joint is sufficient only to form a thin film over the surfaces within an articular capsule.

Many synovial joints also contain **accessory ligaments,** which are called extracapsular ligaments and intracapsular ligaments. *Extracapsular ligaments* are outside of the articular capsule. An example is the fibular collateral ligament of the knee joint (see Figure 8-14). *Intracapsular ligaments* occur within the articular capsule, but are excluded from the joint cavity by reflections of the synovial membrane. Examples are the cruciate ligaments of the knee joint (see Figure 8-14).

Inside some synovial joints, there are pads of fibrocartilage that lie between the articular surfaces of the bones and are attached by their margins to the fibrous capsule. These pads are called **articular discs (menisci).** The discs usually subdivide the joint cavity into two separate spaces, thus providing a mechanism by which each bone can move independently. Articular discs help to maintain the stability of the joint and direct the flow of synovial fluid to areas of greatest friction. A tearing of articular discs in the knee, commonly called **torn cartilage,** occurs frequently among athletes.

The various movements of the body create friction between moving parts. To reduce this friction, saclike structures called **bursae** are situated in the body tissues. These sacs resemble joints in that their walls consist of connective tissue lined by a synovial membrane. They are also filled with a fluid similar to synovial fluid. Bursae are located between the skin and bone in places where skin rubs over bone. They are also found between tendons and bones, muscles and bones, and ligaments and bones. As fluid-filled sacs, they cushion the movement of one part of the body over another. An inflammation of a bursa is called **bursitis.**

The articular surfaces of synovial joints are kept in contact with each other by several factors. One factor is the fit of the articulating bones. This interlocking is very obvious at the hip joint, where the head of the femur articulates with the acetabulum of the coxal bone. Another factor is the strength of the joint ligaments. This is especially important in the hip joint. A third factor is the tension of the muscles around the joint. For example, the fibrous capsule of the knee joint is formed principally from tendinous expansions by muscles acting on the joint.

MOVEMENTS

The movement permitted at synovial joints is limited by several factors. One is the *apposition of soft parts.* For example, during bending of the elbow the anterior surface of the forearm is pressed against the anterior surface of the arm. This apposition limits movement. A second factor is the *tension of ligaments.* The different components of a fibrous capsule are tense only when the joint is in certain positions. Tense ligaments not only restrict the limitation of movement, but also direct the movement of the articulating bones with respect to each other. In the knee joint, for example, the major ligaments are lax when the knee is bent but tense when the knee is straightened. Also, when the knee is straightened the surfaces of the articulating bones are in fullest contact with each other. A third factor that restricts movement at a synovial joint is *muscle tension.* This reinforces the restraint placed on a joint by ligaments. A good example of the effect of muscle tension on a joint is seen at the hip joint. When the thigh is raised, the movement is restricted by the tension of the hamstring muscles on the posterior surface of the thigh if the knee is kept straight. But if the knee is bent, the tension on the hamstring muscles is lessened and the thigh can be raised further.

Following is a description of the specific movements that occur at synovial joints.

Gliding

A **gliding movement** is the simplest kind that can occur at a joint. One surface moves back and forth and from side to side over another surface without angular or rotary motion. Some joints that glide are those between the carpals and between the tarsals. The heads and tubercles of ribs glide on the bodies and transverse processes of vertebrae.

Angular

Angular movements increase or decrease the angle between bones. Among the angular movements are flexion, extension, abduction, and adduction (Figure 8-3). **Flexion** usually involves a decrease in the angle between the anterior surfaces of the articulating bones. An exception to this definition is flexion of the knee and the toe joints in which there is a decrease in the angle between the posterior surfaces of the articulating bones. Examples of flexion include bending the head forward (the joint is between the occipital bone and the atlas), bending the elbow, and bending the knee. Flexion of the foot at the ankle joint is called **dorsiflexion.**

Extension involves an increase in the angle between the anterior surface of the articulating bones, with the same exceptions of knee and toe joints. Extension restores a body part to its anatomical position after it has been flexed. Examples of extension are returning the head to the anatomical position after flexion, straightening the arm after flexion, and straightening the leg after flexion. Continuation of extension beyond the anatomical position, as in bending the head backward, is called **hyperextension.** Extension of the foot at the ankle joint is **plantar flexion.**

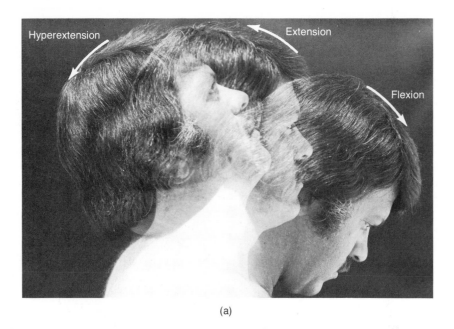

(a)

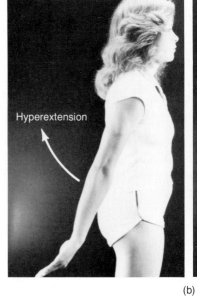

(b)

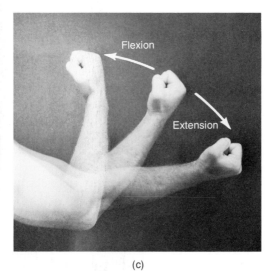

(c)

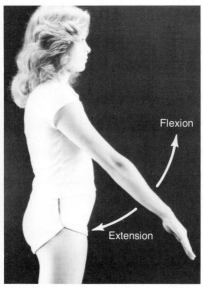

(d)

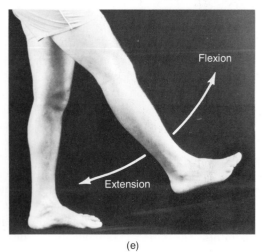

(e)

FIGURE 8-3 Angular movements at synovial joints.

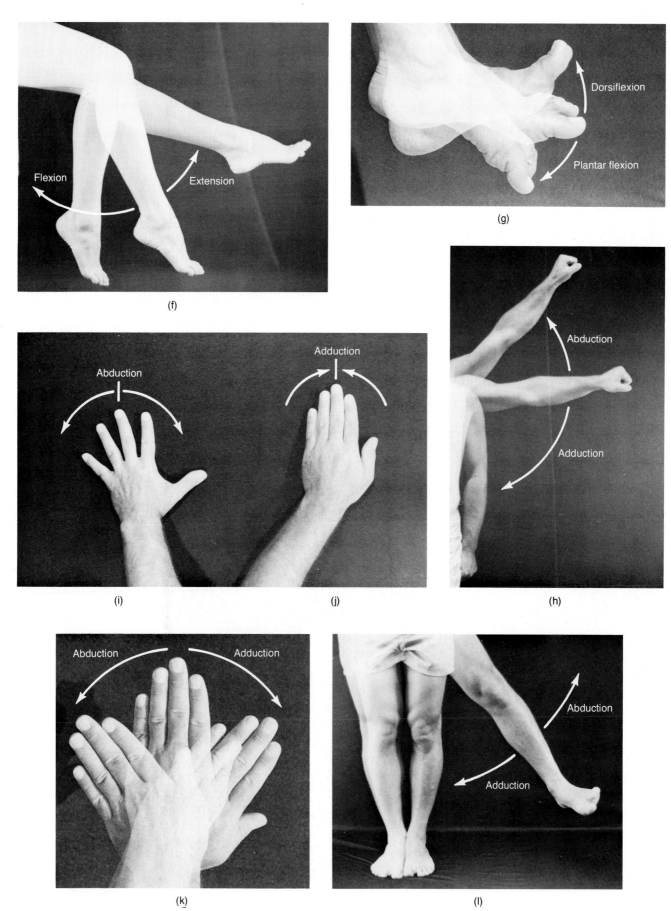

FIGURE 8-3 (*Continued*) Angular movements at synovial joints. (Courtesy of Matt Iacobino and Lynne Tortora.)

Abduction usually means movement of a bone *away from* the midline of the body. An example of abduction is moving the arm upward and away from the body until it is held straight out at right angles to the chest. With the fingers and toes, however, the midline of the body is not used as the line of reference. Abduction of the fingers is a movement away from an imaginary line drawn through the middle finger; in other words, it is spreading the fingers. Abduction of the toes is relative to an imaginary line drawn through the second toe.

Adduction is usually movement of a part *toward* the midline of the body. An example of adduction is returning the arm to the side after abduction. As in abduction, adduction of the fingers is relative to the middle finger, and adduction of the toes is relative to the second toe.

FIGURE 8-4 Rotation and circumduction. (a) Rotation at the atlantoaxial joint (left) and rotation of the humerus (right). (b) Circumduction of the humerus at the shoulder joint. (Courtesy of Matt Iacobino.)

Rotation

Rotation is the movement of a bone around its long axis. During rotation, no other motion is permitted. In *medial rotation* the anterior surface of a bone or extremity moves toward the midline. In *lateral rotation,* the anterior surface of a bone or extremity moves away from the midline. We rotate the atlas around the odontoid process of the axis when we shake the head from side to side. Moving from the shoulder and turning the forearm up, then palm down, and then palm up again is an example of slight lateral and medial rotation of the humerus (Figure 8-4a).

Circumduction

Circumduction is a movement in which the distal end of a bone moves in a circle while the proximal end remains stable. The bone describes a cone in the air. Circumduction typically involves flexion, abduction, adduction, ex-

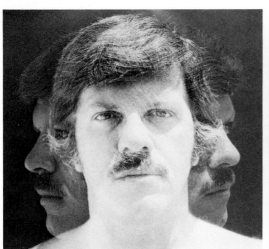
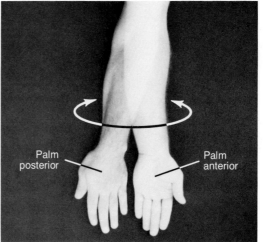

Palm posterior

Palm anterior

(a)

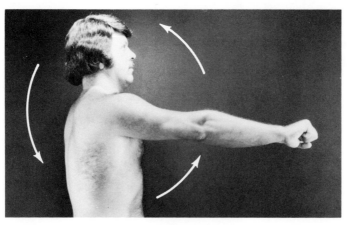

(b)

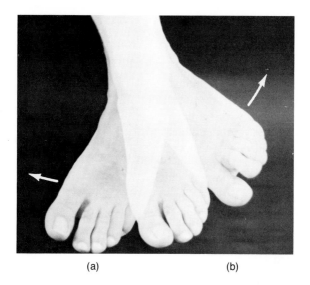

(a)　　　　　　　(b)

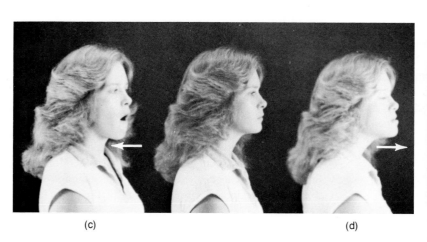

(c)　　　　　　　(d)

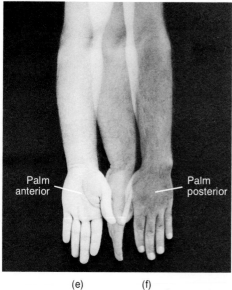

Palm anterior　　　Palm posterior

(e)　　(f)

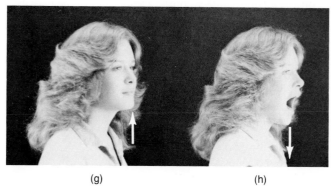

(g)　　　　　　　(h)

FIGURE 8-5 Special movements. (a) Inversion. (b) Eversion. (c) Retraction. (d) Protraction. (e) Supination. (f) Pronation. (g) Elevation. (h) Depression. (Courtesy of Matt Iacobino and Lynne Tortora.)

tension, and rotation. It involves a 360° rotation. An example is moving the outstretched arm in a circle to wind up to pitch a ball (Figure 8-4b).

Special

Special movements are those found only at the joints indicated in Figure 8-5. **Inversion** is the movement of the sole of the foot inward (medially) at the ankle joint. **Eversion** is the movement of the sole outward (laterally) at the ankle joint.

Protraction is the movement of the mandible or clavicle forward on a plane parallel to the ground. Thrusting the jaw outward is protraction of the mandible. Bringing your arms forward until the elbows touch requires protraction of the clavicle. **Retraction** is the movement of a protracted part of the body backward on a plane parallel to the ground. Pulling the lower jaw back in line with the upper jaw is retraction of the mandible.

Supination is a movement of the forearm in which the palm of the hand is turned forward (anterior). To demonstrate supination, flex your arm at the elbow to prevent rotation of the humerus in the shoulder joint. **Pronation** is a movement of the flexed forearm in which the palm is turned backward (posterior).

Elevation is a movement in which a part of the body moves upward. You elevate your mandible when you close your mouth. **Depression** is a movement in which a part of the body moves downward. You depress your mandible when you open your mouth. The shoulders can also be elevated and depressed.

TYPES

Though all synovial joints are similar in structure, variations exist in the shape of the articulating surfaces. Accordingly, synovial joints are divided into six subtypes: gliding, hinge, pivot, ellipsoidal, saddle, and ball-and-socket joints.

Gliding

The articulating surfaces of bones in **gliding joints** or **arthrodia** (ar-THRŌ-dē-a) are usually flat. Only side-to-side and back-and-forth movements are permitted. Since this joint allows movements in two planes, it is called *biaxial*. Twisting and rotation are inhibited at gliding joints generally because ligaments or adjacent bones restrict the range of movement. Examples are the joints between carpal bones, tarsal bones, the sternum and clavicle, and the scapula and clavicle.

Hinge

A **hinge** or **ginglymus** (JIN-gli-mus) **joint** is one in which the convex surface of one bone fits into the concave surface of another bone. Movement is primarily in a single plane, and the joint is therefore known as *monaxial,* or *uniaxial.* The motion is similar to that of a hinged door. Movement is usually flexion and extension. Examples of hinge joints are the elbow, knee, ankle, and interphalangeal joints. The movement allowed by a hinge joint is illustrated by flexion and extension at the elbow and knee (see Figure 8-3c, f).

Pivot

In a **pivot** or **trochoid** (TRŌ-koyd) **joint,** a rounded, pointed, or conical surface of one bone articulates within a ring formed partly by bone and partly by a ligament. The primary movement permitted is rotation, and the joint is therefore monaxial. Examples include the joints between the atlas and axis (atlantoaxial) and between the proximal ends of the radius and ulna. Movement at a pivot joint is illustrated by supination and pronation of the palms and rotation of the head from side to side (see Figure 8-4a).

Ellipsoidal

In an **ellipsoidal** or **condyloid** (KON-di-loyd) **joint,** an oval-shaped condyle of one bone fits into an elliptical cavity of another bone. Since the joint permits side-to-side and back-and-forth movements, it is biaxial. The joint at the wrist between the radius and carpals is ellipsoidal. The movement permitted by such a joint is illustrated when you flex and extend and abduct and adduct the wrist (see Figure 8-3d, f).

Saddle

In a **saddle** or **sellaris** (sel-A-ris) **joint,** the articular surfaces of both bones are saddle-shaped, that is, concave in one direction and convex in the other. Essentially, the saddle joint is a modified ellipsoidal joint in which the movement is somewhat freer. Movements at a saddle joint are side to side and back and forth. Thus the joint is biaxial. The joint between the trapezium and metacarpal of the thumb is an example of a saddle joint.

Ball-and-Socket

A **ball-and-socket** or **spheroid** (SFĒ-royd) **joint** consists of a ball-like surface of one bone fitted into a cuplike depression of another bone. Such a joint permits *triaxial* movement, or movement in three planes of motion: flexion-extension, abduction-adduction, and rotation. Examples of ball-and-socket joints are the shoulder joint and hip joint. The range of movements at a ball-and-socket joint is illustrated by circumduction of the arm (Figure 8-4b).

SUMMARY OF JOINTS

The summary of joints presented in Exhibit 8-1 is based on the anatomy of the joints. Joints can also be classified according to movement. If we rearrange Exhibit 8-1 into a classification based on movement, we arrive at the following.

Synarthroses: immovable joints
1. Suture
2. Synchondrosis
3. Gomphosis

Amphiarthroses: slightly movable joints
1. Syndesmosis
2. Symphysis

Diarthroses: freely movable joints
1. Gliding
2. Hinge
3. Pivot
4. Ellipsoidal
5. Saddle
6. Ball-and-socket

EXHIBIT 8-1 JOINTS

TYPE	DESCRIPTION	MOVEMENT	EXAMPLES
FIBROUS	No joint cavity; bones held together by a thin layer of fibrous tissue or dense fibrous tissue.		
Suture	Found only between bones of the skull; articulating bones separated by a thin layer of fibrous tissue.	None—synarthrotic.	Lambdoidal suture between occipital and parietal bones.
Syndesmosis	Articulating bones united by dense fibrous tissue.	Slight—amphiarthrotic.	Distal ends of tibia and fibula.
Gomphosis	Cone-shaped peg fits into a socket; articulating bones separated by periodontal ligament.	None—synarthrotic.	Roots of teeth in alveolar processes.
CARTILAGINOUS	No joint cavity; articulating bones united by cartilage.		
Synchondrosis	Connecting material is hyaline cartilage.	None—synarthrotic.	Temporary joint between the diaphysis and epiphyses of a long bone.
Symphysis	Connecting material is a broad, flat disc of fibrocartilage.	Slight—amphiarthrotic.	Intervertebral joints and symphysis pubis.
SYNOVIAL	Joint cavity and articular cartilage present; articular capsule composed of an outer fibrous capsule and an inner synovial membrane; may contain accessory ligaments, articular discs (menisci), and bursae.	Freely movable—diarthrotic.	
Gliding	Articulating surfaces usually flat.	Biaxial (flexion-extension, abduction-adduction).	Intercarpal and intertarsal joints.
Hinge	Spool-like surface fits into a concave surface.	Monaxial (flexion-extension).	Elbow, knee, ankle, and interphalangeal joints.
Pivot	Rounded, pointed, or concave surface fits into a ring formed partly by bone and partly by a ligament.	Monaxial (rotation).	Atlantoaxial and radioulnar joints.
Ellipsoidal	Oval-shaped condyle fits into an elliptical cavity.	Biaxial (flexion-extension, abduction-adduction).	Radiocarpal joint.
Saddle	Articular surfaces concave in one direction and convex in opposite direction.	Biaxial (flexion-extension, abduction-adduction).	Carpometacarpal joint of thumb.
Ball-and-socket	Ball-like surface fits into a cuplike depression.	Triaxial (flexion-extension, abduction-adduction, rotation).	Shoulder and hip joints.

SELECTED ARTICULATIONS OF THE BODY

In the next section of the chapter, we will examine in some detail selected articulations of the body. In order to simplify your learning efforts, a series of exhibits has been prepared. Each exhibit considers a specific articulation and contains (1) a definition, that is, a description of the bones that form the joint; (2) the type of joint (its structural classification); (3) its anatomical components—a description of the major connecting ligaments, articular disc, articular capsule, and other distinguishing features of the joint. Each exhibit also refers you to an illustration of the joint.

EXHIBIT 8-2 **TEMPOROMANDIBULAR JOINT (Figure 8-6)**

DEFINITION

Joint formed by mandibular condyle of the mandible and mandibular fossa and articular tubercle of temporal bone. The temporomandibular joint is the only movable joint between skull bones; all other skull joints are sutures and therefore immovable.

TYPE OF JOINT

Synovial; combined hinge (ginglymus) and gliding (arthrodial) type.

ANATOMICAL COMPONENTS

1. **Articular disc (meniscus).** Fibrocartilage disc that separates the joint cavity into a superior and inferior compartment, each of which has a synovial membrane.
2. **Articular capsule.*** Thin, fairly loose envelope around the circumference of the joint.
3. **Lateral ligament.** Two short bands on the lateral surface of the articular capsule that extend inferiorly and posteriorly from the lower border and tubercle of the zygomatic arch to the lateral and posterior aspect of the neck of the mandible. It is covered by the parotid gland and helps prevent displacement of the mandible.
4. **Sphenomandibular ligament.** Thin band that extends inferiorly and anteriorly from the spine of the sphenoid bone to the ramus of the mandible.
5. **Stylomandibular ligament.** Thickened band of deep cervical fascia that extends from the styloid process of the temporal bone to the inferior and posterior border of the ramus of the mandible. The ligament separates the parotid from the submandibular gland.

* Also referred to as the capsular ligament.

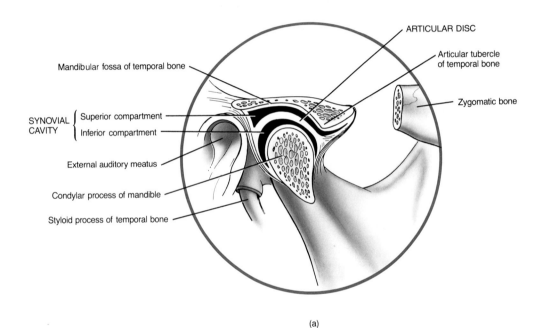

(a)

FIGURE 8-6 Temporomandibular joint. (a) Sagittal section through the temporomandibular joint.

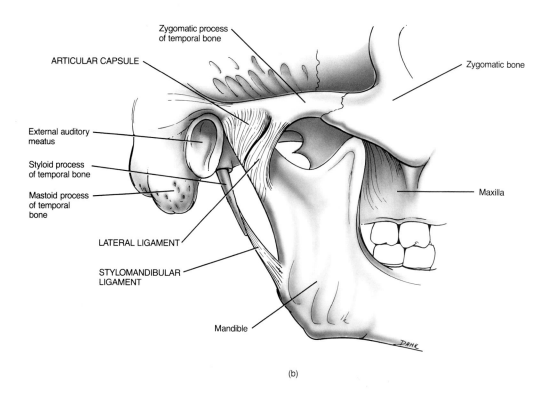

(b)

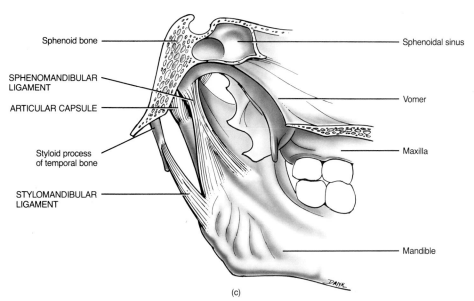

(c)

FIGURE 8-6 (*Continued*) Temporomandibular joint. (b) Right lateral view. (c) Median view.

EXHIBIT 8-3 ATLANTOOCCIPITAL JOINTS (Figure 8-7)

DEFINITION

Joints formed by the superior articular surfaces of the atlas and the occipital condyles of the occipital bone.

TYPE OF JOINT

Synovial; ellipsoidal (condyloid) type.

ANATOMICAL COMPONENTS

1. **Articular capsules.** Thin, loose envelopes that surround the occipital condyles and attach them to the articular processes of the atlas.
2. **Anterior atlantooccipital ligament.** Broad, thick ligament that extends from the anterior surface of the foramen magnum of the occipital bone to the anterior arch of the atlas.
3. **Posterior atlantooccipital ligament.** Broad, thin ligament that extends from the posterior surface of the foramen magnum of the occipital bone to the posterior arch of the atlas. The inferior portion of the ligament is sometimes ossified.
4. **Lateral atlantooccipital ligaments.** Thickened portions of the articular capsules that also contain fibrous tissue bundles and extend from the jugular process of the occipital bone to the transverse process of the atlas.

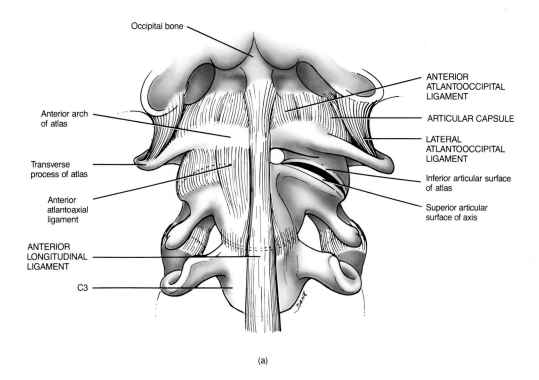

(a)

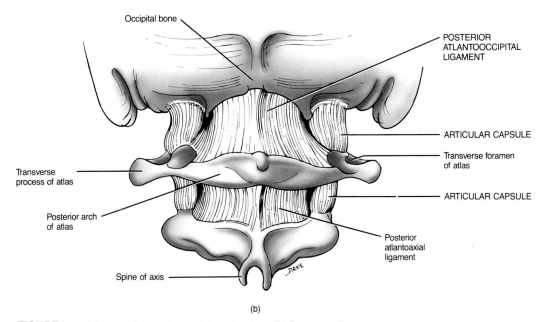

(b)

FIGURE 8-7 Atlantooccipital joints. (a) Anterior view. (b) Posterior view.

EXHIBIT 8-4 INTERVERTEBRAL JOINTS (Figure 8-8)

DEFINITION

Joints formed between (1) vertebral bodies and (2) vertebral arches.

TYPE OF JOINT

Joints between vertebral bodies—cartilaginous, symphysis type. Joints between vertebral arches—synovial, gliding (arthrodial) type.

ANATOMICAL COMPONENTS OF JOINTS BETWEEN VERTEBRAL BODIES

1. **Anterior longitudinal ligament.** Broad, strong band that extends along the anterior surfaces of the vertebral bodies from the axis to the sacrum. It is firmly attached to the intervertebral discs.
2. **Posterior longitudinal ligament.** Extends along the posterior surfaces of the vertebral bodies within the vertebral canal from the axis to the sacrum. The free surface of the ligament is separated from the spinal dura mater by loose (areolar) tissue.
3. **Intervertebral discs.** Unite with adjacent surfaces of the vertebral bodies from the axis to the sacrum. Each disc is composed of a peripheral **anulus fibrosus** consisting of fibrous tissue and fibrocartilage and a central **nucleus pulposus** composed of a soft, pulpy, highly elastic substance.

ANATOMICAL COMPONENTS OF JOINTS BETWEEN VERTEBRAL ARCHES

1. **Articular capsules.** Thin, loose ligaments attached to the margins of the articular processes of adjacent vertebrae.
2. **Ligamenta flava.** Contain elastic tissue and connect the laminae of adjacent vertebrae from the axis to the first segment of the sacrum.
3. **Supraspinous ligament.** Strong fibrous cord that connects the spinous processes from the seventh cervical vertebra to the sacrum.
4. **Ligamentum nuchae.** Fibrous ligament that represents an enlargement of the supraspinous ligament in the neck. It extends from the external occipital protuberance of the occipital bone to the spinous process of the seventh cervical vertebra.
5. **Interspinous ligaments.** Relatively weak bands that run between adjacent spinous processes.
6. **Intertransverse ligaments.** Bands between transverse processes that are readily apparent in the lumbar region.

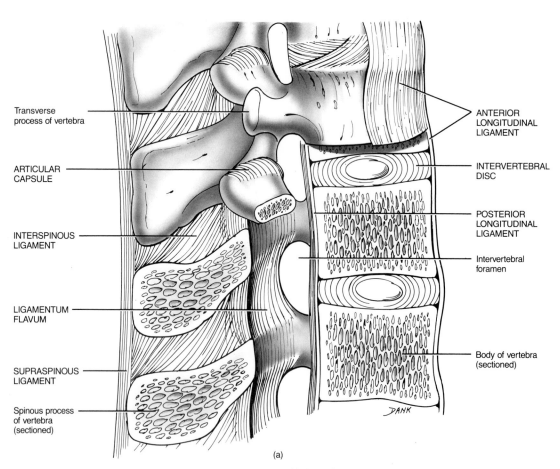

Transverse process of vertebra

ARTICULAR CAPSULE

INTERSPINOUS LIGAMENT

LIGAMENTUM FLAVUM

SUPRASPINOUS LIGAMENT

Spinous process of vertebra (sectioned)

ANTERIOR LONGITUDINAL LIGAMENT

INTERVERTEBRAL DISC

POSTERIOR LONGITUDINAL LIGAMENT

Intervertebral foramen

Body of vertebra (sectioned)

(a)

FIGURE 8-8 Intervertebral joints. (a) Median view.

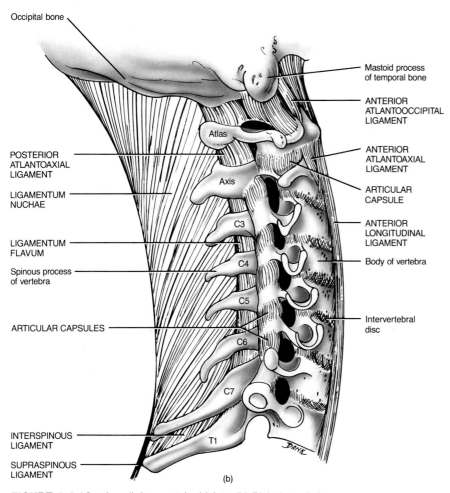

Occipital bone

Mastoid process
of temporal bone

ANTERIOR
ATLANTOOCCIPITAL
LIGAMENT

Atlas

ANTERIOR
ATLANTOAXIAL
LIGAMENT

POSTERIOR
ATLANTOAXIAL
LIGAMENT

Axis

ARTICULAR
CAPSULE

LIGAMENTUM
NUCHAE

C3

ANTERIOR
LONGITUDINAL
LIGAMENT

LIGAMENTUM
FLAVUM

C4

Body of vertebra

Spinous process
of vertebra

C5

ARTICULAR CAPSULES

Intervertebral
disc

C6

C7

INTERSPINOUS
LIGAMENT

T1

SUPRASPINOUS
LIGAMENT

(b)

FIGURE 8-8 (*Continued*) Intervertebral joints. (b) Right lateral view.

EXHIBIT 8-5 **LUMBOSACRAL JOINT (Figure 8-9)**

DEFINITION

Joint formed by the body of the fifth lumbar vertebra and the superior surface of the first sacral vertebra of the sacrum.

TYPE OF JOINT

Joint between the bodies of the fifth lumbar vertebra and the first sacral vertebra—cartilaginous joint, symphysis type. Joint between the articular processes—synovial joint, gliding (arthrodial) type.

ANATOMICAL COMPONENTS

1. The lumbosacral joint is similar to the joints between typical vertebrae, united by an **intervertebral disc, anterior** and **posterior longitudinal ligaments, ligamenta flava, interspinous** and **supraspinous ligaments,** and **articular capsules** between articular processes. The lumbosacral joint also contains an iliolumbar ligament.
2. **Iliolumbar ligament.** Strong ligament that connects the transverse process of the fifth lumbar vertebra with the base of the sacrum and the iliac crest.

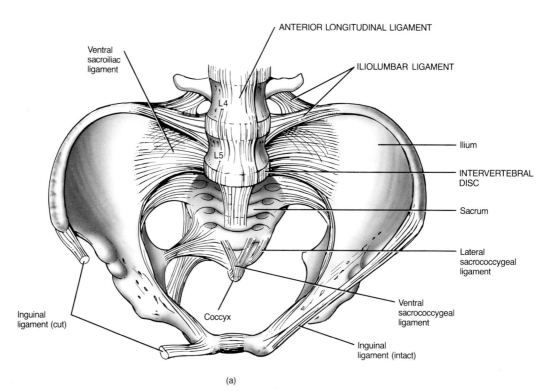

(a)

FIGURE 8-9 Lumbosacral joint. (a) Anterior view.

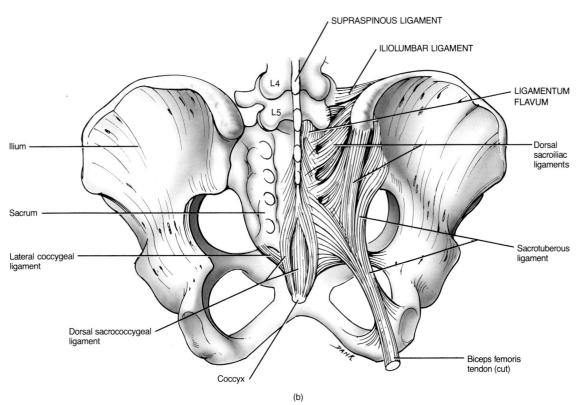

SUPRASPINOUS LIGAMENT

ILIOLUMBAR LIGAMENT

LIGAMENTUM FLAVUM

Dorsal sacroiliac ligaments

Sacrotuberous ligament

Biceps femoris tendon (cut)

Ilium

Sacrum

Lateral coccygeal ligament

Dorsal sacrococcygeal ligament

Coccyx

L4

L5

(b)

FIGURE 8-9 (*Continued*) Lumbosacral joint. (b) Posterior view.

EXHIBIT 8-6 HUMEROSCAPULAR (SHOULDER) JOINT (Figure 8-10)

DEFINITION	Joint formed by the head of the humerus and the glenoid cavity of the scapula.
TYPE OF JOINT	Synovial joint, ball-and-socket (spheroid) type.

ANATOMICAL COMPONENTS

1. **Articular capsule.** Loose sac that completely envelops the joint, extending from the circumference of the glenoid cavity to the anatomical neck of the humerus.
2. **Coracohumeral ligament.** Strong, broad ligament that extends from the coracoid process of the scapula to the greater tubercle of the humerus.
3. **Glenohumeral ligaments.** Three thickenings of the articular capsule over the ventral surface of the joint.
4. **Transverse humeral ligament.** Narrow sheet extending from the greater tubercle to the lesser tubercle of the humerus.
5. **Glenoid labrum.** Narrow rim of fibrocartilage around the edge of the glenoid cavity.
6. Among the bursae associated with the shoulder joint are:
   Subscapular bursa between the tendon of the subscapularis muscle and the underlying joint capsule.
   Subdeltoid bursa between the deltoid muscle and joint capsule.
   Subacromial bursa between the acromion and joint capsule.
   Subcoracoid bursa either lies between the coracoid process and joint capsule or appears as an extension from the subacromial bursa.

CLINICAL APPLICATION

The strength and stability of the shoulder joint are not provided by the shape of the articulating bones or its ligaments. Instead, deep muscles of the shoulder and their tendons—subscapularis, supraspinatus, infraspinatus, and teres minor—strengthen and stabilize the shoulder joint. The muscles and their tendons are so arranged as to form a nearly complete encirclement of the joint. This arrangement is referred to as the **rotator (musculotendinous) cuff** and is a common site of injury to baseball pitchers, especially tearing of the supraspinatus muscle.

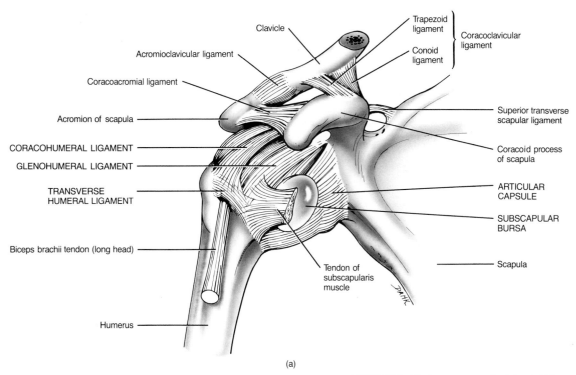

Clavicle

Trapezoid ligament

Conoid ligament

Coracoclavicular ligament

Acromioclavicular ligament

Coracoacromial ligament

Acromion of scapula

CORACOHUMERAL LIGAMENT

GLENOHUMERAL LIGAMENT

TRANSVERSE HUMERAL LIGAMENT

Biceps brachii tendon (long head)

Humerus

Superior transverse scapular ligament

Coracoid process of scapula

ARTICULAR CAPSULE

SUBSCAPULAR BURSA

Scapula

Tendon of subscapularis muscle

(a)

FIGURE 8-10 Humeroscapular joint. (a) Diagram of anterior view. (b) Photograph of anterior view. (Courtesy of C. Yokochi and J. W. Rohen, *Photographic Anatomy of the Human Body,* 2nd ed., 1979, IGAKU-SHOIN, Ltd., Tokyo, New York.)

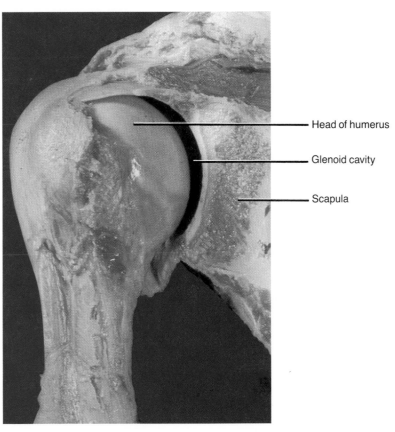

Head of humerus

Glenoid cavity

Scapula

(b)

EXHIBIT 8-7 ELBOW JOINT (Figure 8-11)

DEFINITION	Joint formed by the trochlea of the humerus, the trochlear notch of the ulna, the capitulum of the humerus, and the head of the radius.
TYPE OF JOINT	Synovial joint, hinge (ginglymus) type.
ANATOMICAL COMPONENTS	1. **Articular capsule.** The anterior part covers the anterior part of the joint from the radial and coronoid fossae of the humerus to the coronoid process of the ulna and the annular ligament of the radius. The posterior part extends from the capitulum, olecranon fossa, and lateral epicondyle of the humerus to the annular ligament of the radius, the olecranon of the ulna, and the ulna posterior to the radial notch.
	2. **Ulnar collateral ligament.** Thick, triangular ligament that extends from the medial epicondyle of the humerus to the coronoid process and olecranon of the ulna.
	3. **Radial collateral ligament.** Strong, triangular ligament that extends from the lateral epicondyle of the humerus to the annular ligament of the radius and the radial notch of the ulna.

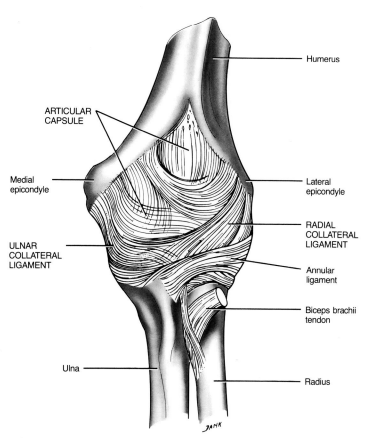

Humerus

ARTICULAR
CAPSULE

Medial
epicondyle

Lateral
epicondyle

RADIAL
COLLATERAL
LIGAMENT

ULNAR
COLLATERAL
LIGAMENT

Annular
ligament

Biceps brachii
tendon

Ulna

Radius

JANK

FIGURE 8-11 Elbow joint. Anterior view.

EXHIBIT 8-8 **RADIOCARPAL (WRIST) JOINT (Figure 8-12)**

DEFINITION	Joint formed by the distal end of the radius, the distal surface of the articular disc separating the carpal and distal radioulnar joint, and the scaphoid, lunate, and triquetral carpal bones.
TYPE OF JOINT	Synovial joint, ellipsoidal (condyloid) type.
ANATOMICAL COMPONENTS	1. **Articular capsule.** Extends from the styloid processes of the ulna and radius to the proximal row of carpal bones.

1. **Articular capsule.** Extends from the styloid processes of the ulna and radius to the proximal row of carpal bones.
2. **Palmar radiocarpal ligament.** Thick, strong ligament that extends from the anterior border of the distal end of the radius to the proximal row of carpals. Some longer fibers extend to the capitate in the distal row.
3. **Dorsal radiocarpal ligament.** Extends from the posterior border of the distal end of the radius to the proximal row of carpals, especially the triquetral.
4. **Ulnar collateral ligament.** Extends from the styloid process of the ulna to the triquetral and pisiform bones and the transverse carpal ligament.
5. **Radial collateral ligament.** Extends from the styloid process of the radius to the scaphoid and pisiform bones.

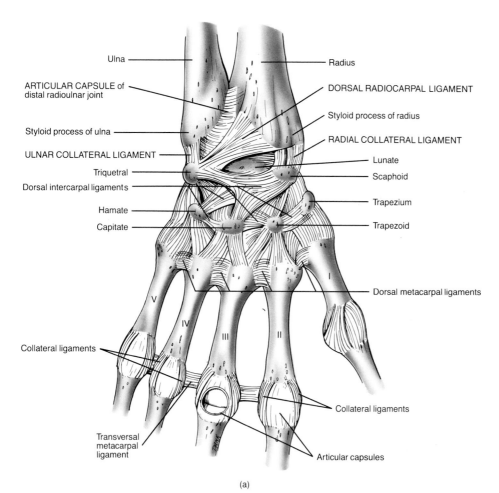

(a)

FIGURE 8-12 Radiocarpal joint. (a) Dorsal view.

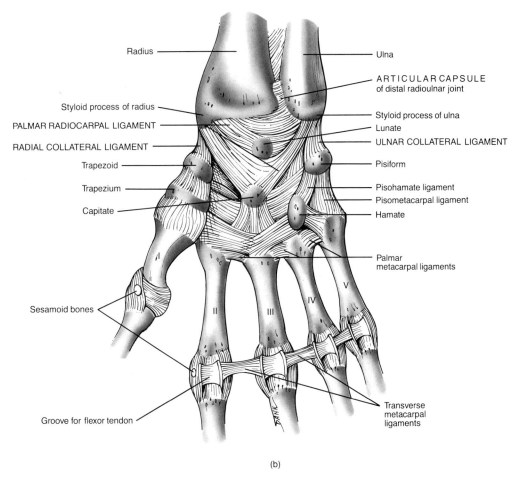

Radius

Ulna

ARTICULAR CAPSULE
of distal radioulnar joint

Styloid process of radius

PALMAR RADIOCARPAL LIGAMENT

RADIAL COLLATERAL LIGAMENT

Styloid process of ulna

Lunate

ULNAR COLLATERAL LIGAMENT

Trapezoid

Trapezium

Capitate

Pisiform

Pisohamate ligament

Pisometacarpal ligament

Hamate

Palmar
metacarpal ligaments

Sesamoid bones

Groove for flexor tendon

Transverse
metacarpal
ligaments

(b)

FIGURE 8-12 (*Continued*) Radiocarpal joint. (b) Palmar view.

EXHIBIT 8-9 COXAL (HIP) JOINT (Figure 8-13)

DEFINITION   Joint formed by the head of the femur and the acetabulum of the coxal bone.

TYPE OF JOINT   Synovial, ball-and-socket (spheroid) type.

ANATOMICAL COMPONENTS

1. **Articular capsule.** Extends from the rim of the acetabulum to the neck of the femur. One of the strongest ligaments of the body, the capsule consists of circular and longitudinal fibers. The circular fibers, called the **zona orbicularis,** form a collar around the neck of the femur. The longitudinal fibers are reinforced by accessory ligaments known as the iliofemoral ligament, the pubofemoral ligament, and the ischiofemoral ligament.
2. **Iliofemoral ligament.** Thickened portion of the articular capsule that extends from the anterior inferior iliac spine of the coxal bone to the intertrochanteric line of the femur.
3. **Pubofemoral ligament.** Thickened portion of the articular capsule that extends from the pubic part of the rim of the acetabulum to the neck of the femur.
4. **Ischiofemoral ligament.** Thickened portion of the articular capsule that extends from the ischial wall of the acetabulum to the neck of the femur.
5. **Ligament of the head of the femur.** Flat, triangular band that extends from the fossa of the acetabulum to the head of the femur.
6. **Acetabular labrum.** Fibrocartilage rim attached to the margin of the acetabulum.
7. **Transverse ligament of the acetabulum.** Strong ligament which crosses over the acetabular notch, converting it to a foramen. It supports part of the acetabular labrum and is connected with the ligament of the head of the femur and the articular capsule.

CLINICAL APPLICATION   **Dislocation of the hip** among adults is quite rare because of (1) the stability of the ball-and-socket joint, (2) the strong, tough, articular capsule, (3) the strength of the intracapsular ligaments, and (4) the extensive musculature over the joint.

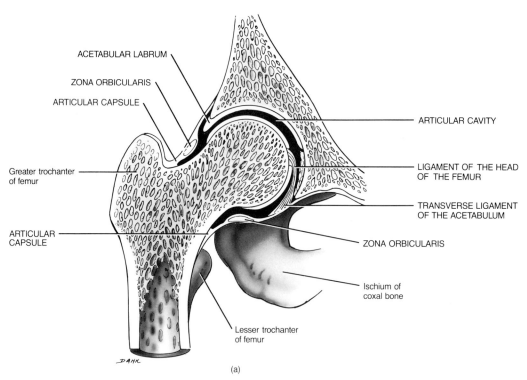

ACETABULAR LABRUM

ZONA ORBICULARIS

ARTICULAR CAPSULE

ARTICULAR CAVITY

Greater trochanter of femur

LIGAMENT OF THE HEAD OF THE FEMUR

TRANSVERSE LIGAMENT OF THE ACETABULUM

ARTICULAR CAPSULE

ZONA ORBICULARIS

Ischium of coxal bone

Lesser trochanter of femur

(a)

FIGURE 8-13 Coxal joint. (a) Frontal section.

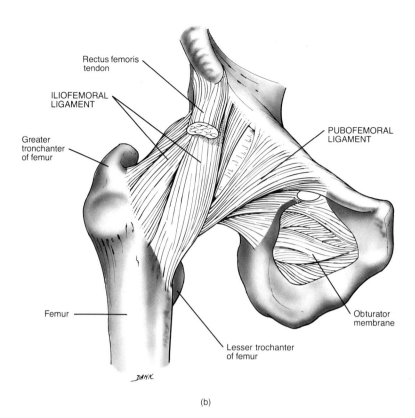

(b)

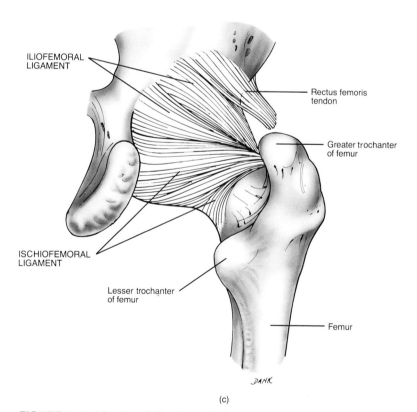

(c)

FIGURE 8-13 (*Continued*) Coxal joint. (b) Anterior view. (c) Posterior view.

EXHIBIT 8-10 **TIBIOFEMORAL (KNEE) JOINT (Figure 8-14)**

DEFINITION

The largest joint of the body, actually consisting of three joints: (1) an intermediate patellofemoral joint between the patella and the patellar surface of the femur; (2) a lateral tibiofemoral joint between the lateral condyle of the femur, lateral meniscus, and lateral condyle of the tibia; and (3) a medial tibiofemoral joint between the medial condyle of the femur, medial meniscus, and medial condyle of the tibia.

TYPE OF JOINT

Patellofemoral joint—partly synovial, gliding (arthrodial) type. Lateral and medial tibiofemoral joints—synovial, hinge (ginglymus) type.

ANATOMICAL COMPONENTS

1. **Articular capsule.** No complete, independent capsule uniting the bones. The ligamentous sheath surrounding the joint consists mostly of muscle tendons or expansions of them. There are, however, some capsular fibers connecting the articulating bones.
2. **Medial and lateral patellar retinacula.** Fused tendons of insertion of the quadriceps femoris muscle and the fascia lata that strengthen the anterior surface of the joint.
3. **Patellar ligament.** Central portion of the common tendon of insertion of the quadriceps femoris muscle that extends from the patella to the tibial tuberosity. This also strengthens the anterior surface of the joint. The posterior surface of the ligament is separated from the synovial membrane of the joint by an **infrapatellar fat pad.**
4. **Oblique popliteal ligament.** Broad, flat ligament that connects the intercondylar fossa of the femur to the head of the tibia. The tendon of the semimembranosus muscle is superficial to the ligament and passes from the medial condyle of the tibia to the lateral condyle of the femur. The ligament and tendon afford strength for the posterior surface of the joint.
5. **Arcuate popliteal ligament.** Extends from the lateral condyle of the femur to the styloid process of the head of the fibula. It strengthens the lower lateral part of the posterior surface of the joint.
6. **Tibial collateral ligament.** Broad, flat ligament on the medial surface of the joint that extends from the medial condyle of the femur to the medial condyle of the tibia. The ligament is crossed by tendons of the sartorius, gracilis, and semitendinosus muscles, all of which strengthen the medial aspect of the joint.
7. **Fibular collateral ligament.** Stong, rounded ligament on the lateral surface of the joint that extends from the lateral condyle of the femur to the lateral side of the head of the fibula. The ligament is covered by the tendon of the biceps femoris muscle. The tendon of the popliteus muscle is deep to the tendon.
8. **Intraarticular ligaments** (ligaments within the capsule) connect the tibia and femur.
    a. **Anterior cruciate ligament.** Extends posteriorly and laterally from the area anterior to the intercondylar eminence of the tibia to the posterior part of the medial surface of the lateral condyle of the femur.
    b. **Posterior cruciate ligament.** Extends anteriorly and medially from the posterior intercondylar fossa of the tibia and lateral meniscus to the anterior part of the medial surface of the medial condyle of the femur.
9. **Menisci** (fibrocartilage discs between the tibial and femoral condyles) help to compensate for the incongruence of the articulating bones.
    a. **Medial meniscus.** Semicircular piece of fibrocartilage. Its anterior end is attached to the anterior intercondylar fossa of the tibia, in front of the anterior cruciate ligament. Its posterior end is attached to the posterior intercondylar fossa of the tibia between the attachments of the posterior cruciate ligament and lateral meniscus.
    b. **Lateral meniscus.** Circular piece of fibrocartilage. Its anterior end is attached anterior to the intercondylar eminence of the tibia and lateral and posterior to the anterior cruciate ligament. Its posterior end is attached posterior to the intercondylar eminence of the tibia and anterior to the posterior end of the medial meniscus. The medial and lateral menisci are connected to each other by the **transverse ligament** and to the margins of the head of the tibia by the **coronary ligaments.**
10. The principal **bursae** of the knee include:
    a. Anterior bursae: (1) between the patella and skin **(prepatellar bursa),** (2) between upper part of tibia and patellar ligament **(infrapatellar bursa),** (3) between lower part of tibial tuberosity and skin, and (4) between lower part of femur and deep surface of quadriceps femoris muscle **(suprapatellar bursa).**
    b. Medial bursae: (1) between medial head of gastrocnemius muscle and the articular capsule, (2) superficial to the tibial collateral ligament between the ligament and tendons of the sartorius, gracilis, and semitendinosus muscles, (3) deep to the tibial collateral ligament between the ligament and the tendon of the semimembranosus muscle, (4) between the tendon of the semimembranosus muscle and the head of the tibia, and (5) between the tendons of the semimembranosus and semitendinosus muscles.
    c. Lateral bursae: (1) between the lateral head of the gastrocnemius muscle and articular capsule, (2) between the tendon of the biceps femoris muscle and fibular collateral ligament, (3) between the tendon of the popliteus muscle and fibular collateral ligament, and (4) between the lateral condyle of the femur and the popliteus muscle.

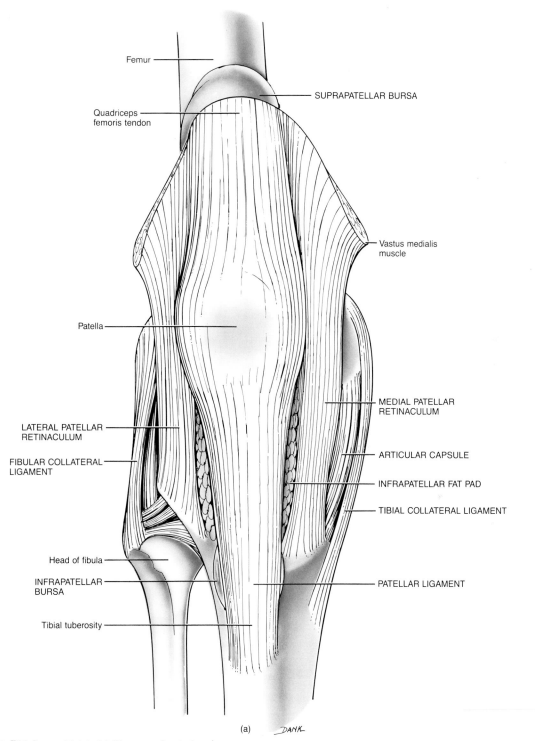

Femur

SUPRAPATELLAR BURSA

Quadriceps
femoris tendon

Vastus medialis
muscle

Patella

MEDIAL PATELLAR
RETINACULUM

LATERAL PATELLAR
RETINACULUM

ARTICULAR CAPSULE

FIBULAR COLLATERAL
LIGAMENT

INFRAPATELLAR FAT PAD

TIBIAL COLLATERAL LIGAMENT

Head of fibula

INFRAPATELLAR
BURSA

PATELLAR LIGAMENT

Tibial tuberosity

(a)   DANK

FIGURE 8-14 Tibiofemoral joint. (a) Diagram of anterior view.

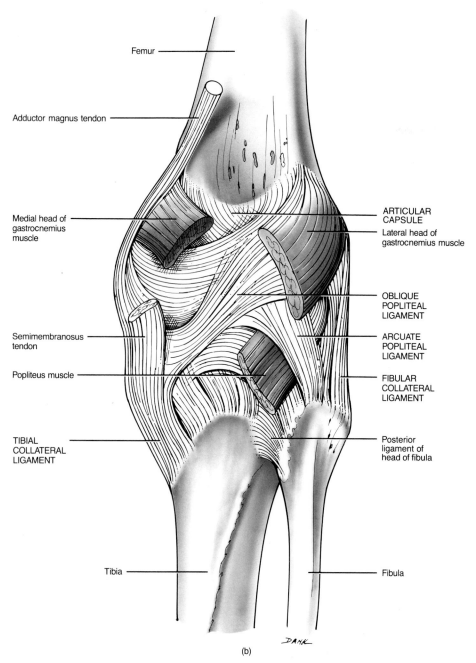

Femur

Adductor magnus tendon

Medial head of gastrocnemius muscle

ARTICULAR CAPSULE

Lateral head of gastrocnemius muscle

OBLIQUE POPLITEAL LIGAMENT

Semimembranosus tendon

ARCUATE POPLITEAL LIGAMENT

Popliteus muscle

FIBULAR COLLATERAL LIGAMENT

TIBIAL COLLATERAL LIGAMENT

Posterior ligament of head of fibula

Tibia

Fibula

(b)

FIGURE 8-14 (*Continued*) Tibiofemoral joint. (b) Diagram of posterior view.

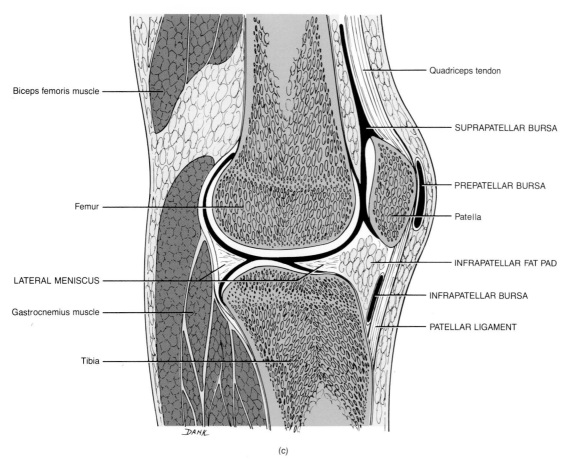

Quadriceps tendon

SUPRAPATELLAR BURSA

PREPATELLAR BURSA

Patella

INFRAPATELLAR FAT PAD

INFRAPATELLAR BURSA

PATELLAR LIGAMENT

Biceps femoris muscle

Femur

LATERAL MENISCUS

Gastrocnemius muscle

Tibia

DANK

(c)

FIGURE 8-14 (*Continued*) Tibiofemoral joint. (c) Diagram of sagittal section.

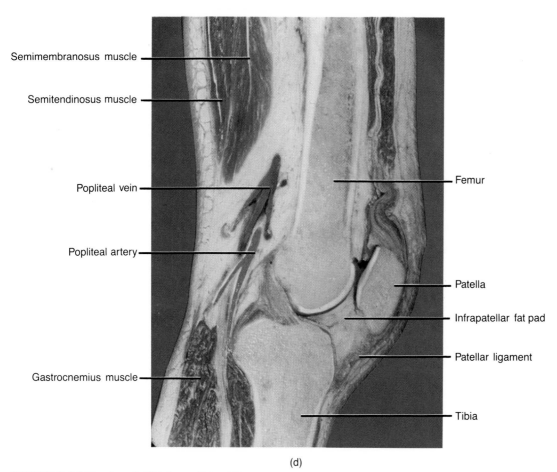

Semimembranosus muscle

Semitendinosus muscle

Popliteal vein

Popliteal artery

Gastrocnemius muscle

Femur

Patella

Infrapatellar fat pad

Patellar ligament

Tibia

(d)

FIGURE 8-14 (*Continued*) Tibiofemoral joint. (d) Photograph of sagittal section. (Courtesy of C. Yokochi and J. W. Rohen, *Photographic Anatomy of the Human Body,* 1st ed., 1969, IGAKU-SHOIN, Ltd., Tokyo, New York.)

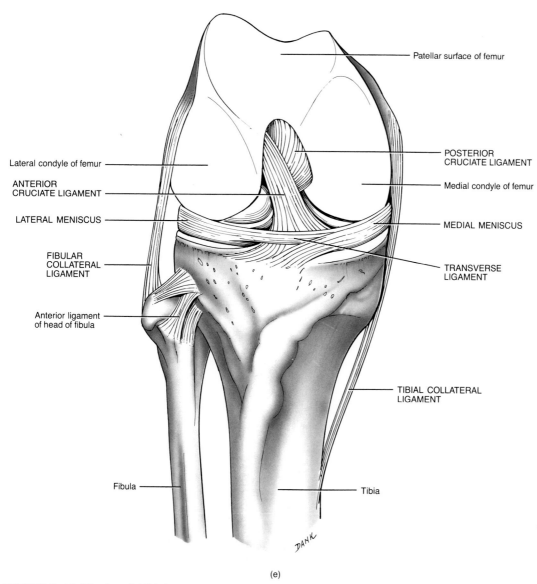

Patellar surface of femur

Lateral condyle of femur

ANTERIOR
CRUCIATE LIGAMENT

LATERAL MENISCUS

FIBULAR
COLLATERAL
LIGAMENT

Anterior ligament
of head of fibula

POSTERIOR
CRUCIATE LIGAMENT

Medial condyle of femur

MEDIAL MENISCUS

TRANSVERSE
LIGAMENT

TIBIAL COLLATERAL
LIGAMENT

Fibula

Tibia

(e)

FIGURE 8-14 (*Continued*) Tibiofemoral joint. (e) Diagram of anterior view (flexed).

EXHIBIT 8-11 TALOCRURAL (ANKLE) JOINTS (Figure 8-15)

DEFINITION   Joints formed between (1) the distal end of the tibia and its medial malleolus and the talus and (2) the lateral malleolus of the fibula and the talus.

TYPE OF JOINT   Both joints—synovial, hinge (ginglymus) type.

ANATOMICAL COMPONENTS

1. **Articular capsule.** Surrounds the joint and extends from the borders of the tibia and malleoli to the talus.
2. **Deltoid (medial) ligament.** Strong, triangular ligament that extends from the medial malleolus of the tibia to the navicular, calcaneus, and talus.
3. **Anterior talofibular ligament.** Extends from the anterior margin of the lateral malleolus of the fibula to the talus.
4. **Posterior talofibular ligament.** Extends from the posterior margin of the lateral malleolus of the fibula to the talus.
5. **Calcaneofibular ligament.** Extends from the apex of the lateral malleolus of the fibula to the talus. Together, the anterior talofibular, posterior talofibular, and calcaneofibular ligaments are referred to as the **lateral ligament.**

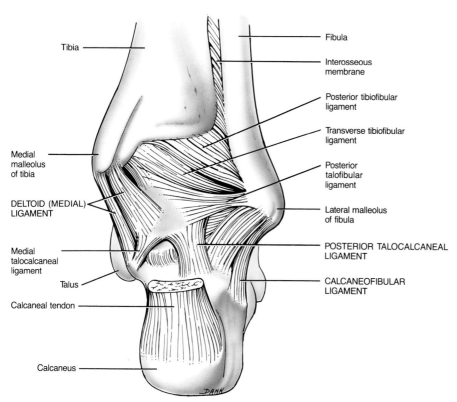

FIGURE 8-15 Talocrural joints. Posterior view.

APPLICATIONS TO HEALTH

RHEUMATISM

Rheumatism (*rheumat* = subject to flux) refers to any painful state of the supporting structures of the body, its bones, ligaments, joints, tendons, or muscles. Arthritis is a form of rheumatism in which the joints have become inflamed.

Rheumatism is not necessarily related to rheumatoid arthritis. About 25 percent of all Americans—most of them over age 40—have rheumatism that is diagnosed as some form of osteoarthritis.

ARTHRITIS

The term **arthritis** refers to at least 25 different diseases, the most common of which are rheumatoid arthritis, osteoarthritis, and gout. All these ailments are characterized by inflammation in one or more joints. Inflammation, pain, and stiffness may also be present in adjacent parts of the body, such as the muscles near the joint.

The causes of arthritis are unknown. In some cases, it has followed the stress of sprains, infections, and joint injury. Some researchers think that the cause is a bacterium or virus, whereas others suspect an allergy. Some believe the nervous system or hormones are involved, whereas others suspect a metabolic disorder. Still others believe that certain types of prolonged psychological stress, such as inhibited hostility, can upset homeostatic balance and bring on arthritic attacks.

RHEUMATOID ARTHRITIS

Rheumatoid (ROO-ma-toyd) **arthritis** is the most common inflammatory form of arthritis. It involves inflammation of the joint, swelling, pain, and a loss of function. Usually this form occurs bilaterally—if your left knee is affected, your right knee may also be affected, although usually not to the same degree.

The disease may afflict at any age, but is especially prevalent between 30 and 50 years old. The frequency of occurrence is approximately 57 percent in females and 43 percent in males. The symptoms in females are more severe than in males. One form, *juvenile rheumatoid arthritis,* afflicts children shortly after birth, and other forms can occur in later childhood.

The primary symptom of rheumatoid arthritis is inflammation of the synovial membrane. If it is completely untreated, the following sequential pathology may occur. The membrane thickens and synovial fluid accumulates. The resulting pressure causes pain and tenderness. The membrane then produces an abnormal tissue called *pannus,* which adheres to the surface of the articular cartilage. The pannus formation sometimes erodes the cartilage completely. When the cartilage is destroyed, fibrous tissue joins the exposed bone ends. The tissue ossifies and fuses the joint so that it is immovable—the ultimate crippling effect of rheumatoid arthritis (Figure 8-16). Most cases do not progress to this stage, but the range of motion of the joint is greatly inhibited by the severe inflammation and swelling.

Damaged joints may be surgically replaced, either partly or entirely, with artificial joints (Figure 8-17). The artificial parts are inserted after removal of the diseased portion of the articulating bone and its cartilage. The new metal or plastic joint is fixed in place with a special acrylic cement. When freshly mixed in the operating room, it hardens as strong as bone in minutes. These new parts function nearly as well as a normal joint—and much better than a diseased joint.

Joint surgery, especially hip and knee replacement, holds much promise for the future. The technique is being extended to other joints—wrist, elbow, shoulder, fingers, ankle—with much optimism.

OSTEOARTHRITIS

A degenerative joint disease far more common than rheumatoid arthritis, and usually less damaging, is **osteoarthritis** (os'-tē-ō-ar-THRĪ-tis). It apparently results from a combination of aging, irritation of the joints, and wear and abrasion.

Degenerative joint disease is a noninflammatory, progressive disorder of movable joints, particularly weight-bearing joints. It is characterized pathologically by the deterioration of articular cartilage and by formation of new bone in the subchondral areas and at the margins of the joint. The cartilage slowly degenerates, and as the bone ends become exposed, small bumps, or *spurs,* of new osseous tissue are deposited on them. These spurs decrease the space of the joint cavity and restrict joint movement. Unlike rheumatoid arthritis, osteoarthritis usually affects only the articular cartilage. The synovial membrane is rarely destroyed, and other tissues are unaffected.

GOUTY ARTHRITIS

Uric acid is a waste product produced during the metabolism of the nucleic acid purine. Normally, all the acid is quickly excreted in the urine. In fact, it gives urine its name. The person who suffers from *gout* either produces excessive amounts of uric acid or is not able to excrete normal amounts. The result is a buildup of uric acid in the blood. This excess acid then reacts with sodium to form a salt called sodium urate. Crystals of this salt are deposited in soft tissues. Typical sites are the kidneys and the cartilage of the ears and joints.

In **gouty** (GOW-tē) **arthritis,** sodium urate crystals are deposited in the soft tissues of the joints. The crystals

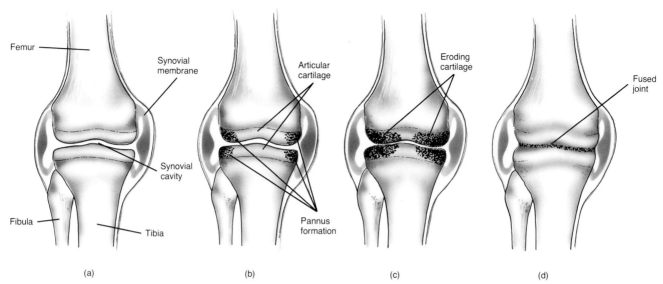

(a)   (b)   (c)   (d)

FIGURE 8-16 Rheumatoid arthritis. (a) Inflammation of the synovial membrane. (b) Early stage of pannus formation and erosion of articular cartilage. (c) Advanced stage of pannus formation and further erosion of articular cartilage. (d) Obliteration of joint cavity and fusion of articulating bones. (e) Anteroposterior projection of the hand and wrist showing changes characteristic of rheumatoid arthritis. The arrows indicate fusion of the bones resulting in obliteration of the joint cavities. (Courtesy of Lester W. Paul and John H. Juhl, *The Essentials of Roentgen Interpretation,* 3d ed., Harper & Row, Publishers, Inc., New York, 1972.)

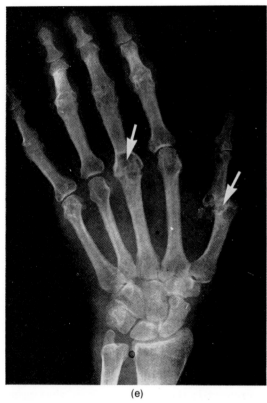

(e)

produces unusually large amounts of uric acid. Diet and environmental factors such as stress and climate are also suspected causes of gout.

Although other forms of arthritis cannot be treated with complete success, the treatment of gout with the use of various drugs has been quite effective. A chemical called colchicine has been utilized periodically since the sixth century to relieve the pain, swelling, and tissue destruction that occurs during attacks of gouty arthritis. This chemical is derived from the variety of crocus plant from which the spice saffron is obtained. Other drugs, which either inhibit uric acid production or assist in the elimination of excess uric acid by the kidneys, are used to prevent further attacks. The drug allopurinal is used for the treatment of gout because it prevents the formation of uric acid without interfering with purine synthesis.

SEPTIC ARTHRITIS

Septic arthritis is caused by the invasion of the synovial-lined joint space by an infectious agent. The invading organisms may be bacteria, viruses, or, rarely, fungi or even parasites. Other synovial-lined spaces such as bursae or tendon sheaths may be similarly infected, often in association with septic arthritis. One of the most common

irritate the cartilage, causing inflammation, swelling, and acute pain. Eventually, the crystals destroy all the joint tissues. If the disorder is not treated, the ends of the articulating bones fuse and the joint becomes immovable.

Gout occurs primarily in males of any age. It is believed to be the cause of 2 to 5 percent of all chronic joint diseases. Numerous studies indicate that gout is sometimes caused by an abnormal gene. As a result of this gene, the body manufactures purine by a mechanism that

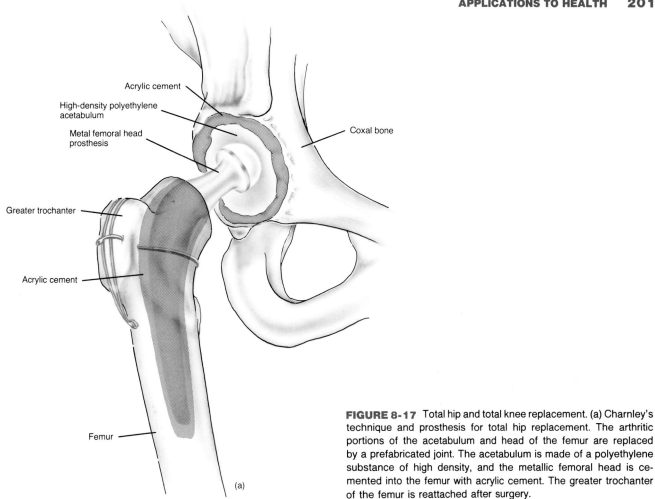

Acrylic cement

High-density polyethylene acetabulum

Metal femoral head prosthesis

Coxal bone

Greater trochanter

Acrylic cement

Femur

(a)

FIGURE 8-17 Total hip and total knee replacement. (a) Charnley's technique and prosthesis for total hip replacement. The arthritic portions of the acetabulum and head of the femur are replaced by a prefabricated joint. The acetabulum is made of a polyethylene substance of high density, and the metallic femoral head is cemented into the femur with acrylic cement. The greater trochanter of the femur is reattached after surgery.

forms of septic arthritis is *gonococcal arthritis. Pyogenic septic arthritis* is caused by invasion of the joints by pyogenic organisms, including staphylococci, streptococci, pneumococci, and pseudomonas. This type of arthritis is most commonly seen in drug addicts.

Various viral infections may cause acute septic arthritis. Perhaps the most common is infectious hepatitis and rubella the second most common. Other viral infections associated with arthritis are mumps, chickenpox, and the Coxsackie virus.

Most of these septic arthritic conditions are treated with antibiotics and drainage of the infected joints.

BURSITIS

An acute chronic inflammation of a bursa is called **bursitis.** The condition may be caused by trauma, by an acute or chronic infection (including syphilis and tuberculosis), or by rheumatoid arthritis. Repeated excessive friction often results in a bursitis with local inflammation and the accumulation of fluid. Bunions are frequently associated with a friction bursitis over the head of the first metatarsal bone. Symptoms include pain, swelling, tender-

ness, and the limitation of motion involving the inflamed bursa. The prepatellar or subcutaneous infrapatellar bursa may become inflamed in individuals who spend a great deal of time kneeling. This bursitis is usually called **"housemaid's knee"** or **"carpet layer's knee."**

Treatment includes rest of the afflicted area and cessation of activity. In some cases, a simple change of occupation may be all that is required in the nonspecific or traumatic forms of bursitis. Aspirin and other analgesic/anti-inflammatory drugs may be used. If this does not suffice, or when the bursitis is very acute, local injection of the bursa with corticosteroids, such as cortisone, is usually effective. When large calcium deposits are present, these may have to be removed.

Aspiration of the knee joint may be necessary to relieve pressure, to evacuate blood, or to obtain a fluid sample for laboratory studies. In this procedure, the needle is inserted somewhat proximal and lateral to the patella through the tendinous part of the vastus lateralis muscle and is directed toward the middle of the joint. Through the same route, the joint cavity may be anesthetized or irrigated, and steroids may be administered in the treatment of knee pathologies.

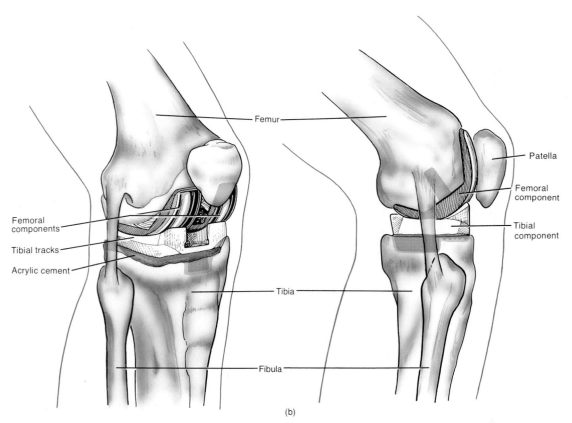

Femur

Patella

Femoral
component

Femoral
components

Tibial
component

Tibial tracks

Acrylic cement

Tibia

Fibula

(b)

FIGURE 8-17 (*Continued*) Total hip and total knee replacement. (b) Illustrated are two total knee replacement prostheses. In the technique on the left, called the polycentric type, there are two femoral and two tibial components. In the technique on the right, the Waldius type, there is one femoral component and one tibial component.

TENDINITIS

Tendinitis or **tenosynovitis** (ten'-ō-sin-ō-VĪ-tis) frequently occurs as inflammation involving the tendon sheaths and synovial membrane surrounding certain joints. The wrists, shoulders, elbows (tennis elbow), finger joints (trigger finger), ankles, and associated tendons are most often affected. The affected sheaths may become visibly swollen because of fluid accumulation or they may remain dry. Local tenderness is variable, and there may be disabling pain with movement of the body part. The condition often follows some form of trauma, strain, or excessive exercise.

Treatment is with analgesic/anti-inflammatory drugs and, if warranted, cortisone injections.

DISLOCATION

A **dislocation** or **luxation** (luks-Ā-shun) is the displacement of a bone from a joint with tearing of ligaments, tendons, and capsules. A partial or incomplete dislocation is called a **subluxation.** The most common dislocations are those involving a finger, thumb, or shoulder. Those of the mandible, elbow, knee, or hip are less common. Symptoms include loss of motion, temporary paralysis of the involved joint, pain, swelling, and occasionally shock. A dislocation is usually caused by a blow or fall, although unusual physical effort may lead to this condition.

A dislocation should be treated as a fracture when first aid is administered. The dislocation should be reduced by a surgeon as soon as possible. Traction, slight flexion, abduction, and rotation often reduce a dislocation. The affected joint is then immobilized to allow for healing of the torn components. In some cases, surgery may be necessary to stabilize the joint.

SPRAIN

A **sprain** is the forcible wrenching or twisting of a joint with partial rupture or other injury to its attachments without luxation. There may be damage to the associated blood vessels, muscles, tendons, ligaments, or nerves. A sprain is more serious than a **strain,** which is simply the overstretching of a muscle. Severe sprains may be so painful that the joint cannot be moved. There is considerable swelling and pain, with reddish to blue discoloration due to hemorrhage from ruptured blood vessels. The ankle joint is most often sprained; the low back area is another frequent location for sprains.

Treatment includes hot or cold compresses, immediate rest, immobilization, and elevation of the joint.

KEY MEDICAL TERMS ASSOCIATED WITH JOINTS

Ankylosis (*agkyle* = stiff joint, *osis* = condition) Severe or complete loss of movement at a joint.

Arthralgia (*algia* = pain) Pain in a joint.

Arthrosis (*arth* = joint) Refers to an articulation; also a disease of a joint.

Bursectomy (*ectomy* = removal of) Removal of a bursa.

Chondritis (*chondro* = cartilage) Inflammation of a cartilage.

Deterioration (*deterior* = worse or poorer) The process or state of growing worse; disintegration or wearing away.

Detritus (*deterere* = to rub away) Particulate matter produced by or remaining after the wearing away

or disintegration of a substance or tissue; scales, crusts, and loosened skin.

Insidious Hidden, not apparent—as a disease that does not exhibit distinct symptoms of its arrival.

Pyogenic (*pyo* = pus) Producing suppuration (pus).

Reduce To replace in normal position—as to reduce a fracture.

Rheumatology (*rheumat* = subject to flux) The medical specialty devoted to arthritis.

Septic (*sept* = poison) Indicating the presence of microorganisms or their toxins.

Synovitis Inflammation of a synovial membrane in a joint.

STUDY OUTLINE

Classification
1. A joint or articulation is a point of contact between two or more bones.
2. Functional classification of joints is based on the degree of movement permitted. Joints may be synarthroses, amphiarthroses, or diarthroses.
3. Structural classification is based on the presence of a joint cavity and type of connecting tissue. Structurally, joints are classified as fibrous, cartilaginous, or synovial.

Fibrous Joints
1. Bones held by fibrous connective tissue, with no joint cavity, are fibrous joints.
2. These joints include immovable sutures (found in the skull), slightly movable syndesmoses (such as the tibiofibular articulation), and immovable gomphoses (roots of teeth in alveolar processes).

Cartilaginous Joints
1. Bones held together by cartilage, with no joint cavity, are cartilaginous joints.
2. These joints include immovable synchondroses united by hyaline cartilage (temporary cartilage between diaphysis and epiphyses) and partially movable symphyses united by fibrocartilage (the symphysis pubis).

Synovial Joints
1. Synovial joints contain a joint cavity, articular cartilage, and a synovial membrane; some also contain ligaments, articular discs, and bursae.
2. All synovial joints are freely movable.
3. Movements at synovial joints are limited by the apposition of soft parts, tension of ligaments, and muscle tension.
4. Types of movements at synovial joints include gliding movements, angular movements, rotation, circumduction, inversion and eversion, protraction and retraction, supination and pronation, and elevation and depression.
5. Types of synovial joints include gliding joints (wrist bones), hinge joints (elbow), pivot joints (radioulnar), ellipsoidal joints (radiocarpal), saddle joints (carpometacarpal), and ball-and-socket joints (shoulder and hip).
6. A joint may be described according to the number of planes of movement it allows as monaxial, biaxial, or triaxial.

Selected Articulations of the Body
1. Temporomandibular is between the mandible and temporal bone.
2. Atlantooccipitals are between the atlas and occipital bones.
3. Intervertebrals are between vertebral bodies and between vertebral arches.
4. Lumbosacral is between the fifth lumbar vertebra and the first sacral vertebra of the sacrum.
5. Humeroscapular is between the humerus and scapula.
6. Elbow is between the humerus and ulna and between the humerus and radius.
7. Radiocarpal is between the radius and scaphoid, lunate, and triquetral carpal bones.
8. Coxal is between the femur and coxal bone.
9. Tibiofemoral is between the patella and femur and between the femur and tibia.

10. Talocrurals are between the tibia and talus and between the fibula and talus.

Applications to Health

1. Rheumatism is a painful state of supporting body structures such as bones, ligaments, tendons, joints, and muscles.
2. Arthritis refers to several disorders characterized by inflammation of joints, often accompanied by stiffness of adjacent structures.
3. Rheumatoid arthritis refers to inflammation of a joint accompanied by pain, swelling, and loss of function.
4. Osteoarthritis is a degenerative joint disease characterized by deterioration of articular cartilage and spur formation.

5. Gouty arthritis is a condition in which sodium urate crystals are deposited in the soft tissues of joints and eventually destroy the tissues.
6. Septic arthritis is the invasion of a joint cavity by infectious microbes.
7. Bursitis is an acute or chronic inflammation of bursae.
8. Tendinitis is an inflammation of tendon sheaths and synovial membranes.
9. A dislocation, or luxation, is a displacement of a bone from its joint; a partial dislocation is called subluxation.
10. A sprain is the forcible wrenching or twisting of a joint with partial rupture to its attachments without dislocation, while a strain is the stretching of a muscle.

REVIEW QUESTIONS

1. Define an articulation. What factors determine the degree of movement at joints?
2. Distinguish among the three kinds of joints on the basis of structure and function. List the subtypes. Be sure to include degree of movement and specific examples.
3. Explain the components of a synovial joint. Indicate the relationship of ligaments and tendons to the strength of the joint and restrictions on movement.
4. Explain how the articulating bones in a synovial joint are held together.
5. What is an accessory ligament? Define the two principal types.
6. What is an articular disc? Why are they important?
7. What are bursae? What is their function?
8. Define the following movements: gliding, angular, rotation, circumduction, and special. Name a joint where each occurs.
9. Have another person assume the anatomical position and execute for you each of the movements at joints discussed in the text. Reverse roles, and see if you can execute the same movements.

10. Contrast monaxial, biaxial, and triaxial planes of movement. Give examples of each, and name a joint at which each occurs.
11. For each joint of the body discussed in Exhibits 8-2 through 8-11, be sure that you can define the bones that form the joint, identify the joint by type, and list the anatomical components of the joint.
12. Define arthritis. What are some suspected causes of arthritis?
13. What is rheumatism?
14. Distinguish between rheumatioid arthritis, osteoarthritis, and gouty arthritis with respect to causes and symptoms.
15. Define bursitis. How is it caused?
16. What is tendinitis? What are some of its causes?
17. Define dislocation. What are the symptoms of dislocation? How is a dislocation caused?
18. Distinguish between a sprain and a strain.
19. Refer to the glossary of key medical terms at the end of the chapter and be sure that you can define each term.

9

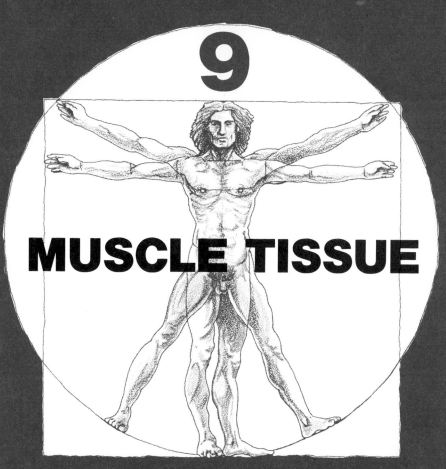

MUSCLE TISSUE

STUDENT OBJECTIVES

- List the characteristics and functions of muscle tissue.

- Compare the location, microscopic appearance, nervous control, and functions of the three kinds of muscle tissue.

- Define fascia, epimysium, perimysium, endomysium, tendons, and aponeuroses, and list their modes of attachment to muscles.

- Describe the relationship of blood vessels and nerves to skeletal muscles.

- Identify the histological characteristics of skeletal muscle tissue.

- Describe the importance of the motor unit.

- Define the all-or-none principle of muscular contraction.

- Compare fast (white) muscle with slow (red) muscle.

- Contrast cardiac muscle tissue with smooth muscle tissue.

- Define such common muscular disorders as fibrosis, fibrositis, "charleyhorse," muscular dystrophy, and myasthenia gravis.

- Compare spasms, cramps, convulsions, tetany, and fibrillation as abnormal muscular contractions.

- Define key medical terms associated with the muscular system.

Although bones and joints provide leverage and form the framework of the body, they are not capable of moving the body by themselves. Motion is an essential body function that results from the contraction and relaxation of muscles. Muscle tissue constitutes about 40 to 50 percent of the total body weight and is composed of highly specialized cells with four striking characteristics.

CHARACTERISTICS

Muscle tissue has four principal characteristics that assume key roles in maintaining homeostasis. (1) **Excitability** is the ability of muscle tissue to receive and respond to stimuli. A stimulus is a change in the internal or external environment strong enough to initiate a nerve impulse. (2) **Contractility** is the ability to shorten and thicken, or contract, when a sufficient stimulus is received. (3) **Extensibility** is the ability of muscle tissue to stretch. Many skeletal muscles are arranged in opposing pairs. While one is contracting, the other is undergoing extension. (4) **Elasticity** is the ability of muscle to return to its original shape after contraction or extension.

FUNCTIONS

Through contraction, muscle performs three important functions:

1. Motion
2. Maintenance of posture
3. Heat production

Motion is obvious in movements involving the whole body, such as walking and running, and in localized movements, such as grasping a pencil or nodding the head. All of these movements rely on the integrated functioning of the bones, joints, and muscles attached to the bones. Less noticeable kinds of motion produced by muscles are the beating of the heart, the churning of food in the stomach, the pushing of food through the intestines, the contraction of the gallbladder to release bile, and the contraction of the urinary bladder to expel urine.

In addition to the movement function, muscle tissue also enables the body to maintain posture. The contraction of skeletal muscles holds the body in stationary positions, such as standing and sitting.

The third function of muscle tissue is heat production. Skeletal muscle contractions produce heat and are thereby important in maintaining normal body temperature.

KINDS

Three kinds of muscle tissue are recognized: **skeletal, visceral,** and **cardiac.** These three types are further categorized by location, microscopic structure, and nervous control. *Skeletal muscle* tissue, which is named for its location, is attached to bones. It is *striated* muscle tissue because striations, or bandlike structures, are visible when the tissue is examined under a microscope. It is a *voluntary* muscle tissue because it can be made to contract by conscious control. *Visceral muscle* is located in the walls of hollow internal structures, such as blood vessels, the stomach, and the intestines. It is *smooth,* or *nonstriated,* that is, it lacks the bandlike structures. It is *involuntary* muscle tissue because its contraction is usually not under conscious control. *Cardiac muscle* tissue forms the walls of the heart. It is striated and involuntary. Thus all muscle tissues are classified in the following ways: (1) skeletal, striated, voluntary muscle, (2) cardiac, striated, involuntary muscle, and (3) visceral, smooth, involuntary muscle.

SKELETAL MUSCLE TISSUE

To understand the fundamental mechanisms of muscle movement, you will need some knowledge of its connective tissue components, its nerve and blood supply, and its histology.

FASCIA

The term **fascia** (FASH-ē-a) is applied to a sheet or broad band of fibrous connective tissue beneath the skin or around muscles and other organs of the body. Fasciae may be divided into three types: superficial, deep, and subserous.

The **superficial fascia,** or **subcutaneous layer,** is immediately deep to the skin. It covers the entire body and varies in thickness in different regions. On the back or dorsum of the hand it is quite thin, whereas over the inferior abdominal wall it is thick. The superficial fascia is composed of adipose tissue and loose connective tissue. The outer layer usually contains fat and varies considerably in thickness, while the inner layer is thin and elastic. Between the two layers are found arteries, veins, lymphatics, nerves, the mammary glands, and the facial muscles. Superficial fascia serves a number of important functions. (1) It serves as a storehouse for water and particularly for fat. Much of the fat of an overweight person is in the superficial fascia. (2) It forms a layer of insulation protecting the body from loss of heat. (3) It provides mechanical protection from blows. (4) It provides a pathway for nerves and vessels.

The **deep fascia** is by far the most extensive of the three types. It is a dense connective tissue. Unlike the superficial fascia, it does not contain fat. The deep fascia lines the body wall and extremities and holds muscles together, separating them into functioning groups. Functionally, deep fascia allows free movement of muscles, carries nerves and blood vessels, fills spaces between muscles, and sometimes provides the origin for muscles.

The **subserous fascia** is located between the internal investing layer of deep fascia and a serous membrane.

CONNECTIVE TISSUE COMPONENTS

Skeletal muscles are further protected, strengthened, and attached to other structures by several connective tissue components (Figure 9-1). The entire muscle is usually wrapped with a substantial quantity of fibrous connective tissue called the **epimysium** (ep'-i-MĪZ-ē-um). The epimysium is an extension of deep fascia. When the muscle is cut in cross section, invaginations of the epimysium are seen to divide the muscle into bundles of fibers (cells) called **fasciculi** (fa-SIK-yoo-lī) or **fascicles** (FAS-i-kuls). These invaginations of the epimysium are called the **perimysium** (per'-i-MĪZ-ē-um). Perimysium, like epimysium, is an extension of deep fascia. In turn, invaginations of the perimysium, called **endomysium** (en'-dō-MĪZ-ē-um), penetrate into the interior of each fascicle and separate each muscle cell. Endomysium is also an extension of deep fascia.

The epimysium, perimysium, and endomysium are all continuous with the connective tissue that attaches the muscle to another structure, such as bone or other muscle. All three elements may be extended beyond the muscle cells as a **tendon**—a cord of connective tissue that attaches a muscle to the periosteum of a bone. In other cases, the connective tissue elements may extend as a broad, flat tendon called an **aponeurosis**. This structure also attaches to the coverings of a bone or another muscle. When a muscle contracts, the tendon and its corresponding bone or muscle are pulled toward the contracting muscle. In this way skeletal muscles produce movement. Certain tendons, especially those of the wrist and ankle, are enclosed by tubes of fibrous connective tissue called **tendon sheaths.** They are lined by a synovial membrane that permits the tendon to slide easily within the sheath. The sheaths also prevent the tendons from slipping out of place.

NERVE AND BLOOD SUPPLY

Skeletal muscles are well supplied with nerves and blood vessels. This innervation and vascularization is directly related to contraction, the chief characteristic of muscle.

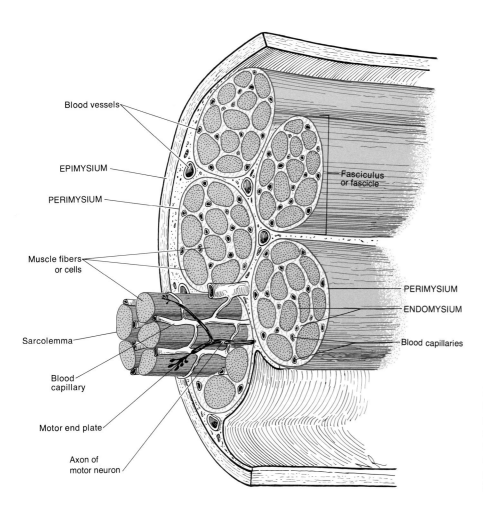

FIGURE 9-1 Relationships of connective tissue to skeletal muscle. Shown is a cross section and longitudinal section of a skeletal muscle indicating the relative positions of the epimysium, perimysium and endomysium. Compare this figure with the photomicrographs in Figure 9-3a, b.

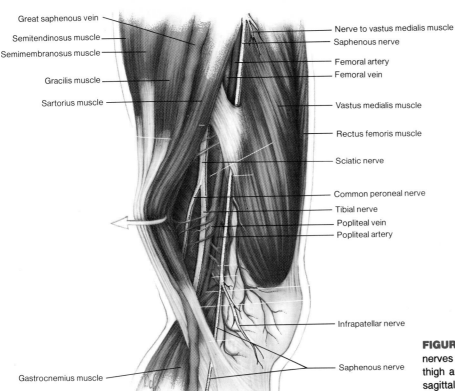

Great saphenous vein
Semitendinosus muscle
Semimembranosus muscle
Gracilis muscle
Sartorius muscle

Nerve to vastus medialis muscle
Saphenous nerve
Femoral artery
Femoral vein
Vastus medialis muscle
Rectus femoris muscle
Sciatic nerve
Common peroneal nerve
Tibial nerve
Popliteal vein
Popliteal artery

Infrapatellar nerve

Gastrocnemius muscle
Saphenous nerve

(a)

FIGURE 9-2 Relationship of blood vessels and nerves to skeletal muscles. (a) Diagram of the left thigh and knee in medial view. (b) Photograph of sagittal section. (Courtesy of C. Yokochi and J. W. Rohen, *Photographic Anatomy of the Human Body,* 1st ed., 1969, IGAKU-SHOIN, Ltd., Tokyo, New York.)

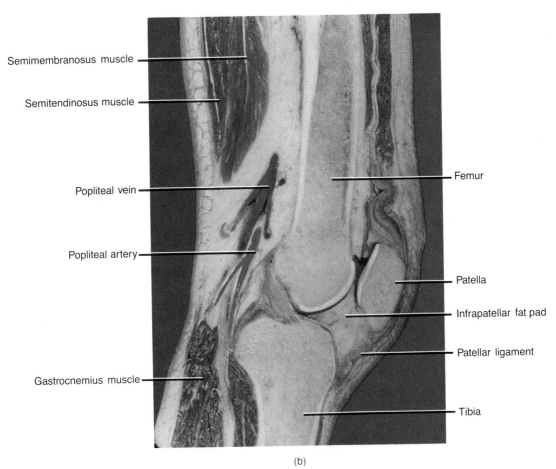

Semimembranosus muscle
Semitendinosus muscle
Popliteal vein
Popliteal artery
Gastrocnemius muscle

Femur
Patella
Infrapatellar fat pad
Patellar ligament
Tibia

(b)

For a skeletal muscle cell to contract, it must first be stimulated by an impulse from a nerve cell. Muscle contraction also requires a good deal of energy—meaning large amounts of nutrients and oxygen. Moreover, the waste products of these energy-producing reactions must be eliminated. Thus muscle action depends on the blood supply.

Generally, an artery and one or two veins accompany each nerve that penetrates a skeletal muscle. The larger branches of the blood vessel accompany the nerve branches through the connective tissue of the muscle (Figure 9-2). Microscopic blood vessels called capillaries are arranged in the endomysium. Each muscle cell is thus in close contact with one or more capillaries. Each skeletal muscle cell usually makes contact with a portion of a nerve cell.

HISTOLOGY

Muscle tissue attached to bones is generally termed **skeletal muscle tissue.** When a typical skeletal muscle is teased apart and viewed microscopically, it can be seen to consist of many elongated, cylindrical cells called **muscle fibers** (Figure 9-3a, b). These fibers lie parallel to one another and range from 10 to 100 μm in diameter. Some fibers may reach lengths of 30 cm (12 inches) or more. Each muscle fiber is surrounded by a plasma membrane called the **sarcolemma** (*sarco* = flesh; *lemma* = sheath). The sarcolemma contains a quantity of cytoplasm called **sarcoplasm.** Within the sarcoplasm of a muscle fiber and lying close to the sarcolemma are many nuclei and a number of mitochondria. Also within a muscle fiber is the **sarcoplasmic reticulum** (sar'-kō-PLAZ-mik rē-TIK-yoo-lum), a network of membrane-enclosed tubules comparable to smooth endoplasmic reticulum (Figure 9-3c). Running transversely through the fiber and perpendicularly to the sarcoplasmic reticulum are **T tubules.** The tubules open to the outside of the fiber. A **triad** consists of a T tubule and the segments of sarcoplasmic reticulum on either side.

A highly magnified view of skeletal muscle fibers reveals threadlike structures, about 1 or 2 μm in diameter, called **myofibrils** (Figure 9-3c, d, e). The prefix *myo* means muscle. The myofibrils run longitudinally through the muscle fiber and consist of two kinds of even smaller structures called **myofilaments.** The **thin myofilaments** are about 6 nm in diameter. The **thick myofilaments** are about 16 nm in diameter.

The myofilaments of a myofibril do not extend the entire length of a muscle fiber—they are stacked in compartments called **sarcomeres.** Sarcomeres are separated from one another by narrow zones of dense material called **Z lines.** Each sarcomere is about 2.6 μm long. In a relaxed muscle fiber, that is, one that is not contracting, the thin and thick myofilaments overlap and form a dark, dense band called the **anisotropic band,** or **A band** (Figure 9-3d). Each A band is about 1.6 μm long. A light-colored, less dense area called the **isotropic band,** or **I band,** is composed of thin myofilaments only. The I bands are about 1 μm long. This combination of alternating dark and light bands gives the muscle fiber its striped appearance. A narrow **H zone** contains thick myofilaments only. The H zone is about 0.5 μm long.

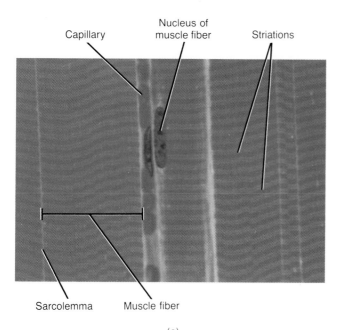

Capillary Nucleus of muscle fiber Striations

Sarcolemma Muscle fiber

(a)

FIGURE 9-3 Histology of skeletal muscle tissue. (a) Photomicrograph of several muscle fibers in longitudinal section at a magnification of 800×. (© 1983 by Michael H. Ross. Used by permission.)

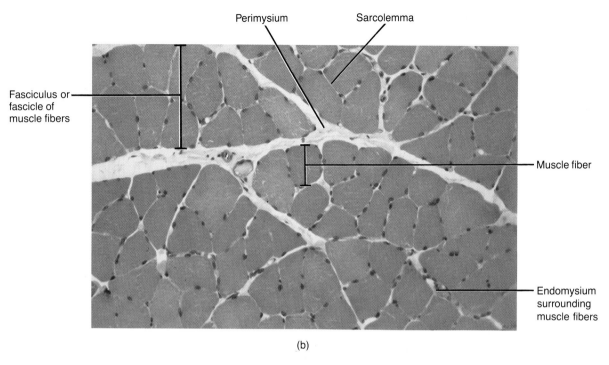

(b)

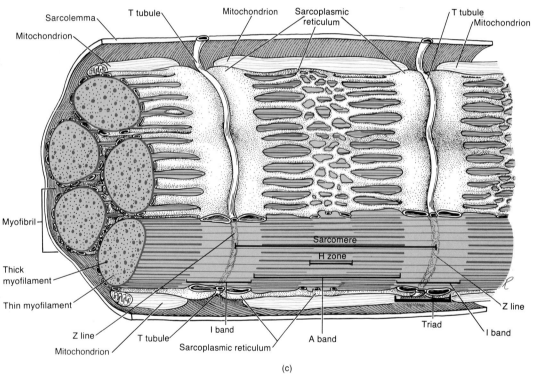

(c)

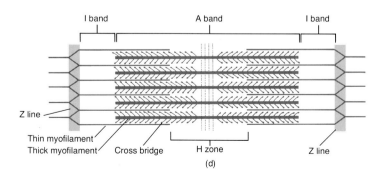

(d)

FIGURE 9-3 (*Continued*) Histology of skeletal muscle tissue. (b) Photomicrograph of several muscle fibers in cross section at a magnification of 500×. (© 1983 by Michael H. Ross. Used by permission.) (c) Enlarged aspect of several myofibrils of a muscle fiber based on an electron micrograph. (d) Enlarged aspect of a sarcomere showing thin and thick myofilaments.

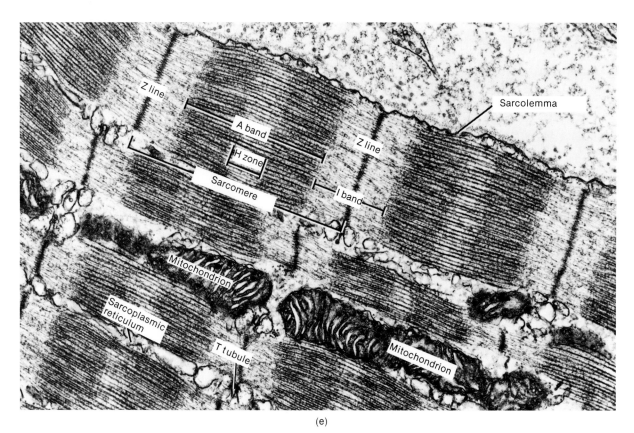

(e)

FIGURE 9-3 (*Continued*) Histology of skeletal muscle tissue. (e) Electron micrograph of several sarcomeres at a magnification of 35,000×. (Courtesy of D. E. Kelly, from *Introduction to the Musculoskeletal System,* by Cornelius Rosse and D. Kay Clawson, Harper & Row, Publishers, Inc., New York, 1970.)

The thin myofilaments are composed mostly of the protein **actin.** The actin molecules are arranged in a double-stranded coil that gives the thin myofilaments their characteristic shape (Figure 9-4a). Besides actin, the thin myofilaments contain two other protein molecules, **tropomyosin** and **troponin.** Together they are referred to as a **tropomyosin-troponin complex.**

The thick myofilaments are composed mostly of the protein **myosin.** A myosin molecule is shaped like a rod with a round head. The rods form the long axis of the thick myofilaments, and the heads form projections called **cross bridges** (Figure 9-4b). The cross bridges are arranged in pairs and seem to spiral around the main axis. The relationship between thin and thick myofilaments is shown in Figure 9-4c.

CONTRACTION

Sliding-Filament Theory

During muscle contraction, the thin myofilaments slide inward toward the H zone. The sarcomere shortens, but the lengths of the thin and thick myofilaments do not change. The cross bridges of the thick myofilaments connect with portions of actin of the thin myofilaments. The myosin cross bridges move like the oars of a boat on the surface of the thin myofilaments, and the thin and thick myofilaments slide past each other (Figure 9-5). As the thin myofilaments move past the thick myofilaments, the width of the H zone gets smaller and may even disappear when the thin myofilaments meet at the center of the sarcomere (Figure 9-6). In fact, the cross bridges may pull the thin myofilaments of each sarcomere so far inward that their ends overlap. As the thin myofilaments slide inward, the Z lines are drawn toward the A band and the sarcomere is shortened. The sliding of myofilaments and shortening of sarcomeres causes the shortening of the muscle fibers. All these events associated with the movement of myofilaments are known as the **sliding-filament theory** of muscle contraction.

Motor Unit

For a skeletal muscle fiber to contract, a stimulus must be applied to it. Such a stimulus is normally transmitted by nerve cells called **neurons.** A neuron has a threadlike process called a fiber, or axon, that may run 91 cm (3 ft) or more to a muscle. A bundle of such fibers from many different neurons composes a nerve. A neuron that transmits a stimulus to muscle tissue is called a **motor neuron.**

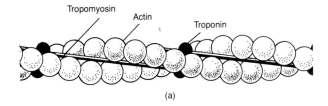

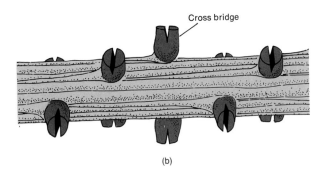

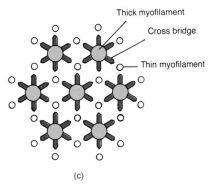

FIGURE 9-4 Detailed structure of myofilaments. (a) Thin myofilament. (b) Thick myofilament. (c) Cross section of several thin and thick myofilaments showing the arrangement of cross bridges. Note that each thick myofilament is surrounded by six thin myofilaments. (Parts (a) and (b) have been redrawn by permission from Arthur W. Ham, *Histology,* 7th ed., Philadelphia: J. B. Lippincott Company, 1974.)

Upon entering a skeletal muscle, the axon of the motor neuron branches into fine endings that come into close approximation at grooves on the muscle membrane. The portion of the muscle membrane directly under the end of the axon is called a **motor end plate** (Figure 9-7). The area of contact between neuron and muscle fiber is called a **neuromuscular junction,** or **myoneural junction.** When a nerve impulse reaches a motor end plate, small vesicles in the terminal branches of the nerve fiber release a chemical called *acetylcholine* (as'-ē-til-KŌ-lēn) or *ACh.* The ACh transmits the nerve impulse from the neuron, across the myoneural junction, to the muscle fiber, thus initiating contraction. Muscle contraction also requires calcium ions and energy (ATP).

A motor neuron, together with all the muscle cells it stimulates, is referred to as a **motor unit.** A single motor neuron may innervate about 150 muscle fibers, depending on the region of the body. This means that stimulation of one neuron will tend to cause the simultaneous contraction of about 150 muscle fibers. In addition, all the muscle fibers of a motor unit that are sufficiently stimulated will contract and relax together. Muscles that control precise movements, such as the eye muscles, have fewer than 10 muscle fibers to each motor unit. Muscles of the body that are responsible for gross movements may have as many as 500 muscle fibers in each motor unit.

All-or-None Principle

According to the **all-or-none principle,** individual muscle fibers of a motor unit will contract to their fullest extent or will not contract at all, providing conditions remain constant. In other words, *muscle fibers* do not partly contract. The principle does not imply that the strength of the contraction is the same every time the fiber is stimulated. The strength of contraction may be decreased by fatigue, lack of nutrients, or lack of oxygen. The weakest stimulus from a neuron that can initiate a contraction is called a **threshold** or **liminal stimulus.** A stimulus of lesser intensity, or one that cannot initiate contraction, is referred to as a **subthreshold** or **subliminal stimulus.** Each muscle fiber in a motor unit has its own threshold level.

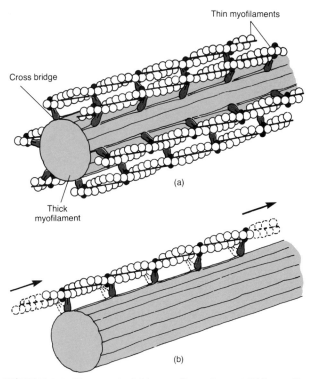

FIGURE 9-5 Movement of thin myofilaments past thick myofilaments. (a) Attachment of myosin cross bridges to actin of thin myofilaments. (b) Mechanism of movement of myosin cross bridges resulting in sliding of thin myofilaments toward H zone.

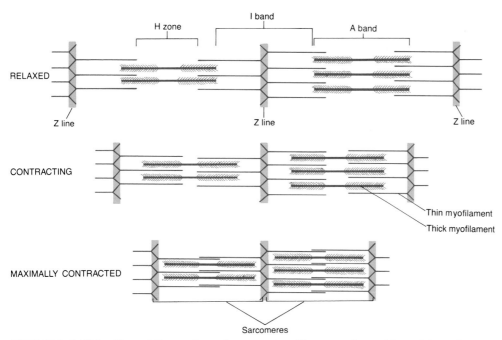

FIGURE 9-6 Sliding-filament theory of muscle contraction. Shown are the positions of the various parts of two sarcomeres in relaxed, contracting, and maximally contracted states. Note the movement of the thin myofilaments and the relative size of the H zone.

MUSCLE TONE

A muscle may be in a state of partial contraction even though contraction of the muscle fibers is always complete. At any given time, some cells in a muscle are contracted while others are relaxed. This contraction tightens a muscle, but there may not be enough fibers contracting at that time to produce movement. The same group of motor units does not function continuously. Instead, there is an asynchronous firing of alternating motor units that relieve one another so smoothly that the contraction can be sustained for long periods of time.

A sustained partial contraction of portions of a skeletal muscle in response to stretch receptors results in **muscle tone.** Tone is essential for maintaining posture. For example, when the muscles in the back of the neck are in tonic contraction, they keep the head in the anatomical position and prevent it from slumping forward onto the chest, but they do not apply enough force to pull the head back into hyperextension.

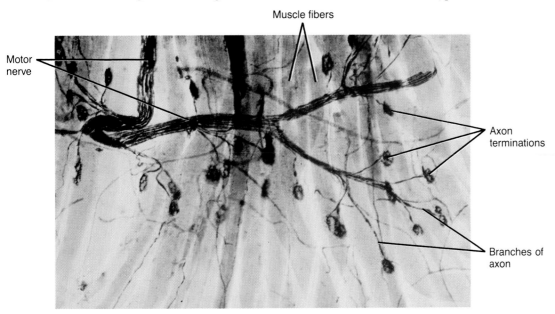

FIGURE 9-7 Motor end plate. (a) Photomicrograph at a magnification of 400×. (© 1983 by Michael H. Ross. Used by permission.)

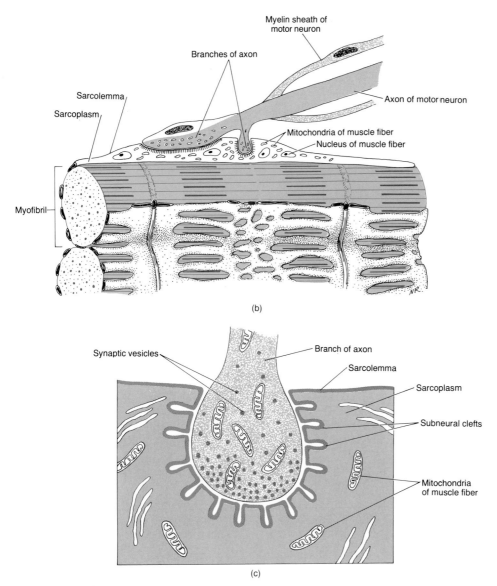

(b)

(c)

FIGURE 9-7 (*Continued*) Motor end plate. (b) Diagram based on a photomicrograph. (c) Enlarged aspect based on an electron micrograph.

CLINICAL APPLICATION

The term **flaccid** (FLAK-sid) is applied to muscles with less than normal tone. Such a loss of tone may be the result of damage or disease of the nerve that conducts a constant flow of impulses to the muscle. If the muscle does not receive impulses for an extended period of time, it may progress from flaccidity to **atrophy** (AT-rō-fē), which is a state of wasting away. Muscles may also become flaccid and atrophied if they are not used. Bedridden individuals and people with casts may experience atrophy because the flow of impulses to the inactive muscle is greatly reduced. If the nerve supply to a muscle is cut, it will undergo complete atrophy. In about six months to two years,

the muscle will be one-quarter its original size and the muscle fibers will be replaced by fibrous tissue. The transition to fibrous tissue, when complete, cannot be reversed.

Muscular hypertrophy (hī-PER-trō-fē) is the reverse of atrophy. It refers to an increase in the diameters of muscle fibers due to the production of more myofibrils, mitochondria, sarcoplasmic reticulum, and nutrients. Hypertrophic muscles are capable of more forceful contractions. Weak muscular activity does not produce significant hypertrophy. It results from very forceful muscular activity or repetitive muscular activity at moderate levels.

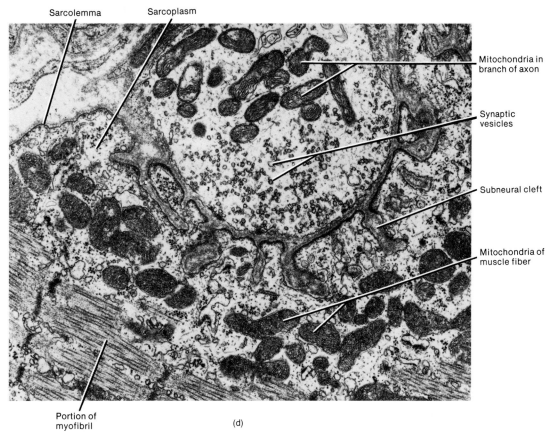

Sarcolemma   Sarcoplasm

Mitochondria in branch of axon

Synaptic vesicles

Subneural cleft

Mitochondria of muscle fiber

Portion of myofibril   (d)

FIGURE 9-7 (*Continued*) Motor end plate. (d) Electron micrograph at a magnification of 30,000×. (Courtesy of Cornelius Rosse and D. Kay Clawson, from *Introduction to the Musculoskeletal System,* Harper & Row, Publishers, Inc., New York, 1970.)

FAST (WHITE) AND SLOW (RED) MUSCLE

The duration of contraction of various muscles of the body varies with the functions of the muscles. For example, the duration of contraction of eye muscles is less than $\frac{1}{100}$ of a second, whereas the duration of contraction of the gastrocnemius muscle at the back of the leg is about $\frac{1}{30}$ of a second. The differences are correlated with the roles of the muscles. Eye movements must be extremely rapid in order to fix the eyes on different objects very quickly. Movements of the legs during walking or running need not be as rapid as eye movements. Usually, **fast muscles,** like those of the eye, have a more extensive sarcoplasmic reticulum for the rapid release and uptake of calcium ions needed for rapid contractions. Generally, **slow muscles,** like the gastrocnemius, are more aerobic; have smaller fibers, more blood capillaries, and more mitochondria; and have a large amount of myoglobin in the sarcoplasm. **Myoglobin** is a substance similar to the hemoglobin in red blood cells that can store oxygen until needed by the mitochondria. Since slow muscle has a reddish tint due to myoglobin and the large amount of red blood cells in capillaries, it is also known as **red muscle.** Fast muscle lacks myoglobin and has fewer capillaries and is thus also known as **white muscle.**

CARDIAC MUSCLE TISSUE

The principal constituent of the heart wall is **cardiac muscle tissue.** It has a striated appearance similar to skeletal muscle tissue, but it is involuntary. The cells of cardiac muscle tissue are roughly quadrangular and have only a single nucleus (Figure 9-8). The individual fibers are covered by a thin, poorly defined sarcolemma, and the internal myofibrils produce the characteristic striations. Cardiac muscle cells have the same basic arrangement of actin, myosin, and sarcoplasmic reticulum that is found in skeletal muscle cells. In addition, they contain a system of transverse tubules similar to the T tubules of skeletal muscle. The nuclei in cardiac cells, however, are centrally located compared to the peripheral location of nuclei in skeletal muscle.

While the groups of skeletal muscle fibers are arranged in a parallel fashion, those of cardiac muscle branch freely with other fibers to form two separate networks. The muscular walls and septum of the upper chambers of the heart (atria) compose one network. The muscular walls and septum of the lower chambers of the heart (ventricles) compose the other network. When a single fiber of either network is stimulated, all the fibers in the network become stimulated as well. Thus each network

Intercalated discs   Muscle fibers

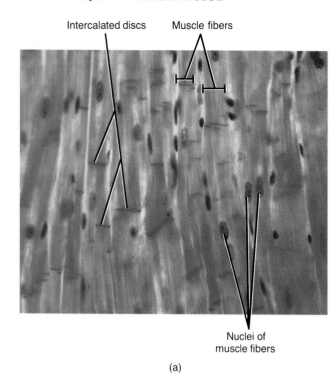

Nuclei of
muscle fibers

(a)

Endomysium surrounding
muscle fibers   Capillaries

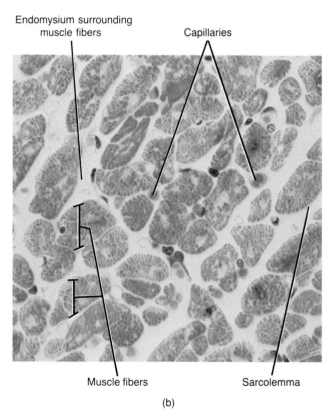

Muscle fibers   Sarcolemma

(b)

FIGURE 9-8 Histology of cardiac muscle tissue. (a) Photomicrograph of several muscle fibers in longitudinal section at a magnification of 400×. (© 1983 by Michael H. Ross. Used by permission.) (b) Photomicrograph of several muscle fibers in cross section at a magnification of 640×. (© 1983 by Michael H. Ross. Used by permission.)

contracts as a functional unit. The fibers of each network were once thought to be fused together into a multinucleated mass called a syncytium. But it is now known that each fiber in a network is separated from the next fiber by an irregular transverse thickening of the sarcolemma called an **intercalated** (in-TER-ka-lāt-ed) **disc.** These discs strengthen the cardiac muscle tissue and aid in impulse conduction.

Under normal conditions, cardiac muscle tissue contracts and relaxes rapidly, continuously, and rhythmically about 72 times a minute without stopping while a person

is at rest. This is a major physiological difference between cardiac and skeletal muscle tissue. Another difference is the source of stimulation. Skeletal muscle tissue ordinarily contracts only when stimulated by a nerve impulse. In contrast, cardiac muscle tissue can contract without nerve stimulation. Its source of stimulation is a conducting tissue of specialized muscle within the heart. About 72 times a minute, this tissue transmits electrical impulses that stimulate cardiac contraction. Nerve stimulation merely causes the conducting tissue to increase or decrease its rate of discharge.

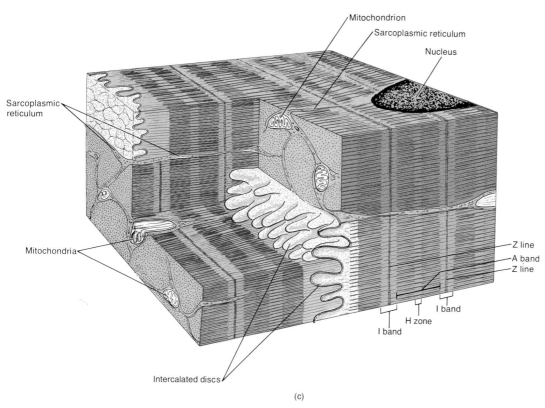

Mitochondrion

Sarcoplasmic reticulum

Nucleus

Sarcoplasmic reticulum

Mitochondria

Z line
A band
Z line

I band
H zone
I band

Intercalated discs

(c)

FIGURE 9-8 (*Continued*) Histology of cardiac muscle tissue. (c) Diagram based on an electron micrograph.

SMOOTH MUSCLE TISSUE

Like cardiac muscle tissue, **smooth muscle tissue** is usually involuntary. However, it is nonstriated. A single fiber of smooth muscle tissue is about 5 to 10 μm in diameter and 30 to 200 μm long. It is spindle-shaped, and within the fiber is a single, oval, centrally located nucleus (Figure 9-9). Smooth muscle cells also contain actin and myosin filaments, but because the filaments are not as orderly as in skeletal and cardiac muscle tissue, the well-differentiated striations do not occur.

Two kinds of smooth muscle tissue, visceral and multiunit, are recognized. The more common type is called visceral muscle tissue. It is found in wraparound sheets that form part of the walls of the hollow viscera such as the stomach, intestines, uterus, and urinary bladder. The terms *smooth muscle tissue* and *visceral muscle tissue* are sometimes used interchangeably. The fibers in visceral muscle tissue are tightly bound together to form a continuous network. When a neuron stimulates one fiber, the impulse travels over the other fibers so that contraction occurs in a wave over many adjacent fibers.

The second kind of smooth muscle tissue, multiunit smooth muscle tissue, consists of individual fibers each with its own motor-nerve endings. Whereas stimulation of a single visceral muscle fiber causes contraction of many adjacent fibers, stimulation of a single multiunit fiber causes contraction of only that fiber. In this respect, multiunit muscle tissue is like skeletal muscle tissue. Multiunit

smooth muscle tissue is found in the walls of blood vessels, in the arrector pili muscles that attach to hair follicles, and in the intrinsic muscles of the eye, such as the iris.

Both kinds of smooth muscle tissue contract and relax more slowly than skeletal muscle tissue. This characteristic is probably due to the arrangement of the thin and

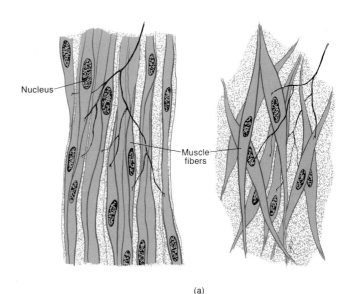

Nucleus

Muscle fibers

(a)

FIGURE 9-9 Histology of smooth muscle tissue. (a) Diagram of visceral smooth muscle tissue (left) and multiunit smooth muscle tissue (right).

Muscle fibers   Nuclei of muscle fibers

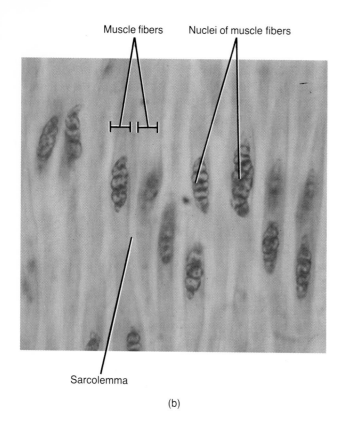

Sarcolemma

(b)

Sarcolemma   Nuclei of muscle fibers

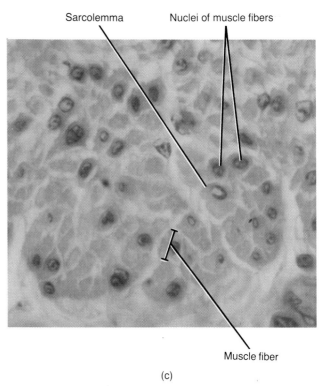

Muscle fiber

(c)

FIGURE 9-9 (*Continued*) Histology of smooth muscle tissue. (b) Photomicrograph of several muscle fibers in longitudinal section at a magnification of 840×. (© 1983 by Michael H. Ross. Used by permission.) (c) Photomicrograph of several muscle fibers in cross section at a magnification of 840×. (© 1983 by Michael H. Ross. Used by permission.)

thick myofilaments of smooth muscle. Whereas skeletal muscle cells contract as individual units, visceral muscle cells contract in sequence as the impulse spreads from one cell to another.

APPLICATIONS TO HEALTH

Disorders of the muscular system are related to disruptions of homeostasis. The disorders may involve a lack of nutrients, the accumulation of toxic products, disease,

injury, disuse, or faulty nervous connections (innervations).

FIBROSIS

The formation of fibrous connective tissue in locations where it normally does not exist is called **fibrosis.** Skeletal and cardiac muscle fibers cannot undergo mitosis, and dead muscle fibers are normally replaced with fibrous connective tissue. Fibrosis, then, is often a consequence of muscle injury or degeneration.

FIBROSITIS

Fibrositis is an inflammation of fibrous tissue. If it occurs in the lumbar region, it is termed **lumbago** (lum-BĀ-gō; *lumbar* = loin). Fibrositis is a common condition characterized by pain, stiffness, or soreness of fibrous tissue, especially in the muscle coverings. It is not destructive or progressive. It may persist for years or spontaneously disappear. Attacks of fibrositis may follow an injury, repeated muscular strain, or prolonged muscular tension.

"CHARLEYHORSE"

Fibromyositis refers to a group of symptoms that include pain, tenderness, and stiffness of the joints, muscles, or adjacent structures. These symptoms ordinarily occur in various combinations. When the thigh is involved, it is usually called **"charleyhorse."** It is characterized by the sudden onset of pain and is aggravated by motion. In some cases, local muscle spasms are noted. The condition is relieved with heat, massage, and rest and completely disappears, although occasionally it may become chronic or recur at frequent intervals.

MUSCULAR DYSTROPHY

The term **muscular dystrophy** (*dystrophy* = degeneration) applies to a number of inherited myopathies, or muscle-destroying diseases. The disease is characterized by degeneration of the individual muscle cells, which leads to a progressive atrophy (reduction in size) of the skeletal muscle. Usually the voluntary skeletal muscles are weakened equally on both sides of the body, whereas the internal muscles, such as the diaphragm, are not affected. Histologically, the changes that occur include the variation in muscle fiber size, degeneration of fibers, and deposition of fat.

Two major varieties of muscular dystrophy can be distinguished clinically. The *Duchenne form* is hereditary and affects only males. Because muscles that attach to the pelvic girdle are weak, the afflicted child shows such symptoms as waddling gait, toe-walking, frequent falls, and difficulty in rising. By early adolescence, the individual is confined to a wheelchair. Advanced cases show abnormal shortening of muscles and scoliosis.

The *fascioscapulohumeral* (*Landouzy-Dejerine*) *form* affects both sexes and is also hereditary. It usually occurs later in life. In early cases, weakness of the shoulder girdle muscles is more prominent than leg weakness and the face is usually affected. Progression is slower than in the Duchenne form.

The cause of muscular dystrophy has been variously attributed to a genetic defect, faulty metabolism of potassium, protein deficiency, and inability of the body to utilize creatine.

There is no specific drug therapy for this disorder. Treatment involves attempts to prolong ambulation by muscle-strengthening exercises, corrective surgical measures, and appropriate braces. Patients are encouraged to keep physically active.

A simple and reliable screening test for a wide variety of skeletal muscle disorders has been developed. Increased levels of an enzyme called *creatine phosphokinase* (*CPK*) in the blood are typical not only of the dystrophies but of other wasting diseases as well. The measurement of CPK, the most consistently elevated enzyme in Duchenne dystrophy, provides a diagnostic test even before clinical symptoms appear. Inflammatory muscle diseases can also be detected with this screening test.

MYASTHENIA GRAVIS

Myasthenia (mī'-as-THĒ-nē-a) **gravis** is a weakness of the skeletal muscles. It is caused by an abnormality at the neuromuscular junction that prevents the muscle fibers from contracting. Recall that motor neurons stimulate the skeletal muscle fibers to contract by releasing acetylcholine. Myasthenia gravis is caused by failure of the neurons to release acetylcholine or by the release from the muscle fibers of an excess amount of cholinesterase, a chemical that destroys acetylcholine (see Chapter 14). As the disease progresses, more neuromuscular junctions become affected. The muscle becomes increasingly weaker and may eventually cease to function altogether.

The cause of myasthenia gravis is unknown. It is more common in females, occurring most frequently between the ages of 20 and 50. The muscles of the face and neck are most apt to be involved. Initial symptoms include a weakness of the eye muscles and difficulty in swallowing. Later, the individual has difficulty chewing and talking. Eventually, the muscles of the limbs may become involved. Death may result from paralysis of the respiratory muscles, but usually the disorder does not progress to this stage.

Anticholinesterase drugs have been the primary treatment for the disease. More recently, steroid drugs, such as prednisone, have been used with great success. It is believed that the immune systems of individuals with myasthenia gravis produce abnormal antibodies that prevent normal muscle contraction. Another recent treatment involves *plasmapheresis,* a procedure that separates blood cells from the plasma that contains the unwanted antibodies. The blood cells are then mixed with a plasma substitute and pumped back into the individual. Preliminary results indicate dramatic improvement in all treated individuals.

ABNORMAL CONTRACTIONS

One kind of abnormal contraction of a muscle is **spasm:** a sudden, involuntary contraction of short duration. A **cramp** is a painful spasmodic contraction of a muscle. It is an involuntary, complete tetanic contraction. **Convul-**

sions are violent, involuntary tetanic contractions of an entire group of muscles. Convulsions occur when motor neurons are stimulated by fever, poisons, hysteria, or changes in body chemistry due to withdrawal of certain drugs. The stimulated neurons send many bursts of seemingly disordered impulses to the muscle fibers. **Fibrillation** is the uncoordinated contraction of individual muscle fibers preventing the smooth contraction of the muscle.

A **tic** is a spasmodic twitching made involuntarily by muscles that are ordinarily under voluntary control. Twitching of the eyelid and face muscles are examples. In general, tics are of psychological origin.

KEY MEDICAL TERMS ASSOCIATED WITH THE MUSCULAR SYSTEM

Gangrene (*gangraena* = an eating sore) Death of tissue that results from interruption of its blood supply.

Myology (*myo* = muscle) Study of muscles.

Myomalacia (*malaco* = soft) Softening of a muscle.

Myopathy (*pathos* = disease) Any disease of muscle tissue.

Myosclerosis (*scler* = hard) Hardening of a muscle.

Myositis (*itis* = inflammation of) Inflammation of muscle fibers (cells).

Myospasm Spasm of a muscle.

Myotonia (*tonia* = tension) Increased muscular excitability and contractility with decreased power of relaxation; tonic spasm of the muscle.

Trichinosis A myositis caused by the parasitic worm *Trichinella spiralis,* which may be found in the muscles of humans, rats, and pigs. People contract the disease by eating insufficiently cooked infected pork.

Volkmann's contracture (*contra* = against) Permanent contraction of a muscle due to replacement of destroyed muscle cells with fibrous tissue that lacks ability to stretch. Destruction of muscle cells may occur from interference with circulation caused by a tight bandage, a piece of elastic, or a cast.

Wryneck or torticollis (*tortus* = twisted; *collam* = neck) Spasmodic contraction of several of the superficial and deep muscles of the neck; produces twisting of the neck and an unnatural position of the head.

STUDY OUTLINE

Characteristics
1. Excitability is the property of receiving and responding to stimuli.
2. Contractility is the ability to shorten and thicken, or contract.
3. Extensibility is the ability to be stretched or extended.
4. Elasticity is the ability to return to original shape after contraction or extension.

Functions
1. Through contraction, muscle tissue performs the three important functions of motion, maintenance of posture, and heat production.

Kinds
1. Skeletal muscle tissue is attached to bones. It is striated and voluntary.
2. Visceral muscle tissue is located in viscera. It is nonstriated (smooth) and involuntary.
3. Cardiac muscle tissue forms the walls of the heart. It is striated and involuntary.

Skeletal Muscle Tissue
Fascia
1. The term fascia is applied to a sheet or broad band of fibrous connective tissue underneath the skin or around muscles and organs of the body.
2. There are three types of fascia: superficial, deep, and subserous.

Connective Tissue Components
1. The entire muscle is covered by the epimysium. Fasciculi are covered by perimysium. Fibers are covered by endomysium.
2. Tendons and aponeuroses attach muscle to bone.

Nerve and Blood Supply
1. Nerves convey impulses for muscular contraction.
2. Blood provides nutrients and oxygen for contraction.

Histology
1. Skeletal muscle consists of fibers covered by a sarcolemma. The fibers contain sarcoplasm, nuclei, sarcoplasmic reticulum, and T tubules.
2. Each fiber contains myofilaments (thin and thick). The myofilaments are compartmentalized into sarcomeres.

Contraction
1. The sliding-filament theory of muscle contraction de-

scribes the movement of myofilaments during contraction.

2. The thin myofilaments of the sarcomere slide inward over the thick myofilaments, thus shortening the sarcomere and the muscle fiber.

3. A motor neuron transmits the stimulus to a skeletal muscle for contraction.

4. The region of the sarcolemma specialized to receive the stimulus is the motor end plate.

5. The area of contact between a motor neuron and muscle fiber is a neuromuscular, or myoneural, junction.

6. A motor neuron and the muscle fibers it stimulates form a motor unit.

7. When a nerve impulse reaches the motor end plate, the neuron releases acetylcholine, which causes an electrical charge.

8. Muscle fibers of a motor unit contract to their fullest extent or not at all.

9. The weakest stimulus capable of causing contraction is a liminal, or threshold, stimulus. A stimulus not capable of inducing contraction is a subliminal, or subthreshold, stimulus.

Muscle Tone

1. A sustained partial contraction of portions of a skeletal muscle results in muscle tone.

2. Tone is essential for maintaining posture.

3. Atrophy is a wasting away or decrease in size; hypertrophy is an enlargement or overgrowth.

Fast (White) and Slow (Red) Muscle

1. Fast or white muscles have an extensive sarcoplasmic reticulum.

2. Slow or red muscles have smaller fibers, more blood capillaries, and a large amount of myoglobin.

Cardiac Muscle Tissue

1. This muscle is found only in the heart. It is striated and involuntary.

2. The cells are quadrangular and contain centrally placed nuclei.

3. The fibers form a continuous, branching network that contracts as a functional unit.

4. Intercalated discs provide strength and aid impulse conduction.

Smooth Muscle Tissue

1. Smooth muscle is found in viscera. It is nonstriated and involuntary.

2. Visceral smooth muscle is found in the walls of viscera. The fibers are arranged in a network.

3. Multiunit smooth muscle is found in blood vessels and the eye. The fibers operate singly rather than as a unit.

Applications to Health

1. Fibrosis is the formation of fibrous tissue where it normally does not exist; it frequently occurs in damaged muscle tissue.

2. Fibrositis is an inflammation of fibrous tissue. If it occurs in the lumbar region, it is called lumbago.

3. "Charleyhorse" refers to pain, tenderness, and stiffness of joints, muscles, and related structures in the thigh.

4. Muscular dystrophy is a hereditary disease of muscles characterized by degeneration of individual muscle cells. It occurs in two principal forms: Duchenne and fascioscapulohumeral.

5. Myasthenia gravis is a disease characterized by great muscular weakness and fatigability resulting from improper neuromuscular transmission.

6. Abnormal contractions include spasms, cramps, convulsions, fibrillation, and tics.

REVIEW QUESTIONS

1. How is the skeletal system related to the muscular system? What are the three basic functions of the muscular system?

2. What are the four characteristics of muscle tissue?

3. How can the three kinds of muscle tissue be distinguished?

4. What is fascia? What are the three different types of fascia and where are they found in the body?

5. Define epimysium, perimysium, endomysium, tendon, and aponeurosis. Describe the nerve and blood supply to a muscle.

6. Discuss the microscopic structure of skeletal muscle.

7. In considering the contraction of skeletal muscle, describe the structure of the motor unit and the sliding-filament theory.

8. What is the all-or-none principle? Relate it to a liminal and subliminal stimulus.

9. What is muscle tone? Why is it important? Distinguish between atrophy and hypertrophy.

10. Compare fast (white) and slow (red) muscle with respect to structure and function.

11. Compare cardiac and smooth muscle with regard to microscopic structure, functions, and locations.

12. What is fibrosis? Define fibrositis.

13. What is a "charleyhorse"?

14. What is muscular dystrophy? Distinguish the two principal forms.

15. What is myasthenia gravis? In this disease, why do the muscles not contract normally?

16. Define the following abnormal muscular contractions: spasm, cramp, convulsion, fibrillation, and tic.

17. Refer to the glossary of key medical terms associated with the muscular system. Be sure that you can define each term.

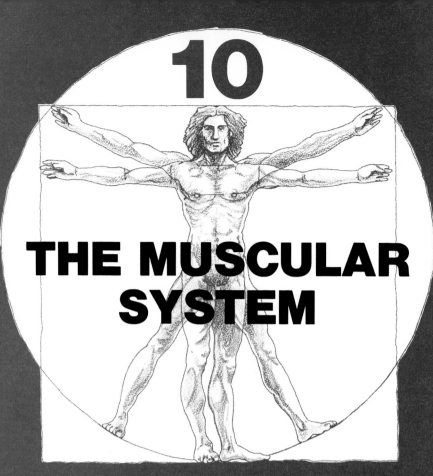

10

THE MUSCULAR SYSTEM

STUDENT OBJECTIVES

■ Describe the relationship between bones and skeletal muscles in producing body movements.

■ Define a lever and fulcrum and compare the three classes of levers on the basis of placement of the fulcrum, effort, and resistance.

■ Describe most body movements as activities of groups of muscles by explaining the roles of the prime mover, antagonist, and synergist.

■ Identify the various arrangements of muscle fibers in a skeletal muscle and relate the arrangements to the strength of contractions and range of movement.

■ Define the criteria employed in naming skeletal muscles.

■ Identify the principal skeletal muscles in different regions of the body by name, origin, insertion, action, and innervation.

■ Identify surface anatomy features of selected skeletal muscles.

■ Compare the common sites for intramuscular injections.

The term **muscle tissue** refers to all the contractile tissues of the body: skeletal, cardiac, and smooth muscle. The **muscular system,** however, refers to the *skeletal* muscle system: the skeletal muscle tissue and connective tissues that make up individual muscle organs, such as the biceps. Cardiac muscle tissue is located in the heart and is therefore considered part of the circulatory system. Smooth muscle tissue of the intestine is part of the digestive system, while smooth muscle tissue of the urinary bladder is part of the urinary system. In this chapter, we discuss only the muscular system. We will see how skeletal muscles produce movement and describe the principal skeletal muscles.

The developmental anatomy of the muscular system is considered in Exhibit 24-3.

HOW SKELETAL MUSCLES PRODUCE MOVEMENT

ORIGIN AND INSERTION

Skeletal muscles produce movements by exerting force on tendons, which in turn pull on bones. Most muscles cross at least one joint and are attached to the articulating bones that form the joint (Figure 10-1). When such a muscle contracts, it draws one articulating bone toward the other. The two articulating bones usually do not move equally in response to the contraction. One is held nearly in its original position because other muscles contract to pull it in the opposite direction or because its structure makes it less movable. Ordinarily, the attachment of a muscle tendon to the stationary bone is called the **origin.** The attachment of the other muscle tendon to the movable bone is the **insertion.** A good analogy is a spring on a door. The part of the spring attached to the door represents the insertion; the part attached to the frame is the origin. The fleshy portion of the muscle between the tendons of the origin and insertion is called the **belly,** or **gaster.** The origin is usually proximal and the insertion distal, especially in the appendages. In addition, muscles that move a body part generally do not cover the moving part. Figure 10-1a shows that although contraction of the biceps muscle moves the forearm, the belly of the muscle lies over the humerus.

LEVER SYSTEMS

In producing a body movement, bones act as levers and joints function as fulcrums of these levers. A **lever** may be defined as a rigid rod that moves about on some fixed point called a **fulcrum.** A fulcrum may be symbolized

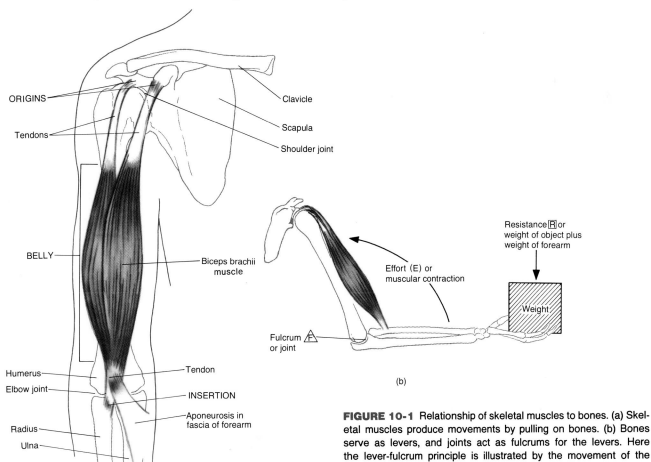

(a)

ORIGINS
Tendons
BELLY
Humerus
Elbow joint
Radius
Ulna

Clavicle
Scapula
Shoulder joint

Biceps brachii muscle

Tendon
INSERTION
Aponeurosis in fascia of forearm

(b)

Effort (E) or muscular contraction

Resistance R or weight of object plus weight of forearm

Weight

Fulcrum F or joint

FIGURE 10-1 Relationship of skeletal muscles to bones. (a) Skeletal muscles produce movements by pulling on bones. (b) Bones serve as levers, and joints act as fulcrums for the levers. Here the lever-fulcrum principle is illustrated by the movement of the forearm lifting a weight. Note where the resistance and effort are applied in this example.

as △. A lever is acted on at two different points by two different forces: the *resistance* ℝ and the *effort* (E). The resistance may be regarded as a force to be overcome, whereas the effort is exerted to overcome the resistance. The resistance may be the weight of the body part that is to be moved. The muscular effort (contraction) is applied to the bone at the insertion of the muscle and produces motion. Consider the biceps flexing the forearm at the elbow as a weight is lifted (Figure 10-1b). When the forearm is raised, the elbow is the fulcrum. The weight of the forearm plus the weight in the hand is the resistance. The shortening of the biceps pulling the forearm up is the effort.

Levers are categorized into three types according to the positions of the fulcrum, the effort, and the resistance.

1. In a **first-class lever,** the fulcrum is between the effort and resistance (Figure 10-2a). An example of a first-class lever is a seesaw. There are not many first-

class levers in the body. One example is the head resting on the vertebral column. When the head is raised, the facial portion of the skull is the resistance. The joint between the atlas and occipital bone (atlantooccipital joint) is the fulcrum. The contraction of the muscles of the back is the effort.

2. Second-class levers have the fulcrum at one end, the effort at the opposite end, and the resistance in between (Figure 10-2b). They operate like a wheelbarrow. Most authorities agree that there are very few examples of second-class levers in the body. One example is raising the body on the toes. The body is the resistance, the ball of the foot is the fulcrum, and the contraction of the calf muscles to pull the heel upward is the effort.

3. Third-class levers consist of the fulcrum at one end, the resistance at the opposite end, and the effort between them (Figure 10-2c). They are the most common levers in the body. A common example is flexing the forearm at the elbow. The weight of the forearm is the resistance,

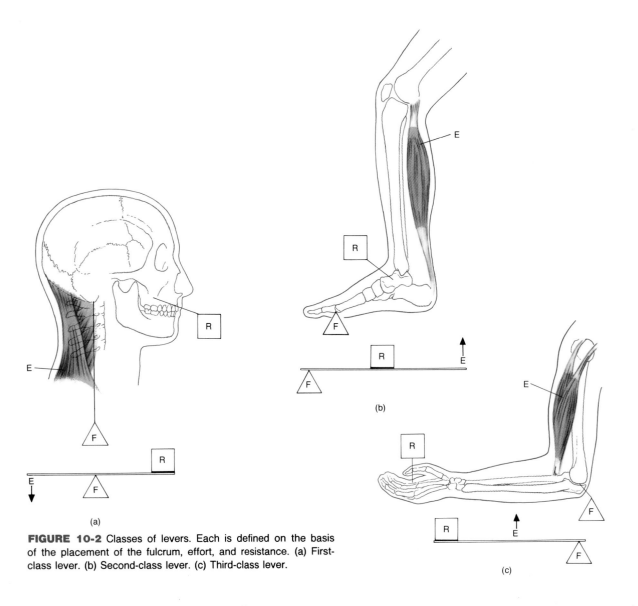

FIGURE 10-2 Classes of levers. Each is defined on the basis of the placement of the fulcrum, effort, and resistance. (a) First-class lever. (b) Second-class lever. (c) Third-class lever.

the contraction of the biceps is the effort, and the elbow joint is the fulcrum.

LEVERAGE

Leverage—the mechanical advantage gained by a lever—is largely responsible for a muscle's strength and range of movement. Consider strength first. Suppose we have two muscles of the same strength crossing and acting on a joint. Assume also that one is attached farther from the joint and one is closer. The muscle attached farther will produce the more powerful movement. Thus strength of movement depends on the placement of muscle attachments.

In considering the range of movement, again assume that we have two muscles of the same strength crossing and acting on a joint and that one is attached farther from the joint than the other. The muscle inserting closer to the joint will produce the greater range of movement. Thus range of movement also depends on the placement of muscle attachments. Since strength increases with distance from the joint and range of movement decreases, maximal strength and maximal range are incompatible, strength and range vary inversely.

ARRANGEMENT OF FASCICULI

Recall from Chapter 9 that skeletal muscle fibers are arranged within the muscle in bundles called fasciculi or fascicles. The muscle fibers are arranged in a parallel fashion within each bundle, but the arrangement of the fasciculi with respect to the tendons may take one of four characteristic patterns.

The first pattern is called **parallel.** The fasciculi are parallel with the longitudinal axis and terminate at either end in flat tendons. The muscle is typically quadrilateral in shape. An example is the stylohyoid muscle (see Figure 10-7). In a modification of the parallel arrangement, called *fusiform,* the fasciculi are nearly parallel with the longitudinal axis and terminate at either end in flat tendons, but the muscle tapers toward the tendons, where the diameter is less than that of the belly. An example is the biceps brachii muscle (see Figure 10-16).

The second distinct pattern is called **convergent.** A broad origin of fasciculi converges to a narrow, restricted insertion. Such a pattern gives the muscle a triangular shape. An example is the deltoid muscle (see Figure 10-15).

The third distinct pattern is referred to as **pennate.** The fasciculi are short in relation to the entire length of the muscle and the tendon extends nearly the entire length of the muscle. The fasciculi are directed obliquely toward the tendon like the plumes of a feather. If the fasciculi are arranged on only one side of a tendon, as in the extensor digitorum muscle, the muscle is referred to as *unipennate* (see Figure 10-17). If the fasciculi are arranged on both sides of a centrally positioned tendon, as in the rectus femoris muscle, the muscle is referred to as *bipennate* (see Figure 10-20).

The final distinct pattern is referred to as **circular.** The fasciculi are arranged in a circular pattern and enclose an orifice. An example is the orbicularis oris muscle (see Figure 10-4).

Fascicular arrangement is correlated with the power of a muscle and range of movement. When a muscle fiber contracts, it shortens to a length just slightly greater than half of its resting length. Thus, the longer the fibers in a muscle, the greater the range of movement it can produce. By contrast, the strength of a muscle depends on the total number of fibers it contains, since a short fiber can contract as forcefully as a long one. Since a given muscle can either contain a small number of long fibers or a large number of short fibers, fascicular arrangement represents a compromise between power and range of movement. Pennate muscles, for example, have a larger number of fasciculi distributed over their tendons, giving them greater power but a smaller range of movement. Parallel muscles, on the other hand, have comparatively few fasciculi that extend the length of the muscle. Thus, they have a greater range of movement, but less power.

GROUP ACTIONS

Most movements are coordinated by several skeletal muscles acting in groups rather than individually. Consider flexing the forearm at the elbow, for example. A muscle that causes a desired action is referred to as the **agonist** or **prime mover.** In this instance, the biceps brachii is the agonist (see Figure 10-17). Simultaneously with the contraction of the biceps brachii, another muscle, called the **antagonist,** is relaxing. In this movement, the triceps brachii serves as the antagonist (see Figure 10-17). The antagonist has an effect opposite to that of the agonist, that is, the antagonist relaxes and yields to the movement of the agonist. If we consider extending the forearm at the elbow, the triceps brachii is the agonist and the biceps brachii is the antagonist. Most joints are operated by antagonistic groups of muscles.

Other muscles called **synergists,** or **fixators,** assist the agonist by reducing undesired action or unnecessary movements in the less mobile articulating bone. While flexing the forearm, the synergists, in this case the deltoid and pectoralis major muscles, hold the arm and shoulder in a suitable position for the flexing action (see Figure 10-14). The deltoid abducts the humerus, and the pectoralis major adducts and medially rotates the humerus. Essentially, synergists contract at the same time as the prime mover and help the prime mover produce an effective movement. Many muscles are, at various times, prime movers, antagonists, or synergists, depending on the action.

NAMING SKELETAL MUSCLES

The names of most of the nearly 700 skeletal muscles are based on several types of characteristics. Learning the terms used to indicate specific characteristics will help you remember the names of muscles.

1. Muscle names may indicate the *direction of the muscle fibers.* *Rectus* fibers usually run parallel to the midline of the body. *Transverse* fibers run perpendicular to the midline. *Oblique* fibers are diagonal to the midline. Muscles named according to these three directions include the rectus abdominis, transversus abdominis, and external oblique.

2. A muscle may be named according to *location.* The temporalis is near the temporal bone. The tibialis anterior is near the tibia.

3. *Size* is another criterion. The term *maximus* means largest, *minimus* means smallest, *longus* means long, and *brevis* means short. Examples include the gluteus maximus, gluteus minimus, adductor longus, and peroneus brevis.

4. Some muscles are named for their *number of origins.* The *bi*ceps has two origins, the *tri*ceps three, and the *quadri*ceps four.

5. Other muscles are named on the basis of *shape.* Common examples include the deltoid (meaning triangular) and trapezius (meaning trapezoid).

6. Muscles may be named after their *origin and insertion.* The sternocleidomastoid originates on the sternum and clavicle and inserts at the mastoid process of the temporal bone; the stylohyoideus originates on the styloid process of the temporal bone and inserts at the hyoid bone.

7. Still another criterion used for naming muscles is *action.* Exhibit 10-1 lists the principal actions of muscles, their definitions, and examples of muscles that perform the actions. For convenience, the actions are grouped as antagonistic pairs where possible.

PRINCIPAL SKELETAL MUSCLES

Exhibits 10-2 through 10-20 list the principal muscles of the body with their origins, insertions, actions, and innervations. (By no means have all the muscles of the body been included.) Refer to Chapters 6 and 7 to review bone markings, since they serve as points of origin and insertion for muscles.

The muscles are divided into groups according to the part of the body on which they act. If you have mastered the naming of the muscles, their actions will have more meaning. Figure 10-3 shows general anterior and posterior views of the muscular system. Do not try to memorize all these muscles yet. As you study groups of muscles in the exhibits, refer to Figure 10-3 to see how each group is related to all others. The figures that accompany the exhibits contain superficial and deep, anterior and posterior, or medial and lateral views to show the muscles' position as clearly as possible. An attempt has been made to show the relationship of the muscles under consideration to other muscles in the area you are studying. Labeled photographs introduce you to surface anatomy.

EXHIBIT 10-1 MUSCLES NAMED ACCORDING TO ACTION

ACTION	DEFINITION	EXAMPLE
Flexor	Usually decreases the anterior angle at a joint; some decrease the posterior angle.	Flexor carpi radialis.
Extensor	Usually increases the anterior angle at a joint; some increase the posterior angle.	Extensor carpi ulnaris.
Abductor	Moves a bone away from the midline.	Abductor hallucis longus.
Adductor	Moves a bone closer to the midline.	Adductor longus.
Levator	Produces an upward movement.	Levator scapulae.
Depressor	Produces a downward movement.	Depressor labii inferioris.
Supinator	Turns the palm upward or anteriorly.	Supinator.
Pronator	Turns the palm downward or posteriorly.	Pronator teres.
Dorsiflexor	Flexes the foot at the ankle joint.	Tibialis anterior.
Plantar flexor	Extends the foot at the ankle joint.	Plantaris.
Invertor	Turns the sole of the foot inward.	Tibialis anterior.
Evertor	Turns the sole of the foot outward.	Peroneus tertius.
Sphincter	Decreases the size of an opening.	Orbicularis oculi.
Tensor	Makes a body part more rigid.	Tensor fasciae latae.
Rotator	Moves a bone around its longitudinal axis.	Obturator.

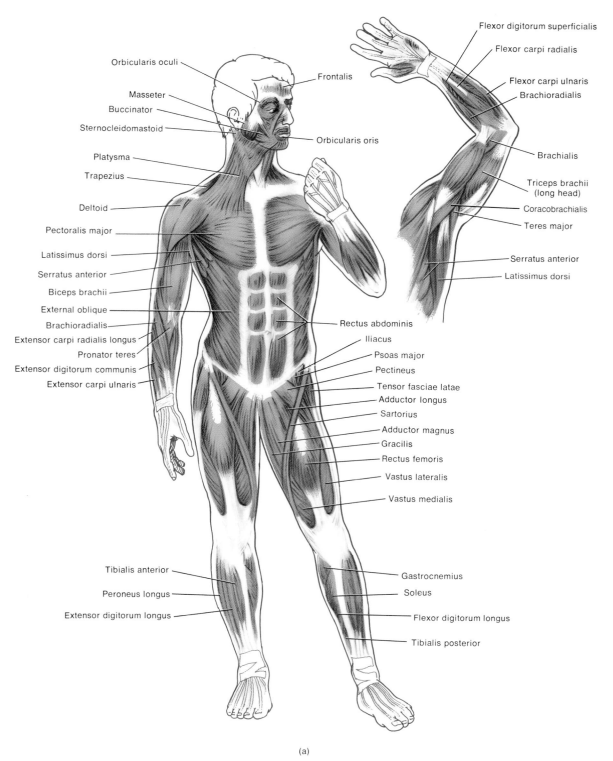

Flexor digitorum superficialis

Flexor carpi radialis

Flexor carpi ulnaris

Brachioradialis

Brachialis

Triceps brachii (long head)

Coracobrachialis

Teres major

Serratus anterior

Latissimus dorsi

Orbicularis oculi

Frontalis

Masseter

Buccinator

Sternocleidomastoid

Orbicularis oris

Platysma

Trapezius

Deltoid

Pectoralis major

Latissimus dorsi

Serratus anterior

Biceps brachii

External oblique

Brachioradialis

Extensor carpi radialis longus

Pronator teres

Extensor digitorum communis

Extensor carpi ulnaris

Rectus abdominis

Iliacus

Psoas major

Pectineus

Tensor fasciae latae

Adductor longus

Sartorius

Adductor magnus

Gracilis

Rectus femoris

Vastus lateralis

Vastus medialis

Tibialis anterior

Peroneus longus

Extensor digitorum longus

Gastrocnemius

Soleus

Flexor digitorum longus

Tibialis posterior

(a)

FIGURE 10-3 Principal superficial muscles. (a) Anterior view.

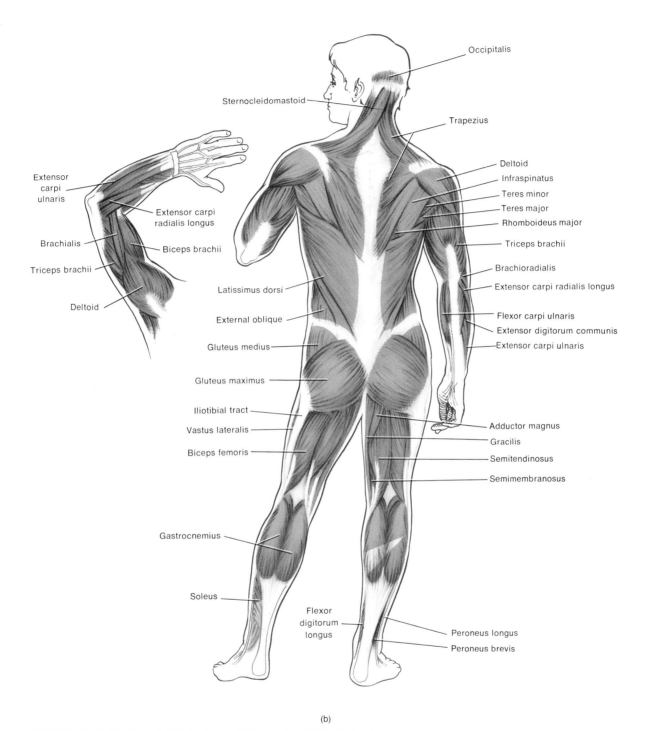

Occipitalis

Sternocleidomastoid

Trapezius

Deltoid
Infraspinatus
Teres minor
Teres major
Rhomboideus major
Triceps brachii
Brachioradialis
Extensor carpi radialis longus
Flexor carpi ulnaris
Extensor digitorum communis
Extensor carpi ulnaris

Extensor
carpi
ulnaris

Extensor carpi
radialis longus

Brachialis
Biceps brachii

Triceps brachii

Deltoid

Latissimus dorsi

External oblique

Gluteus medius

Gluteus maximus

Iliotibial tract
Vastus lateralis
Biceps femoris

Adductor magnus
Gracilis
Semitendinosus
Semimembranosus

Gastrocnemius

Soleus

Flexor
digitorum
longus

Peroneus longus
Peroneus brevis

(b)

FIGURE 10-3 (*Continued*) Principal superficial muscles. (b) Posterior view.

EXHIBIT 10-2 MUSCLES OF FACIAL EXPRESSION (Figure 10-4)

MUSCLE	ORIGIN	INSERTION	ACTION	INNERVATION
Epicranius (*epi* = over; *crani* = skull)	This muscle is divisible into two portions: the frontalis over the frontal bone and the occipitalis over the occipital bone. The two muscles are united by a strong aponeurosis, the galea aponeurotica, which covers the superior and lateral surfaces of the skull.			
Frontalis (*front* = forehead)	Galea aponeurotica.	Skin superior to supraorbital line.	Draws scalp forward, raises eyebrows, and wrinkles forehead horizontally.	Facial nerve (VII).
Occipitalis (*occipito* = base of skull)	Occipital bone and mastoid process of temporal bone.	Galea aponeurotica.	Draws scalp backward.	Facial nerve (VII).
Orbicularis oris (*orb* = circular; *or* = mouth)	Muscle fibers surrounding opening of mouth.	Skin at corner of mouth.	Closes lips, compresses lips against teeth, protrudes lips, and shapes lips during speech.	Facial nerve (VII).
Zygomaticus major (*zygomatic* = cheek bone; *major* = greater)	Zygomatic bone.	Skin at angle of mouth and orbicularis oris.	Draws angle of mouth upward and outward as in smiling or laughing.	Facial nerve (VII).
Levator labii superioris (*levator* = elevates; *labii* = lip; *superioris* = upper)	Superior to infraorbital foramen of maxilla.	Skin at angle of mouth and orbicularis oris.	Elevates upper lip.	Facial nerve (VII).
Depressor labii inferioris (*depressor* = depresses; *inferioris* = lower)	Mandible.	Skin of lower lip.	Depresses lower lip.	Facial nerve (VII).
Buccinator (*bucc* = cheek)	Alveolar processes of maxilla and mandible and pterygomandibular raphe (fibrous band extending from the pterygoid hamulus to the mandible).	Orbicularis oris.	Major cheek muscle; compresses cheek as in blowing air out of mouth and causes the cheeks to cave in, producing the action of sucking.	Facial nerve (VII).
Mentalis (*mentum* = chin)	Mandible.	Skin of chin.	Elevates and protrudes lower lip and pulls skin of chin up as in pouting.	Facial nerve (VII).
Platysma (*platy* = flat, broad)	Fascia over deltoid and pectoralis major muscles.	Mandible, muscles around angle of mouth, and skin of lower face.	Draws outer part of lower lip downward and backward as in pouting; depresses mandible.	Facial nerve (VII).
Risorius (*risor* = laughter)	Fascia over parotid (salivary) gland.	Skin at angle of mouth.	Draws angle of mouth laterally as in tenseness.	Facial nerve (VII).
Orbicularis oculi (*ocul* = eye)	Medial wall of orbit.	Circular path around orbit.	Closes eye.	Facial nerve (VII).
Corrugator supercilii (*corrugo* = to wrinkle; *supercilium* = eyebrow)	Medial end of superciliary arch of frontal bone.	Skin of the eyebrow.	Draws eyebrow downward as in frowning.	Facial nerve (VII).
Levator palpebrae superioris (*palpebrae* = eyelids) (see Figure 10-6b)	Roof of orbit (lesser wing of sphenoid bone).	Skin of the upper eyelid.	Elevates the upper eyelid.	Oculomotor nerve (III).

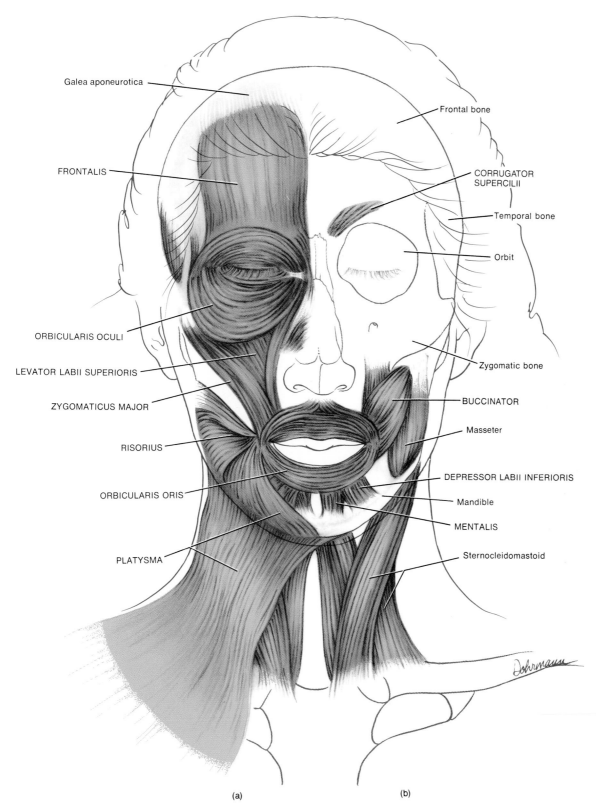

Galea aponeurotica

Frontal bone

FRONTALIS

CORRUGATOR
SUPERCILII

Temporal bone

Orbit

ORBICULARIS OCULI

LEVATOR LABII SUPERIORIS

Zygomatic bone

ZYGOMATICUS MAJOR

BUCCINATOR

RISORIUS

Masseter

ORBICULARIS ORIS

DEPRESSOR LABII INFERIORIS

Mandible

MENTALIS

PLATYSMA

Sternocleidomastoid

(a)

(b)

FIGURE 10-4 Muscles of facial expression. (a) Anterior superficial view. (b) Anterior deep view.

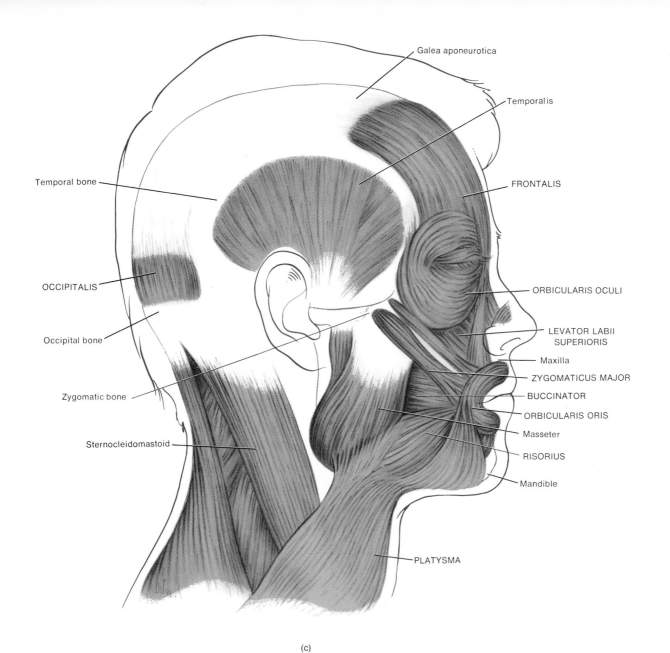

Galea aponeurotica

Temporalis

Temporal bone

FRONTALIS

OCCIPITALIS

ORBICULARIS OCULI

Occipital bone

LEVATOR LABII
SUPERIORIS

Maxilla

ZYGOMATICUS MAJOR

Zygomatic bone

BUCCINATOR

ORBICULARIS ORIS

Masseter

Sternocleidomastoid

RISORIUS

Mandible

PLATYSMA

(c)

LEVATOR LABII
SUPERIORIS

FRONTALIS

ORBICULARIS OCULI

ZYGOMATICUS
MAJOR

ORBICULARIS ORIS

DEPRESSOR
LABII INFERIORIS

(d)

MENTALIS

FIGURE 10-4 (*Continued*)
Muscles of facial expression.
(c) Right lateral superficial
view. (d) Surface anatomy
photograph of the anterior facial muscles. (Courtesy of
Donald Castellaro and Deborah Massimi.)

232

EXHIBIT 10-3 MUSCLES THAT MOVE THE LOWER JAW (Figure 10-5)

MUSCLE	ORIGIN	INSERTION	ACTION	INNERVATION
Masseter (*maseter* = chewer)	Maxilla and zygomatic arch.	Angle and ramus of mandible.	Elevates mandible as in closing the mouth and protracts (protrudes) mandible.	Mandibular branch of trigeminal nerve (V).
Temporalis (*tempora* = temples)	Temporal bone.	Coronoid process of mandible.	Elevates and retracts mandible.	Temporal nerve from mandibular division of trigeminal nerve (V).
Medial pterygoid (*medial* = closer to midline; *pterygoid* = like a wing; pterygoid plate of sphenoid)	Medial surface of lateral pterygoid plate of sphenoid; maxilla.	Angle and ramus of mandible.	Elevates and protracts mandible and moves mandible from side to side.	Mandibular branch of trigeminal nerve (V).
Lateral pterygoid (*lateral* = farther from midline)	Greater wing and lateral surface of lateral pterygoid plate of sphenoid.	Condyle of mandible; temporomandibular articulation.	Protracts mandible, opens mouth, and moves mandible from side to side.	Mandibular branch of trigeminal nerve (V).

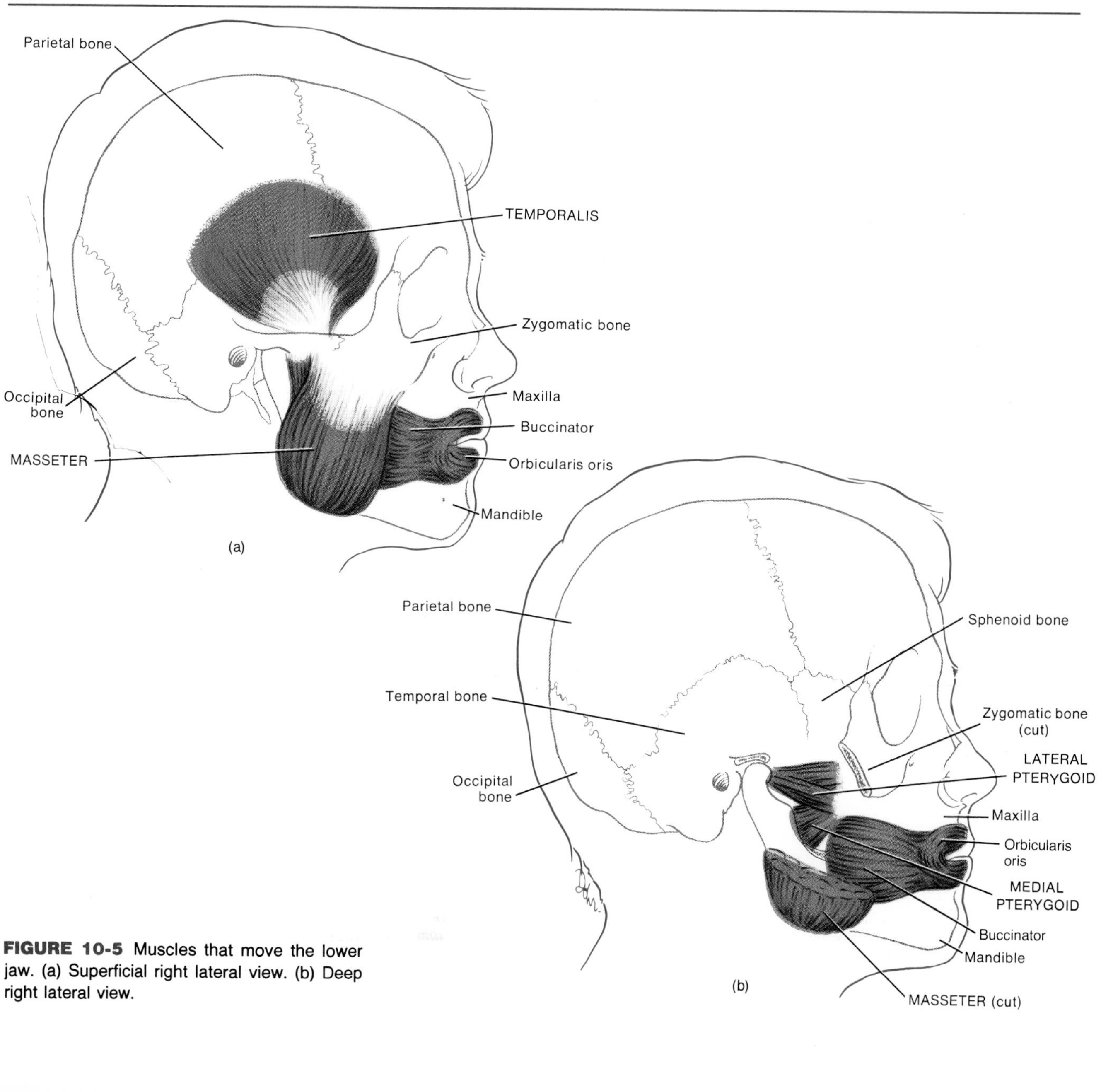

FIGURE 10-5 Muscles that move the lower jaw. (a) Superficial right lateral view. (b) Deep right lateral view.

EXHIBIT 10-4 MUSCLES THAT MOVE THE EYEBALLS—EXTRINSIC MUSCLES* (Figure 10-6)

MUSCLE	ORIGIN	INSERTION	ACTION	INNERVATION
Superior rectus (*superior* = above; *rectus* = in this case, fibers parallel to long axis of eyeball)	Tendinous ring attached to bony orbit around optic foramen.	Superior and central part of eyeball.	Directs eyeball upward.	Oculomotor nerve (III).
Inferior rectus (*inferior* = below)	Same as above.	Inferior and central part of eyeball.	Directs eyeball downward.	Oculomotor nerve (III).
Lateral rectus	Same as above.	Lateral side of eyeball.	Directs eyeball laterally.	Abducens nerve (VI).
Medial rectus	Same as above.	Medial side of eyeball.	Directs eyeball medially.	Oculomotor nerve (III).
Superior oblique (*oblique* = in this case, fibers diagonal to long axis of eyeball)	Same as above.	Eyeball between superior and lateral recti.	Rotates eyeball so that the pupil is directed downward and laterally; note that it moves through a ring of fibrocartilaginous tissue called the trochlea (*trochlea* = pulley).	Trochlear nerve (IV).
Inferior oblique	Floor of the orbit.	Eyeball between superior and lateral recti.	Rotates eyeball so that the pupil is directed upward and laterally.	Oculomotor nerve (III).

* Muscles situated on the outside of the eyeball.

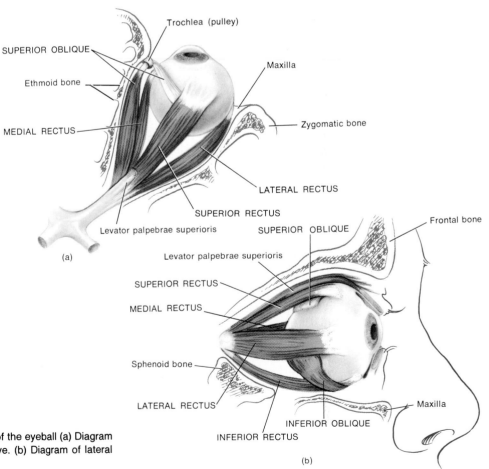

FIGURE 10-6 Extrinsic muscles of the eyeball (a) Diagram of the right eyeball seen from above. (b) Diagram of lateral view of the right eyeball.

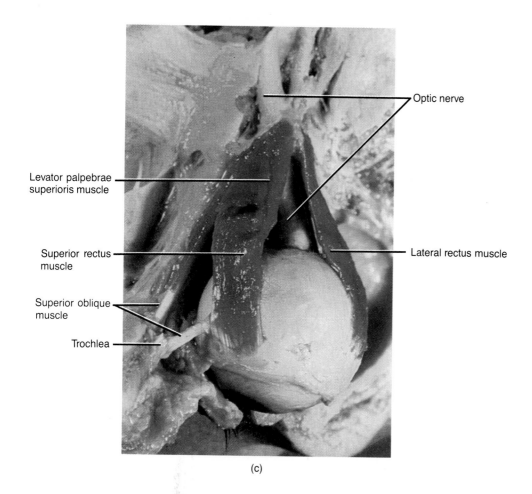

Optic nerve

Levator palpebrae superioris muscle

Superior rectus muscle

Superior oblique muscle

Trochlea

Lateral rectus muscle

(c)

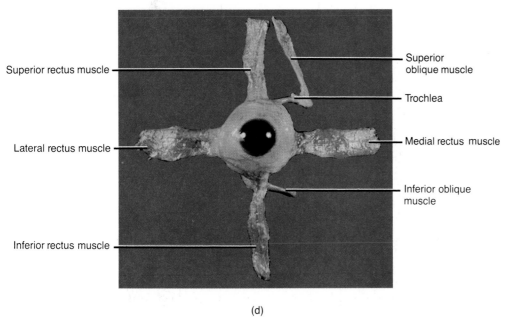

Superior rectus muscle

Lateral rectus muscle

Inferior rectus muscle

Superior oblique muscle

Trochlea

Medial rectus muscle

Inferior oblique muscle

(d)

FIGURE 10-6 (*Continued*) Extrinsic muscles of the eyeball. (c) Photograph of the right eyeball seen from above. (d) Photograph of the right eye muscles extended. (Photographs courtesy of C. Yokochi and J. W. Rohen, *Photographic Anatomy of the Human Body,* 2nd ed., 1979, IGAKU-SHOIN, Ltd., Tokyo, New York.)

EXHIBIT 10-5 MUSCLES THAT MOVE THE TONGUE (Figure 10-7)

MUSCLE	ORIGIN	INSERTION	ACTION	INNERVATION
Genioglossus (*gen-eion* = chin; *glossus* = tongue)	Mandible.	Undersurface of tongue and hyoid bone.	Depresses and pro-tracts tongue.	Hypoglossal nerve (XII).
Styloglossus (*stylo* = stake or pole; styloid process of temporal)	Styloid process of tem-poral bone.	Side and undersurface of tongue.	Elevates and retracts tongue.	Hypoglossal nerve (XII).
Palatoglossus (*palato* = palate)	Anterior surface of soft palate.	Side of tongue.	Elevates tongue and draws soft palate down on tongue.	Hypoglossal nerve (XII).
Hyoglossus	Body of hyoid bone.	Side of tongue.	Depresses tongue and draws down its sides.	Hypoglossal nerve (XII).

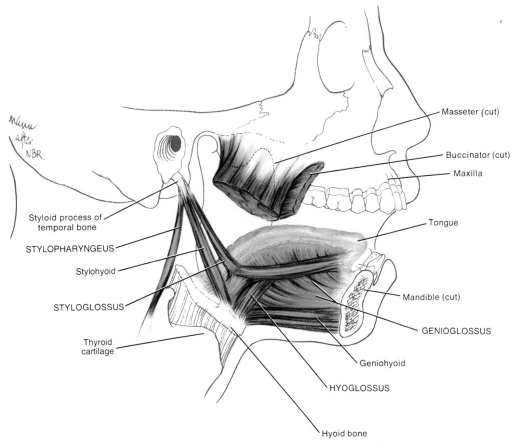

Masseter (cut)

Buccinator (cut)

Maxilla

Tongue

Styloid process of temporal bone

STYLOPHARYNGEUS

Stylohyoid

STYLOGLOSSUS

Thyroid cartilage

Mandible (cut)

GENIOGLOSSUS

Geniohyoid

HYOGLOSSUS

Hyoid bone

FIGURE 10-7 Muscles that move the tongue viewed from the right side.

EXHIBIT 10-6 **MUSCLES OF THE PHARYNX (Figure 10-8)**

MUSCLE	ORIGIN	INSERTION	ACTION	INNERVATION
Inferior constrictor (*inferior* = below; *con-strictor* = decreases diameter of a lumen)	Cricoid and thyroid car-tilages of larynx.	Posterior median raphe of pharynx.	Constricts inferior por-tion of pharynx to pro-pel a bolus into the esophagus.	Pharyngeal plexus.
Middle constrictor	Greater and lesser cornu of hyoid bone and stylohyoid liga-ment.	Posterior median raphe of pharynx.	Constricts middle por-tion of pharynx to pro-pel a bolus into the esophagus.	Pharyngeal plexus.
Superior constric-tor (*superior* = above)	Pterygoid process, pterygomandibular raphe, and mylohyoid line of mandible.	Posterior median raphe of pharynx.	Constricts superior portion of pharynx to propel a bolus into the esophagus.	Pharyngeal plexus.
Stylopharyngeus (*stylo* = stake or pole; styloid process of tem-poral; *pharyngo* = pharynx) (see also Fig-ure 10-7)	Medial side of base of styloid process.	Lateral aspects of pharynx and thyroid cartilage.	Elevates larynx and di-lates pharynx to help bolus descend.	Glossopharyngeal nerve (IX).
Salpingopharyn-geus (*salping* = per-taining to the eusta-chian tube or uterine tube)	Inferior portion of eu-stachian (auditory) tube.	Posterior fibers of pa-latopharyngeus mus-cle.	Elevates superior por-tion of lateral wall of pharynx during swal-lowing and opens ori-fice of eustachian (au-ditory) tube.	Pharyngeal plexus.
Palatopharyngeus (*palato* = palate)	Soft palate.	Posterior border of thy-roid cartilage and lat-eral and posterior wall of pharynx.	Elevates larynx and pharynx and helps close nasopharynx dur-ing swallowing.	Pharyngeal plexus.

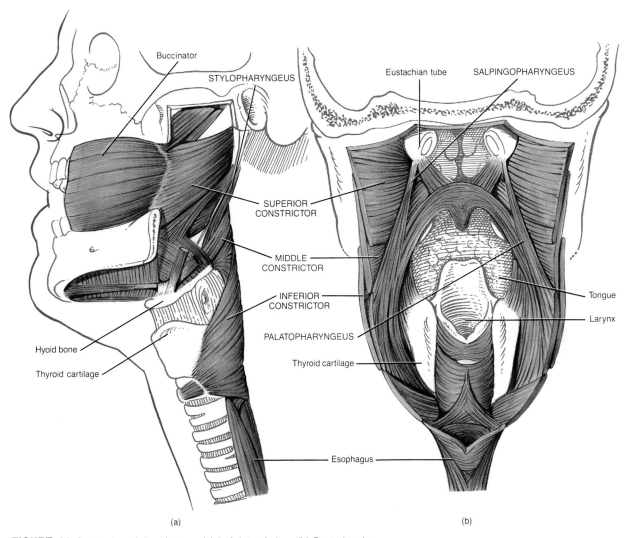

Buccinator

STYLOPHARYNGEUS

Eustachian tube   SALPINGOPHARYNGEUS

SUPERIOR
CONSTRICTOR

MIDDLE
CONSTRICTOR

INFERIOR
CONSTRICTOR

PALATOPHARYNGEUS

Hyoid bone

Thyroid cartilage

Thyroid cartilage

Tongue

Larynx

Esophagus

(a)

(b)

FIGURE 10-8 Muscles of the pharynx. (a) Left lateral view. (b) Posterior view.

EXHIBIT 10-7 MUSCLES OF THE LARYNX (Figure 10-9)

MUSCLE	ORIGIN	INSERTION	ACTION	INNERVATION
EXTRINSIC				
Omohyoid (*omo* = relationship to the shoulder; *hyoedies* = U-shaped, pertaining to the hyoid bone)	Superior border of scapula and superior transverse ligament.	Body of hyoid bone.	Depresses hyoid bone.	Branches of ansa cervicalis nerve (C1–C3).
Sternohyoid (*sterno* = sternum)	Medial end of clavicle and manubrium of sternum.	Body of hyoid bone.	Depresses hyoid bone.	Branches of ansa cervicalis nerve (C1–C3).
Sternothyroid (*thyro* = thyroid gland)	Manubrium of sternum.	Thyroid cartilage of larynx.	Depresses thyroid cartilage.	Branches of ansa cervicalis nerve (C1–C3).
Thyrohyoid	Thyroid cartilage of larynx.	Greater cornu of hyoid bone.	Elevates thyroid cartilage and depresses hyoid bone.	Cervical nerves C1–C2 and descending hypoglossal nerve (XII).
Stylopharyngeus	See Exhibit 10-6.			
Palatopharyngeus	See Exhibit 10-6.			
Inferior constrictor	See Exhibit 10-6.			
Middle constrictor	See Exhibit 10-6.			
INTRINSIC				
Cricothyroid (*crico* = cricoid cartilage of larynx)	Anterior and lateral portion of cricoid cartilage of larynx.	Anterior border of inferior cornu of thyroid cartilage of larynx and posterior part of inferior border of lamina of thyroid cartilage.	Produces tension and elongation of vocal folds.	External laryngeal branch of vagus nerve (X).

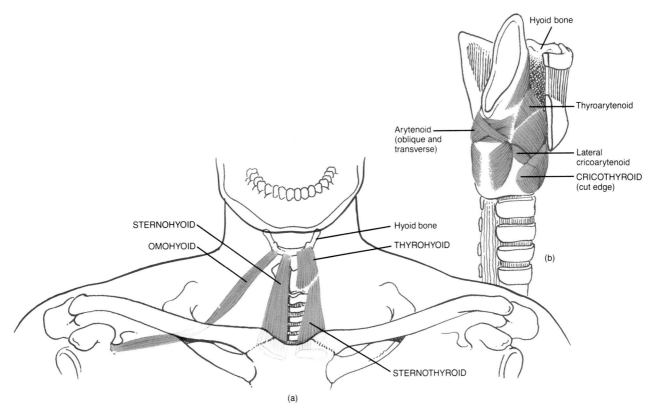

FIGURE 10-9 Muscles of the larynx. (a) Anterior view. (b) Right lateral view.

EXHIBIT 10-8 MUSCLES THAT MOVE THE HEAD

MUSCLE	ORIGIN	INSERTION	ACTION	INNERVATION
Sternocleidomas-toid (*sternum* = breastbone; *cleido* = clavicle; *mastoid* = mastoid process of temporal bone) (see Figure 10-14)	Sternum and clavicle.	Mastoid process of temporal bone.	Contraction of both muscles flexes neck on the chest; contraction of one rotates face toward side opposite contracting muscle.	Accessory nerve (XI) and cervical nerves C2–C3.
Semispinalis capitis (*semi* = half; *spine* = spinous process; *caput* = head) (see Figure 10-18)	Articular process of seventh cervical vertebra and transverse processes of first six thoracic vertebrae.	Occipital bone.	Contraction of both extends head; contraction of one rotates face toward same side as contracting muscle.	Dorsal rami of spinal nerves.
Splenius capitis (*splenion* = bandage) (see Figure 10-18)	Ligamentum nuchae and spines of seventh cervical vertebra and first four thoracic vertebrae.	Occipital bone and mastoid process of temporal bone.	Contraction of both extends head; contraction of one rotates it toward same side as contracting muscle.	Dorsal rami of middle and lower cervical nerves.
Longissimus capitis (*longissimus* = longest) (see Figure 10-18)	Transverse processes of last four cervical vertebrae.	Mastoid process of temporal bone.	Extends head and rotates face toward side opposite contracting muscle.	Dorsal rami of middle and lower cervical nerves.

EXHIBIT 10-9 MUSCLES THAT ACT ON THE ANTERIOR ABDOMINAL WALL (Figure 10-10)

MUSCLE	ORIGIN	INSERTION	ACTION	INNERVATION
Rectus abdominis (*rectus* = fibers parallel to midline; *abdomino* = belly)	Pubic crest and symphysis pubis.	Cartilage of fifth to seventh ribs and xiphoid process.	Flexes vertebral column.	Branches of thoracic nerves T7–T12.
External oblique (*external* = closer to the surface; *oblique* = fibers diagonal to midline)	Lower eight ribs.	Iliac crest; linea alba (midline aponeurosis).	Contraction of both compresses abdomen; contraction of one side alone bends vertebral column laterally.	Branches of thoracic nerves T7–T12, and iliohypogastric nerve.
Internal oblique (*internal* = farther from the surface)	Iliac crest, inguinal ligament, and thoracolumbar fascia.	Cartilage of last three or four ribs.	Compresses abdomen; contraction of one side alone bends vertebral column laterally.	Branches of thoracic nerves T8–T12, iliohypogastric, and ilioinguinal nerves.
Transversus abdominis (*transverse* = fibers perpendicular to midline)	Iliac crest, inguinal ligament, lumbar fascia, and cartilages of last six ribs.	Xiphoid process, linea alba, and pubis.	Compresses abdomen.	Branches of thoracic nerves T8–T12, iliohypogastric, and ilioinguinal nerves.

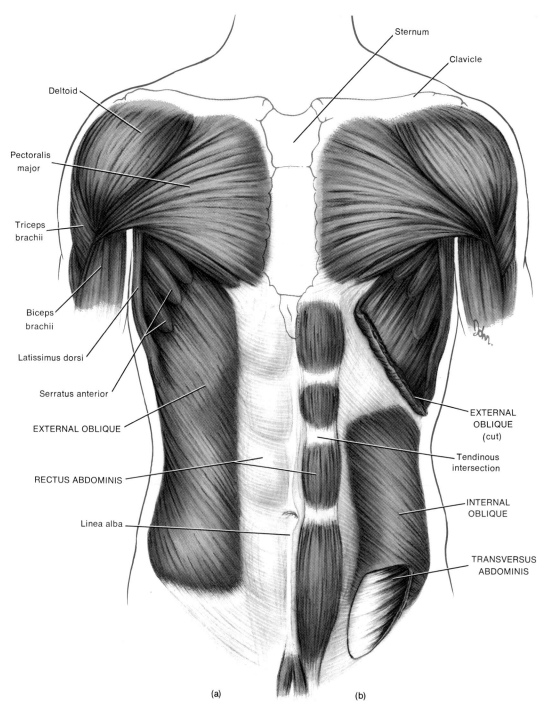

Sternum

Clavicle

Deltoid

Pectoralis major

Triceps brachii

Biceps brachii

Latissimus dorsi

Serratus anterior

EXTERNAL OBLIQUE

RECTUS ABDOMINIS

Linea alba

EXTERNAL OBLIQUE (cut)

Tendinous intersection

INTERNAL OBLIQUE

TRANSVERSUS ABDOMINIS

(a)

(b)

FIGURE 10-10 Muscles of the anterior abdominal wall. (a) Superficial view. (b) Deep view.

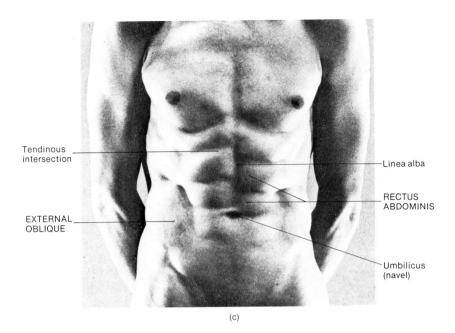

Tendinous intersection

Linea alba

RECTUS ABDOMINIS

EXTERNAL OBLIQUE

Umbilicus (navel)

(c)

FIGURE 10-10 (*Continued*) Muscles of the anterior abdominal wall. (c) Surface anatomy photograph. (Courtesy of R. D. Lockhart, *Living Anatomy,* Faber and Faber, 1974.)

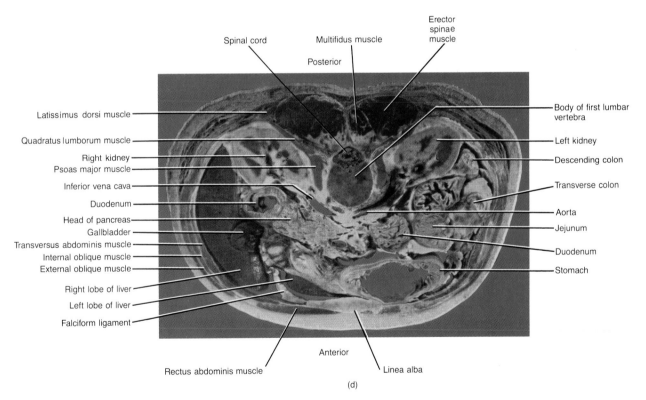

Spinal cord

Multifidus muscle

Erector spinae muscle

Posterior

Latissimus dorsi muscle

Quadratus lumborum muscle

Right kidney

Psoas major muscle

Inferior vena cava

Duodenum

Head of pancreas

Gallbladder

Transversus abdominis muscle

Internal oblique muscle

External oblique muscle

Right lobe of liver

Left lobe of liver

Falciform ligament

Body of first lumbar vertebra

Left kidney

Descending colon

Transverse colon

Aorta

Jejunum

Duodenum

Stomach

Rectus abdominis muscle

Anterior

Linea alba

(d)

FIGURE 10-10 (*Continued*) Muscles of the anterior abdominal wall. (d) Photograph of a cross section through the abdomen showing the musculature and related viscera. (Courtesy of Stephen A. Kieffer and E. Robert Heitzman, *An Atlas of Cross-Sectional Anatomy,* Harper & Row, Publishers, Inc., New York, 1979.)

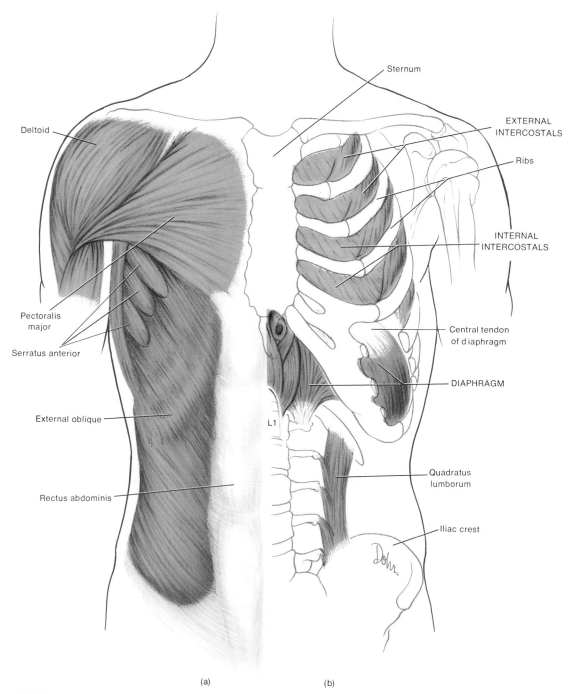

(a)   (b)

FIGURE 10-11 Muscles used in breathing. (a) Diagram of superficial view. (b) Diagram of deep view.

EXHIBIT 10-10 MUSCLES USED IN BREATHING (Figure 10-11)

MUSCLE	ORIGIN	INSERTION	ACTION	INNERVATION
Diaphragm (*dia* = across; *phragma* = wall)	Xiphoid process, costal cartilages of last six ribs, and lumbar vertebrae.	Central tendon.	Forms floor of thoracic cavity; pulls central tendon downward during inspiration and thus increases vertical length of thorax.	Phrenic nerve.
External intercostals (*inter* = between; *costa* = rib)	Inferior border of rib above.	Superior border of rib below.	Elevate ribs during inspiration and thus increase lateral and anteroposterior dimensions of thorax.	Intercostal nerves.
Internal intercostals	Superior border of rib below.	Inferior border of rib above.	Draw adjacent ribs together during forced expiration and thus decrease lateral and anteroposterior dimensions of thorax.	Intercostal nerves.

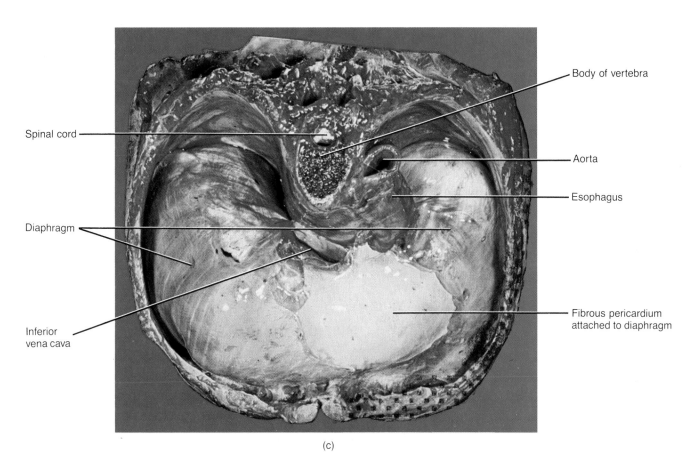

(c)

FIGURE 10-11 (*Continued*) Muscles used in breathing. (c) Photograph of superior view of diaphragm. (Courtesy of C. Yokochi and J. W. Rohen, *Photographic Anatomy of the Human Body,* 1st ed., 1969, IGAKU-SHOIN, Ltd., Tokyo, New York.)

EXHIBIT 10-11 **MUSCLES OF THE PELVIC FLOOR* (Figure 10-12)**

MUSCLE	ORIGIN	INSERTION	ACTION	INNERVATION
Levator ani (*levator* = raises; *ani* = anus)	This muscle is divisible into two parts, the pubococcygeus muscle and the iliococcygeus muscle. It forms the funnel-shaped floor of the pelvic cavity and supports the pelvic structures. It contains openings for the anal canal and urethra in both sexes and the vagina in the female.			
Pubococcygeus (*pubo* = pubis; *coccygeus* = coccyx)	Pubis.	Coccyx, urethra, anal canal, and central tendon of perineum.	Supports and slightly raises pelvic floor, resists increased intraabdominal pressure, and draws anus toward pubis and constricts it.	Sacral nerves S3–S4 or S4 and perineal branch of pudendal nerve.
Iliococcygeus (*ilio* = ilium)	Ischial spine.	Coccyx.	Supports and slightly raises pelvic floor, resists increased intraabdominal pressure, and draws anus toward pubis and constricts it.	Sacral nerves S3–S4 or S4 and perineal branch of pudendal nerve.
Coccygeus	Ischial spine.	Lower sacrum and upper coccyx.	Supports and slightly raises pelvic floor, resists intraabdominal pressure, and pulls coccyx forward following defecation or parturition.	Sacral nerve S3 or S4.

* The muscles of the pelvic floor, together with the fasciae covering their internal and external surfaces, are collectively referred to as the **pelvic diaphragm.**

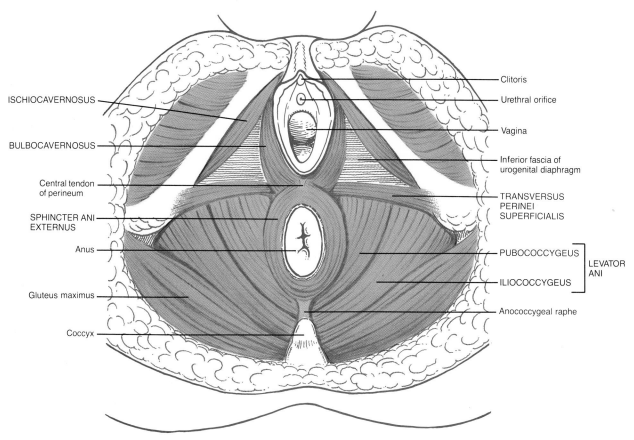

FIGURE 10-12 Muscles of the pelvic floor seen in the female perineum.

EXHIBIT 10-12 MUSCLES OF THE PERINEUM* (Figure 10-13)

MUSCLE	ORIGIN	INSERTION	ACTION	INNERVATION
Transversus perinei superficialis (*transversus* = transverse; *perinei* = perineum; *superficialis* = superficial)	Ischial tuberosity.	Central tendon of perineum.	Helps to stabilize the central tendon of the perineum.	Perineal branch of pudendal nerve.
Bulbocavernosus (*bulbus* = bulb; *caverna* = hollow place)	Central tendon of perineum.	Inferior fascia of the urogenital diaphragm, corpus spongiosum of the penis, and deep fascia on the dorsum of the penis in the male; pubic arch and root and dorsum of clitoris in the female.	Helps expel last drops of urine during micturition, helps propel semen along the urethra, and assists in erection of the penis in the male; decreases vaginal orifice and assists in erection of the clitoris in the female.	Perineal branch of pudendal nerve.
Ischiocavernosus (*ischion* = hip)	Ischial tuberosity and ischial and pubic rami.	Corpus cavernosum of the penis in the male and the clitoris in the female.	Maintains erection of the penis in the male and the clitoris in the female.	Perineal branch of pudendal nerve.
Transversus perinei profundus† (*profundus* = deep)	Ischial rami.	Central tendon of perineum.	Helps eject last drops of urine and semen in the male and urine in the female.	Perineal branch of pudendal nerve.
Sphincter urethrae† (*sphincter* = circular muscle that decreases the size of an opening; *urethrae* = urethra)	Ischial and pubic rami.	Median raphe in the male and vaginal wall in the female.	Helps eject last drops of urine and semen in the male and urine in the female.	Perineal branch of pudendal nerve.
Sphincter ani externus (*externus* = external)	Anococcygeal raphe.	Central tendon of perineum.	Keeps anal canal and orifice closed.	Sacral nerve S4 and inferior rectal branch of pudendal nerve.

* The **perineum** is the entire outlet of the pelvis. It is a diamond-shaped area at the lower end of the trunk between the thighs and buttocks. It is bordered anteriorly by the symphysis pubis, laterally by the ischial tuberosities, and posteriorly by the coccyx. A transverse line drawn between the ischial tuberosities divides the perineum into an anterior **urogenital triangle** that contains the external genitals and a posterior **anal triangle** that contains the anus.

† The transversus perinei profundus, the sphincter urethrae, and a fibrous membrane constitute the **urogenital diaphragm.** It surrounds the urogenital ducts and helps to strengthen the pelvic floor.

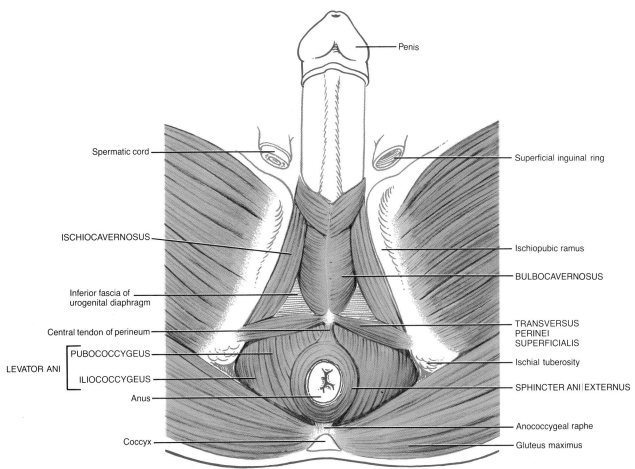

Penis

Spermatic cord

Superficial inguinal ring

ISCHIOCAVERNOSUS

Ischiopubic ramus

BULBOCAVERNOSUS

Inferior fascia of urogenital diaphragm

Central tendon of perineum

TRANSVERSUS PERINEI SUPERFICIALIS

LEVATOR ANI

PUBOCOCCYGEUS

Ischial tuberosity

ILIOCOCCYGEUS

SPHINCTER ANI EXTERNUS

Anus

Coccyx

Anococcygeal raphe

Gluteus maximus

FIGURE 10-13 Muscles of the male perineum.

EXHIBIT 10-13 MUSCLES THAT MOVE THE SHOULDER GIRDLE (Figure 10-14)

MUSCLE	ORIGIN	INSERTION	ACTION	INNERVATION
Subclavius (*sub* = under; *clavius* = clavicle)	First rib.	Clavicle.	Depresses clavicle.	Nerve to subclavius.
Pectoralis minor (*pectus* = breast, chest, thorax; *minor* = lesser)	Third through fifth ribs.	Coracoid process of scapula.	Depresses scapula, rotates shoulder joint anteriorly, and elevates third through fifth ribs during forced inspiration when scapula is fixed.	Medial pectoral nerve.
Serratus anterior (*serratus* = serrated; *anterior* = front)	Upper eight or nine ribs.	Vertebral border and inferior angle of scapula.	Rotates scapula laterally and elevates ribs when scapula is fixed.	Long thoracic nerve.
Trapezius (*trapezoides* = trapezoid-shaped)	Occipital bone, ligamentum nuchae, and spines of seventh cervical and all thoracic vertebrae.	Acromion process of clavicle and spine of scapula.	Elevates clavicle, adducts scapula, elevates or depresses scapula, and extends head.	Acessory nerve (XI) and C3–C4
Levator scapulae (*levator* = raises; *scapulae* = scapula)	Upper four or five cervical vertebrae.	Vertebral border of scapula.	Elevates scapula.	Dorsal scapular nerve and C3–C5.
Rhomboideus major (*rhomboides* = rhomboid or diamond-shaped)	Spines of second to fifth thoracic vertebrae.	Vertebral border of scapula.	Adducts scapula and slightly rotates it upward.	Dorsal scapular nerve.
Rhomboideus minor	Spines of seventh cervical and first thoracic vertebrae.	Superior angle of scapula.	Adducts scapula.	Dorsal scapular nerve.

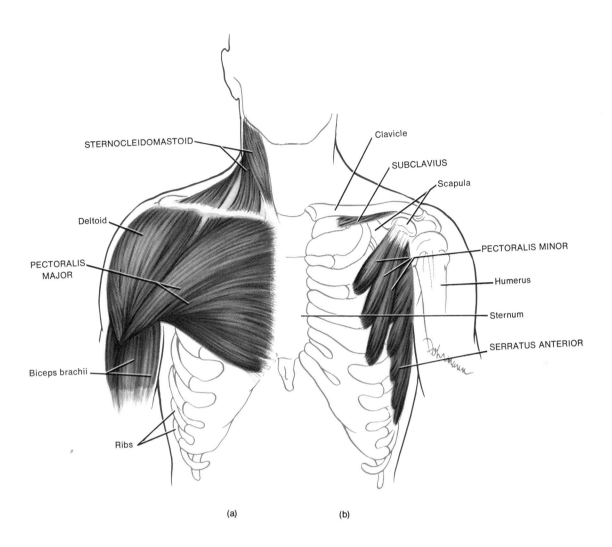

STERNOCLEIDOMASTOID

Deltoid

PECTORALIS
MAJOR

Biceps brachii

Ribs

(a)

Clavicle

SUBCLAVIUS

Scapula

PECTORALIS MINOR

Humerus

Sternum

SERRATUS ANTERIOR

(b)

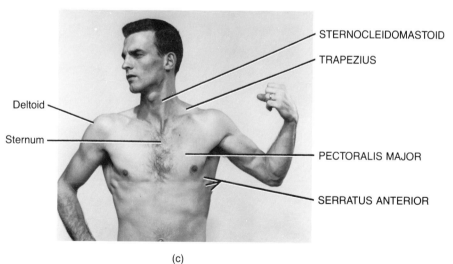

Deltoid

Sternum

STERNOCLEIDOMASTOID

TRAPEZIUS

PECTORALIS MAJOR

SERRATUS ANTERIOR

(c)

FIGURE 10-14 Muscles that move the shoulder girdle. (a) Anterior superficial view. (b) Anterior deep view. (c) Surface anatomy photograph of the neck and chest. (Courtesy of Vincent P. Destro, Mayo Foundation.)

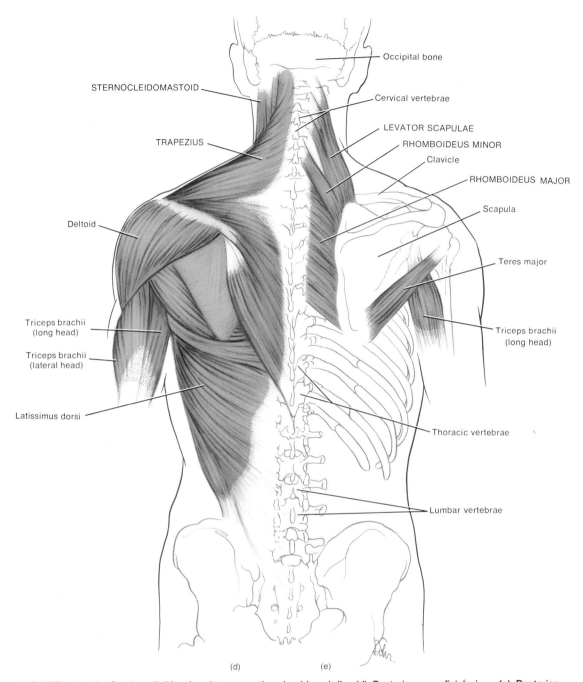

FIGURE 10-14 (*Continued*) Muscles that move the shoulder girdle. (d) Posterior superficial view. (e) Posterior deep view.

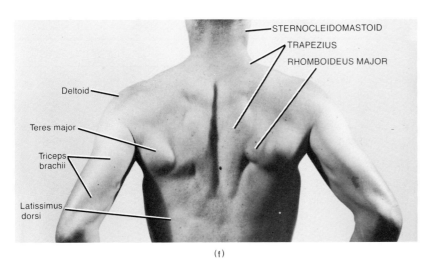

(f)

FIGURE 10-14 (*Continued*) Muscles that move the shoulder girdle. (f) Surface anatomy photograph of the back. (Courtesy of Vincent P. Destro, Mayo Foundation.)

EXHIBIT 10-14 MUSCLES THAT MOVE THE ARM (Figure 10-15)

MUSCLE	ORIGIN	INSERTION	ACTION	INNERVATION
Pectoralis major (see Figure 10-10)	Clavicle, sternum, cartilages of second to sixth ribs.	Greater tubercle of humerus.	Flexes, adducts, and rotates arm medially.	Medial and lateral pectoral nerve.
Latissimus dorsi (*latissimus* = widest; *dorsum* = back)	Spines of lower six thoracic vertebrae, lumbar vertebrae, crests of sacrum and ilium, lower four ribs.	Intertubercular groove of humerus.	Extends, adducts, and rotates arm medially; draws shoulder downward and backward.	Thoracodorsal nerve.
Deltoid (*delta* = triangular)	Clavicle and acromion process and spine of scapula.	Deltoid tuberosity of humerus.	Abducts arm.	Axillary nerve.
Supraspinatus (*supra* = above; *spinatus* = spine of scapula)	Fossa superior to spine of scapula.	Greater tubercle of humerus.	Assists deltoid muscle in abducting arm.	Suprascapular nerve.
Infraspinatus (*infra* = below)	Fossa inferior to spine of scapula.	Greater tubercle of humerus.	Rotates arm laterally.	Suprascapular nerve.
Teres major (*teres* = long and round)	Inferior angle of scapula.	Distal to lesser tubercle of humerus.	Extends arm and draws it down; assists in adduction and medial rotation of arm.	Lower subscapular nerve.
Teres minor	Lateral border of scapula.	Greater tubercle of humerus.	Rotates arm laterally.	Axillary nerve.

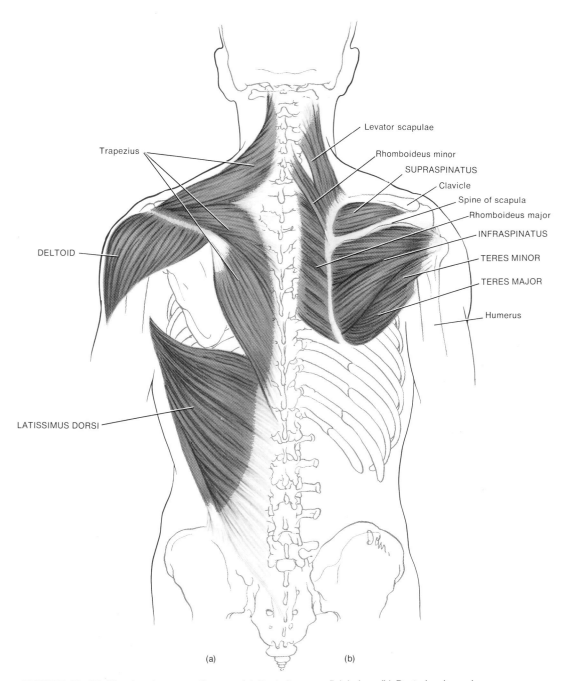

FIGURE 10-15 Muscles that move the arm. (a) Posterior superficial view. (b) Posterior deep view.

DELTOID

Trapezius over
SUPRASPINATUS

INFRASPINATUS

TERES MAJOR

Medial border
of scapula

Trapezius

Inferior angle
of scapula

LATISSIMUS DORSI

(c)

FIGURE 10-15 (*Continued*) Muscles that move the arm. (c) Surface anatomy photograph of the back. (Courtesy of J. Royce, *Surface Anatomy,* Davis, 1965.)

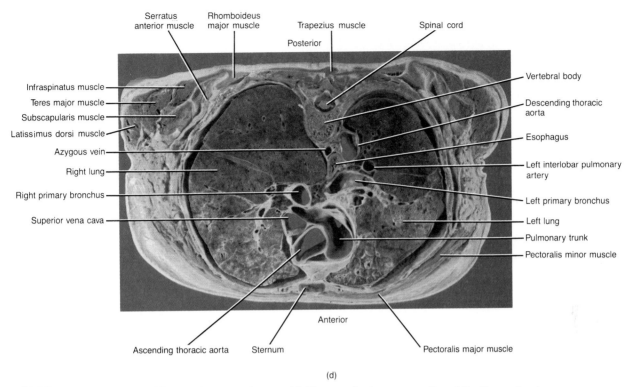

Serratus anterior muscle

Rhomboideus major muscle

Trapezius muscle

Posterior

Spinal cord

Infraspinatus muscle

Teres major muscle

Subscapularis muscle

Latissimus dorsi muscle

Azygous vein

Right lung

Right primary bronchus

Superior vena cava

Vertebral body

Descending thoracic aorta

Esophagus

Left interlobar pulmonary artery

Left primary bronchus

Left lung

Pulmonary trunk

Pectoralis minor muscle

Ascending thoracic aorta

Sternum

Anterior

Pectoralis major muscle

(d)

FIGURE 10-15 (*Continued*) Muscles that move the arm. (d) Photograph of a cross section of the thorax showing the musculature and related viscera. (Courtesy of Stephen A. Kieffer and E. Robert Heitzman, *An Atlas of Cross-Sectional Anatomy*, Harper & Row, Publishers, Inc., New York, 1979.)

Before you move on to Exhibit 10-15 (Muscles That Move the Forearm), refer to Figure 1-4b. This figure is a cross section of the trunk at the level of the heart and lungs. Several of the muscles you have studied up to this point are shown. Figure 1-4b will add to your understanding of how muscles are arranged with respect to each other from external to internal. It will also show you how muscles are oriented with regard to bones and viscera.

EXHIBIT 10-15 MUSCLES THAT MOVE THE FOREARM (Figure 10-16)

MUSCLE	ORIGIN	INSERTION	ACTION	INNERVATION
Biceps brachii (*biceps* = two heads of origin; *brachion* = arm)	Long head originates from tubercle above glenoid cavity; short head originates from coracoid process of scapula.	Radial tuberosity and bicipital aponeurosis.	Flexes and supinates forearm.	Musculocutaneous nerve.
Brachialis	Anterior surface of humerus.	Tuberosity and coronoid process of ulna.	Flexes forearm.	Musculocutaneous, radial, and median nerves.
Brachioradialis (*radialis* = radius) (see also Figure 10-17)	Supracondyloid ridge of humerus.	Superior to styloid process of radius.	Flexes forearm.	Radial nerve.
Triceps brachii (*triceps* = three heads of origin)	Long head originates from infraglenoid tuberosity of scapula; lateral head originates from lateral and posterior surface of humerus superior to radial groove; medial head originates from posterior surface of humerus inferior to radial groove.	Olecranon of ulna.	Extends forearm.	Radial nerve.
Coracobrachialis (*coraco* = coracoid process)	Coracoid process of scapula.	Middle of medial surface of shaft of humerus.	Flexes and adducts forearm.	Musculocutaneous nerve.
Anconeus (*anconeal* = pertaining to the elbow) (see Figure 10-17)	Lateral epicondyle of humerus.	Olecranon and superior portion of shaft of ulna.	Extends forearm.	Radial nerve.
Supinator (*supination* = turning palm upward or anteriorly)	Lateral epicondyle of humerus and ridge on ulna.	Oblique line of radius.	Supinates forearm.	Deep radial nerve.
Pronator teres (*pronation* = turning palm downward or posteriorly)	Medial epicondyle of humerus and coronoid process of ulna.	Midlateral surface of radius.	Pronates forearm.	Median nerve.
Pronator quadratus (*quadratus* = squared, four-sided)	Distal portion of shaft of ulna.	Inferior portion of shaft of radius.	Pronates and rotates forearm.	Median nerve.

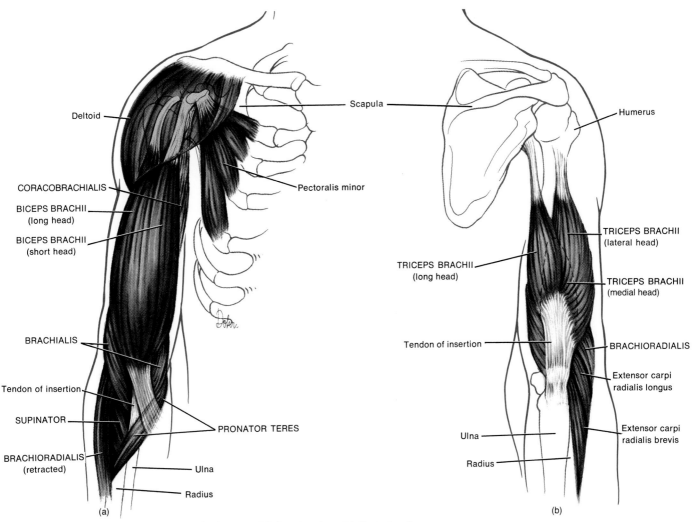

FIGURE 10-16 Muscles that move the forearm. (a) Anterior view. (b) Posterior view.

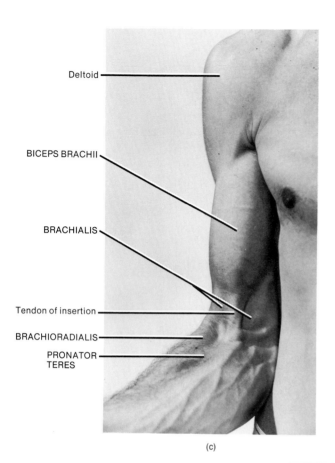

Deltoid

BICEPS BRACHII

BRACHIALIS

Tendon of insertion

BRACHIORADIALIS

PRONATOR
TERES

(c)

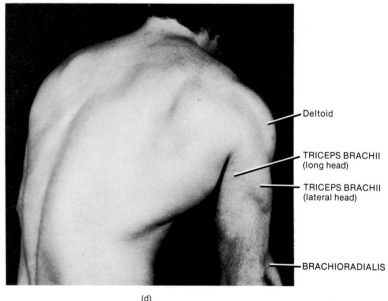

Deltoid

TRICEPS BRACHII
(long head)

TRICEPS BRACHII
(lateral head)

BRACHIORADIALIS

(d)

FIGURE 10-16 (*Continued*) Muscles that move the forearm. (c) Surface anatomy photograph of the anterior arm and upper forearm. (Courtesy of Vincent P. Destro, Mayo Foundation.) (d) Surface anatomy photograph of the posterior arm. (Courtesy of Donald Castellaro and Richard Sollazzo.)

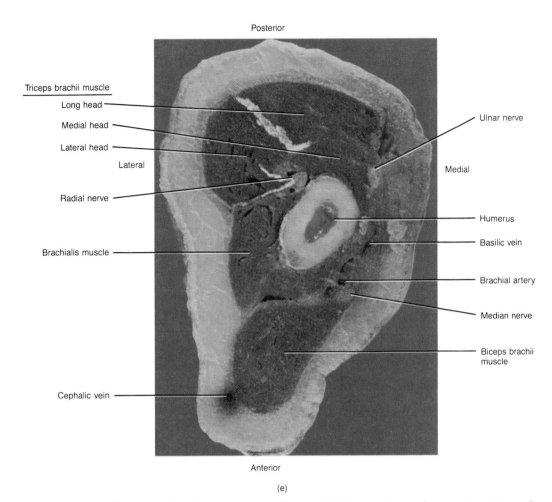

Posterior

Triceps brachii muscle

Long head

Medial head

Lateral head

Lateral

Radial nerve

Brachialis muscle

Cephalic vein

Ulnar nerve

Medial

Humerus

Basilic vein

Brachial artery

Median nerve

Biceps brachii muscle

Anterior

(e)

FIGURE 10-16 (*Continued*) Muscles that move the forearm. (e) Photograph of a cross section of the arm. (Courtesy of Stephen A. Kieffer and E. Robert Heitzman, *An Atlas of Cross-Sectional Anatomy,* Harper & Row, Publishers, Inc., New York, 1979.)

EXHIBIT 10-16 MUSCLES THAT MOVE THE WRIST AND FINGERS (Figure 10-17)

MUSCLE	ORIGIN	INSERTION	ACTION	INNERVATION
Flexor carpi radialis (*flexor* = decreases angle at joint; *carpus* = wrist; *radialis* = radius)	Medial epicondyle of humerus.	Second and third metacarpals.	Flexes and abducts wrist.	Median nerve.
Flexor carpi ulnaris (*ulnaris* = ulna)	Medial epicondyle of humerus and upper dorsal border of ulna.	Pisiform, hamate, and fifth metacarpal.	Flexes and adducts wrist.	Ulnar nerve.
Palmaris longus (*palma*= palm)	Medial epicondyle of humerus.	Transverse carpal ligament and palmar aponeurosis.	Flexes wrist and tenses palmar aponeurosis.	Median nerve.
Extensor carpi radialis longus (*extensor* = increases angle at joint; *longus* = long)	Lateral epicondyle of humerus.	Second metacarpal.	Extends and abducts wrist.	Radial nerve.
Extensor carpi ulnaris	Lateral epicondyle of humerus and dorsal border of ulna.	Fifth metacarpal.	Extends and adducts wrist.	Deep radial nerve.
Flexor digitorum profundus (*digit* = finger or toe; *profundus* = deep)	Anterior medial surface of body of ulna.	Bases of distal phalanges.	Flexes distal phalanges of each finger.	Median and ulnar nerves.
Flexor digitorum superficialis (*superficialis* = superficial)	Medial epicondyle of humerus, coronoid process of ulna, and oblique line of radius.	Middle phalanges.	Flexes middle phalanges of each finger.	Median nerve.
Extensor digitorum	Lateral epicondyle of humerus.	Middle and distal phalanges of each finger.	Extends phalanges.	Deep radial nerve.
Extensor indicis (*indicis* = index)	Dorsal surface of ulna.	Tendon of extensor digitorum of index finger.	Extends index finger.	Deep radial nerve.

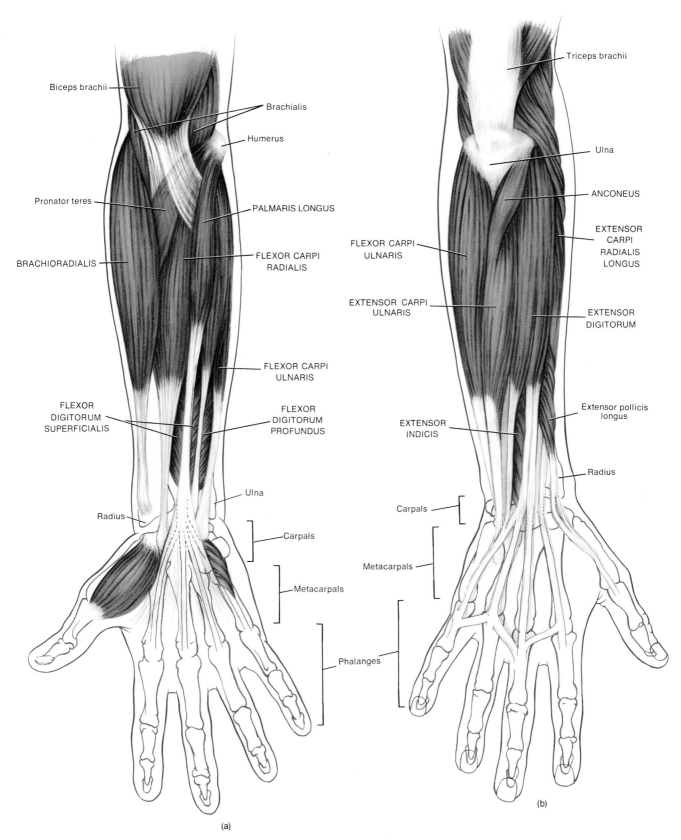

FIGURE 10-17 Muscles that move the wrist and fingers. (a) Anterior view. (b) Posterior view.

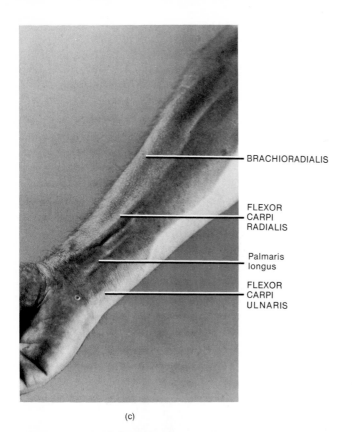

BRACHIORADIALIS

FLEXOR
CARPI
RADIALIS

Palmaris
longus

FLEXOR
CARPI
ULNARIS

(c)

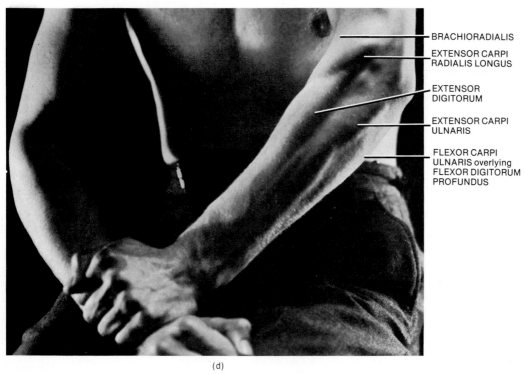

BRACHIORADIALIS

EXTENSOR CARPI
RADIALIS LONGUS

EXTENSOR
DIGITORUM

EXTENSOR CARPI
ULNARIS

FLEXOR CARPI
ULNARIS overlying
FLEXOR DIGITORUM
PROFUNDUS

(d)

FIGURE 10-17 (*Continued*) Muscles that move the wrist and fingers. (c) Surface anatomy photograph of the anterior forearm. (Courtesy of Vincent P. Destro, Mayo Foundation.) (d) Surface anatomy photograph of the posterolateral forearm. (Courtesy of R. D. Lockhart, *Living Anatomy,* Faber and Faber, 1974.)

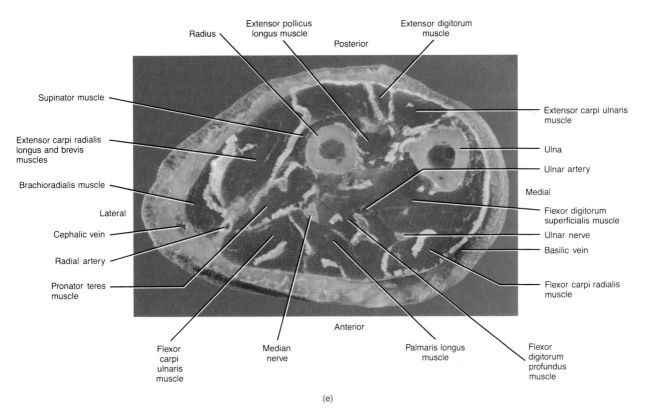

Radius

Extensor pollicus longus muscle

Posterior

Extensor digitorum muscle

Supinator muscle

Extensor carpi radialis longus and brevis muscles

Brachioradialis muscle

Lateral

Cephalic vein

Radial artery

Pronator teres muscle

Extensor carpi ulnaris muscle

Ulna

Ulnar artery

Medial

Flexor digitorum superficialis muscle

Ulnar nerve

Basilic vein

Flexor carpi radialis muscle

Flexor carpi ulnaris muscle

Median nerve

Anterior

Palmaris longus muscle

Flexor digitorum profundus muscle

(e)

FIGURE 10-17 (*Continued*) Muscles that move the wrist and fingers. (e) Photograph of a cross section of the forearm. (Courtesy of Stephen A. Kieffer and E. Robert Heitzman, *An Atlas of Cross-Sectional Anatomy,* Harper & Row, Publishers, Inc., New York, 1979.)

EXHIBIT 10-17 MUSCLES THAT MOVE THE VERTEBRAL COLUMN (Figure 10-18)

MUSCLE	ORIGIN	INSERTION	ACTION	INNERVTION
Rectus abdominis (see Figure 10-10)	Body of pubis of coxal bone.	Cartilages of fifth through seventh ribs.	Flexes vertebral column at lumbar spine and compresses abdomen.	Intercostal nerves T7–T12.
Quadratus lumborum (*quadratus* = squared, four-sided); *lumb* = lumbar region)	Iliac crest.	Twelfth rib and upper four lumbar vertebrae.	Flexes vertebral column laterally.	Intercostal nerve T12 and lumbar nerve L1.
Sacrospinalis (erector spinae) (*sacro* = sacrum; *spine* = spinous process)	This posterior muscle consists of three groupings: iliocostalis, longissimus, and spinalis. These groups, in turn, consist of a series of overlapping muscles. The iliocostalis group is laterally placed, the longissimus group is intermediate in placement, and the spinalis is medially placed.			
Lateral Iliocostalis lumborum (*ilium* = flank; *costa* = rib; *lumbus* = loin)	Iliac crest.	Lower six ribs.	Extends lumbar region of vertebral column.	Dorsal rami of lumbar nerves.
Iliocostalis thoracis (*thorax* = chest)	Lower six ribs.	Upper six ribs.	Maintains erect position of spine.	Dorsal rami of thoracic (intercostal) nerves.
Iliocostalis cervicis (*cervix* = neck)	First six ribs.	Transverse processes of fourth to sixth cervical vertebrae.	Extends cervical region of vertebral column.	Dorsal rami of cervical nerves.
Intermediate Longissimus thoracis (*longissimus* = longest)	Transverse processes of lumbar vertebrae.	Transverse processes of all thoracic and upper lumbar vertebrae and ninth and tenth ribs.	Extends thoracic region of vertebral column.	Dorsal rami of spinal nerves.
Longissimus cervicis	Transverse processes of fourth and fifth thoracic vertebrae.	Transverse processes of second to sixth cervical vertebrae.	Extends cervical region of vertebral column.	Dorsal rami of spinal nerves.
Longissimus capitis (*caput* = head)	Transverse processes of upper four thoracic vertebrae.	Mastoid process of temporal bone.	Extends head and rotates it to opposite side.	Dorsal rami of middle and lower cervical nerves.
Medial Spinalis thoracis	Spines of upper lumbar and lower thoracic vertebrae.	Spines of upper thoracic vertebrae.	Extends vertebral column.	Dorsal rami of spinal nerves.

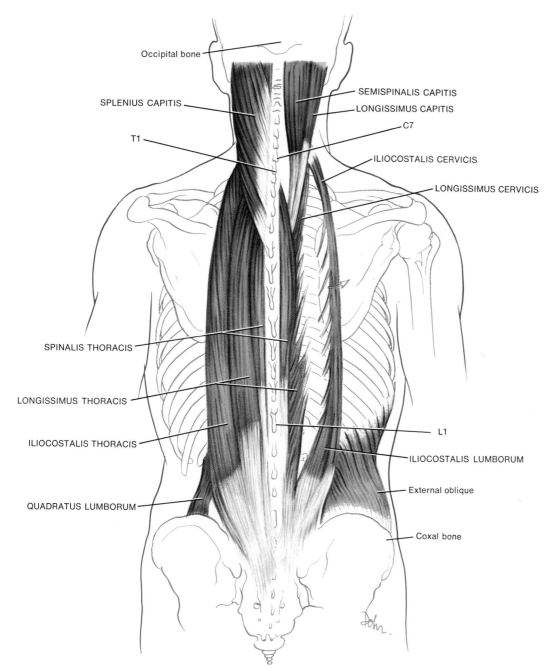

FIGURE 10-18 Muscles that move the vertebral column.

EXHIBIT 10-18 MUSCLES THAT MOVE THE THIGH (Figure 10-19)

MUSCLE	ORIGIN	INSERTION	ACTION	INNERVATION
Psoas major* (*psoa* = muscle of loin)	Transverse processes and bodies of lumbar vertebrae.	Lesser trochanter of femur.	Flexes and rotates thigh laterally; flexes vertebral column.	Lumbar nerves L2-L3.
Iliacus* (*iliac* = ilium)	Iliac fossa.	Tendon of psoas major.	Flexes and rotates thigh laterally; flexes vertebral column slightly.	Femoral nerve.
Gluteus maximus (*gloutos* = buttock; *maximus* = largest)	Iliac crest, sacrum, coccyx, and aponeurosis of sacrospinalis.	Iliotibial tract of fascia lata and gluteal tuberosity of femur.	Extends and rotates thigh laterally.	Inferior gluteal nerve.
Gluteus medius (*media* = middle)	Ilium.	Greater trochanter of femur.	Abducts and rotates thigh medially.	Superior gluteal nerve.
Gluteus minimus (*minimus* = smallest)	Ilium.	Greater trochanter of femur.	Abducts and rotates thigh laterally.	Superior gluteal nerve.
Tensor fasciae latae (*tensor* = makes tense; *fascia* = band; *latus* = wide)	Iliac crest.	Tibia by way of the iliotibial tract.	Flexes and abducts thigh.	Superior gluteal nerve.
Adductor longus (*adductor* = moves part closer to midline; *longus* = long)	Pubic crest and symphysis pubic.	Linea aspera of femur.	Adducts, rotates, and flexes thigh.	Obturator nerve.
Adductor brevis (*brevis* = short)	Inferior ramus of pubis.	Linea aspera of femur.	Adducts, rotates, and flexes thigh.	Obturator nerve.
Adductor magnus (*magnus* = large)	Inferior ramus of pubis and ischium to ischial tuberosity.	Linea aspera of femur.	Adducts, flexes, and extends thigh (anterior part flexes, posterior part extends).	Obturator nerve.
Piriformis (*pirum* = pear; *forma* = shape)	Sacrum.	Greater trochanter of femur.	Rotates thigh laterally and abducts it.	Sacral nerves S2 or S1–S2.
Obturator internus (*obturator* = obturator foramen; *internus* = inside)	Margin of obturator foramen, pubis, and ischium.	Greater trochanter of femur.	Rotates thigh laterally and abducts it.	Obturator nerve.
Pectineus (*pecten* = comb-shaped)	Fascia of pubis.	Pectineal line of femur.	Flexes, adducts, and rotates thigh laterally.	Femoral nerve.

* Together the psoas major and iliacus are sometimes termed the **iliopsoas muscle.**

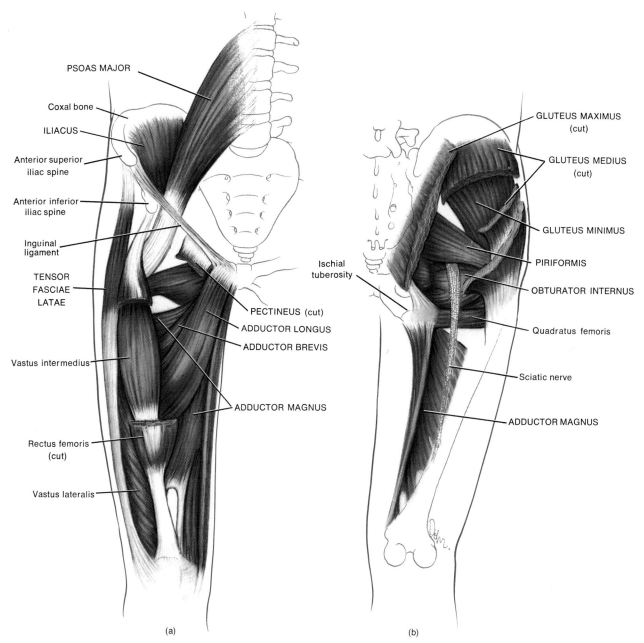

PSOAS MAJOR

Coxal bone

ILIACUS

Anterior superior
iliac spine

Anterior inferior
iliac spine

Inguinal
ligament

TENSOR
FASCIAE
LATAE

Vastus intermedius

Rectus femoris
(cut)

Vastus lateralis

PECTINEUS (cut)

ADDUCTOR LONGUS

ADDUCTOR BREVIS

ADDUCTOR MAGNUS

(a)

GLUTEUS MAXIMUS
(cut)

GLUTEUS MEDIUS
(cut)

GLUTEUS MINIMUS

PIRIFORMIS

OBTURATOR INTERNUS

Quadratus femoris

Sciatic nerve

ADDUCTOR MAGNUS

Ischial
tuberosity

(b)

FIGURE 10-19 Muscles that move the thigh. (a) Anterior view. (b) Posterior view.

EXHIBIT 10-19 MUSCLES THAT ACT ON THE LEG (Figure 10-20)

MUSCLE	ORIGIN	INSERTION	ACTION	INNERVATION
Quadriceps femoris	This composite muscle includes four distinct parts, usually described as four separate muscles. The common tendon that includes the patella and attaches to the tibial tuberosity is known as the **patellar ligament.**			
Rectus femoris (*rectus* = fibers parallel to midline; *femoris* = femur)	Anterior inferior iliac spine.	Upper border of patella.		Femoral nerve.
Vastus lateralis (*vastus* = large; *lateralis* = lateral)	Greater trochanter and linea aspera of femur.	Upper border and sides of patella; tibial tuberosity through patellar ligament (tendon of quadriceps).	All four heads extend leg; rectus portion alone also flexes thigh.	Femoral nerve.
Vastus medialis (*medialis* = medial)	Linea aspera of femur.			Femoral nerve.
Vastus intermedius (*intermedius* = middle)	Anterior and lateral surfaces of body of femur.			Femoral nerve.
Hamstrings	A collective designation for three separate muscles.			
Biceps femoris (*biceps* = two heads of origin)	Long head arises from ischial tuberosity; short head arises from linea aspera of femur.	Head of fibula and lateral condyle of tibia.	Flexes leg and extends thigh.	Tibial nerve from sciatic nerve.
Semitendinosus (*semi* = half; *tendo* = tendon)	Ischial tuberosity.	Proximal part of medial surface of body of tibia.	Flexes leg and extends thigh.	Tibial nerve from sciatic nerve.
Semimembranosus (*membran* = membrane)	Ischial tuberosity.	Medial condyle of tibia.	Flexes leg and extends thigh.	Tibial nerve from sciatic nerve.
Gracilis (*gracilis* = slender)	Symphysis pubis and pubic arch.	Medial surface of body of tibia.	Flexes leg and adducts thigh.	Obturator nerve.
Sartorius (*sartor* = tailor; refers to cross-legged position of tailors)	Anterior superior spine of ilium.	Medial surface of body of tibia.	Flexes leg; flexes thigh and rotates it laterally, thus crossing leg.	Femoral nerve.

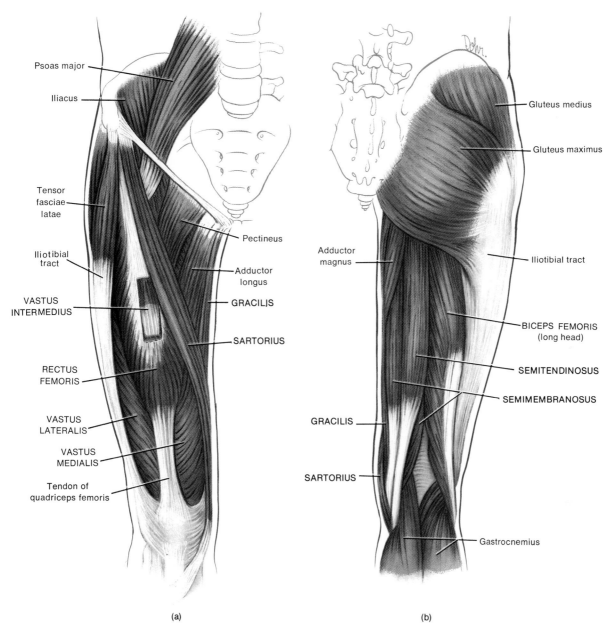

FIGURE 10-20 Muscles that act on the leg. (a) Anterior view. (b) Posterior view.

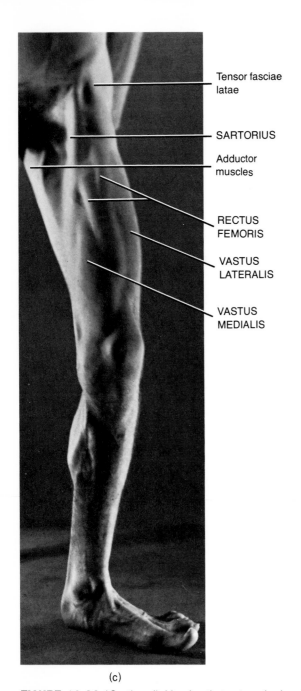

Tensor fasciae latae

SARTORIUS

Adductor muscles

RECTUS FEMORIS

VASTUS LATERALIS

VASTUS MEDIALIS

(c)

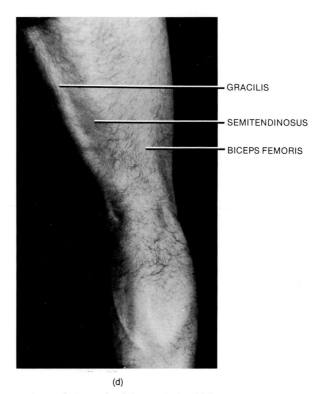

GRACILIS

SEMITENDINOSUS

BICEPS FEMORIS

(d)

FIGURE 10-20 (*Continued*) Muscles that act on the leg. (c) Surface anatomy photograph of the posterior thigh. (Courtesy of R. D. Lockhart, *Living Anatomy,* Faber and Faber, 1974.) (d) Surface anatomy photograph of the posterior thigh and leg. (Courtesy of Donald Castellaro and Richard Sollazzo.)

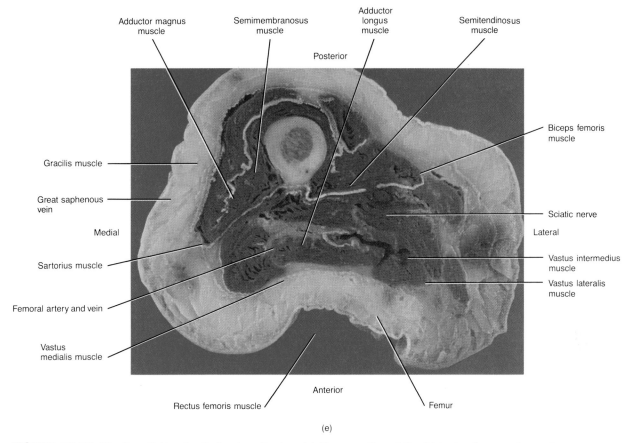

Adductor magnus muscle

Semimembranosus muscle

Adductor longus muscle

Semitendinosus muscle

Posterior

Gracilis muscle

Great saphenous vein

Medial

Sartorius muscle

Femoral artery and vein

Vastus medialis muscle

Rectus femoris muscle

Anterior

Femur

Biceps femoris muscle

Sciatic nerve

Lateral

Vastus intermedius muscle

Vastus lateralis muscle

(e)

FIGURE 10-20 (*Continued*) Muscles that act on the leg. (e) Cross section of the thigh showing muscles and associated structures. (Courtesy of Stephen A. Kieffer and E. Robert Heitzman, *An Atlas of Cross-Sectional Anatomy*, Harper & Row, Publishers, Inc., New York, 1979.)

EXHIBIT 10-20 MUSCLES THAT MOVE THE FOOT AND TOES (Figure 10-21)

MUSCLE	ORIGIN	INSERTION	ACTION	INNERVATION
Gastrocnemius (*gaster* = belly; *kneme* = leg)	Lateral and medial condyles of femur and capsule of knee.	Calcaneus by way of calcaneal ("Achilles") tendon.	Plantar flexes foot.	Tibial nerve.
Soleus (*soleus* = sole of foot)	Head of fibula and medial border of tibia.	Calcaneus by way of calcaneal ("Achilles") tendon	Plantar flexes foot.	Tibial nerve.
Peroneus longus (*perone* = fibula; *longus* = long)	Head and body of fibula and lateral condyle of tibia.	First metatarsal and first cuneiform bone.	Plantar flexes and everts foot.	Superficial peroneal nerve.
Peroneus brevis (*brevis* = short)	Body of fibula.	Fifth metatarsal.	Plantar flexes and everts foot.	Superficial peroneal nerve.
Peroneus tertius (*tertius* = third)	Distal third of fibula.	Fifth metatarsal.	Dorsiflexes and everts foot.	Deep peroneal nerve.
Tibialis anterior (*tibialis* = tibia; *anterior* = front)	Lateral condyle and body of tibia.	First metatarsal and first cuneiform.	Dorsiflexes and inverts foot.	Deep peroneal nerve.
Tibialis posterior (*posterior* = back)	Interosseus membrane between tibia and fibula.	Second, third, and fourth metatarsals; navicular; third cuneiform; and cuboid.	Plantar flexes and inverts foot.	Tibial nerve.
Flexor digitorum longus (*flexor* = decreases angle at joint; *digitorum* = finger or toe)	Tibia.	Distal phalanges of four outer toes.	Flexes toes and plantar flexes and inverts foot.	Tibial nerve.
Extensor digitorum longus (*extensor* = increases angle at joint)	Lateral condyle of tibia and anterior surface of fibula.	Middle and distal phalanges of four outer toes.	Extends toes and dorsiflexes and everts foot.	Deep peroneal nerve.

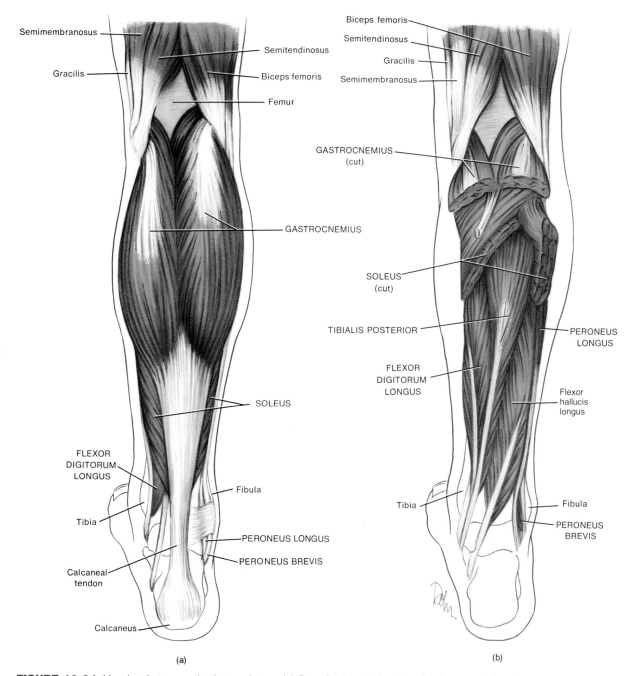

Semimembranosus

Gracilis

Semitendinosus

Biceps femoris

Femur

GASTROCNEMIUS

SOLEUS

FLEXOR
DIGITORUM
LONGUS

Tibia

Fibula

PERONEUS LONGUS

PERONEUS BREVIS

Calcaneal
tendon

Calcaneus

(a)

Biceps femoris

Semitendinosus

Gracilis

Semimembranosus

GASTROCNEMIUS
(cut)

SOLEUS
(cut)

TIBIALIS POSTERIOR

FLEXOR
DIGITORUM
LONGUS

PERONEUS
LONGUS

Flexor
hallucis
longus

Tibia

Fibula

PERONEUS
BREVIS

(b)

FIGURE 10-21 Muscles that move the foot and toes. (a) Superficial posterior view. (b) Deep posterior view.

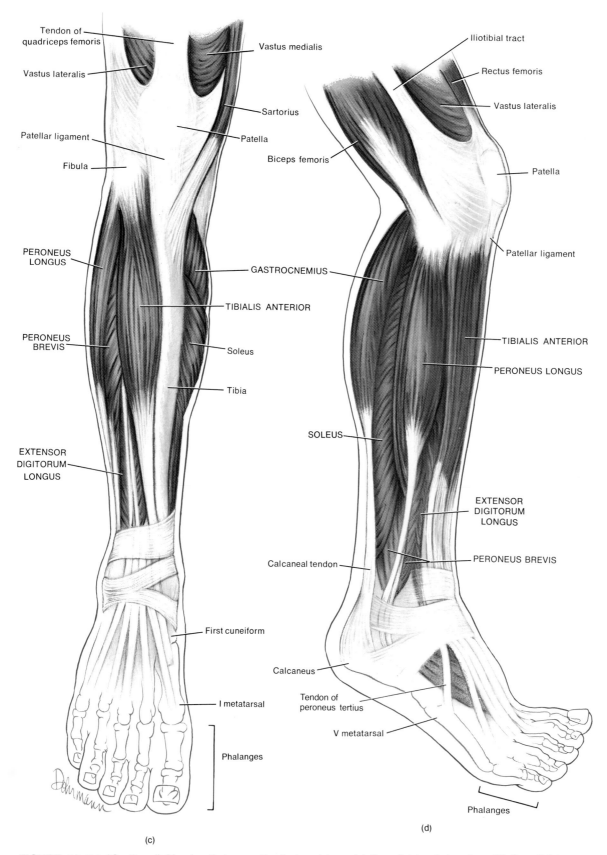

Tendon of quadriceps femoris

Vastus lateralis

Patellar ligament

Fibula

PERONEUS LONGUS

PERONEUS BREVIS

EXTENSOR DIGITORUM LONGUS

Vastus medialis

Sartorius

Patella

GASTROCNEMIUS

TIBIALIS ANTERIOR

Soleus

Tibia

First cuneiform

I metatarsal

Phalanges

(c)

Iliotibial tract

Rectus femoris

Vastus lateralis

Biceps femoris

Patella

Patellar ligament

TIBIALIS ANTERIOR

PERONEUS LONGUS

SOLEUS

EXTENSOR DIGITORUM LONGUS

PERONEUS BREVIS

Calcaneal tendon

Calcaneus

Tendon of peroneus tertius

V metatarsal

Phalanges

(d)

FIGURE 10-21 (*Continued*) Muscles that move the foot and toes. (c) Superficial anterior view. (d) Superficial lateral view.

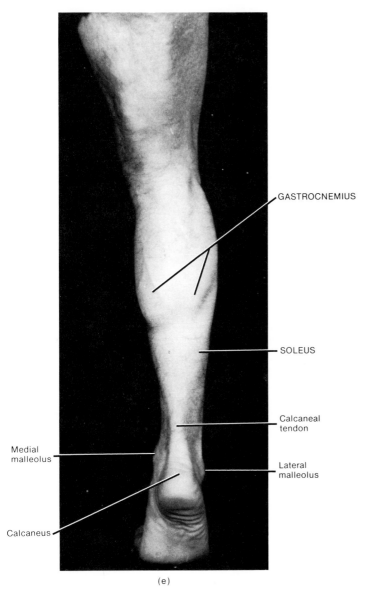

GASTROCNEMIUS

SOLEUS

Calcaneal
tendon

Medial
malleolus

Lateral
malleolus

Calcaneus

(e)

FIGURE 10-21 (*Continued*) Muscles that move the foot and toes.
(e) Surface anatomy of the posterior leg. (Courtesy of Donald Castellaro and Richard Sollazzo.)

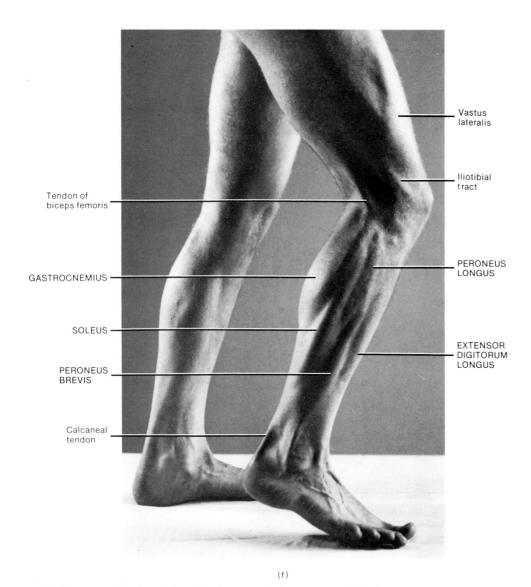

Vastus
lateralis

Iliotibial
tract

Tendon of
biceps femoris

PERONEUS
LONGUS

GASTROCNEMIUS

SOLEUS

EXTENSOR
DIGITORUM
LONGUS

PERONEUS
BREVIS

Calcaneal
tendon

(f)

FIGURE 10-21 (*Continued*) Muscles that move the foot and toes. (f) Surface anatomy photograph of the lateral leg. (Courtesy of R. D. Lockhart, *Living Anatomy,* Faber and Faber, 1974.)

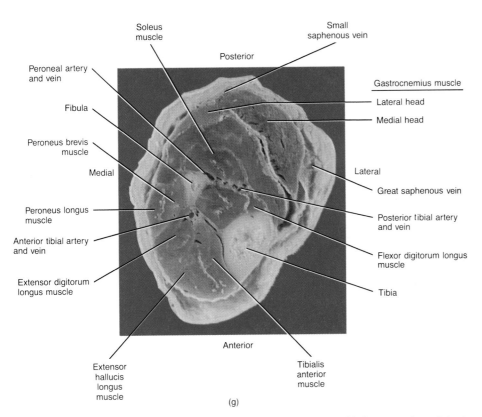

Soleus muscle

Small saphenous vein

Posterior

Peroneal artery and vein

Gastrocnemius muscle

Lateral head

Fibula

Medial head

Peroneus brevis muscle

Medial

Lateral

Great saphenous vein

Peroneus longus muscle

Posterior tibial artery and vein

Anterior tibial artery and vein

Flexor digitorum longus muscle

Extensor digitorum longus muscle

Tibia

Anterior

Extensor hallucis longus muscle

Tibialis anterior muscle

(g)

FIGURE 10-21 (*Continued*) Muscles that move the foot and toes. (g) Cross section of the leg showing muscles and associated structures. (Courtesy of Stephen A. Kieffer and E. Robert Heitzman, *An Atlas of Cross-Sectional Anatomy,* Harper & Row, Publishers, Inc., New York, 1979.)

INTRAMUSCULAR INJECTIONS

Drugs may be introduced into the body by several routes. The term **parenteral** (*par* = aside from; *enteron* = intestine) means administration by a route other than the intestines, that is, by injection. Among the parenteral routes of administration are:

1. Intradermal or **intracutaneous.** The needle tip penetrates the epidermis and is inserted into the dermis.

2. Subcutaneous or **hypodermic.** The needle tip is inserted into the subcutaneous layer.

3. Intramuscular. The needle tip is inserted into muscle.

4. Intravenous. The needle tip is inserted directly into a vein.

5. Intraspinal. The needle tip is inserted into the vertebral canal.

At this point, we will discuss only intramuscular injections.

An intramuscular injection penetrates the skin and subcutaneous tissue to enter the muscle itself. Intramuscular injections are preferred when prompt absorption is desired, when larger doses than can be given cutaneously are indicated, or when the drug is too irritating to give subcutaneously. The common sites for intramuscular injections include the buttock, lateral side of the thigh,

and the deltoid region of the arm. Muscles in these areas, especially the gluteal muscles in the buttock, are fairly thick. Because of the large number of muscle fibers and extensive fascia, the drug has a large surface area for absorption. Absorption is further promoted by the extensive blood supply to muscles. Ideally, intramuscular injections should be given deep within the muscle and away from major nerves and blood vessels.

For many intramuscular injections, the preferred site is the gluteus medius muscle of the buttock (Figure 10-22a). The buttock is divided into four quadrants and the upper outer quadrant is used as the injection site. The iliac crest serves as a landmark for this quadrant. The spot for injection is about 5 to 7.5 cm (2 to 3 inches) below the iliac crest. The upper outer quadrant is chosen because, in this area, the muscle is quite thick and contains few nerves. Thus there is less chance of injury to the sciatic nerve, which can cause paralysis of the lower extremity. The probability of injecting the drug into a blood vessel is also remote in this area. After the needle is inserted into the muscle, the plunger is pulled up for a few seconds. If the syringe fills with blood, the needle is in a blood vessel and a different injection site on the opposite buttock is chosen.

Injections given in the lateral side of the thigh are inserted into the midportion of the vastus lateralis muscle

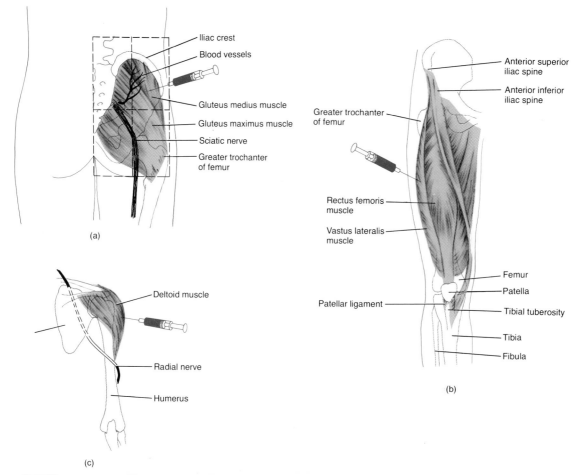

FIGURE 10-22 Intramuscular injections. Shown are the three common sites for intramuscular injections. (a) Buttock. (b) Lateral surface of the thigh. (c) Deltoid region of the arm.

(Figure 10-22b). This site is determined by using the knee and the greater trochanter of the femur as landmarks. The midportion of the muscle is located by measuring a handbreadth above the knee and a handbreadth below the greater trochanter.

The deltoid injection is given in the midportion of the muscle about two to three fingerbreadths below the acromion of the scapula and lateral to the axilla (Figure 10-22c).

STUDY OUTLINE

How Skeletal Muscles Produce Movement
1. Skeletal muscles produce movement by pulling on bones.
2. The stationary attachment is the origin. The movable attachment is the insertion.
3. Bones serve as levers and joints as fulcrums. The lever is acted on by two different forces: resistance and effort.
4. Levers are categorized into three types—first-class, second-class, and third-class—according to the position of the fulcrum, effort, and resistance on the lever.
5. Fascicular arrangements include parallel, convergent, pennate, and circular. Fascicular arrangement is correlated with the power of a muscle and the range of movement.
6. The agonist or prime mover produces the desired action. The antagonist produces an opposite action. The synergist assists the agonist by reducing unnecessary movement.

Naming Skeletal Muscles
Skeletal muscles are named on the basis of distinctive criteria: direction of fibers, location, size, number of

origins (or heads), shape, origin and insertion, and action.

Principal Skeletal Muscles
1. The principal skeletal muscles of the body are grouped according to region in Exhibits 10-2 through 10-20.
2. A knowledge of surface anatomy will help you identify certain superficial muscles by visual inspection or palpation through the skin.

Intramuscular Injections
1. Parenteral routes may be intradermal, subcutaneous, intramuscular, intravenous, or intraspinal.
2. Advantages of intramuscular injections are prompt absorption, use of larger doses than can be given cutaneously, and minimal irritation.
3. Common sites for intramuscular injections are the buttock, lateral side of the thigh, and deltoid region of the shoulder.

REVIEW QUESTIONS

1. What is meant by the muscular system? Explain fully.
2. Using the terms origin, insertion, and belly in your discussion, describe how skeletal muscles produce body movements by pulling on bones.
3. What is a lever? Fulcrum? Apply these terms to the body, and indicate the nature of the forces that act on levers. Describe the three classes of levers, and provide one example for each in the body.
4. Describe the various arrangements of fasciculi. How is fascicular arrangement correlated with the strength of a muscle and its range of movement?
5. Define the role of the agonist, antagonist, and synergist in producing body movements.
6. Select at random several muscles presented in Exhibits 10-2 through 10-20 and see if you can determine the criterion or criteria employed for naming each. In addition, refer to the prefixes, suffixes, roots, and definitions in each exhibit as a guide. Select as many muscles as you wish, as long as you feel you understand the concept involved.
7. Discuss the muscles and their actions involved in facial expression.
8. What muscles would you use to do the following: (a) frown, (b) pout, (c) show surprise, (d) show your upper teeth, (e) pucker your lips, (f) squint, (g) blow up a balloon, (h) smile?
9. What are the principal muscles that move the mandible? Give the function of each.
10. What would happen if you lost tone in the masseter and temporalis muscles?
11. What muscles move the eyeball? In what direction does each muscle move the eyeball?
12. Describe the action of each of the muscles acting on the tongue.
13. What tongue, facial, and mandibular muscles would you use when chewing a piece of gum?
14. What muscles constrict the pharynx? What muscle dilates the pharynx?
15. Describe the actions of the extrinsic and intrinsic muscles of the larynx.
16. What muscles are responsible for moving the head, and how do they move the head?
17. What muscles would you use to signify "yes" and "no" by moving your head?
18. What muscles accomplish compression of the anterior abdominal wall?
19. What are the principal muscles involved in breathing? What are their actions?
20. Describe the actions of the muscles of the pelvic floor.
21. Describe the actions of the muscles of the perineum.
22. In what directions is the shoulder girdle drawn? What muscles accomplish these movements?
23. What muscles are used to (a) raise your shoulders, (b) lower your shoulders, (c) join your hands behind your back, (d) join your hands in front of your chest?
24. What movements are possible at the shoulder joint? What muscles accomplish these movements?
25. What muscles move the arm? In which directions do these movements occur?
26. What muscles move the forearm and what actions are used when striking a match?
27. Discuss the various movements possible at the wrist and fingers. What muscles accomplish these movements?
28. How many muscles and actions of the wrist and fingers used when writing can you list?
29. Discuss the various muscles and movements of the vertebral column.
30. Can you perform an exercise that would involve the use of each of the muscles listed in Exhibit 10-17?
31. What muscles accomplish movements of the femur? What actions are produced by these muscles?
32. Review the various movements involved in your favorite kind of dancing. What muscles listed in Exhibit 10-18 would you be using and what actions would you be performing?
33. What muscles act at the knee joint? What kinds of movements do these muscles perform?
34. Determine the muscles and their actions listed in Exhibit 10-19 that you would use in climbing a ladder

to a diving board, diving into the water, swimming the length of a pool, and then sitting at pool side.

35. Discuss the muscles that plantar flex, evert, pronate, dorsiflex, and supinate the foot.

36. In which directions are the toes moved? What muscles bring about these movements?

37. Define parenteral. What are the various routes of parenteral administration?

38. What are the advantages of intramuscular injections?

39. Describe how you would locate the sites for an intramuscular injection in the buttock, lateral side of the thigh, and deltoid region of the arm.

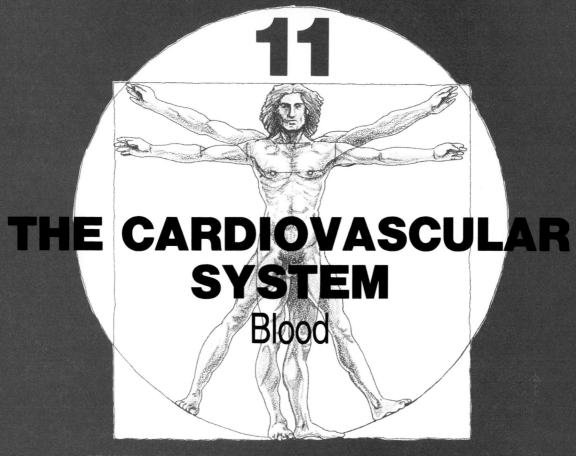

11

THE CARDIOVASCULAR SYSTEM
Blood

STUDENT OBJECTIVES

- Define the principal physical characteristics of blood and its functions in the body.
- Compare the origins of the formed elements in blood and the reticuloendothelial cells.
- Discuss the structure of erythrocytes and their function.
- Define erythropoiesis and describe erythrocyte production and destruction.
- Explain the importance of a reticulocyte count in the diagnosis of abnormal rates of erythrocyte production.
- List the structural features and types of leucocytes.
- Explain the significance of a differential count.
- Discuss the role of leucocytes.
- Discuss the structure and importance of thrombocytes.
- List the components of plasma and explain their importance.
- Compare the location, composition, and functions of interstitial fluid and lymph.
- Contrast the causes and clinical symptoms of hemorrhagic, hemolytic, aplastic, and sickle cell anemia.
- Define polycythemia and describe the importance of hematocrit in its diagnosis.
- Identify the clinical symptoms of infectious mononucleosis and leukemia.
- Define key medical terms associated with blood.

The more specialized a cell becomes, the less capable it is of carrying on an independent existence. For instance, a specialized cell is less capable of protecting itself from extreme temperatures, toxic chemicals, and changes in pH. It may be incapable of devouring whole bits of food. And, if it is firmly implanted in a tissue, it cannot go looking for food or move away from its own wastes. The substance that bathes the cell and carries out these vital functions for it is called **interstitial fluid** (also known as **intercellular** or **tissue fluid**).

The interstitial fluid, in turn, must be serviced by blood and lymph. The blood picks up oxygen from the lungs, nutrients from the digestive tract, hormones from the endocrine glands, and enzymes from still other parts of the body. The blood then transports these substances to all the tissues where they diffuse from the capillaries (the smallest blood vessels) into the interstitial fluid. In the interstitial fluid, the substances are passed on to the cells and exchanged for wastes.

Since the blood must service all the tissues of the body, it can be an important medium for the transport of disease-causing organisms. To protect itself from the spread of disease, the body has a lymphatic system—a collection of vessels containing a fluid called lymph. The lymph picks up materials, including wastes, from the interstitial fluid, cleanses them of bacteria, and returns them to the blood. The blood then carries the wastes to the lungs, kidneys, and sweat glands, where they are eliminated from the body. The blood also takes wastes to the liver, where they are detoxified and recycled.

The blood, heart, and blood vessels constitute the **cardiovascular system.** The lymph, lymph vessels, and lymph glands make up the **lymphatic system.** In this chapter we will take a look at the substance known as blood.

The developmental anatomy of the cardiovascular system is considered in Exhibit 24-4.

PHYSICAL CHARACTERISTICS

The red body fluid that flows through all the vessels except the lymph vessels is called **blood.** Blood is a viscous fluid, that is, it is thicker and more adhesive than water. Water is considered to have a viscosity of 1.0. The viscosity of blood, by comparison, ranges from 4.5 to 5.5. It flows 4½ to 5½ times more slowly than water, at least in part because of its viscosity. The adhesive quality of blood, or its stickiness, may be felt by touching it. Blood is also slightly heavier than water.

Other physical characteristics of blood include a temperature of about 38°C (100.4°F), a pH range of 7.35 to 7.45 (slightly alkaline), and a 0.85 to 0.90 percent concentration of salt (NaCl).

Blood constitutes about 8 percent of the total body weight. The blood volume of an average-sized man is between 5 and 6 liters (about 5 to 6 qt). An average-sized woman has 4 to 5 liters.

FUNCTIONS

Despite its simple appearance, blood is a complex liquid that performs a number of critical functions.

1. It transports oxygen from the lungs to the cells of the body.

2. It transports carbon dioxide from the cells to the lungs.

3. It transports nutrients from the digestive organs to the cells.

4. It transports waste products from the cells to the kidneys, lungs, and sweat glands.

5. It transports hormones from endocrine glands to the cells.

6. It transports enzymes to various cells.

7. It regulates body pH through buffers and amino acids.

8. It regulates normal body temperature through the heat-absorbing and coolant properties of its water content.

9. It regulates the water content of cells, principally through dissolved sodium ions.

10. It prevents body fluid loss through the clotting mechanism.

11. It protects against toxins and foreign microbes through special combat-unit cells.

COMPONENTS

Microscopically, blood is composed of two portions: plasma, which is a liquid containing dissolved substances, and formed elements, which are cells and cell-like bodies suspended in the plasma.

FORMED ELEMENTS

In clinical practice, the most common classification of the **formed elements** of the blood is the following.

Erythrocytes (red blood cells)
Leucocytes (white blood cells)
    Granular leucocytes (granulocytes)
        Neutrophils
        Eosinophils
        Basophils
    Agranular leucocytes (agranulocytes)
        Lymphocytes
        Monocytes
Thrombocytes (platelets)

Origin

The process by which blood cells are formed is called **hemopoiesis** (hē-mō-poy-Ē-sis) or **hematopoiesis.** During embryonic and fetal life, there are several centers for blood cell production. The yolk sac, liver, spleen, thymus gland, lymph nodes, and bone marrow all participate at various times in producing the formed elements. In the adult, however, we can pinpoint the production process to the red bone marrow in the sternum, ribs, vertebrae, and pelvis. Red blood cells, granular leucocytes, and platelets are produced in red bone marrow (myeloid tissue). Agranular leucocytes arise from both myeloid tissue and from lymphoid tissue—spleen, tonsils, lymph nodes. Undiffer- entiated mesenchymal cells in red bone marrow are trans- formed into **hemocytoblasts** (hē'mō-SĪ-tō-blasts), imma- ture cells that are eventually capable of developing into mature blood cells (Figure 11-1). The hemocytoblasts un- dergo differentiation into five types of cells from which the major types of blood cells develop.

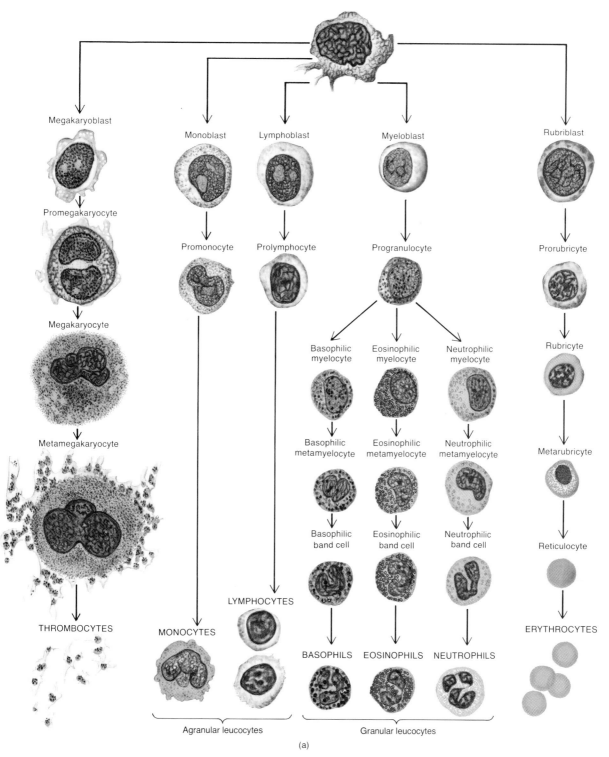

(a)

FIGURE 11-1 Blood cells. (a) Origin, development, and structure.

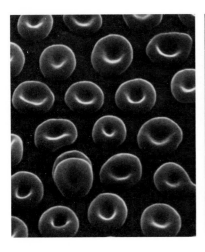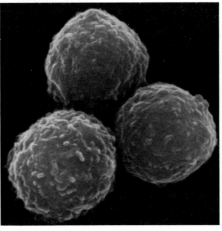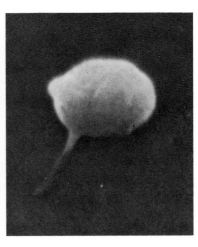

(b)

FIGURE 11-1 (*Continued*) Blood cells. (b) Scanning electron micrographs of erythrocytes (left) at a magnification of 5,000×, leucocytes (center) at a magnification of 10,000×, and a platelet (right) at a magnification of 10,000×. (Courtesy of Fisher Scientific Company and S.T.E.M. Laboratories, Inc., Copyright 1975.)

1. Rubriblasts (*proerythroblasts*) form mature red blood cells.

2. Myeloblasts form mature neutrophils, eosinophils, and basophils.

3. Megakaryoblasts form mature platelets.

4. Lymphoblasts form lymphocytes.

5. Monoblasts form monocytes.

Mature red blood cells survive in the bloodstream for about four months. Their disintegrating bodies pose the danger of clogging small blood vessels, so certain cells clear away their bodies after they die. These cells are called *reticuloendothelial* (re-tik'-yoo-lō-en'-dō-THĒ-lē-al) *cells* and are also formed from primitive reticular cells. Reticuloendothelial cells enter the spleen, tonsils, lymph nodes, liver, lungs, and other organs and become highly specialized for phagocytosis. The reticuloendothelial cells in the lymph nodes are particularly active in destroying microbes and their toxins. The reticuloendothelial cells in the liver and spleen concentrate on ingesting dead blood cells.

Erythrocytes

● *Structure* Microscopically, **red blood cells,** or **erythrocytes** (e-RITH-rō-sīts), appear as biconcave discs averaging about 7.7 μm in diameter (Figure 11-1). Mature red blood cells are quite simple in structure. They lack a nucleus and can neither reproduce nor carry on extensive metabolic activities. The cell contains a network of protein called the *stroma;* some cytoplasm; lipid substances, including cholesterol; and a red pigment called *hemoglobin.* Hemoglobin, which constitutes about 33 percent of the cell weight, is responsible for the red color of blood.

● *Functions* The hemoglobin in erythrocytes combines with oxygen and carbon dioxide and transports them through the blood vessels. The hemoglobin molecule consists of a protein called globin and a pigment called heme, which contains iron. As the erythrocytes pass through the lungs, each of the four iron atoms in the hemoglobin molecules combines with a molecule of oxygen. The oxygen is transported in this state to other tissues of the body. In the tissues, the iron-oxygen reaction reverses, and the oxygen is released to diffuse into the interstitial fluid. On the return trip, the globin portion combines with a molecule of carbon dioxide from the interstitial fluid to form carbaminohemoglobin. This complex is transported to the lungs, where the carbon dioxide is released and then exhaled. Although some carbon dioxide is transported by hemoglobin in this manner (about 23 percent), the greatest portion of carbon dioxide is transported in blood plasma as the bicarbonate ion (HCO_3^-), about 70 percent.

Red blood cells are highly specialized for their transport function. They contain a large number of hemoglobin molecules in order to increase their oxygen-carrying capacity. One estimate is 280 million molecules of hemoglobin per erythrocyte. Hemoglobin is contained in the stroma of a red blood cell because hemoglobin molecules are small and, if they were free in plasma, they would leak through the endothelial membranes of blood vessels and be lost in the urine. The biconcave shape of a red blood cell has a much greater surface area than, say, a sphere or a cube. The erythrocyte thus presents the maximum surface area for the diffusion of gas molecules that pass through the membrane to combine with hemoglobin.

● *Life Span and Number* The cell membrane of a red blood cell becomes fragile and the cell is nonfunctional in about 120 days. A healthy male has about 5.4 million red blood cells per cubic millimeter of blood, and a healthy female has about 4.8 million. The higher value in the male is because of his higher rate of metabolism. To maintain normal quantities of erythrocytes, the body must produce new mature cells at the astonishing rate of 2 million per second. In the adult, production takes place in the red bone marrow in the spongy bone of the cranium, ribs, sternum, bodies of vertebrae, and proximal epiphyses of the humerus and femur.

● *Production*   The process by which erythrocytes are formed is called **erythropoiesis** (e-rith'-rō-poy-Ē-sis). Erythropoiesis starts with the transformation of a hemocytoblast into a rubriblast. The *rubriblast* (proerythroblast) gives rise to a *prorubricyte* (early erythroblast), which then develops into a *rubricyte* (intermediate erythroblast), the first cell in the sequence that begins to synthesize hemoglobin. The rubricyte next develops into a *metarubricyte* (late erythroblast). In the metarubricyte, hemoglobin synthesis is at a maximum and the nucleus is lost by extrusion. In the next stage, the metarubricyte develops into a *reticulocyte,* which in turn becomes an *erythrocyte,* or mature red blood cell. Once the erythrocyte is formed, it leaves the marrow and enters the bloodstream. Aged erythrocytes are destroyed by reticuloendothelial cells in the liver and spleen. The hemoglobin molecules are split apart, the iron is reused, and the rest of the molecule is converted into other substances for reuse or elimination.

Normally erythropoiesis and red cell destruction proceed at the same pace. But if the body suddenly needs more erythrocytes or if erythropoiesis is not keeping up with red blood cell destruction, a homeostatic mechanism steps up erythrocyte production. The mechanism is triggered by the reduced supply of oxygen for body cells which results from a reduced number of erythrocytes. If certain kidney cells become oxygen-deficient, they release a hormone called **erythropoietin** (e-rith'-rō-POY-e-tin). This hormone circulates through the blood to the red bone marrow, where it stimulates more hemocytoblasts to develop into red blood cells. Erythropoietin is also produced by other body tissues, particularly the liver.

Cellular oxygen deficiency, called **hypoxia** (hī-POKS-ē-a), may occur if you do not breathe in enough oxygen. This deficiency commonly occurs at high altitudes, where the air contains less oxygen. Oxygen deficiency may also occur because of anemia. The term **anemia** means that the number of functional red blood cells and/or their hemoglobin content is below normal. Consequently, the erythrocytes are unable to transport enough oxygen from the lungs to the cells. Anemia has many causes: lack of iron, lack of certain amino acids, and lack of vitamin B_{12} are but a few. Iron is needed for the oxygen-carrying part of the hemoglobin molecule. The amino acids are needed for the protein, or globin, part. Vitamin B_{12} helps the red bone marrow to produce erythrocytes. This vitamin is obtained from meat, especially liver, but it cannot be absorbed by the mucosa (lining) of the ileum of the small intestine without the help of another substance—an **intrinsic factor** produced by the mucosal cells of the stomach. The intrinsic factor facilitates the absorption of vitamin B_{12}. Once absorbed, it is stored in the liver. (Several types of anemia are described later in the chapter.)

The rate of erythropoiesis is measured by a procedure called a **reticulocyte** (re-TIK-yoo-lō-sīt) **count.** Some reticulocytes are normally released into the bloodstream before they become mature red blood cells. If the number of reticulocytes in a sample of blood is less than 0.5 percent of the number of mature red blood cells in the sample, erythropoiesis is occurring too slowly. A low reticulocyte count might confirm a diagnosis of nutritional or pernicious anemia. Or it might indicate a kidney disease that prevents the kidney cells from producing erythropoietin. If the reticulocytes number more than 1.5 percent of the mature red blood cells, erythropoiesis is abnormally rapid. Any number of problems may be responsible for a high reticulocyte count: anemia, oxygen deficiency, and uncontrolled red blood cell production caused by a cancer in the bone marrow. A high count may also indicate that treatment for a condition causing anemia has been effective and the bone marrow is making up for lost time.

Another test with important clinical applications is the hematocrit. **Hematocrit** is the percentage of blood that is made up of red blood cells. It is determined by centrifuging blood and noting the ratio of red blood cells to whole blood. The average hematocrit for males is 47 percent. This means that in 100 ml of blood there are 47 ml of cells and 53 ml of plasma. The average hematocrit for females is 42 percent. Anemic blood may have a hematocrit of 15 percent; polycythemic blood (an abnormal increase in the number of functional erythrocytes) may have a hematocrit of 65 percent. Athletes, however, may have higher than normal hematocrits, reflecting constant physical activity rather than a pathological condition.

Leucocytes

● *Structure and Types*   Unlike red blood cells, **leucocytes** (LOO-kō-sīts), or **white blood cells,** have nuclei and do not contain hemoglobin (Figure 11-1). Leucocytes fall into two major groups. The first group is the **granular leucocytes.** They develop from red bone marrow, have granules in the cytoplasm, and possess lobed nuclei. The three kinds of granular leucocytes are **neutrophils (polymorphs), eosinophils,** and **basophils.**

The second principal group of leucocytes is the **agranular leucocytes.** They develop from lymphoid and myeloid tissue, no cytoplasmic granules can be seen under a light microscope, and their nuclei are usually spherical. The two kinds of agranular leucocytes are **lymphocytes** and **monocytes.**

● *Functions*   The general function of the leucocytes is to combat inflammation and infection. Some leucocytes are actively **phagocytotic**—they can ingest bacteria and dispose of dead matter. Others produce chemicals to counteract irritants, toxins, and microorganisms.

Most leucocytes possess, to some degree, the ability to crawl through the minute spaces between the cells that form the walls of capillaries and through connective and epithelial tissue. This movement, like that of amoebas, is called **diapedesis** (dī'-a-pe-DĒ-sis). First, part of the cell membrane stretches out like an arm. Then the cytoplasm and nucleus flow into the projection. Finally, the rest of the membrane snaps up into place. Another projection is made, and so on, until the cell has crawled to its destination.

An increase in the number of white cells present in the blood indicates a state of inflammation or infection. Because each type of white cell plays a different role, determining which types have increased aids in diagnosis of the condition. A **differential count** is the number of each kind of white cell in 100 white blood cells. A normal differential count falls within the following percentages.

Neutrophils	60–70%
Eosinophils	2–4%
Basophils	0.5–1%
Lymphocytes	20–25%
Monocytes	3–8%
	100%

Particular attention is paid to the neutrophils in a differential count. The neutrophils (NOO-trō-fils) are the most active white cells in response to tissue destruction by bacteria. Their major role is phagocytosis. They also release the enzyme lysozyme, which destroys certain bacteria. More often than not, a high neutrophil count indicates damage by invading bacteria.

An increase in the number of monocytes (MON-ō-sīts) generally indicates a chronic infection such as tuberculosis. Apparently monocytes take longer to reach the site of infection than do neutrophils, but once they arrive they do so in larger numbers and destroy more microbes. Monocytes, like neutrophils, are phagocytic. They clean up cellular debris following an infection.

High eosinophil (ē'-ō-SIN-ō-fil) counts indicate allergic conditions, since eosinophils are believed to combat the irritants that cause allergies. Eosinophils leave the capillaries, enter the tissue fluid, and produce antihistamines that destroy antigen-antibody complexes.

Basophils (BĀ-sō-fils) are also believed to be involved in allergic reactions. Basophils leave the capillaries, enter the tissues, and become the mast cells of the tissues, liberating heparin, histamine, and serotonin.

Lymphocytes (LIM-fō-sīts) are involved in the production of antibodies (AN-ti-bod'ēz). **Antibodies** are special proteins that inactivate antigens. An **antigen** (AN-ti-jen) is any substance that will stimulate the production of antibodies and is capable of reacting specifically with the antibody. Most antigens are proteins, and most are not synthesized by the body. Many of the proteins that make up the cell structures and enzymes of bacteria are antigens. The toxins released by bacteria are also antigens. When antigens enter the body, they react chemically with substances in the lymphocytes and stimulate some lymphocytes, called B cells, to become **plasma cells** (Figure 11-2). The plasma cells then produce antibodies, globulin-type proteins that attach to antigens much as enzymes attach to substrates. Like enzymes, a specific antibody will generally attach only to a certain antigen. However, unlike enzymes, which enhance the reactivity of the substrate, antibodies "cover" their antigens so the antigens cannot come in contact with other chemicals in the body. In this way, bacterial poisons can be sealed up and rendered harmless. The bacteria themselves are destroyed by the antibodies. This process is called the **antigen-antibody response.** Eosinophils in tissues phagocytose the antigen-antibody complexes. The antigen-antibody response helps us to combat infection and gives us immunity to some diseases. It is also responsible for blood types, allergies, and the body's rejection of organs transplanted from an individual with a different genetic makeup.

● *Life Span and Number* Foreign bacteria exist everywhere in the environment and have continuous access to the body through the mouth, nose, and pores of the skin. Furthermore, many cells, especially those of epithelial tissue, age and die, and their remains must be disposed of daily. Even when the body is healthy, the leucocytes actively ingest bacteria and debris. However, a leucocyte can phagocytose only a certain number of substances before they interfere with the leucocyte's normal metabolic activities and bring on its death. Consequently, the life span of most leucocytes is very short. In a healthy body, some white blood cells will live only a few days. During a period of infection they may live only a few hours.

Leucocytes are far less numerous than red blood cells, averaging from 5,000 to 9,000 cells per cubic millimeter of blood. Red blood cells, therefore, outnumber white blood cells about 700 to 1. The term **leucocytosis** (loo'-kō-sī-TŌ-sis) refers to an increase in the number of white blood cells. If the increase exceeds 10,000, a pathological condition is usually indicated. An abnormally low level of white blood cells (below 5,000 per cubic millimeter) is termed **leucopenia** (loo-kō-PĒ-nē-a).

Thrombocytes

● *Structure* In addition to the immature cell types that develop into erythrocytes and leucocytes, hemocytoblasts differentiate into still another kind of cell, called a megakaryoblast (Figure 11-1). Megakaryoblasts are transformed into megakaryocytes, large cells that shed frag-

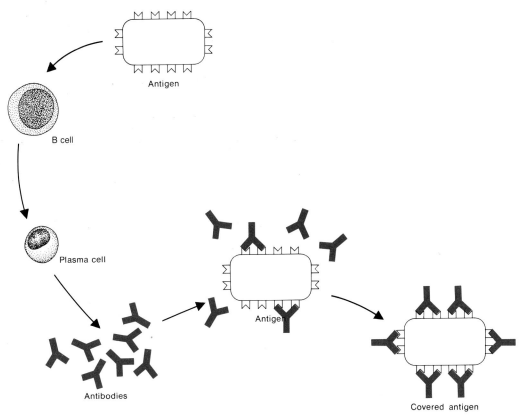

FIGURE 11-2 Antigen-antibody response. An antigen entering the body stimulates a B cell to develop into an antibody-producing plasma cell. The antibodies attach to the antigen, cover it, and render it harmless.

ments of cytoplasm. Each fragment becomes enclosed by a piece of the cell membrane and is called a **thrombocyte** (THROM-bō-sīt) or **platelet.** Platelets are disc-shaped cells without a nucleus. They average from 2 to 4 μm in diameter.

● *Function* Platelets prevent fluid loss by initiating a chain of reactions that results in blood clotting.

● *Life Span and Number* Like the other formed elements of the blood, platelets have a short life, probably only 5 to 9 days. Between 250,000 and 400,000 platelets appear in each cubic millimeter of blood.

PLASMA

When the cellular elements are removed from blood, a straw-colored liquid called **plasma** is left. Plasma consists of about 92 percent water and 8 percent solutes. Among the solutes are proteins (albumins, globulins, and fibrinogen), nonprotein (NPN) substances (urea, uric acid, and creatine), foods (amino acids, glucose, fatty acids, and glycerol), regulatory substances (enzymes and hormones), respiratory gases (oxygen and carbon dioxide), and electrolytes (Na^+, K^+, Ca^{2+}, Mg^{2+}, Cl^-, HCO_3^-, SO_4^{2-}, and PO_4^{3-}).

Seven to 9 percent of the solutes are proteins. Some of these proteins are also found elsewhere in the body, but in blood they are called **plasma proteins. Albumins,** which constitute the majority of plasma proteins, are largely responsible for blood's viscosity. The concentration of albumins is about four times higher in plasma than in interstitial fluid. Along with the electrolytes, albumins also help to regulate blood volume by preventing the water in the blood from diffusing into the interstitial fluid. Recall that water moves by osmosis from an area of low solute (high water) concentration to an area of high solute (low water) concentration. **Globulins,** which are antibody proteins released by plasma cells, form a small component of the plasma proteins. Gamma globulin is especially well known because it is able to form an antigen-antibody complex with the proteins of the hepatitis and measles viruses and the tetanus bacterium, among others. **Fibrinogen,** a third plasma protein, takes part in the blood-clotting mechanism along with the platelets. **Serum** is the term applied to plasma minus fibrinogen.

INTERSTITIAL FLUID AND LYMPH

Whole blood does not flow into the tissue spaces; it remains in closed vessels. However, certain constituents of the plasma do move through the capillary walls, and once they move out of the blood, they are called interstitial fluid. Interstitial fluid and lymph are basically the same. The major difference between the two is location. When the fluid bathes the cells, it is called **interstitial fluid, intercellular fluid,** or **tissue fluid.** When it flows through the lymphatic vessels, it is called **lymph** (Figure 11-3).

Both fluids are similar in composition to plasma. The principal chemical difference is that they contain less protein than plasma because the larger protein molecules are not easily filtered through the cells that form the walls of the capillaries. The transfer of materials between blood and interstitial fluid occurs by osmosis, diffusion, and filtration across the cells that make up the capillary walls.

Interstitial fluid and lymph also differ from plasma in that they contain variable numbers of leucocytes. Leucocytes can enter the tissue fluid by diapedesis, and the lymphoid tissue itself is one of the sites of agranular leucocyte production. Like plasma, interstitial fluid and lymph lack erythrocytes and platelets.

Other substances, especially organic molecules, in interstitial fluid and lymph vary in relation to the location of the sample analyzed. The lymph vessels that drain the organs of the digestive tract, for example, contain a great deal of lipid absorbed from food.

APPLICATIONS TO HEALTH

ANEMIA

Anemia is a sign, not a diagnosis. Many kinds of anemia exist, all characterized by insufficient erythrocytes or hemoglobin. These conditions lead to fatigue and intolerance to cold, both of which are related to lack of oxygen needed for energy and heat production, and to paleness, which is due to low hemoglobin content.

Nutritional Anemia

Nutritional anemia arises from an inadequate diet, one that provides insufficient amounts of iron, the necessary amino acids, or vitamin B_{12}.

Pernicious Anemia

Pernicious anemia is the insufficient production of erythrocytes resulting from an inability of the body to produce intrinsic factor.

Hemorrhagic Anemia

An excessive loss of erythrocytes through bleeding is called **hemorrhagic anemia.** Common causes are large wounds, stomach ulcers, and heavy menstrual bleeding. If bleeding is extraordinarily heavy, the anemia is termed acute. Excessive blood loss can be fatal. Slow, prolonged bleeding is apt to produce a chronic anemia; the chief symptom is fatigue.

Hemolytic Anemia

If erythrocyte cell membranes rupture prematurely, the cells remain as "ghosts" and their hemoglobin pours out into the plasma. A characteristic sign of this condition, called **hemolytic anemia,** is distortions in the shapes of erythrocytes that are progressing toward hemolysis. There may also be a sharp increase in the number of reticulocytes, since the destruction of red blood cells stimulates erythropoiesis.

The premature destruction of red cells may result from inherent defects, such as hemoglobin defects, abnormal red cell enzymes, or defects of the red cell membrane. Agents that may cause hemolytic anemia are parasites, toxins, and antibodies from incompatible blood (Rh^- mother and Rh^+ fetus, for instance). *Erythroblastosis fetalis* is an example of a hemolytic anemia.

The term *thalassemia* (thaḷ'-a-SĒ-mē-a) represents a group of hereditary hemolytic anemias, resulting from a defect in the synthesis of hemoglobin, which produces extremely thin and fragile erythrocytes. It occurs primarily in populations from countries bordering the Mediterranean Sea. Treatment generally consists of blood transfusions.

Aplastic Anemia

Destruction or inhibition of the red bone marrow results in **aplastic anemia.** Typically, the marrow is replaced by fatty tissue, fibrous tissue, or tumor cells. Toxins, gamma radiation, and certain medications are causes. Many of the medications inhibit the enzymes involved in hemopoiesis. Bone marrow transplants can now be done with a reasonable hope of success in patients with aplastic anemia. They are done early, before the victim is sensitized by transfusions. Immunosuppressive drugs are given for 4 days before the transplant and with decreasing frequency for 100 days afterward.

Sickle Cell Anemia

The erythrocytes of a person with **sickle cell anemia** manufacture an abnormal kind of hemoglobin. When an erythrocyte gives up its oxygen to the interstitial fluid, its hemoglobin tends to lose its integrity in places of low oxygen tension and forms long, stiff, rodlike structures that bend the erythrocyte into a sickle shape (Figure 11-4). The sickled cells rupture easily. Even though erythropoiesis is stimulated by the loss of the cells, it cannot keep pace with the hemolysis. The individual consequently suffers from a hemolytic anemia that reduces the amount of oxygen that can be supplied to the tissues. Prolonged oxygen reduction may eventually cause extensive tissue damage.

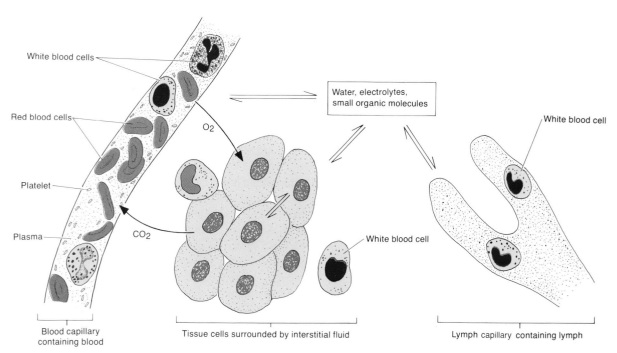

White blood cells

Red blood cells

Platelet

Plasma

O_2

CO_2

Water, electrolytes,
small organic molecules

White blood cell

White blood cell

Blood capillary
containing blood

Tissue cells surrounded by interstitial fluid

Lymph capillary containing lymph

FIGURE 11-3 Composition of blood, interstitial fluid, and lymph.

Furthermore, because of the shape of the sickled cells, they tend to get stuck in blood vessels and can cut off blood supply to an organ altogether.

Sickle cell anemia is inherited. The gene responsible for the tendency of the erythrocytes to sickle during hypo-

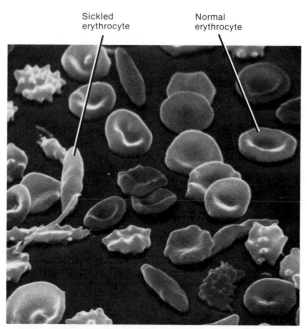

Sickled
erythrocyte

Normal
erythrocyte

FIGURE 11-4 Scanning electron micrograph of erythrocytes as they appear in sickle cell anemia at a magnification of 5,000×. (Courtesy of Fisher Scientific Company and S.T.E.M. Laboratories, Inc., Copyright 1975.)

xia also seems to prevent erythrocytes from rupturing during a malarial crisis. Sickle cell genes are found primarily among populations, or descendants of populations, that live in the malaria belt around the world—including parts of Mediterranean Europe and subtropical Africa and Asia. A person with only one of the sickling genes is said to have sickle cell trait. Such an individual has a high resistance to malaria—a factor that may have tremendous survival value—but does not develop the anemia. Only people who inherit a sickling gene from both parents get sickle cell anemia.

It has recently been learned that sickled cells have lower than normal levels of potassium. The gene responsible for sickling also alters the permeability of the plasma membranes of sickled cells, causing potassium to leak out. Low levels of potassium kill the malarial parasites that infect sickled cells.

POLYCYTHEMIA

The term **polycythemia** (pol'-ē-sī-THĒ-mē-a) refers to an abnormal increase in the number of red blood cells. The hematocrit is an important procedure in diagnosing the condition. Increases of 2 to 3 million cells per cubic millimeter above normal are considered to be polycythemic. The disorder is harmful, since the blood's viscosity is greatly increased because of the extra red blood cells, and viscosity contributes to thrombosis and hemorrhage. It also causes a rise in blood pressure. The thrombosis results from too many red blood cells piling up as they try to enter smaller vessels. The hemorrhage is due to widespread hyperemia (unusually large amount of blood in an organ part).

INFECTIOUS MONONUCLEOSIS

Infectious mononucleosis is a contagious disease caused by a virus. It occurs mainly in children and young adults. Its trademark is an elevated white blood cell count with an abnormally high percentage of lymphocytes and mononucleocytes. An increase in the number of monocytes usually indicates a chronic infection. Symptoms include slight fever, sore throat, brilliant red throat and soft palate, stiff neck, cough, and malaise. The spleen may enlarge. Secondary complications involving the liver, heart, kidneys, and nervous system may develop. There is no cure for mononucleosis, and treatment consists of watching for and treating complications. Usually the disease runs its course in a few weeks, and the individual generally suffers no permanent ill effects.

LEUKEMIA

Also called "cancer of the blood," **leukemia** is an uncontrolled, greatly accelerated production of white blood cells. Many of the cells fail to reach maturity. As with most cancers, the symptoms result not so much from the cancer cells themselves as from their interference with normal body processes. The anemia and bleeding problems commonly seen in leukemia result from the crowding out of normal bone marrow cells, preventing normal production of red blood cells and platelets. The most common cause of death from leukemia is internal hemorrhaging, especially cerebral hemorrhage that destroys the vital centers in the brain. Another frequent cause of death is uncontrolled infection owing to lack of mature or normal white blood cells. The abnormal accumulation of leucocytes may be reduced by using x-rays and antileukemic drugs. Partial or complete remissions may be induced, with some lasting as long as 15 years.

KEY MEDICAL TERMS ASSOCIATED WITH BLOOD

Blood plasma substitute   Substance that mimics the characteristics of plasma and so can be used to maintain blood volume during emergency conditions (such as hemorrhage) until blood can be matched, to prevent dehydration if a patient cannot swallow liquids, or to replace fluid and electrolytes after loss of blood during surgery.

Corpuscles (*corpus* = body)   Cellular elements in the blood such as red and white blood cells.

Direct (immediate) transfusion   Transfer of blood directly from one person to another without exposing the blood to air.

Embolus (*em* = in, *bolus* = a mass)   Any foreign or abnormal particle that is transported through the blood. It may be solid, liquid, or gaseous. Obstruction occurs if the embolus becomes stuck in a small vessel.

Exchange transfusion   Removing blood from the recipient while simultaneously replacing it with donor blood. This method is used for erythroblastosis fetalis and poisoning.

Fibrinogen   A plasma protein; also, a sterile, freeze-dried preparation of this component of normal plasma. In solution the preparation can be converted to insoluble fibrin by adding thrombin. It can be applied to wounds to stop bleeding and hemorrhagic disorders caused by fibrinogen deficiency.

Fractionated blood (*fract* = break)   Blood that has been separated into its components. Only the part needed by the patient is given.

Gamma globulin (immune serum globulin)   Solution of globulins from nonhuman blood consisting of antibodies that react with a specific pathogen, such as measles, epidemic hepatitis, tetanus, and possibly poliomyelitis viruses. It is prepared by injecting the specific virus into animals, removing blood from the animals after antibodies have accumulated, isolating antibodies, and injecting them into a human for short-term immunity.

Hem, Hemo, Hema, Hemato (*heme* = iron)   Various combining forms meaning blood.

Hemolysis (laking)   A swelling and subsequent rupture of erythrocytes in a hypotonic solution with the liberation of hemoglobin into the surrounding fluid.

Hemorrhage (*rrhage* = bursting forth)   Bleeding, either internal (from blood vessels into tissues) or external (from blood vessels directly to the surface of the body).

Heparinized whole blood   Whole blood in a heparin solution, which prevents coagulation.

Indirect (mediate) transfusion   Transfer of blood from a donor to a container and then to the recipient, permitting blood to be stored for an emergency. The blood may be separated into its components so that a patient will receive only a needed part.

Normal plasma   Cell-free plasma containing normal concentrations of all solutes. It is used to bring blood volume up to normal when excessive numbers of blood cells have not been lost.

Plasmolysis (crenation) (*plas* = form; *lysis* = dissolve) Shrinking of the cytoplasm in any cell (in this case, erythrocytes), producing knobbed, starry shapes. It results from a loss of water by osmosis, as when blood is mixed with a 5 percent salt (hypertonic) solution.

Platelet concentrates A preparation of platelets obtained from freshly drawn whole blood and used for transfusions in platelet-deficiency disorders such as hemophilia.

Reciprocal transfusion Transfer of blood from a person who has recovered from a contagious infection into the vessels of a patient suffering with the same infection. An equal amount of blood is returned from the patient to the well person. This method allows the patient to receive antibody-bearing lymphocytes from the recovered person.

Septicemia (*sep* = decay; *emia* = condition of blood) Toxins or disease-causing bacteria in the blood. Also called "blood poisoning."

Thrombus (*thrombo* = clot) A blood clot lodged in a vessel.

Transfusion Transfer of whole blood, blood components (red blood cells only or plasma only), or bone marrow directly into the bloodstream.

Venesection (*veno* = vein) Opening of a vein for withdrawal of blood.

Whole blood Blood containing all formed elements, plasma, and plasma solutes in natural concentration.

STUDY OUTLINE

Physical Characteristics

1. The cardiovascular system consists of blood, the heart, and blood vessels. The lymphatic system consists of lymph, lymph vessels, and lymph glands.
2. Physical characteristics of blood include viscosity, 4.5 to 5.5; temperature, 38°C; pH, 7.35 to 7.45; and salinity, 0.85 to 0.90 NaCl. Blood constitutes about 8 percent of body weight.

Functions

1. Blood transports oxygen, carbon dioxide, nutrients, wastes, hormones, and enzymes.
2. It helps to regulate pH, body temperature, and water content of cells.
3. It prevents excessive fluid loss through clotting.
4. It protects against toxins and microbes.

Components

1. The formed elements in blood include erythrocytes (red blood cells), leucocytes (white blood cells), and thrombocytes (platelets).
2. Blood cells are formed by a process called hemopoiesis.
3. Red bone marrow (myeloid tissue) is responsible for producing red blood cells, granular leucocytes, and platelets; lymphoid tissue and myeloid tissue produce agranular leucocytes.
4. Erythrocytes are biconcave discs without nuclei and containing hemoglobin. Erythrocyte formation, called erythropoiesis, occurs in adult red marrow of certain bones. A reticulocyte count is a diagnostic test that indicates the rate of erythropoiesis. A hematocrit measures the ratio of red blood cells to whole blood.
5. Leucocytes are nucleated cells. Two principal types are granular (neutrophils, eosinophils, basophils) and agranular (lymphocytes and monocytes).
6. The function of leucocytes is to combat inflammation and infection. Neutrophils and monocytes do so through phagocytosis.
7. Eosinophils and basophils are believed to be involved in combating allergic reactions.
8. Lymphocytes, in response to the presence of foreign proteins called antigens, are changed into tissue plasma cells which produce antibodies. Antibodies attach to the antigens and render them harmless. This antigen-antibody response combats infection and provides immunity.
9. Thrombocytes are disc-shaped structures without nuclei. They are formed from megakaryocytes and are involved in clotting.
10. The liquid portion of blood, called plasma, consists of 92 percent water and 8 percent solutes. Principal solutes include proteins (albumins, globulins, fibrinogen), foods, enzymes and hormones, gases, and electrolytes.

Interstitial Fluid and Lymph

1. Interstitial fluid bathes body cells, whereas lymph is found in lymphatic vessels.
2. These fluids are similar in chemical composition. They differ chemically from plasma in that both contain less protein and a variable number of leucocytes. Like plasma, they contain no platelets or erythrocytes.

Applications to Health

1. Anemia is a decreased erythrocyte count or hemoglo-

bin deficiency. Kinds of anemia include nutritional, pernicious, hemorrhagic, hemolytic, aplastic, and sickle cell anemia.

2. Polycythemia is an abnormal increase in the number of erythrocytes.

3. Infectious mononucleosis is characterized by an elevated white cell count, especially the monocytes. The cause is unknown.

4. Leukemia is the uncontrolled production of white blood cells that interferes with normal clotting and vital body activities.

REVIEW QUESTIONS

1. How are blood, interstitial fluid, and lymph related?

2. Distinguish between the cardiovascular system and lymphatic system.

3. Define the principal physical characteristics of blood. List the functions of blood and their relationship to other systems of the body.

4. Distinguish between plasma and formed elements. Where are the formed elements produced?

5. What are reticuloendothelial cells? Describe their function.

6. Describe the microscopic appearance of erythrocytes. What is the essential function of erythrocytes?

7. Define erythropoiesis. Relate erythropoiesis to red blood cell count. What factors accelerate and decelerate erythropoiesis?

8. What is a reticulocyte count? What is its diagnostic significance?

9. Define hematocrit. Compare the hematocrits of anemic and polycythemic blood.

10. Describe the classification of leucocytes. What are their functions?

11. What is the importance of diapedesis and phagocytosis in fighting bacterial invasion?

12. What is a differential count? What is its significance?

13. Distinguish between leucocytosis and leucopenia.

14. Describe the antigen-antibody response. How is it protective?

15. What are the major constituents in plasma? What do they do?

16. What is the difference between plasma and serum?

17. Compare interstitial fluid and lymph with regard to location, chemical composition, and function.

18. Define anemia. Contrast the causes of nutritional, pernicious, hemorrhagic, hemolytic, aplastic, and sickle cell anemias.

19. What is infectious mononucleosis?

20. What is leukemia, and what is the cause of some of its symptoms?

21. Refer to the glossary of key medical terms associated with blood. Be sure that you can define each term.

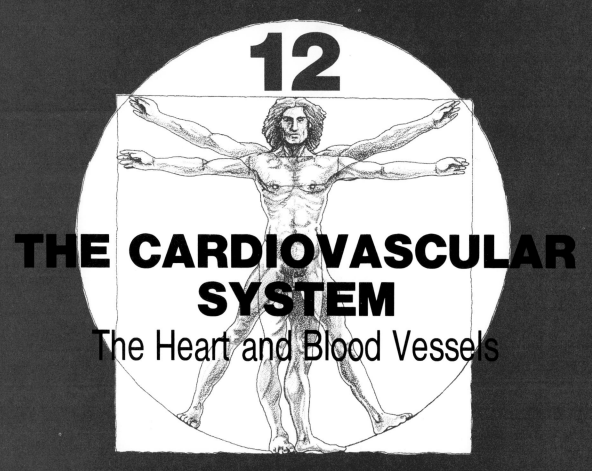

12

THE CARDIOVASCULAR SYSTEM
The Heart and Blood Vessels

STUDENT OBJECTIVES

- Identify the location of the heart in the mediastinum.
- Distinguish between the structure and location of the fibrous and serous pericardium.
- Contrast the structure of the epicardium, myocardium, and endocardium.
- Identify the blood vessels, chambers, and valves of the heart.
- Explain the initiation and conduction of nerve impulses through the electrical conduction system of the heart.
- Contrast the sounds of the heart and their clinical significance.
- Discuss the surface anatomy features of the heart.
- Contrast the effects of sympathetic and parasympathetic stimulation of the heart.
- Contrast the structure and function of arteries, arterioles, capillaries, venules, and veins.
- Identify the principal arteries and veins of systemic circulation.
- Discuss the route of blood in coronary circulation.
- Compare angina pectoris and myocardial infarction as abnormalities of coronary circulation.
- Trace the route of blood involved in hepatic portal circulation and explain its importance.

- Identify the major blood vessels of pulmonary circulation.
- Contrast fetal and adult circulation.
- Describe the importance of cerebral circulation.
- Explain the fate of fetal circulation structures once postnatal circulation is established.
- Discuss heart failure.
- List the risk factors involved in heart attacks.
- Explain why inadequate blood supply, anatomical disorders, and malfunctions of conduction are primary reasons for heart trouble.
- Contrast patent ductus arteriosus, septal defects, and valvular stenosis as congenital heart defects.
- List the four abnormalities of the heart present in tetralogy of Fallot.
- Define atrioventricular block, atrial flutter, atrial fibrillation, and ventricular fibrillation as abnormalities of the conduction system of the heart.
- List the causes and symptoms of aneurysms, atherosclerosis, and hypertension.
- Discuss the diagnosis of atherosclerosis by the use of angiography.
- Define key medical terms associated with the heart and blood vessels.

The heart is the center of the cardiovascular system. It is a hollow, muscular organ that pumps blood through the blood vessels. The blood vessels form a network of tubes that carry blood from the heart to the tissues of the body and then return it to the heart. The specific aspects of the heart that we will consider are its location, major parts, conduction system, surface anatomy, and the cardiac cycle. We will examine the structure of the blood vessels and the circulatory routes through the body. We will also consider several disorders related to the heart and blood vessels.

HEART

LOCATION

The **heart** is situated obliquely between the lungs in the mediastinum. About two-thirds of its mass lies to the left of the body's midline (Figure 12-1a). The heart is shaped like a blunt cone about the size of your closed fist—12 cm (5 inches) long, 9 cm (3½ inches) wide at its broadest point, and 6 cm (2½ inches) thick. Its pointed end, the *apex,* is formed by the tip of the left ventricle, projects inferiorly, anteriorly, and to the left and lies superior to the central depression of the diaphragm. Anteriorly, the apex is in the fifth intercostal space—about 7.5 to 8 cm (3 inches) from the midline of the body. The *left border* is formed almost entirely by the left ventricle, although the left atrium forms part of the upper end of the border. The left border of the heart is indicated approximately as a curved line drawn from the left fifth costochondral junction to the left second costochondral junction. The *superior border,* where the great vessels enter and leave the heart, is formed by both atria. It is represented by a line joining the second left intercostal space to the third right costal cartilage. The *base* of the heart projects superiorly, posteriorly, and to the right. It is formed by the atria, mostly the left atrium. It lies opposite the fifth to ninth thoracic vertebrae. Anteriorly, it lies just inferior to the second rib. The *right border* is formed by the right atrium and corresponds to a curved line drawn from the xiphisternal articulation to about the middle of the right third costal cartilage. The *inferior border* is formed by the right ventricle and slightly by

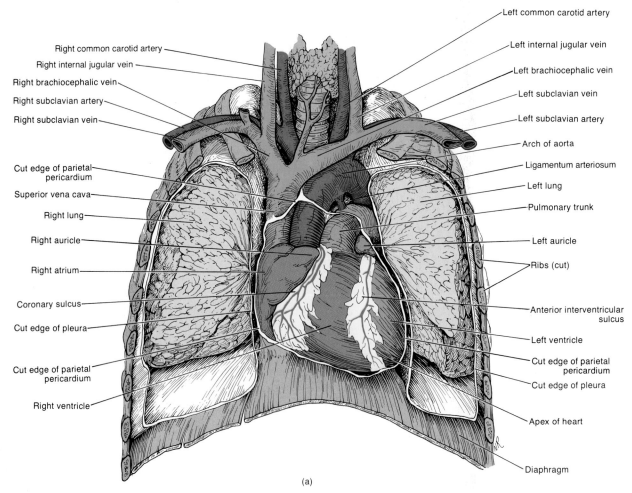

Left common carotid artery

Right common carotid artery

Right internal jugular vein

Right brachiocephalic vein

Right subclavian artery

Right subclavian vein

Left internal jugular vein

Left brachiocephalic vein

Left subclavian vein

Left subclavian artery

Arch of aorta

Ligamentum arteriosum

Cut edge of parietal pericardium

Superior vena cava

Right lung

Right auricle

Right atrium

Coronary sulcus

Cut edge of pleura

Cut edge of parietal pericardium

Right ventricle

Left lung

Pulmonary trunk

Left auricle

Ribs (cut)

Anterior interventricular sulcus

Left ventricle

Cut edge of parietal pericardium

Cut edge of pleura

Apex of heart

Diaphragm

(a)

FIGURE 12-1 Heart. (a) Position of the heart and associated blood vessels in the thoracic cavity.

the left ventricle. It is represented by a line passing from the inferior end of the right border through the xiphisternal joint to the apex of the heart. The *sternocostal (anterior) surface* is formed mainly by the right ventricle and right atrium, whereas the *diaphragmatic (inferior) surface* is formed by the left and right ventricles, mostly the left (see Figure 12-5).

PARIETAL PERICARDIUM (PERICARDIAL SAC)

The heart is enclosed in a loose-fitting serous membrane called the **parietal pericardium** or **pericardial sac** (Figure 12-1a, c). It consists of two layers: the fibrous layer and the serous layer. The *fibrous layer* or *fibrous pericardium* is the outer layer and consists of a tough, fibrous connec-

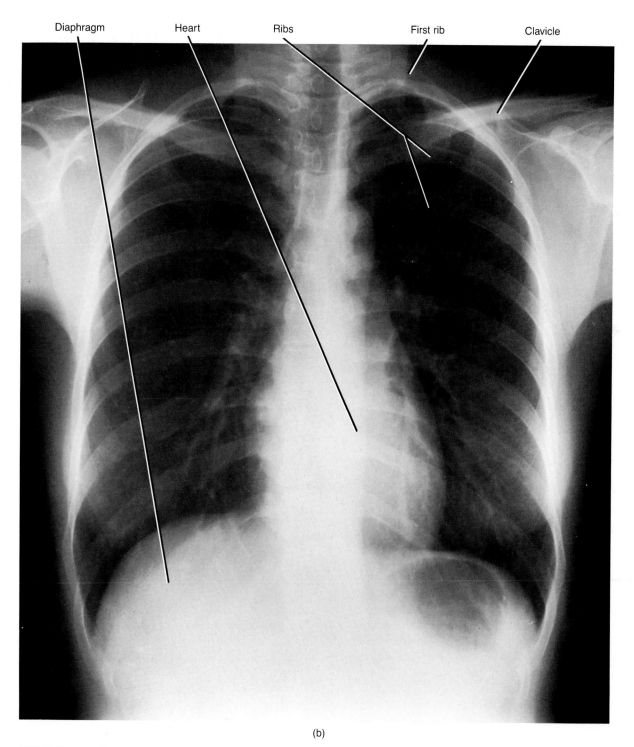

Diaphragm   Heart   Ribs   First rib   Clavicle

(b)

FIGURE 12-1 (*Continued*) Heart. (b) Anteroposterior projection of normal heart. (Courtesy of John C. Bennett, St. Mary's Hospital, San Francisco.)

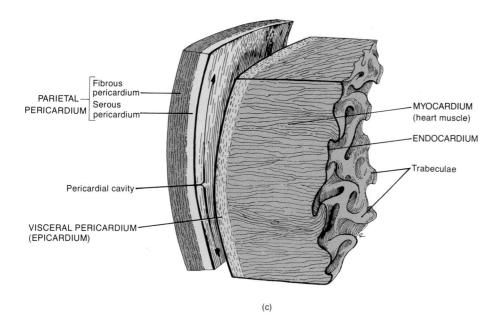

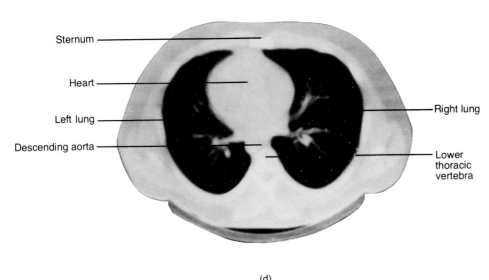

FIGURE 12-1 (*Continued*) Heart. (c) Structure of the parietal pericardium and heart wall. (d) CT scan of a normal heart. (Courtesy of General Electric Co.)

tive tissue. It is attached to the large blood vessels entering and leaving the heart, to the diaphragm, and to the inside of the sternal wall of the thorax. It also adheres to the parietal pleurae. The fibrous pericardium prevents overdistension of the heart, surrounds it with a tough protective membrane, and anchors it in the mediastinum.

The inner layer of the parietal pericardium, known as the *serous layer* or *serous pericardium,* is thinner and more delicate. It is continuous with the visceral pericardium (the outer layer of the wall of the heart) at the base of the heart and around the large blood vessels. An inflammation of the parietal pericardium is known as **pericarditis.**

HEART WALL

The wall of the heart (Figure 12-1c) is divided into three portions: the epicardium (external layer), the myocardium (middle layer), and the endocardium (inner layer). The **epicardium** or **visceral pericardium** is the thin, transparent outer layer of the wall. It is composed of serous tissue and mesothelium. Between the serous pericardium and the epicardium is a potential space called the *pericardial cavity.* The cavity contains a watery fluid, known as *pericardial fluid,* which prevents friction between the membranes as the heart moves.

The **myocardium,** which is cardiac muscle tissue, constitutes the bulk of the heart. Cardiac muscle fibers are

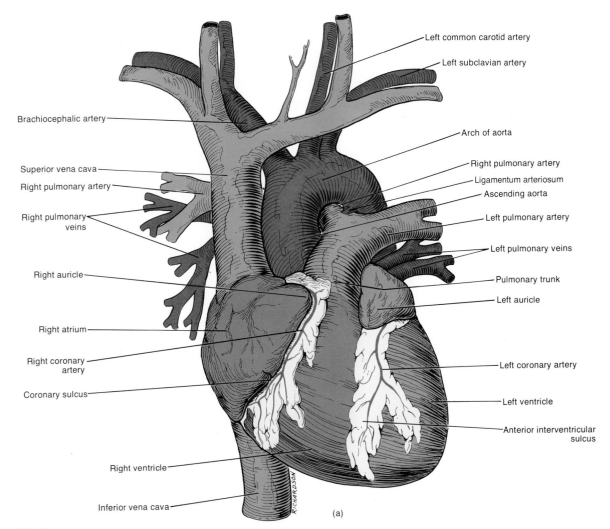

Left common carotid artery
Left subclavian artery
Brachiocephalic artery
Arch of aorta
Superior vena cava
Right pulmonary artery
Right pulmonary artery
Ligamentum arteriosum
Ascending aorta
Right pulmonary veins
Left pulmonary artery
Left pulmonary veins
Pulmonary trunk
Right auricle
Left auricle
Right atrium
Left coronary artery
Right coronary artery
Left ventricle
Coronary sulcus
Anterior interventricular sulcus
Right ventricle
Inferior vena cava
(a)

FIGURE 12-2 Structure of the heart. (a) Anterior external view.

involuntary, striated, and branched, and the tissue is arranged in interlacing bundles of fibers. The myocardium is responsible for the contraction of the heart.

The **endocardium** is a thin layer of endothelium overlying a thin layer of connective tissue pierced by tiny blood vessels and bundles of smooth muscle. It lines the inside of the myocardium and covers the valves of the heart and the tendons that hold them open. It is continuous with the endothelial lining of the large blood vessels of the heart.

Inflammation of the epicardium, myocardium, and endocardium are referred to as **epicarditis, myocarditis,** and **endocarditis** respectively.

CHAMBERS OF THE HEART

The interior of the heart is divided into four spaces, or **chambers,** which receive the circulating blood (Figure 12-2d-f). The two upper chambers are called the right and left **atria.** Each atrium has an appendage called an

auricle (AWR-i-kul; *auris* = ear); so named because its shape resembles a dog's ear. The auricle increases the atrium's surface area. The lining of the atria is smooth, except for the anterior atrial walls and the lining of the auricles, which contain projecting muscle bundles that are parallel to one another and resemble the teeth of a comb: the *musculi pectinati* (MUS-kyoo-lī pek-ti-NA-tē). These bundles give the lining of the auricles a ridged appearance.

The atria are separated by a partition called the *interatrial septum.* A prominent feature of this septum is an oval depression, the *fossa ovalis,* which corresponds to the site of the foramen ovale, an opening in the interatrial septum of the fetal heart. The fossa ovalis faces the opening of the inferior vena cava and is located in the septal wall of the right atrium.

The two lower chambers are the right and left **ventricles.** They are separated by an *interventricular septum.*

The muscle tissue of the atria and ventricles is separated by connective tissue that also forms the valves. This

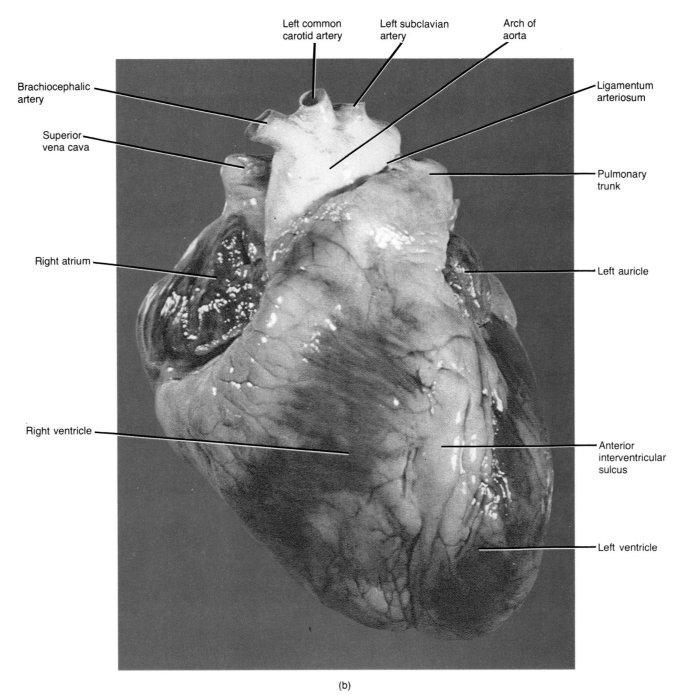

Left common carotid artery

Left subclavian artery

Arch of aorta

Brachiocephalic artery

Superior vena cava

Right atrium

Right ventricle

Ligamentum arteriosum

Pulmonary trunk

Left auricle

Anterior interventricular sulcus

Left ventricle

(b)

FIGURE 12-2 (*Continued*) Structure of the heart. (b) Photograph of anterior external view. (Courtesy of C. Yokochi and J. W. Rohen, *Photographic Anatomy of the Human Body*, 2nd ed., 1979, IGAKU-SHOIN, Ltd., Tokyo, New York.)

"cardiac skeleton" effectively divides the myocardium into two separate muscle masses. Externally, a groove known as the *coronary sulcus* (SUL-kus) separates the atria from the ventricles. It encircles the heart and houses the coronary sinus and circumflex branch of the left coronary artery. The *anterior interventricular sulcus* and *posterior interventricular sulcus* separate the right and left ventricles externally. The sulci contain coronary blood vessels and a variable amount of fat (Figure 12-2a, b).

GREAT VESSELS OF THE HEART

The right atrium receives blood from all parts of the body except the lungs. It receives the blood through three veins. The **superior vena cava** brings blood from parts of the body superior to the heart; the **inferior vena cava** brings blood from parts of the body inferior to the heart; and the **coronary sinus** drains blood from most of the vessels supplying the walls of the heart. The right atrium

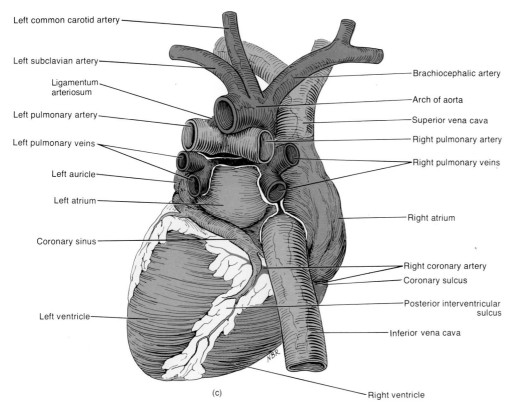

Left common carotid artery

Left subclavian artery

Ligamentum arteriosum

Left pulmonary artery

Left pulmonary veins

Left auricle

Left atrium

Coronary sinus

Left ventricle

Brachiocephalic artery

Arch of aorta

Superior vena cava

Right pulmonary artery

Right pulmonary veins

Right atrium

Right coronary artery

Coronary sulcus

Posterior interventricular sulcus

Inferior vena cava

Right ventricle

(c)

FIGURE 12-2 (*Continued*) Structure of the heart. (c) Posterior external view.

then delivers the blood into the right ventricle, which pumps it into the **pulmonary trunk.** The pulmonary trunk divides into a **right** and **left pulmonary artery,** each of which carries blood to the lungs. In the lungs, the blood releases its carbon dioxide and takes on oxygen. It returns to the heart via four **pulmonary veins** that empty into the left atrium. The blood then passes into the left ventricle, which pumps the blood into the **ascending aorta.** From here blood is passed into the **coronary arteries, arch of the aorta, descending thoracic aorta,** and **abdominal aorta.** These blood vessels transport the blood to all body parts except the lungs (Figure 12-2a-d).

Figure 12-2e, f shows that the sizes of the four chambers vary according to function. The right atrium, which must collect blood coming from almost all parts of the body, is slightly larger than the left atrium, which receives blood only from the lungs. The thickness of the chamber walls varies too. The atria are thin-walled because they need only enough cardiac muscle tissue to deliver the blood into the ventricles with the aid of a reduced pressure created by the expanding ventricles. The right ventricle has a thicker layer of myocardium than the atria, since it must send blood to the lungs and around back to the left atrium. The left ventricle has the thickest walls, since it must pump blood at high pressure through literally thousands of miles of vessels in the head, trunk, and extremities.

VALVES OF THE HEART

As each chamber of the heart contracts, it pushes a portion of blood into a ventricle or out of the heart through an artery. In order to keep the blood from flowing backward, the heart has **valves.**

Atrioventricular Valves

Atrioventricular (AV) valves lie between the atria and ventricles (Figure 12-2d). The right atrioventricular valve, between the right atrium and right ventricle, is also called the **tricuspid valve** because it consists of three flaps, or cusps. These flaps are fibrous tissues that grow out of the walls of the heart and are covered with endocardium. The pointed ends of the cusps project into the ventricle. Cords called *chordae tendineae* (KOR-dē TEN-dī-nē) connect the pointed ends to small conical projections—the *papillary muscles* (muscular columns)—located on the inner surface of the ventricles. The irregular surface of ridges and folds of the myocardium in the ventricles is known as the *trabeculae carneae* (tra-BEK-yoo-lē KAR-nē). The chordae tendineae and their muscles keep the flaps pointing in the direction of the blood flow. As the atrium relaxes and the ventricle pumps the blood out of the heart, any blood driven back toward the atrium is pushed between the flaps and the ventricle walls (Figure

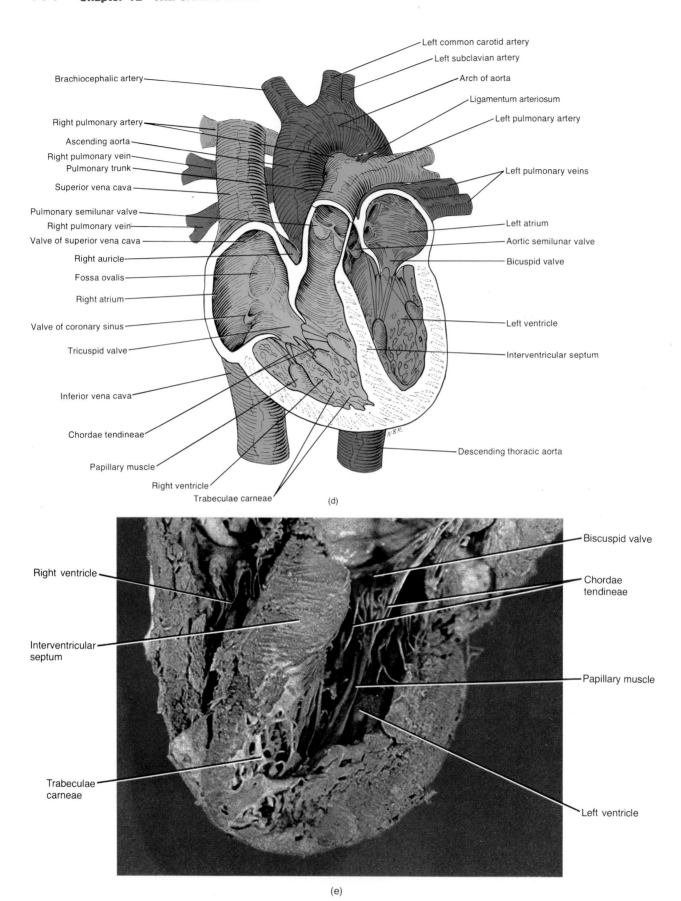

FIGURE 12-2 (*Continued*) Structure of the heart. (d) Anterior internal view. (e) Photograph of a portion of anterior internal view. (Courtesy of J. W. Eads.)

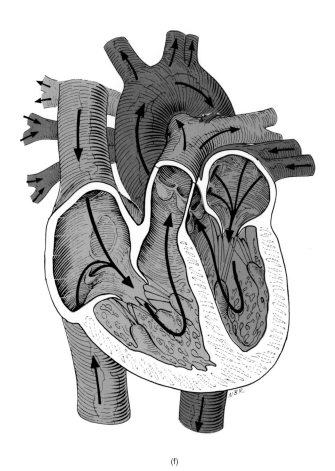

(f)

FIGURE 12-2 (*Continued*) Structure of the heart. (f) Path of blood through the heart.

12-3). This action drives the cusps upward until their edges meet and close the opening. At the same time, contraction of the papillary muscles helps prevent the valve from swinging upward into the atrium.

The atrioventricular valve between the left atrium and left ventricle is called the **bicuspid** or **mitral valve.** It has two cusps that work in the same way as the cusps of the tricuspid valve. The bicuspid valve is also known as the left atrioventricular (left AV) valve. Its cusps are also attached by way of the chordae tendineae to papillary muscles.

Semilunar Valves

Both arteries that leave the heart have a valve that prevents blood from flowing back into the heart. These are the **semilunar valves**—*semilunar* meaning half-moon or crescent-shaped (Figure 12-3b, c). The **pulmonary semilunar valve** lies in the opening where the pulmonary trunk leaves the right ventricle. The **aortic semilunar valve** is situated at the opening between the left ventricle and the aorta.

Both valves consist of three semilunar cusps. Each cusp is attached by its convex margin to the artery wall. The free borders of the cusps curve outward and project into the opening inside the blood vessel. Like the atrioventricular valves, the semilunar valves permit blood to flow in only one direction—in this case, from the ventricles into the arteries.

CONDUCTION SYSTEM

The heart is innervated by the autonomic nervous system (Chapter 17), but the autonomic neurons only increase or decrease the time it takes to complete a cardiac cycle, that is, they do not initiate contraction. The chamber walls can go on contracting and relaxing, contracting and relaxing, without any direct stimulus from the nervous system. This action is possible because the heart has an intrinsic regulating system called the **conduction system.** The conduction system is composed of specialized muscle tissue that generates and distributes the electrical impulses which stimulate the cardiac muscle fibers to contract. These tissues are the sinu-atrial (sinoatrial) node, the atrioventricular node, the atrioventricular bundle, the bundle branches, and the Purkinje fibers. The cells of the conduction system develop during embryological life from certain cardiac muscle cells. These cells lose their ability to contract and become specialized for impulse transmission.

A **node** of the conducting system is a compact mass of conducting cells. The **sinu-atrial (sinoatrial) node,** known as the **SA node** or **pacemaker,** is located in the

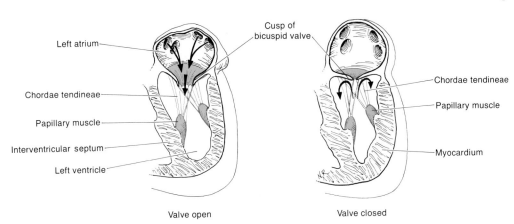

Valve open

Valve closed

(a)

FIGURE 12-3 Valves of the heart. (a) Structure and function of the bicuspid valve.

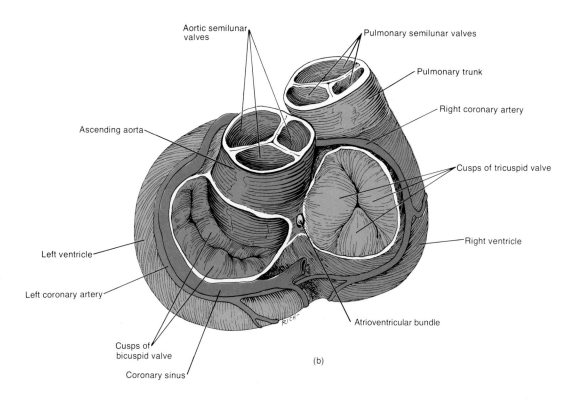

Aortic semilunar valves

Pulmonary semilunar valves

Pulmonary trunk

Right coronary artery

Ascending aorta

Cusps of tricuspid valve

Left ventricle

Right ventricle

Left coronary artery

Atrioventricular bundle

Cusps of bicuspid valve

Right ventricle

Coronary sinus

(b)

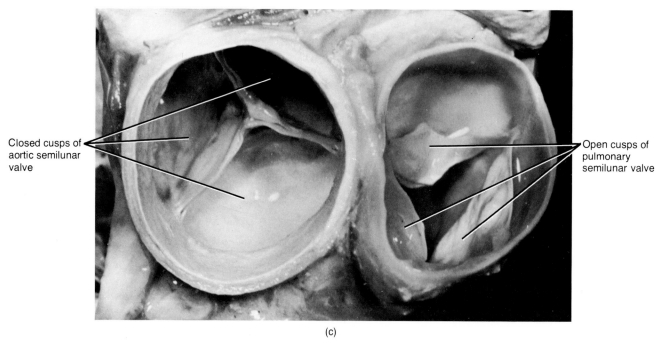

Closed cusps of aortic semilunar valve

Open cusps of pulmonary semilunar valve

(c)

FIGURE 12-3 (*Continued*) Valves of the heart. (b) Valves of the heart viewed from above. The atria have been removed to expose the tricuspid and bicuspid valves. (c) Photograph of a superior view of a closed aortic semilunar valve (left) and an open pulmonary semilunar valve (right). (Courtesy of C. Yokochi and J. W. Rohen, *Photographic Anatomy of the Human Body,* 2nd ed., 1979, IGAKU-SHOIN, Ltd., Tokyo, New York.)

right atrial wall inferior to the opening of the superior vena cava (Figure 12-4a). The SA node initiates each cardiac cycle, and thereby sets the basic pace for the heart rate—hence its common name, pacemaker. The rate set by the SA node may be altered by nervous impulses from the autonomic nervous system or by certain blood-borne chemicals such as thyroid hormone and epinephrine.

Once an electrical impulse is initiated by the SA node, the impulse spreads out over both atria, causing them

Artificial valve

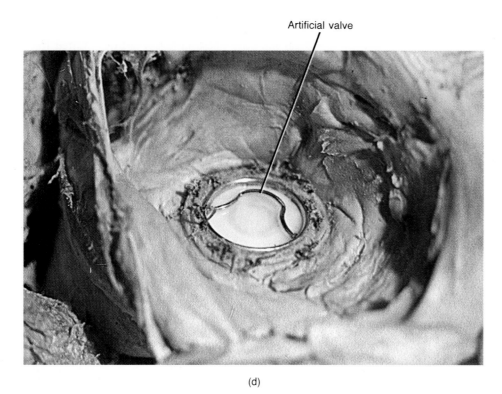

(d)

FIGURE 12-3 (*Continued*) Valves of the heart. (d) Photograph of a superior view of an artificial valve. (Courtesy of John W. Eads.)

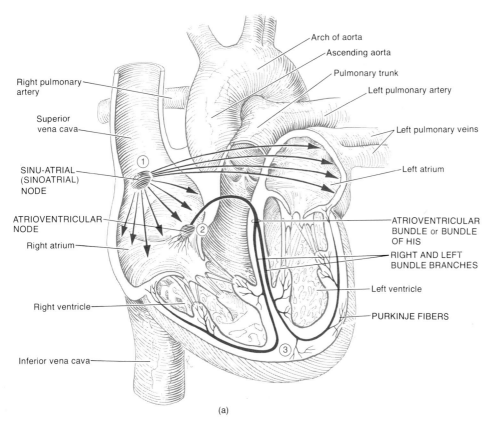

(a)

FIGURE 12-4 Conduction system of the heart. (a) Location of nodes and bundles of the conduction system.

to contract and at the same time depolarizing the **atrioventricular (AV) node.** Because of its location near the inferior portion of the interatrial septum, the AV node is one of the last portions of the atria to be depolarized.

From the AV node, a tract of conducting fibers called the **atrioventricular bundle** or **bundle of His** runs through the cardiac skeleton to the top of the interventricular septum. It then continues down both sides of the septum as the **right** and **left bundle branches.** The bundle of His distributes the charge over the medial surfaces of the ventricles. Actual contraction of the ventricles is stimulated by the **Purkinje** (pur-KIN-jē) **fibers** that emerge from the bundle branches and pass into the cells of the myocardium.

CARDIAC CYCLE

In a normal heartbeat, the two atria contract simultaneously while the two ventricles relax. Then, when the two ventricles contract, the two atria relax. The term **systole** (SIS-tō-lē) refers to the phase of contraction; **diastole** (dī-AS-tō-lē) is the phase of relaxation. A **cardiac cycle,** or complete heartbeat, consists of the systole and diastole of both atria plus the systole and diastole of both ventricles.

Electrocardiogram

Impulse transmission through the conduction system generates electrical currents that may be detected on the body's surface. A recording of the electrical changes that accompany the cardiac cycle is called an **electrocardiogram (ECG).** The instrument used to record the changes is an *electrocardiograph.*

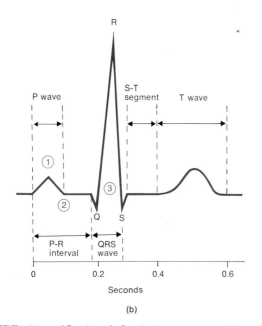

FIGURE 12-4 (*Continued*) Conduction system of the heart. (b) Normal electrocardiogram.

Each portion of the cardiac cycle produces a different electrical impulse. These impulses are transmitted from the electrodes to a recording needle that graphs the impulses as a series of up-and-down waves called *deflection waves.* In a typical record (Figure 12-4b), three clearly recognizable waves accompany each cardiac cycle. The first, called the **P wave,** is a small upward wave. It indicates atrial depolarization—the spread of an impulse from the SA node through the muscle of the two atria. A fraction of a second after the P wave begins, the atria contract. Then there is a deflection wave called the **QRS wave (complex).** It begins as a downward deflection (Q wave), continues as a large, upright, triangular wave (R wave), and ends as a downward wave at its base (S wave). This deflection represents atrial repolarization and ventricular depolarization, that is, the spread of the electrical impulse through the ventricles. The third recognizable deflection is a dome-shaped **T wave.** This wave indicates ventricular repolarization. There is no deflection to show atrial repolarization because the stronger QRS wave masks this event.

In reading an electrocardiogram, it is important to note the size of the deflection waves and certain time intervals. Enlargement of the P wave, for example, indicates enlargement of the atrium, as in mitral stenosis. In this condition, the mitral valve narrows, blood backs up into the atrium, and there is expansion of the atrial wall.

The **P-R interval** is measured from the beginning of the P wave to the beginning of the Q wave. It represents the conduction time from the beginning of atrial excitation to the beginning of ventricular excitation. The P-R interval is the time required for an impulse to travel through the atria and atrioventricular node to the remaining conducting tissues. The lengthening of this interval, as in atherosclerotic heart disease and rheumatic fever, occurs because the heart tissue covered by the P-R interval, namely the atria and atrioventricular node, is scarred or inflamed. Thus the impulse must travel at a slower rate and the interval is lengthened. The normal P-R interval covers no more than 0.20 second.

An enlarged Q wave may indicate a myocardial infarction. An enlarged R wave generally indicates enlarged ventricles.

The **S-T segment** begins at the end of the S wave and terminates at the beginning of the T wave. It represents the time between the end of the spread of the impulse through the ventricles and repolarization of the ventricles. The S-T segment is elevated in acute myocardial infarction and depressed when the heart muscle receives insufficient oxygen.

The T-wave represents ventricular repolarization. It is flat when the heart muscle is receiving insufficient oxygen, as in atherosclerotic heart disease. It may be elevated when the body's potassium level is increased.

Timing

If we assume that the average heart beats 72 times per minute, then each cardiac cycle requires about 0.8 second. During the first 0.1 second, the atria contract and the ventricles relax. The atrioventricular valves are open, and the semilunar valves are closed. For the next 0.3 second, the atria are relaxing and the ventricles are contracting. During the first part of this period, all valves are closed; during the second part, the semilunar valves are open. The last 0.4 second of the cycle is the relaxation, or quiescent, period. All chambers are in diastole. For the first part of the quiescent period, all valves are closed; during the latter part, the atrioventricular valves open and blood starts draining into the ventricles. When the heart beats faster than normal, the quiescent period is shortened accordingly.

Sounds

The sound of the heartbeat comes primarily from turbulence in blood flow created by the closure of the valves, not from the contraction of the heart muscle. The first sound, which can be described as a **lubb** ($\overline{oo}$) sound, is a long, booming sound. The lubb is the sound created by the closure of the atrioventricular valves soon after ventricular systole begins. The second sound, which is heard as a short, sharp sound, can be described as a **dupp** ($\breve{u}$) sound. Dupp is the sound created as the semilunar valves close toward the end of ventricular systole. A pause between the second sound and the first sound of the next cycle is about two times longer than the pause between the first and second sound of each cycle. Thus the cardiac cycle can be heard as a lubb, dupp, pause; lubb, dupp, pause; lubb, dupp, pause.

Although heart sounds are produced in part by the closure of valves, they are not necessarily heard best over these valves. Each sound tends to be clearest in a slightly different location closest to the surface of the body.

SURFACE ANATOMY

The valves of the heart may be identified by surface projection (Figure 12-5). The pulmonary and aortic semilunar valves are represented on the surface by a line about 2.5 cm (1 inch) in length. The pulmonary semilunar valve lies horizontally behind the inner end of the left third costal cartilage and the adjoining part of the sternum. The aortic semilunar valve is placed obliquely behind the left side of the sternum, opposite the third intercostal space. The tricuspid valve lies behind the sternum, extending from the midline at the level of the fourth costal cartilage down toward the right sixth chondrosternal junction. The bicuspid lies behind the left side of the sternum obliquely at the level of the fourth costal cartilage. It is represented by a line about 3 cm in length.

AUTONOMIC CONTROL

The pacemaker initiates contraction and, left to itself, would set an unvarying heart rate. However, the body's need for blood supply varies under different conditions. There are several regulatory mechanisms stimulated by factors such as chemicals present in the body, temperature, emotional state, and age. The most important control of heart rate and strength of contraction is the effect of the autonomic nervous system.

Within the medulla of the brain is a group of neurons called the **cardioacceleratory center.** Arising from this center are sympathetic fibers that travel down a tract in the spinal cord and then pass outward in the *cardiac nerves* and innervate the SA node, AV node, and all chambers (Figure 12-6). When the cardioacceleratory center is stimulated, nerve impulses travel along the sympathetic fibers. This causes them to release norepinephrine, which increases the rate of heartbeat and the strength of contraction.

The medulla also contains a group of neurons that form the **cardioinhibitory center.** Arising from this center are parasympathetic fibers that reach the heart via the *vagus nerve.* These fibers innervate the SA node and atria. When this center is stimulated, nerve impulses transmitted along the parasympathetic fibers cause the release of acetylcholine. This decreases the rate of heartbeat and strength of contraction.

The autonomic control of the heart is therefore the result of complementary action of sympathetic (stimulatory) and parasympathetic (inhibitory) influences. Sensory impulses from nerve cells in different parts of the

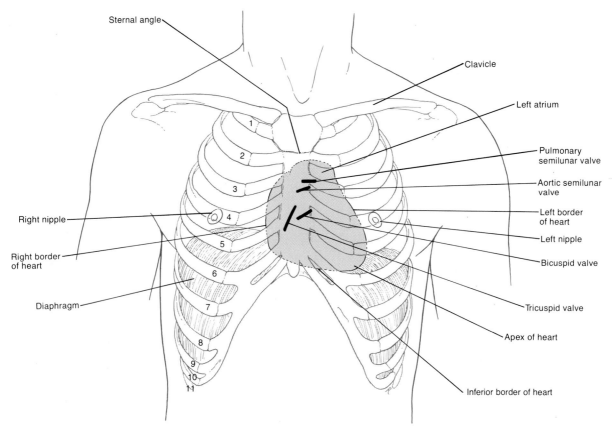

FIGURE 12-5 Surface projection of the heart.

cardiovascular system act upon the centers so that a balance between stimulation and inhibition is maintained. Nerve cells capable of responding to changes in blood pressure are called **pressoreceptors (baroreceptors).** These receptors affect the rate of heartbeat and are involved in three reflex pathways: the carotid sinus reflex, the aortic reflex, and the right heart (atrial) reflex.

The *carotid sinus reflex* is concerned with maintaining normal blood pressure in the brain. The *carotid sinus* is a small widening of the internal carotid artery just above the point where it branches off from the common carotid artery (Figure 12-6). The walls of the carotid sinus contain pressoreceptors. Any increase in blood pressure stretches the walls of the sinus, and the stretching stimulates the pressoreceptors. The impulses then travel from the pressoreceptors over sensory neurons in the glossopharyngeal (IX) nerves. Within the medulla, the impulses stimulate the cardioinhibitory center and inhibit the cardioacceleratory center. Consequently, more parasympathetic impulses pass from the cardioinhibitory center via the vagus (X) nerves to the heart and fewer sympathetic impulses pass from the cardioacceleratory center via cardiac nerves to the heart. The result is a decrease in heart rate and force of contraction. There is a subsequent decrease in cardiac output, a decrease in arterial blood pressure, and

restoration of blood pressure to normal. But if blood pressure falls, reflex acceleration of the heart takes place. The pressoreceptors in the carotid sinus do not stimulate the cardioinhibitory center, and the cardioacceleratory center is free to dominate. The heart then beats faster and more forcefully to restore normal blood pressure. This inverse relationship between blood pressure and heart rate is referred to as **Marey's law of the heart.**

The *aortic reflex* is concerned with general systemic blood pressure. It is initiated by pressoreceptors in the walls of the arch of the aorta (Figure 12-6), and it operates like the carotid sinus reflex.

The *right heart (atrial) reflex* responds to venous blood pressure. It is initiated by pressoreceptors in the superior and inferior venae cavae and in the right atrium. When venous pressure increases, the pressoreceptors send impulses that stimulate the cardioacceleratory center. This action causes the heart rate to increase. This mechanism is called the **Bainbridge reflex.**

BLOOD VESSELS

Blood vessels are called arteries, arterioles, capillaries, venules, and veins. **Arteries** are the vessels that carry blood from the heart to the tissues. Large or elastic ar-

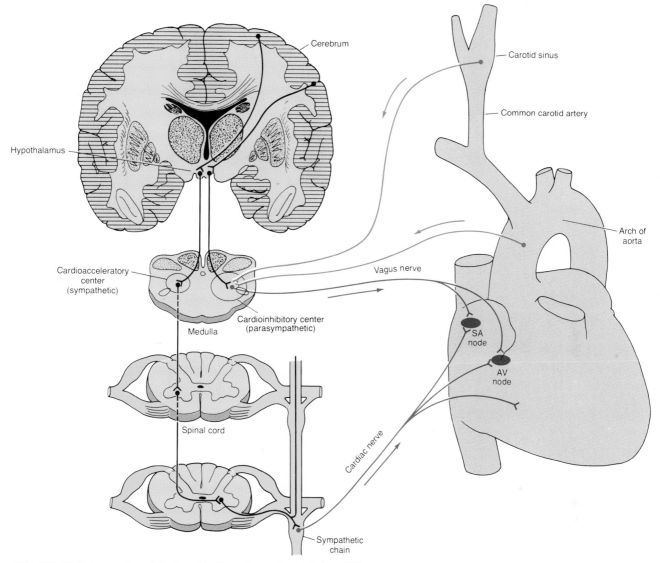

FIGURE 12-6 Innervation of the heart by the autonomic nervous system.

teries leave the heart and divide into medium-sized vessels that deliver blood to various regions of the body. The medium-sized or muscular arteries, in turn, divide into small arteries, which, in turn, divide into smaller arteries called **arterioles.** As the arterioles enter a tissue, they branch into countless microscopic vessels called **capillaries.** Through the walls of the capillaries, substances are exchanged between the blood and body tissues. Before leaving the tissue, groups of capillaries reunite to form small veins called **venules.** These, in turn, merge to form progressively larger tubes—the veins themselves. **Veins,** then, are blood vessels that convey blood from the tissues back to the heart. Since blood vessels require oxygen and nutrients just like other tissues of the body, they also have blood vessels in their own walls called **vasa vasorum.**

ARTERIES

Arteries have walls constructed of three coats or tunics and a hollow core, called a *lumen,* through which the blood flows (Figure 12-7). Veins also have three coats, but the artery walls are considerably thicker and stronger because the pressure in an artery is always greater than in a vein. The inner coat of an arterial wall is called the *tunica interna* or *intima.* It is composed of a lining of endothelium (simple squamous epithelium) that is in contact with the blood. It also has an overlying layer of areolar connective tissue and an outer layer of elastic tissue called the internal elastic membrane. The middle coat, or *tunica media,* is usually the thickest layer. It consists of elastic fibers and smooth muscle. The outer coat, the *tunica externa* or *adventitia,* is composed princi-

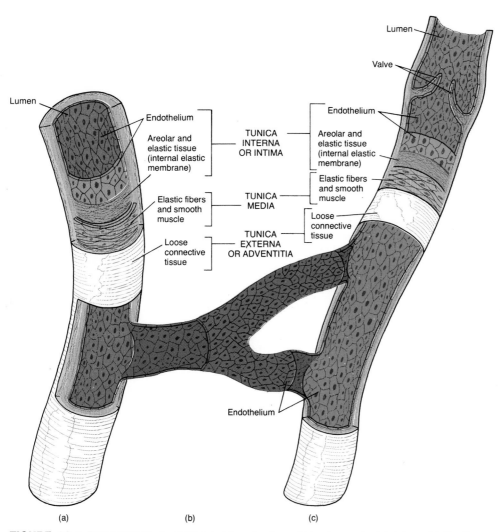

Lumen

Lumen

Valve

Endothelium

Endothelium

Areolar and
elastic tissue
(internal elastic
membrane)

Areolar and
elastic tissue
(internal elastic
membrane)

TUNICA
INTERNA
OR INTIMA

Elastic fibers
and smooth
muscle

TUNICA
MEDIA

Elastic fibers
and smooth
muscle

Loose
connective
tissue

Loose
connective
tissue

TUNICA
EXTERNA
OR ADVENTITIA

Endothelium

(a)   (b)   (c)

FIGURE 12-7 Blood vessels. (a–c) Comparative structure of blood vessels. (a) Artery. (b) Capillary. (c) Vein. The relative size of the capillary is enlarged for emphasis.

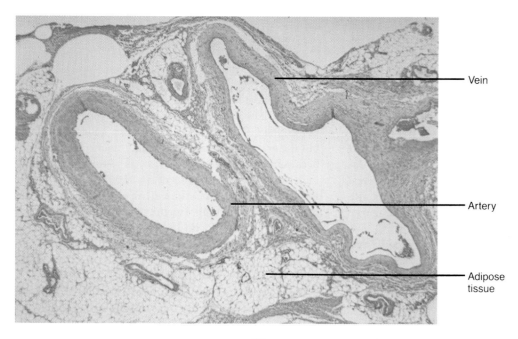

Vein

Artery

Adipose
tissue

(d)

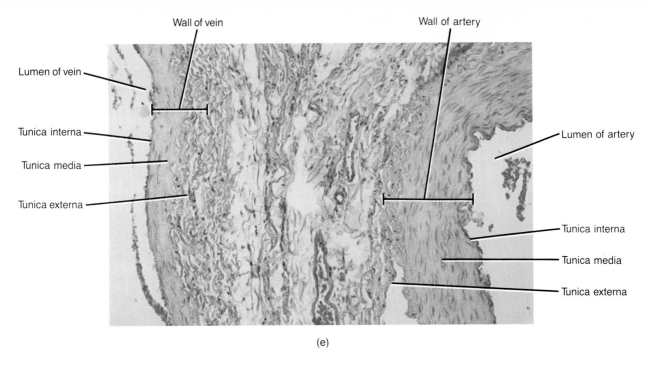

Wall of vein

Lumen of vein

Tunica interna

Tunica media

Tunica externa

Wall of artery

Lumen of artery

Tunica interna

Tunica media

Tunica externa

(e)

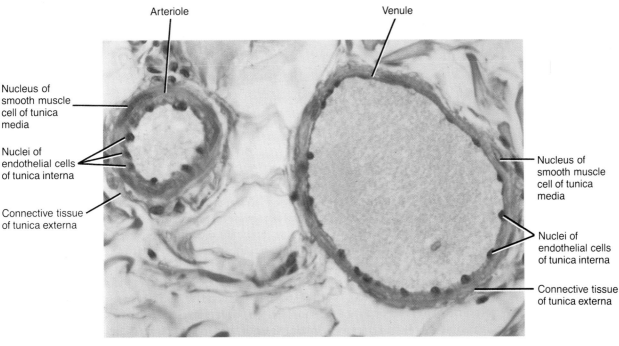

Arteriole

Venule

Nucleus of
smooth muscle
cell of tunica
media

Nuclei of
endothelial cells
of tunica interna

Connective tissue
of tunica externa

Nucleus of
smooth muscle
cell of tunica
media

Nuclei of
endothelial cells
of tunica interna

Connective tissue
of tunica externa

(f)

FIGURE 12-7 (*Continued*) Blood vessels. (d) (*Opposite page*) Photomicrograph of an artery and its accompanying vein at a magnification of 50×. (e) Photomicrograph of a portion of the wall of an artery and its accompanying vein at a magnification of 250×. (f) Photomicrograph of an arteriole and a venule at a magnification of 800×. (Photomicrographs © 1983 by Michael H. Ross. Used by permission.)

pally of loose connective tissue. The tunica externa contains elastic and collagenous fibers and a few smooth muscle fibers. An external elastic membrane may separate the tunica media from the tunica externa.

As a result of the structure of the middle coat especially, arteries have two major properties: elasticity and contractility. When the ventricles of the heart contract and eject blood into the large arteries, the arteries expand to contain the extra blood. Then, as the ventricles relax,

the elastic recoil of the arteries forces the blood onward. The contractility of an artery comes from its smooth muscle. The smooth muscle is arranged longitudinally and in rings around the lumen somewhat like a doughnut. The smooth muscle is innervated by sympathetic branches of the autonomic nervous system. When there is sympathetic stimulation, the smooth muscle contracts, squeezes the wall around the lumen, and narrows the vessel. Such a decrease in the size of the lumen is called **vasoconstric-**

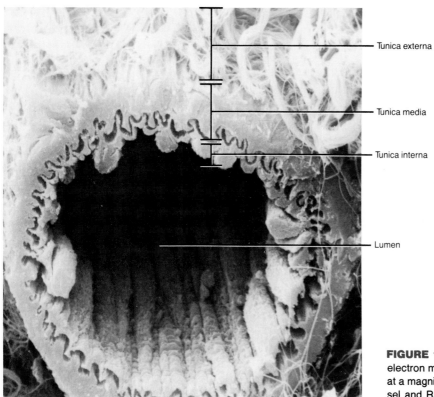

Tunica externa

Tunica media

Tunica interna

Lumen

(g)

FIGURE 12-7 (*Continued*) Blood vessels. (g) Scanning electron micrograph of an artery showing the three tunics at a magnification of 5170×. (Courtesy of Richard G. Kessel and Randy H. Kardon, *Tissues and Organs: A Text-Atlas of Scanning Electron Microscopy,* W. H. Freeman and Company, San Francisco, 1979.)

tion. Conversely, when sympathetic stimulation is removed, the smooth muscle fibers relax and the size of the arterial lumen increases. This increase is called **vasodilation** and is often due to the inhibition of vasoconstriction.

The contractility of arteries also serves a minor function in stopping bleeding. The blood flowing through an artery is under a great deal of pressure. Thus great quantities of blood can be quickly lost from a broken artery. When an artery is cut, its wall constricts so that blood does not escape quite so rapidly. However, there is a limit to how much vasoconstriction can help.

Large arteries are referred to as *elastic* or *conducting arteries.* They include the aorta and brachiocephalic, common carotid, subclavian, vertebral, and common iliac arteries. The wall of elastic arteries is relatively thin in proportion to their diameter and their tunica media contains more elastic fibers and less smooth muscle. As the heart alternately contracts and relaxes, the rate of blood flow tends to be intermittent. When the heart contracts and forces blood into the aorta, the wall of the elastic arteries stretches. During relaxation of the heart, the wall of the elastic arteries recoils, moving the blood forward in a more continuous flow. Elastic arteries are called conducting arteries because they conduct blood from the

heart to medium-sized arteries. Medium-sized arteries are called *muscular* or *distributing arteries.* They include the axillary, brachial, radial, intercostal, splenic, mesenteric, femoral, popliteal, and tibial arteries. Their tunica media contains more smooth muscle than elastic fibers and they are capable of vasoconstriction and vasodilation to adjust the volume of blood to suit the needs of the structure supplied. The wall of muscular arteries is relatively thick, mainly due to the large amounts of smooth muscle. Muscular arteries are called distributing arteries because they distribute blood to various parts of the body.

Most parts of the body receive branches from more than one artery. In such areas the distal ends of the vessels unite. The junction of two or more vessels supplying the same body region is called an **anastomosis** (a-nas-tō-MŌ-sis). Anastomoses may also occur between the origins of veins and between arterioles and venules. Anastomoses between arteries provide alternate routes by which blood can reach a tissue or organ. Thus if a vessel is occluded by disease, injury, or surgery, circulation to a part of the body is not necessarily stopped. The alternate route of blood to a body part through an anastomosis is known as **collateral circulation.** An alternate blood route may also be from nonanastomosing vessels that supply the same region of the body.

ARTERIOLES

An **arteriole** is a small artery that delivers blood to capillaries. Arterioles closer to the arteries from which they branch have a tunica interna like that of arteries, a tunica media composed of smooth muscle and very few elastic fibers, and a tunica externa composed mostly of loose connective tissue. As arterioles get smaller in size as they approach capillaries, the tunics change character so that arterioles closest to capillaries consist of little more than a layer of endothelium surrounded by a few scattered smooth muscle cells.

Arterioles play a key role in regulating blood flow from arteries into capillaries. The smooth muscle of arterioles, like that of arteries, is subject to vasoconstriction and vasodilation. During vasoconstriction, blood flow into capillaries is restricted, but during vasodilation, the flow is significantly increased. The relationship of arterioles to blood flow will be considered in detail later in the chapter.

CAPILLARIES

Capillaries are microscopic vessels that usually connect arterioles and venules (Figure 12-7b). They are found near almost every cell in the body. The distribution of capillaries in the body varies with the activity of the tissue. For example, in places where activity is higher, such as the muscles, the liver, the kidneys, the lungs, and the nervous system, there are rich capillary supplies. In areas where activity is lower, such as tendons and ligaments, the capillary supply is not as extensive. The epidermis, cornea of the eye, and cartilage are devoid of capillaries.

The primary function of capillaries is to permit the exchange of nutrients and wastes between the blood and tissue cells. The structure of the capillaries is admirably suited to this purpose. First, capillary walls are composed of only a single layer of cells (endothelium). They have no tunica media or tunica externa. Thus a substance in the blood must pass through the plasma membrane of just one cell to reach tissue cells. This vital exchange of materials occurs only through capillary walls—the thick walls of arteries and veins present too great a barrier. Although in some places in the body capillaries pass directly from arterioles to venules, in other places they form extensive branching networks. These networks increase the surface area for diffusion and thereby allow a rapid exchange of large quantities of materials. At the point where blood enters capillaries there is a ring of smooth muscle called the *precapillary sphincter.* By its contraction or relaxation, the sphincter can help adjust the amount and force of blood flowing into capillaries.

Some capillaries of the body, such as those found in muscle tissue and other locations, are referred to as *continuous capillaries.* These capillaries are so named because the cytoplasm of the endothelial cells is continuous when viewed in cross section through a microscope; the cytoplasm appears as an uninterrupted ring, except for the endothelial junction. Other capillaries of the body are referred to as *fenestrated capillaries.* They differ from continuous capillaries in that the endothelial cells of fenestrated capillaries have numerous fenestrae (pores) where the cytoplasm is absent. The fenestrae range from 700 to 1000 Å in diameter and are closed by a thin diaphragm, except in the capillaries in the kidneys where they are assumed to be open. Fenestrated capillaries are also found in the villi of the small intestine, choroid plexuses of the ventricles in the brain, ciliary processes of the eyes, and endocrine glands.

Microscopic blood vessels in certain parts of the body, such as the liver, are termed **sinusoids.** They are wider than capillaries and more tortuous. Also, instead of the usual endothelial lining, sinusoids are lined largely by phagocytic cells. In the liver, such cells are called *Kupffer* (KOOP-fer) *cells.* Like capillaries, sinusoids convey blood from arterioles to venules. Other regions containing sinusoids include the spleen, adenohypophysis, parathyroid glands, and adrenal cortex.

VENULES

When several capillaries unite, they form small veins called **venules.** Venules collect blood from capillaries and drain it into veins. The venules closest to the capillaries consist of a tunica interna of endothelium and a tunica externa of connective tissue. As the venules approach the veins, they also contain a tunica media characteristic of veins.

VEINS

Veins are composed of essentially the same three coats as arteries, but they have considerably less elastic tissue and smooth muscle and contain more white fibrous tissue (Figure 12-7). However, they are still distensible enough to adapt to variations in the volume and pressure of blood passing through them.

By the time the blood leaves the capillaries and moves into the veins, it has lost a great deal of pressure. The difference in pressure can be observed in the blood flow from a cut vessel; blood leaves a cut vein in an even flow rather than in the rapid spurts characteristic of arteries. Most of the structural differences between arteries and veins reflect this pressure difference. For example, veins do not need walls as strong as those of arteries. The low pressure in veins, however, has its disadvantages. When you stand, the pressure pushing blood up the veins in your lower extremities is barely enough to balance the force of gravity pushing it back down. For this reason, many veins, especially those in the limbs, contain valves that prevent backflow. Normal valves ensure the flow of blood toward the heart.

CLINICAL APPLICATION

In people with weak venous valves, large quantities of blood are forced by gravity back down into distal parts of the vein. This pressure overloads the vein and pushes the walls outward. After repeated overloading, the walls lose their elasticity and become stretched and flabby. A vein damaged in this way is called a **varicose vein.** Varicose veins may be caused by prolonged standing and pregnancy. Because a varicosed wall is not able to exert a firm resistance against the blood, blood tends to accumulate in the pouched-out area of the vein, causing it to swell and forcing fluid into the surrounding tissue. Veins close to the surface of the legs are highly susceptible to varicosities. Veins that lie deeper are not as vulnerable because surrounding skeletal muscles prevent their walls from overstretching.

Varicosities are also common in the veins that lie in the wall of the anal canal. These varicosities are called **hemorrhoids** (HEM-o-royds). In many people hemorrhoids may be caused by constipation. Repeated straining during defecation forces blood down into the superior hemorrhoidal plexus, increasing pressure in these veins. Constipation is related to low-fiber diets, especially in North America. A recent hypothesis links increased intraabdominal pressure, caused by straining during evacuation of firm feces, directly to hemorrhoids or varicose veins. A new development in the treatment of hemorrhoids is *cryosurgery* (*cryo* = cold), in which external hemorrhoids are destroyed by freezing them with a solution of nitrous oxide or liquid nitrogen.

A **vascular (venous) sinus,** or simply **sinus,** is a vein with a thin endothelial wall that has no smooth muscle to alter its diameter. Surrounding tissue replaces the tunica media and tunica externa to provide support. Intracranial vascular sinuses, which are supported by the dura mater, return cerebrospinal fluid and deoxygenated blood from the brain to the heart. Another example of a vascular sinus is the coronary sinus of the heart.

CIRCULATORY ROUTES

The arteries, arterioles, capillaries, venules, and veins are organized into definite routes in order to circulate the blood throughout the body. We can now look at the basic routes the blood takes as it is transported through its vessels.

Figure 12-8 shows a number of basic **circulatory routes** through which the blood travels. **Systemic circulation** includes all the oxygenated blood that leaves the left ventricle through the aorta and the deoxygenated blood that returns to the right atrium after traveling to all the organs including the nutrient arteries to the lungs. Two of the many subdivisions of the systemic circulation are the **coronary circulation,** which supplies the myocardium of the heart, and the **hepatic portal circulation,** which runs from the digestive tract to the liver. Blood leaving the aorta and traveling through the systemic arteries is a bright red color. As it moves through the capillaries, it loses its oxygen and takes on carbon dioxide, which gives the blood in the systemic veins its dark red color.

When blood returns to the heart from the systemic route, it goes out of the right ventricle through the **pulmonary circulation** to the lungs. In the lungs, it loses its carbon dioxide and takes on oxygen. It is now bright red again. It returns to the left atrium of the heart and reenters the systemic circulation.

Another major route—the **fetal circulation**—exists only in the fetus and contains special structures that allow the developing human being to exchange materials with its mother.

Cerebral circulation (circle of Willis) is discussed in Exhibit 12-3.

SYSTEMIC CIRCULATION

The flow of blood from the left ventricle to all parts of the body except the lungs and back to the right atrium is called the **systemic circulation.** The purpose of systemic circulation is to carry oxygen and nutrients to body tissues and to remove carbon dioxide and other wastes from the tissues. All systemic arteries branch from the *aorta*, which arises from the left ventricle of the heart.

As the aorta emerges from the left ventricle, it passes upward and deep to the pulmonary artery. At this point, it is called the *ascending aorta.* The ascending aorta gives off two coronary branches to the heart muscle. Then it turns to the left, forming the *arch of the aorta* before descending to the level of the fourth thoracic vertebra as the *descending aorta.* The descending aorta lies close to the vertebral bodies, passes through the diaphragm, and terminates at the level of the fourth lumbar vertebra by dividing into two *common iliac arteries,* which carry blood to the lower extremities. The section of the descending aorta between the arch of aorta and the diaphragm is also referred to as the *thoracic aorta.* The section between the diaphragm and the common iliac arteries is termed the *abdominal aorta.* Each section of the aorta gives off arteries that continue to branch into distributing arteries leading to organs and finally into the capillaries that pierce the tissues.

Blood is returned to the heart through the systemic veins. All the veins of the systemic circulation flow into either the *superior* or *inferior venae cavae* or the *coronary sinus.* They in turn empty into the right atrium. The principal arteries and veins of systemic circulation are described and illustrated in Exhibits 12-1 to 12-12 and Figures 12-9 to 12-18.

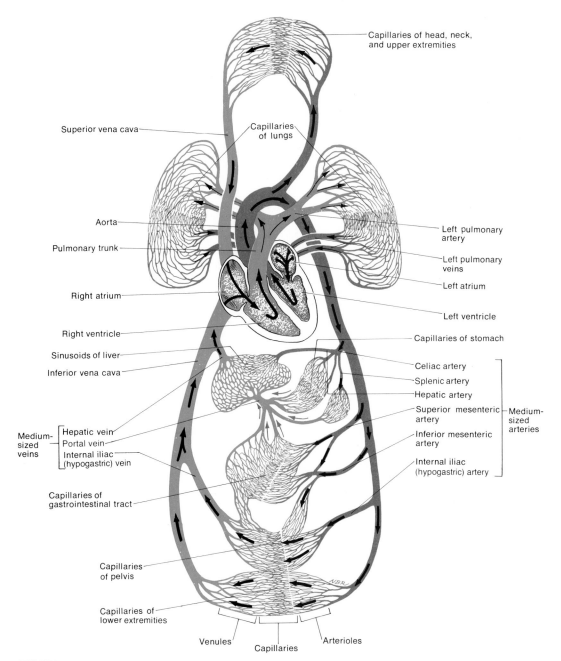

Capillaries of head, neck, and upper extremities

Superior vena cava

Capillaries of lungs

Aorta

Pulmonary trunk

Right atrium

Right ventricle

Sinusoids of liver

Inferior vena cava

Medium-sized veins
{
Hepatic vein
Portal vein
Internal iliac (hypogastric) vein
}

Capillaries of gastrointestinal tract

Capillaries of pelvis

Capillaries of lower extremities

Left pulmonary artery

Left pulmonary veins

Left atrium

Left ventricle

Capillaries of stomach

Celiac artery

Splenic artery

Hepatic artery

Superior mesenteric artery

Inferior mesenteric artery

Internal iliac (hypogastric) artery

}
Medium-sized arteries

Venules   Capillaries   Arterioles

FIGURE 12-8 Circulatory routes. Systemic circulation is indicated by heavy black arrows; pulmonary circulation by thin black arrows; and hepatic portal circulation by thin colored arrows. Refer to Figure 12-19 for the details of coronary circulation and Figure 12-22 for the details of fetal circulation.

EXHIBIT 12-1 AORTA AND ITS BRANCHES (Figure 12-9)

DIVISION OF AORTA	ARTERIAL BRANCH	REGION SUPPLIED
Ascending aorta	Coronaries	Heart
Arch of aorta	Brachiocephalic → Right common carotid	Right side of head and neck.
	Brachiocephalic → Right subclavian	Right upper extremity. Left side of head and neck. Left upper extremity.
	Left common carotid	
	Left subclavian	
Thoracic aorta	Posterior intercostals	Intercostal and chest muscles and pleurae.
	Subcostals	Similar to intercostals.
	Pericardials	Dorsal pericardium.
	Superior phrenics	Posterior surface of diaphragm.
	Mediastinals	Posterior mediastinum.
	Bronchials	Bronchi of lungs.
	Esophageals	Esophagus.
Abdominal aorta	Inferior phrenics	Inferior surface of diaphragm.
	Celiac → Common hepatic	Liver.
	Celiac → Left gastric	Stomach and esophagus.
	Celiac → Splenic	Spleen, pancreas, and stomach.
	Superior mesenteric	Small intestine, cecum, and ascending and transverse colons.
	Suprarenals	Adrenal (suprarenal) glands.
	Renals	Kidneys.
	Gonadals → Testiculars	Testes.
	Gonadals → Ovarians	Ovaries.
	Inferior mesenteric	Transverse, descending, and sigmoid colons and rectum.
	Common iliacs → External iliacs	Lower extremities.
	Common iliacs → Internal iliacs (hypogastrics)	Uterus, prostate, muscles of buttocks, and urinary bladder.

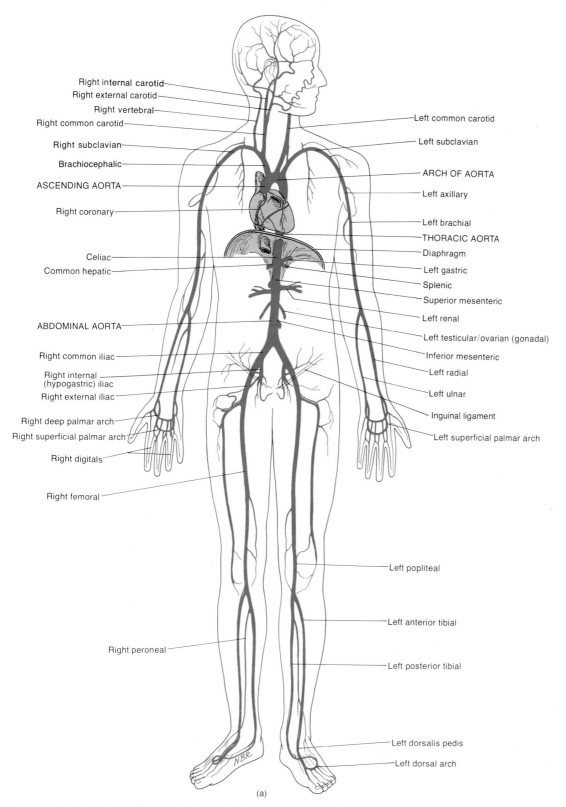

(a)

FIGURE 12-9 Aorta and its principal branches in anterior view. (a) Diagram.

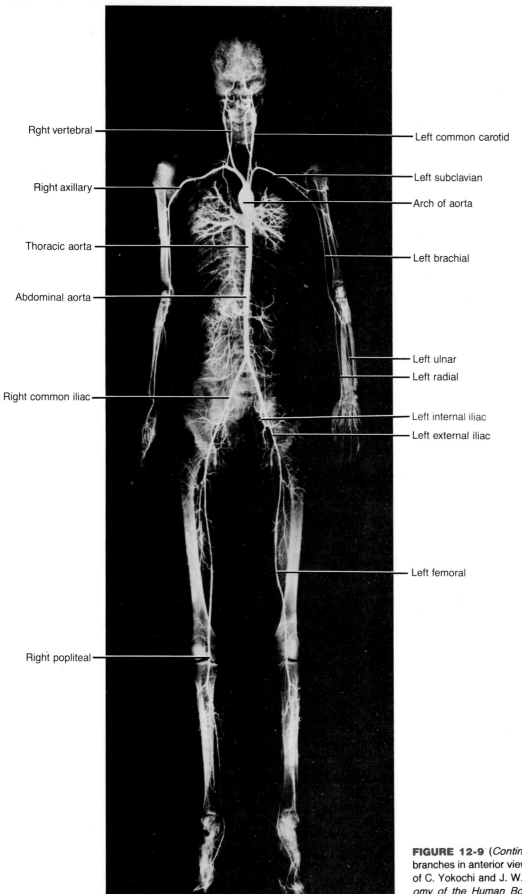

Rght vertebral

Right axillary

Thoracic aorta

Abdominal aorta

Right common iliac

Right popliteal

Left common carotid

Left subclavian

Arch of aorta

Left brachial

Left ulnar
Left radial

Left internal iliac
Left external iliac

Left femoral

(b)

FIGURE 12-9 (*Continued*) Aorta and its principal branches in anterior view. (b) Arteriogram. (Courtesy of C. Yokochi and J. W. Rohen, *Photographic Anatomy of the Human Body,* 2nd ed., 1979, IGAKU-SHOIN, Ltd., Tokyo, New York.)

EXHIBIT 12-2 ASCENDING AORTA (Figure 12-10)

BRANCH	DESCRIPTION AND REGION SUPPLIED
Coronary arteries	These two branches arise from ascending aorta just superior to aortic semilunar valve. They form crown around heart, giving off branches to atrial and ventricular myocardium.

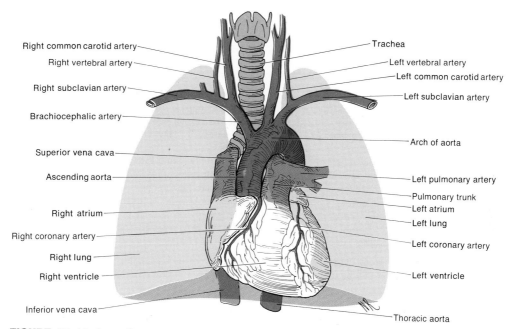

FIGURE 12-10 Ascending aorta and its arterial branches in anterior view.

EXHIBIT 12-3 **ARCH OF AORTA (Figure 12-11)**

BRANCH	DESCRIPTION AND REGION SUPPLIED
Brachiocephalic	**Brachiocephalic artery** is first branch off arch of aorta. It divides to form right subclavian artery and right common carotid artery. **Right subclavian artery** extends from brachiocephalic to first rib and then passes into armpit (axilla) and supplies arm, forearm, and hand. Continuation of right subclavian into axilla is called **axillary artery.*** From here, it continues into upper arm as **brachial artery.** At bend of elbow, brachial artery divides into medial **ulnar** and lateral **radial arteries.** These vessels pass down to palm, one on each side of forearm. In palm, branches of two arteries anastomose to form two palmar arches—**superficial palmar arch** and **deep palmar arch.** From these arches arise **digital arteries,** which supply fingers and thumb.
	Before passing into axilla, right subclavian gives off major branch to brain called **vertebral artery.** Right vertebral artery passes through foramina of transverse processes of cervical vertebrae and enters skull through foramen magnum to reach undersurface of brain. Here it unites with left vertebral artery to form **basilar artery.**
Right common carotid	**Right common carotid artery** passes upward in neck. At upper level of larynx, it divides into **right external** and **right internal carotid arteries.** External carotid supplies right side of thyroid gland, tongue, throat, face, ear, scalp, and dura mater. Internal carotid supplies brain, right eye, and right sides of forehead and nose.
	Anastomoses of left and right internal carotids along with basilar artery form arterial circle at base of brain called **circle of Willis.** From this anastomosis arise arteries supplying brain. Essentially, circle of Willis is formed by union of **anterior cerebral arteries** (branches of internal carotids) and **posterior cerebral arteries** (branches of basilar artery). Posterior cerebral arteries are connected with internal carotids by **posterior communicating arteries.** Anterior cerebral arteries are connected by **anterior communicating arteries.** Circle of Willis equalizes blood pressure to brain and provides alternate routes for blood to brain should arteries become damaged.
Left common carotid	**Left common carotid** is second branch off arch of aorta (see Figure 12-10). Corresponding to right common carotid, it divides into basically same branches with same names, except that arteries are now labeled "left" instead of "right."
Left subclavian	**Left subclavian artery** is third branch off arch of aorta (see Figure 12-10). It distributes blood to left vertebral artery and vessels of left upper extremity. Arteries branching from left subclavian are named like those of right subclavian.

* The right subclavian artery is a good example of the practice of giving the same vessel different names as it passes through different regions.

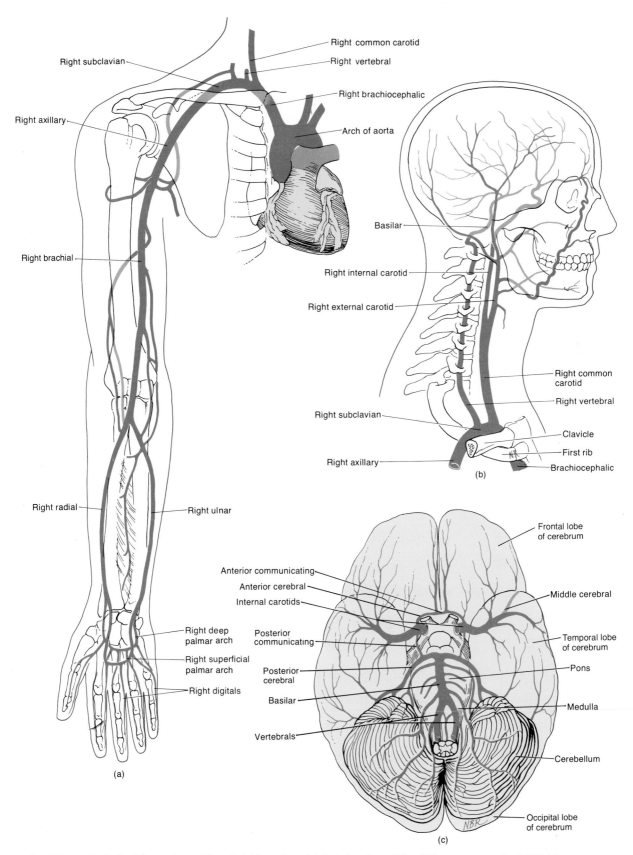

FIGURE 12-11 Arch of the aorta and its arterial branches. (a) Anterior view of the right upper extremity. (b) Right lateral view of the arteries of the neck and head. (c) Arteries of the base of the brain.

EXHIBIT 12-4 THORACIC AORTA (Figure 12-12)

BRANCH	DESCRIPTION AND REGION SUPPLIED
	Thoracic aorta runs from fourth to twelfth thoracic vertebrae. Along its course, it sends off numerous small arteries to viscera and skeletal muscles of the chest. Branches of an artery that supply viscera are called **visceral branches.** Those that supply body wall structures are **parietal branches.**
VISCERAL **Pericardial**	Several minute **pericardial arteries** supply blood to dorsal aspect of pericardium.
Bronchial	One **right** and two **left bronchial arteries** supply the bronchial tubes, areolar tissue of the lungs, bronchial lymph nodes, and esophagus.
Esophageal	Four or five **esophageal arteries** supply the esophagus.
Mediastinal	Numerous small **mediastinal arteries** supply blood to structures in the posterior mediastinum.
PARIETAL **Posterior intercostal**	Nine pairs of **posterior intercostal arteries** supply the intercostal, pectoral, and abdominal muscles; overlying subcutaneous tissue and skin; mammary glands; and vertebral canal and its contents.
Subcostal	The **left** and **right subcostal arteries** have a distribution similar to that of the posterior intercostals.
Superior phrenic	Small **superior phrenic arteries** supply the posterior surface of the diaphragm.

EXHIBIT 12-5 ABDOMINAL AORTA (Figure 12-12)

BRANCH	DESCRIPTION AND REGION SUPPLIED
VISCERAL	
Celiac	**Celiac artery** (trunk) is first visceral aortic branch below diaphragm. It has three branches: (1) **common hepatic artery,** which supplies tissues of liver; (2) **left gastric artery,** which supplies stomach; and (3) **splenic artery,** which supplies spleen, pancreas, and stomach.
Superior mesenteric	**Superior mesenteric artery** distributes blood to small intestine and part of large intestine.
Suprarenals	Right and left **suprarenal arteries** supply blood to adrenal (suprarenal) glands.
Renals	Right and left **renal arteries** carry blood to kidneys.
Testiculars (gonadals)	Right and left **testicular arteries** extend into scrotum and terminate in testes.
Ovarians (gonadals)	Right and left **ovarian arteries** are distributed to ovaries.
Inferior mesenteric	**Inferior mesenteric artery** supplies major part of large intestine and rectum.
PARIETAL	
Inferior phrenics	**Inferior phrenic arteries** are distributed to undersurface of diaphragm.
Lumbars	**Lumbar arteries** supply spinal cord and its meninges and muscles and skin of lumbar region of back.
Middle sacral	**Middle sacral artery** supplies sacrum, coccyx, gluteus maximus muscles, and rectum.

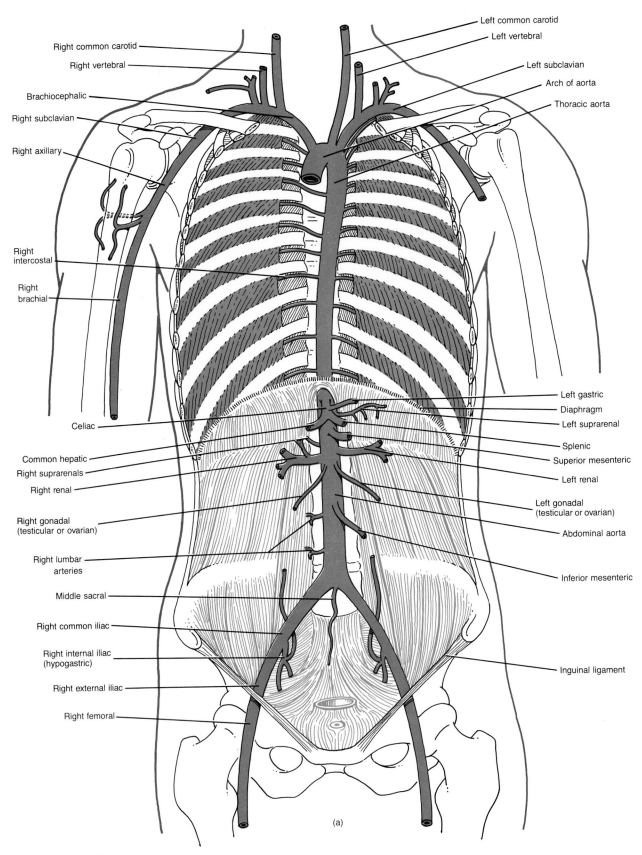

FIGURE 12-12 Abdominal aorta and its principal arterial branches in anterior view. (a) Diagram.

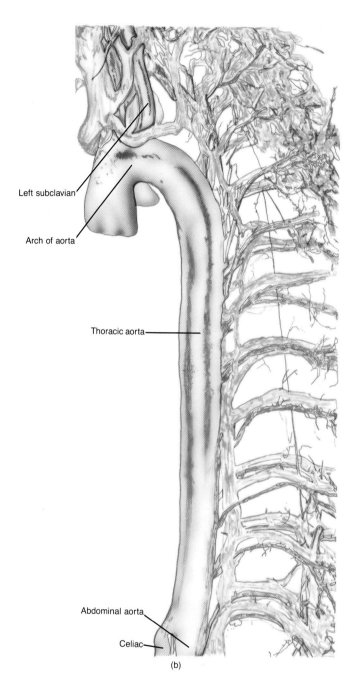

Left subclavian

Arch of aorta

Thoracic aorta

Abdominal aorta

Celiac

(b)

FIGURE 12-12 (*Continued*) Abdominal aorta and its principal arterial branches in anterior view. (b) Diagram of a cast of the aorta and associated vessels.

EXHIBIT 12-6 ARTERIES OF PELVIS AND LOWER EXTREMITIES (Figure 12-13)

BRANCH	DESCRIPTION AND REGION SUPPLIED
Common iliac	At about level of fourth lumbar vertebra, abdominal aorta divides into right and left **common iliac arteries.** Each passes downward about 5 cm (2 inches) and gives rise to two branches: internal iliac and external iliac.
Internal iliacs	**Internal iliac** or **hypogastric arteries** form branches that supply psoas major, quadratus lumborum, and medial side of each thigh, urinary bladder, rectum, prostate gland, ductus deferens, uterus, and vagina.
External iliacs	**External iliac arteries** diverge through pelvis, enter thighs, and here become right and left **femoral arteries.** Both femorals send branches back up to genitals and wall of abdomen. Other branches run to muscles of thigh. Femoral continues down medial and posterior side of thigh at back of knee joint, where it becomes **popliteal artery.** Between knee and ankle, popliteal runs down back of leg and is called **posterior tibial artery.** Below knee, **peroneal artery** branches off posterior tibial to supply structures on medial side of fibula and calcaneus. In calf, **anterior tibial artery** branches off popliteal and runs along front of leg. At ankle, it becomes **dorsalis pedis artery.** At ankle, posterior tibial divides into **medial** and **lateral plantar arteries.** These arteries anastomose with dorsalis pedis and supply blood to foot.

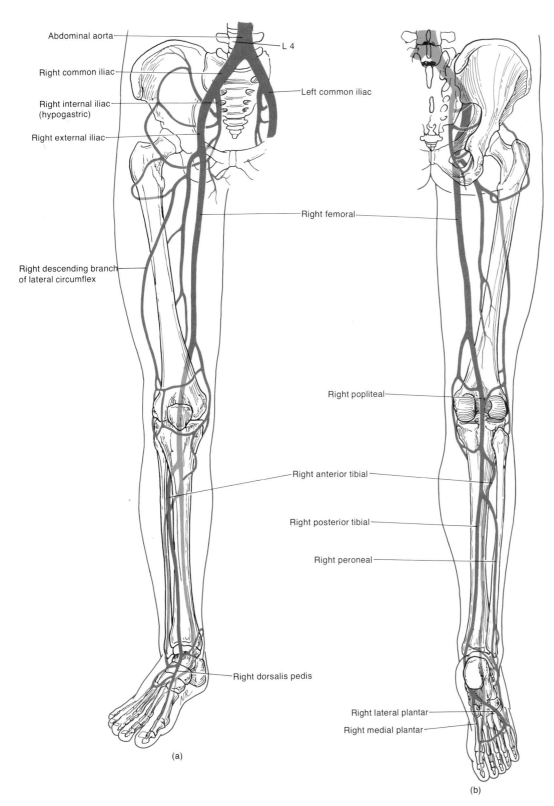

Abdominal aorta

L 4

Right common iliac

Left common iliac

Right internal iliac
(hypogastric)

Right external iliac

Right femoral

Right descending branch
of lateral circumflex

Right popliteal

Right anterior tibial

Right posterior tibial

Right peroneal

Right dorsalis pedis

Right lateral plantar

Right medial plantar

(a)

(b)

FIGURE 12-13 Arteries of the pelvis and right lower extremity. (a) Anterior view. (b) Posterior view.

EXHIBIT 12-7 VEINS OF SYSTEMIC CIRCULATION (Figure 12-14)

VEIN	DESCRIPTION AND REGION DRAINED
	All systemic and cardiac veins return blood to the right atrium of the heart through one of three large vessels. Return flow in coronary circulation is taken up by **cardiac veins,** which empty into the large vein of the heart, the **coronary sinus.** From here, the blood empties into the right atrium of the heart (see Figure 12-3). Return flow in systemic circulation empties into the superior vena cava or inferior vena cava.
Superior vena cava	Veins that empty into the **superior vena cava** are veins of the head and neck, upper extremities, and some from the thorax.
Inferior vena cava	Veins that empty into the **inferior vena cava** are some from the thorax and veins of the abdomen, pelvis, and lower extremities.

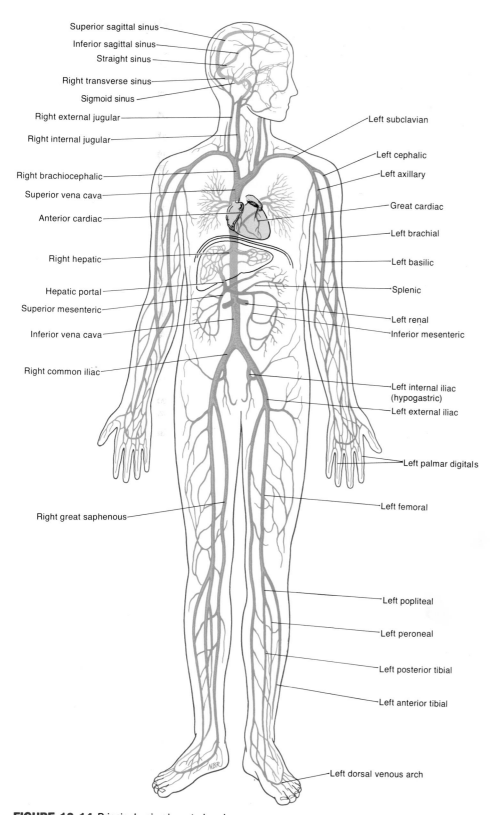

Superior sagittal sinus

Inferior sagittal sinus

Straight sinus

Right transverse sinus

Sigmoid sinus

Right external jugular

Right internal jugular

Right brachiocephalic

Superior vena cava

Anterior cardiac

Right hepatic

Hepatic portal

Superior mesenteric

Inferior vena cava

Right common iliac

Right great saphenous

Left subclavian

Left cephalic

Left axillary

Great cardiac

Left brachial

Left basilic

Splenic

Left renal

Inferior mesenteric

Left internal iliac
(hypogastric)

Left external iliac

Left palmar digitals

Left femoral

Left popliteal

Left peroneal

Left posterior tibial

Left anterior tibial

Left dorsal venous arch

FIGURE 12-14 Principal veins in anterior view.

EXHIBIT 12-8 VEINS OF HEAD AND NECK (Figure 12-15)

VEIN	DESCRIPTION AND REGION DRAINED
Internal jugulars	Right and left **internal jugular veins** receive blood from face and neck. They arise as continuation of **sigmoid sinuses** at base of skull. Intracranial vascular sinuses are located between layers of dura mater and receive blood from brain. Other sinuses that drain into internal jugular include **superior sagittal sinus, inferior sagittal sinus, straight sinus,** and **transverse (lateral) sinuses.** Internal jugulars descend on either side of neck and pass behind clavicles, where they join with right and left **subclavian veins.** Unions of internal jugulars and subclavians form right and left **brachiocephalic veins.** From here blood flows into **superior vena cava.**
External jugulars	Left and right **external jugular veins** run down neck along outside of internal jugulars. They drain blood from parotid (salivary) glands, facial muscles, scalp, and other superficial structures into **subclavian veins.** In cases of heart failure, the venous pressure in the right atrium may rise. In such patients the pressure in the column of blood in the external jugular vein rises so that, even with the patient at rest and sitting in a chair, the external jugular vein will be visibly distended. Temporary distention of the vein is often seen in healthy adults when the intrathoracic pressure is raised in coughing and physical exertion.

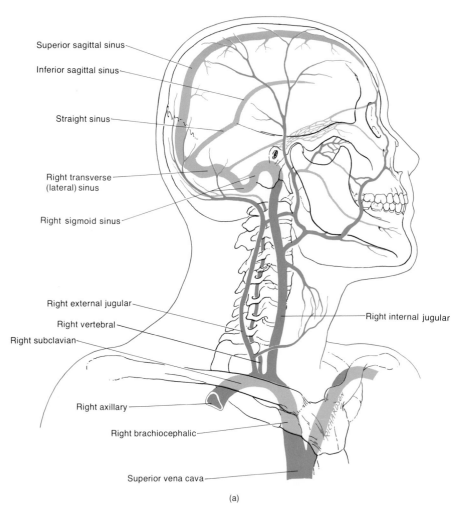

(a)

FIGURE 12-15 Veins of the neck and head in right lateral view. (a) Diagram.

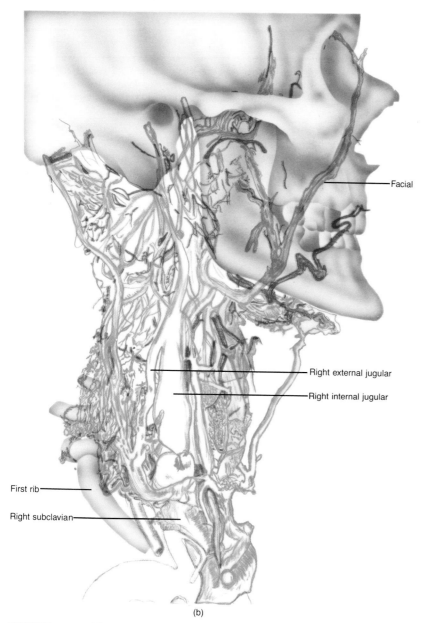

(b)

FIGURE 12-15 (*Continued*) Veins of the neck and head in right lateral view. (b) Diagram of a cast of the vessels of the neck and head.

EXHIBIT 12-9 VEINS OF UPPER EXTREMITIES (Figure 12-16)

VEIN	DESCRIPTION AND REGION DRAINED
	Blood from each upper extremity is returned to the heart by deep and superficial veins. Both sets of veins contain valves. **Deep veins** are located deep in the body. They usually accompany arteries, and many have the same names as corresponding arteries. **Superficial veins** are located just below the skin and are often visible. They anastomose extensively with each other and deep veins.
SUPERFICIAL **Cephalics**	**Cephalic vein** of each upper extremity begins in the medial part of **dorsal arch** and winds upward around radial border of forearm. In front of elbow, it is connected to **basilic vein** by the **median cubital vein.** Just below elbow, cephalic vein unites with **accessory cephalic vein** to form cephalic vein of upper extremity. Ultimately, cephalic vein empties into **axillary vein.**
Basilics	**Basilic vein** of each upper extremity originates in the ulna part of dorsal arch. It extends along posterior surface of ulna to point below elbow where it receives **median cubital vein.** If a vein must be punctured for an injection, transfusion, or removal of a blood sample, median cubitals are preferred. The median cubital vein joins the basilic vein to form the axillary vein.
Median antebrachials	**Median antebrachial veins** drain venous plexus on palmar surface of hand, ascend on ulnar side of anterior forearm, and end in median cubital veins.
DEEP **Radials**	**Radial veins** receive dorsal metacarpal veins.
Ulnars	**Ulnar veins** receive tributaries from deep palmar arch. Radial and ulnar veins unite in bend of elbow to form **brachial vein.**
Brachials	Located on either side of brachial artery, **brachial veins** join into axillary veins.
Axillaries	**Axillary veins** are a continuation of brachials and basilics. Axillaries end at first rib, where they become subclavians.
Subclavians	**Right** and **left subclavian veins** unite with internal jugulars to form **brachiocephalic veins.** Thoracic duct of lymphatic system flows into left subclavian vein at junction with internal jugular. Right lymphatic duct enters right subclavian vein at corresponding junction.

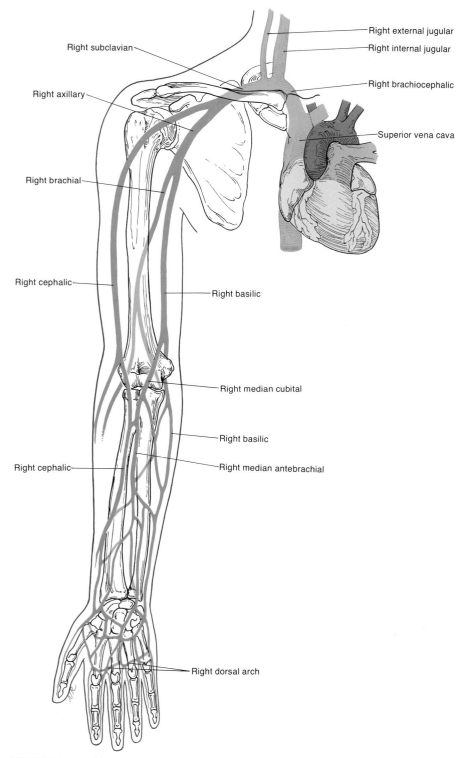

FIGURE 12-16 Veins of the right upper extremity in anterior view.

EXHIBIT 12-10 VEINS OF THORAX (Figure 12-17)

VEIN	DESCRIPTION AND REGION DRAINED
Brachiocephalic	Right and left **brachiocephalic veins,** formed by union of subclavians and internal jugulars, drain blood from head, neck, upper extremities, mammary glands, and upper thorax. Brachiocephalics unite to form **superior vena cava.**
Azygos veins	**Azygos veins,** besides collecting blood from thorax, may serve as bypass for inferior vena cava that drains blood from lower body. Several small veins directly link azygos veins with inferior vena cava. And large veins that drain lower extremities and abdomen dump blood into azygos. If inferior vena cava or hepatic portal vein becomes obstructed, azygos veins can return blood from lower body to superior vena cava.
Azygos	**Azygos vein** lies in front of vertebral column, slightly right of midline. It begins as continuation of **right ascending lumbar vein.** It connects with inferior vena cava, right common iliac, and lumbar veins. Azygos receives blood from right intercostal veins that drain chest muscles; from hemiazygos and accessory hemiazygos veins; from several esophageal, mediastinal, and pericardial veins; and from right bronchial vein. Vein ascends to fourth thoracic vertebra, arches over right lung, and empties into **superior vena cava.**
Hemiazygos	**Hemiazygos vein** is in front of vertebral column and slightly left of midline. It begins as continuation of **left ascending lumbar vein.** It receives blood from lower four or five intercostal veins and some esophageal and mediastinal veins. At level of ninth thoracic vertebra, it joins **azygos vein.**
Accessory hemiazygos	**Accessory hemiazygos vein** is also in front and to left of vertebral column. It receives blood from three or four intercostal veins and left bronchial vein. It joins azygos at level of eighth thoracic vertebra.

EXHIBIT 12-11 VEINS OF THE ABDOMEN AND PELVIS (Figure 12-17)

VEIN	DESCRIPTION AND REGION DRAINED
Inferior vena cava	**Inferior vena cava** is the largest vein of the body; formed by union of two common iliac veins that drain lower extremities and abdomen. Inferior vena cava extends upward through abdomen and thorax to right atrium. Numerous small veins enter the inferior vena cava. Most carry return flow from branches of abdominal aorta and names correspond to names of arteries.
Common iliacs	**Common iliac veins** are formed by union of internal (hypogastric) and external iliac veins and represent distal continuation of inferior vena cava at its bifurcation.
Internal iliacs (hypogastrics)	Tributaries of **internal iliac (hypogastric) veins** basically correspond with branches of external iliac arteries. Internal iliacs drain gluteal muscles, medial side of thigh, urinary bladder, rectum, prostate gland, ductus deferens, uterus, and vagina.
External iliacs	**External iliac veins** are continuation of femoral veins and receive blood from lower extremities and inferior part of anterior abdominal wall.
Renals	**Renal veins** drain kidneys.
Gonadals (testicular and ovarian)	**Testicular veins** drain testes (left testicular vein empties into left renal vein) and **ovarian veins** drain ovaries (left ovarian vein empties into left renal vein).
Suprarenals	**Suprarenal veins** drain suprarenal glands (left suprarenal vein empties into left renal vein).
Inferior phrenics	**Inferior phrenic veins** drain diaphragm (left inferior phrenic vein sends tributary to left renal vein).
Hepatics	**Hepatic veins** drain liver.
Lumbars	A series of parallel **lumbar veins** drain blood from both sides of posterior abdominal wall. Lumbars connect at right angles with right and left ascending lumbar veins, which form origin of corresponding azygos or hemiazygos vein. Lumbars drain blood into ascending lumbars and then run to inferior vena cava, where they release remainder of flow.

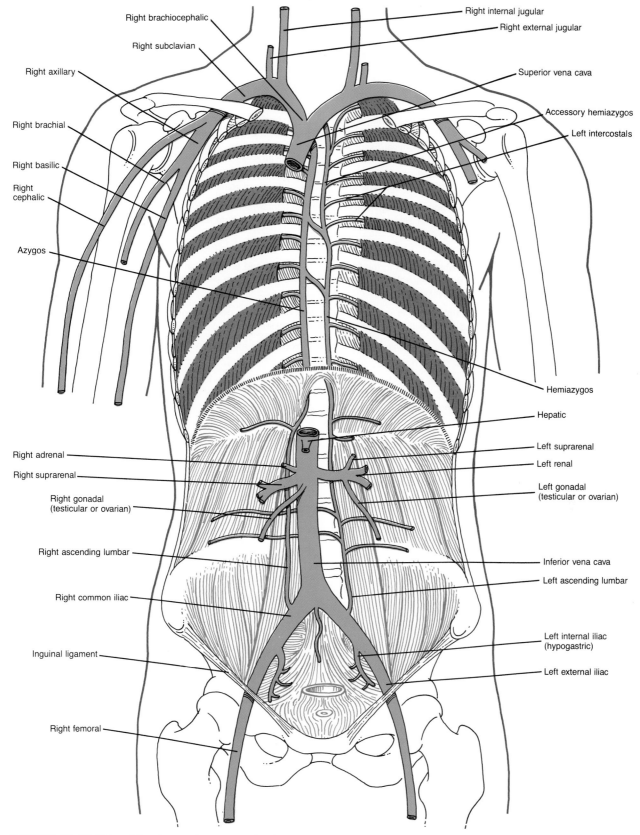

FIGURE 12-17 Veins of the thorax, abdomen, and pelvis in anterior view.

EXHIBIT 12-12 VEINS OF LOWER EXTREMITIES (Figure 12-18)

VEIN	DESCRIPTION AND REGION DRAINED
	Blood from each lower extremity is returned by superficial set and deep set of veins.
SUPERFICIAL VEINS	Main superficial veins are great saphenous and small saphenous. Both, especially great saphenous, frequently become varicosed.
Great saphenous	**Great saphenous vein,** longest vein in body, begins at medial end of **dorsal venous arch** of foot. It passes in front of medial malleolus and then upward along medial aspect of leg and thigh. It receives tributaries from superficial tissues and connects with deep veins as well. It empties into femoral vein in groin. The great saphenous vein is very constant in its position anterior to the medial malleolus. It is frequently used for prolonged administration of intravenous fluids. This is particularly important in very young babies and in patients of any age who are in shock and whose veins are collapsed.
Small saphenous	**Small saphenous vein** begins at lateral end of dorsal venous arch of foot. It passes behind lateral malleolus and ascends under skin of back of leg. It receives blood from foot and posterior portion of leg. It empties into popliteal vein behind knee.
DEEP VEINS **Posterior tibial**	**Posterior tibial vein** is formed by union of **medial** and **lateral plantar veins** behind medial malleolus. It ascends deep in muscle at back of leg, receives blood from **peroneal vein,** and unites with anterior tibial vein just below knee.
Anterior tibial	**Anterior tibial vein** is upward continuation of **dorsalis pedis** veins in foot. It runs between tibia and fibula and unites with posterior tibial to form popliteal vein.
Popliteal	**Popliteal vein,** just behind knee, receives blood from anterior and posterior tibials and small saphenous vein.
Femoral	**Femoral vein** is upward continuation of popliteal just above knee. Femorals run up posterior of thighs and drain deep structures of thighs. After receiving great saphenous veins in groin, they continue as right and left **external iliac veins.** Right and left **internal iliac veins** receive blood from pelvic wall and viscera, external genitals, buttocks, and medial aspect of thigh. Right and left **common iliac veins** are formed by union of internal and external iliacs. Common iliacs unite to form inferior vena cava.

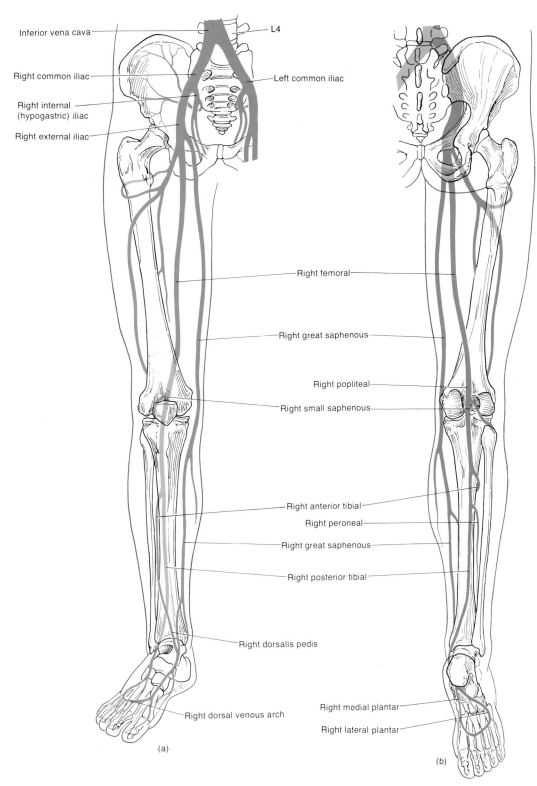

FIGURE 12-18 Veins of the pelvis and right lower extremity. (a) Anterior view. (b) Posterior view.

CORONARY (CARDIAC) CIRCULATION

The walls of the heart, like any other tissue, including large blood vessels, have their own blood vessels. Nutrients could not possibly diffuse through all the layers of cells that make up the heart tissue. And the blood in the left chambers of the heart would never supply enough oxygen. The flow of blood through the numerous vessels that pierce the myocardium is called the **coronary (cardiac) circulation** (Figure 12-19). The *left coronary artery* originates as a branch of the ascending aorta, runs under the left auricle, and divides into the anterior interventricular and circumflex branches. The *anterior interventricular branch* follows the anterior interventricular sulcus and supplies oxygenated blood to the walls of both ventricles. The *circumflex branch* distributes oxygenated blood to the walls of the left ventricle and left atrium. The *right coronary artery* also originates as a branch of

the ascending aorta. It runs under the right auricle and divides into the marginal and posterior interventricular branches. The *posterior interventricular branch* follows the posterior interventricular sulcus and supplies the walls of the two ventricles with oxygenated blood. The *marginal branch* transports oxygenated blood to the myocardium of the right ventricle and right atrium. The left ventricle receives the most abundant blood supply because of the enormous work it must do.

As blood passes through the arterial system of the heart, it delivers oxygen and nutrients and collects carbon dioxide and wastes. Most of the deoxygenated blood, which carries the carbon dioxide and wastes, is collected by a large vein, the *coronary sinus,* which empties into the right atrium. The principal tributaries of the coronary sinus are the *great cardiac vein,* which drains the anterior aspect of the heart, and the *middle cardiac vein,* which drains the posterior aspect of the heart.

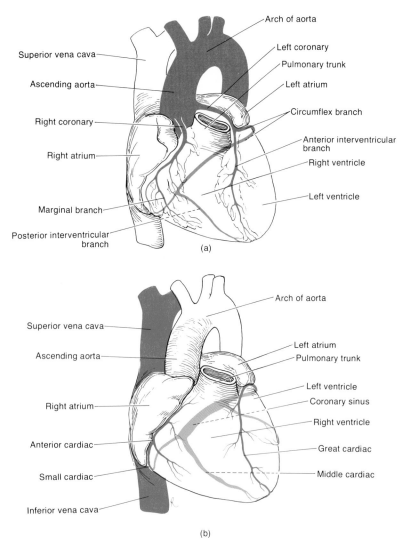

FIGURE 12-19 Coronary (cardiac) circulation. (a) Diagram of anterior view of arterial distribution. (b) Diagram of anterior view of venous drainage.

CLINICAL APPLICATION

Most heart problems result from faulty coronary circulation. If a reduced oxygen supply weakens the cells, but does not actually kill them, the condition is called **ischemia** (is-KĒ-mē-a). **Angina pectoris** (an-JĪ-na PEK-tō-ris), meaning "chest pain," is ischemia of the myocardium. It occurs when coronary circulation is somewhat reduced for some reason. Stress, which produces constriction of vessel walls, is a common cause. Equally common is strenuous exercise after a heavy meal. When any quantity of food enters the stomach, the body increases blood flow to the digestive tract. As a consequence, some blood is diverted away from other organs, including the heart. Exercise, however, increases heart muscle activity and thus the heart's need for oxygen. Doing heavy work while food is in the stomach can therefore lead to oxygen deficiency in the myocardium. Angina pectoris weakens the heart muscle, but it does not produce a full-scale heart attack. The simple remedy of taking nitroglycerin, a drug that dilates coronary vessels and thereby increases blood flow to the area, brings coronary circulation back to normal and stops the pain of angina.

A much more serious problem is **myocardial infarction** (in-FARK-shun), commonly called a "coronary" or "heart attack." **Infarction** means death of an area of tissue because of an interrupted blood supply. Myocardial infarction may result from a thrombus or embolus in one of the coronary arteries. The tissue distal to the obstruction dies and is replaced by noncontractile scar tissue. Thus, the heart muscle loses at least some of its strength. The aftereffects depend partly on the size and location of the infarcted, or dead, area.

New diagnostic tests have been developed that enable physicians to diagnose heart attacks within hours. During a heart attack, certain enzymes are released into the bloodstream and are the strongest indicators that a heart attack has occurred. One test detects creatine kinase MB isoenzyme (CK-MB), which is found only in heart cells. The test detects nearly all the CK-MB released after a heart attack and gives physicians a more accurate measure of the extent of the heart damage. Another test measures lactic dehydrogenase-1 (LDH-1), an enzyme liberated during a myocardial infarction. This test yields results in an hour. Other enzymes that are always elevated after a myocardial infarction are serum glutamic oxaloacetic transaminase (SGOT) and creatine phosphokinase (CPK).

HEPATIC PORTAL CIRCULATION

Blood enters the liver from two sources. The hepatic artery delivers oxygenated blood from the systemic circulation; the hepatic portal vein delivers deoxygenated blood from the digestive organs. The term **hepatic portal circulation** refers to this flow of venous blood from the digestive organs to the liver before returning to the heart (Figure 12-20). Hepatic portal blood is rich with substances absorbed from the digestive tract. The liver monitors these substances before they pass into the general circulation. For example, the liver stores nutrients such as glucose. It modifies other digested substances so they may be used by cells. It detoxifies harmful substances that have been absorbed by the digestive tract and destroys bacteria by phagocytosis.

The hepatic portal system includes veins that drain blood from the pancreas, spleen, stomach, intestines, and gallbladder and transport it to the portal vein of the liver. The *hepatic portal vein* is formed by the union of the superior mesenteric and splenic veins. The *superior mesenteric vein* drains blood from the small intestine and portions of the large intestine and stomach. The *splenic vein* drains the spleen and receives tributaries from the stomach, pancreas, and portions of the colon. The tributaries from the stomach are the *gastric, pyloric,* and *gastroepiploic veins.* The *pancreatic veins* come from the pancreas, and the *inferior mesenteric veins* come from portions of the colon. Before the hepatic portal vein enters the liver, it receives the *cystic vein* from the gallbladder and other veins. Ultimately, blood leaves the liver through the *hepatic veins,* which enter the inferior vena cava.

PULMONARY CIRCULATION

The flow of deoxygenated blood from the right ventricle to the lungs and the return of oxygenated blood from the lungs to the left atrium is called the **pulmonary circulation** (Figure 12-21). The *pulmonary trunk* emerges from the right ventricle and passes upward, backward, and to the left. It then divides into two branches. The *right pulmonary artery* runs to the right lung; the *left pulmonary artery* goes to the left lung. On entering the lungs, the branches divide and subdivide. They get smaller and ultimately form capillaries around the alveoli in the lungs. Carbon dioxide is passed from the blood into the alveoli to be breathed out of the lungs. Oxygen breathed in by the lungs is passed from the alveoli into the blood. The capillaries then unite. They grow larger and become veins. Eventually, two *pulmonary veins* exit from each lung and transport the oxygenated blood to the left atrium. The pulmonary veins are the only postnatal veins that carry oxygenated blood. Contraction of the left ventricle then sends the blood into the systemic circulation.

FETAL CIRCULATION

The circulatory system of a fetus, called **fetal circulation,** differs from an adult's because the lungs, kidneys, and digestive tract of a fetus are nonfunctional. The fetus derives its oxygen and nutrients from the maternal blood and eliminates its carbon dioxide and wastes into the maternal blood (Figure 12-22).

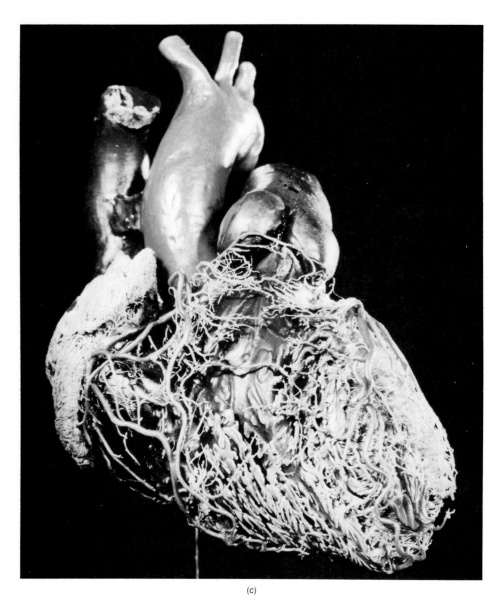

(c)

FIGURE 12-19 (*Continued*) Coronary (cardiac) circulation. (c) Cast of the cardiac blood vessels in anterior view. (© John Watney Photo Library.)

The exchange of materials between fetal and maternal circulation occurs through a structure called the *placenta* (pla-SEN-ta). It is attached to the umbilicus of the fetus by the umbilical (um-BIL-i-kal) cord, and it communicates with the mother through countless small blood vessels that emerge from the uterine wall. The umbilical cord contains blood vessels that branch into capillaries in the placenta. Wastes from the fetal blood diffuse out of the capillaries, into spaces containing maternal blood (intervillous spaces) in the placenta, and finally into the mother's uterine blood vessels. Nutrients travel the opposite route—from the maternal blood vessels to the intervillous spaces to the fetal capillaries. Normally there is no mixing of maternal and fetal blood since all exchanges occur through capillaries.

Blood passes from the fetus to the placenta via two *umbilical arteries*. These branches of the internal iliac arteries are included in the umbilical cord. At the placenta, the blood picks up oxygen and nutrients and eliminates carbon dioxide and wastes. The oxygenated blood returns from the placenta via a single *umbilical vein*. This vein ascends to the liver of the fetus, where it divides into two branches. Some blood flows through the branch that joins the hepatic portal vein and enters the liver. Although the fetal liver manufactures red blood cells, it does not function in digestion. Therefore, most of the blood flows into the second branch: the *ductus venosus* (DUK-tus ve-NŌ-sus). The ductus venosus eventually passes its blood to the inferior vena cava, bypassing the liver.

In general, circulation through other portions of the fetus is not unlike postnatal circulation. Deoxygenated blood returning from the lower regions is mingled with oxygenated blood from the ductus venosus in the inferior

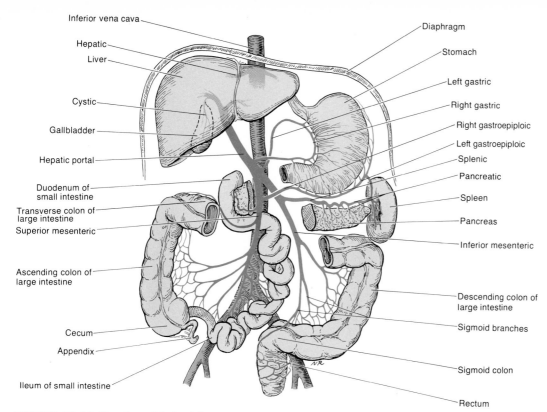

FIGURE 12-20 Hepatic portal circulation.

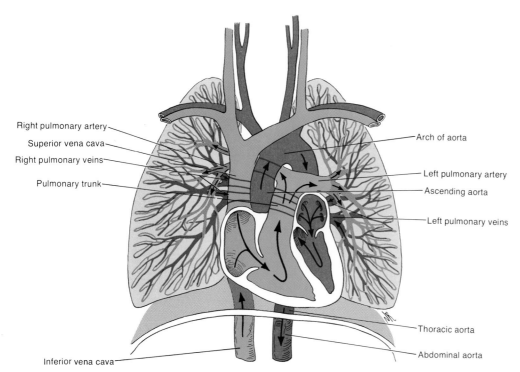

FIGURE 12-21 Pulmonary circulation.

vena cava. This mixed blood then enters the right atrium. The circulation of blood through the upper portion of the fetus is also similar to the postnatal flow. Deoxygenated blood returning from the upper regions of the fetus is collected by the superior vena cava, and it also passes into the right atrium.

Most of the blood does not pass through the right ventricle to the lungs, as it does in postnatal circulation, since the fetal lungs do not operate. In the fetus, an opening called the *foramen ovale* (fō-RĀ-men ō-VAL-ē) exists in the septum between the right and left atria. A valve in the inferior vena cava directs about a third of the

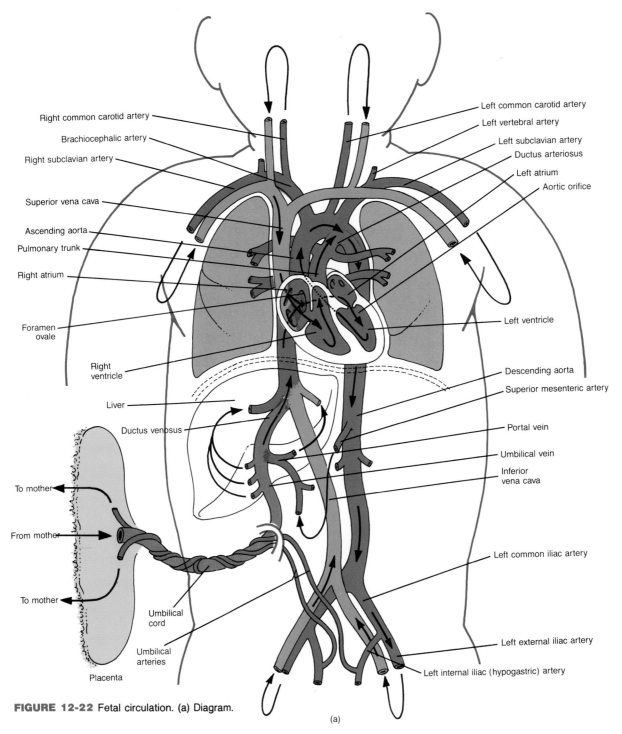

FIGURE 12-22 Fetal circulation. (a) Diagram.

(a)

blood through the foramen ovale so that it may be sent directly into the systemic circulation. The blood that does descend into the right ventricle is pumped into the pulmonary trunk, but little of this blood actually reaches the lungs. Most blood in the pulmonary trunk is sent through the *ductus arteriosus* (ar-tē-rē-Ō-sus). This small vessel connecting the pulmonary trunk with the aorta enables blood in excess of nutrient requirements to bypass the fetal lungs. The blood in the aorta is carried to all parts of the fetus through its systemic branches. When the common iliac arteries branch into the external and internal

iliacs, part of the blood flows into the internal iliacs. It then goes to the umbilical arteries and back to the placenta for another exchange of materials. The only vessel that carries fully oxygenated blood is the umbilical vein.

At birth, when lung, renal, digestive, and liver functions are established, special structures of fetal circulation are no longer needed and the following changes occur.

1. The umbilical arteries atrophy to become the **lateral umbilical ligaments.**

2. The umbilical vein becomes the **round ligament** of the liver.

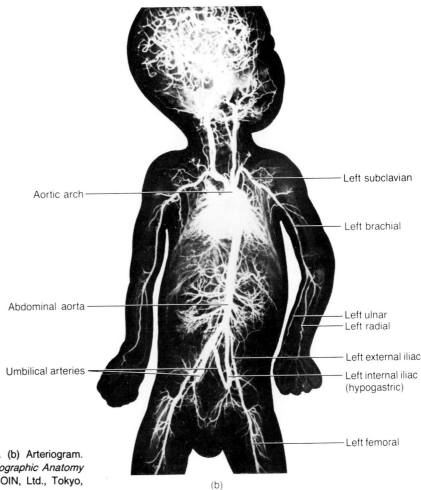

Aortic arch

Abdominal aorta

Umbilical arteries

Left subclavian

Left brachial

Left ulnar
Left radial

Left external iliac

Left internal iliac
(hypogastric)

Left femoral

(b)

FIGURE 12-22 (*Continued*) Fetal circulation. (b) Arteriogram. (Courtesy of C. Yokochi and J. W. Rohen, *Photographic Anatomy of the Human Body,* 1st ed., 1969, IGAKU-SHOIN, Ltd., Tokyo, New York.)

3. The placenta is delivered by the mother as the "afterbirth."

4. The ductus venosus becomes the **ligamentum venosum,** a fibrous cord in the liver.

5. The foramen ovale closes shortly after birth to become the **fossa ovalis,** a depression in the interatrial septum.

6. The ductus arteriosus closes, atrophies, and becomes the **ligamentum arteriosum.**

APPLICATIONS TO HEALTH

HEART FAILURE

Heart failure may be defined as a chronic or acute state that results when the heart is not capable of supplying the oxygen demands of the body. Symptoms and signs of heart failure are produced by diminished blood flow to the various tissues of the body and by accumulation of excess blood in the various organs because the heart is unable to pump out the blood returned to it by the great veins. Both diminished blood flow and accumulation of excess blood in the organs occur together, but certain symptoms result from congestion, while others are produced by poor tissue nutrition.

Although the treatment of heart failure should be directed toward correcting the cause, its medical management is basically the same in all cases. Heart failure occurs when the heart pumps less blood than the body requires. Consequently, the aim of therapy is to increase the output of the heart, improve tissue oxygenation, remove excess fluid, and prevent fluid reaccumulation. Medical management of heart failure requires digitalis (a drug that strengthens the contraction of heart muscles), diuretics, dietary salt restriction, possibly oxygen, and antiarrhythmic medications (drugs that help maintain normal cardiac rhythm).

RISK FACTORS

It is estimated that one in every five persons who reaches 60 will have a **heart attack.** One in every four persons between 30 and 60 has the potential to be stricken. Heart disease is epidemic in this country, despite the fact that some of the causes can be foreseen and prevented.

The Framingham, Massachusetts, Heart Study, which began in 1950, is the longest and most famous study of the susceptibility of a community to heart disease. Approximately 13,000 people in the town have participated by receiving examinations every 2 years since the study began. The results of this research indicate that people who develop combinations of certain risk factors eventually have heart attacks. These factors are:

1. High blood cholesterol level.
2. High blood pressure.
3. Cigarette smoking.
4. Obesity.
5. Lack of exercise.
6. Diabetes mellitus.
7. Genetic predisposition.

The first five risk factors all contribute to increasing the heart's work load. High blood cholesterol and hypertension are discussed later in this chapter. Cigarette smoking, through the effects of nicotine, stimulates the adrenal gland to oversecrete aldosterone, epinephrine, and norepinephrine—powerful vasoconstrictors. Overweight people develop miles of extra capillaries to nourish fat tissue. The heart has to work harder to pump the blood through more vessels. Without exercise, venous return gets less help from contracting skeletal muscles. In addition, regular exercise strengthens the smooth muscle of blood vessels and enables them to assist general circulation. Exercise also increases cardiac efficiency and output. In diabetes mellitus, fat metabolism dominates glucose metabolism. As a result, cholesterol levels get progressively

higher and result in plaque formation, a situation that may lead to high blood pressure. High blood pressure drives fat into the vessel wall, encouraging atherosclerosis.

There are significant differences between males and females in terms of coronary heart disease. For example, white males have more severe coronary artery atherosclerosis and more frequent myocardial infarction and sudden death than white females. By contrast, white females have a greater incidence of angina pectoris than white males. These differences are not as great in nonwhites.

Current research done with rhesus monkeys and baboons indicates that muscle cells of the atria and ventricles contain receptors for androgens, steroid hormones produced by the adrenal cortex. Although the significance of this is unknown, it does suggest that the sex steroid hormones may affect heart function directly and may explain the differences between males and females in terms of coronary heart disease.

Generally, the immediate cause of heart trouble is one of the following: inadequate coronary blood supply, anatomical disorders, or faulty electrical conduction in the heart.

INADEQUATE CORONARY BLOOD SUPPLY

Angina pectoris and myocardial infarction result from insufficient oxygen supply to the myocardium. Coronary artery disease kills about 1 in 12 of all Americans who die between the ages of 25 and 34 and almost 1 in 4 of all those who die between 35 and 44. It has been reported that 50 to 65 percent of all sudden deaths are due to coronary heart disease.

ANATOMICAL DISORDERS

Less than 1 percent of all new babies have a **congenital, or inborn, heart defect.** Even so, the total number in this country each year is estimated to be 30,000 to 40,000. Some of these infants may live quite healthy and long lives without any need for repairing their hearts. But sometimes an inborn heart defect is so severe that an infant lives only a few hours. A common anatomical defect is **patent ductus arteriosus.** The connection between the aorta and the pulmonary artery remains open instead of closing completely after birth. As a result, aortic blood flows into the lower-pressure pulmonary trunk, thus increasing the pulmonary trunk blood pressure and overworking both ventricles and the heart.

A **septal defect** is an opening in the septum that separates the interior of the heart into a left and right side. **Interatrial septal defect** is failure of the fetal foramen ovale between the two atria to close after birth. Because pressure in the right atrium is low, interatrial septal defect generally allows a good deal of blood to flow from the left atrium to the right without going through systemic circulation. This defect, an example of left-to-right shunt,

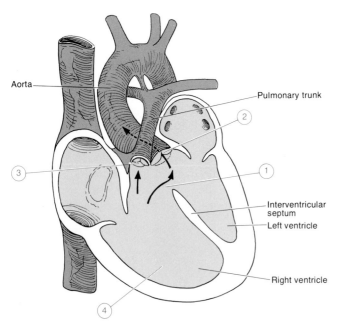

Aorta

Pulmonary trunk

③

②

①

Interventricular septum

Left ventricle

Right ventricle

④

FIGURE 12-23 Tetralogy of Fallot. The four abnormalities associated with this condition are indicated by numbers. (1) Opening in the interventricular septum. (2) Origin of the aorta in both ventricles. (3) Stenosed pulmonary semilunar valve. (4) Enlarged right ventricle.

overloads the pulmonary circulation, produces fatigue, and increases respiratory infections. If it occurs early in life, it inhibits growth because the systemic circulation may be deprived of a considerable portion of the blood destined for the organs and tissues of the body.

Interventricular septal defect is caused by an incomplete closure of the interventricular septum. Owing to the higher pressure in the left ventricle, blood is shunted directly from the left ventricle into the right ventricle. This is another example of a left-to-right shunt. Septal openings can now be sewn shut or covered with synthetic patches.

Valvular stenosis is a narrowing, or *stenosis,* of one of the valves regulating blood flow in the heart. Stenosis may occur in the valve itself—most commonly in the mitral valve from rheumatic heart disease or in the aortic valve from sclerosis or rheumatic fever. Or it may occur near a valve. All stenoses are serious because they all place a severe work load on the heart by making it work harder to push the blood through the abnormally narrow valve openings. As a result of mitral stenosis, blood pressure is increased. Angina pectoris and heart failure may accompany this disorder. Most stenosed valves are totally replaced with artificial valves.

Tetralogy of Fallot (tet-RAL-ō-jē fal-Ō) is a combination of four defects: an interventricular septal opening, an aorta that emerges from both ventricles instead of from the left ventricle, a stenosed pulmonary semilunar valve, and an enlarged right ventricle (Figure 12-23). The condition is an example of a right-to-left shunt, in which blood is shunted from the right ventricle to the left ventricle without going through pulmonary circulation. Because there is stenosis of the pulmonary semilunar valve, the increased right ventricular pressure forces deoxygenated blood from the right ventricle to enter the left ventricle through the interventricular septum. As a result, deoxygenated blood gets mixed with the oxygenated blood that is pumped into systemic circulation. Also, because the aorta emerges from the right ventricle and the pulmonary artery is stenosed, very little blood ever gets to the lungs and pulmonary circulation is bypassed almost completely. Insufficient amount of oxygenated blood in systemic circulation results in *cyanosis* (si-a-NO-sis), a blue or dark purple discoloration of skin. For this reason the tetralogy of Fallot is one of the conditions that causes a "blue baby." Lung disorders and suffocation also result in cyanosis.

Today it is possible to correct cases of tetralogy of Fallot when the patient is of proper age and condition. Open-heart operations are performed in which the narrowed pulmonary valve is cut open and the interventricular septal defect is sealed with a Dacron patch.

FAULTY CONDUCTION

At least half the deaths from myocardial infarction occur before the patient reaches the hospital. These early deaths could result from an irregular heart rhythm—an *arrhythmia* (a-RITH-mē-a). Sometimes this condition progresses to the stage called *cardiac arrest* or ventricular fibrillation, in which the heart stops functioning. An arrhythmia is caused by disturbances in the conduction system. This abnormal rhythm of the heartbeat can result in cardiac arrest if the heart cannot supply its own oxygen demands, as well as those of the rest of the body. Serious arrhythmias can be controlled, and the normal heart rhythm can be reestablished, if they are detected and treated early enough. Coronary care units have reduced hospital mortality rates from acute myocardial infarctions to about 30 to 20 percent or less by preventing or controlling serious arrhythmias.

Arrhythmia arises when electrical impulses through the heart are blocked at critical points in the conduction system. One such arrhythmia is called a **heart block.** Perhaps the most common blockage is in the atrioventricular node, which conducts impulses from the atria to the ventricles. This disturbance is called *atrioventricular (AV) block.* It usually indicates a myocardial infarction, atherosclerosis, rheumatic heart disease, diphtheria, or syphilis. In a *first-degree AV block,* which can be detected only with an electrocardiograph, the conduction of impulses from the atria to the ventricles through the AV node is delayed. As a result, the P-R interval is longer than normal.

Occasionally, the delay at the AV node will progressively increase from beat to beat until conduction to the ventricles is blocked. This causes the P-R interval to become longer and longer until finally a P wave occurs without a following QRS complex and T wave. By the time the sinu-atrial node fires again, AV conduction has had time to recover and the sequence starts over. This condition represents one variation of a *second-degree AV block.*

In another variation of a second-degree AV block, the delay at the AV node is sufficiently long that occasionally an atrial impulse fails to reach the ventricles. This usually occurs in a regular sequence such as 2:1, 3:2, or 4:1, the ratios being that of atrial to ventricular beats. When ventricular contraction does not occur (dropped beat), oxygenated blood is not pumped efficiently to all parts of the body. The patient may feel faint or may collapse if there are many dropped ventricular beats.

In a *third-degree* or *complete AV block,* there is practically no conduction of impulses between the atria and ventricles. Atrial and ventricular rates get out of synchronization (Figure 12-24a). The ventricles may go into systole at any time. This condition could occur when the atria are in systole or just before. Or the ventricles may rest for a few cardiac cycles. With complete AV block, patients may have vertigo, unconsciousness, or convulsions. These symptoms result from a decreased cardiac output with diminished cerebral blood flow and cerebral hypoxia or lack of sufficient oxygen.

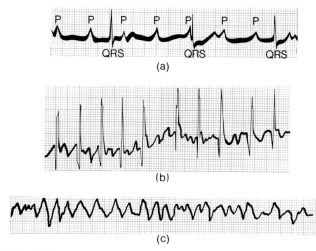

FIGURE 12-24 Abnormal electrocardiograms. (a) Complete AV block. (b) Atrial fibrillation. (c) Ventricular fibrillation.

Among the causes of AV block are excessive stimulation by the vagus nerves that depresses conductivity of the junctional fibers, destruction of the AV bundle as a result of coronary infarct, atherosclerosis, myocarditis, or depression caused by various drugs. Other heart blocks include *intraatrial (IA) block, interventricular (IV) block,* and *bundle branch block (BBB).* In the latter condition, the ventricles do not contract together because of the delayed impulse in the blocked branch.

Two abnormal rhythms that indicate heart trouble are **flutter** and **fibrillation.** In **atrial flutter** the atrial rhythm averages between 240 and 360 beats per minute. The condition is essentially rapid atrial contractions accompanied by a second-degree AV block. It generally indicates severe damage to heart muscle. Atrial flutter usually becomes fibrillation after a few hours, days, or weeks. **Atrial fibrillation** is asynchronous contraction of the atrial muscles that causes the atria to contract irregularly and still faster (Figure 12-24b). Atrial flutter and fibrillation occur in myocardial infarction, acute and chronic rheumatic heart disease, and hyperthyroidism. Atrial fibrillation results in complete uncoordination of atrial contraction so that atrial pumping ceases altogether. When the muscle fibrillates, the muscle fibers of the atrium quiver individually instead of contracting together. The quivering cancels out the pumping of the atrium. In a strong heart, atrial fibrillation reduces the pumping effectiveness of the heart by only 25 to 30 percent.

Ventricular fibrillation is another abnormality that indicates imminent cardiac arrest and death. It is characterized by asynchronous, haphazard, ventricular muscle contractions. The rate may be rapid or slow. The impulse travels to the different parts of the ventricles at different rates. Thus part of the ventricle may be contracting while other parts are still unstimulated. Ventricular contraction becomes ineffective and circulatory failure and death occur immediately unless the arrhythmia is reversed quickly

(Figure 12-24c). Ventricular fibrillation may be caused by coronary occlusion. It sometimes occurs during surgical procedures on the heart or pericardium. It may be the cause of death in electrocution.

Another form of arrhythmia arises when a small region of the heart becomes more excitable than normal, causing an occasional abnormal impulse to be generated between normal impulses. The region from which the abnormal impulse is generated is called an **ectopic focus.** As a wave of depolarization spreads outward from the ectopic focus, it causes a *premature beat.* The beat occurs early in diastole before the SA node is normally scheduled to discharge its impulses. Ectopic foci may be caused by emotional stress, excessive intake of stimulants such as caffeine or nicotine, lack of sleep, and local ischemic areas of cardiac muscle.

ANEURYSM

A blood-filled sac formed by an outpouching in an arterial or venous wall is called an **aneurysm** (AN-yoo-rizm) (Figure 12-25). Aneurysms may occur in any major blood vessel and include the following types.

1. Berry. A small dilation of a vessel. This type occurs frequently in the cerebral artery. If it ruptures, it may cause a hemorrhage below the dura mater. Hemorrhaging is one cause of a stroke.

2. Ventricular. A dilation of a ventricle of the heart.

3. Aortic. A dilation of the aorta. The aorta has a higher incidence of aneurysms than any other artery, probably because of its curved shape, large size, and high pressure. In the thorax, aneurysms usually occur in the ascending or descending aorta but seldom in the aortic arch. Symptoms of thoracic aortic aneurysm depend on pressure exerted on adjacent structures. For example, pressure on the inferior (recurrent) laryngeal nerve causes hoarseness and a brassy cough. Pressure on the esophagus may cause dysphagia (difficulty in swallowing). Dyspnea (difficulty in breathing) may follow pressure on the trachea, the root of the lung, or the phrenic nerve.

ATHEROSCLEROSIS

Atherosclerosis (ath'-er-ō-skle-RŌ-sis) is a form of arteriosclerosis, which includes many diseases of the arterial wall. It is the major disease responsible for the principal clinical complications. In this disorder, the tunica intima of an artery becomes thickened with soft fatty deposits called *atheromatous* (ath'-er-Ō-ma-tus) *plaques* (Figure 12-26a).

An *atheroma* (ath'-er-Ō-ma) is an abnormal mass of fatty or lipid material deposited in an arterial wall. Atheromas involve the abdominal aorta and major leg arteries more extensively than the thoracic aorta. They may also be found in coronary, cerebral, and peripheral arteries. The atheroma looks like a pearly gray or yellow mound

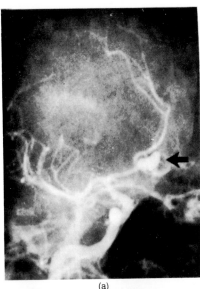

(a)

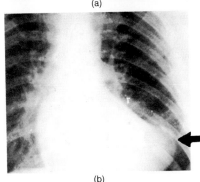

(b)

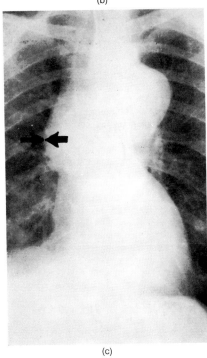

(c)

FIGURE 12-25 Aneurysms. (a) Aneurysm of the anterior cerebral artery. (b) Ventricular aneurysm. (c) Aortic aneurysm. (Courtesy of Lester W. Paul and John H. Juhl, *The Essentials of Roentgen Interpretation,* 3d ed., Harper & Row, Publishers, Inc., New York, 1972.)

of tissue. As it grows, it may impede blood flow in the affected artery and damage the tissues the artery supplies. An additional danger is that the plaque may provide a roughened surface for clot formation, or it may rupture and form a thrombus (a blood clot lodged in a vessel). If the clot breaks off and forms an embolus (an abnormal particle transported through the blood), it may obstruct small arteries, capillaries, and veins quite a distance from the site of formation.

The degree of arterial stenosis necessary to produce symptoms depends on the site of stenosis and the artery. The process causes a collateral circulation to develop around the stenotic block. These collaterals may prevent ischemic necrosis. When they do not, ischemia may progress to gangrene.

Atherosclerosis is generally a slow, progressive disease. It may start in childhood, and its development may produce absolutely no symptoms for 40 years or longer. Even if it reaches the advanced stages, the individual may feel no symptoms and the condition may be discovered only at postmortem examination. Diagnosis is possible by injecting radiopaque substances into the blood and then taking x-rays of the arteries. This technique is called *angiography* (an'-jē-OG-ra-fē) or *arteriography* (ar'-te-rē-OG-ra-fē). The film is called an *arteriogram* (Figure 12-26b).

Animal experiments have given us considerable scientific information about the plaques. It is possible to produce the streaks in many animals by feeding them a diet high in fat and cholesterol. This diet raises the blood lipid levels—a condition called *hyperlipidemia* (*hyper* = above; *lipo* = fat). Hyperlipidemia increases the risk of atherosclerosis. Patients with a high blood level of cholesterol should be treated with diet and drug therapy.

Currently, the preferred treatment for most individuals with atheromatous plaques in coronary arteries is a bypass operation. In this procedure, blood vessels from the patient's body or synthetic tubes are used to reroute blood around the diseased area. A relatively new treatment for coronary artery disease was introduced in Germany near the end of 1977. This procedure consists of inserting a balloon into the obstructed coronary artery, inflating the balloon, and squashing the plaque against the arterial wall. Then the balloon is deflated and removed. Unfortunately, the procedure has several drawbacks. For example, calcified plaques are not crushed by the balloon. Also, the balloon can damage the arterial wall. Despite these drawbacks, it is hoped that the procedure can be refined so that it will come into general use.

HYPERTENSION

Hypertension, or high blood pressure, is the most common disease affecting the heart and blood vessels. Statistics from a recent National Health Survey indicate that hypertension afflicts at least 17 million American adults and perhaps as many as 22 million.

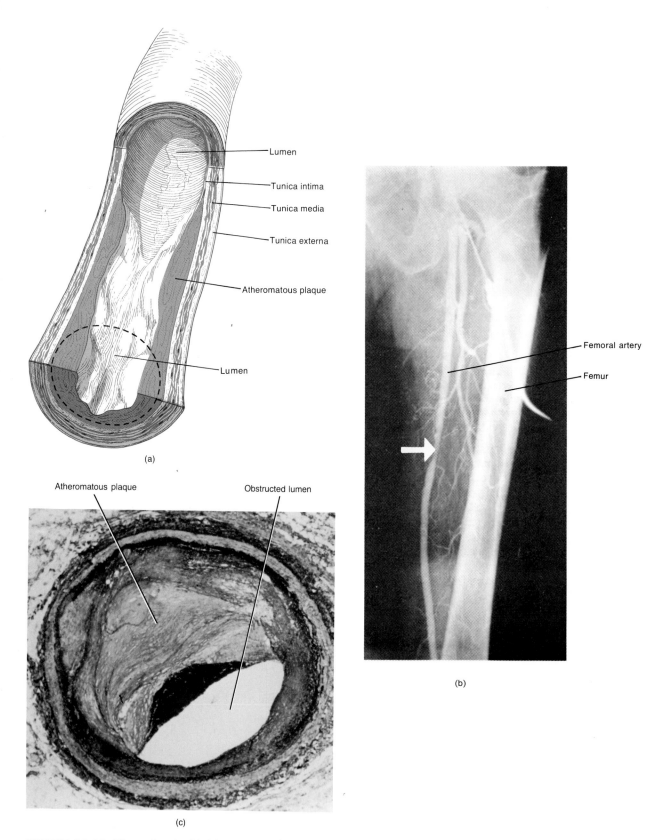

FIGURE 12-26 Atherosclerosis. (a) Atheromatous plaque formation. The broken circular line indicates the approximate size of a normal lumen. (b) Femoral anteriogram showing an atheromatous plaque (arrow) in the middle third of the thigh. (Courtesy of Lester W. Paul and John H. Juhl, *The Essentials of Roentgen Interpretation,* 3d ed., Harper & Row, Publishers, Inc., New York, 1972.) (c) Photomicrograph of a cross section of an artery partially obstructed by an atheromatous plaque. (Courtesy of the National Heart, Lung and Blood Inst., NIH.) (Photomicrographs from *The Healthy Heart* by Arthur Fisher, Time-Life Books, Alexandria, Va., © 1981.)

Primary hypertension, or essential hypertension, is a persistently elevated blood pressure that cannot be attributed to any particular organic cause. Specifically, the diastolic pressure continually exceeds 95 mm Hg and the systolic pressure continually exceeds 140 mm Hg. Approximately 85 percent of all hypertension cases fit this definition. The other 15 percent have *secondary hypertension.* Secondary hypertension is caused by disorders such as atherosclerosis, kidney disease, and adrenal hypersecretion. Atherosclerosis increases blood pressure by reducing the elasticity of the arterial walls and narrowing the space through which the blood can flow. Kidney disease and obstruction of blood flow may cause the kidney to release renin into the blood. This enzyme catalyzes the formation of angiotensin II from a plasma protein. Angiotensin II is a powerful blood-vessel constrictor—and the most potent agent known for raising blood pressure. It also stimulates aldosterone release. Aldosteronism, the hypersecretion of aldosterone, may also cause an increase in blood pressure. Aldosterone is the adrenal cortex hormone that promotes the retention of salt and water by the kidneys. It thus tends to increase plasma volume. Pheochromocytoma is a tumor of the adrenal medulla. It produces and releases into the blood large quantities of norepinephrine and epinephrine. These hormones also raise blood pressure. Epinephrine causes an increase in heart rate and norepinephrine causes vasoconstriction.

High blood pressure is of considerable concern because of the harm it can do to the heart, brain, and kidneys if it remains uncontrolled. The heart is most commonly affected by high blood pressure. When pressure is high, the heart uses more energy in pumping. Because of the increased effort, the heart muscle thickens and the heart becomes enlarged. The heart also needs more oxygen. If it cannot meet the demands put on it, angina pectoris or even myocardial infarction may develop. Continued high blood pressure may produce a cerebral vascular accident, or "stroke." In this case, severe strain has been imposed on the cerebral arteries that supply the brain. These arteries are usually less protected by surrounding tissues than are the major arteries in other parts of the body. These weakened cerebral arteries may finally rupture, and a brain hemorrhage follows.

The kidney is another prime target of hypertension. The principal site of damage is in the arterioles that supply these vital organs. The continual high blood pressure pushing against the walls of the arterioles causes them to thicken, thus narrowing the lumen. The blood supply to the kidney is thereby gradually reduced. In response, the kidney may secrete renin, which raises the blood pressure even higher and complicates the problem. The reduced blood flow to the kidney cells may eventually lead to the death of the cells.

The causes of primary hypertension are unknown. Medical science cannot cure it. However, almost all cases of hypertension, whether mild or severe, can be controlled by a variety of drugs that reduce blood pressure. In addition to drugs, treatment often involves the restriction of sodium intake. Sodium restriction curbs fluid retention by the body and tends to reduce blood volume. The overweight person with hypertension will usually be placed on a reducing diet, because blood pressure often falls with weight loss.

KEY MEDICAL TERMS ASSOCIATED WITH THE HEART AND BLOOD VESSELS

Angiocardiography (*angio* = vessel; *cardio* = heart; *graph* = writing)   X-ray examination of the heart and great blood vessels after injection of a radiopaque dye into the bloodstream.

Aortic insufficiency   An improper closing of the aortic semilunar valve that permits a backflow of blood.

Aortography   X-ray examination of the aorta and its main branches after injection of a radiopaque dye.

Arteriography   X-ray examination of arteries after injection of a radiopaque dye.

Cardiac arrest (*cardio* = heart)   Complete stoppage of the heartbeat.

Cardiomegaly (*mega* = large)   Heart enlargement.

Carditis (*itis* = inflammation of)   Inflammation of the heart.

Catheterization   In cardiology, the process of examining the heart by means of introducing a thin tube (catheter) into a vein or artery and passing it into the heart.

Claudication   Pain and lameness or limping caused by defective circulation of the blood in the vessels of the limbs.

Commissurotomy   An operation that is performed to widen the opening in a heart valve which has become narrowed by scar tissue.

Compliance   The passive or diastolic stiffness properties of the left ventricle. A hypertrophied or fibrosed heart with a stiff wall, for example, has decreased compliance.

Constrictive pericarditis.   A shrinking and thickening of the pericardium which prevents the heart muscle from expanding and contracting normally.

Coronary endarterectomy   The removal of the obstructing area within the lumen of the vessel.

Cor pulmonale   Heart disease resulting from disease of the lungs or the blood vessels in the lungs. This is due to resistance to the passage of blood through the lungs.

Cyanosis (*cyano* = blue) Slightly bluish, dark purple skin coloration due to oxygen deficiency in systemic blood.

Defibrillator A mechanical device for applying electrical shock to the heart to terminate abnormal cardiac rhythms.

Hematoma (*hemato* = blood; *oma* = tumor) Leakage of blood from a vessel, which clots to form a solid mass or swelling in tissue.

Hemorrhage (*rrhage* = bursting forth) The escape of blood from a ruptured vessel.

Occlusion The closure or obstruction of the lumen of a structure such as a blood vessel.

Palpitation A fluttering of the heart or abnormal rate or rhythm of the heart.

Pancarditis (*pan* = all) Inflammation of the whole heart including inner layer (endocardium), heart muscle (myocardium), and outer sac (pericardium).

Paroxysmal tachycardia A period of rapid heartbeats which begins and ends suddenly.

Phlebitis (*phleb* = vein) Inflammation of a vein.

Shunt A passage between two blood vessels or between the two sides of the heart.

Stokes-Adams syndrome Sudden attacks of unconsciousness, sometimes with convulsions, which may accompany heart block.

Thrombectomy (*thrombo* = clot) An operation to remove a blood clot from a blood vessel.

Thrombophlebitis Inflammation of a vein with clot formation.

STUDY OUTLINE

Heart
Location
1. The heart is situated obliquely between the lungs in the mediastinum.
2. About two-thirds of its mass is to the left of the midline.

Parietal Pericardium (Pericardial Sac)
1. The parietal pericardium, consisting of an outer fibrous layer and an inner serous layer, encloses the heart.
2. Between the serous pericardium and epicardium is a space called the pericardial cavity.
3. The pericardial cavity contains fluid that prevents friction between the membranes.

Wall, Chambers, Vessels, and Valves
1. The wall of the heart has three layers: epicardium, myocardium, and endocardium.
2. Between the serous pericardium and the epicardium is the pericardial cavity, a space filled with pericardial fluid that prevents friction between the two membranes.
3. The chambers include two upper atria and two lower ventricles.
4. The blood flows through the heart from the superior and inferior venae cavae and the coronary sinus to the right atrium. It then goes through the tricuspid valve to the right ventricle and through the pulmonary trunk to the lungs. It returns to the heart through the pulmonary veins into the left atrium, through the bicuspid valve to the left ventricle, and out through the aorta.
5. Valves prevent backflow of blood in the heart.
6. Atrioventricular valves, between the atria and their ventricles, are the tricuspid valve on the right side of the heart and the bicuspid valve on the left.

7. The chordae tendineae and their muscles keep the flaps of the valves pointing in the direction of blood flow.
8. The two arteries that leave the heart both have a semilunar valve.

Conduction System and Cardiac Cycle
1. The conduction system consists of tissue specialized for impulse conduction. Components of the system are the sinu-atrial node (pacemaker), atrioventricular node, atrioventricular bundle, and Purkinje fibers.
2. The record of electrical changes during each cardiac cycle is referred to as an electrocardiogram (ECG). The ECG is invaluable in diagnosing abnormal cardiac rhythms and conduction patterns, detecting the presence of fetal life, determining the presence of several fetuses, and following the course of recovery from a heart attack.
3. A normal ECG consists of a P wave (spread of impulse from SA node over atria), QRS wave (spread of impulse through ventricles), and T wave (ventricular repolarization). The P-R interval represents the conduction time from the beginning of atrial excitation to the beginning of ventricular excitation. The S-T segment represents the time between the end of the spread of the impulse through the ventricles and repolarization of the ventricles.
4. The cardiac cycle consists of the systole (contraction) and diastole (relaxation) of both atria plus the systole and diastole of both ventricles followed by a short pause.
5. With an average heartbeat of 72/minute, a complete cardiac cycle requires 0.8 second.
6. The first sound (lubb) represents the closing of the atrioventricular valves. The second sound (dupp) represents the closing of semilunar valves.

Autonomic Control

1. The pacemaker may be accelerated by sympathetic stimulation from the cardioacceleratory center in the medulla and slowed down by parasympathetic stimulation from the cardioinhibitory center in the medulla.
2. Pressoreceptors are nerve cells that respond to changes in blood pressure. They act on the cardiac centers in the medulla through three reflex pathways: carotid sinus reflex, aortic reflex, and right heart reflex.
3. Other influences on heart rate include chemicals (epinephrine, sodium, potassium), temperature, emotion, sex, and age.

Blood Vessels

1. Arteries carry blood away from the heart. They are stronger and thicker than veins. Arteries consist of a tunica interna, tunica media (which maintains elasticity and contractility), and tunica externa.
2. Many arteries anastomose—the distal ends of two or more vessels unite. An alternate blood route from an anastomosis is called collateral circulation.
3. Large arteries are referred to as elastic (conducting) arteries and medium-sized arteries are called muscular (distributing) arteries.
4. Arterioles are small arteries that deliver blood to capillaries. Through constriction and dilation they assume a key role in regulating blood flow from arteries into capillaries.
5. Capillaries are microscopic blood vessels through which materials are exchanged between blood and tissue cells; some capillaries are continuous, others are fenestrated.
6. Capillaries branch to form an extensive capillary network throughout the tissue. This network increases the surface area, allowing a rapid exchange of large quantities of materials.
7. Microscopic blood vessels in the liver are called sinusoids.
8. Venules are small vessels that continue from capillaries and merge to form veins. They drain blood from capillaries into veins.
9. Veins consist of three tunics, but have less elastic tissue and smooth muscle than arteries. They contain valves to prevent backflow of blood.
10. Weak valves can lead to varicose veins or hemorrhoids.
11. Vascular (venous) sinuses or simply sinuses are veins with very thin walls.

Circulatory Routes

Systemic Circulation

1. The systemic circulation takes oxygenated blood from the left ventricle through the aorta to all parts of the body including lung tissue.
2. The aorta is divided into the ascending aorta, the arch of the aorta, and the descending aorta.
3. Blood is returned to the heart through the systemic veins. All the veins of the systemic circulation flow into either the superior or inferior venae cavae or the coronary sinus. They in turn empty into the right atrium.

Coronary (Cardiac) Circulation

1. The coronary circulation, a subdivision of the systemic circulation, takes oxygenated blood through the arterial system of the myocardium.
2. Deoxygenated blood returns to the right atrium via the coronary sinus.
3. Complications of this system are angina pectoris and myocardial infarction.

Hepatic Portal Circulation

1. The hepatic portal circulation, another subdivision of the systemic circulation, collects blood from the veins of the pancreas, spleen, stomach, intestines, and gallbladder and directs it into the hepatic portal vein of the liver.
2. This circulation enables the liver to utilize nutrients and detoxify harmful substances in the blood.

Pulmonary Circulation

1. The pulmonary circulation takes deoxygenated blood from the right ventricle to the lungs and returns oxygenated blood from the lungs to the left atrium.
2. It allows blood to be oxygenated for systemic circulation.

Fetal Circulation

1. The fetal circulation involves the exchange of materials between fetus and mother.
2. The fetus derives its oxygen and nutrients and eliminates its carbon dioxide and wastes through the maternal blood supply by means of a structure called the placenta.
3. At birth, when lung, digestive, and liver functions are established, the special structures of fetal circulation are no longer needed.

Applications to Health

1. Heart disorders related to inadequate blood supply are angina pectoris and myocardial infarction.
2. Congenital heart defects include septal defects, valvular stenosis, and tetralogy of Fallot.
3. Heart conditions relative to conduction problems include heart blocks, atrial flutter and fibrillation, ventricular fibrillation, and ectopic foci.
4. An aneurysm is a sac formed by an outpocketing of a portion of an arterial or venous wall.
5. Atherosclerosis is caused by formation of plaques in the wall of an artery.
6. Hypertension is high blood pressure and may damage the heart, brain, and kidneys.

REVIEW QUESTIONS

1. Describe the location of the heart in the mediastinum. Distinguish the subdivisions of the pericardium. What is the purpose of this structure?
2. Compare the three portions of the heart wall. Define atria and ventricles. What vessels enter or exit the atria and ventricles?
3. Discuss the principal valves in the heart and how they operate.
4. Describe the path of a nerve impulse through the heart's conducting system.
5. Define and label the deflection waves of a normal electrocardiogram. Explain why the ECG is an important diagnostic tool.
6. Define systole and diastole and their relationship to the cardiac cycle.
7. Describe the various sounds of the heart and the importance of each.
8. Prepare a diagram to illustrate the surface landmarks used to locate the borders and valves of the heart.
9. Describe the structural and functional differences among arteries, arterioles, capillaries, venules, and veins.
10. Distinguish between elastic and muscular arteries in terms of location, histology, and function. What is an anastomosis?
11. Define varicose veins and hemorrhoids.
12. What is meant by a circulatory route? Define systemic circulation.
13. By means of a diagram, indicate the major divisions of the aorta, their principal arterial branches, and the regions supplied.
14. Trace a drop of blood from the arch of the aorta through its systemic circulatory route and back to the heart again. Remember that the major branches of the arch are the brachiocephalic artery, the left common carotid artery, and the left subclavian artery. In giving your answer, be sure to indicate which veins return the blood to the heart.
15. What is the circle of Willis? Why is it important?
16. What are visceral branches of an artery? Parietal branches?
17. What major organs are supplied by branches of the thoracic aorta? How is blood returned from these organs to the heart?
18. What organs are supplied by the celiac, superior mesenteric, renal, inferior mesenteric, inferior phrenic, and middle sacral arteries? How is blood returned to the heart?
19. Trace a drop of blood from the common iliac arteries through their branches to the big toe on your left foot and back to the heart again.
20. What is a deep vein? A superficial vein? Define a venous sinus in relation to blood vessels. What are the three major groups of systemic veins?
21. Describe the route of blood in the coronary (cardiac) circulation. Distinguish between angina pectoris and myocardial infarction.
22. What is hepatic portal circulation? Describe the route by means of a diagram. Why is this route significant?
23. Define pulmonary circulation. Prepare a diagram to indicate the route. What is the purpose of the route?
24. Discuss in detail the anatomy and physiology of fetal circulation. Be sure to indicate the function of the umbilical arteries, umbilical vein, ductus venosus, foramen ovale, and ductus arteriosus.
25. What is the fate of the special structures involved in fetal circulation once postnatal circulation is established?
26. Describe the cause and treatment of patent ductus arteriosus.
27. Describe the risk factors involved in heart attacks.
28. Describe the various types of septal defects. What are the common causes of valvular stenosis? What are the consequences of this disorder?
29. What is heart block? Describe the various types of atrioventricular block.
30. Distinguish between atrial flutter, atrial fibrillation, and ventricular fibrillation.
31. What is an aneurysm? Distinguish three types on the basis of location.
32. Discuss the causes, symptoms, and diagnosis of atherosclerosis.
33. Compare primary and secondary hypertension with regard to cause. How does hypertension affect the body?
34. Refer to the glossary of medical terms associated with the heart and blood vessels. Be sure that you can define each term.

13

THE LYMPHATIC SYSTEM

STUDENT OBJECTIVES

- Define the lymphatic system and list its functions.

- Describe the structure and origin of lymphatics and contrast them with veins.

- Describe the histological aspects of lymph nodes and explain their functions.

- Trace the general plan of lymph circulation from lymphatics into the thoracic duct or right lymphatic duct.

- Describe the principal lymph nodes of the head and neck, extremities, and trunk, their location, and the areas they drain.

- Contrast the locations and functions of the tonsils, spleen, and thymus gland as lymphatic organs.

- Explain the forces responsible for maintaining the circulation of lymph.

- Discuss how edema develops.

- Define key medical terms associated with the lymphatic system.

ymph, lymph vessels, a series of small masses of lymphoid tissue called lymph nodes, and three organs—tonsils, thymus, and spleen—make up the **lymphatic** (lim-FAT-ik) **system.** The primary function of the lymphatic system is to drain from the tissue spaces protein-containing fluid that escapes from the blood capillaries. Such proteins cannot be directly reabsorbed. Other functions of the lymphatic system are to transport fats from the digestive tract to the blood, to produce lymphocytes, and to develop antibodies.

The developmental anatomy of the lymphatic system is considered in Exhibit 24-5.

LYMPHATIC VESSELS

Lymphatic vessels originate as blind-end tubes that begin in spaces between cells (see Figure 11-3). The tubes, called **lymph capillaries,** occur singly or in extensive plexuses (*plexus* = braid; a network of interlacing or anastomosing vessels or nerves). Lymph capillaries originate throughout the body, but not in avascular tissue, the central nervous system, splenic pulp, and bone marrow. They are slightly larger and more permeable than blood capillaries.

Just as blood capillaries converge to form venules and veins, lymph capillaries unite to form larger and larger lymph vessels called **lymphatics** (Figure 13-1). Lymphatics resemble veins in structure, but have thinner walls and more valves and contain lymph nodes at various intervals. Lymphatics of the skin travel in loose subcutaneous tissue and generally follow veins. Lymphatics of the viscera generally follow arteries, forming plexuses around them. Ultimately, lymphatics converge into two main channels—the thoracic duct and the right lymphatic duct. This will be described shortly.

CLINICAL APPLICATION

Lymphangiography (lim-fan'-jē-OG-ra-fē) is the x-ray examination of lymphatic vessels and lymph organs after they are filled with a radiopaque substance. Such an x-ray is called a **lymphangiogram** (lim-FAN-jē-ō-gram). Lymphangiograms are useful in detecting edema and carcinomas and in localizing lymph nodes for surgical or radiotherapeutic treatment. A normal lymphangiogram of lymphatic vessels and a few nodes in the upper thighs and pelvis is shown in Figure 13-2.

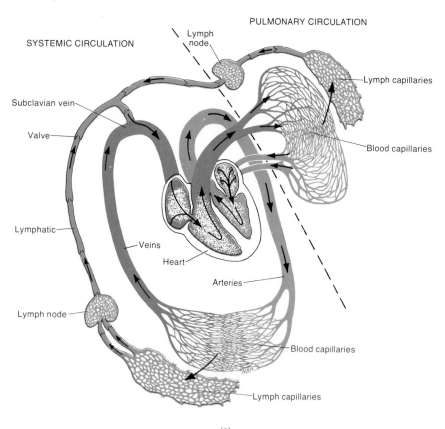

(a)

FIGURE 13-1 Lymphatic system. (a) Schematic representation of the relationship of the lymphatic system to the cardiovascular system.

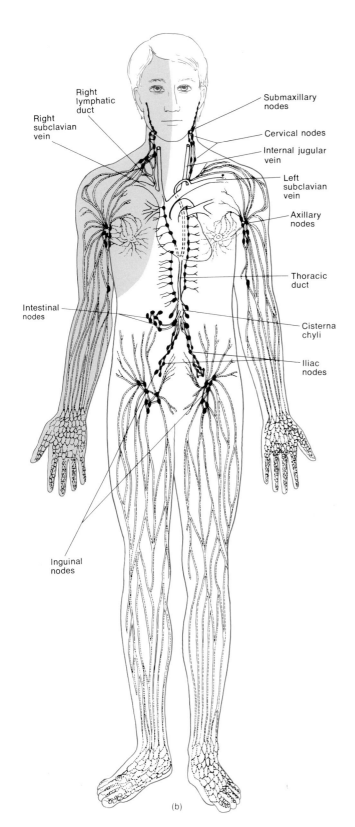

Right
lymphatic
duct

Right
subclavian
vein

Submaxillary
nodes

Cervical nodes

Internal jugular
vein

Left
subclavian
vein

Axillary
nodes

Thoracic
duct

Intestinal
nodes

Cisterna
chyli

Iliac
nodes

Inguinal
nodes

(b)

FIGURE 13-1 (*Continued*) Lymphatic system. (b) Location of the principal lymphatics and lymph nodes. The light gray area indicates those portions of the body drained by the right lymphatic duct. All other areas of the body are drained by the thoracic duct.

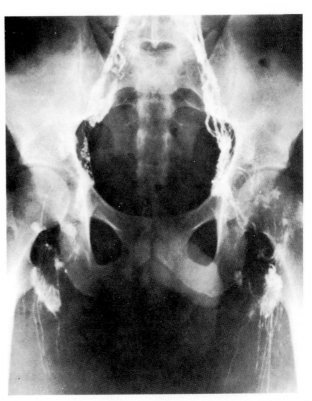

FIGURE 13-2 Normal lymphangiogram of the upper thighs and pelvis. Can you identify the lymphatics and lymph nodes? (Courtesy of Lester W. Paul and John H. Juhl, *The Essentials of Roentgen Interpretation,* 3d ed., Harper & Row, Publishers, Inc., New York, 1972.)

STRUCTURE OF LYMPH NODES

The oval or bean-shaped structures located along the length of lymphatics are called **lymph nodes** or **lymph glands.** They range from 1 to 25 mm (0.04 to 1 inch) in length. A lymph node contains a slight depression on one side called a *hilus* (HĪ-lus) or *hilum,* where blood vessels and an afferent lymphatic vessel leave the node (Figure 13-3). Each node is covered by a *capsule* of fibrous connective tissue that extends into the node. The capsular extensions are called *trabeculae* (tra-BEK-yoo-lē). The capsule, trabeculae, and hilum constitute the stroma (framework) of a lymph node. The parenchyma of a lymph node is specialized into two regions. The outer *cortex* contains densely packed lymphocytes arranged in masses called *lymph nodules.* The nodules often contain lighter-staining central areas, the *germinal centers,* where lymphocytes are produced. The inner region of a lymph node is called the *medulla.* In the medulla, the lymphocytes are arranged in strands called *medullary cords.*

The circulation of lymph through a node involves afferent (to convey toward a center) lymphatic vessels, sinuses in the node, and efferent (to convey away from a center) lymphatic vessels. *Afferent lymphatic vessels* enter the convex surface of the node at several points. They contain

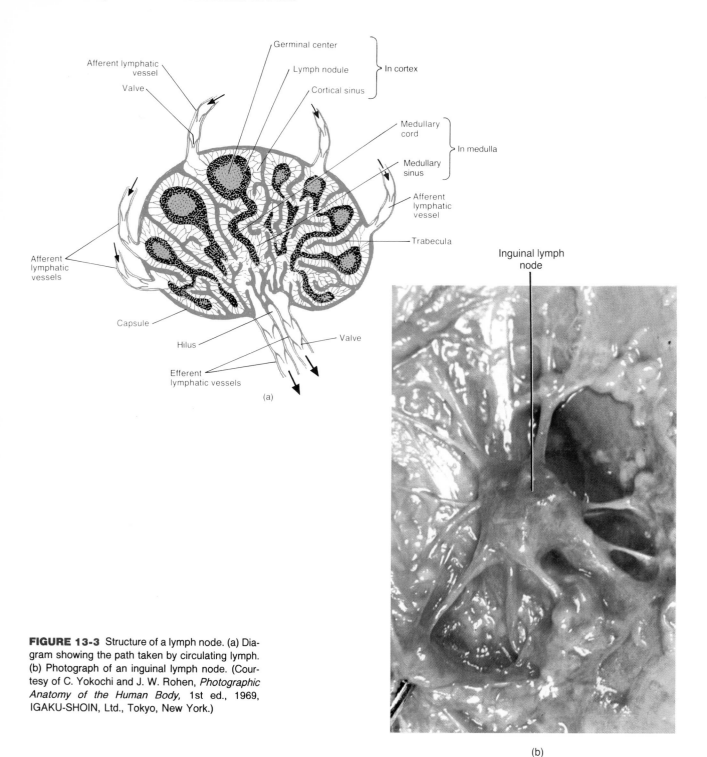

FIGURE 13-3 Structure of a lymph node. (a) Diagram showing the path taken by circulating lymph. (b) Photograph of an inguinal lymph node. (Courtesy of C. Yokochi and J. W. Rohen, *Photographic Anatomy of the Human Body,* 1st ed., 1969, IGAKU-SHOIN, Ltd., Tokyo, New York.)

valves that open toward the node so that the circulation of lymph through the afferent lymphatic vessels is into the lymph node. Once inside the lymph node, the lymph enters the sinuses of the node, which are a series of irregular channels. Lymph from the afferent lymphatic vessels enters the *cortical sinuses* under the capsule. From here the lymph circulates to the *medullary sinuses* between the medullary cords. From these sinuses the lymph circulates into a usually single *efferent lymphatic vessel*. This

vessel is located at the hilus of the lymph node. The efferent vessel is wider than the afferent vessels and contains valves that open away from the lymph node. Thus it conveys lymph out of the node.

As the lymph circulates through the nodes, it is processed by macrophages. The macrophages are fixed phagocytic cells of the reticuloendothelial system that line the sinuses. They remove bacteria, foreign material, and cell debris from the lymph. Lymph nodes also give rise

Valve

Wall of
lymphatic

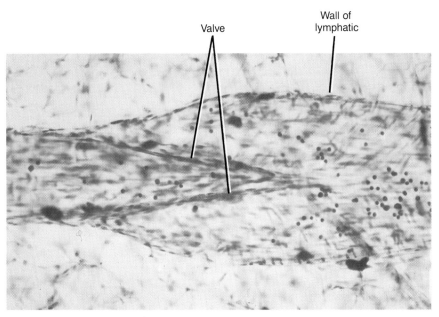

(a)

Blood vessel   Hilum

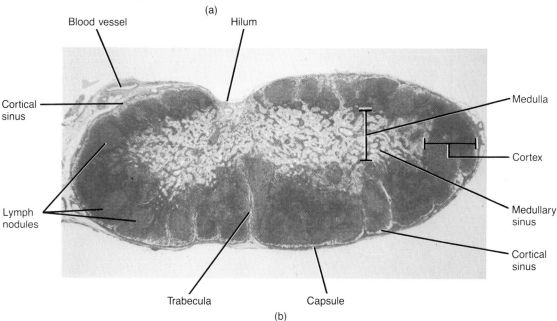

Cortical
sinus

Medulla

Cortex

Lymph
nodules

Medullary
sinus

Cortical
sinus

Trabecula   Capsule

(b)

FIGURE 13-4 Histology of lymphatics and lymph nodes. (a) Photomicrograph of a lymphatic at a magnification of 180×. (b) Photomicrograph of a lymph node at a magnification of 25×. (Photomicrographs © 1983 by Michael H. Ross. Used by permission.)

to lymphocytes and plasma cells. The plasma cells produce antibodies.

Histological features of lymphatics and lymph nodes are shown in Figure 13-4.

CLINICAL APPLICATION

At times the number of microbes entering the nodes in the lymph is so great that the macrophages and lymphocytes cannot remove or detoxify them. The nodes become infected. **Infected lymph nodes** become enlarged and tender.

LYMPH CIRCULATION

When plasma is filtered by blood capillaries it passes into the interstitial spaces; it is then known as interstitial fluid. When this fluid passes from the interstitial spaces into lymph capillaries it is called lymph. Lymph from lymph capillary plexuses is then passed to lymphatics that run toward lymph nodes. At the nodes, afferent vessels penetrate the capsules at numerous points and the lymph passes through the sinuses of the nodes. Efferent vessels from the nodes either run with afferent vessels into another node of the same group or pass on to another group

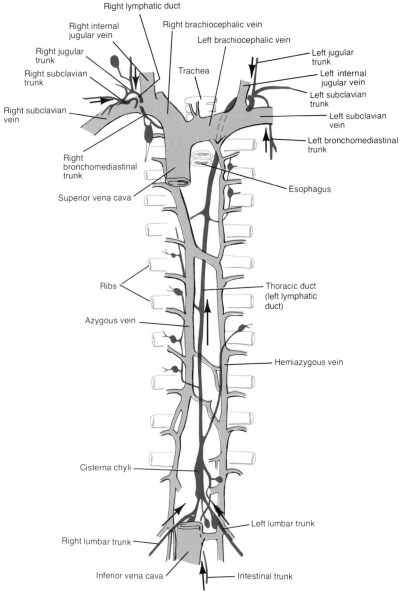

Right lymphatic duct

Right internal jugular vein

Right brachiocephalic vein

Left brachiocephalic vein

Right jugular trunk

Left jugular trunk

Right subclavian trunk

Trachea

Left internal jugular vein

Right subclavian vein

Left subclavian trunk

Left subclavian vein

Right bronchomediastinal trunk

Left bronchomediastinal trunk

Superior vena cava

Esophagus

Ribs

Thoracic duct (left lymphatic duct)

Azygous vein

Hemiazygous vein

Cisterna chyli

Right lumbar trunk

Left lumbar trunk

Inferior vena cava

Intestinal trunk

FIGURE 13-5 Relationship of lymph trunks to the thoracic duct and right lymphatic duct.

of nodes. From the most proximal group of each chain of nodes, the efferent vessels unite to form **lymph trunks.** The principal trunks are the *lumbar, intestinal, bronchomediastinal, subclavian,* and *jugular trunks.*

The principal trunks pass their lymph into two main channels, the thoracic duct and the right lymphatic duct. The **thoracic duct** or **left lymphatic duct** is about 38 to 45 cm (15 to 18 in) in length and begins as a dilation in front of the second lumbar vertebra called the *cisterna chyli* (sis-TER-na KĪ-lē). The thoracic duct receives lymph from the left side of the head, neck, and chest, the left upper extremity, and the entire body below the ribs (see Figure 13-1b). It is the main collecting duct of the lymphatic system. Here is how the thoracic duct collects lymph from its trunks (Figure 13-5). The cisterna chyli receives lymph from the right and left lumbar trunks and from the intestinal trunk. The lumbar trunks drain

lymph from the lower extremities, walls and viscera of the pelvis, kidneys, adrenals, and the deep lymphatics from most of the abdominal wall. The intestinal trunk drains lymph from the stomach, intestines, pancreas, spleen, and visceral surface of the liver. In the neck, the thoracic duct also receives lymph from the left jugular, left subclavian, and left bronchomediastinal trunks. The left jugular trunk drains lymph from the left side of the head and neck; the left subclavian trunk drains lymph from the upper left extremity; and the left bronchomediastinal trunk drains lymph from the left side of the deeper parts of the anterior thoracic wall, upper part of the anterior abdominal wall, anterior part of the diaphragm, left lung, and left side of the heart.

The **right lymphatic duct** is about 1.25 cm (0.5 inches) long and drains lymph from the upper right side of the body (see Figure 13-1b). The right lymphatic duct collects lymph from its trunks as follows (Figure 13-5). It receives lymph from the right jugular trunk, which drains the right side of the head and neck, from the right subclavian trunk, which drains the right upper extremity, and from the right bronchomediastinal trunk, which drains the right side of the thorax, right lung, right side of the heart, and part of the convex surface of the liver.

Ultimately, the thoracic duct empties all of its lymph into the junction of the left internal jugular vein and left subclavian vein and the right lymphatic duct empties all of its lymph into the junction of the right internal jugular vein and right subclavian vein. Thus, lymph is drained back into the blood and the cycle repeats itself continuously.

The flow of lymph from tissue spaces to the large lymphatic ducts to the subclavian veins is maintained primarily by the milking action of muscle tissue. Skeletal muscle contractions compress lymph vessels and force lymph toward the subclavian veins. Moreover, lymph vessels, like veins, contain valves, and the valves ensure the movement of lymph toward the subclavian veins (see Figure 13-4a). Another factor that maintains lymph flow is respiratory movements. These movements create a pressure gradient between the two ends of the lymphatic system. Lymph flows from the tissue spaces, where the pressure is higher, toward the thoracic region, where it is lower.

CLINICAL APPLICATION

Edema, an excessive accumulation of lymph in tissue spaces, may be caused by an obstruction, such as an infected node or a blockage of vessels, in the pathway between the lymphatic capillaries and the subclavian veins. Another cause is excessive lymph formation and increased permeability of blood capillary walls. A rise in capillary blood pressure, in which interstitial fluid is formed faster then it is passed into lymphatics, also may result in edema.

PRINCIPAL GROUPS OF LYMPH NODES

Lymph nodes are scattered throughout the body, usually in groups. Typically, these groups are arranged in two sets: *superficial* and *deep*. Since lymph nodes may become enlarged and tender, an understanding of the regions drained by the nodes may be helpful in diagnosing the site of an infection.

Exhibits 13-1 through 13-5 list the principal groups of lymph nodes of the body by region and the general areas of the body they drain.

EXHIBIT 13-1 PRINCIPAL LYMPH NODES OF THE HEAD AND NECK (Figure 13-6)

LYMPH NODES	LOCATION AND AREAS DRAINED
LYMPH NODES OF THE HEAD	
Occipital	Near trapezius and semispinalis capitis muscles. They drain the occipital portion of scalp and upper neck.
Retroauricular	Behind ear. They drain skin of ear and posterior parietal region of scalp.
Preauricular	Anterior to tragus. They drain pinna and temporal region of the scalp.
Parotid	Embedded in and below parotid gland. They drain root of nose, eyelids, anterior temporal region, external auditory meatus, tympanic cavity, nasopharynx, and posterior portions of nasal cavity.
Facial	Consist of three groups: infraorbital, buccal, and mandibular.
Infraorbital	Below the orbit. They drain eyelids and conjunctiva.
Buccal	At angle of mouth. They drain the skin and mucous membrane of nose and cheek.
Mandibular	Over mandible. They drain the skin and mucous membrane of nose and cheek.
LYMPH NODES OF THE NECK	
Submandibular	Along inferior border of mandible. They drain chin, lips, nose, nasal cavity, cheeks, gums, lower surface of palate, and anterior portion of tongue.
Submental	Between digastric muscles. They drain chin, lower lip, cheeks, tip of tongue, and floor of mouth.
Superficial cervical	Along external jugular vein. They drain lower part of ear and parotid region.
Deep cervical	Largest group of nodes in neck, consisting of numerous large nodes forming a chain extending from base of skull to root of neck. They are arbitrarily divided into superior deep cervical nodes and inferior deep cervical nodes.
Superior deep cervical	Under sternocleidomastoid muscle. They drain posterior head and neck, pinna, tongue, larynx, esophagus, thyroid gland, nasopharynx, nasal cavity, and palate.
Inferior deep cervical	Near subclavian vein. They drain posterior scalp and neck, superficial pectoral region, and part of arm.

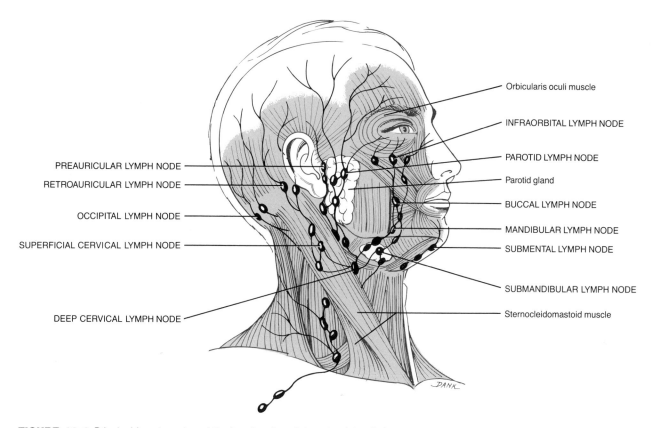

Orbicularis oculi muscle

INFRAORBITAL LYMPH NODE

PAROTID LYMPH NODE

Parotid gland

BUCCAL LYMPH NODE

MANDIBULAR LYMPH NODE

SUBMENTAL LYMPH NODE

SUBMANDIBULAR LYMPH NODE

Sternocleidomastoid muscle

PREAURICULAR LYMPH NODE

RETROAURICULAR LYMPH NODE

OCCIPITAL LYMPH NODE

SUPERFICIAL CERVICAL LYMPH NODE

DEEP CERVICAL LYMPH NODE

FIGURE 13-6 Principal lymph nodes of the head and neck in anterolateral view.

EXHIBIT 13-2 PRINCIPAL LYMPH NODES OF THE UPPER EXTREMITIES (Figure 13-7)

LYMPH NODES	LOCATION AND AREAS DRAINED
Supratrochlear	Above medial epicondyle of humerus. They drain medial fingers, palm, and forearm.
Deltopectoral	Below clavicle. They drain lymphatic vessels on radial side of upper extremity.
Axillary	Most deep lymph nodes of the upper extremities are in the axilla and are called the axillary nodes. They are large in size and may be grouped as follows.
Lateral	Medial and posterior aspects of axillary artery. They drain most of whole upper extremity. Since infection or malignancy of upper extremity may cause tenderness and swelling in axilla, the axillary nodes, especially the lateral group, are clinically important since they filter lymph from much of upper extremity.
Pectoral (anterior)	Along inferior border of the pectoralis minor muscle. They drain skin and muscles of anterior and lateral thoracic walls and central and lateral portions of mammary gland.
Subscapular (posterior)	Along subscapular artery. They drain skin and muscles of posterior part of neck and thoracic wall.
Central (intermediate)	Base of axilla embedded in adipose tissue. They drain lateral, pectoral (anterior), and subscapular (posterior) nodes.
Subclavicular (medial)	Posterior and superior to pectoralis minor muscle. They drain deltopectoral nodes.

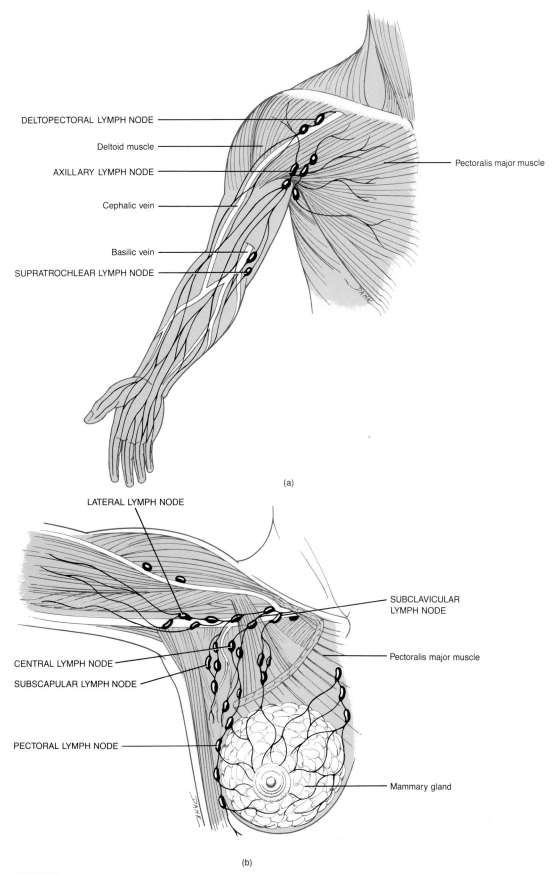

DELTOPECTORAL LYMPH NODE

Deltoid muscle

AXILLARY LYMPH NODE

Cephalic vein

Basilic vein

SUPRATROCHLEAR LYMPH NODE

Pectoralis major muscle

(a)

LATERAL LYMPH NODE

SUBCLAVICULAR LYMPH NODE

CENTRAL LYMPH NODE

SUBSCAPULAR LYMPH NODE

Pectoralis major muscle

PECTORAL LYMPH NODE

Mammary gland

(b)

FIGURE 13-7 Principal lymph nodes of the upper extremities. (a) Anterior view. (b) Anterior view.

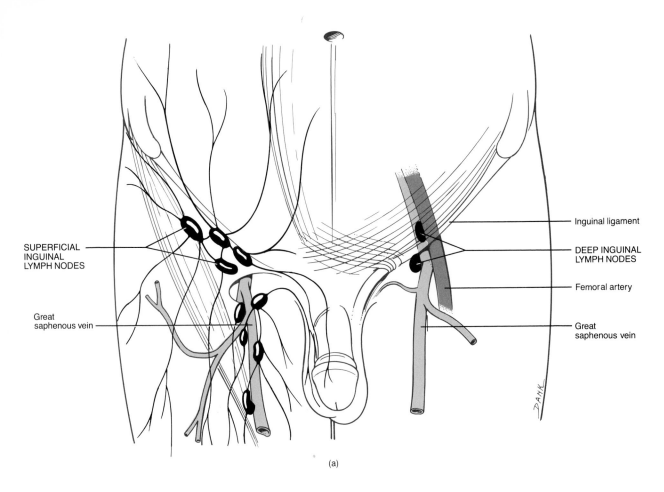

SUPERFICIAL
INGUINAL
LYMPH NODES

Great
saphenous vein

Inguinal ligament

DEEP INGUINAL
LYMPH NODES

Femoral artery

Great
saphenous vein

(a)

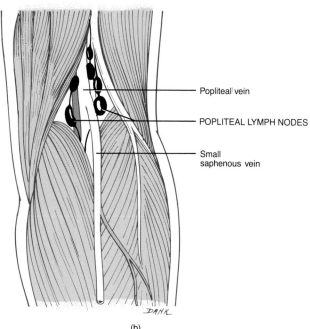

Popliteal vein

POPLITEAL LYMPH NODES

Small
saphenous vein

(b)

FIGURE 13-8 Principal lymph nodes of the lower extremities. (a) Anterior view. (b) Posterior view.

EXHIBIT 13-3 PRINCIPAL LYMPH NODES OF THE LOWER EXTREMITIES (Figure 13-8)

LYMPH NODES	LOCATION AND AREAS DRAINED
Popliteal	In adipose tissue in popliteal fossa. They drain knee.
Superficial inguinal	Parallel to saphenous vein. They drain anterior and lateral abdominal wall to level of umbilicus, gluteal region, external genitals, perineal region, and entire superficial lymphatics of lower extremity.
Deep inguinal	Medial to femoral vein. They drain deep lymphatics of lower extremity, penis, and clitoris.

EXHIBIT 13-4 PRINCIPAL LYMPH NODES OF THE ABDOMEN AND PELVIS (Figure 13-9)

LYMPH NODES	LOCATION AND AREAS DRAINED
	Lymph nodes of the abdomen and pelvis are divided into *parietal lymph nodes* that are retroperitoneal (behind the parietal peritoneum) and in close association with larger blood vessels and *visceral lymph nodes* found in association with visceral arteries.
PARIETAL	
External iliac	Arranged about external iliac vessels. They drain the deep lymphatics of abdominal wall below umbilicus, adductor region of thigh, urinary bladder, prostate gland, ductus deferens, seminal vesicles, prostatic and membranous urethra, uterine tubes, uterus, and vagina.
Common iliac	Arranged along course of common iliac vessels. They drain pelvic viscera.
Internal iliac	Near internal iliac artery. They drain pelvic viscera, perineum, gluteal region, and posterior surface of thigh.
Sacral	In hollow of sacrum. They drain rectum, prostate gland, and posterior pelvic wall.
Lumbar	From aortic bifurcation to diaphragm; arranged around aorta and designated as **right lateral aortic nodes, left lateral aortic nodes, preaortic nodes,** and **retroaortic nodes.** They drain the efferents from testes, ovaries, uterine tubes, uterus, kidneys, suprarenal glands, abdominal surface of diaphragm, and lateral abdominal wall.
VISCERAL	
Celiac	Consist of three groups of nodes: gastric, hepatic, and pancreaticosplenic.
Gastric	Lie along lesser and greater curvatures of stomach. They drain lesser curvature of stomach, inferior, anterior, and posterior aspects of stomach, and esophagus.
Hepatic	Along the hepatic artery. They drain stomach, duodenum, liver, gallbladder, and pancreas.
Pancreaticosplenic	Along splenic artery. They drain stomach, spleen, and pancreas.
Superior mesenteric	These nodes are divided into mesenteric, ileocolic, and mesocolic groups.
Mesenteric	Along superior mesenteric artery. They drain jejunum and all parts of ileum, except for terminal portion.
Ileocolic	Along ileocolic artery. They drain terminal portion of ileum, appendix, cecum, and ascending colon.
Mesocolic	Between layers of transverse mesocolon. They drain descending iliac and sigmoid parts of colon.
Inferior mesenteric	Near left colic, sigmoid, and superior rectal arteries. They drain descending, iliac, and sigmoid parts of colon, superior part of rectum, and superior anal canal.

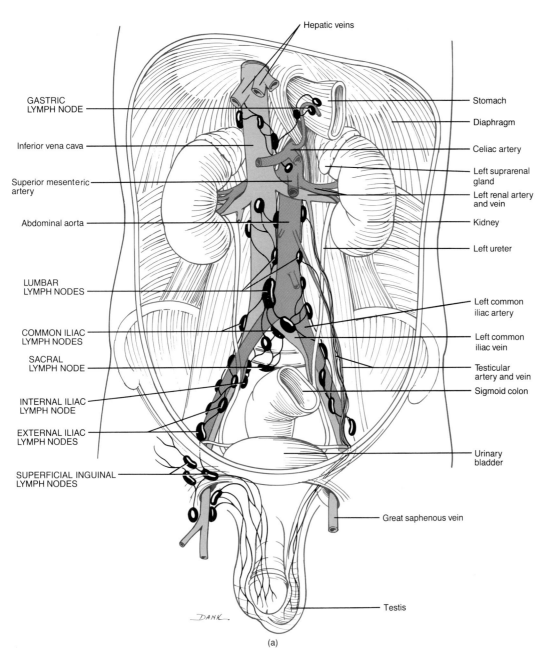

Hepatic veins

GASTRIC LYMPH NODE

Inferior vena cava

Superior mesenteric artery

Abdominal aorta

LUMBAR LYMPH NODES

COMMON ILIAC LYMPH NODES

SACRAL LYMPH NODE

INTERNAL ILIAC LYMPH NODE

EXTERNAL ILIAC LYMPH NODES

SUPERFICIAL INGUINAL LYMPH NODES

Stomach

Diaphragm

Celiac artery

Left suprarenal gland

Left renal artery and vein

Kidney

Left ureter

Left common iliac artery

Left common iliac vein

Testicular artery and vein

Sigmoid colon

Urinary bladder

Great saphenous vein

Testis

DANK

(a)

FIGURE 13-9 Principal lymph nodes of the abdomen and pelvis. (a) Anterior view.

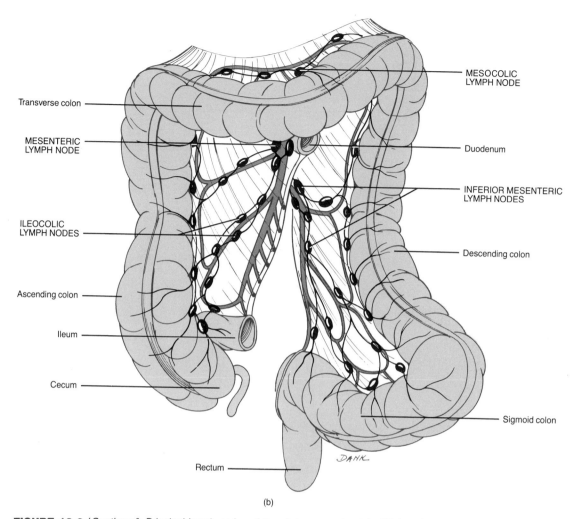

Transverse colon

MESENTERIC
LYMPH NODE

ILEOCOLIC
LYMPH NODES

Ascending colon

Ileum

Cecum

Rectum

MESOCOLIC
LYMPH NODE

Duodenum

INFERIOR MESENTERIC
LYMPH NODES

Descending colon

Sigmoid colon

(b)

FIGURE 13-9 (*Continued*) Principal lymph nodes of the abdomen and pelvis. (b) Anterior view.

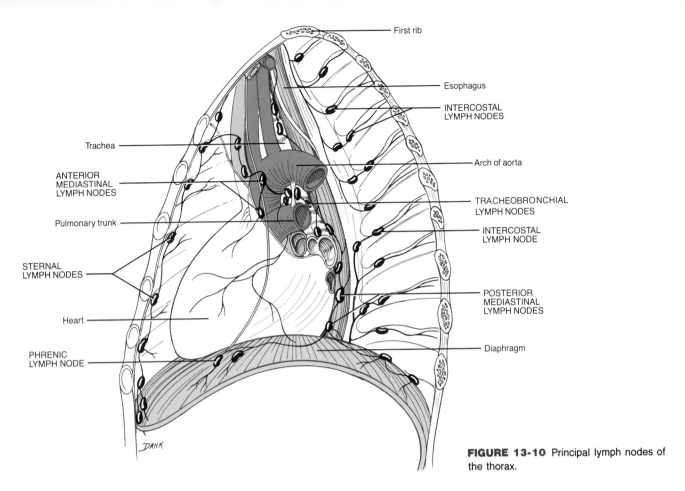

Labels on figure:
- First rib
- Esophagus
- INTERCOSTAL LYMPH NODES
- Trachea
- ANTERIOR MEDIASTINAL LYMPH NODES
- Pulmonary trunk
- STERNAL LYMPH NODES
- Heart
- PHRENIC LYMPH NODE
- Arch of aorta
- TRACHEOBRONCHIAL LYMPH NODES
- INTERCOSTAL LYMPH NODE
- POSTERIOR MEDIASTINAL LYMPH NODES
- Diaphragm
- DANK

FIGURE 13-10 Principal lymph nodes of the thorax.

EXHIBIT 13-5 PRINCIPAL LYMPH NODES OF THE THORAX (Figure 13-10)

LYMPH NODES	LOCATION AND AREAS DRAINED
	Lymph nodes of the thorax are divided into *parietal lymph nodes,* which drain the wall of the thorax, and *visceral lymph nodes,* which drain the viscera.
PARIETAL **Sternal**	Alongside internal thoracic artery. They drain central and lateral parts of mammary gland, deeper structures of anterior abdominal wall above umbilicus, diaphragmatic surface of liver, and deeper parts of anterior portion of thoracic wall.
Intercostal	Near heads of ribs at posterior parts of intercostal spaces. They drain posterolateral aspect of thoracic wall.
Phrenic (diaphragmatic)	Located on thoracic aspect of the diaphragm and divisible into three sets called anterior phrenic, middle phrenic, and posterior phrenic.
Anterior phrenic	Behind base of xiphoid process. They drain convex surface of liver, diaphragm, and anterior abdominal wall.
Middle phrenic	Close to phrenic nerves where they pierce diaphragm. They drain middle part of diaphragm and convex surface of liver.
Posterior phrenic	Back of diaphragm near aorta. They drain posterior part of diaphragm.
VISCERAL **Anterior mediastinal**	Anterior part of superior mediastinum anterior to arch of aorta. They drain thymus gland and pericardium.
Posterior mediastinal	Posterior to pericardium. They drain esophagus, posterior aspect of the pericardium, diaphragm, and convex surface of liver.
Tracheobronchial	The tracheobronchial nodes drain lungs, bronchi, thoracic part of trachea and heart and are divisible into four groups:
Tracheal	On either side of trachea.
Bronchial	At inferior part of trachea and in angle between two bronchi.
Bronchopulmonary	In the hilus of each lung.
Pulmonary	Within lungs on larger bronchial branches.

LYMPHATIC ORGANS

TONSILS

Tonsils are basically masses of lymphoid tissue embedded in mucous membrane. The *pharyngeal* (fa-RIN-jē-al) *tonsil* is embedded in the posterior wall of the nasopharynx (see Figure 20-3a). When it becomes enlarged, it is known as the adenoids. The *palatine* (PAL-a-tīn) *tonsils* are situated in the tonsillar fossae between the pharyngopalatine and glossopalatine arches (see Figure 21-4). These are the ones commonly removed by a tonsillectomy. The *lingual* (LIN-gwal) *tonsil* is located at the base of the tongue and may also have to be removed by a tonsillectomy (see Figure 21-5a). The tonsils are supplied with reticuloendothelial cells.

SPLEEN

The oval **spleen** is the largest mass of lymphatic tissue in the body, measuring about 12 cm (5 inches) in length. It is situated in the left hypochondriac region between the fundus of the stomach and diaphragm (see Figure 1-5d). Its *visceral surface* (Figure 13-11a) contains the contours of the organs adjacent to it—the gastric impression (stomach), renal impression (left kidney), and colic impression (left flexure of colon). The *diaphragmatic* (dī-a-fra-MAT-ik) *surface* (Figure 13-11b) is smooth and convex and conforms to the concave surface of the diaphragm to which it is adjacent.

The spleen is surrounded by a capsule of fibroelastic tissue and scattered smooth muscle (Figure 13-11c). The

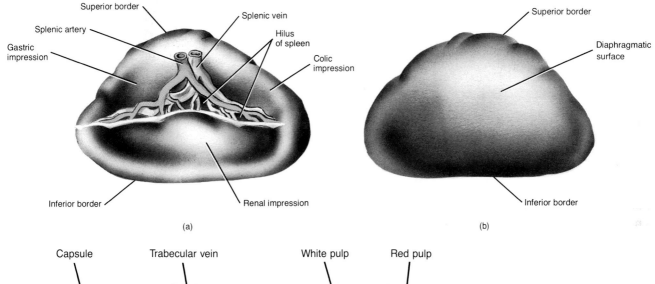

(a)

(b)

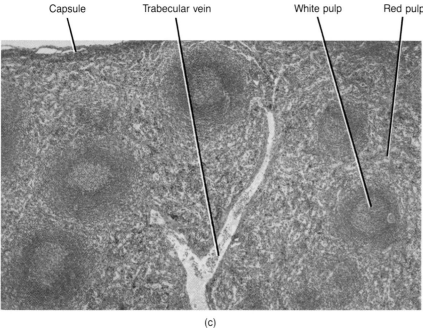

(c)

FIGURE 13-11 Spleen. (a) Gross structure of visceral surface. (b) Gross structure of diaphragmatic surface. (c) Photomicrograph of a portion of the spleen at a magnification of 60×. (© 1983 by Michael H. Ross. Used by permission.)

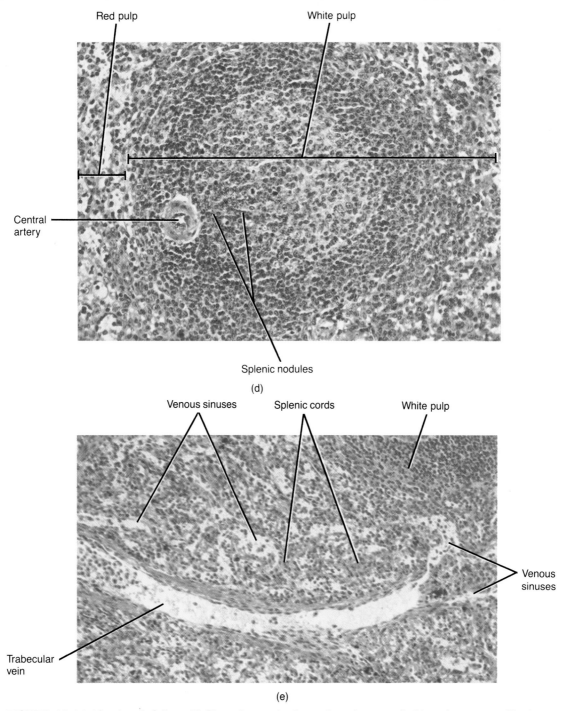

Red pulp

White pulp

Central artery

Splenic nodules

(d)

Venous sinuses

Splenic cords

White pulp

Venous sinuses

Trabecular vein

(e)

FIGURE 13-11 (*Continued*) Spleen. (d) Photomicrograph of an enlarged aspect of white pulp at a magnification of 150×. (e) Photomicrograph of an enlarged aspect of red pulp at a magnification of 150×. (Photomicrographs © 1983 by Michael H. Ross. Used by permission.)

capsule, in turn, is covered by a serous membrane, the peritoneum. Like lymph nodes, the spleen contains trabeculae and a hilus. The capsule, trabeculae, and hilus constitute the stroma of the spleen.

The parenchyma of the spleen consists of two different kinds of tissue called white pulp and red pulp. *White pulp* is essentially lymphoid tissue arranged around arteries. The clusters of lymphocytes surrounding the arteries at intervals and the expansions are referred to as *splenic nodules* or *Malpighian* (mal-PIG-ē-an) *corpuscles* (*bodies*). The *red pulp* consists of venous sinuses filled

with blood and cords of splenic tissue called *splenic* or *Billroth's* (BIL-rōts) *cords*. Veins are closely associated with the red pulp.

The splenic artery and vein and the efferent lymphatics pass through the hilus. Since the spleen has no afferent lymphatic vessels or lymph sinuses, it does not filter lymph. The spleen phagocytoses bacteria and worn-out red blood cells and platelets. It also produces lymphocytes and antibody-producing plasma cells. In addition, the spleen stores and releases blood in case of demand, such as during hemorrhage. The release seems to be purely

CLINICAL APPLICATION

Although the spleen is protected from traumatic injuries, it is the most frequently damaged organ in cases of abdominal trauma, particularly those involving severe blows over the lower left chest or upper abdomen that fracture the protecting ribs. Such a crushing injury may **rupture the spleen,** which causes severe intraperitoneal hemorrhage and shock. This requires prompt removal of the spleen, called a **splenectomy,** to prevent the patient from bleeding to death. The functions of the spleen are then assumed by other reticuloendothelial organs.

sympathetic (see Chapter 17). The sympathetic impulses cause the smooth muscle of the spleen to contract.

THYMUS GLAND

The **thymus gland** is a bilobed mass of lymphatic tissue in the upper thoracic cavity. It is found along the trachea behind the sternum (see Figure 19-13). The thymus is relatively large in children. It reaches its maximum size at puberty and then undergoes involution. Eventually, it is replaced by fat and connective tissue. Its role in immunity is to help produce cells that destroy invading microbes directly or cells that manufacture antibodies against invading microbes.

KEY MEDICAL TERMS ASSOCIATED WITH THE LYMPHATIC SYSTEM

Adenitis (*adeno* = gland) Enlarged, tender, and inflamed lymph nodes resulting from an infection.

Elephantiasis Great enlargement of a limb (especially lower limbs) and scrotum resulting from obstruction of lymph glands or vessels by a parasitic worm.

Hypersplenism (*hyper* = over) Abnormal splenic activity involving highly increased blood cell destruction.

Lymphadenectomy (*ectomy* = removal) Removal of a lymph node.

Lymphadenopathy (*patho* = disease) Enlarged, sometimes tender lymph glands.

Lymphangioma A benign tumor of the lymph vessels.

Lymphangitis (*itis* = inflammation of; *angio* = vessel) Inflammation of the lymphatic vessels.

Lymphedema (*oidema* = swelling) Accumulation of lymph fluid producing subcutaneous tissue swelling.

Lymphoma (*oma* = tumor) Any tumor composed of lymph tissue. Malignancy of reticuloendothelial cells of lymph nodes is called Hodgkin's disease.

Lymphostasis (*stasis* = halt) A lymph flow stoppage.

Splenectomy Surgical removal of the spleen.

Splenomegaly (*mega* = large) Enlarged spleen.

STUDY OUTLINE

Lymphatic Vessels
1. Lymphatic vessels are similar in structure to veins.
2. Lymph capillaries merge to form larger vessels, called lymphatics, which ultimately converge into the thoracic duct or right lymphatic duct.

Structure of Lymph Nodes
1. Lymph nodes are oval structures located along lymphatics.
2. Lymph passing through the nodes is filtered by macrophages. Lymphoid tissue is a site of production of agranular leucocytes and plasma cells, from which antibodies arise.

Lymph Circulation
1. The passage of lymph is from interstitial fluid, to lymph capillaries, to lymphatics, to lymph trunks, to the thoracic duct or right lymphatic trunk, to the subclavian veins.

2. Lymph flows as a result of skeletal muscle contractions and respiratory movements. It is also aided by valves in the lymphatics.

Principal Groups of Lymph Nodes
1. Lymph nodes are scattered throughout the body as superficial and deep groups.
2. The principal groups of lymph nodes are found in the head and neck, upper extremities, lower extremities, abdomen and pelvis, and thorax.

Lymphatic Organs
1. Tonsils are masses of lymphoid tissue embedded in mucous membrane.
2. The spleen functions as a lymphatic organ in phagocytosis of bacteria and worn-out cells and production of lymphocytes and plasma cells. It also acts as a reservoir for blood.
3. The thymus gland functions in immunity.

REVIEW QUESTIONS

1. Identify the components and functions of the lymphatic system.
2. How do lymphatic vessels originate? Compare veins and lymphatics with regard to structure.
3. What is a lymphangiogram? What is its diagnostic value?
4. Describe the structure of a lymph node. What functions do lymph nodes serve?
5. Construct a diagram to indicate the plan of lymph circulation.
6. List and explain the various factors involved in the maintenance of lymph circulation.
7. Define edema. What are some of its causes?
8. For each of the following regions of the body, list the major lymph nodes and the areas of the body they drain: head, neck, upper extremities, lower extremities, abdomen and pelvis, and thorax.
9. Identify the tonsils by location.
10. Describe the location, gross anatomy, histology, and functions of the spleen. What is a splenectomy?
11. Where is the thymus gland located?

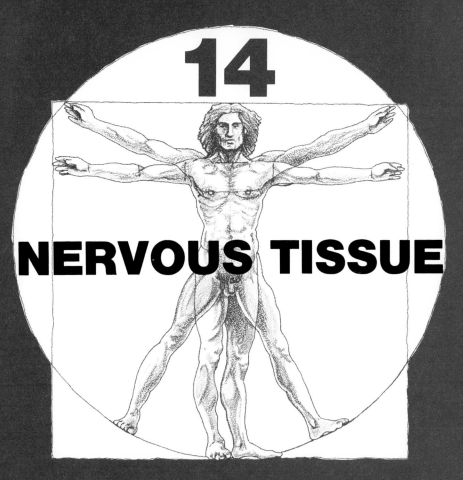

14

NERVOUS TISSUE

STUDENT OBJECTIVES

- Classify the organs of the nervous system into central and peripheral divisions.

- Contrast the histological characteristics and functions of neuroglia and neurons.

- Classify neurons by shape and function.

- List the necessary conditions for the regeneration of nervous tissue.

- Define a nerve impulse and describe its importance in the body.

- Explain how a nerve impulse is transmitted across a synapse.

The **nervous system** is the body's control center and communications network. In human beings it performs two broad functions. First, it stimulates movements. Second, it shares the maintenance of homeostasis with the endocrine system. Human life cannot exist without a functioning nervous system. Skeletal and most smooth muscle cells cannot contract until stimulated by a nerve impulse. If the intercostal muscles and diaphragm do not contract, we cannot breathe. If the digestive glands are not stimulated to release their secretions, we cannot digest our food. It is obvious, then, that our cells cannot receive nutrients unless the digestive system is connected to a functioning nervous system. But suppose our muscles and glands could stimulate themselves. Even then, we could not live very long without our nervous system. The nervous system senses changes within the body and in the outside environment. It then interprets these changes and may initiate action to maintain homeostasis.

The developmental anatomy of the nervous system is considered in Exhibit 24-6.

ORGANIZATION

The nervous system may be divided into two principal divisions, the central nervous system and the peripheral nervous system, and several subdivisions (Figure 14-1).

The **central nervous system (CNS)** is the control center for the entire system and consists of the brain and spinal cord. All body sensations must be relayed from receptors to the central nervous system if they are to be interpreted

and acted on. All the nerve impulses that stimulate muscles to contract and glands to secrete must also pass from the central system.

The various nerve processes that connect the brain and spinal cord with receptors, muscles, and glands constitute the **peripheral (pe-RIF-er-al) nervous system (PNS).** The peripheral nervous system may be divided into an afferent system and an efferent system. The **afferent system** consists of nerve cells that convey information from receptors in the periphery of the body to the central nervous system. These nerve cells, called *afferent (sensory) neurons,* are the first cells to pick up incoming information. The **efferent system** consists of nerve cells that convey information from the central nervous system to muscles and glands. These nerve cells are called *efferent (motor) neurons.*

The efferent system is subdivided into a somatic nervous system and an autonomic nervous system. The **somatic nervous system** or **SNS** (*soma* = body) consists of efferent neurons that conduct impulses from the central nervous system to skeletal muscle tissue. Since the somatic nervous system produces movement only in skeletal muscle tissue, it is under conscious control and therefore voluntary. The **autonomic nervous system** or **ANS** (*auto* = self; *nomos* = law), by contrast, contains efferent neurons that convey impulses from the central nervous system to smooth muscle tissue, cardiac muscle tissue, and glands. Since it produces responses only in involuntary muscles and glands, it is usually considered to be involuntary.

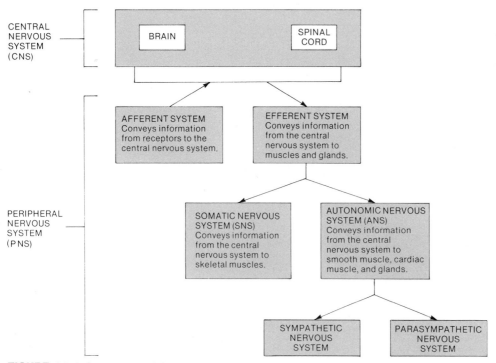

FIGURE 14-1 Organization of the nervous system.

With few exceptions, the viscera receive nerve fibers from the two divisions of the autonomic nervous system: the **sympathetic division** and the **parasympathetic division.**

HISTOLOGY

Despite the organizational complexity of the nervous system, it consists of only two principal kinds of cells. The first of these, the neurons, make up the nervous tissue that forms the structural and functional portion of the system. Neurons are highly specialized for impulse conduction and for all special functions attributed to the nervous system: thinking, controlling muscle activity, regulating glands. The second type of cell, the neuroglia, serves as a special supporting and protective component of the nervous system.

NEUROGLIA

The cells of the nervous system that perform the functions of support and protection are called **neuroglia** (noo-ROG-lē-a) or **glial cells** (*neuro* = nerve; *glia* = glue). About 50 percent of all brain cells are neuroglial cells. Many of the glial cells form a supporting network by twining around the nerve cells in the brain and spinal cord. Others bind nervous tissue to supporting structures and attach the neurons to their blood vessels. A few types of glial cells also serve specialized protective functions. For example, many nerve fibers are coated with a thick, fatty sheath produced by one type of neuroglia. Certain small glial cells are phagocytotic; they protect the central nervous system from disease by engulfing invading microbes and clearing away debris. Neuroglia are of clinical interest because they are a common source of tumors of the nervous system. Exhibit 14-1 lists the neuroglial cells and summarizes their functions.

NEURONS

Nerve cells, called **neurons,** are responsible for conducting impulses from one part of the body to another. They are the structural and functional units of the nervous system.

Structure

A neuron consists of three distinct portions: (1) cell body, (2) dendrites, and (3) axon (Figure 14-2a). The **cell body,** or **perikaryon** (per'-i-KAR-ē-on), contains a well-defined nucleus and nucleolus surrounded by a granular cytoplasm. Within the cytoplasm are typical organelles such

EXHIBIT 14-1 **NEUROGLIA OF CENTRAL NERVOUS SYSTEM**

TYPE	DESCRIPTION	MICROSCOPIC APPEARANCE	FUNCTION
Astrocytes (*astro* = star; *cyte* = cell)	Star-shaped cells with numerous processes. *Protoplasmic astrocytes* are found in the gray matter of the CNS, and *fibrous astrocytes* are found in the white matter of the CNS.		Twine around nerve cells to form supporting network in brain and spinal cord; attach neurons to their blood vessels.
Oligodendrocytes (*oligo* = few; *dendro* = tree)	Resemble astrocytes in some ways, but processes are fewer and shorter.		Give support by forming semirigid connective tissue rows between neurons in brain and spinal cord; produce a myelin sheath around axons of neurons of central nervous system.
Microglia (*micro* = small; *glia* = glue)	Small cells with few processes; normally stationary, but may migrate to injured area if nervous tissue is damaged.		Engulf and destroy microbes and cellular debris.

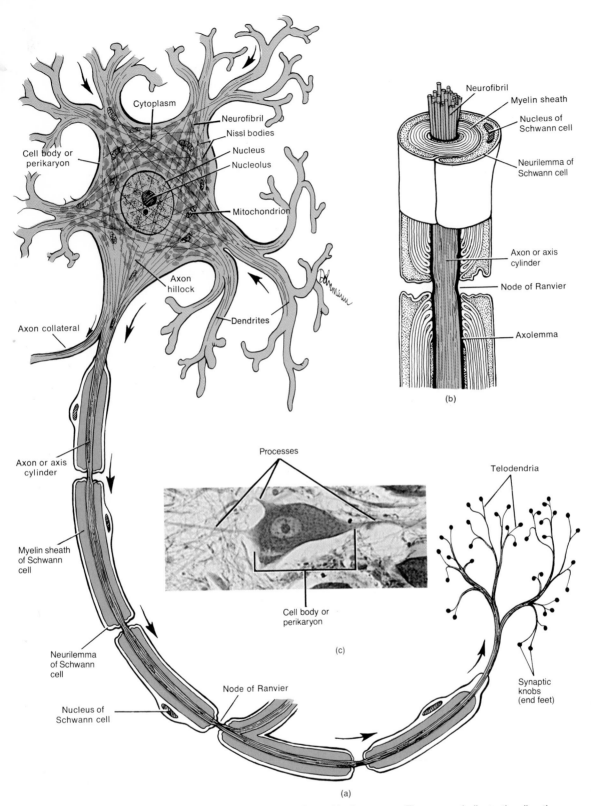

Neurofibril

Myelin sheath

Nucleus of
Schwann cell

Neurilemma of
Schwann cell

Axon or axis
cylinder

Node of Ranvier

Axolemma

(b)

Cytoplasm

Neurofibril

Nissl bodies

Nucleus

Nucleolus

Mitochondrion

Cell body or
perikaryon

Axon
hillock

Dendrites

Axon collateral

Axon or axis
cylinder

Myelin sheath
of Schwann
cell

Neurilemma
of Schwann
cell

Nucleus of
Schwann cell

Node of Ranvier

Processes

Cell body or
perikaryon

(c)

Telodendria

Synaptic
knobs
(end feet)

(a)

FIGURE 14-2 Structure of a neuron. (a) Diagram of an entire multipolar neuron. The arrows indicate the direction in which the nerve impulse travels. (b) Sections through a myelinated fiber. (c) Photomicrograph of a multipolar neuron. (Courtesy of Biophoto Associates.)

as mitochondria and Golgi apparatus. Also located in the cytoplasm are structures characteristic of neurons: Nissl bodies and neurofibrils. *Nissl bodies* are orderly arrangements of granular (rough) endoplasmic reticulum whose function is protein synthesis. Newly synthesized proteins pass from the perikaryon into the neuronal processes, mainly the axon, at the rate of about 1 mm (0.04 inch)/day. These proteins replace those lost during metabolism and are used for growth of neurons and regeneration of peripheral nerve fibers. *Neurofibrils* are long, thin fibrils composed of microtubules. They may assume a function in support.

The cytoplasmic processes of neurons generally depend on the direction in which they conduct impulses. There are two kinds: dendrites and axons. **Dendrites** (*dendron* = related to a tree) are highly branched thick extensions of the cytoplasm of the cell body. They typically contain Nissl bodies, mitochondria, and other cytoplasmic organelles. A neuron usually has several main dendrites. The function of dendrites is to conduct an impulse toward the cell body.

The second type of cytoplasmic process, called an **axon** or **axis cylinder**, is a single, highly specialized, and long thin process that conducts impulses away from the cell body to another neuron or tissue.

An axon usually originates from the cell body as a small conical elevation called the *axon hillock.* An axon contains mitochondria and neurofibrils but no Nissl bodies, thus it does not carry on protein synthesis. Its cytoplasm, called *axoplasm,* is surrounded by a plasma membrane known as the *axolemma* (*lemma* = sheath or husk). Axons vary in length from a few millimeters (1 mm = 0.04 inch) in the brain to a meter (3.28 ft) or more between the spinal cord and toes. Along the course of an axon, there may be side branches called *axon collaterals.* The axon and its collaterals terminate by branching into many fine filaments called *telodendria.* The distal ends of telodendria are expanded into bulblike structures called *synaptic knobs* or *end feet,* which are important in nerve impulse conduction.

The term **nerve fiber** is applied to an axon and its sheaths. Figure 14-2b shows two sections of a nerve fiber of the peripheral nervous system. Many axons, especially large, peripheral axons, are surrounded by a multilayered white, phospholipid, segmented covering called the **myelin sheath.** Axons containing such a covering are *myelinated,* while those without it are *unmyelinated.* The function of the myelin sheath is to increase the speed of nerve impulse conduction and to insulate and maintain the axon. Myelin is responsible for the color of the white matter in the nerves, brain, and spinal cord.

The myelin sheath is produced by flattened cells, called **Schwann cells,** located along the axon. They are neuroglial cells of the peripheral nervous system. In the formation of a sheath, a developing Schwann cell encircles the axon until its ends meet and overlap (Figure 14-3). The cell then winds around the axon many times and, as it does so, the cytoplasm and nucleus are pushed to the outside layer. The inner portion, consisting of several layers of Schwann cell membrane, is the myelin sheath. The peripheral nucleated cytoplasmic layer of the Schwann cell (the outer layer that encloses the sheath) is called the **neurilemma** or **sheath of Schwann.**

The neurilemma is found only around fibers of the peripheral nervous system. Its function is to assist in the regeneration of injured axons. Between the segments of the myelin sheath are unmyelinated gaps called **nodes of Ranvier** (ron-VĒ-ā). Unmyelinated fibers are also enclosed by Schwann cells, but without multiple wrappings.

Nerve fibers of the central nervous system may be myelinated or unmyelinated. Myelination of central nervous system axons is accomplished by oligodendrocytes in somewhat the same manner that Schwann cells myelinate peripheral nervous system axons. Myelinated axons of the central nervous system also contain nodes of Ranvier, but they are not so numerous.

Classification

The different neurons in the body may be classified by structure and function.

The structural classification is based on the number of processes extending from the cell body. **Multipolar neurons** have several dendrites and one axon. Most neurons in the brain and spinal cord are of this type. **Bipolar neurons** have one dendrite and one axon and are found in the retina of the eye, the inner ear, and the olfactory area. **Unipolar neurons** have only one process extending from the cell body. The single process divides into a central branch, which functions as an axon, and a peripheral branch, which functions as a dendrite. Unipolar neurons originate in the embryo as bipolar neurons. During development, the axon and dendrite fuse into a single process and become a functional axon. Unipolar neurons are found in posterior (sensory) root ganglia of spinal nerves.

The functional classification of neurons is based on the direction in which they transmit impulses. **Sensory neurons,** called **afferent neurons,** transmit impulses from receptors in the skin, sense organs, and viscera to the brain and spinal cord. They are usually unipolar. **Motor neurons,** called **efferent neurons,** convey impulses from the brain and spinal cord to effectors, which may be either muscles or glands. Other neurons, called **association (connecting** or **internuncial) neurons,** carry impulses from sensory neurons to motor neurons and are located in the brain and spinal cord.

The processes of afferent and efferent neurons are arranged into bundles called *nerves.* Since nerves lie outside

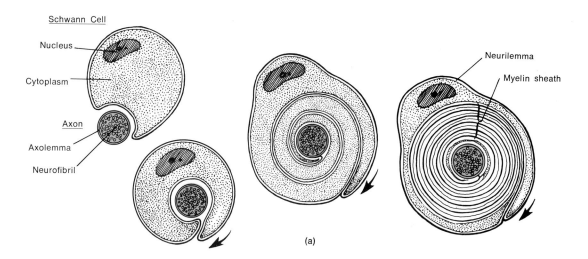

Schwann Cell
Nucleus
Cytoplasm
Axon
Axolemma
Neurofibril

Neurilemma
Myelin sheath

(a)

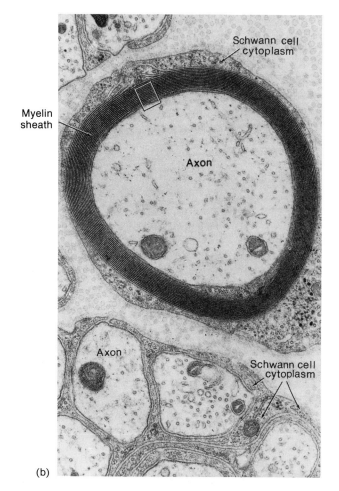

Schwann cell cytoplasm

Myelin sheath

Axon

Axon

Schwann cell cytoplasm

(b)

FIGURE 14-3 Myelin sheath. (a) Stages in the formation of a myelin sheath by a Schwann cell. (b) Electron micrograph of a nerve in cross section showing a myelinated nerve axon (above) and several unmyelinated nerve axons (below) at a magnification of 12,000×. (Courtesy of William Bloom and Don W. Fawcett, *A Textbook of Histology,* W. B. Saunders, Philadelphia, 1968.)

2. General somatic efferent fibers conduct impulses from the central nervous system to skeletal muscles. Impulses over these fibers cause the contraction of skeletal muscles.

3. General visceral afferent fibers convey impulses from the viscera and blood vessels to the central nervous system.

4. General visceral efferent fibers belong to the autonomic nervous system and are also called *autonomic fibers.* They convey impulses from the central nervous system to cause contractions of smooth and cardiac muscle and secretion by glands.

Structural Variation

Although all neurons conform to the general plan described, there are considerable differences in structure. For example, cell bodies range in diameter from 5 μm for the smallest cells to 135 μm for large motor neurons. The pattern of dendritic branching is varied and distinctive for neurons in different parts of the body. The axons of very small neurons are only a fraction of a millimeter in length and lack a myelin sheath, while axons of large neurons are over a meter long and are usually enclosed in a myelin sheath.

A few patterns of diversity are shown in Figure 14-4. Note the structure of a typical afferent (sensory) neuron. Compare it to the typical efferent (motor) neuron shown in Figure 14-2a. What structural differences do you observe? A few examples of association neurons are also shown: a *stellate cell, cell of Martinotti,* and a *horizontal cell of Cajal.* All are found in the cerebral cortex,

the central nervous system, they belong to the peripheral nervous system. The functional components of nerves are the nerve fibers, which may be grouped according to the following scheme.

1. General somatic afferent fibers conduct impulses from the skin, skeletal muscles, and joints to the central nervous system.

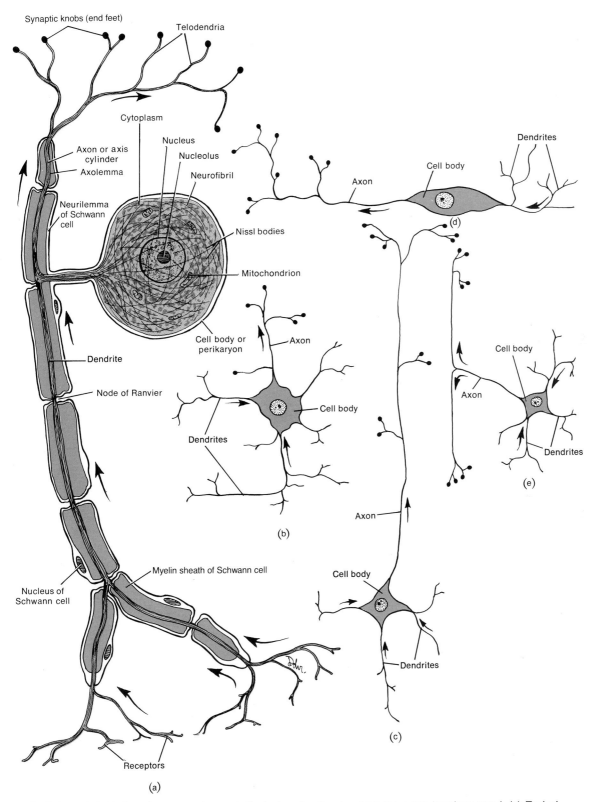

FIGURE 14-4 Varieties of neurons. Arrows indicate the direction in which the nerve impulses travel. (a) Typical afferent neuron. (b–e) Representative association neurons. (b) Stellate cell. (c) Cell of Martinotti. (d) Horizontal cell of Cajal. (e) Granule cell.

the outer layer of the cerebrum. Note the *granule cell,* an association neuron in the cortex of the cerebellum.

PHYSIOLOGY

Two striking features of nervous tissue are its limited ability to regenerate and its highly developed ability to produce and transmit electrical messages called nerve impulses.

REGENERATION

Unlike the cells of epithelial tissue, neurons have only limited powers for regeneration. Around the time of birth, the cell bodies of most developing nerve cells lose their mitotic apparatus (centrioles and spindles) and their ability to reproduce. Thus, when a neuron is damaged or destroyed it cannot be replaced by the daughter cells of other neurons. A neuron destroyed is permanently lost and only some types of damage may be repaired.

Damage to some types of myelinated axons can often be repaired if the cell body remains intact. Whereas axons in the peripheral nervous system are myelinated by Schwann cells, axons in the brain and spinal cord are myelinated by oligodendroglial cells. Scar tissue is formed by neuroglial cells other than oligodendrocytes faster than the axon can repair itself. An injury to a nerve in the arm may repair itself before scar tissue forms and so some nerve function may be restored. However, an injury to the brain or spinal cord is permanent because axonal regeneration is blocked by the rapid scar tissue formation.

NERVE IMPULSE

Although it is beyond the scope of this text to describe the details of a nerve impulse, certain concepts must be understood in order to know how the nervous system works. Very simply, a **nerve impulse** is a wave of electrical negativity that travels along the surface of the membrane of a neuron. Among other things, a nerve impulse depends upon the movement of sodium, potassium, and other ions between interstitial fluid and the inside of a neuron. For a nerve impulse to begin, a stimulus of adequate strength must be applied to the neuron. A stimulus is a change in the environment of sufficient strength to initiate an impulse. The ability of a neuron to respond to a stimulus and convert it into an impulse is known as **excitability.**

The nerve impulse is the most rapid way that the body can respond to environmental changes. It provides the quickest means for achieving homeostasis. The speed of a nerve impulse is determined by the size, type, and physiological condition of the nerve fiber. For example, myelinated fibers with the largest diameters can transmit impulses at speeds up to about 100 m (328 ft)/sec. Unmyelinated fibers with the smallest diameters can transmit impulses at the rate of about 0.5 m (1.5 ft)/sec.

In addition to excitability, neurons are also characterized by **conductivity,** the ability to transmit an impulse to another neuron or another tissue, such as a muscle or a gland. The junction between two neurons is called a **synapse** (*synapse* = to join). The two neurons do not touch, but rather the impulse is conducted across a gap, the **synaptic cleft,** that measures about 200 Å (Figure 14-5). The neuron before the synapse is called the **presynaptic neuron,** and the neuron after the synapse is the **postsynaptic neuron.** The synaptic knobs of a presynaptic neuron may synapse with the dendrites, cell body, or axon of postsynaptic neurons.

When an impulse is transmitted between a neuron and a muscle cell, the junction is called a **neuromuscular junction. A neuroglandular junction** is between a neuron and a glandular cell. Together, they are called **neuroeffector junctions.**

Whether an impulse is conducted across a synapse, neuromuscular junction, or neuroglandular junction depends on the presence of chemicals called **transmitter substances,** or **neurotransmitters.** These chemicals are made by the neuron, usually from amino acids. Each neuron can make only one kind of transmitter substance. Following its production and transport to the synaptic knob, the transmitter substance is stored in small membrane-enclosed sacs called **synaptic vesicles** (Figure 14-

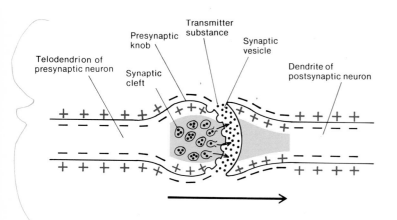

FIGURE 14-5 Impulse conduction at a synapse. The negative charges on the outside of the membrane indicate impulse conduction; the positive charges on the outside of the membrane indicate that no impulse is conducted. The arrow indicates the direction of impulse conduction.

5). Each of the thousands of synaptic vesicles in a synaptic knob may contain between 10,000 and 100,000 transmitter molecules. When a nerve impulse arrives at a presynaptic knob, it causes the discharge of exact quantities of transmitter molecules into the synaptic cleft. Some scientists believe that the synaptic vesicles fuse with the plasma membrane of the presynaptic neuron, form openings, and release the transmitter through the openings. Other scientists think that the transmitter leaves through small channels. In either case, the transmitter enters the synaptic cleft and, depending on the chemical nature of the transmitter and the interaction of the transmitter with the postsynaptic plasma membrane, several things can happen.

An *excitatory transmitter* initiates a nerve impulse in the postsynaptic neuron, while an *inhibitory transmitter* inhibits impulse formation. Perhaps the best studied transmitter substance is **acetylcholine** (as'-ē-til-KŌ-lēn) or **ACh,** an excitatory transmitter released by many neurons outside the brain and spinal cord and by some neurons inside the brain and cord. At neuromuscular junctions, ACh binds to receptor sites on the muscle fiber membrane and increases its permeability to Na^+ ions. This leads to a sequence of events that generates a nerve impulse, causing the muscle fiber to contract. As long as acetylcholine is present in the neuromuscular junction or synapse, it can stimulate a muscle fiber almost indefinitely. The transmission of a continuous succession of impulses by acetylcholine is normally prevented by an enzyme called **acetylcholinesterase (AChE)** or simply **cholinesterase** (kō'-lin-ES-ter-ās). Cholinesterase is released into the neuromuscular junction by the muscle cell. Within $\frac{1}{500}$ second, cholinesterase inactivates acetylcholine. This action permits the membrane of the muscle fiber to discontinue impulse transmission so that another impulse may be generated. When the next impulse comes through, the synaptic vesicles release more acetylcholine, the impulse is conducted, and cholinesterase again inactivates acetylcholine. This cycle is repeated over and over again.

ACh is released at some neuromuscular junctions that have cardiac and smooth muscle, as well as those that have skeletal muscle. It is also released at some neuroglandular junctions. Although ACh is an excitatory transmitter in many parts of the body, it is inhibitory with respect to the heart (vagus nerve).

CLINICAL APPLICATION

Certain drugs have pronounced effects on synaptic transmission. *Hypnotics* and *anesthetics,* for example, **depress** synaptic transmission. Stimulants like *caffeine* and *benzedrine* **facilitate** synaptic transmission. Pressure has an effect on impulse transmission and conduction also. If excessive or prolonged **pressure** is applied to a nerve, impulse transmission is interrupted and part of the body may "go to sleep," producing a prickly sensation. This sensation is caused by an accumulation of waste products and a depressed circulation of blood.

Impulse conduction at a synapse is one-way. The direction is from the presynaptic neuron to the postsynaptic neuron. In other words, impulses must move forward over their pathways. They cannot back up into another presynaptic neuron. Such a mechanism is important in preventing impulse conduction along wrong pathways. Imagine the result if impulses transmitted along the motor neurons that move your hand could move back and stimulate the sensory neuron which relays information about heat. You would feel heat, cry in pain, and go through all the emotions of being burned every time you simply wanted to move your hand.

STUDY OUTLINE

Organization

1. The nervous system controls and integrates all body activities by sensing changes, interpreting them, and reacting to them.
2. The central nervous system consists of the brain and spinal cord.
3. The peripheral nervous system is classified into an afferent system and an efferent system.
4. The efferent system is subdivided into a somatic nervous system and an autonomic nervous system.
5. The somatic nervous system consists of efferent neurons that conduct impulses from the central nervous system to skeletal muscle tissue.
6. The autonomic nervous system contains efferent neurons that convey impulses from the central nervous system to smooth muscle tissue, cardiac muscle tissue, and glands.

Histology

1. Neuroglia are specialized tissue cells that support neurons, attach neurons to blood vessels, produce the myelin sheath, and carry out phagocytosis.
2. Neurons, or nerve cells, consist of a perikaryon or cell body, dendrites that pick up stimuli and convey impulses to the cell body, and usually a single axon. The axon transmits impulses from the neuron to the dendrites or cell body of another neuron or to an effector organ of the body.
3. On the basis of structure, neurons may be classified

as multipolar, bipolar, or unipolar, according to the number of processes extending from the cell body.

4. On the basis of function, neurons are classified according to the direction in which they transmit impulses. Sensory (afferent) neurons transmit impulses to the central nervous system; association neurons transmit impulses to other neurons including motor neurons; and motor (efferent) neurons transmit impulses to effectors.

Physiology
Regeneration

1. Around the time of birth, the cell body loses its mitotic apparatus and is no longer able to divide.
2. Nerve fibers (axis cylinders) that have a neurilemma are capable of regeneration.

Nerve Impulse

1. A nerve impulse is a wave of negativity that travels along the surface of the membrane of a neuron.
2. The ability of a neuron to respond to a stimulus and convert it into a nerve impulse is called excitability.
3. The speed of a nerve impulse is determined by the size, type, and physiological condition of the nerve fiber.
4. Conductivity is the ability of a neuron to transmit a nerve impulse to another neuron or another tissue.
5. The junction between neurons is called a synapse. Impulse conduction across a synapse requires the release of chemical transmitters by presynaptic neurons.
6. Impulse conduction at a synapse is one-way.

REVIEW QUESTIONS

1. What is the overall function of the nervous system?
2. Distinguish between the central and peripheral nervous systems. Relate the terms *voluntary* and *involuntary* to the nervous system.
3. What are neuroglia? List the principal types and their functions.
4. Define a neuron. Diagram and label a neuron. Next to each part list its function.
5. What is the myelin sheath? How is it formed?
6. Define the neurilemma. Why is it important?
7. Discuss the structural classification of neurons. Give an example of each.
8. Describe the functional classification of neurons.
9. Distinguish among the following kinds of fibers: general somatic afferent, general somatic efferent, general visceral afferent, and general visceral efferent.
10. What are the structural differences between a typical afferent and a typical efferent neuron?
11. What determines neuron regeneration?
12. Define a nerve impulse.
13. Compare excitability and conductivity.
14. What factors determine the rate of impulse conduction?
15. What is a synapse? How does an impulse cross a synapse?

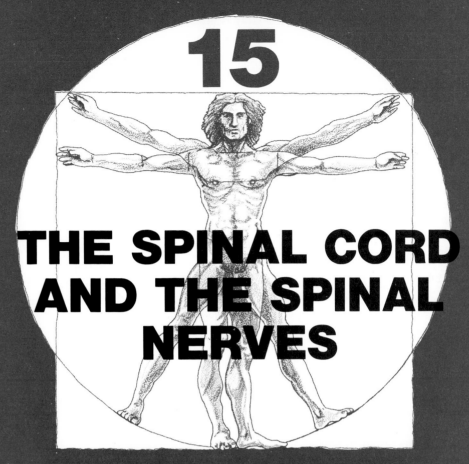

15

THE SPINAL CORD AND THE SPINAL NERVES

STUDENT OBJECTIVES

- Define white matter, gray matter, nerve, ganglion, tract, nucleus, and horn.

- Describe the gross anatomical features of the spinal cord.

- Explain how the spinal cord is protected by the meninges.

- Describe the structure of the spinal cord in cross section.

- Explain the functions of the spinal cord as a conduction pathway and a reflex center.

- List the location, origin, termination, and function of the principal ascending and descending tracts of the spinal cord.

- Describe the components of a reflex arc.

- Compare the functional anatomy of a stretch reflex, flexor reflex, and crossed extensor reflex.

- Describe the composition and coverings of a spinal nerve.

- Name the 31 pairs of spinal nerves.

- Explain how a spinal nerve branches upon leaving the intervertebral foramen.

- Explain the composition and distribution of the cervical, brachial, lumbar, sacral, and coccygeal plexuses.

- Define an intercostal nerve.

- Define a dermatome and its clinical importance.

- Describe spinal cord injury and list the immediate and long-range effects.

- Identify the effects of peripheral nerve damage and conditions necessary for its regeneration.

- Distinguish between sciatica and neuritis.

- Explain the cause and symptoms of shingles.

I n this chapter, our main concern will be a study of the structure and function of the spinal cord and the nerves that originate from it.

GROUPING OF NEURAL TISSUE

The term **white matter** refers to aggregations of myelinated axons from many neurons supported by neuroglia. The lipid substance, myelin, has a whitish color that gives white matter its name. The gray areas of the nervous system are called **gray matter.** They contain either nerve cell bodies and dendrites or bundles of unmyelinated axons and neuroglia.

A **nerve** is a bundle of fibers located outside the central nervous system. Since the dendrites of somatic afferent neurons and axons of somatic efferent neurons of the peripheral nervous system are myelinated, most nerves are white matter. Nerve cell bodies that lie outside the central nervous system are generally grouped with other nerve cell bodies to form **ganglia** (GANG-lē-a; *ganglion* = knot). Ganglia, since they are made up principally of nerve cell bodies, are masses of gray matter.

A **tract** is a bundle of fibers in the central nervous system. Tracts may run long distances up and down the spinal cord. Tracts also exist in the brain and connect parts of the brain with each other and with the spinal cord. The chief spinal tracts that conduct impulses up the cord are concerned with sensory impulses and are called *ascending tracts.* Spinal tracts that carry impulses down the cord are motor tracts and are called *descending tracts.* The major tracts consist of myelinated fibers and are therefore white matter. A **nucleus** is a mass of nerve cell bodies and dendrites in the central nervous system. It forms gray matter. **Horns** or **columns** are the chief areas of gray matter in the spinal cord. The term *horn* describes the two-dimensional appearance of the organization of gray matter in the spinal cord as seen in cross section. The term *column* describes the three-dimensional appearance of the gray matter in longitudinal columns. Since the white matter of the spinal cord is also arranged in columns, we will refer to the gray matter as being arranged in horns.

SPINAL CORD
GENERAL FEATURES

The **spinal cord** is a cylindrical structure that is slightly flattened anteriorly and posteriorly. It begins as a continuation of the medulla oblongata, the inferior part of the brain stem, and extends from the foramen magnum of the occipital bone to the level of the second lumbar vertebra (Figure 15-1). The length of the adult spinal cord ranges from 42 to 45 cm (16 to 18 inches). The diameter of the cord varies at different levels.

When the cord is viewed externally, two conspicuous enlargements can be seen. The superior enlargement, the **cervical enlargement,** extends from the fourth cervical to the first thoracic vertebra. Nerves that supply the upper extremities arise from the cervical enlargement. The inferior enlargement, called the **lumbar enlargement,** extends from the ninth to the twelfth thoracic vertebra. Nerves that supply the lower extremities arise from the lumbar enlargement.

Below the lumbar enlargement, the spinal cord tapers to a conical portion known as the **conus medullaris** (KŌ-nus med-yoo-LAR-is). The conus medullaris ends at the level of the intervertebral disc between the first and second lumbar vertebra. Arising from the conus medullaris is the **filum terminale** (FĪ-lum ter-mi-NAL-ē), a nonnervous fibrous tissue of the spinal cord that extends inferiorly to attach to the coccyx. The filum terminale consists mostly of pia mater, the innermost of three membranes that cover and protect the spinal cord and brain. Some nerves that arise from the lower portion of the cord do not leave the vertebral column immediately. They angle inferiorly in the vertebral canal like wisps of coarse hair flowing from the end of the cord. They are appropriately named the **cauda equina** (KAW-da ē-KWĪ-na), meaning horse's tail.

The spinal cord is a series of 31 segments, each giving rise to a pair of spinal nerves. **Spinal segment** refers to a region of the spinal cord from which a pair of spinal nerves arises. Figure 15-3 shows that the cord is divided into right and left sides by two grooves. One of these, the **anterior median fissure,** is a deep, wide groove on the anterior (ventral) surface. The other is the **posterior median sulcus,** a shallower, narrow groove on the posterior (dorsal) surface.

PROTECTION AND COVERINGS
Vertebral Canal

The spinal cord is located in the vertebral canal of the vertebral column. The vertebral canal is formed by the foramina of all the vertebra arranged on top of each other. Since the wall of the vertebral canal is essentially a ring of bone surrounding the spinal cord, the cord is well protected. A certain degree of protection is also provided by the meninges, the cerebrospinal fluid, and the vertebral ligaments.

Meninges

The **meninges** (me-NIN-jēz) are coverings that run continuously around the spinal cord and brain. Those associated specifically with the cord are known as spinal meninges (Figure 15-2). The outer spinal meninx is called the **dura mater** (DYOO-ra MĀ-ter), meaning tough

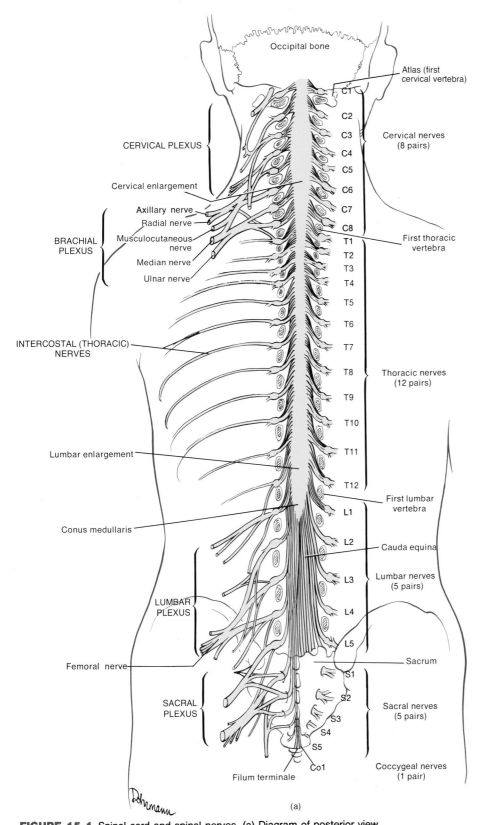

Occipital bone

Atlas (first
cervical vertebra)

C1
C2
C3
C4
C5
C6
C7
C8

Cervical nerves
(8 pairs)

CERVICAL PLEXUS

Cervical enlargement

Axillary nerve
Radial nerve
Musculocutaneous
nerve
Median nerve
Ulnar nerve

BRACHIAL
PLEXUS

First thoracic
vertebra

T1
T2
T3
T4
T5
T6
T7
T8
T9
T10
T11
T12

Thoracic nerves
(12 pairs)

INTERCOSTAL (THORACIC)
NERVES

Lumbar enlargement

Conus medullaris

First lumbar
vertebra

L1
L2
L3
L4
L5

Cauda equina

Lumbar nerves
(5 pairs)

LUMBAR
PLEXUS

Femoral nerve

Sacrum

S1
S2
S3
S4
S5

Sacral nerves
(5 pairs)

SACRAL
PLEXUS

Co1

Filum terminale

Coccygeal nerves
(1 pair)

(a)

FIGURE 15-1 Spinal cord and spinal nerves. (a) Diagram of posterior view.

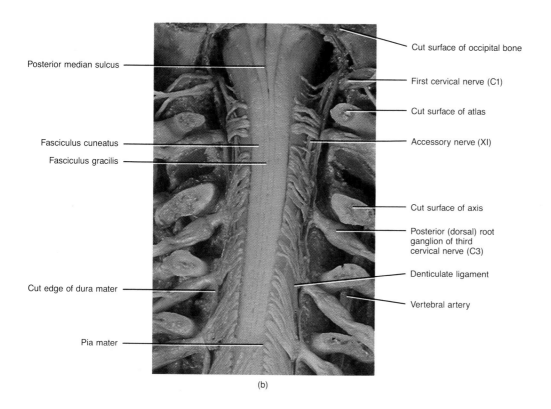

Posterior median sulcus

Fasciculus cuneatus

Fasciculus gracilis

Cut edge of dura mater

Pia mater

Cut surface of occipital bone

First cervical nerve (C1)

Cut surface of atlas

Accessory nerve (XI)

Cut surface of axis

Posterior (dorsal) root ganglion of third cervical nerve (C3)

Denticulate ligament

Vertebral artery

(b)

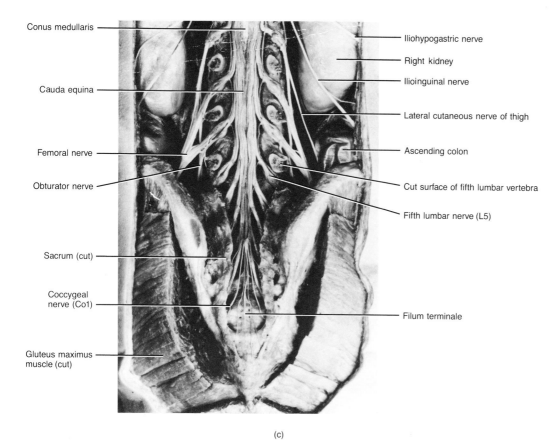

Conus medullaris

Cauda equina

Femoral nerve

Obturator nerve

Sacrum (cut)

Coccygeal nerve (Co1)

Gluteus maximus muscle (cut)

Iliohypogastric nerve

Right kidney

Ilioinguinal nerve

Lateral cutaneous nerve of thigh

Ascending colon

Cut surface of fifth lumbar vertebra

Fifth lumbar nerve (L5)

Filum terminale

(c)

FIGURE 15-1 (*Continued*) Spinal cord and spinal nerves. (b) Photograph of the posterior aspect of the inferior portion of the medulla and the superior six segments of the cervical portion of the spinal cord. (c) Photograph of the posterior aspect of the conus medullaris and cauda equina. (Courtesy of N. Gluhbegovic and T. H. Williams, *The Human Brain: A Photographic Guide*, Harper & Row, New York, 1980.)

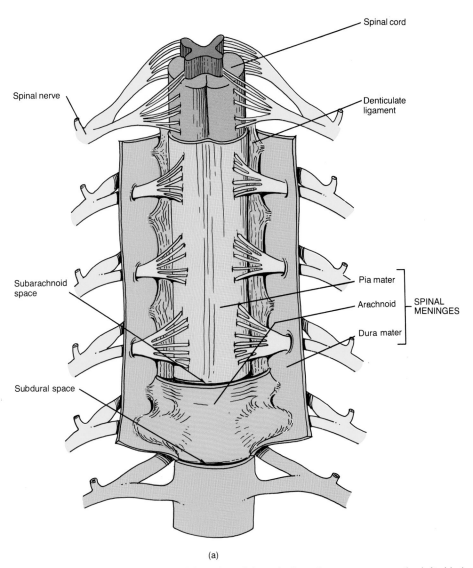

Spinal cord

Spinal nerve

Denticulate
ligament

Subarachnoid
space

Pia mater

Arachnoid   SPINAL
MENINGES

Dura mater

Subdural space

(a)

FIGURE 15-2 Spinal meninges. (a) Location of the spinal meninges as seen on the left side in a cross section of the spinal cord.

mother, and forms a tube from the level of the second sacral vertebra, where it is fused with the filum terminale, to the foramen magnum, where it is continuous with the dura mater of the brain. The dura mater is composed of dense, fibrous connective tissue. Between the dura mater and the wall of the vertebral canal is the *epidural space,* which is filled with fat, connective tissue, and blood vessels. It serves as padding around the cord. The epidural space inferior to the second lumbar vertebra is the site for the injection of anesthetics, such as a saddle-block for childbirth.

The middle spinal meninx is called the **arachnoid** (a-RAK-noyd), or spider layer. It is a delicate connective tissue membrane that forms a tube inside the dura mater. It is also continuous with the arachnoid of the brain. Between the dura mater and the arachnoid is the *subdural space,* which contains serous fluid.

The inner meninx is known as the **pia mater** (PĒ-a MĀ-ter), or delicate mother. It is a transparent fibrous membrane that forms a tube around and adheres to the surface of the spinal cord and brain. It contains numerous blood vessels. Between the arachnoid and the pia mater is the *subarachnoid space,* where the cerebrospinal fluid circulates.

All three spinal meninges cover the spinal nerves up to the point of exit from the spinal column through the intervertebral foramina. The spinal cord is suspended in the middle of its dural sheath by membranous extensions of the pia mater. These extensions, called the *denticulate* (den-TIK-yoo-lāt) *ligaments,* are attached laterally to the dura mater along the length of the cord between the ventral and dorsal nerve roots on either side. The ligaments protect the spinal cord against shock and sudden displacement.

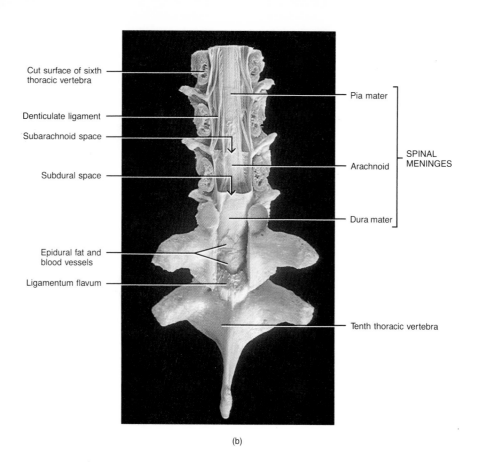

(b)

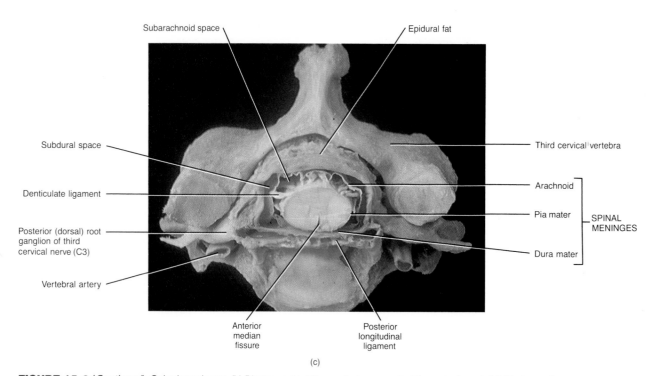

(c)

FIGURE 15-2 (*Continued*) Spinal meninges. (b) Photograph of the posterior aspect of the spinal cord. (c) Photograph of a cross section through the spinal cord between the second and third thoracic vertebrae. (Courtesy of N. Gluhbegovic and T. H. Williams, *The Human Brain: A Photographic Guide,* Harper & Row, New York, 1980.)

CLINICAL APPLICATION

Inflammation of the meninges is known as **meningitis.** If only the dura mater becomes inflamed, the condition is **pachymeningitis.** Inflammation of the arachnoid and pia mater is **leptomeningitis,** the most common form of meningitis.

Cerebrospinal fluid is removed from the subarachnoid space in the inferior lumbar region of the spinal cord by a **spinal (lumbar) puncture,** or **tap.** The procedure is normally performed between the third and fourth or fourth and fifth lumbar vertebrae. The spinous process of the fourth lumbar vertebra is easily located by drawing a line across the highest points of the iliac crests. This line will pass right through the spinous process of the fourth lumbar vertebra. A lumbar puncture is below the spinal cord and thus poses little danger to it. If the patient lies on one side, drawing the knees and chest together, the vertebrae separate slightly so that a needle can be conveniently inserted. In its course the needle pierces the skin, superficial fascia, supraspinous ligament, interspinous ligament, epidural space, and arachnoid, to enter the subarachnoid space. Lumbar punctures are used to withdraw fluid for diagnostic purposes and to introduce antibiotics (as in the case of meningitis) and radiopaque dyes.

STRUCTURE IN CROSS SECTION

The spinal cord consists of both gray and white matter. Figure 15-3 shows that the gray matter lies in an area shaped like an H. The gray matter consists primarily of nerve cell bodies and unmyelinated axons and dendrites of association and motor neurons. The white matter surrounds the gray matter and consists of bundles of myelinated axons of motor and sensory neurons.

In the center of the gray matter the **gray commissure** (KOM-mi-shyoor), forms the cross bar of the H connecting the right and left portions. In the center of the gray commissure is a small space called the **central canal.** This canal runs the length of the spinal cord and is continuous with the fourth ventricle of the medulla. It contains cerebrospinal fluid. Anterior to the gray commissure is the **anterior (ventral) white commissure,** which connects the white matter of the right and left sides of the spinal cord. The upright portions of the H are further subdivided into regions. Those closer to the front of the cord are called **anterior (ventral) gray horns.** They represent the motor part of the gray matter. The regions closer to the back of the cord are referred to as **posterior (dorsal) gray horns.** They represent the sensory part of the gray matter. The regions between the anterior and posterior gray horns are intermediate **lateral gray horns.** The lateral gray horns are present in the thoracic, upper lumbar, and sacral segments of the cord.

The gray matter of the cord also contains several nuclei that serve as relay stations for impulses and origins for certain nerves. Nuclei are clusters of nerve cell bodies and dendrites in the spinal cord and brain.

The white matter, like the gray matter, is organized into regions. The anterior and posterior gray horns divide the white matter on each side into three broad areas: **anterior (ventral) white columns, posterior (dorsal) white columns,** and **lateral white columns.** Each column (or *funiculus*) in turn consists of distinct bundles of myelinated fibers that run the length of the cord. These bundles are called **fasciculi** (fa-SIK-yoo-lī) or **tracts.** The longer **ascending tracts** consist of sensory axons that conduct

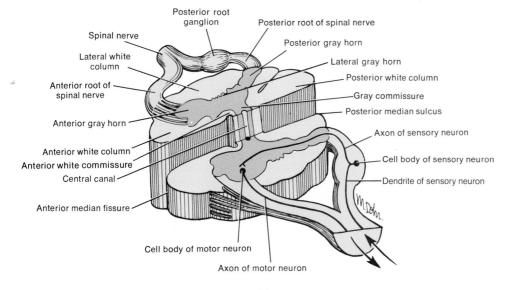

Posterior root
ganglion
Spinal nerve
Lateral white
column
Anterior root of
spinal nerve
Anterior gray horn
Anterior white column
Anterior white commissure
Central canal
Anterior median fissure

Posterior root of spinal nerve
Posterior gray horn
Lateral gray horn
Posterior white column
Gray commissure
Posterior median sulcus
Axon of sensory neuron
Cell body of sensory neuron
Dendrite of sensory neuron

Cell body of motor neuron
Axon of motor neuron

(a)

FIGURE 15-3 Spinal cord. (a) The organization of gray and white matter in the spinal cord as seen in cross section. The front of the figure has been sectioned at a lower level than the back so that you can see what is inside the posterior root ganglion, posterior root of the spinal nerve, anterior root of the spinal nerve, and the spinal nerve.

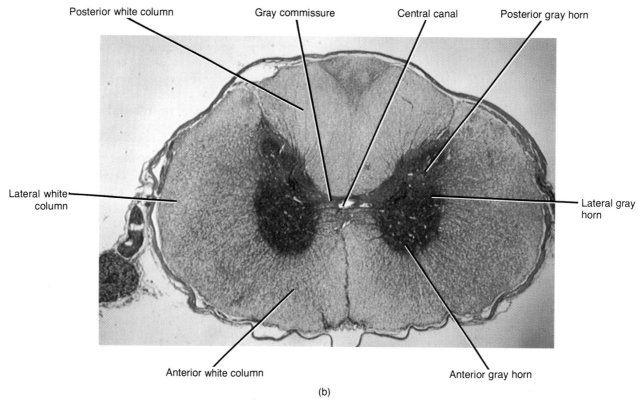

Posterior white column   Gray commissure   Central canal   Posterior gray horn

Lateral white column

Lateral gray horn

Anterior white column

Anterior gray horn

(b)

FIGURE 15-3 (*Continued*) Spinal cord. (b) Photograph of the spinal cord at the seventh cervical segment. (Courtesy of Victor B. Eichler, Wichita State University.)

impulses which enter the spinal cord upward to the brain. The longer **descending tracts** consist of motor axons that conduct impulses from the brain downward into the spinal cord where they synapse with other neurons whose axons pass out to muscles and glands. Thus the ascending tracts are sensory tracts and the descending tracts are motor tracts. Still other short tracts contain ascending or descending axons that convey impulses from one level of the cord to another.

FUNCTIONS

A major function of the spinal cord is to convey sensory impulses from the periphery to the brain and to conduct motor impulses from the brain to the periphery. A second principal function is to provide a means of integrating reflexes. Both functions are essential to maintaining homeostasis.

Conduction Pathway

The vital function of conveying sensory and motor information is carried out by the ascending and descending tracts of the cord. The names of the tracts indicate the white column (funiculus) in which the tract travels, where the cell bodies of the tract originate, and where the axons

of the tract terminate. Since the origin and termination are specified, the direction of impulse conduction is also indicated by the name. For example, the anterior spinothalamic tract is located in the *anterior* white column, it originates in the *spinal* cord, and it terminates in the *thalamus* (a region of the brain). It is an ascending (sensory) tract since it conveys impulses from the cord upward to the brain.

The principal ascending and descending tracts are listed in Exhibit 15-1 and shown in Figure 15-4.

Reflex Center

The second principal function of the spinal cord is to provide reflexes. Spinal nerves are the paths of communication between the spinal cord tracts and the periphery. Figure 15-3 reveals that each pair of spinal nerves is connected to a segment of the cord by two points of attachment called roots. The **posterior** or **dorsal (sensory) root** contains sensory nerve fibers only and conducts impulses from the periphery to the spinal cord. These fibers extend into the posterior (dorsal) gray horn. Each dorsal root also has a swelling, the **posterior** or **dorsal (sensory) root ganglion,** which contains the cell bodies of the sensory neurons from the periphery. The other point of attachment of a spinal nerve to the cord is the **anterior** or

ventral (motor) root. It contains motor nerve fibers only and conducts impulses from the spinal cord to the periphery.

The cell bodies of the motor neurons are located in the gray matter of the cord. If the motor impulse supplies a skeletal muscle, the cell bodies are located in the anterior (ventral) gray horn. If, however, the impulse supplies smooth muscle, cardiac muscle, or a gland through the autonomic nervous system, the cell bodies are located in the lateral gray horn.

Reflex Arc

The path an impulse follows from its origin in the dendrites or cell body of a neuron in one part of the body to its termination elsewhere in the body is called a **conduction pathway.** All conduction pathways consist of circuits of neurons. One pathway is known as a **reflex arc,** the functional unit of the nervous system. A reflex arc contains two or more neurons over which impulses are conducted from a receptor to the brain or spinal cord and then to an effector. The basic components of a reflex arc are as follows (see Figure 15-5).

1. Receptor. The distal end of a dendrite or a sensory structure associated with the distal end of a dendrite. Its role in the reflex arc is to respond to a change in the internal or external environment by initiating a nerve impulse in a sensory neuron.

2. Sensory neuron. Passes the impulse from the receptor to its axonal termination in the central nervous system.

3. Center. A region, usually in the central nervous system, where an incoming sensory impulse generates an outgoing motor impulse. In the center, the impulse may be inhibited, transmitted, or rerouted. In the center of some reflex arcs, the sensory neuron directly generates the impulse in the motor neuron. The center may also contain an association neuron between the sensory neuron and the motor neuron leading to a muscle or a gland.

4. Motor neuron. Transmits the impulse generated by the sensory or association neuron in the center to the organ of the body that will respond.

5. Effector. The organ of the body, either muscle or gland, that responds to the motor impulse. This response is called a **reflex action** or **reflex.**

Reflexes are fast responses to changes in the internal or external environment that allow the body to maintain homeostasis. Reflexes carried out by the spinal cord alone are called **spinal reflexes.** Reflexes that result in the contraction of skeletal muscles are known as **somatic reflexes.** Those that cause the contraction of smooth or cardiac muscle or secretion by glands are **visceral (autonomic) reflexes.** Our concern at this point is to examine a few somatic spinal reflexes: the stretch reflex, the flexor reflex, and the crossed extensor reflex.

● *Stretch Reflex* The **stretch reflex** is based on a *two-neuron,* or *monosynaptic, reflex arc.* Only two neurons are involved, and there is only one synapse in the pathway (Figure 15-5). This reflex results in the contraction of a muscle when it is stretched. Slight stretching of a muscle stimulates receptors in the muscle called *neuromuscular spindles.* The spindles monitor changes in the length of the muscle. Once the spindle is stimulated, an impulse is sent along a sensory neuron to the spinal cord. The sensory neuron lies in the posterior root of a spinal nerve and synapses with a motor neuron in the anterior gray horn. The sensory neuron generates an impulse at the synapse that is transmitted along the motor neuron. The motor neuron lies in the anterior root of the spinal nerve and terminates in a skeletal muscle. Once the impulse reaches the stretched muscle, it contracts. Thus the stretch is counteracted by contraction.

Since the sensory impulse enters the spinal cord on the same side that the motor impulse leaves the spinal cord, the reflex arc is called an *ipsilateral* (ip'-si-LAT-er-al) *reflex arc.* All monosynaptic reflex arcs are ipsilateral and are deep tendon reflexes.

The stretch reflex is essential in maintaining muscle tone. Moreover, it is the basis for several tests used in neurological examinations. One such reflex is the *knee jerk,* or *patellar reflex.* This reflex is tested by tapping the patellar ligament (stimulus). Neuromuscular spindles in the quadriceps femoris muscle attached to the ligament send the sensory impulse to the spinal cord, and the returning motor impulse causes contraction of the muscle. The response is extension of the leg at the knee, or a knee jerk.

● *Flexor Reflex and Crossed Extensor Reflex* Reflexes other than stretch reflexes involve association neurons in addition to the sensory and motor neurons. Since more than two neurons are involved, there is more than one synapse, and so these reflexes are *polysynaptic reflex arcs.* One example of a reflex based on a polysynaptic reflex arc is the **flexor reflex,** or **withdrawal reflex** (Figure 15-6). Suppose you step on a tack. As a result of the painful stimulus, you immediately withdraw your foot. What has happened? A sensory neuron transmits an impulse from the receptor to the spinal cord. A second impulse is generated in an association neuron, which generates a third impulse in a motor neuron. The motor neuron stimulates the muscles of your foot and you withdraw it. Thus a flexor reflex is protective in that it results in the movement of an extremity to avoid pain.

The flexor reflex, like the stretch reflex, is also ipsilateral. The incoming and outgoing impulses are on the same side of the spinal cord. The flexor reflex also illustrates another feature of polysynaptic reflex arcs. In the monosynaptic stretch reflex, the returning motor impulse affects only the quadriceps muscle of the thigh. When

EXHIBIT 15-1 **SELECTED ASCENDING AND DESCENDING TRACTS OF SPINAL CORD (Figure 15-4)**

TRACT	LOCATION (WHITE COLUMN)	ORIGIN	TERMINATION	FUNCTION
ASCENDING TRACTS				
Anterior (ventral) spinothalamic	Anterior (ventral) column.	Posterior (dorsal) gray horn on one side of cord, but crosses to opposite site of thalamus.	Thalamus. Impulse eventually conveyed to cerebral cortex.	Conveys sensations for crude touch and pressure from one side of body to opposite side of thalamus. Eventually sensations reach cerebral cortex.
Lateral spinothalamic	Lateral column.	Posterior (dorsal) gray horn on one side of cord, but crosses to opposite side of thalamus.	Thalamus. Impulse eventually conveyed to cerebral cortex.	Conveys sensations for pain and temperature from one side of body to opposite side of thalamus. Eventually sensations reach cerebral cortex.
Fasciculus gracilis and fasciculus cuneatus	Posterior (dorsal) column.	Axons of afferent neurons from periphery that enter posterior (dorsal) column on one side of cord and rise to same side of medulla.	Nucleus gracilis and nucleus cuneatus of medulla. Impulse eventually conveyed to cerebral cortex.	Convey sensations from one side of body to same side of medulla for fine touch; two-point discrimination (ability to distinguish that two points on skin are touched even though close together); proprioception (awareness of precise position of body parts and their direction of movement); stereognosis (ability to recognize size, shape, and texture of object); weight discrimination (ability to assess weight of an object); and vibrations. Eventually sensations reach cerebral cortex.
Posterior (dorsal) spinocerebellar	Posterior (dorsal) portion of lateral column.	Posterior (dorsal) gray horn on one side of cord and rises on same side.	Cerebellum.	Conveys sensations from one side of body to same side of cerebellum for subconscious proprioception.
Anterior (ventral) spinocerebellar	Anterior (ventral) column.	Posterior (dorsal) gray horn on one side of cord; tract contains both crossed and uncrossed fibers.	Cerebellum.	Conveys sensations from both sides of body to cerebellum for subconscious proprioception.

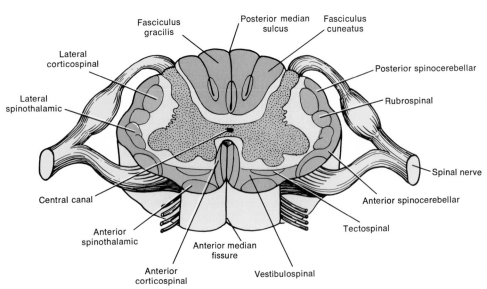

FIGURE 15-4 Selected tracts of the spinal cord. Ascending (sensory) tracts are indicated in orange; descending (motor) tracts are shown in purple.

EXHIBIT 15-1 (*Continued*)

TRACT	LOCATION (WHITE COLUMN)	ORIGIN	TERMINATION	FUNCTION
DESCENDING TRACTS				
Lateral corticospinal	Lateral column.	Cerebral cortex on one side of brain, but crosses in base of medulla to opposite side of cord.	Anterior (ventral) gray horn.	Conveys motor impulses from one side of cortex to anterior gray horn of opposite side. Eventually impulses reach skeletal muscles on opposite side of body that coordinate precise, discrete movements.
Anterior (ventral) corticospinal	Anterior (ventral) column.	Cerebral cortex on one side of brain, uncrossed in medulla, but crosses to opposite side of cord.	Anterior (ventral) gray horn.	Conveys motor impulses from one side of cortex to anterior gray horn of opposite side. Eventually impulses reach skeletal muscles on opposite side of body that coordinate precise, discrete movements.
Rubrospinal	Lateral column.	Midbrain (red nucleus) on one side of brain, but crosses to opposite side of cord.	Anterior (ventral) gray horn.	Conveys motor impulses from one side of midbrain to skeletal muscles on opposite side of body that are concerned with muscle tone and posture.
Tectospinal	Anterior (ventral) column.	Midbrain on one side of brain, but crosses to opposite side of cord.	Anterior (ventral) gray horn.	Conveys motor impulses from one side of midbrain to skeletal muscles on opposite side of body that control movements of head in response to auditory, visual, and cutaneous stimuli.
Vestibulospinal	Anterior (ventral) column.	Medulla on one side of brain and descends on same side of cord.	Anterior (ventral) gray horn.	Conveys motor impulses from one side of medulla to skeletal muscles on same side of body that regulate body tone in response to movements of head (equilibrium).

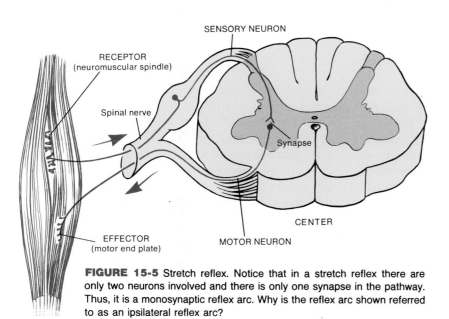

FIGURE 15-5 Stretch reflex. Notice that in a stretch reflex there are only two neurons involved and there is only one synapse in the pathway. Thus, it is a monosynaptic reflex arc. Why is the reflex arc shown referred to as an ipsilateral reflex arc?

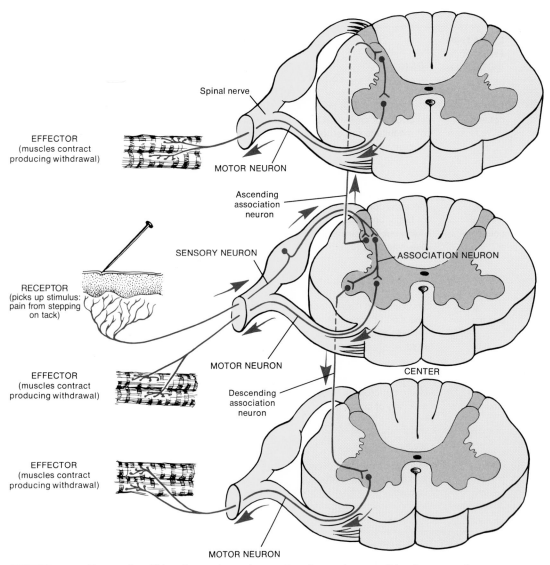

EFFECTOR
(muscles contract
producing withdrawal)

Spinal nerve

MOTOR NEURON

Ascending
association
neuron

SENSORY NEURON

ASSOCIATION NEURON

RECEPTOR
(picks up stimulus:
pain from stepping
on tack)

MOTOR NEURON

EFFECTOR
(muscles contract
producing withdrawal)

Descending
association
neuron

CENTER

EFFECTOR
(muscles contract
producing withdrawal)

MOTOR NEURON

FIGURE 15-6 Flexor reflex. This reflex arc is a polysynaptic reflex arc because it involves more than one synapse; it contains association neurons as well as sensory and motor neurons. It is also ipsilateral. Why is the reflex arc shown also an intersegmental reflex arc?

you withdraw your entire lower or upper extremity from a noxious stimulus, more than one muscle is involved, and several motor neurons are simultaneously returning impulses to several upper and lower extremity muscles at the same time. Thus a single sensory impulse causes several motor responses. This kind of reflex arc, in which a single sensory neuron splits into ascending and descending branches, each forming a synapse with association neurons at different segments of the cord, is called an *intersegmental reflex arc.* Because of intersegmental reflex arcs, a single sensory neuron can activate several motor neurons and thereby cause stimulation of more than one effector.

Something else may happen when you step on a tack. You may lose your balance as your body weight shifts to the other foot. Then you do whatever you can to regain

your balance so you do not fall. This means motor impulses are also sent to your unstimulated foot and both upper extremities. The motor impulses that travel to your unaffected foot cause extension at the knee so you can place your entire body weight on the foot. These impulses cross the spinal cord as shown in Figure 15-7. The incoming sensory impulse not only initiates the flexor reflex that causes you to withdraw, it also initiates an extensor reflex. The incoming sensory impulse crosses to the opposite side of the spinal cord through association neurons at that level and several levels above and below the point of sensory stimulation. From these levels, the motor neurons cause extension of the knee, thus maintaining balance. Unlike the flexor reflex, which passes over an ipsilateral reflex arc, the extensor reflex passes over a *contralateral* (kon'-tra-LAT-er-al) *reflex arc*—the impulse

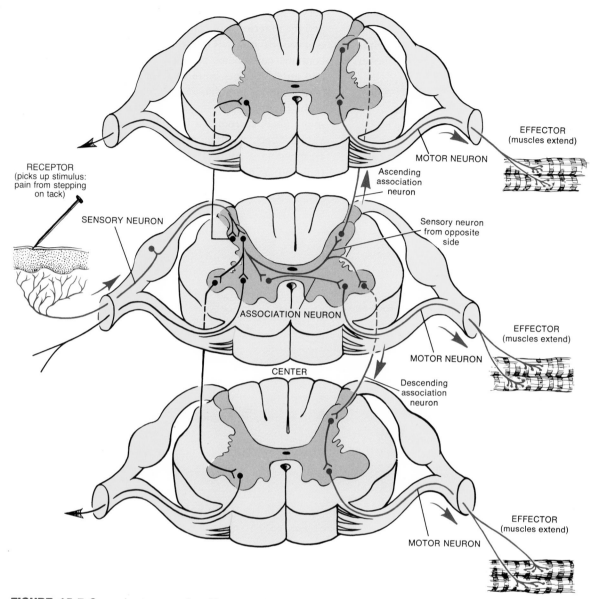

EFFECTOR
(muscles extend)

MOTOR NEURON

Ascending
association
neuron

RECEPTOR
(picks up stimulus:
pain from stepping
on tack)

SENSORY NEURON

Sensory neuron
from opposite
side

ASSOCIATION NEURON

EFFECTOR
(muscles extend)

CENTER

MOTOR NEURON

Descending
association
neuron

EFFECTOR
(muscles extend)

MOTOR NEURON

FIGURE 15-7 Crossed extensor reflex. The crossed extensor reflex is shown on the right side of the diagram. The flexor reflex is shown on the left so that you can correlate it with the crossed extensor reflex. Why is the crossed extensor reflex classified as a contralateral reflex arc?

enters one side of the spinal cord and exits on the opposite side. The reflex just described, in which extension of the muscles in one limb occurs as a result of contraction of the muscles of the opposite limb, is called a **crossed extensor reflex.**

The flexor reflex and crossed extensor reflex also illustrate *reciprocal inhibition,* another feature of many reflexes. Reciprocal inhibition occurs when a reflex excites a muscle to cause its contraction and also inhibits another muscle to allow its extension. Thus, in this reflex, excitation and inhibition occur simultaneously.

In the flexor reflex, when the flexor muscles of your lower extremity are contracting, the extensor muscles of the same extremity are being extended. If both sets of

muscles contracted at the same time, you would not be able to flex your limb because both sets of muscles would pull on the limb bones. But because of reciprocal inhibition, one set of muscles contracts while the other is being extended.

In the crossed extensor reflex, reciprocal inhibition also occurs. While you are flexing the muscles of the limb that has been stimulated by the tack, the muscles of your other limb are producing extension to help maintain balance. Reciprocal inhibition is vital in coordinating body movements. You may recall from Chapter 10 that skeletal muscles act in groups rather than alone and that each muscle in the group has a specific role in bringing about the movement. In flexing the forearm at the elbow, there is a prime mover, an antagonist, and a synergist. The prime mover (biceps) contracts to cause flexion, the antagonist (triceps) extends to yield to the action of the prime mover, and the synergist (deltoid) helps the prime mover perform its role efficiently.

SPINAL NERVES

COMPOSITION AND COVERINGS

A **spinal nerve** has two points of attachment to the cord: a posterior root and an anterior root. A short distance from the spinal cord, the roots unite to form a spinal nerve. Since the posterior root contains sensory fibers and the anterior root contains motor fibers, a spinal nerve is a *mixed nerve*. The posterior (dorsal) root ganglion contains cell bodies of sensory neurons. The posterior and anterior roots unite to form the spinal nerve at the intervertebral foramen.

In Figure 15-8, you can see that the nerve contains many fibers surrounded by different coverings. The individual fibers, whether myelinated or unmyelinated, are wrapped in a connective tissue called the **endoneurium** (en'-dō-NYOO-rē-um). Groups of fibers with their endoneurium are arranged in bundles called fascicles, and each

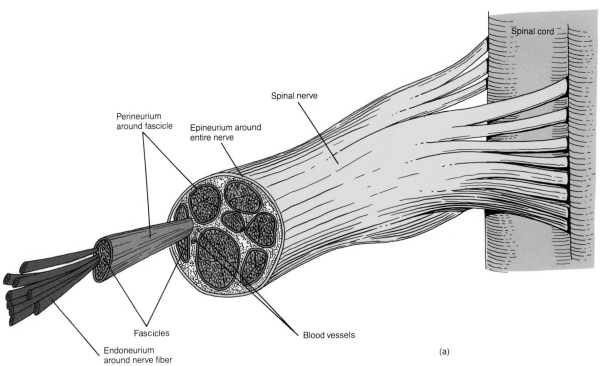

Spinal cord

Spinal nerve

Perineurium around fascicle

Epineurium around entire nerve

Fascicles

Blood vessels

Endoneurium around nerve fiber

(a)

FIGURE 15-8 Coverings of a spinal nerve. (a) Diagram.

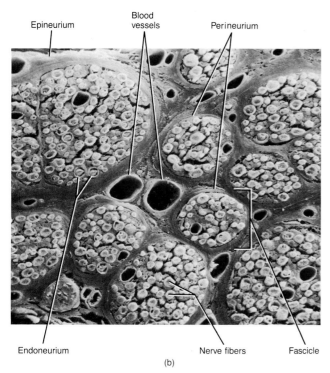

Epineurium   Blood vessels   Perineurium

Endoneurium   Nerve fibers   Fascicle

(b)

FIGURE 15-8 (*Continued*) Coverings of a spinal nerve. (b) Scanning electron micrograph at a magnification of 900×. (Courtesy of Richard G. Kessel and Randy H. Kardon, from *Tissues and Organs: A Text-Atlas of Scanning Electron Microscopy,* W. H. Freeman and Company, San Francisco, 1979.)

bundle is wrapped in connective tissue called the **perineurium** (per'-i-NYOO-rē-um). The outermost covering around the entire nerve is the **epineurium** (ep'-i-NYOO-rē-um). The spinal meninges fuse with the epineurium as the nerve exits from the vertebral canal.

NAMES

The 31 pairs of spinal nerves are named and numbered according to the region and level of the spinal cord from which they emerge (see Figure 15-1). The first cervical pair emerges between the atlas and the occipital bone. All other spinal nerves leave the vertebral column from the intervertebral foramina between adjoining vertebrae. There are 8 pairs of cervical nerves, 12 pairs of thoracic nerves, 5 pairs of lumbar nerves, 5 pairs of sacral nerves, and 1 pair of coccygeal nerves.

During fetal life, the spinal cord and vertebral column grow at different rates, the cord growing more slowly. Thus not all the spinal cord segments are in line with their corresponding vertebrae. Remember that the spinal cord terminates near the level of the first or second lumbar vertebra. Thus the lower lumbar, sacral, and coccygeal nerves must descend more and more to reach their foramina before emerging from the vertebral column. This arrangement constitutes the cauda equina.

BRANCHES

Shortly after a spinal nerve leaves its intervertebral foramen, it divides into several branches (Figure 15-9). These branches are known as **rami** (RĀ-mī). The **dorsal ramus** (RĀ-mus) innervates (supplies) the deep muscles and skin of the dorsal surface of the back. The **ventral ramus** of a spinal nerve innervates the superficial back muscles and all the structures of the extremities and the lateral and ventral trunk. Except for thoracic nerves T2 to T12, the ventral rami of other spinal nerves enter into the formation of plexuses before supplying a part of the body. In addition to dorsal and ventral rami, spinal nerves also give off a **meningeal branch.** This branch reenters the spinal canal through the intervertebral foramen and supplies the vertebrae, vertebral ligaments, blood vessels of the spinal cord, and the meninges. Other branches of a spinal nerve are the **rami communicantes** (ko-myoo-nē-KAN-tēz), components of the autonomic nervous system whose structure and function are discussed in Chapter 16.

PLEXUSES

The ventral rami of spinal nerves, except for T2 to T12, do not go directly to the structures of the body they supply. Instead, they form networks by joining with adjacent nerves on either side of the body. Such a network is called a **plexus.** The principal plexuses are the cervical plexus, the brachial plexus, the lumbar plexus, and the sacral plexus. Emerging from the plexuses are nerves bearing names that are often descriptive of the general regions they supply or the course they take. Each of the nerves, in turn, may have several branches named for the specific structures they innervate.

Cervical Plexus

The **cervical plexus** is formed by the ventral rami of the first four cervical nerves (C1 to C4) with contributions from C5. There is one on each side of the neck alongside the first four cervical vertebrae (Figure 15-10). The *roots* of the plexus indicated in the diagram are the ventral rami. The cervical plexus supplies the skin and muscles of the head, neck, and upper part of the shoulders. Branches of the cervical plexus also connect with cranial nerves XI (accessory) and XII (hypoglossal). A major pair of nerves arising from the cervical plexuses are the phrenic nerves supplying motor fibers to the diaphragm. Damage to the spinal cord above the origin of the phrenic nerves results in paralysis of the diaphragm since the phrenic nerves no longer send impulses to the diaphragm. Contractions of the diaphragm are essential for normal breathing.

Exhibit 15-2 summarizes the nerves and distributions of the cervical plexus. The relationship of the cervical plexus to the other plexuses is shown in Figure 15-1.

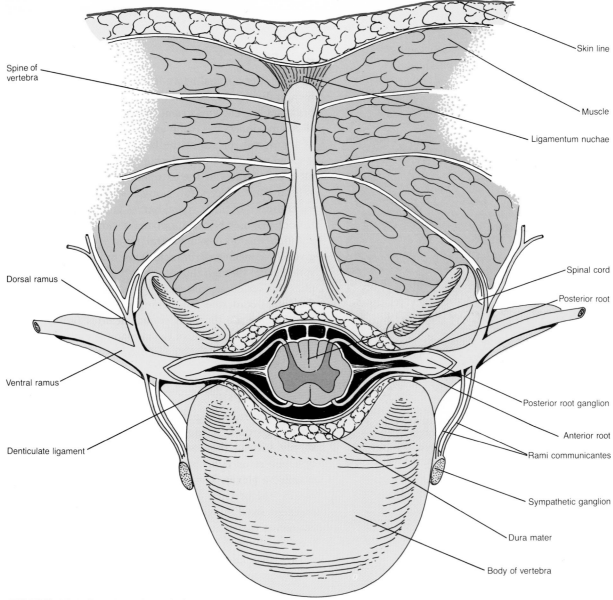

Spine of vertebra

Skin line

Muscle

Ligamentum nuchae

Dorsal ramus

Spinal cord

Posterior root

Ventral ramus

Posterior root ganglion

Anterior root

Rami communicantes

Denticulate ligament

Sympathetic ganglion

Dura mater

Body of vertebra

FIGURE 15-9 Branches of a typical spinal nerve.

EXHIBIT 15-2 CERVICAL PLEXUS

NERVE	ORIGIN	DISTRIBUTION
SUPERFICIAL OR CUTANEOUS BRANCHES		
Lesser occipital	C2–C3.	Skin of scalp behind and above ear.
Greater auricular	C2–C3.	Skin in front, below, and over ear and over parotid glands.
Transverse cervical	C2–C3.	Skin over anterior aspect of neck.
Supraclavicular	C3–C4.	Skin over upper portion of chest and shoulder.
DEEP OR LARGELY MOTOR BRANCHES		
Ansa cervicalis		This nerve is divided into a superior root and an inferior root.
Superior root	C1–C2.	Infrahyoid, thyrohyoid, and geniohyoid muscles of neck.
Inferior root	C3–C4.	Omohyoid, sternohyoid, and sternothyroid muscles of neck.
Phrenic	C3–C5.	Diaphragm between thorax and abdomen.
Segmental branches	C1–C5.	Prevertebral (deep) muscles of neck, levator scapulae, and middle scalene muscles.

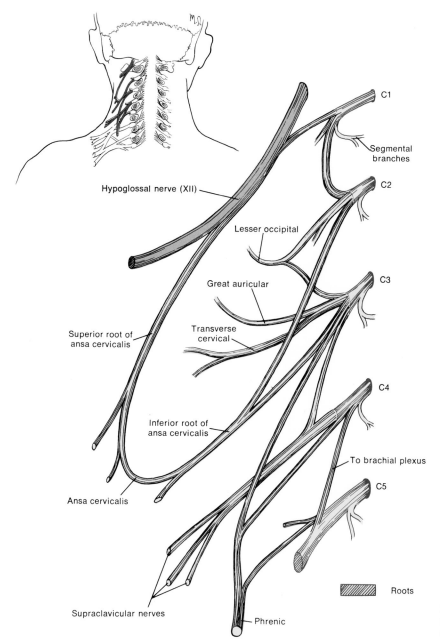

Segmental branches

C1

Hypoglossal nerve (XII)

C2

Lesser occipital

Great auricular

C3

Superior root of ansa cervicalis

Transverse cervical

C4

Inferior root of ansa cervicalis

To brachial plexus

C5

Ansa cervicalis

Supraclavicular nerves

Phrenic

Roots

FIGURE 15-10 Cervical plexus. Consult Exhibit 15-2 so that you can determine the distribution of each of the nerves of the plexus.

Brachial Plexus

The **brachial plexus** is formed by the ventral rami of spinal nerves C5 to C8 and T1 with contributions from C4 and T2. On either side of the last four cervical and first thoracic vertebrae, the brachial plexus extends downward and laterally, passes over the first rib behind the clavicle, and then enters the axilla (Figure 15-11). The brachial plexus constitutes the entire nerve supply for the upper extremities and shoulder region.

The *roots* of the brachial plexus, like those of the cervical plexus, are the ventral rami of the spinal nerves. The roots of C5 and C6 unite to form the *superior trunk,* C7 becomes the *middle trunk,* and C8 and T1 form the *inferior trunk.* Each trunk, in turn, divides into an *anterior*

division and a *posterior division.* The divisions then unite to form cords. The *posterior cord* is formed by the union of the posterior divisions of the superior, middle, and inferior trunks. The *medial cord* is formed as a continuation of the anterior division of the inferior trunk. The *lateral cord* is formed by the union of the anterior divisions of the superior and middle trunk. The peripheral nerves arise from the cords. Thus the brachial plexus begins as roots that unite to form trunks, the trunks branch into divisions, the divisions form cords, and the cords give rise to the peripheral nerves.

Three important nerves arising from the brachial plexus are the radial, median, and ulnar. The radial nerve supplies the muscles on the posterior aspect of the arm

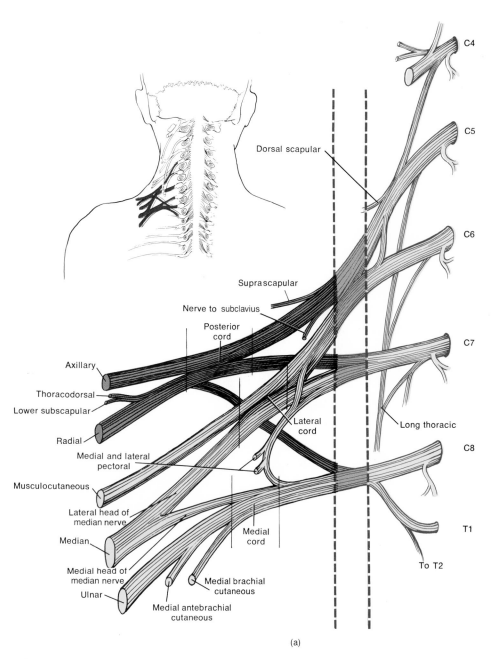

Roots

Trunks

Anterior divisions

Posterior divisions

FIGURE 15-11 Brachial plexus. (a) Diagram.

and forearm. The median nerve supplies most of the muscles of the anterior forearm and some of the muscles in the palm. The ulnar nerve supplies the anteromedial muscles of the forearm and most of the muscles of the palm.

A summary of the nerves and distributions of the brachial plexus is given in Exhibit 15-3. The relationship of the brachial plexus to the other plexuses is shown in Figure 15-1.

Lumbar Plexus

The **lumbar plexus** is formed by the ventral rami of spinal nerves L1 to L4. It differs from the brachial plexus in that there is no intricate interlacing of fibers. It also consists of *roots* and an *anterior* and *posterior division*. On either side of the first four lumbar vertebrae, the lumbar plexus passes obliquely outward behind the psoas major

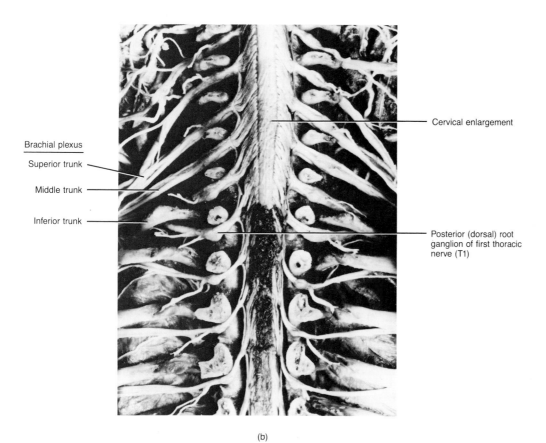

Cervical enlargement

Brachial plexus

Superior trunk

Middle trunk

Inferior trunk

Posterior (dorsal) root
ganglion of first thoracic
nerve (T1)

(b)

FIGURE 15-11 (*Continued*) Brachial plexus. (b) Photograph of the posterior aspect of the lower cervical and upper thoracic segments of the spinal cord showing the trunks of the brachial plexus. (Courtesy of N. Gluhbegovic and T. H. Williams, *The Human Brain: A Photographic Guide,* Harper & Row, New York, 1980.)

muscle (posterior division) and anterior to the quadratus lumborum muscle (anterior division) and then gives rise to its peripheral nerves (Figure 15-12). The lumbar plexus supplies the anterolateral abdominal wall, external genitals, and part of the lower extremity. The largest nerve arising from the lumbar plexus is the femoral nerve.

CLINICAL APPLICATION

Prolonged use of a crutch that presses into the axilla may result in injury to a portion of the brachial plexus. The usual **crutch palsy** involves the posterior cord of the brachial plexus or, more frequently, just the radial nerve, which, in general, supplies extensors.

Radial nerve damage is indicated by wrist drop: inability to extend the hand at the wrist. Care must be taken not to injure the radial nerve when deltoid intramuscular injections are given. **Median nerve damage** is indicated by inability to pronate the forearm, flex the wrist properly, and flex the fingers and thumb properly. **Ulnar nerve damage** is indicated by an inability to flex and adduct the wrist and difficulty in spreading the fingers.

CLINICAL APPLICATION

Injury to the femoral nerve is indicated by an inability to extend the leg and by loss of sensation in the skin over the anteromedial aspect of the thigh.

A summary of the nerves and distributions of the lumbar plexus is presented in Exhibit 15-4. The relationship of the lumbar plexus to the other plexuses is shown in Figure 15-1.

Sacral Plexus

The **sacral plexus** is formed by the ventral rami of spinal nerves L4 to L5 and S1 to S4. It is situated largely in front of the sacrum (Figure 15-13). Like the lumbar plexus, it contains *roots* and an *anterior* and *posterior division.* The sacral plexus supplies the buttocks, perineum, and lower extremities. The largest nerve arising from the sacral plexus—and, in fact, the largest nerve in the body—is the sciatic nerve. The sciatic nerve supplies the entire musculature of the leg and foot.

EXHIBIT 15-3 BRACHIAL PLEXUS

NERVE	ORIGIN	DISTRIBUTION
ROOT NERVES		
Dorsal scapular	C5.	Levator scapulae, rhomboideus major, and rhomboideus minor muscles.
Long thoracic	C5–C7.	Serratus anterior muscle.
TRUNK NERVES		
Nerve to subclavius	C5–C6.	Subclavius muscle.
Suprascapular	C5–C6.	Supraspinatus and infraspinatus muscles.
LATERAL CORD NERVES		
Musculocutaneous	C5–C7.	Coracobrachialis, biceps brachii, and brachialis muscles.
Median (lateral head)	C5–C7.	See distribution for Median (medial head).
Lateral pectoral	C5–C7.	Pectoralis major muscle.
POSTERIOR CORD NERVES		
Upper subscapular	C5–C6.	Subscapularis muscle.
Thoracodorsal	C6–C8.	Latissimus dorsi muscle.
Lower subscapular	C5–C6.	Subscapularis and teres major muscles.
Axillary (circumflex)	C5–C6.	Deltoid and teres minor muscles; skin over deltoid and upper posterior aspect of arm.
Radial	C5–C8 and T1.	Extensor muscles of arm and forearm (triceps brachii, brachioradialis, extensor carpi radialis longus, extensor digitorum, extensor carpi ulnaris, extensor indicis); skin of posterior arm and forearm, lateral two-thirds of dorsum of hand, and fingers over proximal and middle phalanges.
MEDIAL CORD NERVES		
Medial pectoral	C8–T1.	Pectoralis major and pectoralis minor muscles.
Medial brachial cutaneous	C8–T1.	Skin of medial and posterior aspects of lower third of arm.
Medial antebrachial cutaneous	C8–T1.	Skin of medial and posterior aspects of forearm.
Median (medial head)	C5–C8 and T1.	Medial and lateral heads of median nerve form median nerve. Distributed to flexors of forearm (pronator teres, flexor carpi radialis, flexor digitorum superficialis), except flexor carpi ulnaris and flexor digitorum profundus; skin of lateral two-thirds of palm of hand and fingers.
Ulnar	C8–T1.	Flexor carpi ulnaris and flexor digitorum profundus muscles; skin of medial side of hand, little finger, and medial half of ring finger.
OTHER CUTANEOUS DISTRIBUTIONS		
Intercostobrachial	Second intercostal nerve.	Skin over medial side of arm.
Upper lateral brachial cutaneous	Axillary.	Skin over deltoid muscle and down to elbow.
Posterior brachial cutaneous	Radial.	Skin over posterior aspect of arm.
Lower lateral brachial cutaneous	Radial.	Skin over lateral aspect of elbow.
Lateral antebrachial cutaneous	Musculocutaneous.	Skin over lateral aspect of forearm.
Posterior antebrachial cutaneous	Radial.	Skin over posterior aspect of forearm.

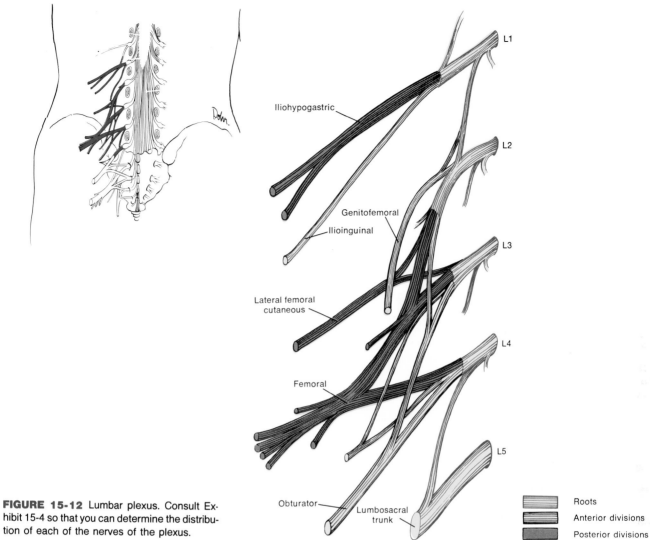

FIGURE 15-12 Lumbar plexus. Consult Exhibit 15-4 so that you can determine the distribution of each of the nerves of the plexus.

Iliohypogastric

Genitofemoral
Ilioinguinal

Lateral femoral cutaneous

Femoral

Obturator   Lumbosacral trunk

L1
L2
L3
L4
L5

Roots
Anterior divisions
Posterior divisions

EXHIBIT 15-4 LUMBAR PLEXUS

NERVE	ORIGIN	DISTRIBUTION
Iliohypogastric	T12–L1.	Muscles of anterolateral abdominal wall (external oblique, internal oblique, transversus abdominis); skin of lower abdomen and buttock.
Ilioinguinal	L1.	Muscles of anterolateral abdominal wall as indicated above; skin of upper medial aspect of thigh, root of penis and scrotum in male, and labia majora and mons pubis in female.
Genitofemoral	L1–L2.	Cremaster muscle; skin over middle anterior surface of thigh, scrotum in male, and labia majora in female.
Lateral femoral cutaneous	L2–L3.	Skin over lateral, anterior, and posterior aspects of thigh.
Femoral	L2–L4.	Flexor muscles of thigh (iliacus, psoas major, pectineus, rectus femoris, sartorius); extensor muscles of leg (rectus femoris, vastus lateralis, vastus medialis, vastus intermedius); skin on front and over medial aspect of thigh and medial side of leg and foot.
Obturator	L2–L4.	Adductor muscles of leg (obturator externus, pectineus, adductor longus, adductor brevis, adductor magnus, gracilis); skin over medial aspect of thigh.

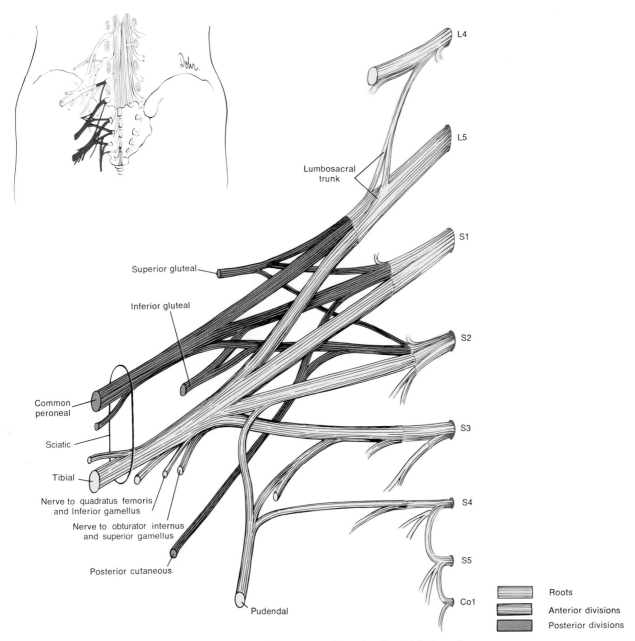

FIGURE 15-13 Sacral plexus. Consult Exhibit 15-5 so that you can determine the distribution of each of the nerves of the plexus.

Labels in figure:

L4
L5
Lumbosacral trunk
S1
Superior gluteal
Inferior gluteal
S2
Common peroneal
Sciatic
S3
Tibial
Nerve to quadratus femoris and Inferior gamellus
Nerve to obturator internus and superior gamellus
S4
Posterior cutaneous
S5
Co1
Pudendal

Roots
Anterior divisions
Posterior divisions

CLINICAL APPLICATION

Damage to the sciatic nerve and its branches results in foot drop, an inability to dorsiflex the foot. This nerve may be injured because of a slipped disc, dislocated hip, pressure from the uterus during pregnancy, or an improperly given gluteal intramuscular injection.

A summary of the nerves and distributions of the sacral plexus is given in Exhibit 15-5. The relationship of the sacral plexus to the other plexuses is shown in Figure 15-1.

INTERCOSTAL (THORACIC) NERVES

Spinal nerves T2 to T12 do not enter into the formation of plexuses. Nerves T2-T11 are known as **intercostal (thoracic) nerves** and are distributed directly to the structures they supply in intercostal spaces (see Figure 15-1). After leaving the vertebral foramina, the ventral ramus of nerve T2 supplies the intercostal muscles of the second intercostal space and the skin of the axilla and posteromedial aspect of the arm. Nerves T3 to T6 pass in the costal grooves of the ribs and are distributed to the intercostal muscles and skin of the anterior and lateral chest wall.

EXHIBIT 15-5 SACRAL PLEXUS

NERVE	ORIGIN	DISTRIBUTION
Superior gluteal	L4–L5 and S1.	Gluteus minimus and gluteus medius muscles and tensor fasciae latae.
Inferior gluteal	L5–S2.	Gluteus maximus muscle.
Nerve to piriformis	S1–S2.	Piriformis muscle.
Nerve to quadratus femoris	L4–L5 and S1.	Inferior gemellus and quadratus femoris muscles.
Nerve to obturator internus	L5–S2.	Superior gemellus and obturator internus muscles.
Perforating cutaneous	S2–S3.	Skin over lower medial aspect of buttock.
Posterior cutaneous	S1–S3.	Skin over anal region, lower lateral aspect of buttock, upper posterior aspect of thigh, upper part of calf, scrotum in male, and labia majora in female.
Sciatic	L4–S3.	Actually two nerves: tibial and common peroneal, bound together by common sheath of connective tissue. It splits into its two divisions, usually at knee. (See below for distributions.) As sciatic nerve descends through thigh, it sends branches to hamstring muscles (biceps femoris, semitendinosus, semimembranosus) and adductor magnus.
Tibial (medial popliteal)	L4–S3.	Gastrocnemius, plantaris, soleus, popliteus, tibialis posterior, flexor digitorum, and hallucis muscles. Branches of tibial nerve in foot are medial plantar nerve and lateral plantar nerve.
Medial plantar		Abductor hallucis, flexor digitorum brevis, and flexor hallucis muscles; skin over medial two-thirds of plantar surface of foot.
Lateral plantar		Remaining muscles of foot not supplied by medial plantar nerve; skin over lateral third of plantar surface of foot.
Common peroneal (lateral popliteal)	L4–S2.	Divides into a superficial peroneal and a deep peroneal branch.
Superficial peroneal		Peroneus longus and peroneus brevis muscles; skin over distal third of anterior aspect of leg and dorsum of foot.
Deep peroneal		Tibialis anterior, extensor hallucis longus, peroneus tertius, and extensor digitorum brevis muscles; skin over great and second toes.
Pudendal	S2–S4.	Muscles of perineum; skin of penis and scrotum in male and clitoris, labia majora, labia minora, and lower vagina in female.

Nerves T7 to T11 supply the intercostal muscles and the abdominal muscles and overlying skin. The dorsal rami of the intercostal nerves supply the deep back muscles and skin of the dorsal aspect of the thorax.

DERMATOMES

The skin over the entire body is supplied segmentally by spinal nerves. This means that the spinal nerves innervate specific, constant segments of the skin. With the exception of spinal nerve C1, all other spinal nerves supply branches to the skin. The skin segment supplied by the dorsal root of a spinal nerve is a **dermatome** (Figure 15-14).

In the neck and trunk, the dermatomes form consecutive bands of skin. In the trunk, there is an overlap of adjacent dermatome nerve supply. Thus there is little loss of sensation if only a single nerve supply to a dermatome is interrupted. Most of the skin of the face and scalp is supplied by the trigeminal (V) cranial nerve.

Since a physician knows that a particular dermatome is associated with particular spinal nerves, it is possible to determine which segment of the spinal cord or spinal nerve is malfunctioning. If a dermatome is stimulated and the sensation is not perceived, it can be assumed that the nerves supplying the dermatome are involved.

APPLICATIONS TO HEALTH
SPINAL CORD INJURY

The spinal cord may be damaged by fracture or dislocation of the vertebrae enclosing it or by wounds. All can result in **transection**—partial or complete severing of the spinal cord. Complete transection means that all ascend-

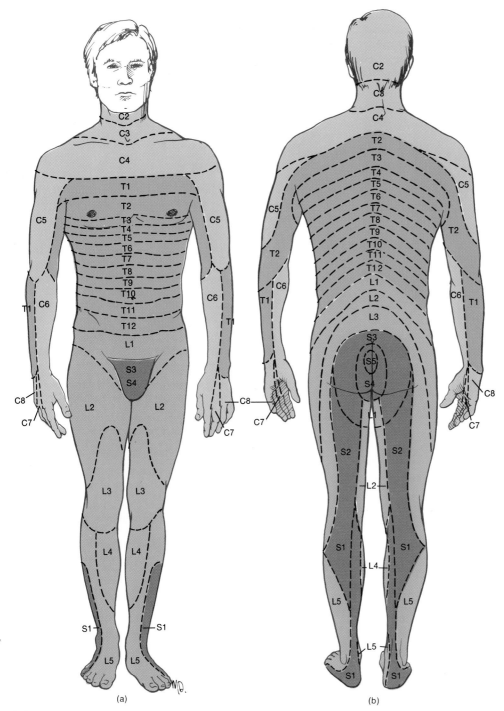

FIGURE 15-14 Distribution of spinal nerves to dermatomes. The lines are not perfectly aligned so that there is often considerable overlap. (a) Anterior view. (b) Posterior view.

ing and descending pathways are cut. It results in loss of all sensation and voluntary muscular movement below the level of transection. A complete cervical transection close to the base of the skull usually leads to death by asphyxiation before treatment can be administered because impulses from the phrenic nerves to the breathing muscles are interrupted. If the upper cervical cord is partially transected, both the upper and lower extremities are paralyzed and the patient is classified as *quadriplegic* (*plege* = stroke). Partial transection between the cervical and lumbar enlargements results in paralysis of the lower extremities only, and the patient is classified as *paraplegic*.

In the case of partial transection, **spinal shock** lasts from a few days to several weeks. During this period, all reflex activity is abolished, a condition called *areflexia* (a'-rē-FLEK-sē-a). In time, however, there is a return of reflex activity. The first reflex to return is the knee jerk. Its reappearance may take several days. Next the flexion reflexes return. This may take up to several months. Then the crossed extensor reflexes return. Visceral reflexes such as erection and ejaculation are also affected by transection. Moreover, urinary bladder and bowel functions are no longer under voluntary control.

Until recently, severe damage resulting from transection was thought to be irreversible. However, a team of researchers has developed a technique for regenerating severed spinal cords in animals. The technique, called *delayed nerve grafting,* involves cutting away the crushed or injured section of the spinal cord and bridging the gap with nerve segments from the arm or leg. The original severed axons in the cord can then grow through the bridge. If the animal research is successful, the procedure may be attempted on humans. Delayed nerve grafting offers hope for paraplegics who have lost voluntary movements of the lower extremities, as well as urinary, bowel, and sexual functions.

PERIPHERAL NERVE DAMAGE AND REPAIR

As we have seen, axons that have a neurilemma can be repaired as long as the cell body is intact and fibers are in association with Schwann cells. Most nerves that lie outside the brain and spinal cord consist of axons that are covered with a neurilemma. A person who injures a nerve in the upper extremity, for example, has a good chance of regaining nerve function. Axons in the brain and spinal cord do not have a neurilemma. Injury there is permanent.

When there is damage to an axon (or to dendrites of somatic afferent neurons), there are usually changes in the cell body and always changes in the portions of the nerve processes distal to the site of damage. These changes associated with the cell body are referred to as the **axon reaction** or **retrograde degeneration.** Those associated with the distal portion of the cut fiber are called **Wallerian** (wal-LE-rē-an) **degeneration.** The axon reaction occurs in essentially the same way, whether the damaged fiber is in the central or peripheral nervous system. The Wallerian reaction, however, depends on whether the fiber is central or peripheral. We will consider the axon reaction first.

Axon Reaction

When there is damage to an axon of a central or peripheral neuron, certain structural changes occur in the cell body. One of the most significant features of the axon reaction occurs 24 to 48 hours after damage. The Nissl bodies, arranged in an orderly fashion in an uninjured cell body, break down into finely granular masses. This alteration is called *chromatolysis* (krō'-ma-TOL-i-sis; *chromo* = color, *lysis* = dissolution). It begins between the axon

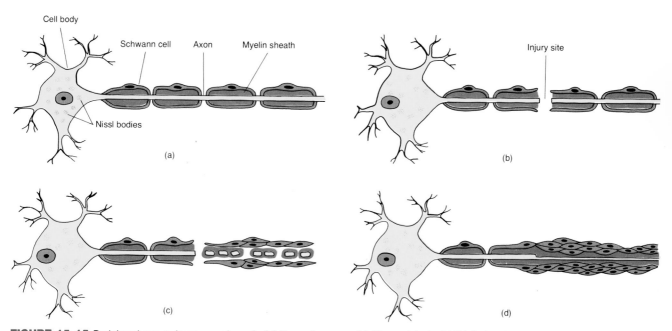

FIGURE 15-15 Peripheral nerve damage and repair. (a) Normal neuron. (b) Chromatolysis. (c) Wallerian degeneration. (d) Regeneration.

hillock and nucleus, but spreads throughout the cell body. As a result of chromatolysis, the cell body swells and the swelling reaches its maximum between 10 and 20 days after injury (Figure 15-15b). Chromatolysis results in a loss of ribosomes by the rough endoplasmic reticulum and an increase in the number of free ribosomes. Another sign of the axon reaction is the off-center position of the nucleus in the cell body. This change makes it possible to identify the cell bodies of damaged fibers through a microscope.

Wallerian Degeneration

The part of the axon distal to the damage becomes slightly swollen and then breaks up into fragments by the third to fifth day. The myelin sheath around the axon also undergoes degeneration (Figure 15-15c). Degeneration of the distal portion of the axon and myelin sheath is called Wallerian degeneration. Following degeneration, there is phagocytosis of the remains by macrophages.

Even though there is degeneration of the axon and myelin sheath, the neurilemma of the Schwann cells remains. The Schwann cells on either side of the site of injury multiply by mitosis and grow toward each other. This growth results in the formation of a tube. This "tunnel" provides a means for new axons to grow from the proximal area, across the injured area, and into the distal area previously occupied by the original nerve fiber (Figure 15-15d). The growth of the new axons will not occur if the gap at the site of injury is too large or if the gap becomes filled with dense collagenous fibers.

Regeneration

Following chromatolysis, there are signs of recovery in the cell body. There is an acceleration of RNA and protein synthesis, which favors regeneration of the axon. Recovery often takes several months and involves the restoration of normal levels of RNA and proteins and the Nissl bodies to their usual, uninjured patterns.

Accelerated protein synthesis is required for repair of the damaged axon. The proteins synthesized in the cell body pass into the axon at about the rate of 1 mm (0.04 inch)/day. The proteins assist in regenerating the damaged axon. During the first few days following damage, regenerating axons begin to invade the tube formed by the Schwann cells. Axons from the proximal area grow at the rate of about 1.5 mm (0.06 inch)/day across the area of damage, find their way into the distal neurilemmal tubes, and grow toward the distally located receptors and effectors. Thus sensory and motor connections are reestablished. In time, a new myelin sheath is also produced by the Schwann cells. However, function is never completely restored after a nerve is severed.

SCIATICA

Sciatica (sī-AT-i-ka) is a type of neuritis characterized by severe pain along the path of the sciatic nerve or its branches. The term is commonly applied to a number of disorders affecting this nerve. Because of its length and size, the sciatic nerve is exposed to many kinds of injury. Inflammation of or injury to the nerve causes pain that passes from the back or thigh down its length into the leg, foot, and toes.

Probably the most common cause of sciatica is a slipped or herniated disc. Other causes include irritation from osteoarthritis, back injuries, or pressure on the nerve from certain types of exertion. Other cases are idiopathic and sciatica may be associated with diabetes mellitus, gout, or vitamin deficiencies.

NEURITIS

Neuritis is inflammation of a single nerve, two or more nerves in separate areas, or many nerves simultaneously. It may result from irritation to the nerve produced by direct blows, bone fractures, contusions, or penetrating injuries. Additional causes include vitamin deficiency (usually thiamine) and poisons such as carbon monoxide, carbon tetrachloride, heavy metals, and some drugs.

Neuritis exists in many forms and is usually considered a symptom rather than a disease. A thorough physical examination, along with laboratory studies, is necessary to discover its exact cause.

SHINGLES

Shingles is an acute infection of the peripheral nervous system. It is caused by a virus called herpes zoster (HER-pēz ZOS-ter), the chickenpox virus that attacks the cell bodies in dorsal root ganglia. The inflammation spreads peripherally along the spinal nerves and infiltrates the epidermis and dermis over the nerves, frequently producing intense pain, a characteristic line of skin blisters, and discoloration of the skin along the line. The intercostal nerves and thoracic spinal nerves in the waist area are most commonly affected. Occasionally, the infection involves the first branch of the trigeminal (V) nerve, causing eye pain and corneal ulcerations and opacities.

Shingles occurs most frequently in individuals over age 50. Most recover uneventfully except for skin scarring and possible neuralgic pain for months or years, especially in the elderly.

A new treatment for shingles consists of injecting a local anesthetic into nerve junctions. In many cases, this treatment has brought about complete relief of pain and healing of skin lesions in 4 to 10 days. In addition, injections prevent the development of neuralgia or chronic pain.

STUDY OUTLINE

Grouping of Neural Tissue

1. White matter is an aggregation of myelinated axons and associated neuroglia.
2. Gray matter is a collection of nerve cell bodies and dendrites or unmyelinated axons along with associated neuroglia.
3. A nerve is a bundle of nerve fibers outside the central nervous system.
4. A ganglion is a collection of cell bodies outside the central nervous system.
5. A tract is a bundle of fibers of similar function in the central nervous system.
6. A nucleus is a mass of nerve cell bodies and dendrites in the gray matter of the brain.
7. A horn or column is an area of gray matter in the spinal cord.

Spinal Cord
General Features

1. The spinal cord begins as a continuation of the medulla oblongata and terminates at about the second lumbar vertebra.
2. It contains a cervical and lumbar enlargement, which serve as points of origin for nerves to the extremities.
3. The tapered portion of the spinal cord is the conus medullaris, from which arise the filum terminale and cauda equina.
4. The spinal cord is partially divided into right and left sides by the anterior median fissure and posterior median sulcus.
5. The gray matter in the spinal cord is divided into horns and the white matter into funiculi or columns.
6. In the center of the spinal cord is the central canal, which runs the length of the spinal cord and contains cerebrospinal fluid.
7. There are ascending (sensory) tracts and descending (motor) tracts.

Protection and Coverings

1. The spinal cord is protected by the vertebral canal, the meninges, the cerebrospinal fluid, and the vertebral ligaments.
2. The meninges are three coverings that run continuously around the spinal cord and brain: dura mater, arachnoid, and pia mater.
3. Removal of cerebrospinal fluid from the subarachnoid space or ventricle is called a spinal (lumbar) puncture. The procedure is used to diagnose pathologies and to introduce antibiotics.

Structure in Cross Section

1. Parts of the spinal cord observed in cross section are the gray commissure; central canal; anterior, posterior, and lateral gray horns; anterior, posterior, and lateral white columns; and ascending and descending tracts.
2. The spinal cord conveys sensory and motor information by way of the ascending and descending tracts, respectively.

Functions

1. A major function of the spinal cord is to convey sensory impulses from the periphery to the brain and to conduct motor impulses from the brain to the periphery.
2. Another function is to serve as a reflex center. The posterior root, posterior root ganglion, and anterior root are involved in conveying an impulse.
3. A reflex arc is the shortest route that can be taken by an impulse from a receptor to an effector. Its basic components are a receptor, a sensory neuron, a center, a motor neuron, and an effector.
4. A reflex is a quick, involuntary response to a stimulus that passes along a reflex arc. Reflexes represent the body's principal mechanisms for responding to changes in the internal and external environment.
5. Somatic spinal reflexes include the stretch reflex, flexor reflex, and crossed extensor reflex.
6. A two-neuron reflex arc contains one sensory and one motor neuron. Stretch reflexes such as the patellar reflex are all monosynaptic.
7. A polysynaptic reflex arc contains a sensory, association, and motor neuron. A withdrawal or flexor reflex and a crossed extensor reflex are examples.
8. Stretch and flexor reflexes are ipsilateral. The crossed extensor reflex is contralateral.
9. The flexor and crossed extensor reflexes illustrate reciprocal inhibition of muscles.

Spinal Nerves
Composition and Coverings

1. Spinal nerves are attached to the spinal cord by means of a posterior root and an anterior root. All spinal nerves are mixed.
2. Spinal nerves are covered by endoneurium, perineurium, and epineurium.

Names and Branches

1. The 31 pairs of spinal nerves are named and numbered according to the region and level of the spinal cord from which they emerge.
2. Branches of a spinal nerve include the dorsal ramus, ventral ramus, meningeal branch, and rami communicantes.

Plexuses

1. The ventral rami of spinal nerves, except for T2 to T12, form networks of nerves called plexuses.
2. Emerging from the plexuses are nerves bearing names that are often descriptive of the general regions they supply or the course they take.
3. The principal plexuses are called the cervical, brachial, lumbar, sacral, and coccygeal plexuses.
4. Nerves that do not form plexuses are called intercostal nerves (T2–T11). They are distributed directly to the structures they supply in intercostal spaces.

Dermatomes

1. With the exception of spinal nerve C1, all spinal nerves supply branches to the skin. The skin segment supplied by spinal nerves is called a dermatome.

2. Knowledge of dermatomes helps a physician to determine which segment of the spinal cord or spinal nerve is malfunctioning.

Applications to Health

1. Complete or partial severing of the spinal cord is called transection. It may result in quadriplegia or paraplegia. Partial transection is followed by a period of loss of reflex activity called areflexia.
2. Following peripheral nerve damage, repair is accomplished by an axon reaction, Wallerian degeneration, and regeneration.
3. Neuritis of the sciatic nerve and its branches is called sciatica.
4. Inflammation of nerves is known as neuritis.
5. Shingles is acute infection of peripheral nerves.

REVIEW QUESTIONS

1. Define the following terms: white matter, gray matter, nerve, ganglion, tract, nucleus, and horn.
2. Describe the location of the spinal cord. What are the cervical and lumbar enlargements?
3. Define conus medullaris, filum terminale, and cauda equina. What is a spinal segment? How is the spinal cord partially divided into a right and left side?
4. Describe the bony covering of the spinal cord.
5. Explain the location and composition of the spinal meninges. Describe the location of the epidural, subdural, and subarachnoid spaces. What are the denticulate ligaments?
6. Define meningitis.
7. What is a spinal puncture? Give several purposes served by a spinal puncture.
8. Based upon your knowledge of the structure of the spinal cord in cross section, define the following: gray commissure, central canal, anterior gray horn, lateral gray horn, posterior gray horn, anterior white column, lateral white column, posterior white column, ascending tract, and descending tract.
9. Describe the function of the spinal cord as a conduction pathway.
10. Using Exhibit 15-1 as a guide, be sure that you can list the location, origin, termination, and function of the principal ascending and descending tracts.
11. Describe how the spinal cord serves as a reflex center.
12. What is a reflex arc? List and define the components of a reflex arc.

13. Describe the mechanism of a stretch reflex, a flexor reflex, and a crossed extensor reflex.
14. Define the following terms: monosynaptic reflex arc, ipsilateral reflex arc, polysynaptic reflex arc, intersegmental reflex arc, contralateral reflex arc, and reciprocal inhibition.
15. Define a spinal nerve. Why are all spinal nerves classified as mixed nerves?
16. Describe how a spinal nerve is attached to the spinal cord.
17. Explain how a spinal nerve is enveloped by its several different coverings.
18. How are spinal nerves named and numbered?
19. Describe the branches and innervations of a typical spinal nerve.
20. What is a plexus?
21. Explain the location, origin, nerves, and distributions of the following plexuses: cervical, brachial, lumbar, and sacral.
22. What are intercostal nerves?
23. Define a dermatome. Why is a knowledge of dermatomes important?
24. What is transection? Distinguish between quadriplegia and paraplegia.
25. What is meant by spinal shock?
26. Outline the principal events that occur as part of the axon reaction, Wallerian degeneration, and regeneration following peripheral nerve damage.
27. Distinguish between sciatica and neuritis.
28. What is shingles? How is it treated?

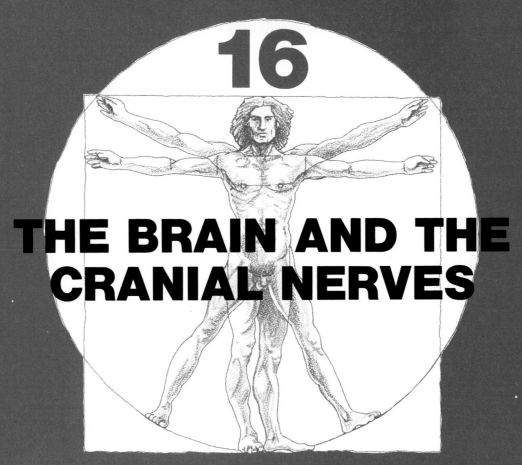

16

THE BRAIN AND THE CRANIAL NERVES

STUDENT OBJECTIVES

- Identify the principal areas of the brain.

- Describe the location of the cranial meninges.

- Explain the formation and circulation of cerebrospinal fluid.

- Describe the blood supply to the brain and the concept of the blood–brain barrier.

- Compare the components of the brain stem with regard to structure and function.

- Identify the structure and functions of the diencephalon.

- Identify the structural features of the cerebrum.

- Describe the lobes, tracts, and basal ganglia of the cerebrum.

- Describe the structure and functions of the limbic system.

- Compare the motor, sensory, and association areas of the cerebrum.

- Describe the principle of the electroencephalograph and its significance in the diagnosis of certain disorders.

- Explain brain lateralization and the split-brain concept.

- Describe the anatomical characteristics and functions of the cerebellum.

- Define a cranial nerve.

- Identify the 12 pairs of cranial nerves by name, number, type, location, and function.

- Explain the effects of injury on cranial nerves.

- List the clinical symptoms of these disorders of the nervous system: poliomyelitis, syphilis, cerebral palsy, Parkinsonism, epilepsy, multiple sclerosis, cerebrovascular accidents, dyslexia, Tay-Sachs disease, headache, and trigeminal neuralgia.

- Define key medical terms associated with the central nervous system.

Now we will consider the principal parts of the brain, how the brain is protected, how it is related to the spinal cord, and how it is related to the 12 pairs of cranial nerves.

BRAIN

PRINCIPAL PARTS

The **brain** of an average adult is one of the largest organs of the body, weighing about 1,300 g (3 lb). Figure 16-1 shows that the brain is mushroom-shaped. It is divided into four principal parts: brain stem, diencephalon, cerebrum, and cerebellum. The **brain stem,** the stalk of the mushroom, consists of the medulla oblongata, pons varolii, and midbrain. The lower end of the brain stem is a continuation of the spinal cord. Above the brain stem is the **diencephalon** (dī-en-SEF-a-lon) consisting primarily of the thalamus and hypothalamus. The **cerebrum** spreads over the diencephalon. The cerebrum constitutes about seven-eighths of the total weight of the brain and occupies most of the cranium. Inferior to the cerebrum and posterior to the brain stem is the **cerebellum.**

PROTECTION AND COVERINGS

The brain is protected by the cranial bones (see Figure 6-2). Like the spinal cord, the brain is also protected by meninges. The **cranial meninges** surround the brain, are continuous with the spinal meninges, and have the same basic structure and bear the same names as the spinal meninges: the outermost **dura mater,** middle **arachnoid,** and innermost **pia mater** (Figure 16-2a).

The cranial dura mater consists of two layers. The thicker, outer layer (periosteal layer) tightly adheres to the cranial bones and serves as a periosteum. The thinner, inner layer (meningeal layer) includes a mesothelial layer on its smooth surface. The spinal dura mater corresponds to the meningeal layer of the cranial dura mater.

CLINICAL APPLICATION

The middle meningeal artery, running on the inner surface of the temporal bone between dura mater and the skull, is closely adherent to the bones of the skull. It, and the meningeal veins, may be ruptured by a blow on the temple, especially if it fractures the bone. This rupture produces an **extradural hemorrhage,** which results in gradually increasing cranial pressure, drowsiness, unconsciousness, and death unless there is surgical intervention.

CEREBROSPINAL FLUID

The brain, as well as the rest of the central nervous system, is further protected against injury by **cerebrospinal fluid.**

This fluid circulates through the subarachnoid space around the brain and spinal cord and through the ventricles of the brain. The subarachnoid space is the area between the arachnoid and pia mater.

The **ventricles** (VEN-tri-kuls) are cavities in the brain that communicate with each other, with the central canal of the spinal cord, and with the subarachnoid space (Figure 16-2a–c). Each of the two **lateral ventricles** is located in a hemisphere (side) of the cerebrum under the corpus callosum (Figure 16-2a). The **third ventricle** is a slit between and inferior to the right and left halves of the thalamus and between the lateral ventricles. Each lateral ventricle communicates with the third ventricle by a narrow, oval opening: the **interventricular foramen (foramen of Monro).** The **fourth ventricle** lies between the inferior brain stem and the cerebellum. It communicates with the third ventricle via the **cerebral aqueduct (aqueduct of Sylvius),** which passes through the midbrain. The roof of the fourth ventricle has three openings: a **median aperture (foramen of Magendie)** and two **lateral apertures (foramina of Luschka).** Through these openings, the fourth ventricle also communicates with the subarachnoid space of the brain and cord.

The entire central nervous system contains about 125 ml (4 oz) of cerebrospinal fluid. It is a clear, colorless fluid of watery consistency. Chemically, it contains proteins, glucose, urea, and salts. It also contains some white blood cells. The fluid serves as a shock absorber for the central nervous system. It also circulates nutritive substances filtered from the blood.

Cerebrospinal fluid is formed primarily by filtration from networks of capillaries, called **choroid** (KO-royd; *chorion* = delicate) **plexuses,** located in the ventricles (Figure 16-2d). The fluid formed in the choroid plexuses of the lateral ventricles circulates through the interventricular foramen to the third ventricle, where more fluid is added by the choroid plexus in the third ventricle. It then flows through the cerebral aqueduct into the fourth ventricle. Here there are contributions from the choroid plexus in the fourth ventricle. The fluid then circulates through the apertures of the fourth ventricle into the subarachnoid space around the back of the brain. It also passes downward to the subarachnoid space around the posterior surface of the spinal cord, up the anterior surface of the spinal cord, and around the anterior part of the brain. From here it is gradually reabsorbed into veins. Some cerebrospinal fluid may be formed by ependymal (neuroglial) cells lining the central canal of the spinal cord. This small quantity of fluid ascends to reach the fourth ventricle. Most of the fluid is absorbed into the superior sagittal sinus (Figure 16-2e, f). The absorption actually occurs through **arachnoid villi**—fingerlike projections of the arachnoid that push into the superior sagittal sinus. Normally, cerebrospinal fluid is absorbed as rapidly as it is formed.

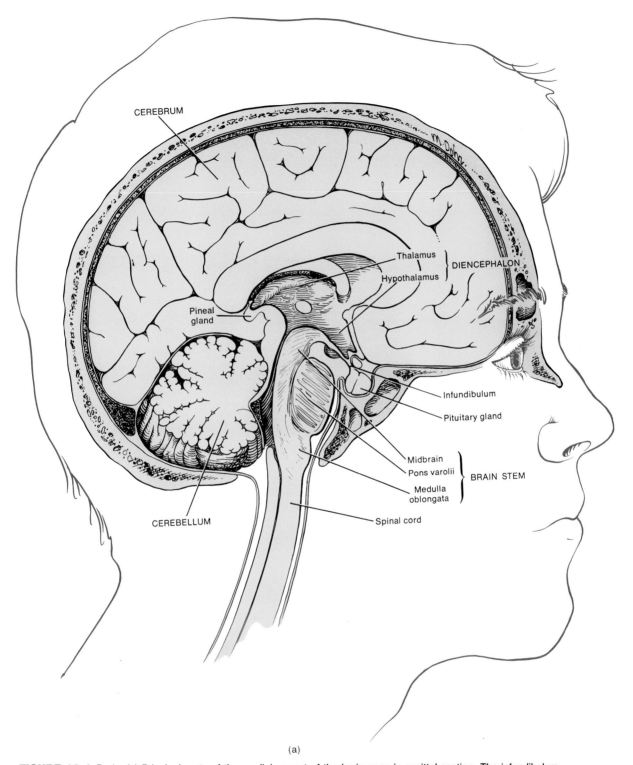

(a)

FIGURE 16-1 Brain. (a) Principal parts of the medial aspect of the brain seen in sagittal section. The infundibulum and pituitary gland are discussed in conjunction with the endocrine system in Chapter 18.

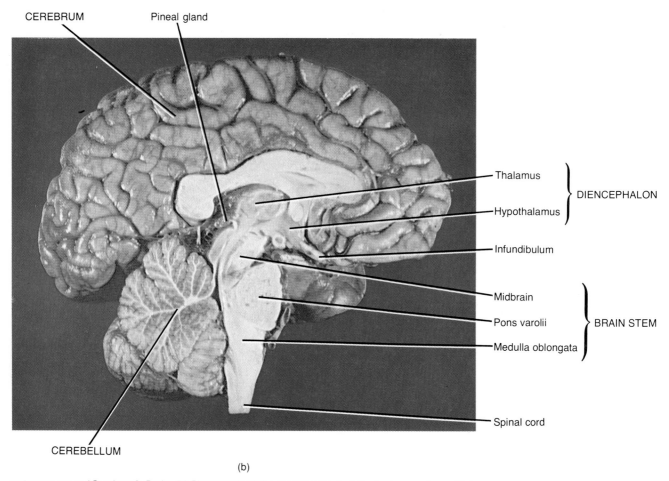

CEREBRUM Pineal gland

Thalamus
Hypothalamus
} DIENCEPHALON

Infundibulum

Midbrain
Pons varolii
Medulla oblongata
} BRAIN STEM

Spinal cord

CEREBELLUM

(b)

FIGURE 16-1 (*Continued*) Brain. (b) Photograph of the medial aspect of the brain seen in sagittal section. (Courtesy of C. Yokochi and J. W. Rohen, *Photographic Anatomy of the Human Body,* 2nd ed., 1979, IGAKU-SHOIN, Ltd., Tokyo, New York.)

CLINICAL APPLICATION

If an obstruction, such as a tumor, arises in the brain and interferes with the drainage of fluid from the ventricles into the subarachnoid space, large amounts of fluid accumulate in the ventricles. Fluid pressure inside the brain increases, and, if the fontanels have not yet closed, the head bulges to relieve the pressure. This condition is called **internal hydrocephalus** (*hydro* = water; *enkephalos* = brain). If an obstruction interferes with drainage somewhere in the subarachnoid space and cerebrospinal fluid accumulates inside the space, the condition is termed **external hydrocephalus.**

BLOOD SUPPLY

The brain is well supplied with blood vessels, which supply oxygen and nutrients (Figure 16-3; see also Figure 12-11c). If the blood flow to the brain is interrupted for even a few moments, unconsciousness may result. A one or two minute interruption may weaken the brain cells by starving them of oxygen. If the cells are totally deprived of oxygen for four minutes, many are permanently injured. Occasionally during childbirth the oxygen supply from the mother's blood is interrupted before the baby leaves the birth canal and can breathe. Often such babies are stillborn or suffer permanent brain damage that may result in mental retardation, epilepsy, and paralysis.

Blood supplying the brain also contains glucose, the principal source of energy for brain cells. Because carbohydrate storage in the brain is limited, the supply of glucose must be continuous. If blood entering the brain has a low glucose level, mental confusion, dizziness, convulsions, and loss of consciousness may occur.

Glucose, oxygen, and certain ions pass rapidly from the circulating blood into brain cells. Other substances, such as creatinine, urea, chloride, insulin, and sucrose, enter quite slowly. Still other substances—proteins and most antibiotics—do not pass at all from the blood into brain cells. The differential rates of passage of certain materials from the blood into the brain suggest a concept called the **blood–brain barrier.** Electron micrograph stud-

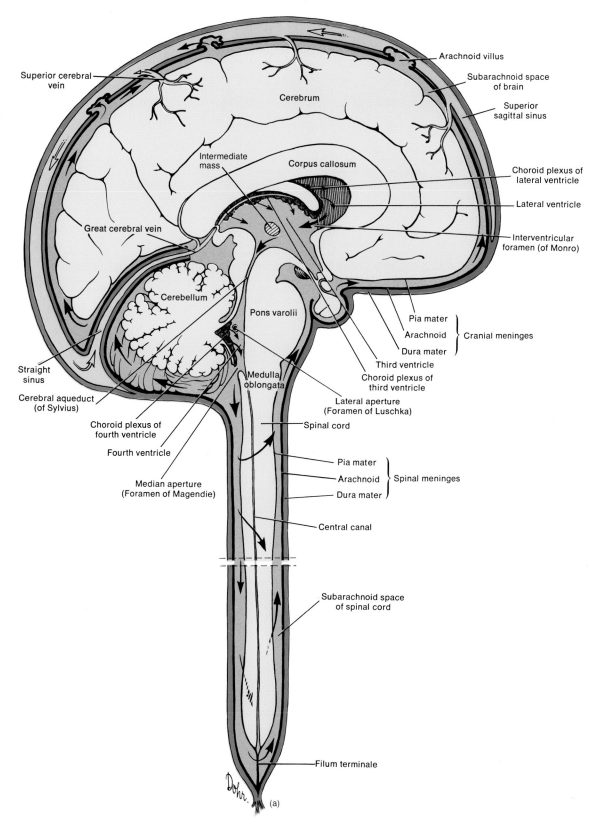

FIGURE 16-2 Meninges and ventricles of the brain. (a) Brain and meninges seen in sagittal section. The direction of flow of cerebrospinal fluid is indicated by black arrows.

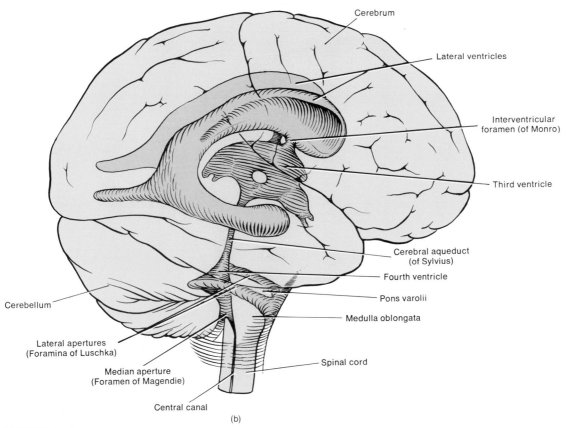

(b)

FIGURE 16-2 (*Continued*) Meninges and ventricles of the brain. (b) Diagrammatic lateral projection of the ventricles.

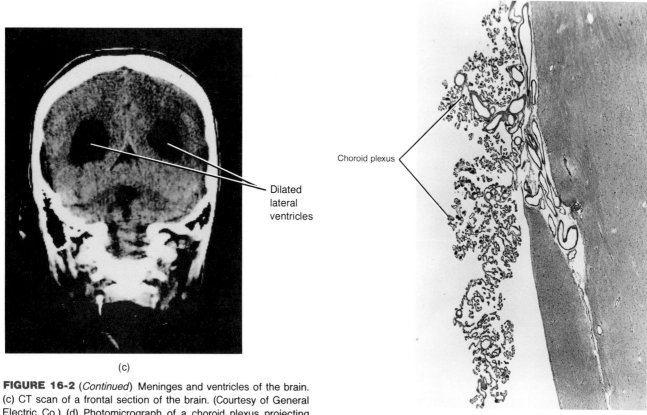

(c)

FIGURE 16-2 (*Continued*) Meninges and ventricles of the brain. (c) CT scan of a frontal section of the brain. (Courtesy of General Electric Co.) (d) Photomicrograph of a choroid plexus projecting into a ventricle. (Courtesy of Carolina Biological Supply Company.)

(d)

Superior sagittal sinus

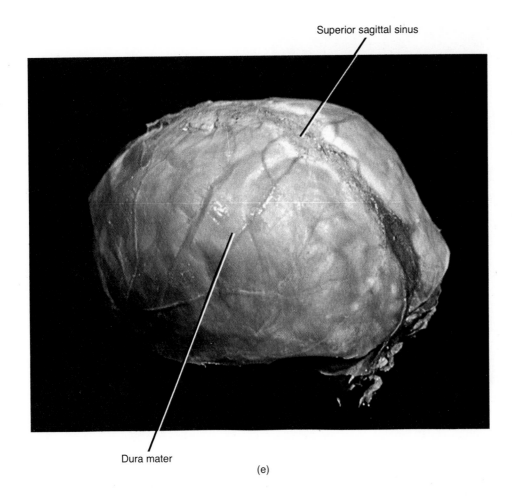

Dura mater

(e)

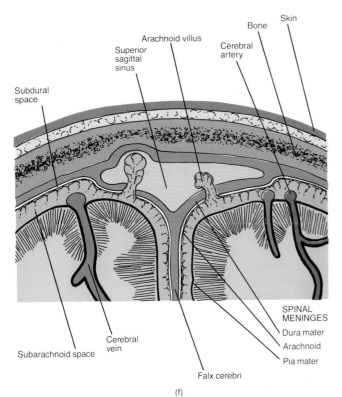

(f)

FIGURE 16-2 (*Continued*) Meninges and ventricles of the brain. (e) Photograph of a brain with dura mater and showing the superior sagittal sinus. (Courtesy of J. W. Eads.) (f) Frontal section through the brain showing the relationship of the superior sagittal sinus to the arachnoid villi.

ies of the capillaries of the brain reveal that they differ structurally from other capillaries. Brain capillaries are constructed of more densely packed cells and are surrounded by large numbers of glial cells and a continuous basement membrane. These features form a barrier to the passage of certain materials. Substances that cross the barrier are either very small molecules or require the assistance of a carrier molecule to cross by active transport. The function of the blood–brain barrier is not known. It may protect brain cells from harmful substances.

BRAIN STEM

Medulla Oblongata

The **medulla oblongata** (me-DULL-la ob'-long-GA-ta), or simply **medulla,** is a continuation of the upper portion of the spinal cord and forms the inferior part of the brain

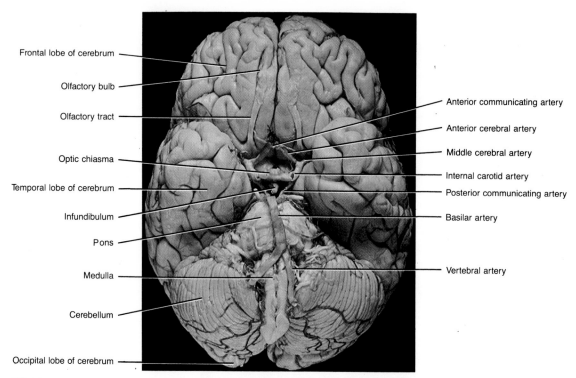

Frontal lobe of cerebrum

Olfactory bulb

Olfactory tract

Optic chiasma

Temporal lobe of cerebrum

Infundibulum

Pons

Medulla

Cerebellum

Occipital lobe of cerebrum

Anterior communicating artery

Anterior cerebral artery

Middle cerebral artery

Internal carotid artery

Posterior communicating artery

Basilar artery

Vertebral artery

FIGURE 16-3 Photograph of the ventral surface of the brain showing the principal arteries and associated structures. (Courtesy of N. Gluhbegovic and T. H. Williams, *The Human Brain: A Photographic Guide*, Harper & Row, New York, 1980.)

stem (Figure 16-4). Its position in relation to the other parts of the brain may be noted in Figure 16-1. It lies just superior to the level of the foramen magnum and extends upward to the inferior portion of the pons varolii. The medulla measures 3 cm (about 1 inch) in length.

The medulla contains all ascending and descending tracts that communicate between the spinal cord and various parts of the brain. These tracts constitute the white matter of the medulla. Some tracts cross as they pass through the medulla. Let us see how this crossing occurs and what it means.

On the ventral side of the medulla are two roughly triangular structures called **pyramids** (Figures 16-4 and 16-5). The pyramids are composed of the largest motor tracts that pass from the outer region of the cerebrum (cerebral cortex) to the spinal cord. Just above the junction of the medulla with the spinal cord, most of the fibers in the left pyramid cross to the right side, and most of the fibers in the right pyramid cross to the left. This crossing is called the **decussation** (de'ku-SA-shun) **of pyramids.** The adaptive value, if any, of this phenomenon is unknown. The principal motor fibers that undergo decussation belong to the lateral corticospinal tracts. These tracts originate in the cerebral cortex and pass inferiorly to the medulla. The fibers cross in the pyramids and descend in the lateral columns of the spinal cord, terminating in the anterior gray horns. Here synapses occur with motor neurons that terminate in skeletal mus-

cles. As a result of the crossing, fibers that originate in the left cerebral cortex activate muscles on the right side of the body, and fibers that originate in the right cerebral cortex activate muscles on the left side. Decussation explains why motor areas of one side of the cerebral cortex control muscular movements on the opposite side of the body.

The dorsal side of the medulla contains two pairs of prominent nuclei: the right and left **nucleus gracilis** (gras-I-lis; *gracilis* = slender) and **nucleus cuneatus** (kyoo-ne-A-tus; *cuneus* = wedge). These nuclei receive sensory fibers from ascending tracts (right and left fasciculus gracilis and fasciculus cuneatus) of the spinal cord and relay the sensory information to the opposite side of the medulla. This information is conveyed to the thalamus and then to the sensory areas of the cerebral cortex. Nearly all sensory impulses received on one side of the body cross in the medulla or spinal cord and are perceived in the opposite side of the cerebral cortex.

In addition to its function as a conduction pathway for motor and sensory impulses between the brain and spinal cord, the medulla also contains an area of dispersed gray matter containing some white fibers. This region is called the **reticular formation.** Actually, portions of the reticular formation are located in the spinal cord, pons, midbrain, and diencephalon. The reticular formation functions in consciousness and arousal. Within the medulla are three vital reflex centers of the reticular system.

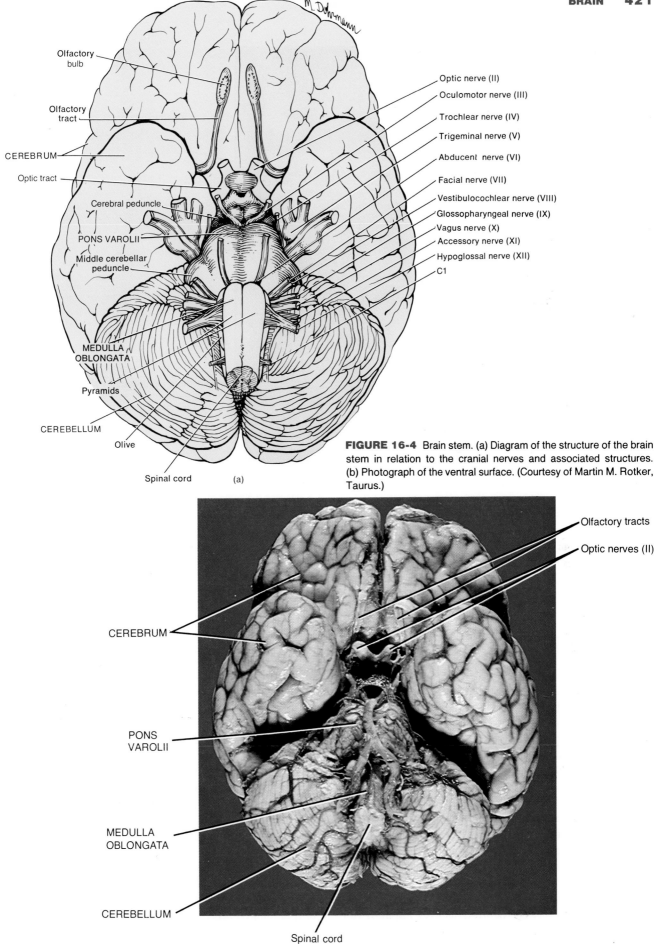

Olfactory bulb
Olfactory tract
CEREBRUM
Optic tract
Cerebral peduncle
PONS VAROLII
Middle cerebellar peduncle
MEDULLA OBLONGATA
Pyramids
CEREBELLUM
Olive
Spinal cord
(a)

Optic nerve (II)
Oculomotor nerve (III)
Trochlear nerve (IV)
Trigeminal nerve (V)
Abducent nerve (VI)
Facial nerve (VII)
Vestibulocochlear nerve (VIII)
Glossopharyngeal nerve (IX)
Vagus nerve (X)
Accessory nerve (XI)
Hypoglossal nerve (XII)
C1

FIGURE 16-4 Brain stem. (a) Diagram of the structure of the brain stem in relation to the cranial nerves and associated structures. (b) Photograph of the ventral surface. (Courtesy of Martin M. Rotker, Taurus.)

Olfactory tracts
Optic nerves (II)
CEREBRUM
PONS VAROLII
MEDULLA OBLONGATA
CEREBELLUM
Spinal cord
(b)

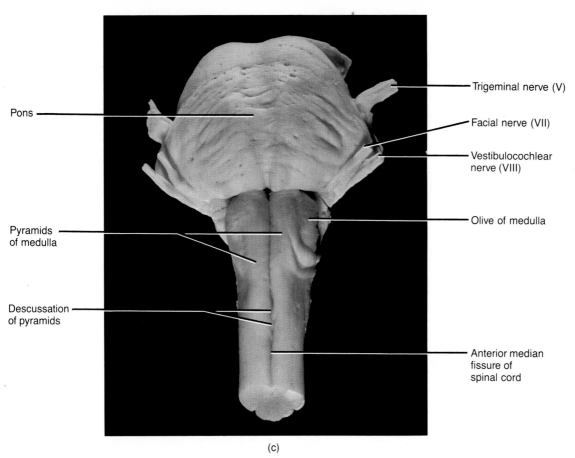

Pons

Pyramids
of medulla

Descussation
of pyramids

Trigeminal nerve (V)

Facial nerve (VII)

Vestibulocochlear
nerve (VIII)

Olive of medulla

Anterior median
fissure of
spinal cord

(c)

FIGURE 16-4 (*Continued*) Brain stem. (c) Photograph of the ventral surface of the brain stem. (Courtesy of N. Gluhbegovic and T. H. Williams, *The Human Brain: A Photographic Guide,* Harper & Row, New York, 1980.)

The **cardiac center** regulates heartbeat and force of contraction, the **medullary rhythmicity area** adjusts the basic rhythm of breathing, and the **vasomotor (vasoconstrictor) center** regulates the diameter of blood vessels. Other centers in the medulla are considered nonvital and coordinate swallowing, vomiting, coughing, sneezing, and hiccuping.

The medulla also contains the nuclei of origin for several pairs of cranial nerves (Figure 16-4a). These are the cochlear and vestibular branches of the vestibulocochlear nerves (VIII), which are concerned with hearing and equilibrium (there is also a nucleus for the vestibular branches in the pons); the glossopharyngeal nerves (IX), which relay impulses related to swallowing, salivation, and taste; the vagus nerves (X), which relay impulses to and from many thoracic and abdominal viscera; the cranial portions of the accessory nerves (XI), which convey impulses re-

lated to head and shoulder movements (a part of this nerve also arises from the first five segments of the spinal cord); and the hypoglossal nerves (XII), which convey impulses that involve tongue movements.

On each lateral surface of the medulla is an oval projection called the **olive** (Figure 16-4c, e). The olive contains an inferior olivary nucleus and two accessory olivary nuclei. The nuclei are connected to the cerebellum by fibers.

Also associated with the medulla is the greater part of the **vestibular nuclear complex.** This nuclear group consists of the *lateral, medial,* and *inferior vestibular nuclei* in the medulla and the *superior vestibular nucleus* in the pons. As you will see later (Chapter 18), the vestibular nuclei assume an important role in helping the body to maintain its sense of equilibrium.

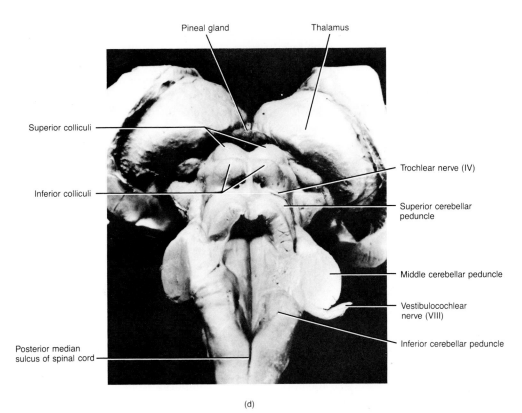

Pineal gland   Thalamus

Superior colliculi

Trochlear nerve (IV)

Inferior colliculi

Superior cerebellar
peduncle

Middle cerebellar peduncle

Vestibulocochlear
nerve (VIII)

Posterior median
sulcus of spinal cord

Inferior cerebellar peduncle

(d)

FIGURE 16-4 (*Continued*) Brain stem. (d) Photograph of the posterior surface of the brain stem. (Courtesy of N. Gluhbegovic and T. H. Williams, *The Human Brain: A Photographic Guide,* Harper & Row, New York, 1980.)

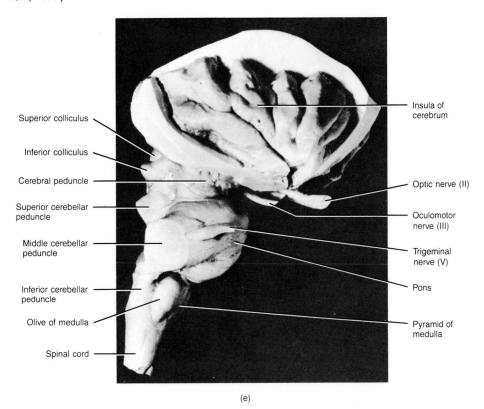

Superior colliculus

Inferior colliculus

Cerebral peduncle

Superior cerebellar
peduncle

Middle cerebellar
peduncle

Inferior cerebellar
peduncle

Olive of medulla

Spinal cord

Insula of
cerebrum

Optic nerve (II)

Oculomotor
nerve (III)

Trigeminal
nerve (V)

Pons

Pyramid of
medulla

(e)

FIGURE 16-4 (*Continued*) Brain stem. (e) Photograph of the brain stem in right lateral view. (Courtesy of N. Gluhbegovic and T. H. Williams, *The Human Brain: A Photographic Guide,* Harper & Row, New York, 1980.)

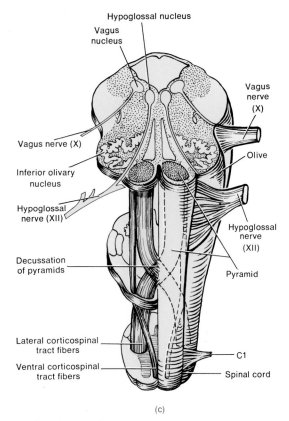

Hypoglossal nucleus

Vagus nucleus

Vagus nerve (X)

Vagus nerve (X)

Inferior olivary nucleus

Olive

Hypoglossal nerve (XII)

Hypoglossal nerve (XII)

Decussation of pyramids

Pyramid

Lateral corticospinal tract fibers

Ventral corticospinal tract fibers

C1

Spinal cord

(c)

FIGURE 16-5 Details of the medulla.

In view of the many vital activities controlled by the medulla, it is not surprising that a hard blow to the base of the skull can be fatal. Nonfatal medullary injury may be indicated by cranial nerve malfunctions on the same side of the body as the area of medullary injury, paralysis and loss of sensation on the opposite side of the body, and irregularities in respiratory control.

Pons Varolii

The relationship of the **pons varolii** (va-RŌ-lē-ī) or **pons** to other parts of the brain can be seen in Figures 16-1 and 16-4. The pons, which means bridge, lies directly above the medulla and anterior to the cerebellum. It measures about 2.5 cm (1 inch) in length. Like the medulla, the pons consists of white fibers scattered throughout with nuclei. As the name implies, the pons is a bridge connecting the spinal cord with the brain and parts of the brain with each other. These connections are provided by fibers that run in two principal directions. The transverse fibers connect with the cerebellum through the *middle cerebellar peduncles.* The longitudinal fibers of the pons belong to the motor and sensory tracts that connect the spinal cord or medulla with the upper parts of the brain stem.

The nuclei for certain paired cranial nerves are also contained in the pons (Figure 16-4a). These include the trigeminal nerves (V), which relay impulses for chewing and for sensations of the head and face; the abducens

nerves (VI), which regulate certain eyeball movements; the facial nerves (VII), which conduct impulses related to taste, salivation, and facial expression; and the vestibular branches of the vestibulocochlear nerves (VIII), which are concerned with equilibrium.

Other important nuclei in the reticular formation of the pons are the **pneumotaxic** (noo-mō-TAK-sik) **area** and the **apneustic** (ap-NOO-stik) **area.** Together with the medullary rhythmicity area in the medulla, they help control respiration.

Midbrain

The **midbrain,** or **mesencephalon** (*meso* = middle; *enkephalos* = brain), extends from the pons to the lower portion of the diencephalon (Figure 16-1 and 16-4). It is about 2.5 cm (1 inch) in length. The cerebral aqueduct passes through the midbrain and connects the third ventricle above with the fourth ventricle below.

The ventral portion of the midbrain contains a pair of fiber bundles referred to as **cerebral peduncles** (peDUNG-kulz). The cerebral peduncles contain many motor fibers that convey impulses from the cerebral cortex to the pons and spinal cord. They also contain sensory fibers that pass from the spinal cord to the thalamus. The cerebral peduncles constitute the main connection for tracts between upper parts of the brain and lower parts of the brain and the spinal cord.

The dorsal portion of the midbrain is called the **tectum** (*tectum* = roof) and contains four rounded eminences: the **corpora quadrigemina** (KOR-po-ra kwad-ri-JEM-in-a). Two of the eminences are known as the **superior colliculi** (ko-LIK-yoo-lī). These serve as a reflex center for movements of the eyeballs and head in response to visual and other stimuli. The other two eminences are the **inferior colliculi.** They serve as reflex centers for movements of the head and trunk in response to auditory stimuli. The midbrain also contains the **substantia** (sub-STAN-shē-a) **nigra** (NĪ-gra), a large, heavily pigmented nucleus near the cerebral peduncles.

A major nucleus in the reticular formation of the midbrain is the **red nucleus.** Fibers from the cerebellum and cerebral cortex terminate in the red nucleus. The red nucleus is also the origin of cell bodies of the descending rubrospinal tract. Other nuclei in the midbrain are associated with cranial nerves (Figure 16-4a). These include the oculomotor nerves (III), which mediate some movements of the eyeballs and changes in pupil size and lens shape, and the trochlear nerves (IV), which conduct impulses that move the eyeballs.

A structure called the **medial lemniscus** (*lemniskos* = ribbon or band) is common to the medulla, pons, and midbrain. The medial lemniscus is a band of white fibers containing axons that convey impulses for fine touch, proprioception, and vibrations from the medulla to the thalamus.

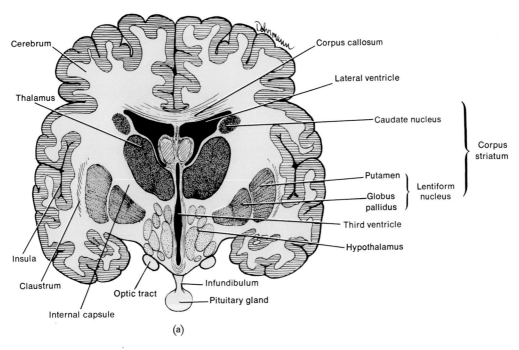

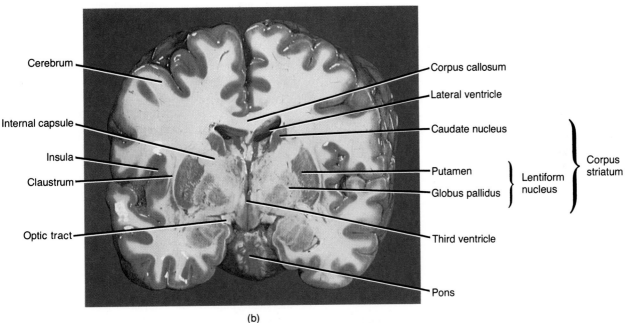

FIGURE 16-6 Thalamus. (a) Frontal section showing the thalamus and associated structures. (b) Photograph of a frontal section of the cerebrum. (Courtesy of C. Yokochi and J. W. Rohen, *Photographic Anatomy of the Human Body,* 2nd ed., 1979, IGAKU-SHOIN, Ltd., Tokyo, New York.)

DIENCEPHALON

The **diencephalon** (*dia* = through; *enkephalos* = brain) consists principally of the thalamus and hypothalamus. The relationship of these structures to the rest of the brain is shown in Figure 16-1.

Thalamus

The **thalamus** (THAL-a-mus; *thalamos* = inner chamber) is an oval structure above the midbrain that measures about 3 cm (1 inch) in length and constitutes four-fifths

of the diencephalon. It consists of two oval masses of mostly gray matter organized into nuclei that form the lateral walls of the third ventricle. The masses are joined by a bridge of gray matter called the **intermediate mass** (Figure 16-6d). Each mass is deeply embedded in a cerebral hemisphere and is bounded laterally by the **internal capsule** (Figure 16-6a).

Although the thalamic masses are primarily gray matter, some portions are white matter. Among the white matter portions are the **stratum zonale,** which covers the dorsal surface; the **external medullary lamina,** covering

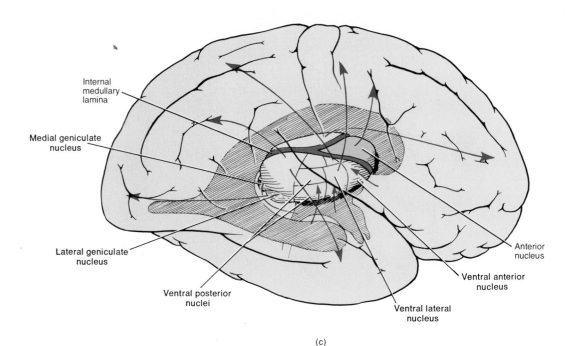

(c)

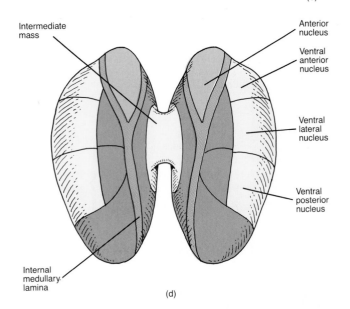

(d)

FIGURE 16-6 (*Continued*) Thalamus. (c) Right lateral view of the thalamic nuclei. (d) Dorsal view of the thalamic nuclei.

relay station for sensory impulses that reach the cerebral cortex from the spinal cord, brain stem, cerebellum, and parts of the cerebrum.

The thalamus also functions as an interpretation center for some sensory impulses, such as pain, temperature, crude touch, and pressure. The thalamus also contains a **reticular nucleus** in its reticular formation, which in some way seems to modify neuronal activity in the thalamus, and an **anterior nucleus** in the floor of the lateral ventricle, which is concerned with certain emotions and memory.

Hypothalamus

The **hypothalamus** (*hypo* = under) is a small portion of the diencephalon. Its relationship to other parts of the brain is shown in Figures 16-1 and 16-6a. The hypothalamus forms the floor and part of the lateral walls of the third ventricle. It is partially protected by the sella turcica of the sphenoid bone.

Despite its small size, nuclei in the hypothalamus control many body activities, most of them related to homeostasis. Although differentiation of the hypothalamic nuclei is far from precise, it is possible to identify certain nuclei. Some of these nuclei are more readily identified in lower animals and are more distinct in fetuses than adults. Also, within a given nucleus there may be several kinds of cells that can be differentiated histologically. The localization of function, with a few exceptions, is not specific to the individual nuclei; certain functions tend to overlap nu-

the lateral surface; and the **internal medullary lamina,** which divides the gray matter masses into an anterior nuclear group, a medial nuclear group, and a lateral nuclear group.

Within each group are nuclei that assume various roles. Some nuclei in the thalamus serve as relay stations for all sensory impulses, except smell, to the cerebral cortex. These include the **medial geniculate** (je-NIK-yoo-lāt) **nuclei** (hearing), the **lateral geniculate nuclei** (vision), and the **ventral posterior nuclei** (general sensations and taste). Other nuclei are centers for synapses in the somatic motor system. These include the **ventral lateral nuclei** (voluntary motor actions) and **ventral anterior nuclei** (voluntary motor actions and arousal). The thalamus is the principal

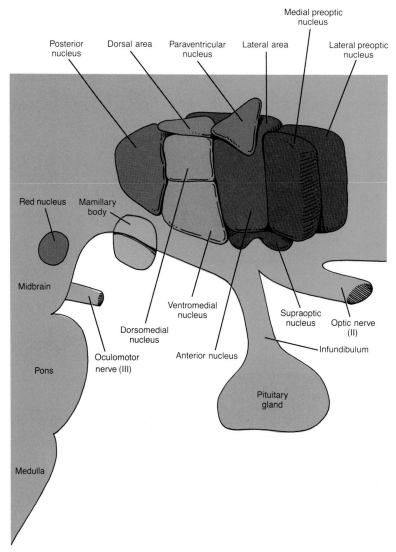

Medial preoptic nucleus

Posterior nucleus   Dorsal area   Paraventricular nucleus   Lateral area   Lateral preoptic nucleus

Red nucleus   Mamillary body

Midbrain

Oculomotor nerve (III)

Dorsomedial nucleus

Ventromedial nucleus

Anterior nucleus

Supraoptic nucleus   Optic nerve (II)

Infundibulum

Pons

Medulla

Pituitary gland

FIGURE 16-7 Hypothalamus. Shown here is a three-dimensional view of the hypothalamic nuclei.

clear boundaries. For this reason, functions are attributed to regions rather than specific nuclei. Since the hypothalamic nuclei are useful landmarks in understanding subsequent discussion of functions, a three-dimensional view of the hypothalamic nuclei is shown in Figure 16-7. The chief functions of the hypothalamus are as follows.

1. It controls and integrates the autonomic nervous system, which stimulates smooth muscle, regulates the rate of contraction of cardiac muscle, and controls the secretions of many glands. This is accomplished by axons of neurons whose dendrites and cell bodies are in hypothalamic nuclei. The axons form tracts from the hypothalamus to sympathetic and parasympathetic centers in the brain stem and spinal cord. Through the autonomic nervous system, the hypothalamus is the main regulator

of visceral activities. It regulates heart rate, movement of food through the digestive tract, and contraction of the urinary bladder.

2. It is involved in the reception of sensory impulses from the viscera.

3. It is the principal intermediary between the nervous system and the endocrine system—the two major control systems of the body. The hypothalamus lies just above the pituitary, the main endocrine gland. When the hypothalamus detects certain changes in the body, it releases chemicals called regulating factors that stimulate or inhibit the anterior pituitary gland. The anterior pituitary then releases or holds back hormones that regulate various physiological activities of the body. The hypothalamus also produces two hormones, antidiuretic hormone (ADH) and oxytocin, which are transported to and stored

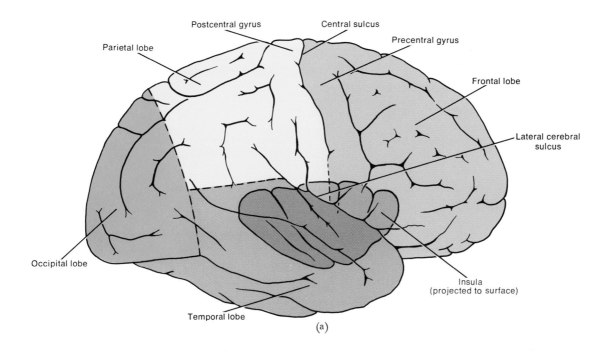

(a)

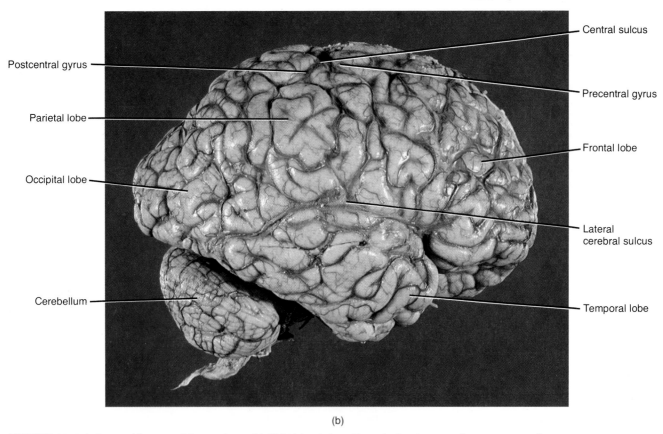

(b)

FIGURE 16-8 Lobes and fissures of the cerebrum. (a) Right lateral view. Since the insula cannot be seen externally, it has been projected to the surface. It can be seen in Figure 16-6a, b. (b) Photograph of right lateral view. (Courtesy of Martin M. Rotker, Taurus.)

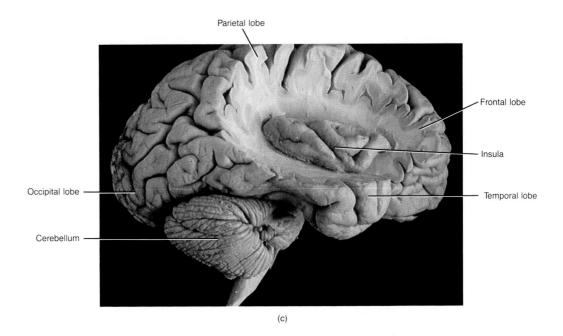

Parietal lobe

Frontal lobe

Insula

Temporal lobe

Occipital lobe

Cerebellum

(c)

FIGURE 16-8 (*Continued*) Lobes and fissures of the cerebrum. (c) Photograph of right lateral view showing the insula after removal of a portion of the cerebrum. (Courtesy of N. Gluhbegovic and T. H. Williams, *The Human Brain: A Photographic Guide,* Harper & Row, New York, 1980.)

in the posterior pituitary gland. The hormones are released from storage when needed by the body.

4. It is the center for the mind-over-body phenomenon. When the cerebral cortex interprets strong emotions, it often sends impulses along the tracts that connect the cortex with the hypothalamus. The hypothalamus then directs impulses via the autonomic nervous system and also releases chemicals that stimulate the anterior pituitary gland. The result can be a wide range of changes in body activities. For instance, when you panic, impulses leave the hypothalamus to stimulate your heart to beat faster. Likewise, continued psychological stress can produce long-term abnormalities in body function that result in serious illness. These so-called psychosomatic disorders are definitely real.

5. It is associated with feelings of rage and aggression.

6. It controls normal body temperature. Certain cells of the hypothalamus serve as a thermostat. If blood flowing through the hypothalamus is above normal temperature, the hypothalamus directs impulses along the autonomic nervous system to stimulate activities that promote heat loss. Heat can be lost through relaxation of the smooth muscle in the blood vessels, causing vasodilation of cutaneous vessels and increased heat loss from the skin. Heat loss also occurs by sweating. Conversely, if the temperature of the blood is below normal, the hypothalamus generates impulses that promote heat retention. Heat can be retained through the contraction of cutaneous blood vessels, cessation of sweating, and shivering.

7. It regulates food intake through two centers. The **feeding center** is stimulated by hunger sensations from an empty stomach. When sufficient food has been ingested, the **satiety** (sa-TĪ-e-tē) **center** is stimulated and sends out impulses that inhibit the feeding center.

8. It contains a **thirst center.** Certain cells in the hypothalamus are stimulated when the extracellular fluid volume is reduced. The stimulated cells produce the sensation of thirst.

9. It is one of the centers that maintain the waking state and sleep patterns.

CEREBRUM

Supported on the brain stem and forming the bulk of the brain is the **cerebrum** (see Figure 16-1). The surface of the cerebrum is composed of gray matter 2 to 4 mm (0.08 to 0.16 inch) thick and is referred to as the **cerebral cortex** (*cortex* = rind or bark). The cortex, containing roughly 15 billion cells, consists of six layers of nerve cell bodies. Beneath the cortex lies the cerebral white matter.

During embryonic development, when there is a rapid increase in brain size, the gray matter of the cortex enlarges out of proportion to the underlying white matter. As a result, the cortical region rolls and folds upon itself. The upfolds are called **gyri** (JĪ-rī) or **convolutions** (Figure 16-8). The deep downfolds are referred to as **fissures;** the shallow downfolds are **sulci** (SUL-sī). The most prom-

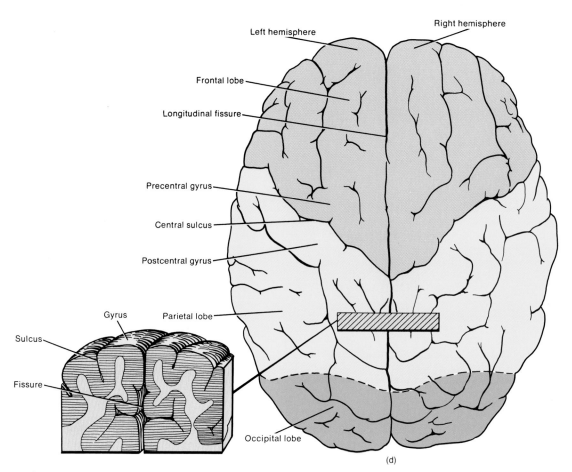

FIGURE 16-8 (*Continued*) Lobes and fissures of the cerebrum. (d) Superior view. The insert indicates the relative differences between a gyrus, sulcus, and fissure.

inent fissure, the **longitudinal fissure,** nearly separates the cerebrum into right and left halves, or **hemispheres** (Figure 16-8d, e). The hemispheres, however, are connected internally by a large bundle of transverse fibers composed of white matter called the **corpus callosum** (kal-LŌ-sum; *corpus* = body; *callosus* = hard). Between the hemispheres is an extension of the cranial dura mater called the **falx** (FALKS) **cerebri** (Figure 16-8f).

Lobes

Each cerebral hemisphere is further subdivided into four lobes by deep sulci or fissures (Figure 16-8a–e). The **central sulcus (fissure of Rolando)** separates the **frontal lobe** from the **parietal lobe.** A major gyrus, the **precentral gyrus,** is located immediately anterior to the central sulcus. The **lateral cerebral sulcus (fissure of Sylvius)** separates the **frontal lobe** from the **temporal lobe.** The **parieto-occipital sulcus** separates the **parietal lobe** from the **occipital lobe.** Another prominent fissure, the **transverse fissure,** separates the cerebrum from the cerebellum. The frontal lobe, parietal lobe, temporal lobe, and occipital

lobe are named after the bones that cover them. A fifth part of the cerebrum, the **insula (island of Reil),** lies deep within the lateral cerebral fissure, under the parietal, frontal, and temporal lobes. It cannot be seen in an external view of the brain (Figure 16-8a, c).

White Matter

The white matter underlying the cortex consists of myelinated axons running in three principal directions (Figure 16-9).

1. Association fibers connect and transmit impulses between gyri in the same hemisphere.

2. Commissural fibers transmit impulses from the gyri in one cerebral hemisphere to the corresponding gyri in the opposite cerebral hemisphere. Three important groups of commissural fibers are the *corpus callosum, anterior commissure,* and *posterior commissure.*

3. Projection fibers form ascending and descending tracts that transmit impulses from the cerebrum to other parts of the brain and spinal cord.

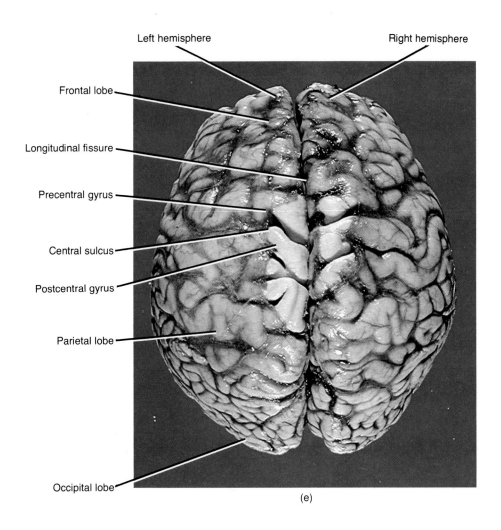

Left hemisphere

Right hemisphere

Frontal lobe

Longitudinal fissure

Precentral gyrus

Central sulcus

Postcentral gyrus

Parietal lobe

Occipital lobe

(e)

FIGURE 16-8 (*Continued*) Lobes and fissures of the cerebrum. (e) Photograph of a superior view of the cerebrum. (Courtesy of Martin M. Rotker, Taurus.) (f) Photograph of the skull showing the tentorium cerebelli and falx cerebri. (Courtesy of J. W. Eads.)

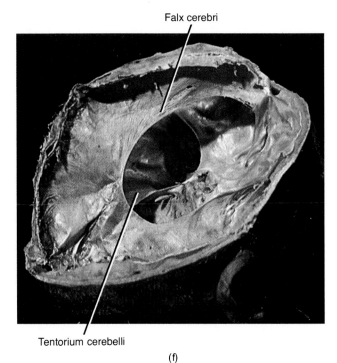

Falx cerebri

Tentorium cerebelli

(f)

Basal Ganglia (Cerebral Nuclei)

The **basal ganglia,** or **cerebral nuclei,** are paired masses of gray matter in each cerebral hemisphere (Figures 16-6 and 16-10). The largest of the basal ganglia of each hemisphere is the **corpus striatum** (strī-Ā-tum; *corpus* = body; *striatus* = striped). It consists of the **caudate nucleus** (*cauda* = tail) and the **lentiform nucleus** (*lenticula* = shaped like a lentil or lens). The lentiform nucleus, in turn, is subdivided into a lateral portion called the **putamen** (pu-TĀ-men; *putamen* = shell) and a medial portion called the **globus pallidus** (*globus* = ball; *pallid* = pale).

The portion of the *internal capsule* passing between the lentiform nucleus and the caudate nucleus and between the lentiform nucleus and thalamus is sometimes considered part of the corpus striatum. The internal capsule is made up of a group of sensory and motor white matter tracts that connect the cerebral cortex with the brain stem and spinal cord.

Other structures frequently considered part of the basal ganglia are the claustrum and amygdaloid nucleus. The **claustrum** (KLAWS-trum) is a thin sheet of gray matter

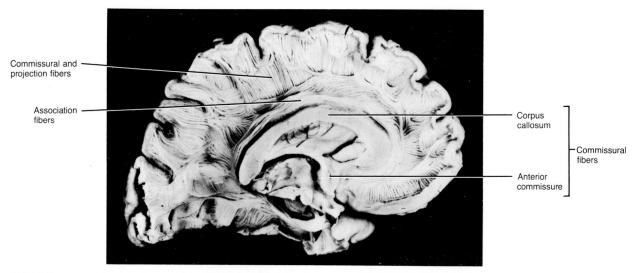

FIGURE 16-9 White matter tracts of the left cerebral hemisphere seen in sagittal section. (Courtesy of N. Gluhbegovic and T. H. Williams, *The Human Brain: A Photographic Guide,* Harper & Row, New York, 1980.)

lateral to the putamen. The **amygdaloid** (a-MIG-da-loyd; *amygda* = almond) **nucleus** is located at the tail end of the caudate nucleus. Some authorities also consider the **substantia nigra,** the **subthalamic nucleus,** and the **red nucleus** to be part of the basal ganglia. The substantia nigra is a large nucleus in the midbrain. The subthalamic nucleus lies against the internal capsule. Its major connection is with the globus pallidus.

The basal ganglia are interconnected by many fibers. They are also connected to the cerebral cortex, thalamus, and hypothalamus. The caudate nucleus and the putamen control large subconscious movements of the skeletal muscles—such as swinging the arms while walking. Such gross movements are also consciously controlled by the cerebral cortex. The globus pallidus is concerned with the regulation of muscle tone required for specific body movements.

CLINICAL APPLICATION

Damage to the basal ganglia results in abnormal body movements, such as uncontrollable shaking, called **tremor,** and **involuntary movements of skeletal muscles.** Moreover, destruction of a substantial portion of the caudate nucleus almost totally **paralyzes** the side of the body opposite to the damage. The caudate nucleus is an area often affected by a stroke.

A lesion in the subthalamic nucleus results in a motor disturbance on the opposite side of the body called **hemiballismus** (*hemi* = half; *ballismos* = jumping), which is characterized by involuntary movements occurring suddenly with great force and rapidity. The movements are purposeless and generally of the withdrawal type, although they may be jerky. The spontaneous movements affect the proximal portions of the extremities most severely, especially the arms.

Limbic System

Certain components of the cerebral hemispheres and diencephalon constitute the **limbic** (*limbus* = border) **system.** Among its components are the following regions of gray matter.

1. Limbic lobe. Formed by two gyri of the cerebral hemisphere: cingulate gyrus and hippocampal gyrus.

2. Hippocampus. An extension of the hippocampal gyrus that extends into the floor of the lateral ventricle.

3. Amygdaloid nucleus. Located at the tail end of the caudate nucleus.

4. Mammillary bodies of the hypothalamus. Two round masses close to the midline near the cerebral peduncles.

5. Anterior nucleus of the thalamus. Located in the floor of the lateral ventricle.

The limbic system functions in the emotional aspects of behavior related to survival. It also functions in memory. Although behavior is a function of the entire nervous system, the limbic system controls most of its involuntary aspects. Experiments on the limbic system of monkeys and other animals indicate that the amygdaloid nucleus assumes a major role in controlling the overall pattern of behavior.

Other experiments have shown that the limbic system is associated with pleasure and pain. When certain areas of the limbic system of the hypothalamus, thalamus, and midbrain are stimulated, the reactions of experimental animals indicate they are experiencing intense punishment. When other areas are stimulated, the animals' reactions indicate they are experiencing extreme pleasure. In still other studies, stimulation of the perifornical nuclei of the hypothalamus results in a behavioral pattern called *rage.* The animal assumes a defensive posture—extending

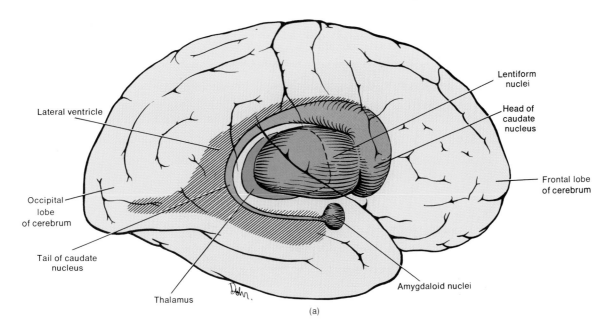

FIGURE 16-10 Basal ganglia. (a) In this right lateral view of the cerebrum, the basal ganglia have been projected to the surface. Refer to Figure 16-6a, b to note the positions of the basal ganglia in the frontal section of the cerebrum. (b) Photograph of the medial surface of the left cerebral hemisphere showing portions of the basal ganglia. (Courtesy of N. Gluhbegovic and T. H. Williams, *The Human Brain: A Photographic Guide,* Harper & Row, New York, 1980.)

CLINICAL APPLICATION

Lesions of the cerebrum are many and varied, but their symptoms and signs are relatively few—for example, vomiting, dizziness, headaches, convulsions, and partial or complete paralysis. Specific symptoms are related to the location of the lesion more than to the pathological origin of the lesion. For instance, any lesion that blocks the motor pathway between the motor cortex of the cerebrum and the skeletal muscles where the nerve terminates will cause paralysis whether the obstruction is due to a tumor, blood clot, inflammation, compression, or scar.

its claws, raising its tail, hissing, spitting, growling, and opening its eyes wide. Stimulating other areas of the limbic system results in an opposite behavioral pattern: docility, tameness, and affection. Because the limbic system assumes a primary function in emotions such as pain, pleasure, anger, rage, fear, sorrow, sexual feelings, docility, and affection, it is sometimes called the "visceral" or "emotional" brain.

Functional Areas of Cerebral Cortex

The functions of the cerebrum are numerous and complex. In a general way, the cerebral cortex is divided into sensory, motor, and association areas. The **sensory areas**

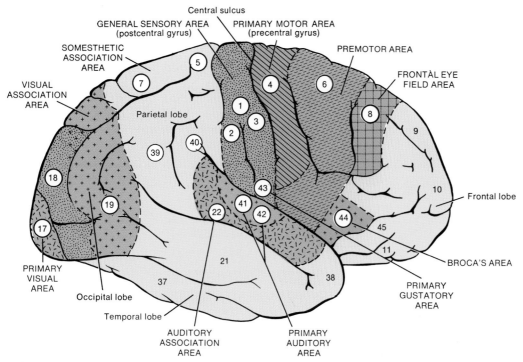

FIGURE 16-11 Functional areas of the cerebrum. The sensory and motor areas are indicated on a right lateral view. Although the right hemisphere is illustrated, Broca's area is in the left hemisphere of most people.

interpret sensory impulses, the **motor areas** control muscular movement, and the **association areas** are concerned with emotional and intellectual processes.

• *Sensory Areas* The **general sensory area** or **somesthetic** (sō'-mes-THET-ik; *soma* = body; *aisthesis* = perception) **area** is located directly posterior to the central sulcus of the cerebrum in the postcentral gyrus of the parietal lobe. It extends from the longitudinal fissure on the top of the cerebrum to the lateral cerebral sulcus. In Figure 16-11 the general sensory area is designated by the areas numbered 1, 2, and 3.*

The general sensory area receives sensations from cutaneous, muscular, and visceral receptors in various parts of the body. Each point of the area receives sensations from specific parts of the body, and essentially the entire body is spatially represented in it. The size of the portion of the sensory area receiving stimuli from body parts is not dependent on the size of the part, but on the number of receptors the part contains. For example, a larger portion of the sensory area receives impulses from the lips than from the thorax. The major function of the general sensory area is to localize exactly the points of the body where the sensations originate. The thalamus is capable of localizing sensations in a general way, that is, it receives sensations from large areas of the body but cannot distin-

guish between specific areas of stimulation. This ability is reserved to the general sensory area of the cortex.

Posterior to the general sensory area is the **somesthetic association area.** It corresponds to the areas numbered 5 and 7 in Figure 16-11. The somesthetic association area receives input from the thalamus, other lower portions of the brain, and the general sensory area. Its role is to integrate and interpret sensations. This area permits you to determine the exact shape and texture of an object without looking at it, to determine the orientation of one object to another as they are felt, and to sense the relationship of one body part to another. Another role of the somesthetic association area is the storage of memories of past sensory experiences. Thus you can compare sensations with previous experiences.

Other sensory areas of the cortex include:

1. Primary visual area (area 17). Located on the medial surface of the occipital lobe and occasionally extends around to the lateral surface. It receives sensory impulses from the eyes and interprets shape and color.

2. Visual association area (areas 18 and 19). Located in the occipital lobe. It receives sensory signals from the primary visual area and the thalamus. It relates present to past visual experiences with recognition and evaluation of what is seen.

3. Primary auditory area (areas 41 and 42). Located in the superior part of the temporal lobe near the lateral cerebral sulcus. It interprets the basic characteristics of sound such as pitch and rhythm.

* These numbers, as well as most of the others shown, are based on K. Brodmann's cytoarchitectural map of the cerebral cortex. His map, first published in 1909, is an attempt to correlate structure and function.

4. Auditory association area (area 22). Inferior to the primary auditory area in the temporal cortex. It determines if a sound is speech, music, or noise. It also interprets the meaning of speech by translating words into thoughts.

5. Primary gustatory area (area 43). Located at the base of the postcentral gyrus above the lateral cerebral sulcus in the parietal cortex. It interprets sensations related to taste.

6. Primary olfactory area. Located in the temporal lobe on the medial aspect. It interprets sensations related to smell.

7. Gnostic (NOS-tik; *gnosis* = knowledge) **area** (areas 5, 7, 39, and 40). This *common integrative area* is located between the somesthetic, visual, and auditory association areas. The gnostic area receives impulses from these areas, as well as from the taste and smell areas, the thalamus, and lower portions of the brain stem. It integrates sensory interpretations from the association areas and impulses from other areas so that a common thought can be formed from the various sensory inputs. It then transmits signals to other parts of the brain to cause the appropriate response to the sensory signal.

● *Motor Areas*   The **primary motor area** (area 4) is located in the precentral gyrus of the frontal lobe (Figure 16-11). Like the general sensory area, the primary motor area consists of regions that control specific muscles or groups of muscles. Stimulation of a specific point of the primary motor area results in a muscular contraction, usually on the opposite side of the body.

The **premotor area** (area 6) is anterior to the primary motor area. It is concerned with learned motor activities of a complex and sequential nature. It generates impulses that cause a specific group of muscles to contract in a specific sequence. An example of this is writing. Thus the premotor area controls skilled movements.

The **frontal eye field area** (area 8) in the frontal cortex is sometimes included in the premotor area. This area controls voluntary scanning movements of the eyes—searching for a word in a dictionary, for instance.

The **language areas** are also significant parts of the motor cortex. When you listen to someone speaking, sounds are relayed to the primary auditory area of the cortex. The sounds are then interpreted as words in the auditory association area. The words are interpreted as thoughts in the gnostic area. Written words are interpreted by the visual association area and converted into thoughts by the gnostic area. Thus you can translate speech or written words into thoughts.

The translation of thoughts into speech involves **Broca's** (BRŌ-kaz) **area** or the **motor speech area** (area 44), located in the frontal lobe just superior to the lateral cerebral sulcus. From this area, a sequence of signals is sent to the premotor regions that control the muscles of the larynx, pharynx, and mouth. The impulses from the premotor area to the muscles result in specific, coordinated contractions that enable you to speak. Simultaneously, impulses are sent from Broca's area to the primary motor area. From here, impulses reach your breathing muscles to regulate the proper flow of air past the vocal cords. The coordinated contractions of your speech and breathing muscles enable you to translate your thoughts into speech.

CLINICAL APPLICATION

Broca's area is usually located in the left cerebral hemisphere of most individuals regardless of whether they are left-handed or right-handed. Injury to the sensory or motor speech areas results in **aphasia** (a-FĀ-zē-a; *a* = without; *phasis* = speech), an inability to speak; **agraphia** (*a* = without; *graph* = write), an inability to write; **word deafness,** an inability to understand spoken words; or **word blindness,** an inability to understand written words.

● *Association Areas*   The **association areas** of the cerebrum are made up of association tracts that connect motor and sensory areas (see Figure 16-9). The association region of the cortex occupies the greater portion of the lateral surfaces of the occipital, parietal, and temporal lobes and the frontal lobes anterior to the motor areas. The association areas are concerned with memory, emotions, reasoning, will, judgment, personality traits, and intelligence.

Brain Waves

Brain cells can generate electrical activity as a result of literally millions of action potentials of individual neurons. These electrical potentials are called **brain waves** and indicate activity of the cerebral cortex. Brain waves pass through the skull easily and can be detected by sensors called electrodes. A record of such waves is called an **electroencephalogram (EEG). Distinct EEG patterns** appear in certain abnormalities. In fact, the EEG is used clinically in the diagnosis of epilepsy, infectious diseases, tumors, trauma, and hematomas. Electroencephalograms also furnish information regarding sleep and wakefulness.

BRAIN LATERALIZATION (SPLIT-BRAIN CONCEPT)

A distinctive characteristic of the human brain is the allocation of functions to the two cerebral hemispheres. At a glance the brain appears to have perfect bilateral symmetry, like most other organs of the body. It might therefore be expected that the two halves of the brain would also be functionally equivalent. Actually many of the more specialized functions are found in only one hemisphere or the other. In fact, careful examination indicates that even the apparent anatomical symmetry is an illusion, since brain asymmetries have been detected by CT scans. In these images a peculiar departure from bilateral symmetry is observed. In right-handed people the right frontal lobe is usually wider than the left, but the left occipital and parietal lobes are wider than the right. The inner surfaces of the skull itself bulge at the right front and the left rear to accommodate the protuberances.

The distribution of the more specialized functions is profoundly asymmetrical. For example, linguistic ability is dependent primarily on the left hemisphere. The right side of the brain is more important for melodies, as evidenced by the fact that aphasic patients with left-hemisphere damage can sing. The perception and analysis of nonverbal visual patterns such as perspective drawings is mostly a function of the right hemisphere, although the left hemisphere contributes.

In everyday life this lateralization of function can seldom be detected because information is readily passed between the hemispheres through several commissures, including the corpus callosum.

One of the most recent findings is that different emotional reactions result from damage to the right and left sides of the brain. Lesions in most areas on the left side are accompanied by the feelings of loss that might be expected as a result of any serious injury, and the patient is disturbed and depressed. Damage in much of the right hemisphere sometimes leaves the patient unconcerned with his condition.

A common cerebral dominance and also one of the most puzzling is the phenomenon of handedness. In the human population no more than 9 percent are left-handed. This considerable bias toward right-handedness may represent a unique specialization of the human brain. Studies indicate that the distribution of brain asymmetries in left-handed people is different from that in right-handed people.

Scientists are testing the possibility that faulty connections between the hemispheres may be a cause of dyslexia and that other peculiarities in communication between the hemispheres may sometimes be clues to psychiatric symptoms and may also be involved in creativity and invention.

CEREBELLUM

The **cerebellum** is the second-largest portion of the brain and occupies the inferior and posterior aspects of the cranial cavity. Specifically, it is posterior to the medulla and pons below the occipital lobes of the cerebrum and is separated from the cerebrum by the **transverse fissure** (see Figure 16-1). The cerebellum is also separated from the cerebrum by an extension of the cranial dura mater called the **tentorium** (*tentorium* = tent) **cerebelli** (see Figure 16-8f).

The cerebellum is shaped somewhat like a butterfly. The central constricted area is the **vermis,** which means worm-shaped, and the lateral "wings" or lobes are referred to as **hemispheres** (Figure 16-12). Each hemisphere consists of lobes that are separated by deep and distinct fissures. The **anterior lobe** and **posterior lobe** are concerned with subconscious movements of skeletal muscles. The **flocculonodular lobe** is concerned with the sense of equilibrium (Chapter 18). Between the hemispheres is another extension of the cranial dura mater: the **falx cerebelli** (see Figure 16-8f). It passes only a short distance between the cerebellar hemispheres.

The surface of the cerebellum, called the **cortex,** consists of gray matter in a series of slender, parallel ridges called **folia.** They are less prominent than the gyri of the cerebral cortex. Beneath the gray matter are **white matter tracts (arbor vitae)** that resemble branches of a tree. Deep within the white matter are masses of gray matter, the **cerebellar nuclei.**

The cerebellum is attached to the brain stem by three paired bundles of fibers called **cerebellar peduncles** (see Figure 16-4d, e). These are as follows:

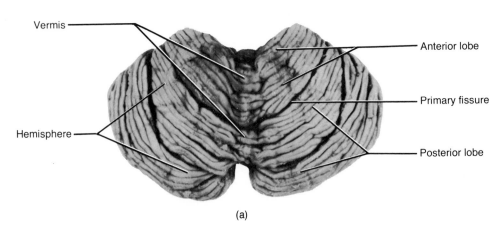

Vermis

Hemisphere

Anterior lobe

Primary fissure

Posterior lobe

(a)

FIGURE 16-12 Cerebellum. (a) Viewed from below.

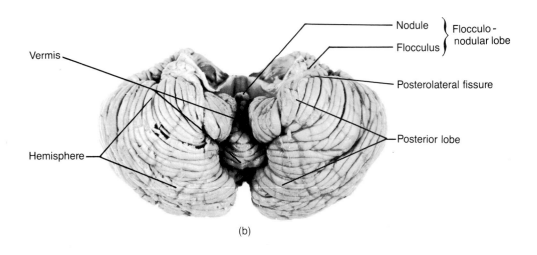

(b)

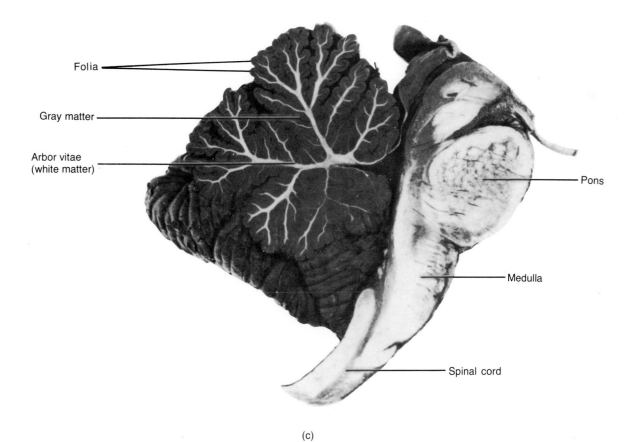

(c)

FIGURE 16-12 (*Continued*) Cerebellum. (b) Viewed from below. (c) Viewed in sagittal section. (Courtesy of M. L. Barr, *The Human Nervous System,* Second Edition, Harper & Row, New York, 1974.)

1. Inferior cerebellar peduncles connect the cerebellum with the medulla at the base of the brain stem and with the spinal cord.

2. Middle cerebellar peduncles connect the cerebellum with the pons.

3. Superior cerebellar peduncles connect the cerebellum with the midbrain.

The cerebellum is a motor area of the brain concerned with certain subconscious movements in the skeletal muscles. These movements are required for coordination, for maintenance of posture, and for balance. The cerebellar peduncles are the fiber tracts that allow the cerebellum to perform its functions.

Let us now see how the cerebellum produces coordinated movement. Motor areas of the cerebral cortex voluntarily initiate muscle contraction. Once the movement has begun, the sensory areas of the cortex receive impulses from nerves in the joints. The impulses provide information about the extent of muscle contraction and the amount of joint movement. The term *proprioception* is applied to this sense of the position of one body part relative to another. The cerebral cortex uses the proprioceptive sensations to determine which muscles are required to contract next and with what strength they are to contract in order to continue moving in the desired direction. Then a pattern of impulses is generated by the cerebral cortex along tracts to the pons and midbrain, which relay the impulses over the middle and superior cerebellar peduncles to the cerebellum. The cerebellum then generates subconscious motor impulses along the inferior cerebellar peduncles to the medulla and spinal cord. The impulses pass downward along the spinal cord and out the nerves that stimulate the prime movers and synergists to contract and that inhibit the contraction of the antagonists. The result is a smooth, coordinated movement. A well-functioning cerebellum is essential for delicate movements such as playing the piano.

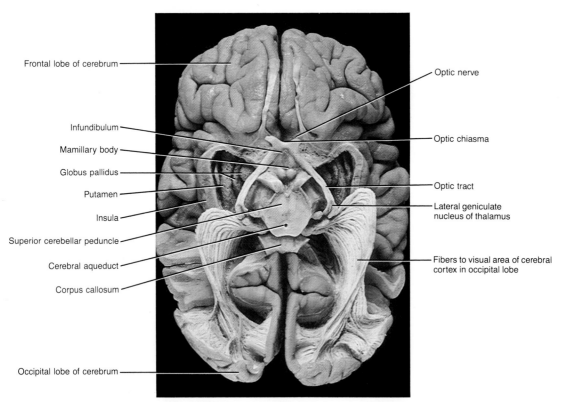

FIGURE 16-13 Photograph of the visual pathway as seen from the ventral aspect of the brain. (Courtesy of N. Gluhbegovic and T. H. Williams, *The Human Brain: A Photographic Guide,* Harper & Row, New York, 1980.)

The cerebellum also transmits impulses that control postural muscles, that is, it maintains normal muscle tone. The cerebellum also maintains body equilibrium. The inner ear contains structures that sense balance. Information such as whether the body is leaning to the left or right is transmitted from the inner ear to the cerebellum. The cerebellum then discharges impulses that cause the contraction of the muscles necessary for maintaining equilibrium.

CLINICAL APPLICATION
Damage to the cerebellum through trauma or disease is characterized by certain symptoms involving skeletal muscles on the same side of the body as the damage. There may be lack of muscle coordination, called *ataxia* (*a* = without; *taxis* = order). Blindfolded people with ataxia cannot touch the tip of their nose with a finger because they cannot coordinate movement with their sense of where a body part is located. Another sign of ataxia is a change in the speech pattern due to a lack of coordination of speech muscles. Cerebellar damage may also result in *disturbances of gait,* in which the subject staggers or cannot coordinate normal walking movements, and *severe dizziness.*

CRANIAL NERVES

Of the 12 pairs of **cranial nerves,** 10 originate from the brain stem, but all leave the skull through foramina of the skull (see Figure 16-4a). The cranial nerves are designated with Roman numerals and with names. The Roman numerals indicate the order in which the nerves arise from the brain (front to back). The names indicate the distribution or function of the nerves. Cranial nerves that contain both sensory and motor fibers are termed *mixed nerves.* Other cranial nerves contain sensory fibers only. The cell bodies of sensory fibers are located in ganglia outside the brain. The cell bodies of motor fibers lie in nuclei within the brain.

Some motor fibers control subconscious movements, yet the somatic nervous system has been defined as a *conscious* system. The reason for this apparent contradiction is that some fibers of the autonomic nervous system leave the brain bundled together with somatic fibers of the cranial nerves, as is the case for spinal nerves. Therefore, subconscious functions transmitted by the autonomic fibers are described along with the conscious functions of the somatic fibers of the cranial nerves.

Although the cranial nerves are mentioned singly in the following description of their type, location, and function, remember that they are paired structures. A number of clinical applications related to the cranial nerves are presented in Exhibit 16-1.

I. OLFACTORY

The **olfactory nerve** is entirely sensory and conveys impulses related to smell. It arises as bipolar neurons from the olfactory mucosa of the nasal cavity (see Figure 18-7b). The dendrites and cell bodies of these neurons are generally limited to the mucosa covering the superior nasal conchae and the adjacent nasal septum. Axons from the neurons pass through the cribriform plate of the ethmoid bone and synapse with other olfactory neurons in the *olfactory bulb,* an extension of the brain lying above the cribriform plate. The axons of these neurons make up the *olfactory tract.* The fibers from the tract terminate in the primary olfactory area in the cerebral cortex.

II. OPTIC

The **optic nerve** is entirely sensory and conveys impulses related to vision (Figure 16-13). Impulses initiated by rods and cones of the retina are relayed by bipolar neurons to ganglion cells. Axons of the ganglion cells, the optic nerve fibers, exit the optic foramen, after which the two optic nerves unite to form the *optic chiasma* (kī-AZ-ma) (see also Figure 18-11). Within the chiasma, fibers from the medial half of each retina cross to the opposite side; those from the lateral half remain on the same side. From the chiasma, the regrouped fibers pass posteriorly to the *optic tracts.* From the optic tracts, the majority of fibers terminate in a nucleus (lateral geniculate) in the thalamus. They then synapse with neurons that pass to the visual areas of the cerebral cortex. Some fibers from the optic chiasma terminate in the superior colliculi of the midbrain. They synapse with neurons whose fibers terminate in the nuclei that convey impulses to the oculomotor (III), trochlear (IV), and abducens (VI) nerves—nerves that control the extrinsic (external) and intrinsic (internal) eye muscles. Through this relay, there are widespread motor responses to light stimuli.

III. OCULOMOTOR

The **oculomotor nerve** is a mixed cranial nerve (Figure 16-14a). It originates from neurons in a nucleus in the ventral portion of the midbrain. It runs forward, divides into superior and inferior divisions, both of which pass through the superior orbital fissure into the orbit. The superior branch is distributed to the superior rectus (an extrinsic eyeball muscle) and the levator palpebrae superioris (the muscle of the upper eyelid). The inferior branch is distributed to the medial rectus, inferior rectus, and inferior oblique muscles—all extrinsic eyeball muscles. These distributions to the levator palpebrae superioris and extrinsic eyeball muscles constitute the motor portion of the oculomotor nerve. Through these distributions, impulses are sent that control movements of the eyeball and upper eyelid.

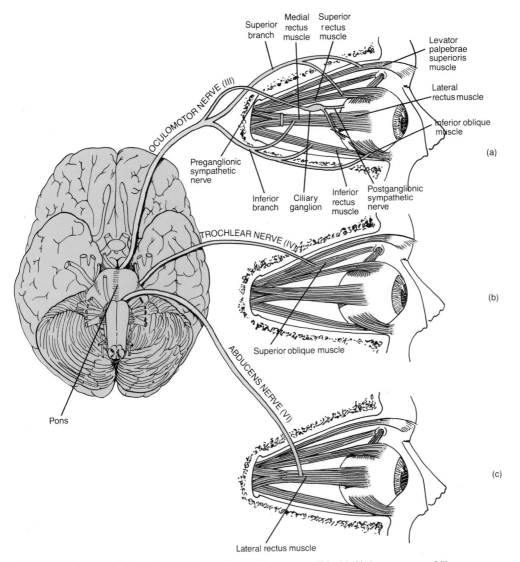

FIGURE 16-14 (a) Oculomotor nerve (III). (b) Trochlear nerve (IV). (c) Abducens nerve (VI).

The inferior branch of the oculomotor nerve also sends a branch to the *ciliary ganglion,* a relay center of the autonomic nervous system that connects a nucleus in the midbrain with the intrinsic eyeball muscles. These intrinsic muscles include the ciliary muscle of the eyeball and the sphincter muscle of the iris. Through the ciliary ganglion, the oculomotor nerve controls the smooth muscle (ciliary muscle) responsible for accommodation of the lens for near vision and the smooth muscle (sphincter muscle of iris) responsible for constriction of the pupil.

The sensory portion of the oculomotor nerve consists of afferent fibers from proprioceptors in the eyeball muscles supplied by the nerve to the midbrain. These fibers convey impulses related to muscle sense (proprioception).

IV. TROCHLEAR

The **trochlear** (TROK-lē-ar) **nerve** is a mixed cranial nerve (Figure 16-14b). It is the smallest of the 12 cranial nerves. The motor portion originates in a nucleus in the midbrain, and axons from the nucleus pass through the superior orbital fissure of the orbit. The motor fibers innervate the superior oblique muscle of the eyeball, another extrinsic eyeball muscle. It controls movement of the eyeball.

The sensory portion of the trochlear nerve consists of afferent fibers that run from proprioceptors in the superior oblique muscle to the nucleus of the nerve in the midbrain. The sensory portion is responsible for muscle sense.

V. TRIGEMINAL

The **trigeminal nerve** is a mixed cranial nerve and the largest of the cranial nerves (Figure 16-15). As indicated by its name, the trigeminal nerve has three branches: ophthalmic (of-THAL-mik), maxillary, and mandibular. The trigeminal nerve contains two roots on the ventrolateral surface of the pons. The large sensory root has a swelling called the *semilunar (gasserian) ganglion* lo-

EXHIBIT 16-1 CLINICAL APPLICATIONS RELATED TO THE CRANIAL NERVES

NERVE	DISORDER
I. Olfactory	Loss of the sense of smell, called *anosmia*, may result from head injuries in which the cribriform plate of the ethmoid bone is fractured and from lesions along the olfactory pathway.
II. Optic	Fractures in the orbit, lesions along the visual pathway, and diseases of the nervous system may result in visual field defects and loss of visual acuity. A defect of vision is called *anopsia*.
III. Oculomotor	A lesion in the nerve causes *strabismus* (squinting), *ptosis* (drooping) of the upper eyelid, pupil dilation, the movement of the eyeball downward and outward on the damaged side, a loss of accommodation for near vision, and double vision (*diplopia*).
IV. Trochlear	In trochlear nerve paralysis, the head is tilted to the affected side and diplopia and strabismus occur.
V. Trigeminal	Injury results in paralysis of the muscles of mastication and a loss of sensation of touch and temperature. Neuralgia (pain) of one or more branches of trigeminal nerve is called *trigeminal neuralgia* (*tic douloureux*).
VI. Abducens	With damage to this nerve, the affected eyeball cannot move laterally beyond the midpoint and the eye is usually directed medially.
VII. Facial	Injury produces paralysis of the facial muscles, called *Bell's palsy,* loss of taste, and the eyes remain open, even during sleep.
VIII. Vestibulocochlear	Injuries to the vestibular branch are *vertigo* (a subjective feeling of rotation) *ataxia,* and *nystagmus* (involuntary rapid movement of the eyeball). Injury to the cochlear branch may cause *tinnitus* (ringing) or deafness.
IX. Glossopharyngeal	Injury results in pain during swallowing, reduced secretion of saliva, loss of sensation in the throat, and loss of taste.
X. Vagus	Severing of both nerves in the upper body interferes with swallowing, paralyzes vocal cords, and interrupts sensations from many organs. Injury to both nerves in the abdominal area has little effect, since the abdominal organs are also supplied by autonomic fibers from the spinal cord.
XI. Accessory	Injury results in paralysis of the sternocleidomastoid and trapezius muscles, with resulting inability to turn the head or raise the shoulders.
XII. Hypoglossal	Injury results in difficulty in chewing, speaking, and swallowing. The tongue, when protruded, curls toward the affected side and the affected side becomes atrophied, shrunken, and deeply furrowed.

cated in a fossa on the inner surface of the petrous portion of the temporal bone. From this ganglion, the **ophthalmic branch** enters the orbit via the superior orbital fissure, the **maxillary branch** enters the foramen rotundum, and the **mandibular branch** exits through the foramen ovale. The smaller motor root originates in a nucleus in the pons. The motor fibers join the mandibular branch and supply the muscles of mastication. These motor fibers, which control chewing movements, constitute the motor portion of the trigeminal nerve.

The sensory portion of the trigeminal nerve delivers impulses related to touch, pain, and temperature and consists of the ophthalmic, maxillary, and mandibular branches. The ophthalmic branch receives sensory fibers from the skin over the upper eyelid, eyeball, lacrimal glands, upper part of the nasal cavity, side of the nose, forehead, and anterior half of the scalp. The maxillary branch receives sensory fibers from the mucosa of the nose, palate, parts of the pharynx, upper teeth, upper lip, cheek, and lower eyelid. The mandibular branch transmits sensory fibers from the anterior two-thirds of the tongue (not taste), lower teeth, skin over the mandible and side of the head in front of the ear, and mucosa of the floor of the mouth. Sensory fibers from the three branches of the trigeminal nerve enter the semilunar ganglion and terminate in a nucleus in the pons. There are

also sensory fibers from proprioceptors in the muscles of mastication.

CLINICAL APPLICATION

The inferior alveolar nerve, a branch of the mandibular nerve, supplies all the teeth in one half of the mandible and is frequently **anesthetized in dental procedures.** The same procedure will anesthetize the lower lip because the mental nerve is a branch of the inferior alveolar nerve. Since the lingual nerve runs very close to the inferior alveolar near the mental foramen, it too is often anesthetized at the same time. For anesthesia to the upper teeth, the superior alveolar nerve endings, branches of the maxillary branch, are blocked by inserting the needle beneath the mucous membrane, and the anesthetic solution is then infiltrated slowly throughout the area of the roots of the teeth to be treated.

VI. ABDUCENS

The **abducens** (ab-DOO-sens) **nerve** is a mixed cranial nerve that originates from a nucleus in the pons (see Figure 16-14c). The motor fibers extend from the nucleus to the lateral rectus muscle of the eyeball, an extrinsic

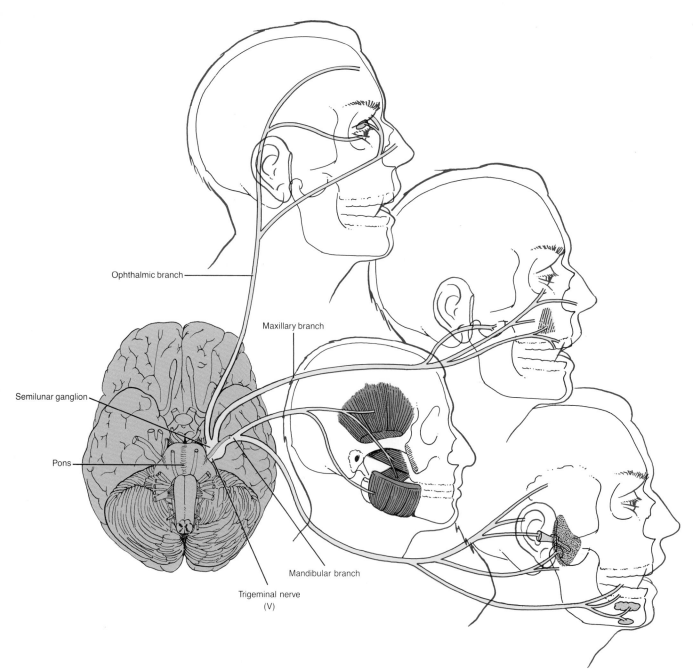

FIGURE 16-15 Trigeminal nerve (V).

eyeball muscle. Impulses over the fibers bring about movement of the eyeball. The sensory fibers run from proprioceptors in the lateral rectus muscle to the pons and mediate muscle sense. The abducens nerve reaches the lateral rectus muscle through the superior orbital fissure of the orbit.

VII. FACIAL

The **facial nerve** is a mixed cranial nerve (Figure 16-16). Its motor fibers originate from a nucleus in the pons and enter the petrous portion of the temporal bone. The motor fibers are distributed to facial, scalp, and neck mus-

cles. Impulses along these fibers cause contraction of the muscles of facial expression. Some motor fibers are also distributed to the lacrimal, sublingual, submandibular, nasal, and palatine glands.

The sensory fibers extend from the taste buds of the anterior two-thirds of the tongue to the *geniculate ganglion,* a swelling of the facial nerve. From here, the fibers pass to a nucleus in the pons, which sends fibers to the thalamus for relay to the gustatory area of the cerebral cortex. The sensory portion of the facial nerve also conveys deep general sensations from the face. There are also sensory fibers from proprioceptors in the muscles of the face and scalp.

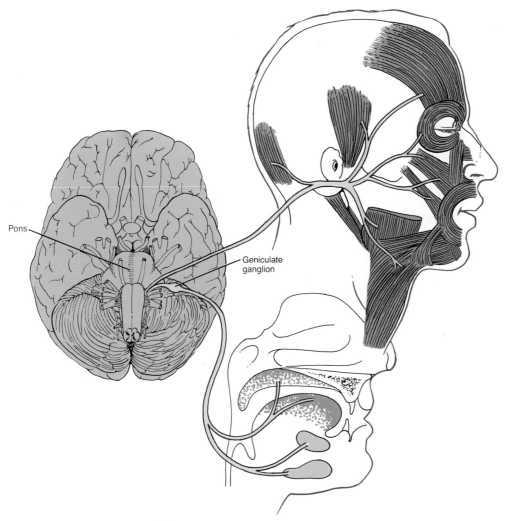

Pons

Geniculate
ganglion

FIGURE 16-16 Facial nerve (VII).

VIII. VESTIBULOCOCHLEAR

The **vestibulocochlear** (ves-tib'-yoo-lō-KŌK-lē-ar) **nerve** is another sensory cranial nerve. It consists of two branches: the cochlear (auditory) branch and the vestibular branch (see Figure 18-16). The **cochlear branch,** which conveys impulses associated with hearing, arises in the spiral organ of Corti in the cochlea of the internal ear. The cell bodies of the cochlear branch are located in the *spiral ganglion* of the cochlea. From here the axons pass through a nucleus in the medulla and terminate in the medial geniculate nucleus in the thalamus. Ultimately, the fibers synapse with neurons that relay the impulses to the auditory areas of the cerebral cortex.

The **vestibular branch** arises in the semicircular canals, the saccule, and the utricle of the inner ear. Fibers from the semicircular canals, saccule, and utricle extend to the *vestibular ganglion,* where the cell bodies are contained. The cell bodies of the fibers synapse in the ganglion with fibers that extend to a nucleus in the medulla and

pons and terminate in the thalamus. Some fibers also enter the cerebellum. The vestibular branch transmits impulses related to equilibrium.

IX. GLOSSOPHARYNGEAL

The **glossopharyngeal** (glos'-ō-fa-RIN-jē-al) **nerve** is a mixed cranial nerve (Figure 16-17). Its motor fibers originate in a nucleus in the medulla. The nerve exits the skull through the jugular foramen. The motor fibers are distributed to the swallowing muscles of the pharynx and the parotid gland to mediate swallowing movements and the secretion of saliva.

The sensory fibers of the glossopharyngeal nerve supply the pharynx and taste buds of the posterior third of the tongue. Some sensory fibers also originate from receptors in the carotid sinus, which assumes a major role in blood pressure regulation. The sensory fibers terminate in a nucleus in the thalamus. There are also sensory fibers from proprioceptors in the muscles innervated by this nerve.

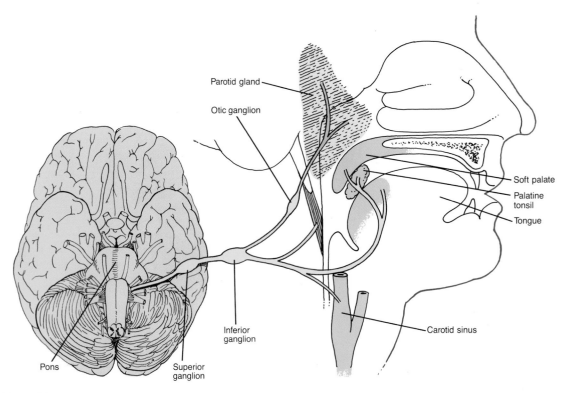

Parotid gland

Otic ganglion

Soft palate

Palatine tonsil

Tongue

Carotid sinus

Inferior ganglion

Pons

Superior ganglion

FIGURE 16-17 Glossopharyngeal nerve (IX).

X. VAGUS

The **vagus nerve** is a mixed cranial nerve that is widely distributed from the head and neck into the thorax and abdomen (Figure 16-18). Its motor fibers originate in a nucleus of the medulla and terminate in the muscles of the pharynx, larynx, respiratory passageways, lungs, heart, esophagus, stomach, small intestine, most of the large intestine, and gallbladder. Impulses along the motor fibers generate visceral, cardiac, and skeletal muscle movement.

Sensory fibers of the vagus nerve supply essentially the same structures as the motor fibers. They convey impulses for various sensations from the larynx, the viscera, and the ear. The fibers terminate in the medulla and pons. There are also sensory fibers from proprioceptors in the muscles supplied by this nerve.

XI. ACCESSORY

The **accessory nerve** (formerly the spinal accessory nerve) is a mixed cranial nerve (Figure 16-19). It differs from all other cranial nerves in that it originates from both the brain stem and the spinal cord. The **bulbar (medullary) portion** originates from nuclei in the medulla, passes through the jugular foramen, and supplies the voluntary muscles of the pharynx, larynx, and soft palate that are used in swallowing. The **spinal portion** originates in the anterior gray horn of the first five segments of the cervical portion of the spinal cord. The fibers from the segments

join, enter the foramen magnum, and exit through the jugular foramen along with the bulbar portion.

The spinal portion conveys motor impulses to the sternocleidomastoid and trapezius muscles to coordinate head movements. The sensory fibers originate from proprioceptors in the muscles supplied by its motor neurons and terminate in upper cervical posterior root ganglia.

XII. HYPOGLOSSAL

The **hypoglossal nerve** is a mixed cranial nerve (Figure 16-20). The motor fibers originate in a nucleus in the medulla, pass through the hypoglossal canal, and supply the muscles of the tongue. These fibers conduct impulses related to speech and swallowing.

The sensory portion of the hypoglossal nerve consists of fibers originating from proprioceptors in the tongue muscles and terminating in the medulla. The sensory fibers conduct impulses for muscle sense.

APPLICATIONS TO HEALTH

Many disorders can affect the central nervous system. Some are caused by viruses or bacteria. Others are caused by damage to the nervous system during birth. The origins of many conditions, however, are unknown. Here we discuss the origins and symptoms of some common central nervous system disorders.

Computerized transaxial tomography is a diagnostic technique used to study the structure of the brain. A

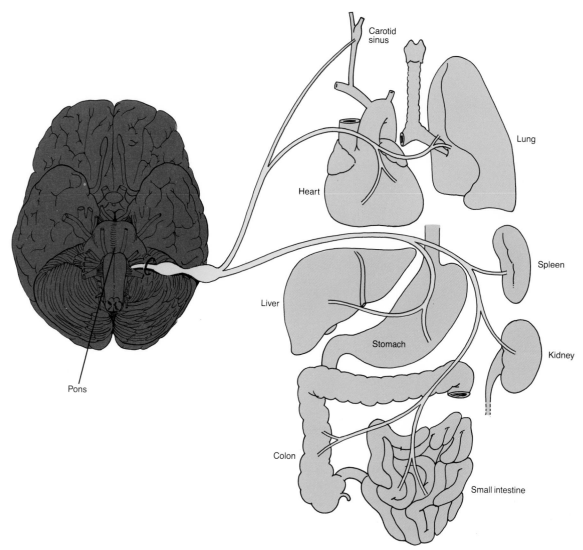

FIGURE 16-18 Vagus nerve (X).

form of radiologic imaging, it provides three-dimensional views of the cranial contents. Computerized transaxial tomography is a safe diagnostic tool for evaluation of neurological problems. It can differentiate between intracranial pathologies such as tumors, cysts, edema, and hemorrhage.

POLIOMYELITIS

Poliomyelitis, also known as **infantile paralysis** or simply **polio,** is a viral infection that is most common during childhood. The onset of the disease is marked by fever, severe headache, a stiff neck and back, deep muscle pain and weakness, and loss of certain somatic reflexes. The virus may affect almost all parts of the body. In its most serious form, called **bulbar polio,** the virus spreads via the respiratory passages and blood to the central nervous system, where it destroys the motor nerve cell bodies, specifically those in the anterior horns of the spinal cord and in the nuclei of the cranial nerves. Injury to the

spinal gray matter is the basis for the name of this disease (*polio* = gray matter; *myel* = spinal cord). Destruction of the anterior horns produces paralysis. The first sign of bulbar polio is difficulty in swallowing, breathing, and speaking. Poliomyelitis can cause death from respiratory or heart failure if the virus invades the brain cells of the vital medullary centers. In recent years, an immunization against the disease has been used.

SYPHILIS

Syphilis is a venereal disease caused by the *Treponema pallidum* bacterium. Venereal diseases are infectious disorders that can be spread through sexual contact. The disease progresses through several stages: primary, secondary, latent, and sometimes tertiary. During the *primary stage,* the chief symptom is an open sore, called a chancre, at the point of contact. The chancre heals within 1 to 5 weeks. From 6 to 24 weeks later, symptoms such as a skin rash, fever, and aches in the joints and muscles

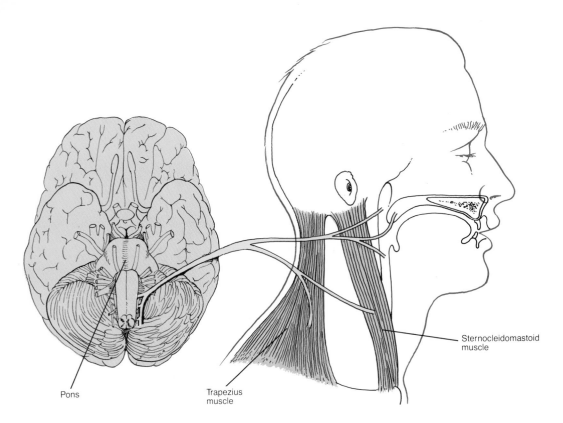

Sternocleidomastoid muscle

Pons

Trapezius muscle

FIGURE 16-19 Accessory nerve (XI).

usher in the *secondary stage.* These symptoms also eventually disappear (in about 4 to 12 weeks) and the disease ceases to be infectious, but a blood test for the presence of the bacteria generally remains positive. During this "symptomless" period, called the *latent stage,* the bacteria may invade body organs. When signs of organ degeneration appear, the disease is said to be in the *tertiary stage.*

If the syphilis bacteria attack the organs of the nervous system, the tertiary stage is called *neurosyphilis.* Neurosyphilis may take different forms, depending on the tissue involved. For instance, about two years after the onset of the disease, the bacteria may attack the meninges, producing meningitis. The blood vessels that supply the brain may also become infected. In this case, symptoms depend on the parts of the brain destroyed by oxygen and glucose starvation. Cerebellar damage is manifested by uncoordinated movements in such activities as writing. As the motor areas become extensively damaged, victims may be unable to control urine and bowel movements. Eventually, they may become bedridden, unable even to feed themselves. Damage to the cerebral cortex produces memory loss and personality changes that range from irritability to hallucinations.

A common form of neurosyphilis is *tabes dorsalis,* a progressive degeneration of the posterior columns of the spinal cord. The sensory ganglia and sensory nerve roots are also affected. Tabes dorsalis forms an interesting contrast with polio. The polio virus attacks the anterior col-umns and destroys motor neurons. Polio victims are unable to move the affected muscles voluntarily, but they retain sensory function. Individuals with tabes dorsalis suffer from the reverse condition. Motor control is maintained, but sensory functions are lost. Tingling and numbness are often experienced in the limbs and trunk.

Syphilis can be treated with antibiotics during the primary, secondary, and latent periods. Certain forms of neurosyphilis may also be successfully treated. However, the prognosis for tabes dorsalis is very poor. Noticeable symptoms do not always appear during the first two stages of the disease. Syphilis, however, is usually diagnosed through a blood test whether noticeable symptoms appear or not. The importance of these blood tests and follow-up treatment cannot be overemphasized.

CEREBRAL PALSY

The term **cerebral palsy** refers to a group of motor disorders caused by damage to the motor areas of the brain during fetal life, birth, or infancy. One cause is infection of the mother with German measles during the first three months of pregnancy. During early pregnancy, certain cells in the fetus are dividing and differentiating in order to lay down the basic structures of the brain. These cells can be abnormally changed by toxin from the measles virus. Radiation during fetal life, temporary oxygen star-

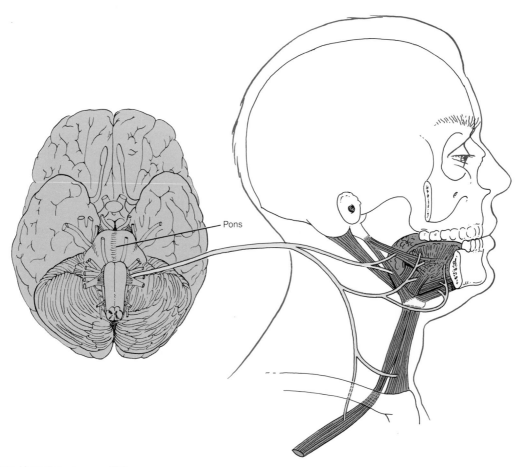

FIGURE 16-20 Hypoglossal nerve (XII).

vation during birth, and hydrocephalus during infancy may also damage brain cells.

Cases of cerebral palsy are categorized into three groups depending on whether the cortex, the basal ganglia of the cerebrum, or the cerebellum is affected most severely. Most cerebral palsy victims have at least some damage in all three areas. The location and extent of motor damage determine the symptoms. The victim may be deaf or partially blind. About 70 percent of cerebral palsy victims appear to be mentally retarded. The apparent mental slowness, however, is often due to the person's inability to speak or hear well. Such individuals are often more mentally acute than they appear.

Cerebral palsy is not a progressive disease; it does not worsen as time elapses. Once the damage is done, however, it is irreversible.

PARKINSONISM

This disorder, also called **Parkinson's disease,** is a progressive degeneration of the basal ganglia of the cerebrum. The basal ganglia regulate subconscious contractions of skeletal muscles that aid activities desired by the motor areas of the cerebral cortex—swinging the arms when walking, for example. In Parkinsonism, neurons that re-

lease the excitatory transmitter dopamine degenerate, causing unnecessary skeletal movements that often interfere with voluntary movement. For instance, the muscles of the upper extremities may alternately contract and relax, causing the hands to shake. This shaking is called *tremor.* Other muscles may contract continuously, causing rigidity of the involved body part. *Rigidity* of the facial muscles gives the face a masklike appearance. The expression is characterized by a wide-eyed, unblinking stare and a slightly open mouth with uncontrolled drooling. Vision, hearing, and intelligence are unaffected by the disorder, indicating that Parkinsonism does not attack the cerebral cortex.

Although people with Parkinsonism do not manufacture enough dopamine, injections of dopamine are useless; the blood–brain barrier stops it. However, symptoms are somewhat relieved by a drug developed a few years ago, levodopa, and its successors carbidopa and bromocriptine—none without distressing side effects.

MULTIPLE SCLEROSIS

Multiple sclerosis is the progressive destruction of the myelin sheaths of neurons in the central nervous system. The sheaths deteriorate to *scleroses,* which are hardened scars or plaques, in multiple regions—hence the name. The destruction of myelin sheaths interferes with the transmission of impulses from one neuron to another, literally short-circuiting conduction pathways. Multiple sclerosis is one of the most common disorders of the central nervous system. Usually the first symptoms occur between the ages of 20 and 40. Early symptoms are generally produced by the formation of a few plaques and are, consequently, mild. Plaque formation in the cerebellum may produce lack of coordination in one hand. The patient's handwriting becomes strained and irregular. A short-circuiting of pathways in the corticospinal tract may partially paralyze the leg muscles so that the patient drags a foot when walking. Other early symptoms include double vision and urinary tract infections. Following a period of remission during which the symptoms temporarily disappear, a new series of plaques develop and the victim suffers a second attack. One attack follows another over the years. Each time the plaques form, certain neurons are damaged by the hardening of their sheaths. Other neurons are uninjured by their plaques. The result is a progressive loss of function interspersed with remission periods during which the undamaged neurons regain their ability to transmit impulses.

The symptoms of multiple sclerosis depend on the areas of the central nervous system most heavily laden with plaques. Sclerosis of the white matter of the spinal cord is common. As the sheaths of the neurons in the corticospinal tract deteriorate, the patient loses the ability to contract skeletal muscles. Damage to the ascending tracts produces numbness and short-circuits impulses related to position of body parts and flexion of joints. Damage to either set of tracts also destroys spinal cord reflexes.

As the disease progresses, most voluntary motor control is eventually lost and the patient becomes bedridden. Death occurs anywhere from 7 to 30 years after the first symptoms appear. The usual cause of death is a severe infection resulting from the loss of motor activity. Without the constricting action of the urinary bladder wall, for example, the bladder never totally empties and stagnant urine provides an environment for bacterial growth. Bladder infection may then spread to the kidney, damaging kidney cells.

Multiple sclerosis may be caused by a virus. The occasional appearance of more than one case in a family suggests such an infectious agent, but the same circumstances also suggest a genetic predisposition. Like other demyelinating diseases, multiple sclerosis is incurable. Electrical stimulation of the spinal cord can improve function in certain patients, however.

EPILEPSY

Epilepsy is a disorder characterized by short, recurrent, periodic attacks of motor, sensory, or psychological malfunction. The attacks, called *epileptic seizures,* are initi-

ated by abnormal and irregular discharges of electricity from millions of neurons in the brain. The discharges stimulate many of the neurons to send impulses over their conduction pathways. As a result, a person undergoing an attack may contract skeletal muscles involuntarily. Lights, noise, or smells may be sensed when the eyes, ears, and nose actually have not been stimulated. The electrical discharges may also inhibit certain brain centers. For instance, the waking center in the brain may be depressed so that the person loses consciousness.

Many different types of epileptic seizures exist. The particular type of seizure depends on the area of the brain that is electrically stimulated and whether the stimulation is restricted to a small area or spreads throughout the brain. *Grand mal* seizures are initiated by a burst of electrical discharges that travel throughout the motor areas and spread to the areas of consciousness in the brain. The person loses consciousness, has spasms of the voluntary muscles, and may also lose urinary and bowel control. Sensory and intellectual areas may also be involved. For instance, just as the attack begins, the person may sense a peculiar taste in the mouth or see flashes of light or have olfactory hallucinations. This is called an aura and is a warning that may allow the person to lie down and avoid injury. The unconsciousness and motor activity last a few minutes. Then the muscles relax, and the person awakens. Afterward, the individual may be mentally confused for a short period of time. Studies with EEGs show that grand mal attacks are characterized by a rapid rate of 25 to 30 brain waves per second. The normal adult rate is 10 waves per second.

Many epileptics suffer from electrical discharges that are restricted to one or several relatively small areas of the brain. An example is the *petit mal* form, which apparently involves the thalamus and hypothalamus. Petit mal seizures are characterized by an abnormally slow brain wave pattern (3 waves/second). The person may lose contact with the environment for about 5 to 30 seconds, but does not undergo the loss of motor control that is typical of a grand mal seizure. The victim merely seems to be daydreaming. A few people experience several hundred petit mal seizures each day. For them, the chief problems are a loss of productivity in school or work and periodic inattentiveness while driving a car.

A form of epilepsy that is sometimes confused with mental illness is *psychomotor epilepsy.* The electrical outburst occurs in the temporal lobe, where it causes the person to lose contact with reality.

The causes of epilepsy are varied. Many conditions can cause nerve cells to produce periodic bursts of impulses. These causes include head injuries, tumors and abscesses of the brain, and childhood infections, such as mumps, whooping cough, and measles. Epilepsy may also be *idiopathic,* that is, have no demonstrable cause. It should be noted that epilepsy almost never affects intelligence. If frequent severe seizures are allowed to occur over a long period of time, however, some cerebral damage may occasionally result. Damage can be prevented by controlling the seizures with drug therapy.

Epileptic seizures can be eliminated or alleviated by drugs that make neurons more difficult to stimulate. Many of these drugs change the permeability of the neuron cell membrane so that it does not depolarize as easily.

CEREBROVASCULAR ACCIDENTS

The most common brain disorder is a **cerebrovascular accident (CVA),** also called a **stroke** or **cerebral apoplexy.** A CVA is the destruction of brain tissue (infarction) resulting from disorders in the vessels that supply the brain. Common causes of CVAs are intracerebral hemorrhage from aneurysms, embolism, and atherosclerosis of the cerebral arteries. An *intracerebral hemorrhage* is a rupture of a vessel in the pia mater or brain. Blood seeps into the brain and damages neurons by increasing intracranial fluid pressure. An *embolus* is a blood clot, air bubble, or bit of foreign material, most often debris from an inflammation, that becomes lodged in an artery and blocks circulation. *Atherosclerosis* is the formation of plaques in the artery walls. The plaques may slow down circulation by constricting the vessel. Both emboli and atherosclerosis cause brain damage by reducing the supply of oxygen and glucose needed by brain cells.

Many elderly people suffer mild CVAs as a result of short periods of reduced blood supply. One cause is atherosclerosis. Another is *arteriosclerosis,* or hardening of the arteries, which occurs with aging. Damage is generally undetectable or very mild. During these mild CVAs the individual may have a short blackout, blurred vision, or dizziness and does not realize anything serious has occurred. A CVA can also cause sudden, massive damage, however. Severe CVAs cause about 21 percent of all deaths from cardiovascular disease. The person who recovers may suffer partial paralysis and mental disorders such as speech difficulty. The malfunction depends on the parts of the brain that were injured. Vascular disorders are more common after age 40.

DYSLEXIA

Dyslexia (dis-LEK-sē-a; *dys* = difficulty; *lexis* = words) is unrelated to basic intellectual capacity, but it causes a mysterious difficulty in handling words and symbols. Apparently some peculiarity in the brain's organizational pattern distorts the ability to read, write, and count. Letters in words seem transposed, reversed, or even topsy-turvy—*dog* becomes *god; b* changes identity with *d;* a sign saying "OIL" inverts into "710." Many dyslexics cannot orient themselves in the three dimensions of space and may show bodily awkwardness.

The cause of dyslexia is unknown, since it is unaccompanied by outward scars of detectable neurological dam-

age and its symptoms vary from victim to victim. It occurs three times as often among boys as among girls. It has been variously attributed to defective vision, brain damage, lead in the air, physical trauma, or oxygen deprivation during birth, but it remains an unsolved problem.

TAY-SACHS DISEASE

Tay-Sachs disease is a central nervous system affliction that brings death before age 5. The Tay-Sachs gene is carried mostly by normal-appearing individuals descended from the Ashkenazi Jews of Eastern Europe. Approximately 1 in 3,600 of their offspring will be afflicted with Tay-Sachs disease. The disease involves the neuronal degeneration of the central nervous system because of excessive amounts of a substance known as ganglioside G_{m2} in the nerve cells of the brain. The afflicted child will develop normally until the age of 4 to 8 months. Then the symptoms follow a course of progressive degeneration: paralysis, blindness, inability to eat, decubitus ulcers, and death from infection. There is no known cure.

HEADACHE

One of the most common human afflictions is **headache (cephalagia).** Based upon origin, two general types are distinguished: intracranial and extracranial. Serious headaches of intracranial origin are caused by brain tumors, blood vessel abnormalities, inflammation of the brain or meninges, decrease in oxygen supply to the brain, and damage to brain cells. Extracranial headaches are related to infections of the eyes, ears, nose, and sinuses and are commonly felt as headaches because of the location of these structures. The sinuses are really pain sensitive in only one or two areas and do not usually cause pain except in acute sinusitis, or with some obvious chronic sinus problem as with tumor or cyst formation or some pressure on the conchae. Headaches that accompany sore throats and influenza may arise when the infection produces chemical changes in the blood that irritate pain sensors in the brain. Other extracranial causes include oral complications, ocular disease such as glaucoma, and orthopedic disorders such as cervical spine disease.

Most people with headaches usually have a tension headache, a migraine headache, or a cluster headache. *Tension headaches* or *muscle-contraction headaches* are the most frequent headache complaint. They commonly develop toward the end of a workday and are most frequently caused by anxiety or anger. Substances such as norepinephrine, serotonin, and dopamine are normally found in considerable concentration in the limbic system. It is believed that a depletion of these substances in the limbic system is responsible for depressive headache symptoms. This is the rationale for using antidepressant drugs to treat depressed people with tension headaches.

Migraine headaches are more often associated with a throbbing pain around the eye that spreads to one or both sides. Classically, migraine headaches produce vasoconstriction first, to be followed by a period of vasodilation. While the vasoconstriction phase is painless, it does cause such sensory manifestations as visual field defects or spots before the eyes. The pain of migraine begins with the onset of the vasodilation phase and is aching or throbbing in type, with the pain throbbing in rhythm with the individual's pulse. Nausea and vomiting are also commonly experienced. Approximately 6 to 8 percent of the adult population suffers from migraine headaches, and its incidence is greater in adult women. One type of migraine headache seems to run in some families. The cause of migraine headaches is unknown. Sleeping habits, emotional factors, work problems, food allergies, and menstrual and obstetric factors are some possible causes.

Cluster headache is largely a disease of men, often follows the drinking of alcohol, and is associated with tearing, photophobia (an abnormal intolerance of light), and nasal congestion and drainage. Foods that contain tyramine (a vasoactive amine) may precipitate cluster headaches. Tyramine-rich foods include strong or aged cheese, pickled herring, chicken livers, and canned figs.

Most headaches require no special treatment. Analgesic and tranquilizing compounds are generally effective for tension headaches, but not for migraine headaches. Drugs that constrict the blood vessels can be helpful for migraine. Biofeedback training and dietary changes may also be of some help. Taking a careful and complete history is an important procedure in determining the cause or causes of a patient's headaches.

TRIGEMINAL NEURALGIA (TIC DOULOUREUX)

As noted earlier, pain arising from irritation of the trigeminal nerve (V) is known as **trigeminal neuralgia** or **tic douloureux** (doo-loo-ROO). The disorder is characterized by brief but extreme pain in the face and forehead on the affected side. Many patients describe sensitive regions around the mouth and nose that can cause an attack when touched. Eating, drinking, washing the face, and exposure to cold may also bring on an attack.

Treatment may be palliative (relieving symptoms without curing the disease) or surgical. One technique involves alcohol injections directly into the semilunar (gasserian) ganglion, which controls the trigeminal nerve. This method is superior to open surgery because it is safer. Moreover, it can bring lasting pain relief with preservation of touch sensation in the face and release of the patient 24 hours after treatment.

KEY MEDICAL TERMS ASSOCIATED WITH THE CENTRAL NERVOUS SYSTEM

Agnosia (*a* = without; *gnosis* = knowledge)   Inability to recognize the significance of sensory stimuli such as auditory, visual, olfactory, gustatory, and tactile.

Analgesia (*an* = without; *algia* = painful condition)   Insensibility to pain.

Anesthesia (*esthesia* = feeling)   Loss of feeling.

Aphasia (*phasis* = speech; *ia* = condition)   Diminished or complete loss of ability to comprehend or express spoken or written words due to injury or disease of the brain centers. The most common cause is CVA.

Apraxia (*pratto* = to do)   Inability to carry out purposeful movements in the absence of paralysis.

Bacterial meningitis (*itis* = inflammation of)   Acute inflammation of the meninges caused by bacteria.

Bradykinesia (*brady* = slow; *kinesis* = motion)   Abnormal slowness of movement.

Coma   Abnormally deep unconsciousness with an absence of voluntary response to stimuli and with varying degrees of reflex activity. It may be due to illness or to an injury.

Epidural (*epi* = above)   External to the dura mater.

Idiopathic (*idios* = own; *patho* = disease)   Self-originated; occurring without known cause or due to some other condition already present.

Lethargy   A condition of functional torpor or sluggishness.

Nerve block   Loss of sensation in a region, such as in local dental anesthesia.

Neuralgia (*neur* = nerve)   Attacks of pain along the entire course or branch of a peripheral sensory nerve.

Neuritis   Inflammation of a nerve. It can result from irritation to the nerve produced by trauma, bone fractures, nutritional deficiency (usually thiamine), poisons such as carbon monoxide and carbon tetrachloride, heavy metals such as lead, and some drugs.

Paralysis   Diminished or total loss of motor function resulting from damage to nervous tissue or a muscle.

Spastic (*spas* = draw or pull)   Resembling spasms or convulsions.

Stupor   Condition of unconsciousness, torpor, or lethargy with suppression of sense or feeling.

Torpor Abnormal inactivity or lack of response to normal stimuli.

Viral encephalitis An acute inflammation of the brain caused by a direct attack by various viruses or by an allergic reaction to any of the many viruses that are normally harmless to the central nervous system. If the virus affects the spinal cord as well, it is called *encephalomyelitis.*

STUDY OUTLINE

Brain
Principal Parts
1. The principal parts of the brain are the brain stem, the diencephalon, the cerebrum, and the cerebellum.
2. The brain is protected by the cranial bones, the cranial meninges, and the cerebrospinal fluid.

Cerebrospinal Fluid
1. Cerebrospinal fluid is formed in the choroid plexuses and circulates through the subarachnoid space, ventricles, and central canal. Most of the fluid is absorbed by the arachnoid villi of the superior sagittal sinus.
2. Cerebrospinal fluid protects by serving as a shock absorber. It also circulates nutritive substances from the blood.
3. The accumulation of cerebrospinal fluid in the head is called hydrocephalus. If the fluid accumulates in the ventricles, it is called internal hydrocephalus. If it accumulates in the subarachnoid space, it is called external hydrocephalus.

Blood Supply
1. The blood supply to the brain is via the circle of Willis.
2. Interruption of the oxygen supply to the brain can result in paralysis, mental retardation, epilepsy, or death.
3. Glucose deficiency may produce dizziness, convulsions, and unconsciousness.
4. The blood–brain barrier is a concept that explains the differential rates of passage of certain materials from the blood into the brain.

Brain Stem
1. The medulla oblongata is continuous with the upper part of the spinal cord. It contains nuclei that are reflex centers for regulation of heart rate, respiratory rate, vasoconstriction, swallowing, coughing, vomiting, sneezing, and hiccuping. It also contains the nuclei of origin of cranial nerves VIII (cochlear and vestibular branches) to XII.
2. The pons is superior to the medulla. It connects the spinal cord with the brain and links parts of the brain with one another. It relays impulses from the cerebral cortex to the cerebellum related to voluntary skeletal movements. It contains the nuclei for certain cranial nerves (V to VII and the vestibular branch of VIII). The reticular formation of the pons contains the pneumotaxic center, which helps control respiration.
3. The midbrain connects the pons and the diencephalon. It conveys motor impulses from the cerebrum to the cerebellum and cord and conveys sensory impulses from cord to thalamus. It regulates auditory and visual reflexes.

Diencephalon
1. The diencephalon consists of the thalamus and hypothalamus.
2. The thalamus is superior to the midbrain and contains nuclei that serve as relay stations for all sensory impulses, except smell, to the cerebral cortex. It also registers conscious recognition of pain and temperature and some awareness of crude touch and pressure.
3. The hypothalamus is inferior to the thalamus. It controls the autonomic nervous system, connects the nervous and endocrine systems, controls body temperature, food and fluid intake, the waking state, and sleep.

Cerebrum
1. The cerebrum is the largest part of the brain. Its cortex contains convolutions, fissures, and sulci.
2. The cerebral lobes are named the frontal, parietal, temporal, and occipital.
3. The white matter is under the cortex and consists of myelinated axons running in three principal directions.
4. The basal ganglia are paired masses of gray matter in the cerebral hemispheres. They help to control muscular movements.
5. The limbic system is found in the cerebral hemispheres and diencephalon. It functions in emotional aspects of behavior and memory.
6. The motor areas of the cerebral cortex are the regions that govern muscular movement. The sensory areas are concerned with the interpretation of sensory impulses. The association areas are concerned with emotional and intellectual processes.
7. Brain waves generated by the cerebral cortex are recorded as an EGG. They may be used to diagnose epilepsy, infections, and tumors.

8. Recent research indicates that the two hemispheres of the brain are not bilaterally symmetrical, either in function or anatomically.
9. The left hemisphere controls linguistic ability, while the right hemisphere controls melodies and most of our perceptions and analyses of nonverbal visual patterns.
10. Recent findings show that different emotional reactions follow damage to the right and left sides of the brain.
11. A unique specialization of the brain might be the phenomenon of handedness. In the human population no more than 9 percent are left-handed.
12. Scientists are testing the possibility that faulty connection between the hemispheres in some children may be a cause of dyslexia and other pathologies.

Cerebellum
1. The cerebellum occupies the inferior and posterior aspects of the cranial cavity. It consists of two hemispheres and a central constricted vermis.
2. It is attached to the brain stem by three pairs of cerebellar peduncles.
3. The cerebellum functions in the coordination of skeletal muscles, the maintenance of posture, and keeping the body balanced.

Cranial Nerves
1. Twelve pairs of cranial nerves originate from the brain.
2. The pairs are named primarily on the basis of distribution and numbered by order of attachment to the brain.

Applications to Health
1. Poliomyelitis is a viral infection that results in paralysis.
2. Syphilis is caused by the bacterium *Treponema pallidum* and may result in blindness, memory defects, abnormal behavior, and loss of sensory functions in trunk and limbs.
3. Cerebral palsy includes a group of central nervous system disorders that primarily involve the cerebral cortex, cerebellum, and basal ganglia. The disorders damage motor centers.
4. Parkinsonism is a progressive degeneration of the basal ganglia of the cerebrum resulting in insufficient dopamine.
5. Epilepsy results from irregular electrical discharges of brain cells and may be diagnosed by an EEG. The victim experiences convulsive seizures.
6. Multiple sclerosis is the destruction of myelin sheaths of the neurons of the central nervous system. Impulse transmission is interrupted.
7. Cerebrovascular accidents are also called strokes. Brain tissue is destroyed due to hemorrhage, thrombosis, and arteriosclerosis.
8. Dyslexia involves an inability of an individual to comprehend written language.
9. Tay-Sachs disease is an inherited disorder that involves neurological degeneration of the CNS because of excessive amounts of ganglioside G_{m2}.
10. Headaches are of two types: intracranial and extracranial.
11. Irritation of the trigeminal nerve is known as trigeminal neuralgia.

REVIEW QUESTIONS

1. Identify the four principal parts of the brain and the components of each, where applicable.
2. Describe the location of the cranial meninges.
3. Where is cerebrospinal fluid formed? Describe its circulation. Where is cerebrospinal fluid absorbed?
4. Distinguish between internal and external hydrocephalus.
5. Describe the blood supply to the brain. Explain the importance of oxygen and glucose to brain cells.
6. What is the blood–brain barrier? Is it of any advantage?
7. Describe the location and structure of the medulla. Define decussation of pyramids. Why is it important?
8. List the principal functions of the medulla.
9. Describe the location and structure of the pons. What are its functions?
10. Describe the location and structure of the midbrain. What are some of its functions?
11. Describe the location and structure of the thalamus. List some of the functions of the thalamus.
12. Where is the hypothalamus located? Explain some of the major functions of the hypothalamus.
13. Where is the cerebrum located? Describe the cortex, convolutions, fissures, and sulci of the cerebrum.
14. List and locate the lobes of the cerebrum. How are they separated from one another? What is the insula?
15. Describe the organization of cerebral white matter. Be sure to indicate the function of each group of fibers.
16. What are basal ganglia? Name the important basal ganglia and list the function of each.
17. Describe the effects of damage on the basal ganglia.
18. Define the limbic system. Explain several of its functions.
19. What is meant by a sensory area of the cerebral cortex? List, locate, and give the function of each sensory area.
20. What is meant by a motor area of the cerebral cortex? List, locate, and give the function of each motor area.
21. What is an association area of the cerebral cortex?

What are its functions?

22. Define an electroencephalogram. What is the diagnostic value of an EEG?

23. Describe brain lateralization and the split-brain concept.

24. What evidence is there that lateralization exists in the human brain?

25. Discuss the phenomenon of handedness in the human population.

26. Describe the location of the cerebellum. List the principal parts of the cerebellum.

27. Describe the relationship of the dural extensions to the cerebellum.

28. What are cerebellar peduncles? List and explain the function of each.

29. Explain the functions of the cerebellum. What is ataxia?

30. Define a cranial nerve. How are cranial nerves named and numbered? Distinguish between a mixed and sensory cranial nerve.

31. For each of the 12 pairs of cranial nerves, list (a) its name, number, and type, (b) its location, and (c) its function. In addition, list the effects of damage, where applicable.

32. Define each of the following: poliomyelitis, syphilis, tabes dorsalis, cerebral palsy, Parkinsonism, multiple sclerosis, epilepsy, cerebrovascular accidents, dyslexia, Tay-Sachs disease, tension headache, migraine headache, cluster headache, and trigeminal neuralgia.

33. Refer to the glossary of medical terminology associated with the nervous system. Be sure that you can define each term.

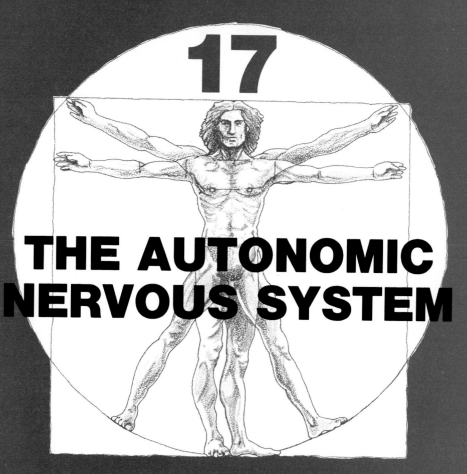

17

THE AUTONOMIC NERVOUS SYSTEM

STUDENT OBJECTIVES

■ Compare the structural and functional differences between the somatic efferent and autonomic portions of the nervous system.

■ Identify the structural features of the autonomic nervous system.

■ Compare the sympathetic and parasympathetic divisions of the autonomic nervous system in terms of structure, physiology, and chemical transmitters released.

■ Describe a visceral autonomic reflex and its components.

■ Explain the role of the hypothalamus and its relationship to the sympathetic and parasympathetic division.

■ Explain the relationship between biofeedback and the autonomic nervous system.

■ Describe the relationship between meditation and the autonomic nervous system.

The portion of the nervous system that regulates the activities of smooth muscle, cardiac muscle, and glands is the **autonomic nervous system.** Structurally, the system consists of visceral efferent neurons organized into nerves, ganglia, and plexuses. Functionally, it usually operates without conscious control. The system was originally named *autonomic* because physiologists thought it functioned with no control from the central nervous system, that it was autonomous or self-governing. It is now known that the autonomic system is neither structurally nor functionally independent of the central nervous system. It is regulated by centers in the brain, in particular by the cerebral cortex, hypothalamus, and medulla oblongata. However, the old terminology has been retained, and since the autonomic nervous system does differ from the somatic nervous system in some ways, the two are separated for convenience of study.

SOMATIC EFFERENT AND AUTONOMIC NERVOUS SYSTEMS

Whereas the somatic efferent nervous system produces conscious movement in skeletal muscles, the autonomic nervous system (visceral efferent nervous system) regulates visceral activities, and it generally does so involuntarily and automatically. Examples of visceral activities regulated by the autonomic nervous system are changes in the size of the pupil, accommodation for near vision, dilation and constriction of blood vessels, adjustment of the rate and force of the heartbeat, movements of the gastrointestinal tract, formation of gooseflesh, and secretion by most glands. These activities usually lie beyond conscious control. They are automatic.

The autonomic nervous system is entirely motor. All its axons are efferent fibers, which transmit impulses from the central nervous system to visceral effectors. Autonomic fibers are called **visceral efferent fibers. Visceral effectors** include cardiac muscle, smooth muscle, and glandular epithelium. This does not mean there are no afferent (sensory) impulses from visceral effectors, however. Impulses that give rise to visceral sensations pass over visceral afferent neurons that have cell bodies located in the posterior (dorsal) root ganglia of spinal nerves. Some functions of these afferent neurons were described with the cranial and spinal nerves. The hypothalamus, which largely controls the autonomic nervous system, also receives impulses from the visceral sensory fibers.

The autonomic nervous system consists of two principal divisions: the **sympathetic** and the **parasympathetic.** Many organs innervated by the autonomic nervous system receive visceral efferent neurons from both components of the autonomic system—one set from the sympathetic division, another from the parasympathetic division. In general, impulses transmitted by the fibers of one division stimulate the organ to start or increase activity, whereas impulses from the other division decrease the organ's activity. Organs that receive impulses from both sympathetic and parasympathetic fibers are said to have *dual innervation.* In the somatic efferent nervous system, only one kind of motor neuron innervates an organ, which is always a skeletal muscle. When the somatic neurons stimulate the cells of the skeletal muscle, the muscle becomes active. When the neuron ceases to stimulate the muscle, contraction stops altogether. Skeletal muscle cells of each motor unit contract only when stimulated by their motor neuron. When the impulse stops, contraction stops.

STRUCTURE OF THE AUTONOMIC NERVOUS SYSTEM

VISCERAL EFFERENT PATHWAYS

Autonomic visceral efferent pathways always consist of two neurons. One extends from the central nervous system to a ganglion. The other extends directly from the ganglion to the effector (muscle or gland).

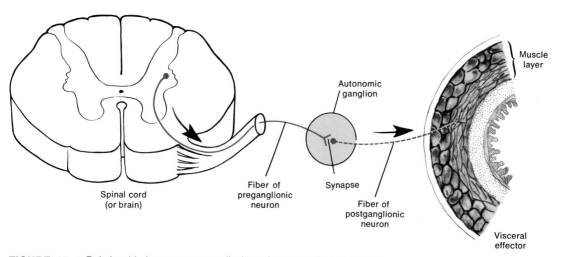

FIGURE 17-1 Relationship between preganglionic and postganglionic neurons.

The first of the visceral efferent neurons in an autonomic pathway is called a **preganglionic neuron** (Figure 17-1). Its cell body is in the brain or spinal cord. Its myelinated axon, called a **preganglionic fiber,** passes out of the central nervous system as part of a cranial or spinal nerve. At some point, the fiber separates from the nerve and courses to autonomic ganglia, where it synapses with the dendrites or cell body of the postganglionic neuron, the second neuron in the visceral efferent pathway.

The **postganglionic neuron** lies entirely outside the central nervous system. Its cell body and dendrites (if it has dendrites) are located in the autonomic ganglia, where the synapse with the preganglionic fibers occurs. The axon of a postganglionic neuron, called a **postganglionic fiber,** is unmyelinated and terminates in a visceral effector.

Thus preganglionic neurons convey efferent impulses from the central nervous system to autonomic ganglia. Postganglionic neurons relay the impulses from the autonomic ganglia to visceral effectors.

Preganglionic Neurons

In the sympathetic division, the preganglionic neurons have their cell bodies in the lateral gray horns of the twelve thoracic segments and first two lumbar segments of the spinal cord (Figure 17-2). It is for this reason that the sympathetic division is also called the **thoracolumbar** (thō'-ra-kō-LUM-bar) **division** and the fibers of the sympathetic preganglionic neurons are known as the **thoracolumbar outflow.**

The cell bodies of the preganglionic neurons of the parasympathetic division are located in the nuclei of cranial nerves III, VII, IX, and X in the brain stem and in the lateral gray horns of the second through fourth sacral segments of the spinal cord. Hence the parasympathetic division is also known as the **craniosacral division,** and the fibers of the parasympathetic preganglionic neurons are referred to as the **craniosacral outflow.**

Autonomic Ganglia

Autonomic pathways also include **autonomic ganglia,** where synapses between visceral efferent neurons occur. Autonomic ganglia differ from posterior root ganglia. The latter contain cell bodies of sensory neurons and no synapses occur in them.

The autonomic ganglia may be divided into three general groups. The **sympathetic trunk** or **vertebral chain ganglia** are a series of ganglia that lie in a vertical row on either side of the vertebral column, extending from the base of the skull to the coccyx (Figure 17-3). They are also known as **paravertebral** or **lateral ganglia.** They receive preganglionic fibers only from the thoracolumbar (sympathetic) division (Figure 17-2).

The second kind of autonomic ganglion also belongs to the sympathetic division. It is called a **prevertebral**

or **collateral ganglion** (Figure 17-3). The ganglia of this group lie anterior to the spinal column and close to the large abdominal arteries from which their names are derived. Examples of prevertebral ganglia so named are the celiac ganglion, on either side of the celiac artery just below the diaphragm; the superior mesenteric ganglion, near the beginning of the superior mesenteric artery in the upper abdomen; and the inferior mesenteric ganglion, located near the beginning of the inferior mesenteric artery in the middle of the abdomen (Figure 17-2). Prevertebral ganglia receive preganglionic fibers from the thoracolumbar (sympathetic) division.

The third kind of autonomic ganglion belongs to the parasympathetic division and is called a **terminal** or **intramural ganglion.** The ganglia of this group are located at the end of a visceral efferent pathway very close to visceral effectors or within the walls of visceral effectors. Terminal ganglia receive preganglionic fibers from the craniosacral (parasympathetic) division. The preganglionic fibers do not pass through sympathetic trunk ganglia (Figure 17-2).

In addition to autonomic ganglia, the autonomic nervous system also contains **autonomic plexuses.** Slender nerve fibers from ganglia containing postganglionic nerve cell bodies arranged in a branching network constitute an autonomic plexus.

Postganglionic Neurons

Axons from preganglionic neurons of the sympathetic division pass to ganglia of the sympathetic trunk. They can either synapse in the sympathetic chain ganglia with postganglionic sympathetics or they can continue, without synapsing, through the chain ganglia to end at a prevertebral ganglion where synapses with the postganglionic sympathetics can take place. Each sympathetic preganglionic fiber synapses with several postganglionic fibers in the ganglion, and the postganglionic fibers pass to several visceral effectors. Upon exiting their ganglia, the postsynaptic fibers innervate their visceral effectors.

Axons from preganglionic neurons of the parasympathetic division pass to terminal ganglia near or within a visceral effector. In the ganglion, the presynaptic neuron usually synapses with only four or five postsynaptic neurons to a single visceral effector. Upon exiting their ganglia, the postsynaptic fibers supply their visceral effectors.

With this background in mind, we can now examine some specific structural features of the sympathetic and parasympathetic divisions of the autonomic nervous system.

SYMPATHETIC DIVISION

The preganglionic fibers of the sympathetic division have their cell bodies located in the lateral gray horn of the spinal cord in the thoracic and first two lumbar segments

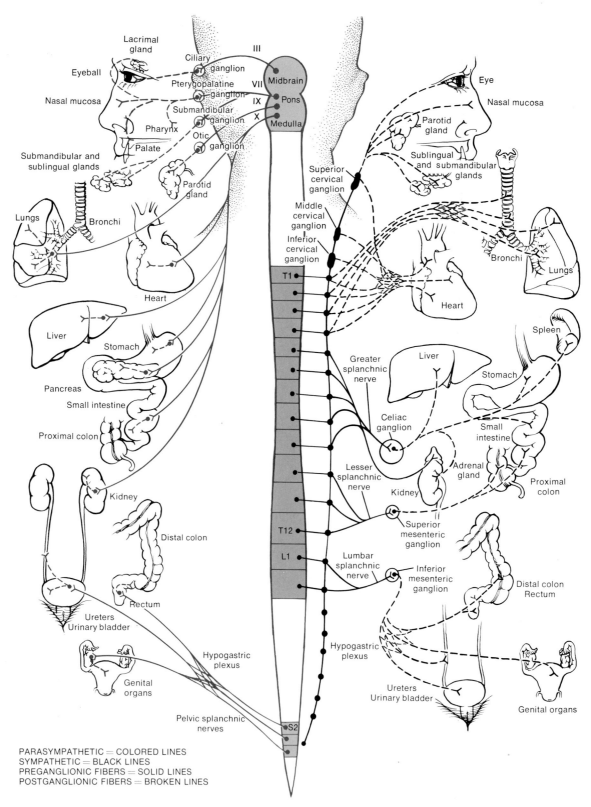

FIGURE 17-2 Structure of the autonomic nervous system.

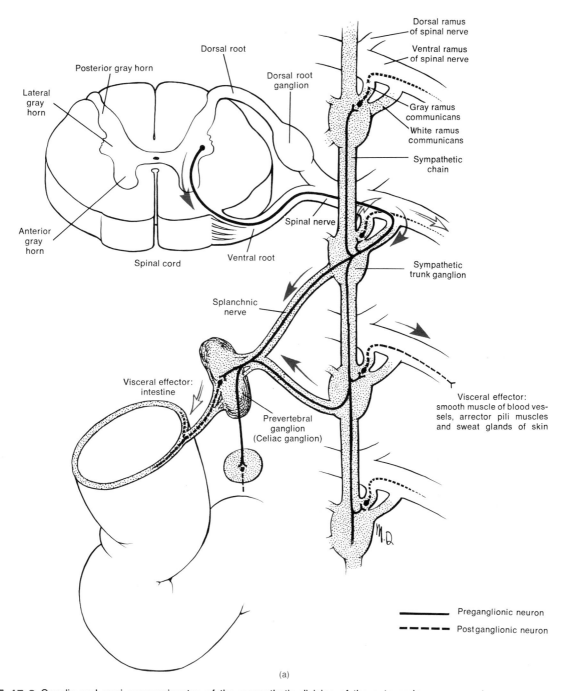

Posterior gray horn

Lateral gray horn

Dorsal root

Dorsal root ganglion

Dorsal ramus of spinal nerve

Ventral ramus of spinal nerve

Gray ramus communicans

White ramus communicans

Sympathetic chain

Spinal nerve

Sympathetic trunk ganglion

Anterior gray horn

Spinal cord

Ventral root

Splanchnic nerve

Visceral effector: intestine

Prevertebral ganglion (Celiac ganglion)

Visceral effector: smooth muscle of blood vessels, arrector pili muscles and sweat glands of skin

———— Preganglionic neuron

– – – – Postganglionic neuron

(a)

FIGURE 17-3 Ganglia and rami communicantes of the sympathetic division of the autonomic nervous system. (a) Diagram.

(Figure 17-2). The preganglionic fibers are myelinated and leave the spinal cord through the ventral root of a spinal nerve along with the somatic efferent fibers at the same segmental levels. After exiting through the intervertebral foramina, the preganglionic sympathetic fibers enter a white ramus to pass to the nearest sympathetic trunk ganglion on the same side. Collectively, the white rami are called the **white rami communicantes** (kō-myoo-ne-KAN-tēz). Their name indicates that they contain myelinated fibers. Only thoracic and upper lumbar nerves have white rami communicantes. The white rami communicantes connect the ventral ramus of the spinal nerve with the ganglia of the sympathetic trunk.

The paired sympathetic trunks are situated anterolaterally to the spinal cord, one on either side. Each consists of a series of ganglia arranged more or less segmentally. The divisions of the sympathetic trunk are named on the basis of location. Typically, there are 22 ganglia in each chain: 3 cervical, 11 thoracic, 4 lumbar, and 4 sacral. Although the trunk extends downward from the neck,

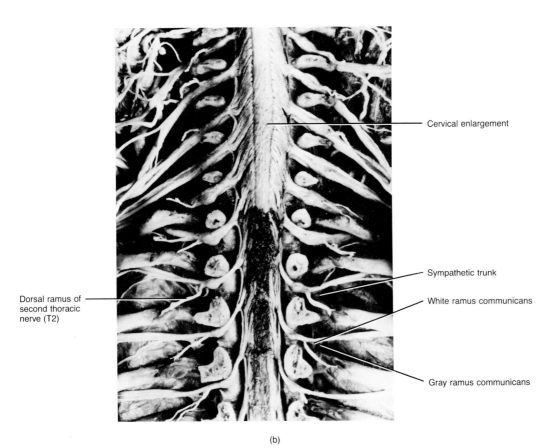

Cervical enlargement

Sympathetic trunk

White ramus communicans

Dorsal ramus of
second thoracic
nerve (T2)

Gray ramus communicans

(b)

FIGURE 17-3 (*Continued*) Ganglia and rami communicantes of the sympathetic division of the autonomic
nervous system. (b) Photograph of the posterior aspect of the lower cervical and upper thoracic segments of
the spinal cord showing the trunks of the brachial plexus. (Courtesy of N. Gluhbegovic and T. H. Williams, *The
Human Brain: A Photographic Guide*, Harper & Row, New York, 1980.)

thorax, and abdomen to the coccyx, it receives preganglionic fibers only from the thoracic and lumbar segments
of the spinal cord (Figure 17-2).

The cervical portion of each sympathetic trunk is located in the neck anterior to the prevertebral muscles.
It is subdivided into a superior, middle, and inferior ganglion (Figure 17-2). The *superior cervical ganglion* is posterior to the internal carotid artery and anterior to the
transverse processes of the second cervical vertebra. Postganglionic fibers leaving the ganglion serve the head,
where they are distributed to the sweat glands, the smooth
muscle of the eye and the smooth muscle of the blood
vessels of the face, the nasal mucosa, and the submandibular, sublingual, and parotid salivary glands. Gray rami
communicantes from the ganglion also pass to the upper
two to four cervical spinal nerves. The *middle cervical
ganglion* is situated near the sixth cervical vertebra at
the level of the cricoid cartilage. Postganglionic fibers
from it innervate the heart. The *inferior cervical ganglion*
is located near the first rib, anterior to the transverse
processes of the seventh cervical vertebra. Its postganglionic fibers also supply the heart.

The thoracic portion of each sympathetic trunk usually
consists of 11 segmentally arranged ganglia, lying ventral

to the necks of the corresponding ribs. This portion of
the sympathetic trunk receives most of the sympathetic
preganglionic fibers. Postganglionic fibers from the thoracic sympathetic trunk innervate the heart, lungs, bronchi,
and other thoracic viscera.

The lumbar portion of each sympathetic trunk is found
on either side of the corresponding lumbar vertebrae. The
sacral portion of the sympathetic trunk lies in the pelvic
cavity on the medial side of the sacral foramina. Postganglionic fibers from the lumbar and sacral sympathetic
chain ganglia are distributed with the respective spinal
nerves via gray rami, or they may join the hypogastric
plexus via direct visceral branches.

When a preganglionic fiber of a white ramus communicans enters the sympathetic trunk, it may terminate
(synapse) in several ways. Some fibers synapse in the first
ganglion at the level of entry. Others pass up or down
the sympathetic trunk for a variable distance to form
the fibers on which the ganglia are strung. These fibers,
known as **sympathetic chains** (Figure 17-3), may not
synapse until they reach a ganglion in the cervical or
sacral area. Some postganglionic fibers leaving the sympathetic trunk ganglia pass directly to visceral effectors of
the head, neck, chest, and abdomen. Most, however, re-

join the spinal nerves before supplying peripheral visceral effectors such as sweat glands and the smooth muscle in blood vessels and around hair follicles in the extremities. The **gray ramus communicans** (kō-MYOO-ne-kanz) is the structure containing the postganglionic fibers that connect the ganglion of the sympathetic trunk to the spinal nerve (Figure 17-3). The term *gray* refers to the fact that the fiber is unmyelinated. All spinal nerves have gray rami communicantes. Gray rami communicantes outnumber the white rami, since there is a gray ramus leading to each of the 31 pairs of spinal nerves.

In most cases, a sympathetic preganglionic fiber terminates by synapsing with a large number of postganglionic cell bodies in a ganglion, usually 20 or more. Often the postganglionic fibers then terminate in widely separated organs of the body. Thus an impulse that starts in a single preganglionic neuron may affect several visceral effectors. For this reason, most sympathetic responses have widespread effects on the body.

Some preganglionic fibers pass through the sympathetic trunk without terminating in the trunk. Beyond the trunk, they form nerves known as **splanchnic** (SPLANK-nik) **nerves** (Figure 17-2). After passing through the trunk of ganglia, the splanchnic nerves terminate in the *celiac* or *solar plexus.* In the plexus, the preganglionic fibers synapse in ganglia with postganglionic cell bodies. These ganglia are prevertebral ganglia. The greater splanchnic nerve passes to the celiac ganglion of the celiac plexus. From here, postganglionic fibers are distributed to the stomach, spleen, liver, kidney, and small intestine. The lesser splanchnic nerve passes through the celiac plexus to the superior mesenteric ganglion of the superior mesenteric plexus. Postganglionic fibers from this ganglion innervate the small intestine and colon. The lowest splanchnic nerve, not always present, enters the renal plexus. Postganglionics supply the renal artery and ureter. The lumbar splanchnic nerve enters the inferior mesenteric plexus. In the plexus, the preganglionic fibers synapse with postganglionic fibers in the inferior mesenteric ganglion. These fibers pass through the hypogastric plexus and supply the distal colon and rectum, urinary bladder, and genital organs. As noted earlier, the postganglionic fibers leaving the prevertebral ganglia follow the course of various arteries to abdominal and pelvic visceral effectors.

PARASYMPATHETIC DIVISION

The preganglionic cell bodies of the parasympathetic division are found in nuclei in the brain stem and the lateral gray horn of the second through fourth sacral segments of the spinal cord (Figure 17-2). Their fibers emerge as part of a cranial nerve or as part of the ventral root of a spinal nerve. The **cranial parasympathetic outflow** consists of preganglionic fibers that leave the brain stem by way of the oculomotor nerves (III), facial nerves (VII), glossopharyngeal nerves (IX), and vagus nerves (X). The

sacral parasympathetic outflow consists of preganglionic fibers that leave the ventral roots of the second through fourth sacral nerves. The preganglionic fibers of both the cranial and sacral outflows end in terminal ganglia, where they synapse with postganglionic neurons. We will first look at the cranial outflow.

The cranial outflow has five components: four pairs of ganglia and the plexuses associated with the vagus nerve. The four pairs of cranial parasympathetic ganglia innervate structures in the head and are located close to the organs they innervate. The *ciliary ganglion* is near the back of an orbit lateral to each optic nerve. Preganglionic fibers pass with the oculomotor nerve (III) to the ciliary ganglion. Postganglionic fibers from the ganglion innervate smooth muscle cells in the eyeball. Each *pterygopalatine* (ter'-i-gō-PAL-a-tin) *ganglion* is situated lateral to a sphenopalatine foramen. It receives preganglionic fibers from the facial nerve (VII) and transmits postganglionic fibers to the nasal mucosa, palate, pharynx, and lacrimal gland. Each *submandibular ganglion* is found near the duct of a submandibular salivary gland. It receives preganglionic fibers from the facial nerve (VII) and transmits postganglionic fibers that innervate the submandibular and sublingual salivary glands. The *otic ganglia* are situated just below each foramen ovale. The otic ganglion receives preganglionic fibers from the glossopharyngeal nerve (IX) and transmits postganglionic fibers that innervate the parotid salivary gland. Ganglia associated with the cranial outflow are classified as terminal ganglia. Since the terminal ganglia are close to their visceral effectors, postganglionic parasympathetic fibers are short. Postganglionic sympathetic fibers are relatively long.

The last component of the cranial outflow consists of the preganglionic fibers that leave the brain via the vagus nerves (X). This component has the most extensive distribution of the parasympathetic fibers. It provides about 80 percent of the craniosacral outflow. Each vagus nerve enters into the formation of several plexuses in the thorax and abdomen. As it passes through the thorax, it sends fibers to the *superficial cardiac plexus* in the arch of the aorta and the *deep cardiac plexus* anterior to the branching of the trachea. These plexuses contain terminal ganglia, and the postganglionic parasympathetic fibers emerging from them supply the heart. Also in the thorax is the *pulmonary plexus,* in front of and behind the roots of the lungs and within the lungs themselves. It receives preganglionic fibers from the vagus and transmits postganglionic parasympathetic fibers to the lungs and bronchi. Other plexuses associated with the vagus nerve are described in later chapters in conjunction with the appropriate thoracic, abdominal, and pelvic viscera. Postganglionic fibers from these plexuses innervate viscera such as the liver, pancreas, stomach, kidneys, small intestine, and part of the colon.

The sacral parasympathetic outflow consists of preganglionic fibers from the ventral roots of the second through fourth sacral nerves. Collectively, they form the *pelvic splanchnic nerves*. They pass into the hypogastric plexus. From ganglia in the plexus, parasympathetic postganglionic fibers are distributed to the colon, ureters, urinary bladder, and reproductive organs.

Salient structural features of the sympathetic and parasympathetic divisions are compared in Exhibit 17-1.

EXHIBIT 17-1 STRUCTURAL FEATURES OF SYMPATHETIC AND PARASYMPATHETIC DIVISIONS

SYMPATHETIC	PARASYMPATHETIC
Forms thoracolumbar outflow.	Forms craniosacral outflow.
Contains sympathetic trunk and prevertebral ganglia.	Contains terminal ganglia.
Ganglia are close to the CNS and distant from visceral effectors.	Ganglia are near or within visceral effectors.
Each preganglionic fiber synapses with many postganglionic neurons that pass to many visceral effectors.	Each preganglionic fiber usually synapses with four or five postganglionic neurons that pass to a single visceral effector.
Distributed throughout the body, including the skin.	Distribution limited primarily to head and viscera of thorax, abdomen, and pelvis.

PHYSIOLOGY

TRANSMITTER SUBSTANCES

Autonomic fibers, like other axons of the nervous system, release transmitter substances at synapses as well as at points of contact between autonomic fibers and visceral effectors. These latter points are called **neuroeffector junctions.** Neuroeffector junctions may be either neuromuscular junctions or neuroglandular junctions. On the basis of the transmitter substance produced, autonomic fibers may be classified as either cholinergic (kō'-lin-ER-jik) or adrenergic (ad'-ren-ER-jik).

Cholinergic fibers release acetylcholine (ACh) and include the following: (1) all sympathetic and parasympathetic preganglionic axons, (2) all parasympathetic postganglionic axons, and (3) some sympathetic postganglionic axons. The cholinergic sympathetic postganglionic axons include those to sweat glands and blood vessels in skeletal muscles, to the skin, and to the external genitalia. Since acetylcholine is quickly inactivated by the enzyme cholinesterase (AChE), the effects of cholinergic fibers are short-lived and local.

Adrenergic fibers produce the transmitter substance norepinephrine (NE), also called noradrenaline or sympathin. Most sympathetic postganglionic axons are adrenergic. Since norepinephrine is inactivated much more slowly by catechol-o-methyltransferase (COMT) or monamine oxidase (MAO) than acetylcholine is, and since norepinephrine may enter the bloodstream, the effects of sympathetic stimulation are longer lasting and more widespread than parasympathetic stimulation.

ACTIVITIES

Most visceral effectors have dual innervation, that is, they receive fibers from both the sympathetic and the parasympathetic divisions. In these cases, impulses from one division stimulate the organ's activities, whereas impulses from the other division inhibit the organ's activities. The stimulating division may be either the sympathetic or the parasympathetic, depending on the organ. For example, sympathetic impulses increase heart activity, whereas parasympathetic impulses decrease it. On the other hand, parasympathetic impulses increase digestive activities, whereas sympathetic impulses inhibit them. The actions of the two systems are carefully integrated to help maintain homeostasis. A summary of the activities of the autonomic nervous system is presented in Exhibit 17-2.

The parasympathetic division is primarily concerned with activities that restore and conserve body energy. It is a **rest-repose system.** Under normal body conditions, for instance, parasympathetic impulses to the digestive glands and the smooth muscle of the digestive system dominate over sympathetic impulses. Thus energy-supplying food can be digested and absorbed by the body.

The sympathetic division, by contrast, is primarily concerned with processes involving the expenditure of energy. When the body is in homeostasis, the main function of the sympathetic division is to counteract the parasympathetic effects just enough to carry out normal processes requiring energy. During extreme stress, however, the sympathetic dominates the parasympathetic. When people are confronted with a stress condition, for example, their bodies become alert and they sometimes perform feats of unusual strength. Fear stimulates the sympathetic division.

Activation of the sympathetic division sets into operation a series of physiological responses collectively called the **fight-or-flight response.** It produces the following effects.

1. The pupils of the eyes dilate.
2. The heart rate increases.
3. The blood vessels of the skin and viscera constrict.
4. The remainder of the blood vessels dilate. This reaction causes a rise in blood pressure and a faster flow of blood into the dilated blood vessels of skeletal muscles, cardiac muscle, lungs, and brain—organs involved in fighting off danger.
5. Rapid breathing occurs as the bronchioles dilate to allow faster movement of air in and out of the lungs.
6. Blood sugar level rises as liver glycogen is converted

EXHIBIT 17-2 ACTIVITIES OF AUTONOMIC NERVOUS SYSTEM

VISCERAL EFFECTOR	EFFECT OF SYMPATHETIC STIMULATION	EFFECT OF PARASYMPATHETIC STIMULATION
Eye		
Iris	Contracts dilator muscle of iris and brings about dilation of pupil.	Contracts sphincter muscle of iris and brings about constriction of pupil.
Ciliary muscle	No innervation.	Contracts ciliary muscle and accommodates lens for near vision.
Glands		
Sweat	Stimulates secretion.	No innervation.
Lacrimal (tear)	Vasoconstriction, which inhibits secretion.	Normal or excessive secretion.
Salivary	Vasoconstriction, which decreases salivary secretion.	Stimulation of salivary secretion and vasodilation.
Gastric	Vasoconstriction, which inhibits secretion.	Secretion stimulated.
Intestinal	Vasoconstriction, which inhibits secretion.	Secretion stimulated.
Adrenal medulla	Promotes epinephrine and norepinephrine secretion.	No innervation.
Adrenal cortex	Promotes glucocorticoid secretion.	No innervation.
Lungs (bronchial tubes)	Dilation.	Constriction.
Heart	Increases rate and strength of contraction; dilates coronary vessels that supply blood to heart muscle cells.	Decreases rate and strength of contraction; constricts coronary vessels.
Blood vessels		
Skin	Constriction.	No innervation for most.
Skeletal muscle	Dilation.	No innervation.
Visceral organs (except heart and lungs)	Constriction.	No innervation for most.
Liver	Promotes glycogenolysis; decreases bile secretion.	Promotes glycogenesis; increases bile secretion.
Stomach	Decreases motility.	Increases motility.
Intestines	Decreases motility.	Increases motility.
Kidney	Constriction of blood vessels that results in decreased urine volume.	No innervation.
Pancreas	Inhibits secretion.	Promotes secretion.
Spleen	Contraction and discharge of stored blood into general circulation.	No innervation.
Urinary bladder	Relaxes muscular wall; increases tone in internal sphincter.	Contracts muscular wall; relaxes internal sphincter.
Arrector pili of hair follicles	Contraction results in erection of hairs ("goose pimples").	No innervation.
Uterus	Inhibits contraction if nonpregnant; stimulates contraction if pregnant.	Minimal effect.
Sex organs	In male, vasoconstriction of ductus deferens, seminal vesicle, prostate; results in ejaculation. In female, reverse uterine peristalsis.	Vasodilation and erection in both sexes; secretion in female.

to glucose to supply the body's additional energy needs.

7. The medulla of the adrenal gland is stimulated to produce epinephrine and norepinephrine, hormones that intensify and prolong the sympathetic effects noted above.

8. Processes that are not essential for meeting the stress situation are inhibited. For example, muscular movements of the gastrointestinal tract and digestive secretions are slowed down or even stopped.

CLINICAL APPLICATION

If the sympathetic trunk is cut on one side, the sympathetic supply to that side of the head is removed and the result is **Horner's syndrome,** in which the patient exhibits (on the affected side): anhidrosis (lack of sweating), enophthalmos (the eye appears sunken), a constricted pupil, and flushing of the skin.

VISCERAL AUTONOMIC REFLEXES

A **visceral autonomic reflex** adjusts the activity of a visceral effector. In other words, it results in the contraction of smooth or cardiac muscle or secretion by a gland. Such reflexes assume a key role in activities such as regulating heart action, blood pressure, respiration, digestion, defecation, and urinary bladder functions.

A visceral autonomic reflex arc consists of the following components.

1. Receptor. The receptor is the distal end of an afferent neuron in an exteroceptor or enteroceptor.

2. Afferent neuron. This neuron, either a somatic afferent or visceral afferent neuron, conducts the sensory impulse to the spinal cord or brain.

3. Association neurons. These neurons are found in the central nervous system.

4. Visceral efferent preganglionic neuron. In the thoracic and abdominal regions, this neuron is in the lateral gray horn of the spinal cord. The axon passes through the ventral root of the spinal nerve, the spinal nerve, and the white ramus communicans. It then enters a sympathetic trunk or prevertebral ganglion, where it synapses with a postganglionic neuron. In the cranial and sacral regions, the visceral efferent preganglionic axon leaves the central nervous system and passes to a terminal ganglion, where it synapses with a postganglionic neuron. The role of the visceral efferent preganglionic neuron is to convey a motor impulse from the brain or spinal cord to an autonomic ganglion.

5. Visceral efferent postganglionic neuron. This neuron conducts a motor impulse from a visceral efferent preganglionic neuron to the visceral effector.

6. Visceral effector. A visceral effector is smooth muscle, cardiac muscle, or a gland.

The basic difference between a somatic reflex arc and a visceral autonomic reflex arc is that in a somatic reflex arc, only one efferent neuron is involved. In a visceral autonomic reflex arc, two efferent neurons are involved.

Visceral sensations do not always reach the cerebral cortex. Most remain at subconscious levels. Under normal conditions, you are not aware of muscular contractions of the digestive organs, heartbeat, changes in the diameter of blood vessels, and pupil dilation and constriction. When your body is making adjustments in such visceral activities, they are handled by visceral reflex arcs whose centers are in the spinal cord or lower regions of the brain. Among such centers are the cardiac, respiratory, vasomotor, swallowing, and vomiting centers in the medulla and the temperature control center in the hypothalamus. Stimuli delivered by somatic or visceral afferent neurons synapse in these centers, and the returning motor impulses conducted by visceral efferent neurons bring about an adjustment in the visceral effector without conscious recognition. The impulses are interpreted and acted on subconsciously. Some visceral sensations do give rise to conscious recognition: hunger, nausea, and fullness of the urinary bladder and rectum.

CONTROL BY HIGHER CENTERS

The autonomic nervous system is not a separate nervous system. Axons from many parts of the central nervous system are connected to both the sympathetic and the parasympathetic divisions of the autonomic nervous system and thus exert considerable control over it. Autonomic centers in the cerebral cortex are connected to autonomic centers of the thalamus, for example. These, in turn, are connected to the hypothalamus. In this hierarchy of command, the thalamus sorts incoming impulses before they reach the cerebral cortex. The cerebral cortex then turns over control and integration of visceral activities to the hypothalamus. It is at the level of the hypothalamus that the major control and integration of the autonomic nervous system is exerted.

The hypothalamus is connected to both the sympathetic and the parasympathetic divisions of the autonomic nervous system. The posterior and lateral portions of the hypothalamus appear to control the sympathetic division. When these areas are stimulated, there is an increase in visceral activities—an increase in heart rate, a rise in blood pressure due to vasoconstriction of blood vessels, an increase in the rate and depth of respiration, dilation of the pupils, and inhibition of the digestive tract. On the other hand, the anterior and medial portions of the hypothalamus seem to control the parasympathetic division. Stimulation of these areas results in a decrease in heart rate, lowering of blood pressure, constriction of the pupils, and increased motility of the digestive tract.

Control of the autonomic nervous system by the cerebral cortex occurs primarily during emotional stress. In extreme anxiety, the cerebral cortex can stimulate the hypothalamus. This stimulation, in turn, increases heart rate and blood pressure. If the cortex is stimulated by an extremely unpleasant sight, the stimulation causes vasodilation of blood vessels, a lowering of blood pressure, and fainting.

Evidence of even more direct control of visceral responses is provided by data gathered from studies of biofeedback and meditation.

BIOFEEDBACK

In the simplest terms, **biofeedback** is a process in which people get constant signals, or feedback, about visceral body functions such as blood pressure, heart rate, and muscle tension. By using special monitoring devices, they can control these visceral functions consciously.

Suppose you are connected to a monitor that informs you of your heartbeat by means of lights. A red light indicates a fast heart beat, an amber light a normal rate, and a green light a slow rate. When you see the red light flash, you know your heart is beating too fast. You have been informed of a visceral response. This is the biofeedback. According to some researchers, you can be taught to slow down the heart rate by thinking of something pleasant and thus relaxing the body. The green light flashes and your reward is a slower heart rate. In similar experiments, some individuals have learned to control heart rhythm.

Researchers estimate that between 5 and 10 percent of the American population suffers from migraine headaches. Moreover, no effective treatment has been developed that does not have significant side effects and serious risks. One approach to alleviating migraine headaches is biofeedback.* The first clue that biofeedback could be used in this way came when a patient in the voluntary control laboratory of the Menninger Foundation demonstrated that with a 10°F rise in hand skin temperature (as a result of vasodilation and increased blood flow) she could spontaneously recover from a migraine headache.

In a study conducted at the Menninger Foundation, subjects suffering from migraine headaches received instructions in the use of a monitor that registers the skin temperature of the right index finger. Subjects were also given a typewritten sheet containing two sets of phrases. The first set was designed to help them relax the entire body. The second set was designed to bring about an increased flow of blood in the hands. The subjects practiced raising their skin temperature at home for 5 to 15 minutes a day. When skin temperature increased, the monitor emitted a high-pitched sound. In time, the monitor was abandoned.

Once the subjects learned how to vasodilate their blood vessels, the migraine headaches lessened. Since migraine headaches are believed to involve a distension of blood vessels in the head, the shunting of blood from head to hands relieved the distension and thus the pain.

Other experiments have shown that biofeedback can be applied to childbirth. Women were given monitors hooked up to their fingers and arms to measure electrical conductivity of the skin and skeletal muscle tension. Both conductivity and tension increase with nervousness and make labor difficult. Muscle tension was recorded as a sirenlike sound that became louder with nervousness. Skin conductivity was recorded as a crackling noise that also increased with nervousness. The monitors kept the women informed of their nervousness. This was the biofeedback. Having pleasant thoughts reduced the sound levels. The reward was less nervousness. The results of the study indicate that the women needed less medication during labor and labor time itself was shortened.

There is no way to determine where biofeedback will lead. Perhaps the outstanding contribution of biofeedback research has been to demonstrate that the autonomic nervous system is not autonomous. Visceral responses can be controlled. Strong supporters point out that possible applications of biofeedback in medicine are legion. They envision its use in lowering blood pressure in patients with hypertension, altering heart rates and rhythms, relieving pain from migraine headaches, making delivery easier, and controlling anxiety related to a host of illnesses that may be linked to stress. One researcher concludes that "an important trend is beginning to take place in the areas of psychosomatic disorders and medicine. This is the increasing involvement of the patient in his own treatment. The traditional doctor-patient relationship is giving way slowly to a shared responsibility."

MEDITATION

Yoga, which literally means union, is defined as a higher consciousness achieved through a fully rested and relaxed body and a fully awake and relaxed mind. One widely practiced technique for achieving higher consciousness is called **transcendental meditation.** One sits in a comfortable position with the eyes closed and concentrates on a suitable sound or thought.

Research indicates that transcendental meditation can alter physiological responses. Oxygen consumption decreases drastically along with carbon dioxide elimination. Subjects have experienced a reduction in metabolic rate and blood pressure. Researchers have also observed a decrease in heart rate, an increase in the intensity of alpha brain waves, a sharp decrease in the amount of lactic acid in the blood, and an increase in the skin's electrical resistance. These last four responses are characteristic of a highly relaxed state of mind. Alpha waves are found in the EEGs of almost all individuals in a resting, but awake state; they disappear during sleep.

These responses have been called an **integrated response**—essentially, a hypometabolic state due to inactivation of the sympathetic division of the autonomic nervous system. The response is the exact opposite of the fight-or-flight response, which is a hyperactive state of the sympathetic division. The existence of the integrated response suggests that the central nervous system does exert some control over the autonomic nervous system.

* Much of the following discussion of the use of biofeedback for the treatment of migraine headaches is based upon the information provided by Dr. Joseph D. Sargent of the Menninger Foundation, Topeka, Kansas.

STUDY OUTLINE

Somatic Efferent and Autonomic Nervous Systems

1. The autonomic nervous system, or visceral efferent nervous system, regulates visceral activities, that is, activities of smooth muscle, cardiac muscle, and glands.
2. It usually operates without conscious control.
3. It is regulated by centers in the brain, in particular by the cerebral cortex, the hypothalamus, and the medulla oblongata.
4. The somatic efferent nervous system produces conscious movement in skeletal muscles.

Structure of the Autonomic Nervous System

1. The autonomic nervous system consists of visceral efferent neurons organized into nerves, ganglia, and plexuses.
2. It is entirely motor. All autonomic axons are efferent fibers.
3. Efferent neurons are preganglionic (with myelinated axons) and postganglionic (with unmyelinated axons).
4. The autonomic system consists of two principal divisions: sympathetic (thoracolumbar) and parasympathetic (craniosacral).
5. Autonomic ganglia are classified as sympathetic trunk ganglia (on sides of spinal column), prevertebral ganglia (anterior to spinal column), and terminal ganglia (near or inside visceral effectors).

Physiology

1. Autonomic fibers release chemical transmitters at synapses. On the basis of the transmitter produced, these fibers may be classified as cholinergic or adrenergic.
2. Cholinergic fibers release acetylcholine. Adrenergic fibers produce norepinephrine (noradrenaline), also called sympathin.
3. Sympathetic responses are widespread and, in general, concerned with energy expenditure. Parasympathetic responses are restricted and are typically concerned with energy restoration and conservation.

Visceral Autonomic Reflexes

1. A visceral autonomic reflex adjusts the activity of a visceral effector.
2. A visceral autonomic reflex arc consists of a receptor, afferent neuron, association neuron, visceral efferent preganglionic neuron, visceral efferent postganglionic neuron, and visceral effector.

Control by Higher Centers

1. The hypothalamus controls and integrates the autonomic nervous system. It is connected to both the sympathetic and the parasympathetic divisions.
2. Biofeedback is a process in which people learn to monitor visceral functions and to control them consciously. It has been used to control heart rate, to alleviate migraine headaches, and to make childbirth easier.
3. Yoga is a higher consciousness achieved through a fully rested and relaxed body and a fully awake and relaxed mind.
4. Transcendental meditation produces the following physiological responses: decreased oxygen consumption and carbon dioxide elimination, reduced metabolic rate, decrease in heart rate, increase in the intensity of alpha brain waves, a sharp decrease in the amount of lactic acid in the blood, and an increase in the skin's electrical resistance.

REVIEW QUESTIONS

1. What are the principal components of the autonomic nervous system? What is its general function? Why is it called involuntary?
2. What is the principal anatomical difference between the voluntary nervous system and the autonomic nervous system?
3. Relate the role of visceral efferent fibers and visceral effectors to the autonomic nervous system.
4. Distinguish between preganglionic neurons and postganglionic neurons with respect to location and function.
5. What is an autonomic ganglion? Describe the location and function of the three types of autonomic ganglia. Define white and gray rami communicantes.
6. On what basis are the sympathetic and parasympathetic divisions of the autonomic nervous system differentiated anatomically and functionally?
7. Discuss the distinction between cholinergic and adrenergic fibers of the autonomic nervous system.
8. Give examples of the antagonistic effects of the sympathetic and parasympathetic divisions of the autonomic nervous system.
9. Summarize the principal functional differences between the voluntary nervous system and the autonomic nervous system.
10. See page 467 for a diagram of the human body as it might appear in a fear situation. Specific parts of the body have been labeled. Write the *sympathetic response* for each of the labeled parts.

11. Define a visceral autonomic reflex and give three examples.
12. Describe a complete visceral autonomic reflex in proper sequence.
13. Describe how the hypothalamus controls and integrates the autonomic nervous system.
14. Define biofeedback. Explain how it could be useful.
15. What is transcendental meditation? How is the integrated response related to the autonomic nervous system?

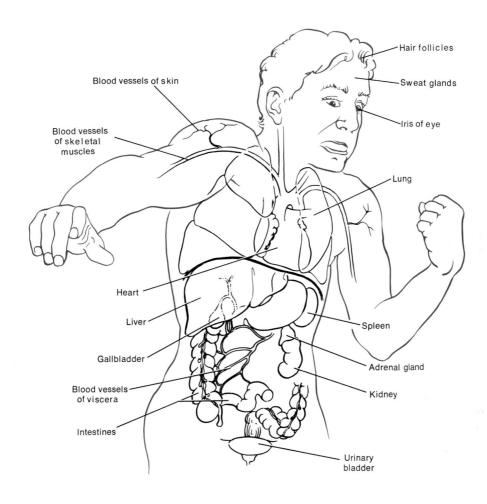

Blood vessels of skin

Blood vessels of skeletal muscles

Hair follicles

Sweat glands

Iris of eye

Lung

Heart

Liver

Gallbladder

Blood vessels of viscera

Intestines

Spleen

Adrenal gland

Kidney

Urinary bladder

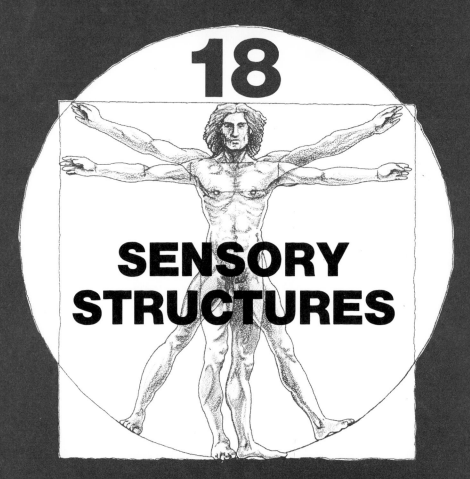

18

SENSORY STRUCTURES

STUDENT OBJECTIVES

- ■ Define a sensation and list the four prerequisites for its transmission.

- ■ Compare the location and function of exteroceptors, visceroceptors, and proprioceptors.

- ■ Describe the distribution of cutaneous receptors.

- ■ List the location and function of the receptors for tactile sensations (touch, pressure, vibration), thermoreceptive sensations (heat and cold), pain, and proprioception.

- ■ Distinguish among somatic, visceral, referred, and phantom pain.

- ■ Describe how acupuncture is believed to relieve pain.

- ■ Identify the locations and components of the principal sensory and motor pathways.

- ■ Locate the receptors for olfaction and describe the neural pathway for smell.

- ■ Identify the gustatory receptors and describe the neural pathway for taste.

- ■ Describe the structure of the accessory visual organs.

- ■ List the structural divisions of the eye.

- ■ Describe the afferent pathway of light impulses to the brain.

- ■ Define the anatomical subdivisions of the ear.

- ■ List the principal events in hearing.

- ■ Identify the receptor organs for static and dynamic equilibrium.

- ■ Contrast the causes and symptoms of cataract, glaucoma, conjunctivitis, trachoma, labyrinthine disease, Ménière's disease, impacted cerumen, otitis media, and motion sickness.

- ■ Define key medical terms associated with the sense organs.

SENSATIONS

Your ability to sense stimuli is vital to your survival. If pain could not be sensed, burns would be common and an inflamed appendix or stomach ulcer would progress unnoticed. A lack of sight would increase the risk of injury from obstacles, a loss of smell would allow a harmful gas to be inhaled, a loss of hearing would prevent recognition of warning sounds, and a lack of taste would allow toxic substances to be ingested. In short, if you could not "sense" your environment and make the necessary homeostatic adjustments, you could not survive on your own.

DEFINITION

In its broadest context, **sensation** refers to a state of awareness of external or internal conditions of the body. **Perception** refers to the conscious registration of a sensory stimulus. For a sensation to occur, four prerequisites must be fulfilled.

1. A **stimulus,** or change in the environment, capable of initiating a response by the nervous system must be present.

2. A **receptor** or **sense organ** must pick up the stimulus and convert it to a nerve impulse. A sense receptor or sense organ may be viewed as specialized nervous tissue that is extremely sensitive to internal or external conditions.

3. The impulse must be **conducted** along a nervous pathway from the receptor or sense organ to the brain.

4. A region of the brain must **translate** the impulse into a sensation.

Receptors are capable of converting a specific stimulus into a nerve impulse. The stimulus may be light, heat, pressure, mechanical energy, or chemical energy. Each stimulus is capable of causing the membrane of the receptor to depolarize. This depolarization is called a **generator potential (receptor potential).**

The generator potential is a graded response, that is, within limits, the magnitude increases with stimulus strength and frequency. When the generator potential reaches the threshold level, it initiates an action potential (nerve impulse). Once initiated, the action potential is propagated along the nerve fiber. Whereas a generator potential is a local, graded response, an action potential is propagated and obeys the all-or-none principle. The function of a generator potential is to initiate an action potential by transducing (converting) a stimulus into a nerve impulse.

A receptor may be quite simple. It may consist of the dendrites of a single neuron in the skin that are sensitive to pain stimuli. Or it may be contained in a complex organ such as the eye. Regardless of complexity, all sense receptors contain the dendrites of sensory neurons. The dendrites occur either alone or in close association with specialized cells of other tissues.

Receptors are at the same time very excitable and very specialized. Except for pain receptors, each has a low threshold of response to its specific stimulus and a high threshold to all others.

Once a stimulus is received by a receptor and converted into an impulse, the impulse is conducted along an afferent pathway that enters either the spinal cord or the brain. Many sensory impulses are conducted to the sensory areas of the cerebral cortex. Only in this region can a stimulus produce conscious sensations. Sensory impulses that terminate in the spinal cord or brain stem can initiate motor activities, but they do not produce conscious sensations.

CLASSIFICATION

One convenient method of classifying sensations is by location of the receptor. Receptors may be classified as exteroceptors (eks'-ter-ō-SEP-tors), visceroceptors (vis'-er-ō-SEP-tors), and proprioceptors (prō'-prē-ō-SEP-tors). **Exteroceptors** provide information about the external environment. They are sensitive to stimuli outside the body and transmit sensations of hearing, sight, smell, touch, pressure, temperature, and pain. Exteroceptors are located near the surface of the body.

Visceroceptors or **enteroceptors** provide information about the internal environment. These sensations arise from within the body and may be felt as pain, pressure, taste, fatigue, hunger, thirst, and nausea. Visceroceptors are located in blood vessels and viscera.

Proprioceptors provide information about body position and movement. Such sensations give us information about muscle tension, the position and tension of our joints, and equilibrium. These receptors are located in muscles, tendons, joints, and the internal ear.

Sensations may also be classified according to the simplicity or complexity of the receptor and neural pathway involved. **General senses** involve a simple receptor and neural pathway. The receptors for general sensations are numerous and widespread. They are located in the skin, subcutaneous tissue, muscles, tendons, joints, and viscera. Examples include cutaneous sensations, such as touch, pressure, vibration, heat, and cold; pain sensations; and proprioceptive sensations. **Special senses,** by contrast, involve complex receptors and neural pathways. The receptors for each special sense are found in only one or two specific areas of the body. Among the special senses are smell, taste, sight, hearing, and equilibrium.

GENERAL SENSES
CUTANEOUS SENSATIONS

Cutaneous sensations include tactile sensations (touch, pressure, vibration), thermoreceptive sensations (cold and heat), and pain. The receptors for these sensations are in the skin, connective tissue, and the ends of the gastrointestinal tract. The cutaneous receptors are randomly distributed over the body surface so that certain parts of the skin are densely populated with receptors and other parts contain only a few. Areas of the body that have few cutaneous receptors are insensitive; those containing many are very sensitive. The order for the following receptors, from greatest sensitivity to least, has been established: tip of tongue, tip of finger, side of nose, back of hand, and back of neck.

Cutaneous receptors have simple structures. They consist of the dendrites of sensory neurons that may or may not be enclosed in a capsule of epithelial or connective tissue. Impulses generated by cutaneous receptors pass along somatic afferent neurons in spinal and cranial nerves, through the thalamus, to the general sensory area of the parietal lobe of the cortex.

Tactile Sensations

Even though the tactile sensations—touch, pressure, and vibration—are classified as separate sensations, they are all detected by the same types of receptors, receptors subject to deformation. **Touch sensations** generally result from stimulation of tactile receptors in the skin or tissues immediately beneath the skin. The term *light touch* refers to the ability to recognize exactly what point of the body is touched and to recognize the texture of the objects touched; *crude touch* refers to the ability to perceive that something has touched the skin, although its exact location, shape, size, or texture cannot be determined. **Pressure sensations** generally result from stimulation of tactile receptors in deeper tissues and are longer lasting and have less variation in intensity than touch sensations. Moreover, pressure is felt over a larger area than touch. **Vibration sensations** result from rapidly repetitive sensory signals from tactile receptors.

Tactile receptors for light touch include Meissner's (MĪS-nerz) corpuscles, Merkel's (MER-kelz) discs, root hair plexuses, and free nerve endings (Figure 18-1). **Meissner's corpuscles** are egg-shaped receptors containing a mass of dendrites enclosed by connective tissue. They are located in the dermal papillae of the skin and enable us to detect touch in the skin. Meissner's corpuscles are most numerous in the fingertips, palms of the hand, and soles of the feet. They are also abundant in the eyelids,

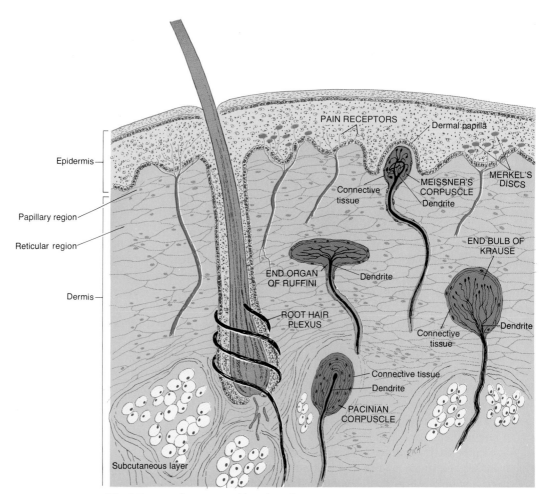

FIGURE 18-1 Structure and location of cutaneous receptors.

tip of the tongue, lips, nipples, clitoris, and tip of the penis.

Merkel's discs are receptors for touch that consist of disclike formations of dendrites attached to deeper layers of epidermal cells. They are distributed in many of the same locations as Meissner's corpuscles.

Root hair plexuses are dendrites arranged in networks around the roots of hairs. If a hair shaft is moved, the dendrites are stimulated.

Receptors that are not surrounded by supportive or protective structures are called **free, or naked, nerve endings.** Free nerve endings are found everywhere in the skin and many other tissues and are receptors for touch. Root hair plexuses are free nerve endings.

Pressure receptors are free nerve endings and Pacinian corpuscles. **Pacinian (pa-SIN-ē-an) corpuscles** (Figure 18-1) are oval structures composed of a capsule resembling an onion and consist of connective tissue layers enclosing dendrites. Pacinian corpuscles are located in the subcutaneous tissue under the skin, the deep subcutaneous tissues that lie under mucous membranes, in serous membranes, around joints and tendons, in the perimysium of muscles, in the mammary glands, in the external genitalia of both sexes, and in certain viscera.

The receptors for vibration sensations are Meissner's corpuscles and Pacinian corpuscles. Whereas Meissner's corpuscles detect low frequency vibration, Pacinian corpuscles detect higher frequency vibration.

Thermoreceptive Sensations

The cutaneous receptors for **cold** are not known, but were once thought to be **end bulbs of Krause** (Figure 18-1). The commonest form of these receptors is an oval connective tissue capsule containing dendrites. They are widely distributed in the dermis and subcutaneous connective tissue and are also located in the conjunctiva of the eye, the tip of the tongue, and external genitals.

The cutaneous receptors for **heat** are also not known, but were once thought to be **end organs of Ruffini** (Figure 18-1). These receptors are deeply embedded in the dermis and are less abundant than end bulbs of Krause.

Pain Sensations

The receptors for **pain** are simply the branching ends of the dendrites of certain sensory neurons (Figure 18-1). Pain receptors are found in practically every tissue of the body. They may be stimulated by any type of stimulus. Excessive stimulation of a sense organ causes pain. When stimuli for other sensations, such as touch, pressure, heat, and cold, reach a certain threshold, they stimulate the sensation of pain as well. Additional stimuli for pain receptors include excessive distension or dilation of an organ, prolonged muscular contractions, muscle

spasms, inadequate blood flow to an organ, or the presence of certain chemical substances. Pain receptors, because of their sensitivity to all stimuli, perform a protective function by identifying changes that may endanger the body. Pain receptors adapt only slightly or not at all. Adaptation is the decrease or disappearance of the perception of a sensation even though the stimulus is still present. If there were adaptation to pain, it would cease to be sensed and irreparable damage could result.

Sensory impulses for pain are conducted to the central nervous system along spinal and cranial nerves. The lateral spinothalamic tracts of the spinal cord relay impulses to the thalamus. From here the impulses may be relayed to the postcentral gyrus of the parietal lobe. Recognition of the kind and intensity of most pain is ultimately localized in the cerebral cortex. Some awareness of pain occurs at subcortical levels.

Pain may be divided into two types: somatic and visceral. **Somatic pain** arises from stimulation of receptors in the skin, in which case it is called *superficial somatic pain,* or from stimulation of receptors in skeletal muscles, joints, tendons, and fascia, then called *deep somatic pain.* **Visceral pain** results from stimulation of receptors in the viscera.

The ability of the cerebral cortex to locate the origin of pain is related to past experience. In most instances of somatic pain and in some instances of visceral pain, the cortex accurately projects the pain back to the stimulated area. If you burn your finger, you feel the pain in your finger. If the lining of your pleural cavity is inflamed, you experience pain there. In most instances of visceral pain, however, the sensation is not projected back to the point of stimulation. Rather, the pain may be felt in or just under the skin that overlies the stimulated organ. The pain may also be felt in a surface area far from the stimulated organ. This phenomenon is called **referred pain.** In general, the area to which the pain is referred and the visceral organ involved receive their innervation from the same segment of the spinal cord. Consider the following example. Afferent fibers from the heart as well as from the skin over the heart and along the medial aspect of the left upper extremity enter spinal cord segments T1 to T4. Thus the pain of a heart attack is typically felt in the skin over the heart and along the left arm. Figure 18-2 illustrates cutaneous regions to which visceral pain may be referred.

Inhibition of Pain

Pain sensations may be controlled by interrupting the pain impulse between the receptors and the interpretation centers of the brain. This may be done chemically, surgically, or by other means. Most pain sensations respond to pain-reducing drugs, which, in general, act to inhibit nerve impulse conduction at synapses.

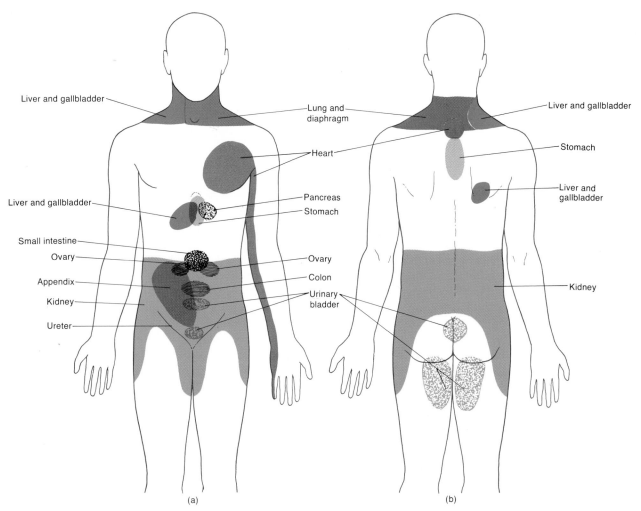

Liver and gallbladder

Lung and diaphragm

Heart

Liver and gallbladder

Liver and gallbladder

Stomach

Pancreas
Stomach

Liver and gallbladder

Small intestine
Ovary
Appendix
Kidney
Ureter

Ovary
Colon
Urinary bladder

Kidney

(a)   (b)

FIGURE 18-2 Referred pain. The color parts of the diagrams indicate cutaneous areas to which visceral pain is referred. (a) Anterior view. (b) Posterior view.

Occasionally, however, pain may be controlled only by surgery. The purpose of surgical treatment is to interrupt the pain impulse somewhere between the receptors and the interpretation centers of the brain by severing the sensory nerve, its spinal root, or certain tracts in the spinal cord or brain. *Sympathectomy* is excision of portions of the neural tissue from the autonomic nervous system; *cordotomy* is severing of a spinal cord tract; *prefrontal lobotomy* is the destruction of tracts that connect the thalamus with the prefrontal and frontal lobes of the cerebral cortex. In each instance, the pathway for pain is severed so that pain impulses are no longer conducted to the cortex.

Another method of inhibiting pain impulses is **acupuncture** (*acus* = needle; *pungere* = sting). Needles are inserted through selected areas of the skin and then twirled by the acupuncturist or by a mechanical device. After 20 to 30 minutes, pain is deadened for 6 to 8 hours. The location of needle insertion depends on the part of the body the acupuncturist wishes to anesthetize. To pull a tooth, a needle is inserted in the web between thumb and index finger. For a tonsillectomy, a needle is inserted approximately 5 cm (2 inches) above the wrist. For removal of a lung, a needle is placed in the forearm midway between wrist and elbow.

There is no satisfactory explanation of how acupuncture works. According to the "gate control" theory,

CLINICAL APPLICATION

A kind of pain frequently experienced by patients who have had a limb amputated is called **phantom pain.** They still experience pain or other sensations in the extremity as if the limb were still there. An explanation for this phenomenon is that the remaining proximal portions of the sensory nerves that previously received impulses from the limb are being stimulated by the trauma of the amputation. Stimuli from these nerves are interpreted by the brain as coming from the nonexistent (phantom) limb.

twirling the acupuncture needle stimulates two sets of nerves that eventually enter the spinal cord and synapse with the same association neurons. One nerve contains nerve fibers of small diameter (C fibers) for pain, and the other nerve contains nerve fibers of larger diameter (A fibers) for touch. The impulse passing along the touch-conducting neurons is faster than that passing along the pain-conducting neurons—fibers with large diameters conduct impulses faster than those with small diameters. Because the touch impulse reaches the posterior gray horn of the cord first, it has priority over the pain impulse. It thus "closes the gate" to the brain before the pain impulse reaches the cord. Since the pain impulse does not pass to the brain, no pain is felt. Currently, acupuncture is used in the United States mostly for childbirth, tic douloureux, arthritis, and other nonsurgical conditions.

PROPRIOCEPTIVE SENSATIONS

An awareness of the activities of muscles, tendons, and joints is provided by the **proprioceptive** or **kinesthetic** (kin'-es-THET-ik) **sense.** It informs us of the degree to which muscles are contracted and the amount of tension created in the tendons. The proprioceptive sense enables us to recognize the location and rate of movement of one body part in relation to others. It also allows us to estimate weight and determine the muscular work necessary to perform a task. With the proprioceptive sense, we can judge the position and movements of our limbs without using our eyes when we walk, type, or dress in the dark.

Proprioceptive receptors are located in muscles, tendons, joints, and the internal ear. The **joint kinesthetic receptors** are located in the capsules of joints. These receptors provide feedback information on the degree and rate of angulation (change of position) of a joint.

Neuromuscular spindles consist of the endings of sensory neurons that are wrapped around specialized muscle fibers. They are located in nearly all skeletal muscles and are numerous in the muscles of the extremities. Neuromuscular spindles provide feedback information on the degree of muscle stretch. This information is relayed to the central nervous system to assist in the coordination and efficiency of muscle contraction. Neuromuscular spindles are involved in the stretch and extensor reflexes.

Golgi tendon organs are also proprioceptive receptors that provide information about skeletal muscles. These organs are located at the junction of muscle and tendon. They function by sensing the tension applied to a tendon. The degree of tension is related to the degree of muscle contraction and is translated by the central nervous system.

The proprioceptors in the internal ear are the macula of the saccule and the utricle and cristae in the semicircular ducts. Their function in equilibrium is discussed later in the chapter.

Proprioceptors adapt only slightly. This feature is advantageous since the brain must be apprised of the status of different parts of the body at all times so that adjustments can be made to ensure coordination.

The afferent pathway for muscle sense consists of impulses generated by proprioceptors via cranial and spinal nerves to the central nervous system. Impulses for conscious proprioception pass along ascending tracts in the cord, where they are relayed to the thalamus and cerebral cortex. The sensation is registered in the general sensory area in the parietal lobe of the cortex posterior to the central sulcus. Proprioceptive impulses that have resulted in reflex action pass to the cerebellum along spinocerebellar tracts.

LEVELS OF SENSATION

As we have said, a receptor converts a stimulus into a nerve impulse, and only after that impulse has been conducted to a region of the spinal cord or brain can it be translated into a sensation. The nature of the sensation and the type of reaction generated vary with the level of the central nervous system at which the sensation is translated.

Sensory fibers terminating in the spinal cord can generate spinal reflexes without immediate action by the brain. Sensory fibers terminating in the lower brain stem bring about far more complex motor reactions than simple spinal reflexes. When sensory impulses reach the lower brain stem, they cause subconscious motor reactions. Sensory impulses that reach the thalamus can be localized crudely in the body. At the thalamic levels sensations are sorted by *modality,* that is, identified as the specific sensation of touch, pressure, pain, position, hearing, vision, smell, or taste. When sensory information reaches the cerebral cortex, we experience precise localization. It is at this level that memories of previous sensory information are stored and the perception of sensation occurs on the basis of past experience.

Let us now examine how sensory information is transmitted from receptors to the central nervous system.

SENSORY PATHWAYS

Before reading this section, you will find it helpful to review the principal ascending and descending tracts of the spinal cord (see Exhibit 15-1 and Figure 15-4). Sensory information transmitted from the spinal cord to the brain is conducted along two general pathways: the posterior column pathway and the spinothalamic pathway.

In the **posterior column pathway** to the cerebral cortex there are three separate sensory neurons. The **first-order neuron** connects the receptor with the spinal cord and medulla on the same side of the body. The cell body of the first-order neuron is in the posterior root ganglion of a spinal or cranial nerve. The first-order neuron synapses with a **second-order neuron.** The second-order neuron passes from the medulla upward to the thalamus. The cell body of the second-order neuron is located in the nuclei cuneatus and gracilis of the medulla. Before passing into the thalamus, the second-order neuron crosses to the opposite side of the medulla and enters the medial lemniscus, a projection tract that terminates at the thalamus. In the thalamus, the second-order neuron synapses with a **third-order neuron.** The third-order neuron terminates in the somesthetic sensory area of the cerebral cortex. The posterior column pathway conducts impulses related to proprioception, fine touch, two-point discrimination, and vibrations.

The **spinothalamic pathway** is also composed of three orders of sensory neurons. The first-order neuron connects a receptor of the neck, trunk, and extremities with the spinal cord. The cell body of the first-order neuron is in the posterior root ganglion also. The first-order neuron synapses with the second-order neuron, which has its cell body in the posterior gray horn of the spinal cord. The fiber of the second-order neuron crosses to the opposite side of the spinal cord and passes upward to the brain stem in the lateral spinothalamic tract or ventral spinothalamic tract. The fibers from the second-order neuron terminate in the thalamus. There the second-order neuron synapses with a third-order neuron. The third-order neuron terminates in the somesthetic sensory area of the cerebral cortex. The spinothalamic pathway conveys sensory impulses for pain and temperature as well as crude touch and pressure.

The second-order neurons of the spinothalamic pathway enter the medulla, pons, and midbrain. Thus the spinothalamic pathway conducts sensory signals that result in subconscious motor reactions. By contrast, second-order neurons of the posterior column pathway have a direct connection with the thalamus and cerebral cortex. Thus the posterior column pathway conducts sensory information primarily into the conscious area of the brain.

Now we can examine the anatomy of specific sensory pathways—for pain and temperature; for crude touch and pressure; and for fine touch, proprioception, and vibration.

PAIN AND TEMPERATURE

The sensory pathway for pain and temperature is called the **lateral spinothalamic pathway** (Figure 18-3). The first-order neuron conveys the impulse for pain or temperature from the appropriate receptor to the posterior gray horn

on the same side of the spinal cord. In the horn, the first-order neuron synapses with a second-order neuron. The axon of the second-order neuron crosses to the opposite side of the cord. Here it becomes a component of the *lateral spinothalamic tract* in the lateral white column. The second-order neuron passes upward in the tract through the brain stem to a nucleus in the thalamus called the ventral posterolateral nucleus. In the thalamus, conscious recognition of pain and temperature occurs. The sensory impulse is then conveyed from the thalamus through the internal capsule to the somesthetic area of the cortex by a third-order neuron. The cortex analyzes the sensory information for the precise source, severity, and quality of the pain and heat stimuli.

CRUDE TOUCH AND PRESSURE

The neural pathway that conducts impulses for crude touch and pressure is the **anterior (ventral) spinothalamic pathway** (Figure 18-4). The first-order neuron conveys

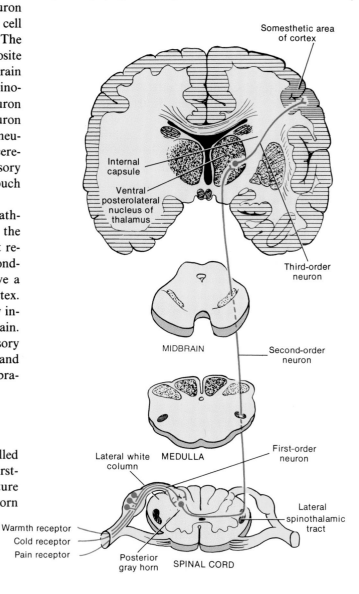

Somesthetic area of cortex

Internal capsule

Ventral posterolateral nucleus of thalamus

Third-order neuron

MIDBRAIN

Second-order neuron

Lateral white column   MEDULLA

First-order neuron

Lateral spinothalamic tract

Warmth receptor
Cold receptor
Pain receptor

Posterior gray horn   SPINAL CORD

FIGURE 18-3 Sensory pathway for pain and temperature—the lateral spinothalamic pathway.

the impulse from a crude touch or pressure receptor to the posterior gray horn on the same side of the spinal cord. In the horn, the first-order neuron synapses with a second-order neuron. The axon of the second-order neuron crosses to the opposite side of the cord and becomes a component of the *anterior spinothalamic tract* in the anterior white column. The second-order neuron passes upward in the tract through the brain stem to the ventral posterolateral nucleus of the thalamus. The sensory impulse is then relayed from the thalamus through the internal capsule to the somesthetic area of the cerebral cortex by a third-order neuron. Although there is some awareness of crude touch and pressure at the thalamic level, it is not fully perceived until the impulses reach the cortex.

FINE TOUCH, PROPRIOCEPTION, AND VIBRATION

The neural pathway for fine touch, proprioception, and vibration is called the **posterior column pathway** (Figure 18-5). This pathway conducts impulses that give rise to several discriminating senses.

1. Fine touch: the ability to recognize the exact location of stimulation and to distinguish that two points are touched, even though they are close together (two-point discrimination).

2. Stereognosis (ste'-rē-og-NŌ-sis): recognizing, by "feel," the size, shape, and texture of an object.

3. Weight discrimination: the ability to assess the weight of an object.

4. Proprioception: the awareness of the precise position of body parts and directions of movement.

5. The ability to sense vibrations.

First-order neurons for the discriminating senses just noted follow a pathway different from those for pain and temperature and crude touch and pressure. Instead of terminating in the posterior gray horn, the first-order neurons from appropriate receptors pass upward in the fasciculus gracilis or fasciculus cuneatus in the posterior white column of the cord. From here the first-order neurons enter either the nucleus gracilis or nucleus cuneatus in the medulla, where they synapse with second-order neurons. The axons of the second-order neurons cross to the opposite side of the medulla and ascend to the thalamus through the medial lemniscus, a projection tract of white fibers passing through the medulla, pons, and midbrain. The second-order neuron axons synapse with third-order neurons in the ventral posterior nucleus in the thalamus. In the thalamus, there is no conscious awareness of the discriminating senses, except for a possible crude awareness of vibrations. The third-order neu-

rons convey the sensory impulses to the somesthetic area of the cerebral cortex. It is here that you perceive your sense of position and movement and fine touch.

CEREBELLAR TRACTS

The *posterior spinocerebellar tract* is an uncrossed tract that conveys impulses concerned with subconscious muscle sense and thus assumes a role in reflex adjustments for posture and muscle tone. The nerve impulses originate in neurons that run between proprioceptors in muscles, tendons, and joints and the posterior gray horn of the spinal cord. Here the neurons synapse with afferent neurons that pass to the ipsilateral lateral white column of the cord to enter the posterior cerebellar tract. The tract enters the inferior cerebellar peduncles from the medulla and ends at the cerebellar cortex. In the cerebellum, synapses are made that ultimately result in the transmission of impulses back to the spinal cord to the anterior

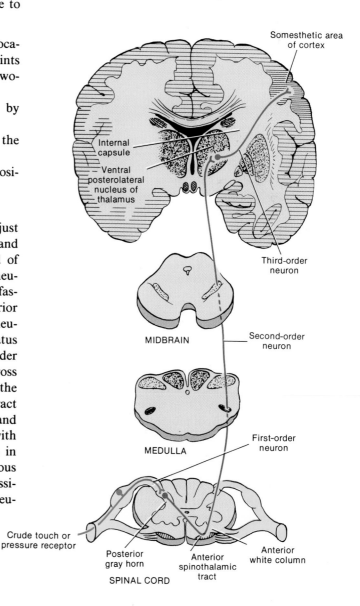

FIGURE 18-4 Sensory pathway for crude touch and pressure—the anterior spinothalamic pathway.

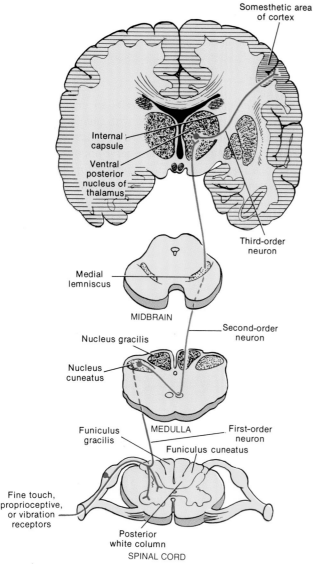

FIGURE 18-5 Sensory pathway for fine touch, proprioception, and vibration—the posterior column pathway.

gray horn to synapse with the lower motor neurons leading to skeletal muscles.

The *anterior spinocerebellar tract* also conveys impulses for subconscious muscle sense. It, however, is made up of both crossed and uncrossed nerve fibers. Sensory neurons deliver impulses from proprioceptors to the posterior gray horn of the spinal cord. Here a synapse occurs with neurons that make up the anterior spinocerebellar tracts. Some fibers cross to the opposite side of the spinal cord in the anterior white commissure. Others pass laterally to the ipsilateral anterior spinocerebellar tract and move upward, through the brain stem, to the pons to enter the cerebellum through the superior cerebellar peduncles. Here again the impulses for subconscious muscle sense are registered.

MOTOR PATHWAYS

After receiving and interpreting sensations, the central nervous system generates impulses to direct responses to that sensory input. We have already considered somatic reflex arcs and visceral efferent pathways. Now we will look at the transmission of motor impulses that result in the movement of skeletal muscles.

The principal parts of the brain concerned with skeletal muscle control are the cerebral motor cortex, basal ganglia, reticular formation, and cerebellum. The motor cortex assumes the major role for controlling precise, discrete muscular movements. The basal ganglia largely integrate semivoluntary movements like walking, swimming, and laughing. The cerebellum, although not a control center, assists the motor cortex and basal ganglia by making body movements smooth and coordinated. Voluntary motor impulses are conveyed from the brain through the spinal cord by way of two major pathways: the pyramidal pathways and the extrapyramidal pathways.

PYRAMIDAL PATHWAYS

Voluntary motor impulses are conveyed from the motor areas of the brain to somatic efferent neurons leading to skeletal muscles via the **pyramidal** (pi-RAM-i-dal) **pathways.** Most pyramidal fibers originate from cell bodies in the precentral gyrus. They descend through the internal capsule of the cerebrum and cross to the opposite side of the brain. They terminate in nuclei of cranial nerves that innervate voluntary muscles or in the anterior gray horn of the spinal cord. A short connecting neuron probably completes the connection of the pyramidal fibers with the motor neurons that activate voluntary muscles.

The pathways over which the impulses travel from the motor cortex to skeletal muscles have two components: **upper motor neurons (pyramidal fibers)** in the brain and **lower motor neurons (peripheral fibers)** in the spinal cord. Here we consider three tracts of the pyramidal system.

1. Lateral corticospinal tract (pyramidal tract proper). This tract begins in the motor cortex and descends through the internal capsule of the cerebrum, the cerebral peduncle of the midbrain, and then the pons on the same side as the point of origin (Figure 18-6). In the medulla, the fibers decussate to the opposite side to descend through the spinal cord in the lateral white column in the lateral corticospinal tract. Thus the motor cortex of the right side of the brain controls muscles on the left side of the body and vice versa. Most upper motor neuron fibers of the lateral corticospinal tract synapse with short association neurons in the anterior gray horn of the cord. These then synapse in the anterior gray horn with lower motor neurons that exit all levels of the cord via the ventral roots of spinal nerves. The lower motor neurons terminate in skeletal muscles.

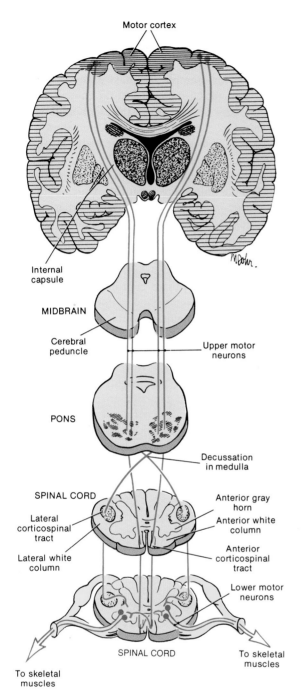

Motor cortex

Internal
capsule

MIDBRAIN

Cerebral
peduncle

Upper motor
neurons

PONS

Decussation
in medulla

SPINAL CORD

Lateral
corticospinal
tract

Anterior gray
horn

Anterior white
column

Anterior
corticospinal
tract

Lateral white
column

Lower motor
neurons

SPINAL CORD

To skeletal
muscles

To skeletal
muscles

FIGURE 18-6 Pyramidal pathways.

2. Anterior corticospinal tract. About 15 percent of the upper motor neurons from the motor cortex do not cross in the medulla. These pass through the medulla and continue to descend on the same side to the anterior white column to become part of the anterior (straight or uncrossed) corticospinal tract. The fibers of these upper motor neurons decussate and synapse with association neurons in the anterior gray horn of the spinal cord of the side opposite the origin of the anterior corticospinal tract. The association neurons in the horn synapse with

lower motor neurons that exit the cervical and upper thoracic segments of the cord via the ventral roots of spinal nerves. The lower motor neurons terminate in skeletal muscles that control muscles of the neck and part of the trunk.

3. Corticobulbar tract. The fibers of this tract arise from upper motor neurons in the motor cortex. They accompany the corticospinal tracts through the internal capsule to the brain stem, where they decussate and terminate in the nuclei of cranial nerves in the pons and medulla. These cranial nerves include the oculomotor (III), trochlear (IV), trigeminal (V), abducens (VI), facial (VII), glosspharyngeal (IX), vagus (X), accessory (XI), and hypoglossal (XII). The corticobulbar tract conveys impulses that largely control voluntary movements of the head and neck.

The various tracts of the pyramidal system convey impulses from the cortex that result in precise muscular movements.

EXTRAPYRAMIDAL PATHWAYS

The **extrapyramidal pathways** include all descending tracts other than the pyramidal tracts. Generally, these include tracts that begin in the basal ganglia and reticular formation. The main extrapyramidal tracts are as follows.

1. Rubrospinal tract. This tract originates in the red nucleus of the midbrain (after receiving fibers from the cerebellum), crosses over to descend in the lateral white column of the opposite side, and extends through the entire length of the spinal cord. The tract transmits impulses to skeletal muscles concerned with tone and posture.

2. Tectospinal tract. This tract originates in the superior colliculus of the midbrain, crosses to the opposite side, descends in the anterior white column, and enters the anterior gray horns in the cervical segments of the cord. Its function is to transmit impulses that control movements of the head in response to visual stimuli.

3. Vestibulospinal tract. This tract originates in the vestibular nucleus of the medulla, descends on the same side of the cord in the anterior white column, and terminates in the anterior gray horns, mostly in the cervical and lumbrosacral segments of the cord. It conveys impulses that regulate muscle tone in response to movements of the head. This tract, therefore, plays a major role in equilibrium.

Only one motor neuron carries the impulse from the cerebral cortex to the cranial nerve nuclei or spinal cord: an **upper motor neuron.** Only one motor neuron in the pathway actually terminates in a skeletal muscle: the **lower motor neuron.** This neuron, a somatic efferent neuron, always extends from the central nervous system to the skeletal muscle. Since it is the final transmitting neuron in the pathway, it is also called the **final common pathway.**

CLINICAL APPLICATION

The lower motor and upper motor neurons are important clinically. If the lower motor neuron is damaged or diseased, there is neither voluntary nor reflex action of the muscle it innervates, and the muscle remains in a relaxed state, a condition called **flaccid paralysis.** Injury or disease of upper motor neurons in a motor pathway is characterized by varying degrees of continued contraction of the muscle, a condition called **spasticity,** and exaggerated reflexes. Another characteristic is the **sign of Babinski,** which is a slow dorsiflexion of the great toe accompanied by fanning of the lateral toes in response to stroking the plantar surface along the outer border of the foot. The normal response is plantar flexion of the toes.

Lower motor neurons are subjected to stimulation by many other presynaptic neurons. Some signals are excitatory; others are inhibitory. The algebraic sum of the opposing signals determines the final response of the lower motor neuron. It is not just a simple matter of the brain sending an impulse and the muscle always contracting.

Association neurons are of considerable importance in the motor pathways. Impulses from the brain are conveyed to association neurons before being received by lower motor neurons. These association neurons are integrators. They integrate the pattern of muscle contraction.

The basal ganglia have many connections with other parts of the brain. Through these connections, they help to control subconscious movements. The caudate nucleus controls gross intentional movements. The caudate nucleus and putamen, together with the cortex, control patterns of movement. The globus pallidus controls positioning of the body for performing a complex movement. The subthalamic nucleus is thought to control walking and possibly rhythmic movements. Many potential functions of the basal ganglia are held in check by the cerebrum. Thus if the cerebral cortex is damaged early in life, a person can still perform many gross muscular movements.

The role of the cerebellum is significant also. The cerebellum is connected to other parts of the brain that are concerned with movement. The vestibulocerebellar tract transmits impulses from the equilibrium apparatus in the ear to the cerebellum. The olivocerebellar tract transmits impulses from the basal ganglia to the cerebellum. The corticopontocerebellar tract conveys impulses from the cerebrum to the cerebellum. The spinocerebellar tracts relay proprioceptive information to the cerebellum. Thus the cerebellum receives considerable information regarding the overall physical status of the body. Using this information, the cerebellum generates impulses that integrate body responses.

Take tennis, for example. To make a good serve, you must bring your racket forward just far enough to make solid contact. How do you stop at the exact point without swinging too far? This is where the cerebellum comes in. It receives information about your body status while you are serving. Before you even hit the ball, the cerebellum has already sent information to the cerebral cortex and basal ganglia informing them that your swing must stop at an exact point. In response to cerebellar stimulation, the cortex and basal ganglia transmit motor impulses to your opposing body muscles to stop the swing. The cerebellar function of stopping overshoot when you want to zero in on a target is called its *damping function.* The cerebellum also helps you to coordinate different body parts while walking, running, and swimming. Finally, the cerebellum helps you maintain equilibrium.

SPECIAL SENSES

The special senses—smell, taste, sight, hearing, and equilibrium—have receptor organs that are structurally more complex than receptors for general sensations. The sense of smell is the least specialized, as opposed to the sense of sight, which is the most specialized. Like the general senses, however, the special senses allow us to detect changes in our environment.

OLFACTORY SENSATIONS

The receptors for the **olfactory** (ol-FAK-tō-rē) **sense** are located in the nasal epithelium in the superior portion of the nasal cavity on either side of the nasal septum (Figure 18-7). The nasal epithelium consists of two principal kinds of cells: supporting and olfactory. The **supporting cells** are columnar epithelial cells of the mucous membrane lining the nose. Olfactory glands in the mucosa keep the mucous membrane moist. The **olfactory cells** are bipolar neurons whose cell bodies lie between the supporting cells. The distal (free) end of each olfactory cell contains six to eight dendrites, called **olfactory hairs.** The unmyelinated axons of the olfactory cells unite to form the **olfactory nerves (I),** which pass through foramina in the cribriform plate of the ethmoid bone. The olfactory nerves terminate in paired masses of gray matter called the **olfactory bulbs.** The olfactory bulbs lie beneath the frontal lobes of the cerebrum on either side of the crista galli of the ethmoid bone. The first synapse of the olfactory neural pathway occurs in the olfactory bulbs between the axons of the olfactory nerves and the dendrites of neurons inside the olfactory bulbs. Axons of these neurons run posteriorly to form the **olfactory tract.** From here, impulses are conveyed to the olfactory portion

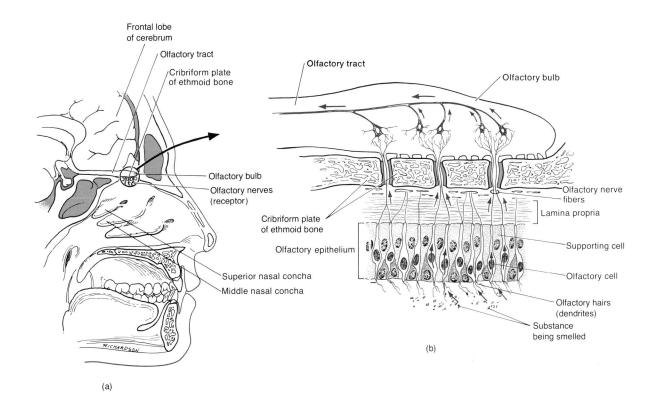

Frontal lobe
of cerebrum

Olfactory tract

Cribriform plate
of ethmoid bone

Olfactory bulb

Olfactory nerves
(receptor)

Superior nasal concha

Middle nasal concha

RICHARDSON

(a)

Olfactory tract

Olfactory bulb

Cribriform plate
of ethmoid bone

Olfactory epithelium

Olfactory nerve
fibers

Lamina propria

Supporting cell

Olfactory cell

Olfactory hairs
(dendrites)

Substance
being smelled

(b)

FIGURE 18-7 Olfactory receptors. (a) Location of receptors in nasal cavity. (b) Enlarged aspect of olfactory receptors. (c) Photomicrograph of the olfactory mucosa at a magnification of 400×. (Courtesy of Donald I. Patt, from *Comparative Vertebrate Histology,* by Donald I. Patt and Gail R. Patt, Harper & Row, Publishers, Inc., New York, 1969.) (d) Scanning electron micrograph of the ciliated nasal mucosa at a magnification of 5,000×. (Courtesy of Fisher Scientific Company and S.T.E.M. Laboratories, Inc., Copyright, 1975.)

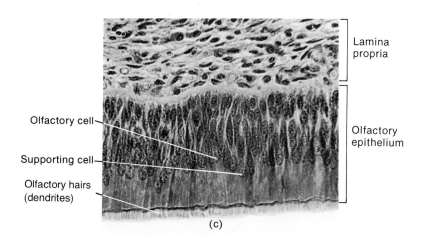

Lamina
propria

Olfactory
epithelium

Olfactory cell

Supporting cell

Olfactory hairs
(dendrites)

(c)

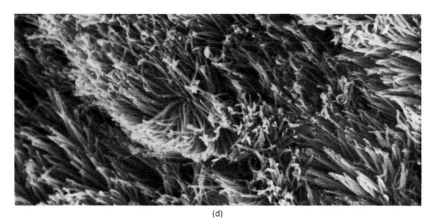

(d)

of the cerebral cortex. In the cerebral cortex, the impulses are interpreted as odor and give rise to the sensation of smell. A few of the surface anatomy features of the nose are shown in Exhibit 25-1.

Both the supporting cells of the nasal epithelium and tear glands are innervated by branches of the trigeminal nerve (V). The nerve receives stimuli of pain, cold, heat, tickling, and pressure. Olfactory stimuli such as pepper, ammonia, and chloroform are irritating and may cause tearing because they stimulate the lacrimal and nasal mucosal receptors of the trigeminal nerve as well as the olfactory neurons.

GUSTATORY SENSATIONS

The receptors for **gustatory** (GUS-ta-tō'-rē) **sensations,** or sensations of taste, are located in the taste buds (Figure 18-8). Taste buds are most numerous on the tongue, but they are also found on the soft palate and in the throat. The **taste buds** are oval bodies consisting of two kinds of cells. The **supporting cells** are a specialized epithelium that forms a capsule. Inside each capsule are 4 to 20 **gustatory cells.** Each gustatory cell contains a hairlike process **(gustatory hair)** that projects to the external surface through an opening in the taste bud called the **taste**

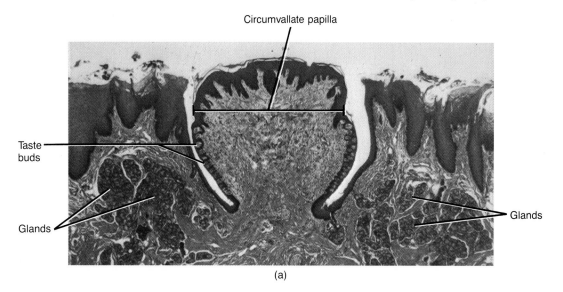

Circumvallate papilla

Taste buds

Glands

Glands

(a)

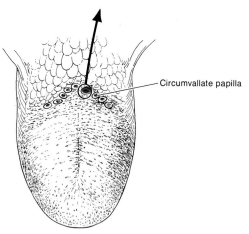

Circumvallate papilla

FIGURE 18-8 Gustatory receptors. (a) Photomicrograph of a circumvallate papilla containing taste buds at a magnification of 50×.

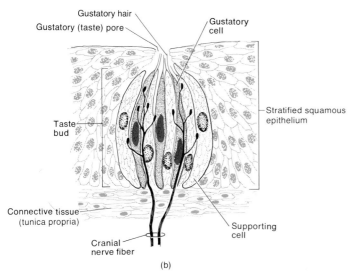

Gustatory hair

Gustatory (taste) pore

Gustatory cell

Stratified squamous epithelium

Taste bud

Connective tissue (tunica propria)

Supporting cell

Cranial nerve fiber

(b)

FIGURE 18-8 (*Continued*) Gustatory receptors. (b) Diagram of the structure of a taste bud.

Gustatory (taste) pore

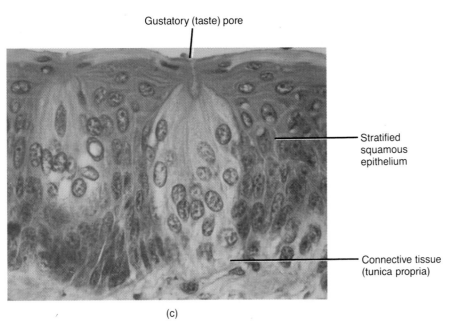

Stratified squamous epithelium

Connective tissue (tunica propria)

(c)

FIGURE 18-8 (*Continued*) Gustatory receptors. (c) Photomicrograph of a taste bud at a magnification of 600×. (© 1983 by Michael H. Ross. Used by permission.)

pore. Gustatory cells make contact with taste stimuli through the taste pore.

Taste buds are found in some connective tissue elevations on the tongue called **papillae.** The papillae give the upper surface of the tongue its rough appearance. **Circumvallate** (ser-kum-VAL-āt) **papillae,** the largest type, are circular and form an inverted V-shaped row at the poste-rior portion of the tongue. **Fungiform** (FUN-ji-form; meaning mushroom-shaped) **papillae** are knoblike elevations found primarily on the tip and sides of the tongue. All circumvallate and most fungiform papillae contain taste buds. **Filiform** (FIL-i-form) **papillae** are pointed threadlike structures that cover the anterior two-thirds of the tongue (see Figure 21-5b).

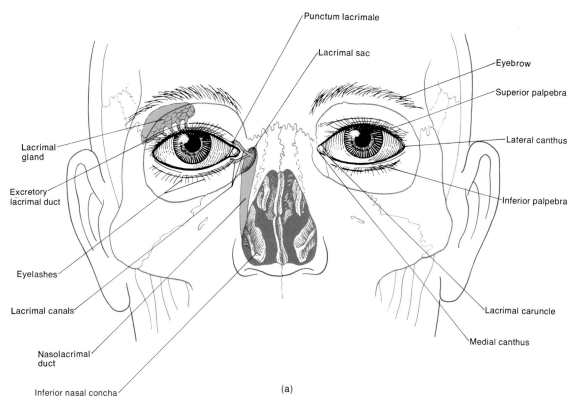

FIGURE 18-9 Accessory structures of the eye. (a) Anterior view.

For gustatory cells to be stimulated, the substances we taste must be in solution in the saliva so they can enter the taste pores. Despite the many substances we seem to taste, there are basically only four taste sensations: sour, salt, bitter, and sweet. All other "tastes," such as chocolate, pepper, and coffee, are actually odors. Each of the four tastes is due to a different response to different chemicals. Certain regions of the tongue react more strongly than others to certain taste sensations. Although the tip of the tongue reacts to all four taste sensations, it is highly sensitive to sweet and salty substances. The posterior portion of the tongue is highly sensitive to bitter substances. The lateral edges of the tongue are more sensitive to sour substances (see Figure 21-5a).

The cranial nerves that supply afferent fibers to taste buds are the facial nerve (VII), which supplies the anterior two-thirds of the tongue; the glossopharyngeal (IX), which supplies the posterior one-third of the tongue; and the vagus (X), which supplies the epiglottis. Taste impulses are conveyed from the gustatory cells in taste buds along the three nerves just cited to the medulla and then to the thalamus. They terminate in the parietal lobe of the cerebral cortex.

VISUAL SENSATIONS

The structures related to vision are the eyeball, the optic nerve, the brain, and a number of accessory structures.

The developmental anatomy of the eye is considered in Exhibit 24-7.

Accessory Structures of Eye

Among the **accessory structures** are the eyebrows, eyelids, eyelashes, and the lacrimal apparatus (Figure 18-9). The **eyebrows** form a transverse arch at the junction of the upper eyelid and forehead. Structurally, they resemble the hairy scalp. The skin of the eyebrows is richly supplied with sebaceous glands. The hairs are generally coarse and directed laterally. Deep to the skin of the eyebrows are the fibers of the orbicularis oculi muscles. The eyebrows help to protect the eyeballs from falling objects, perspiration, and the direct rays of the sun.

The upper and lower **eyelids,** or **palpebrae** (PAL-pe-brē), have several important roles. They shade the eyes during sleep, protect the eyes from excessive light and foreign objects, and spread lubricating secretions over the eyeballs. The upper eyelid is more movable than the lower and contains in its superior region a special levator muscle know as the **levator palpebrae superioris.** The space between the upper and lower eyelids that exposes the eyeball is called the **palpebral fissure.** Its angles are known as the **lateral canthus** (KAN-thus), which is narrower and closer to the temporal bone, and the **medial canthus,** which is broader and nearer the nasal bone. In the medial canthus there is a small reddish elevation,

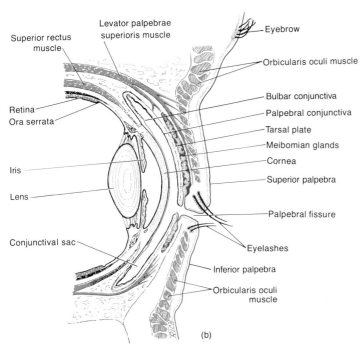

FIGURE 18-9 (*Continued*) Accessory structures of the eye. (b) Sagittal section of the eyelids and anterior portion of the eyeball.

the **lacrimal caruncle** (KAR-ung-kul), containing sebaceous and sudoriferous glands. A whitish material secreted by the caruncle collects in the medial canthus.

From superficial to deep, each eyelid consists of epidermis, dermis, subcutaneous areolar connective tissue, fibers of the orbicularis oculi muscle, a tarsal plate, tarsal glands, and a conjunctiva. The **tarsal plate** is a thick fold of connective tissue that forms much of the inner wall of each eyelid and gives form and support to the eyelids. Embedded in grooves on the deep surface of each tarsal plate is a row of elongated tarsal glands known as **Meibomian** (mī-BŌ-mē-an) **glands.** These are modified sebaceous glands, and their oily secretion helps keep the eyelids from adhering to each other. Infection of the Meibomian glands produces a tumor or cyst on the eyelid called a **chalazion** (ka-LĀ-zē-on). The **conjunctiva** (kon'-junk-TĪ-va) is a thin mucous membrane. It is called the **palpebral conjunctiva** when it lines the inner aspect of the eyelids. It is called the **bulbar** or **ocular conjunctiva** when it is reflected from the eyelids onto the eyeball to the periphery of the cornea.

Projecting from the border of each eyelid, anterior to the Meibomian glands, is a row of short, thick hairs, the **eyelashes.** In the upper lid, they are long and turn upward; in the lower lid, they are short and turn downward. Sebaceous glands at the base of the hair follicles of the eyelashes called **glands of Zeis** (ZĪS) pour a lubricating fluid into the follicles. Infection of these glands is called a **sty.**

The **lacrimal apparatus** is a term used for a group of structures that manufactures and drains away tears. These structures are the lacrimal glands, the excretory lacrimal ducts, the lacrimal canals, the lacrimal sacs, and the nasolacrimal ducts. A **lacrimal gland** is a compound tubuloacinar gland located at the superior lateral portion of each orbit. Each is about the size and shape of an almond. Leading from the lacrimal glands are 6 to 12 **excretory lacrimal ducts** that empty lacrimal fluid, or tears, onto the surface of the conjunctiva of the upper lid. From here the lacrimal fluid passes medially and enters two small openings (**puncta lacrimalia**) that appear as two small pores, one in each papilla of the eyelid, at the medial canthus of the eye. The lacrimal secretion then passes into two ducts, the **lacrimal canals,** and is next conveyed into the lacrimal sac. The lacrimal canals are located in the lacrimal grooves of the lacrimal bones. The **lacrimal sac** is the superior expanded portion of the **nasolacrimal duct,** a canal that transports the lacrimal secretion into the inferior meatus of the nose.

The **lacrimal secretion** is a watery solution containing salts, some mucus, and a bactericidal enzyme called **lysozyme.** It cleans, lubricates, and moistens the eyeball. After being secreted by the lacrimal glands, it is spread over the surface of the eyeball by the blinking of the eyelids. Usually, 1 ml per day is produced. Normally, the secretion is carried away by evaporation or by passing into the lacrimal canals and then into the nasal cavities as fast as it is produced. If, however, an irritating substance

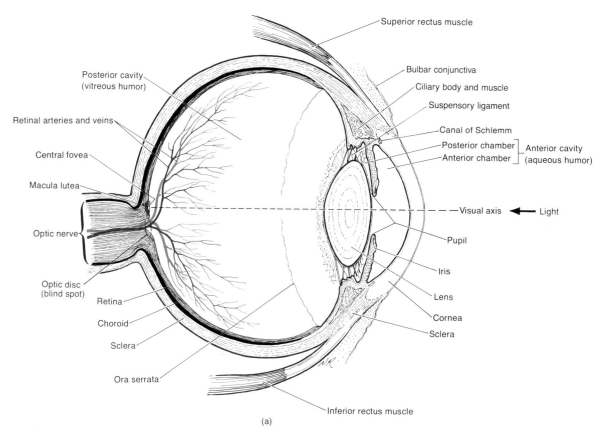

(a)

FIGURE 18-10 Structure of the eyeball. (a) Gross structure in sagittal section.

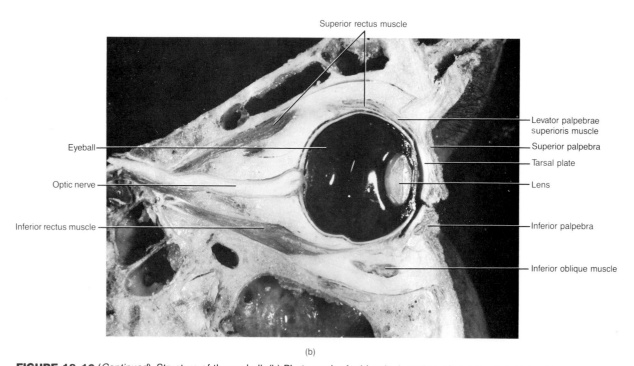

(b)

FIGURE 18-10 (*Continued*) Structure of the eyeball. (b) Photograph of midsagittal section. (Courtesy of C. Yokochi and J. W. Rohen, *Photographic Anatomy of the Human Body,* 1st ed., 1969, IGAKU-SHOIN, Tokyo, New York.)

makes contact with the conjunctiva, the lacrimal glands are stimulated to oversecrete. Tears then accumulate more rapidly than they can be carried away. This is a protective mechanism since the tears dilute and wash away the irritating substance. "Watery" eyes also occur when an inflammation of the nasal mucosa, such as a cold, obstructs the nasolacrimal ducts so that drainage of tears is blocked.

Structure of Eyeball

The adult **eyeball** measures about 2.5 cm (1 inch) in diameter. Of its total surface area, only the anterior one-sixth is exposed. The remainder is recessed and protected by the orbit into which it fits.

Anatomically, the eyeball can be divided into three layers: fibrous tunic, vascular tunic, and retina or nervous tunic (Figure 18-10). The **fibrous tunic** is the outer coat of the eyeball. It can be divided into two regions: the posterior portion is the sclera (SKLE-ra) and the anterior portion is the cornea (KOR-nē-a). The **sclera,** the "white of the eye," is a white coat of fibrous tissue that covers

all the eyeball except the anterior colored portion. The sclera gives shape to the eyeball and protects its inner parts. Its posterior surface is pierced by the optic nerve. The **cornea** is a nonvascular, nervous, transparent fibrous coat that covers the iris, the colored part of the eye. The cornea's outer surface is covered by an epithelial layer continuous with the epithelium of the bulbar conjunctiva. At the junction of the sclera and cornea is a venous sinus known as the **canal of Schlemm.**

The **vascular tunic** is the middle layer of the eyeball and is composed of three portions: the choroid, the ciliary body, and the iris. Collectively, these three structures are called the **uvea** (YOO-vē-a). The **choroid** (KŌ-royd), the posterior portion of the vascular tunic, is a thin, dark-brown membrane that lines most of the internal surface of the sclera. It contains numerous blood vessels and a large amount of pigment. The choroid absorbs light rays so they are not reflected back out of the eyeball. Through its blood supply, it nourishes the retina. Like the sclera, it is pierced by the optic nerve at the back of the eyeball. In the anterior portion of the vascular tunic, the choroid becomes the **ciliary** (SIL-ē-ar'-ē) **body.** It is the thickest

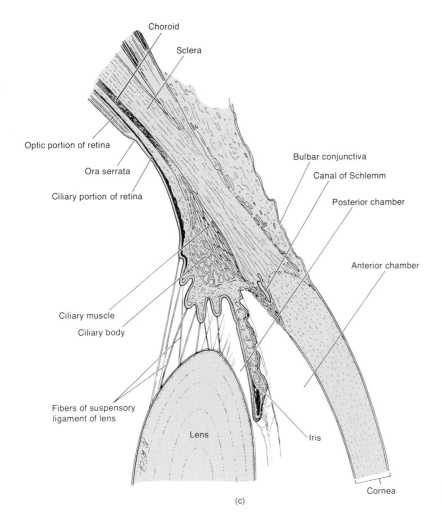

Choroid
Sclera
Optic portion of retina
Ora serrata
Ciliary portion of retina
Ciliary muscle
Ciliary body
Fibers of suspensory ligament of lens
Lens
Iris
Bulbar conjunctiva
Canal of Schlemm
Posterior chamber
Anterior chamber
Cornea
(c)

FIGURE 18-10 (*Continued*) Structure of the eyeball. (c) Section through the anterior portion of the eyeball at the sclerocorneal junction.

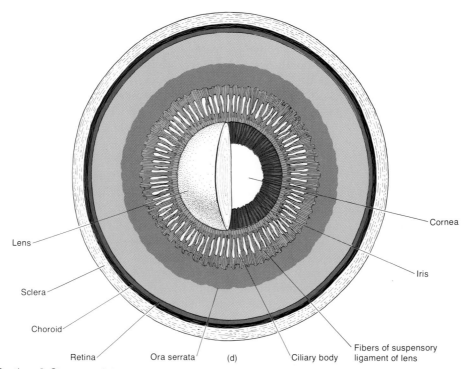

Lens

Sclera

Choroid

Retina   Ora serrata   (d)   Ciliary body

Cornea

Iris

Fibers of suspensory
ligament of lens

FIGURE 18-10 (*Continued*) Structure of the eyeball. (d) Anterior portion of the eyeball viewed from behind.

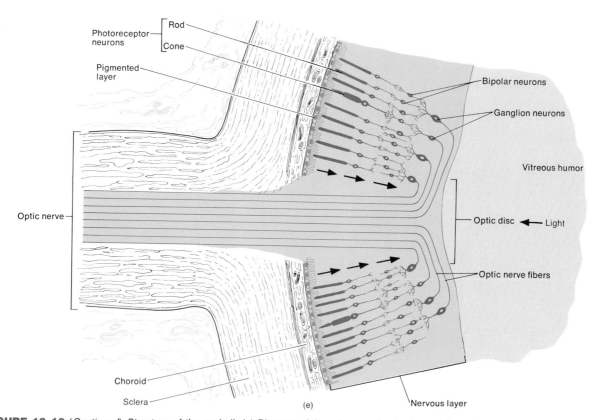

Photoreceptor
neurons   Rod

Cone

Pigmented
layer

Optic nerve

Choroid

Sclera

(e)

Bipolar neurons

Ganglion neurons

Vitreous humor

Optic disc ◄— Light

Optic nerve fibers

Nervous layer

FIGURE 18-10 (*Continued*) Structure of the eyeball. (e) Diagram of the microscopic structure of the retina.

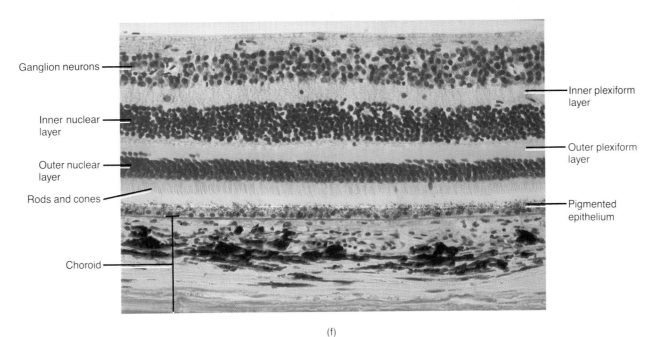

Ganglion neurons

Inner nuclear layer

Outer nuclear layer

Rods and cones

Choroid

Inner plexiform layer

Outer plexiform layer

Pigmented epithelium

(f)

FIGURE 18-10 (*Continued*) Structure of the eyeball. (f) Photomicrograph of a portion of the retina at a magnification of 300×. The outer nuclear layer contains nuclei of photoreceptor neurons; the outer plexiform layer is where photoreceptor and bipolar neurons synapse; the inner nuclear layer contains the nuclei of bipolar neurons; and the inner plexiform layer is where bipolar and ganglion neurons synapse. (© 1983 by Michael H. Ross. Used by permission.)

portion of the vascular tunic. It extends from the **ora serrata** (Ō-ra ser-RĀ-ta) of the retina (inner tunic) to a point just behind the sclerocorneal junction. The ora serrata is simply the jagged margin of the retina. This second portion of the vascular tunic contains the **ciliary muscle**—a smooth muscle that alters the shape of the lens for near or far vision. The **iris** is the third portion of the vascular tunic. It consists of circular and radial smooth muscle fibers arranged to form a doughnut-shaped structure. The black hole in the center of the iris is the **pupil,** the area through which light enters the eyeball. The iris is suspended between the cornea and the lens and is attached at its outer margin to the ciliary body. A principal function of the iris is to regulate the amount of light entering the eyeball. When the eye is stimulated by bright light, the circular muscles of the iris contract and decrease the size of the pupil. When the eye must adjust to dim light, the radial muscles of the iris contract and increase the pupil's size.

The third and inner coat of the eye, the **retina** or **nervous tunic,** lies only in the posterior portion of the eye. Its primary function is image formation. It consists of a nervous tissue layer and a pigmented layer. The retina covers the choroid. At the edge of the ciliary body, it ends in a scalloped border called the ora serrata. This is where the outer nervous layer or visual portion of the retina ends. The outer pigmented layer extends anteriorly over the back of the ciliary body and the iris as the nonvisual portion of the retina. The inner nervous layer contains three zones of neurons. These three zones, named in the order in which they conduct impulses, are the

photoreceptor neurons, the **bipolar neurons,** and the **ganglion neurons.**

CLINICAL APPLICATION

Detachment of the retina may occur in trauma, such as a blow to the head. The tear in the retina causes blindness in the corresponding field of vision. The actual detachment occurs between the sensory part of the retina and the underlying pigmented layer. Fluid accumulates between these layers, forcing the thin, pliable retina to billow out toward the vitreous humor. The retina may be reattached by photocoagulation by laser beam, cryosurgery, or scleral resection.

The dendrites of the photoreceptor neurons are called rods and cones because of their shapes. They are visual receptors highly specialized for stimulation by light rays. **Rods** are specialized for vision in dim light. They also allow us to discriminate between different shades of dark and light and permit us to see shapes and movement. **Cones,** by contrast, are specialized for color vision and sharpness of vision (*visual acuity*). Cones are stimulated only by bright light. This is why we cannot see color by moonlight. It is estimated that there are 7 million cones and somewhere between 10 and 20 times as many rods. Cones are most densely concentrated in the **central fovea,** a small depression in the center of the macula lutea. The **macula lutea** (MAK-yoo-la LOO-tē-a), or yellow spot, is in the exact center of the retina. The fovea is the area of sharpest vision because of the high concentration of cones. Rods are absent from the fovea and macula, but they increase in density toward the periphery of the retina.

When impulses for sight have passed through the photoreceptor neurons, they are conducted across synapses to the bipolar neurons in the intermediate zone of the nervous layer of the retina. From here the impulses are passed to the ganglion neurons.

The axons of the ganglion neurons extend posteriorly to a small area of the retina called the **optic disc,** or **blind spot.** This region contains openings through which the fibers of the ganglion neurons exit as the optic nerve. Since this area contains neither rods nor cones, and only nerve fibers, no image striking it is perceived. Thus it is called the blind spot.

In addition to the fibrous tunic, vascular tunic, and retina, the eyeball itself contains the lens, just behind the pupil and iris. The **lens** is constructed of numerous layers of protein fibers arranged like the layers of an onion. Normally, the lens is perfectly transparent. It is enclosed by a clear connective tissue capsule and held in position by the **suspensory ligament.** A loss of transparency of the lens is known as a **cataract.**

The interior of the eyeball is a large cavity divided into two smaller ones by the lens. The **anterior cavity,** the division anterior to the lens, is further divided into the **anterior chamber,** which lies behind the cornea and in front of the iris, and the **posterior chamber,** which lies behind the iris and in front of the suspensory ligament and lens. The anterior cavity is filled with a watery fluid, similar to cerebrospinal fluid, called the **aqueous humor.** The fluid is believed to be secreted into the posterior chamber by choroid plexuses of the ciliary processes of the ciliary bodies behind the iris. Once the fluid is formed, it permeates the posterior chamber and then passes forward between the iris and the lens, through the pupil, into the anterior chamber. From the anterior chamber, the aqueous humor, which is continually produced, is drained off into the **canal of Schlemm** and then into the blood. The anterior chamber thus serves a function similar to the subarachnoid space around the brain and spinal cord. The canal of Schlemm is analogous to a venous sinus of the dura mater. The pressure in the eye, called **intraocular pressure,** is produced mainly by the aqueous humor. The intraocular pressure keeps the retina smoothly applied to the choroid so the retina will form clear images. Normal intraocular pressure (about 24 mm Hg) is maintained by drainage of the aqueous humor through the canal of Schlemm and into the blood. Excessive intraocular pressure, called **glaucoma** (glow-KŌ-ma), results in degeneration of the retina and blindness. Besides maintaining normal intraocular pressure, the aqueous humor is also the principal link between the circulatory system and the lens and cornea. Neither the lens nor the cornea has blood vessels.

The second, and larger, cavity of the eyeball is the **posterior cavity.** It lies between the lens and the retina and contains a jellylike substance called the **vitreous humor.** This substance contributes to intraocular pressure, helps to prevent the eyeball from collapsing, and holds the retina flush against the internal portions of the eyeball. The vitreous humor, unlike the aqueous humor, does not undergo constant replacement. It is formed during embryonic life and is not replaced thereafter.

Surface anatomy features of the eye are presented in Exhibit 25-1.

Afferent Pathway to Brain

From the rods and cones, impulses are transmitted through bipolar neurons to ganglion cells. The cell bodies of the ganglion cells lie in the retina, and their axons leave the eyeball via the **optic nerve (II)** (Figure 18-11). The axons pass through the **optic chiasma** (kī-AZ-ma), a crossing point of the optic nerves. Some fibers cross to the opposite side. Others remain uncrossed. On passing through the optic chiasma, the fibers, now part of the **optic tract,** enter the brain and terminate in the thalamus. Here the fibers synapse with third-order neurons whose axons pass to the visual centers located in the occipital lobes of the cerebral cortex.

Analysis of the afferent pathway to the brain reveals that the visual field of each eye is divided into two regions: the **nasal,** or **medial, half** and the **temporal,** or **lateral, half.** For each eye, light rays from an object in the nasal half of the visual field fall on the temporal half of the retina. Light rays from an object in the temporal half of the vision field fall on the nasal half of the retina. Note that in the optic chiasma nerve fibers from the nasal halves of the retinas cross and continue on to the thalamus. Note also that nerve fibers from the temporal halves of the retinas do not cross but continue directly on to the thalamus. As a result, the visual center in the cortex of the right occipital lobe "sees" the left side of an object via impulses from the temporal half of the retina of the right eye and the nasal half of the retina of the left eye. The cortex of the left occipital lobe interprets visual sensations from the right side of an object via impulses from the nasal half of the right eye and the temporal half of the left eye.

CLINICAL APPLICATION

Blind spots in the field of vision (other than the normal blind spot of the optic disc) may indicate a brain tumor along one of the afferent pathways. For instance, a symptom of a tumor in the right optic tract might be an inability to see the left side of a normal field of vision without moving the eyeball.

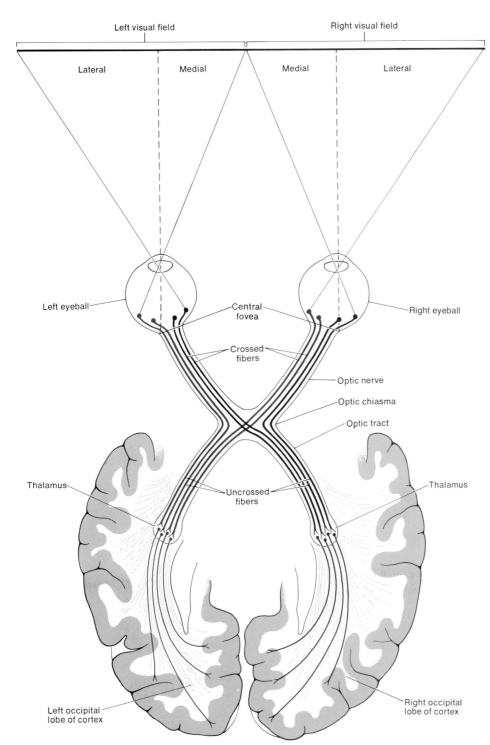

FIGURE 18-11 Afferent pathway for visual impulses. See Figure 16-13 also.

AUDITORY SENSATIONS AND EQUILIBRIUM

In addition to containing receptors for sound waves, the **ear** also contains receptors for equilibrium. Anatomically, the ear is divided into three principal regions: the external or outer ear, the middle ear, and the internal or inner ear.

The developmental anatomy of the ear is considered in Exhibit 24-7.

External or Outer Ear

The **external** or **outer ear** is structurally designed to collect sound waves and direct them inward (Figure 18-12a).

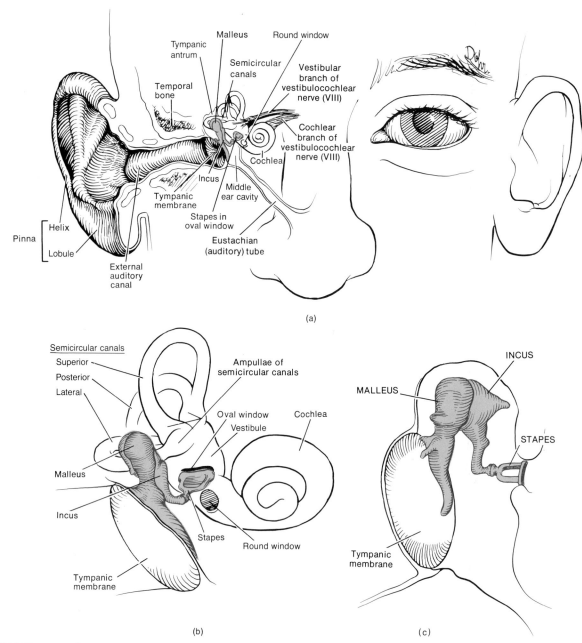

FIGURE 18-12 Structure of the auditory apparatus. (a) Divisions of the right ear into external, middle, and internal portions seen in a coronal section through the right side of the skull. The middle ear is shown in color. (b) Details of the middle ear and the bony labyrinth of the internal ear. (c) Ossicles of the middle ear.

It consists of the pinna, the external auditory canal, and the tympanic membrane, also called the eardrum.

The **pinna,** or **auricle,** is a trumpet-shaped flap of elastic cartilage covered by thick skin. The rim of the pinna is called the helix; the inferior portion is the lobe. (See Exhibit 25-1 for surface anatomy features of the external ear.) The pinna is attached to the head by ligaments and muscles.

The **external auditory canal** or **meatus** is a tube about 2.5 cm (1 inch) in length that lies in the external auditory

meatus of the temporal bone. It leads from the pinna to the eardrum. The walls of the canal consist of bone lined with cartilage that is continuous with the cartilage of the pinna. The cartilage in the external auditory canal is covered with thin, highly sensitive skin. Near the exterior opening, the canal contains a few hairs and specialized sebaceous glands called **ceruminous** (se-ROO-mi-nus) **glands,** which secrete **cerumen** (se-ROO-min) (earwax). The combination of hairs and cerumen helps to prevent foreign objects from entering the ear.

The **tympanic** (tim-PAN-ik) **membrane,** or **eardrum,** is a thin, semitransparent partition of fibrous connective tissue between the external auditory meatus and the middle ear. Its external surface is concave and covered with skin. Its internal surface is convex and covered with a mucous membrane.

Middle Ear

The **middle ear,** also called the **tympanic cavity,** is a small, epithelial-lined, air-filled cavity hollowed out of the temporal bone (Figure 18-12a-c). The cavity is separated from the external ear by the eardrum and from the internal ear by a thin bony partition that contains two small openings: the oval window and the round window.

The posterior wall of the cavity communicates with the mastoid air cells of the temporal bone through a chamber called the **tympanic antrum.** This anatomical fact explains why a middle ear infection may spread to the temporal bone, causing mastoiditis, or even to the brain.

The anterior wall of the cavity contains an opening that leads directly into the **eustachian** (yoo-STĀ-kē-an) **tube,** also called the **auditory tube** or **meatus.** The eustachian tube connects the middle ear with the nasopharynx of the throat. Through this passageway, infections may travel from the throat and nose to the ear. The function of the tube is to equalize air pressure on both sides of the tympanic membrane. Abrupt changes in external or internal air pressure might otherwise cause the eardrum to rupture. Since the tube opens during swallowing and yawning, these activities allow atmospheric air to enter or leave the middle ear until the internal pressure equals the external pressure. Any sudden pressure changes against the eardrum may be equalized by deliberately swallowing.

Extending across the middle ear are three exceedingly small bones called **auditory ossicles** (OS-si-kuls). The bones, named for their shape, are the malleus, incus, and stapes, commonly called the hammer, anvil, and stirrup, respectively (Figure 18-12b, c). They are connected by synovial joints. The "handle" of the **malleus** is attached to the internal surface of the tympanic membrane. Its head articulates with the body of the incus. The **incus** is the intermediate bone in the series and articulates with the head of the stapes. The base or footplate of the **stapes** fits into a small opening between the middle and inner ear called the **fenestra vestibuli** (fe-NES-tra ves-TIB-yoo-lī), or **oval window.** Directly below the oval window is another opening, the **fenestra cochlea** (fe-NES-tra KŌK-lē-a), or **round window.** This opening, which also separates the middle and inner ears, is enclosed by a membrane called the secondary tympanic membrane. The auditory ossicles are attached to the tympanic cavity by means of ligaments.

Three ligaments are associated with the malleus (Figure 18-13). The *anterior ligament of the malleus* is attached by one end to the anterior process of the malleus and by the other to the anterior wall of the tympanic cavity. The *superior ligament of the malleus* extends from the head of the malleus to the roof of the tympanic cavity. The *lateral ligament of the malleus* extends from the neck of the malleus to the lateral wall of the tympanic cavity. Two ligaments are associated with the incus. The *posterior ligament of the incus* extends from the short crus of the incus to the posterior wall of the tympanic cavity and the *superior ligament of the incus* extends from the body of the incus to the roof of the tympanic cavity. A single ligament, the *annular ligament of the base of the stapes,* is associated with the stapes. It extends from the base of the stapes to the fenestra vestibuli.

In addition to the ligaments, there are also two muscles attached to the ossicles. The *tensor tympani muscle* originates from the cartilaginous portion of the eustachian tube, the greater wing of the sphenoid bone, as well as from the osseous canal in which it is contained. The muscle inserts into the manubrium of the malleus. It is supplied by the mandibular nerve via fibers that pass through the otic ganglion. The action of the muscle is to draw the malleus medially, thus increasing tension on the tympanic membrane and reducing the amplitude of oscillations. This action tends to prevent damage to the inner ear when exposed to loud sounds.

The *stapedius muscle* is the smallest of all skeletal muscles. It originates from a projection (pyramid) in the posterior wall of the tympanic cavity and inserts into the neck of the stapes. The muscle is innervated by the

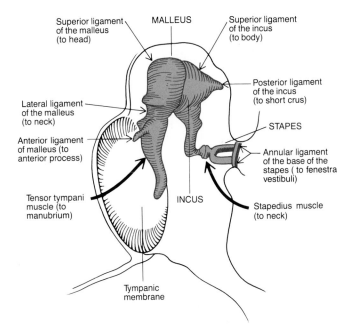

FIGURE 18-13 Areas of attachment of the ligaments and muscles of the auditory ossicles.

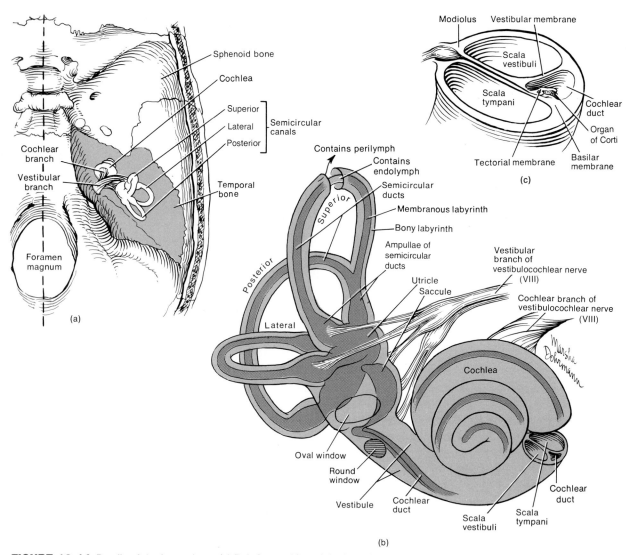

FIGURE 18-14 Details of the internal ear. (a) Relative position of the bony labyrinth projected to the inner surface of the floor of the skull. (b) The outer, gray area belongs to the bony labyrinth. The inner, colored area belongs to the membranous labyrinth. (c) Cross section through the cochlea.

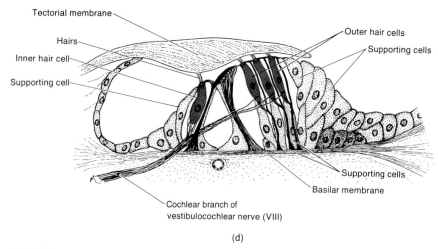

FIGURE 18-14 (*Continued*) Details of the internal ear. (d) Diagram of the organ of Corti.

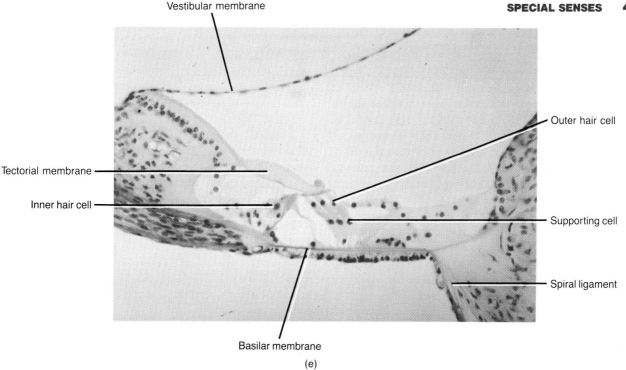

Vestibular membrane

Tectorial membrane

Inner hair cell

Outer hair cell

Supporting cell

Spiral ligament

Basilar membrane

(e)

FIGURE 18-14 (*Continued*) Details of the internal ear. (e) Photomicrograph of the organ of Corti at a magnification of 300×. (© 1983 by Michael H. Ross. Used by permission.)

facial nerve. The action of the muscle is to draw the stapes posteriorly, thereby tightening the annular ligament of the base of the stapes and reducing the oscillatory range of the the stapes (prevents excessive movement). Like the tensor tympani muscle, the stapedius muscle has a protective function in that it dampens (checks) large vibrations that result from loud noises. It is for this reason that paralysis of the stapedius muscle is associated with *hyperacusia* (acuteness of hearing).

Internal or Inner Ear

The **internal** or **inner ear** is also called the **labyrinth** (LAB-i-rinth) because of its complicated series of canals (Figure 18-14). Structurally, it consists of two main divisions: a bony labyrinth and a membranous labyrinth that fits in the bony labyrinth. The **bony labyrinth** is a series of cavities in the petrous portion of the temporal bone. It can be divided into three areas named on the basis of shape: the vestibule, cochlea, and semicircular canals. The bony labyrinth is lined with periosteum and contains a fluid called **perilymph.** This fluid surrounds the **membranous labyrinth,** a series of sacs and tubes lying inside and having the same general form as the bony labyrinth. Epithelium lines the membranous labyrinth and contains a fluid called **endolymph.**

The **vestibule** constitutes the oval central portion of the bony labyrinth. The membranous labyrinth in the vestibule consists of two sacs called the **utricle** and **saccule.** These sacs are connected to each other by a small duct.

Projecting upward and posteriorly from the vestibule are the three bony **semicircular canals.** Each is arranged at approximately right angles to the other two. On the basis of their positions, they are called the superior, posterior, and lateral canals. One end of each canal enlarges into a swelling called the **ampulla** (am-POOL-la). Inside the bony semicircular canals lie portions of the membranous labyrinth: the **semicircular ducts** or **membranous semicircular canals.** These structures are almost identical in shape to the bony semicircular canals and communicate with the utricle of the vestibule.

Lying in front of the vestibule is the **cochlea** (KŌK-lē-a), so designated because of its resemblance to a snail's shell. The cochlea consists of a bony spiral canal that makes about 2¾ turns around a central bony core called the **modiolus.** A cross section through the cochlea shows the canal is divided by partitions into three separate channels resembling the letter Y on its side. The stem of the Y is a bony shelf that protrudes into the canal. The wings of the Y are composed mainly of membranous labyrinth. The channel above the bony partition is the **scala vestibuli;** the channel below is the **scala tympani.** The cochlea adjoins the wall of the vestibule, into which the scala vestibuli opens. The scala tympani terminates at the round window. The perilymph of the vestibule is continuous with that of the scala vestibuli. The third channel (between the wings of the Y) is the membranous labyrinth: the **cochlear duct.** The cochlear duct is separated from the scala vestibuli by the **vestibular membrane,** also called **Reissner's membrane.** It is separated from the scala tympani by the **basilar membrane.**

Resting on the basilar membrane is the **organ of Corti (spiral organ),** the organ of hearing. The organ of Corti is a series of epithelial cells on the inner surface of the basilar membrane. It consists of a number of supporting cells and hair cells, which are the receptors for auditory sensations. The inner hair cells are medially placed in a single row and extend the entire length of the cochlea. The outer hair cells are arranged in several rows throughout the cochlea. The hair cells have long hairlike processes at their free ends that extend into the endolymph of the cochlear duct. The basal ends of the hair cells are in contact with fibers of the cochlear branch of the vestibulocochlear nerve (VIII). Projecting over and in contact with the hair cells of the organ of Corti is the **tectorial membrane,** a delicate and flexible gelatinous membrane.

Mechanism of Hearing

Sound waves are the alternate compression and decompression of air originating from a vibrating object. They travel through air much as waves travel on water. The events involved in the physiology of hearing sound waves are illustrated in Figure 18-15 and listed below.

1. Sound waves that reach the ear are directed by the pinna into the external auditory canal.

2. When the waves strike the tympanic membrane, the alternate compression and decompression of the air causes the membrane to vibrate.

3. The central area of the tympanic membrane is connected to the malleus, which also starts to vibrate. The vibration is then picked up by the incus, which transmits the vibration to the stapes.

4. As the stapes moves back and forth, it pushes the oval window in and out.

5. The movement of the oval window sets up waves in the perilymph.

6. As the window bulges inward, it pushes the perilymph of the scala vestibuli and waves are propagated through the scala vestibuli into the fluid of the scala tympani.

7. This pressure pushes the vestibular membrane inward and increases the pressure of the endolymph inside the cochlear duct.

8. The basilar membrane gives under the pressure and bulges out into the scala tympani.

9. The sudden pressure in the scala tympani pushes

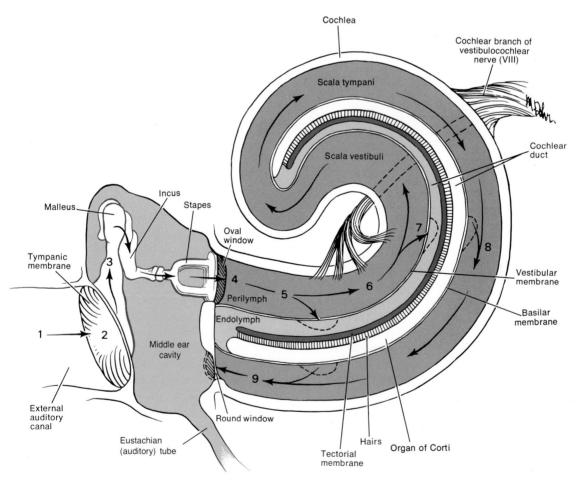

FIGURE 18-15 Physiology of hearing. Follow the text carefully to understand the meaning of the numbers.

the perilymph toward the round window, causing it to bulge back into the middle ear. Conversely, as the sound waves subside, the stapes moves backward and the procedure is reversed. That is, the fluid moves in the opposite direction along the same pathway and the basilar membrane bulges into the cochlear duct.

10. When the basilar membrane vibrates, the hair cells of the organ of Corti are moved against the tectorial membrane. In some unknown manner, the movement of the hairs stimulates the dendrites of neurons at their base, and sound waves are converted into nerve impulses.

11. The impulses are then passed on to the cochlear branch of the vestibulocochlear nerve (Figure 18-16) and the medulla. Here some impulses cross to the opposite side and finally travel to the auditory area of the temporal lobe of the cerebral cortex.

If sound waves passed directly to the oval window without passing through the tympanic membrane and auditory bones, hearing would be inadequate. A minimal amount of sound energy is required to transmit sound waves through the perilymph of the cochlea. Since the tympanic membrane has a surface area about 22 times larger than that of the oval window, it can collect about 22 times more sound energy. This energy is sufficient to transmit sound waves through the perilymph.

Mechanism of Equilibrium

The term **equilibrium** has two meanings. One kind of equilibrium, called **static equilibrium,** refers to the orientation of the body (mainly the head) relative to the ground (gravity). The second kind, **dynamic equilibrium,** is the maintenance of body position (mainly the head) in response to sudden movements. The receptor organs for equilibrium are the maculae of the saccule and the utricle and cristae in the semicircular ducts.

● *Static Equilibrium*   The utricle and saccule are the receptors that are concerned with static equilibrium. They provide sensory information regarding the orientation of the head in space and are essential for maintaining posture.

The walls of both the utricle and saccule contain a small, flat, plaquelike region called a **macula** (Figure 18-17). Microscopically, the two maculae resemble the organ of Corti. They consist of differentiated neuroepithelial cells that are innervated by the vestibular branch of the vestibulocochlear nerve. The maculae are anatomically located in planes perpendicular to one another and possess two kinds of cells: **hair (receptor) cells** and **supporting cells.** Two shapes of receptor cells have been identified: one is more flask-shaped and the other is more cylindrical.

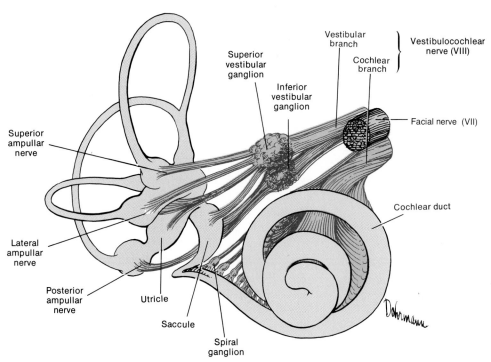

FIGURE 18-16 Principal nerves of the internal ear. Shown here is the right membranous labyrinth with the bony labyrinth removed. Nerve branches from each of the ampullae of the semicircular ducts and from the utricle and saccule form the vestibular branch of the vestibulocochlear nerve. The vestibular ganglia contain the cell bodies of the vestibular branch. The cochlear branch of the vestibulocochlear nerve arises in the organ of Corti. The spiral ganglion contains the cell bodies of the cochlear branch.

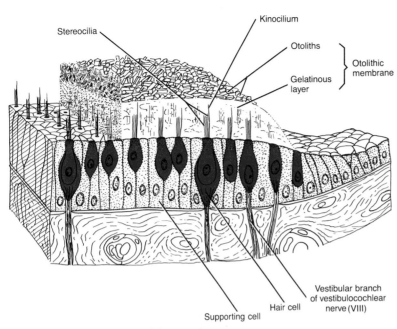

Stereocilia

Kinocilium

Otoliths

Gelatinous layer

Otolithic membrane

Vestibular branch of vestibulocochlear nerve (VIII)

Hair cell

Supporting cell

FIGURE 18-17 Structure of the macula.

Both show long extensions of the cell membrane consisting of many *stereocilia* (they are actually microvilli) and one *kinocilium* (a conventional cilium) anchored firmly to its basal body and extending beyond the longest microvilli. Some of the microvilli reach lengths of over 100 μm, and some receptor cells have over 80 such projections. It is easy to understand why these are also named hair cells.

The columnar supporting cells of the maculae are scattered between the hair cells. Floating directly over the hair cells is a thick, gelatinous, glycoprotein layer, probably secreted by the supporting cells, called the **otolithic membrane.** A layer of calcium carbonate crystals, called **otoliths** (*oto* = ear; *lithos* = stone), extends over the entire surface of the otolithic membrane. The specific gravity of these otoliths is about 3, which makes them much denser than the endolymph fluid that fills the rest of the utricle.

The otolithic membrane sits on top of the macula like a discus on a greased cookie sheet. If you tilt your head forward, the otolithic membrane (the discus in our analogy) slides downhill over the hair cells in the direction determined by the tilt of your head. In the same manner, if you are sitting upright in a car that suddenly jerks forward, the otolithic membrane, due to its inertia, slides backward and stimulates the hair cells that, in turn, would signal the forward movement. As the otoliths move, they pull on the gelatinous layer which pulls on the stereocilia and makes them bend. The movement of the stereocilia initiates a nerve impulse that is then transmitted to the vestibular branch of the vestibulocochlear nerve (see Figure 18-16).

Most of the vestibular branch fibers enter the brain stem and terminate in the vestibular nuclear complex in the medulla. The remaining fibers enter the flocculonodular lobe of the cerebellum through the inferior cerebellar peduncle. Bidirectional pathways connect the vestibular nuclei and cerebellum. Fibers from all the vestibular nuclei form the medial longitudinal fasciculus which extends from the brain stem into the cervical portion of the spinal cord. The fasciculus sends impulses to the nuclei of cranial nerves that control eye movements (oculomotor, trochlear, and abducens) and to the accessory nerve nucleus that helps control movements of the head and neck. In addition, fibers from the lateral vestibular nucleus form the vestibulospinal tract which conveys impulses to skeletal muscles that regulate body tone in response to head movements. Various pathways between the vestibular nuclei, cerebellum, and cerebrum enable the cerebellum to assume a key role in helping the body to maintain static equilibrium. The cerebellum continuously receives updated sensory information from the utricle and saccule concerning static equilibrium. Using this information, the cerebellum monitors and makes corrective adjustments in the motor activities that originate in the cerebral cortex. Essentially, the cerebellum sends continuous impulses to the motor areas of the cerebrum, in response to input from the utricle and saccule, causing the motor system to increase or decrease its impulses to specific skeletal muscles in order to maintain static equilibrium.

● *Dynamic Equilibrium* The three semicircular ducts maintain dynamic equilibrium. They are positioned at right angles to one another in three planes: frontal (the

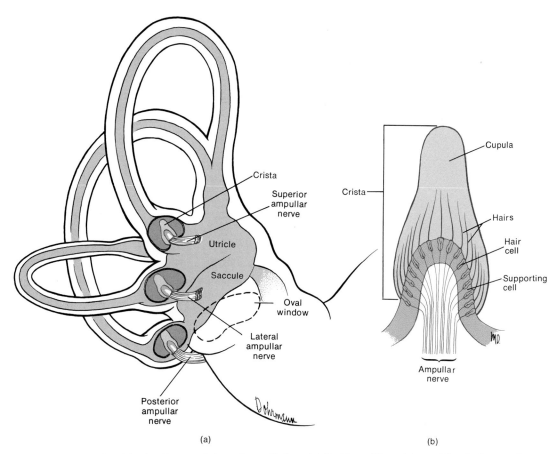

FIGURE 18-18 Semicircular ducts and dynamic equilibrium. (a) Position of the cristae relative to the membranous ampullae. (b) Enlarged aspect of a crista.

superior duct), sagittal (the posterior duct), and lateral (the lateral duct). This positioning permits detection of an imbalance in three planes. In the ampulla, the dilated portion of each duct, there is a small elevation called the **crista** (Figure 18-18a). Each crista is composed of a group of **hair (receptor) cells** and **supporting cells** covered by a mass of gelatinous material called the **cupula** (Figure 18-18b). When the head moves, the endolymph in the semicircular ducts flows over the hairs and bends them. The movement of the hairs stimulates sensory neurons, and the impulses pass over the vestibular branch of the vestibulocochlear nerve. The impulses follow the same pathways as those involved in static equilibrium and are eventually sent to the muscles that must contract to maintain body balance in the new position.

APPLICATIONS TO HEALTH

The special sense organs can be altered or damaged by numerous disorders. The causes of disorder can range from congenital origins to the effects of old age. Here we discuss a few common disorders of the eyes and ears.

CATARACT

The most prevalent disorder resulting in blindness is cataract formation. A **cataract** is a clouding of the lens or its capsule so that it becomes opaque or milk-white (Figure 18-19). As a result, the light from an object, which normally passes directly through the lens to produce a sharp image, produces only a degraded image. If the cataract is severe enough, no image at all is produced.

Cataracts can occur at any age, but we will discuss the type that develops with old age. As a person gets older the cells in the lenses may degenerate and be replaced with nontransparent fibrous protein. Or the lens may start to manufacture nontransparent protein. The cataract formation results in a progressive, painless loss of vision. The degree of loss depends on the location and extent of the opacity. If vision loss is gradual, frequent changes in glasses may help maintain useful vision for a while. However, when the changes become so extensive that light rays are blocked out altogether, surgery is indicated. Essentially, the surgical procedure consists of removing the opaque lens and substituting an artificial lens by means of eyeglasses or by implanting a plastic lens inside the eyeball to replace the natural one.

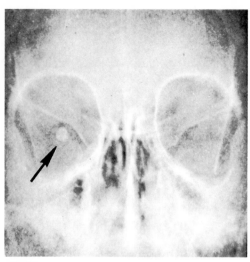

FIGURE 18-19 Anteroposterior projection of a cataract (arrow) in the lens of the right eye. (Courtesy of Lester W. Paul and John H. Juhl, *The Essentials of Roentgen Interpretation,* Harper & Row, Publishers, Inc., New York, 1972.)

GLAUCOMA

The second most common cause of blindness, especially in the elderly, is **glaucoma.** This disorder is characterized by an abnormally high pressure of fluid inside the eyeball. The aqueous humor does not return into the bloodstream through the canal of Schlemm as quickly as it is formed. The fluid accumulates and, by compressing the lens into the vitreous humor, puts pressure on the neurons of the retina. If the pressure continues over a long period of time, it destroys the neurons and brings about blindness. It can affect a person of any age, but 95 percent of the victims are over 40. Glaucoma affects the eyesight of more than 1 million people in the United States.

Treatment is through drugs or surgery. Drug treatment involves reducing the pressure in the eye by giving cholinergic compounds that cause the sphincter muscles of the iris and ciliary body to contract. These contractions prevent the obstruction of the anterior chamber and the canal of Schlemm, resulting in improved fluid outflow from the eye. Surgery consists of removing a small piece of the iris (iridectomy) to allow drainage and lessen the interior pressure.

CONJUNCTIVITIS

Many different eye inflammations exist, but the most common type is **conjunctivitis (pinkeye),** an inflammation of the membrane that lines the insides of the eyelids and covers the cornea. Conjunctivitis can be caused by microorganisms, most often the pneumococci or staphylococci bacteria. In such cases, the inflammation is very contagious. This epidemic type, common in children, is rarely serious. Conjuntivitis may also be caused by a number of irritants, in which case the inflammation is not contagious. Irritants include dust, smoke, wind, pollutants in the air, and excessive glare. The condition may be acute or chronic.

TRACHOMA

A form of chronic contagious conjunctivitis that is serious is **trachoma** (tra-KŌ-ma). It is caused by an organism, called the TRIC agent, that has characteristics of both viruses and bacteria. Trachoma is characterized by many granulations or fleshy projections on the eyelids. If untreated, these projections can irritate and inflame the cornea and reduce vision. The disease produces an excessive growth of subconjunctival tissue and the invasion of blood vessels into the upper half of the front of the cornea. The disease progresses until it covers the entire cornea, bringing about a loss of vision because of corneal opacity.

Antibiotics, such as tetracycline and the sulfa drugs, kill the organisms that cause trachoma and have reduced the seriousness of this infection.

LABYRINTHINE DISEASE

Labyrinthine (lab'-i-RIN-thēn) **disease** refers to a malfunction of the inner ear that is characterized by deafness, tinnitus (ringing in the ears), vertigo (hallucination of movement), nausea, and vomiting. There may also be a blurring of vision, nystagmus (rapid, involuntary movement of the eyeballs), and a tendency to fall in a certain direction.

Among the causes of labyrinthine disease are:

1. Infection of the middle ear.
2. Trauma from brain concussion producing hemorrhage or splitting of the labyrinth.
3. Cardiovascular diseases such as arteriosclerosis and blood vessel disturbances.
4. Congenital malformation of the labyrinth.
5. Excessive formation of endolymph.
6. Allergy.
7. Blood abnormalities.
8. Aging.

In severe cases of labyrinthine disease, surgical treatment is considered. Treatment with ultrasound, which consists of beaming ultrasound waves into the labyrinth, destroys the semicircular ducts (equilibrium) while preserving the cochlear duct (hearing). This treatment has produced excellent results. Another successful procedure involves electrical coagulation of the labyrinth, which changes the fluid to a jelly or a solid.

MÉNIÈRE'S DISEASE

Ménière's (men-YAIRZ) **disease** is one type of labyrinthine disorder that is characterized by fluctuating hearing loss, attacks of vertigo, and roaring tinnitus. Etiology of Ménière's disease is unknown. It is now thought that there is either an overproduction or underabsorption of endolymph in the cochlear duct. The hearing loss is caused by distortions in the basilar membrane of the cochlea. The classic type of Ménière's disease involves both the semicircular canals and the cochlea. The natural course of Ménière's disease may stretch out over a period of years, with the end result being almost total destruction of the patient's hearing.

The treatment for Ménière's disease may be either medical or surgical. Medical treatment is aimed at reducing the excess amount of endolymph in order to lessen the amount of distention and distortion of the basilar membrane. To accomplish this, the overall extracellular volume is diminished by means of salt restriction and the administration of diuretics. A popular treatment for the acute cases is the inhalation of 5 percent CO_2 with 95 percent O_2 for 10 minutes, four times a day. High levels of CO_2 in the cerebral circulation are powerful dilators of the cerebral vessels. For the 20 to 30 percent of Ménière's patients who do not respond to medical treatment, some type of vestibular surgery is indicated, for relief from vertigo. There are two basic procedures available to the otologic surgeon: those that destroy residual hearing and those that preserve it. As a rule, labyrinthectomy is reserved for advanced cases in which the hearing is poor.

IMPACTED CERUMEN

Some people produce an abnormal amount of cerumen, or earwax, in the external auditory canal. It becomes impacted and prevents sound waves from reaching the tympanic membrane. The treatment for **impacted cerumen** is usually periodic ear irrigation or removal of wax with a blunt instrument.

OTITIS MEDIA

Otitis media is an acute infection of the middle ear cavity with a reddening and outward bulging of the eardrum, which may rupture. Second only to the common cold as a disease of childhood, otitis media is the most frequent diagnosis made by physicians who care for children. Infants and young children are at highest risk for otitis media; the peak prevalence is between 6 and 24 months of age.

Abnormal function of the eustachian tube appears to be the most important mechanism in the pathogenesis of middle ear disease. This allows bacteria from the nasopharynx, which are a primary cause of middle ear infection, to enter the middle ear. Otitis media can be acute or chronic, and the primary etiology of both forms is bacterial infection. *Streptococcus pneumoniae* is the most common agent found in all age groups, followed by *Hemophilus influenzae*. Hearing loss is by far the most prevalent complication of otitis media.

Antimicrobial therapy has practically eliminated such complications of middle ear infections as mastoiditis and brain abscess, but perforation of the tympanic membrane and labyrinthitis still occur.

MOTION SICKNESS

Motion sickness is a functional disorder brought on by repetitive angular, linear, or vertical motion and characterized by various symptoms, primarily nausea and vomiting. Warning symptoms include yawning, salivation, pallor, hyperventilation, profuse cold sweating, and prolonged drowsiness. Specific kinds of motion sickness are seasickness or air-, car-, space-, train-, or swing sickness.

Etiology is excessive stimulation of the vestibular apparatus by motion. The pathways taken by the afferent impulses from the labyrinth to the vomiting center in the medulla have not been defined. Visual stimuli and emotional factors like fear and anxiety can also contribute to motion sickness. Ideally, treatment with drugs of susceptible individuals should be instituted prior to their entering into conditions that would produce motion sickness, since prevention is more successful than treatment of symptoms once they have developed.

KEY MEDICAL TERMS ASSOCIATED WITH SENSE ORGANS

Achromatopsia (*a* = without; *chrom* = color) Complete color blindness.

Ametropia (*ametro* = disproportionate; *ops* = eye) Refractive defect of the eye resulting in an inability to focus images properly on the retina.

Anopsia (*opsis* = vision) A defect of vision.

Astereognosis (*stereos* = solid; *gnosis* = knowledge) Loss of ability to recognize objects or to appreciate their form by touching them.

Blepharitis (*blepharo* = eyelid; *itis* = inflammation of) An inflammation of the eyelid.

Chordotomy (*chord* = cord; *tome* = a cutting) Division of the lateral spinothalamic tract for intractable pain.

Dyskinesia (*dys* = difficult; *kinesis* = movement) Abnormality of motor function characterized by involuntary, purposeless movements.

Eustachitis An inflammation or infection of the eustachian tube.

Keratitis (*kerato* = cornea) An inflammation or infection of the cornea.

Keratoplasty (*plasty* = form) Corneal graft or transplant. An opaque cornea is removed and replaced with a normal transparent cornea to restore vision.

Kinesthesis (*aisthesis* = sensation) The sense of perception of movement.

Labyrinthitis An inflammation of the inner ear or labyrinth.

Myringitis (*myringa* = eardrum) An inflammation of the eardrum; also called tympanitis.

Nystagmus A constant, rapid involuntary movement of the eyeballs, possibly caused by a disease of the central nervous system.

Otalgia (*oto* = ear; *algia* = pain) Earache.

Presbyopia (*presby* = old) Inability to focus on nearby objects due to loss of elasticity of the crystalline lens. The loss is usually caused by aging.

Proprioceptor (*proprius* = one's own; *receptor* = receiver) A sensory ending in muscles, tendons, and joints that provides information concerning movement and position of body parts (proprioception).

Ptosis (*ptosis* = fall) Falling or drooping of the eyelid. (This term is also used for the slipping of any organ below its normal position.)

Retinoblastoma (*blast* = bud; *oma* = tumor) A common tumor arising from immature retinal cells and accounting for 2 percent of childhood malignancies.

Strabismus An eye muscle disorder, commonly called "crossed eyes," in which the eyeballs do not move in unison. It may be caused by lack of coordination of the extrinsic eye muscles.

Tinnitus A ringing in the ears.

STUDY OUTLINE

Sensations

1. Sensation is a state of awareness of external and internal conditions of the body.
2. The prerequisites for sensation are reception of a stimulus, conversion of the stimulus into a nerve impulse by a receptor, conduction of the impulse to the brain, and translation of the impulse into a sensation by a region of the brain.
3. Exteroceptors receive stimuli from the external environment, visceroceptors from blood vessels and viscera, and proprioceptors from muscles, tendons, and joints for body position and movement.

General Senses

Cutaneous Sensations

1. Cutaneous sensations include tactile sensations (touch, pressure, vibration), thermoreceptive sensations (heat and cold), and pain. Receptors for these sensations are located in the skin, connective tissues, and the ends of the gastrointestinal tract.
2. Receptors for fine touch are Meissner's corpuscles, Merkel's discs, root hair plexuses, and free nerve endings. Pacinian corpuscles and free nerve endings are receptors for pressure. Krause's end bulbs and Ruffini's end organs were once thought to be receptors for cold and heat, respectively.

Pain Sensations

1. Pain receptors are located in nearly every body tissue.
2. Two kinds of pain recognized in the parietal lobe of the cortex are somatic and visceral.

3. Referred pain is felt in the skin near or away from the organ sending pain impulses.
4. Phantom pain is the sensation of pain in a limb that has been amputated.
5. Pain impulses may be inhibited by drugs, surgery, and acupuncture.

Proprioceptive Sensations

1. Receptors located in muscles, tendons, and joints convey impulses related to muscle tone, movement of body parts, and body position.
2. The receptors include joint kinesthetic receptors, neuromuscular spindles, and Golgi tendon organs.

Levels of Sensation

1. Sensory fibers terminating in the lower brain stem bring about far more complex motor reactions than simple spinal reflexes.
2. When sensory impulses reach the lower brain stem, they cause subconscious motor reactions.
3. Sensory impulses that reach the thalamus can be localized crudely in the body.
4. When sensory impulses reach the cerebral cortex, we experience precise localization.

Sensory Pathways

1. In the posterior column pathway and the spinothalamic pathway there are first-order, second-order, and third-order neurons.
2. The sensory pathway for pain and temperature is the lateral spinothalamic pathway.

3. The neural pathway that conducts impulses for crude touch and pressure is the anterior spinothalamic pathway.
4. The neural pathway for fine touch, proprioception, and vibration is the posterior column pathway.
5. The pathways to the cerebellum are the anterior and posterior spinocerebellar tracts.

Motor Pathways

1. Voluntary motor impulses are conveyed from the brain through the spinal cord along the pyramidal pathways and the extrapyramidal pathways.
2. Pyramidal pathways include the lateral corticospinal, anterior corticospinal, and corticobulbar tracts.
3. Major extrapyramidal tracts are the rubrospinal, tectospinal, and vestibulospinal tracts.

Special Senses

Olfactory Sensations
Receptor cells in the nasal epithelium send impulses to the olfactory bulbs, olfactory tracts, and cortex.

Gustatory Sensations
Receptors in the taste buds send impulses to the cranial nerves, thalamus, and cortex.

Visual Sensations

1. Accessory structures of the eyes include the eyebrows, eyelids, eyelashes, and the lacrimal apparatus.
2. The eye is constructed of three coats: (a) fibrous tunic (sclera and cornea), (b) vascular tunic (choroid, ciliary body, and iris), and (c) retina, which contains rods and cones.
3. The anterior cavity contains aqueous humor; the posterior cavity contains viterous humor.
4. Rods and cones convert light rays into visual nerve impulses.
5. Impulses from rods and cones are conveyed through retina to optic nerve, optic chiasma, optic tract, thalamus, and cortex.

Auditory Sensations and Equilibrium
1. The ear consists of three anatomical subdivisions: (a) the external or outer ear (pinna, external auditory canal, and tympanic membrane), (b) the middle ear (eustachian tube, ossicles, oval window, and round window), and (c) the internal or inner ear (bony labyrinth and membranous labyrinth). The internal ear contains the organ of Corti, the organ of hearing.
2. Sound waves enter the external auditory canal, strike the tympanic membrane, pass through the ossicles, strike the oval window, set up waves in the perilymph, strike the vestibular membrane and scala tympani, increase pressure in the endolymph, strike the basilar membrane, and stimulate hairs on the organ of Corti. A sound impulse is then initiated.
3. Static equilibrium is the relationship of the body relative to the pull of gravity. The maculae of the utricle and saccule are the sense organs of static equilibrium.
4. Dynamic equilibrium is equilibrium in response to movement of the body. The cristae in the semicircular ducts are the sense organs of dynamic equilibrium.

Applications to Health

1. Cataract is the loss of transparency of the lens or capsule.
2. Glaucoma is abnormally high intraocular pressure, which destroys neurons of the retina.
3. Conjunctivitis is an inflammation of the conjunctiva.
4. Trachoma is a chronic, contagious inflammation of the conjunctiva.
5. Labyrinthine disease is basically a malfunction of the inner ear that has a variety of causes.
6. Ménière's disease is the malfunction of the inner ear that may cause deafness and loss of equilibrium.
7. Impacted cerumen is an abnormal amount of earwax in the external auditory canal.
8. Otitis media is an acute infection of the middle ear cavity.
9. Motion sickness is a functional disorder precipitated by repetitive angular, linear, or vertical motion.

REVIEW QUESTIONS

1. Define a sensation and a sense receptor. What prerequisites are necessary for the perception of a sensation?
2. Compare the location and function of exteroceptors, visceroceptors, and proprioceptors.
3. Distinguish between a general sense and a special sense.
4. What is a cutaneous sensation? Distinguish tactile, thermoreceptive, and pain sensations.
5. For each of the following cutaneous sensations, describe the receptor involved in terms of structure, function, and location: touch, pressure, vibration, cold, heat, and pain.
6. How do cutaneous sensations help maintain homeostasis?
7. Why are pain receptors important?
8. Differentiate somatic pain, visceral pain, referred pain, and phantom pain.
9. Why is the concept of referred pain useful to a physician diagnosing internal disorders?
10. What is acupuncture? Describe the "gate control" theory of acupuncture.

11. What is the proprioceptive sense? Where are the receptors for this sense located?
12. What is the sensory pathway for pain and temperature? How does it function?
13. Which pathway controls crude touch and pressure? How does it function?
14. Which pathway is responsible for fine touch, proprioception, and vibration? How does it function?
15. Which pathways control voluntary motor impulses from the brain through the spinal cord? How are they distinguished?
16. Discuss the origin and path of an impulse that results in smelling.
17. How are papillae related to taste buds? Describe the structure and location of the papillae. Discuss how an impulse for taste travels from a taste bud to the brain.
18. Describe the structure and importance of the following accessory structures of the eye: eyelids, eyelashes, eyebrows, and lacrimal apparatus.
19. By means of a labeled diagram, indicate the principal anatomical structures of the eye. How is the retina adapted to its function?
20. Distinguish a sty from a chalazion.
21. How do extrinsic and intrinsic eye muscles differ?
22. Describe the location and contents of the chambers of the eye. What is intraocular pressure? How is the canal of Schlemm related to this pressure?
23. Describe the path of a visual impulse from the optic nerve to the brain.
24. Define visual field. Relate the visual field to image formation on the retina.
25. Diagram the principal parts of the outer, middle, and inner ear. Describe the function of each part labeled.
26. List and describe the points of attachment of the ligaments of the auditory ossicles.
27. Explain the events involved in the transmission of sound from the pinna to the organ of Corti.
28. What is the afferent pathway for sound impulses from the vestibulocochlear nerve to the brain?
29. Compare the function of the saccule and utricle in maintaining static equilibrium with the role of the semicircular ducts in maintaining dynamic equilibrium.
30. Define each of the following: cataract, glaucoma, conjunctivitis, trachoma, labyrinthine disease, Ménière's disease, impacted cerumen, otitis media, and motion sickness.
31. Refer to the key medical terms associated with the sense organs. Be sure that you can define each term listed.

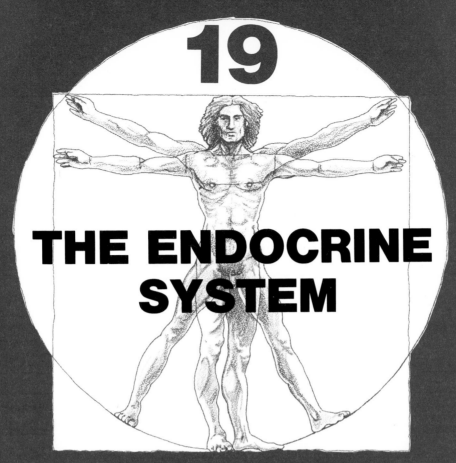

19

THE ENDOCRINE SYSTEM

STUDENT OBJECTIVES

■ Describe the relationship between the endocrine system and the nervous system in maintaining homeostasis.

■ Define an endocrine gland and an exocrine gland.

■ Describe the location, histology, and blood and nerve supply of the pituitary gland.

■ Explain how the pituitary gland is structurally and functionally divided into an adenohypophysis and neurohypophysis.

■ Discuss how the pituitary gland and hypothalamus are structurally and functionally related.

■ Describe the location, histology, and blood and nerve supply of the thyroid gland.

■ Describe the location, histology, and blood and nerve supply of the parathyroid glands.

■ Describe the location, histology, and blood and nerve supply of the adrenal glands.

■ Explain the subdivisions of the adrenal glands into cortical and medullary portions.

■ Describe the location, histology, and blood and nerve supply of the pancreas.

■ Explain why the ovaries and testes are classified as endocrine glands.

■ Describe the location, histology, and blood and nerve supply of the pineal gland.

■ Describe the location, histology, and blood and nerve supply of the thymus gland.

■ Discuss the symptoms of the following endocrine disorders: pituitary dwarfism, giantism, acromegaly, diabetes insipidus, cretinism, myxedema, exophthalmic goiter, simple goiter, tetany, osteitis fibrosa cystica, Addison's disease, Cushing's syndrome, aldosteronism, adrenogenital syndrome, gynecomastia, pheochromocytomas, diabetes mellitus, and hyperinsulinism.

■ Define key medical terms associated with the endocrine system.

The nervous system controls the body by means of electrical impulses delivered over neurons. The body's other control system, the **endocrine system,** affects bodily activities by releasing chemical messengers, called hormones, into the bloodstream. Because the body could not function if the two great control systems were to pull in opposite directions, their activities are coordinated into an interlocking supersystem. Certain parts of the nervous system stimulate or inhibit the release of hormones, and the hormones, in turn, are capable of stimulating or inhibiting the flow of nerve impulses.

The developmental anatomy of the endocrine system is considered in Exhibit 24-8.

ENDOCRINE GLANDS

The body contains two kinds of glands: exocrine and endocrine. **Exocrine glands** secrete their products onto a free surface or into ducts. The ducts then carry the secretions into body cavities, into the lumens of various organs, or to the body's surface. They include sweat, sebaceous, mucous, and digestive glands. **Endocrine glands,** by contrast, secrete their products into the extracellular

space around the secretory cells (*endo* = within). The secretion passes into the capillaries to be transported in the blood. Since they have no ducts, endocrine glands are also called *ductless glands.* The endocrine glands of the body are the pituitary (hypophysis), thyroid, parathyroids, adrenals (suprarenals), pancreas, ovaries, testes, pineal (epiphysis cerebri), and thymus. The endocrine glands make up the **endocrine system.** The location of many organs of the endocrine system is illustrated in Figure 19-1.

The secretions of endocrine glands are called **hormones** (*hormone* = set in motion). The one thing all hormones have in common is the function of maintaining homeostasis by changing the physiological activities of cells. A hormone may stimulate changes in specific cells or organs, called *target cells* or *organs,* or it directly affects the activities of all the cells in the body.

PITUITARY (HYPOPHYSIS)

The hormones of the **pituitary gland,** also called the **hypophysis** (hī-POF-i-sis), regulate so many body activities that the pituitary has been nicknamed the "master gland."

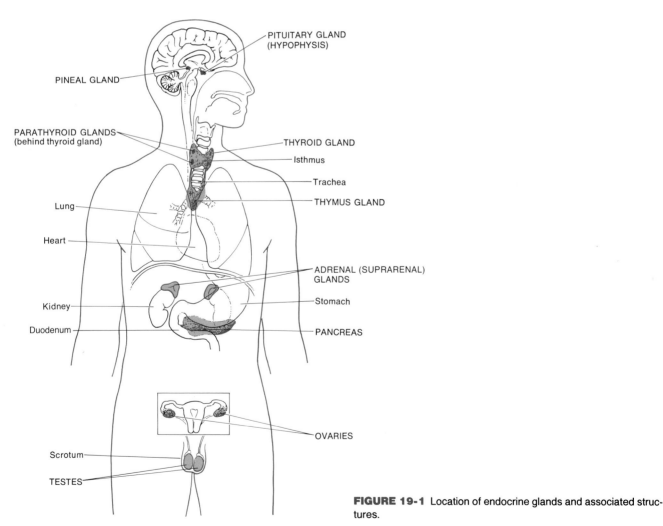

FIGURE 19-1 Location of endocrine glands and associated structures.

504

The hypophysis is a round structure and surprisingly small, measuring about 1.3 cm (0.5 inch) in diameter. It lies in the sella turcica of the sphenoid bone and is attached to the hypothalamus of the brain via a stalklike structure, the **infundibulum** (see Figure 16-1).

The pituitary is divided structurally and functionally into an anterior lobe and a posterior lobe. Both are connected to the hypothalamus. Between the lobes is a small, relatively avascular zone, the **pars intermedia.** Although much larger and more clearly defined in structure and function in some lower animals, its role in humans is obscure. The **anterior lobe** is derived from the ectoderm from an embryological invagination of pharyngeal epithelium called Rathke's pouch. Accordingly, the anterior lobe contains many glandular epithelium cells and forms the glandular part of the pituitary. A system of blood vessels connects the anterior lobe with the hypothalamus. The **posterior lobe** is also derived from the ectoderm from an outgrowth of the hypothalamus of the brain. Accordingly, the posterior lobe contains axonic ends of neurons whose cell bodies are located in the hypothalamus. The nerve fibers that terminate in the posterior lobe are supported by neuroglial cells called pituicytes. Other nerve fibers connect the posterior lobe directly with the hypothalamus.

ADENOHYPOPHYSIS

The anterior lobe of the pituitary is also called the **adenohypophysis** (ad'-i-nō-hī-POF-i-sis). It releases hormones that regulate a whole range of bodily activities from growth to reproduction. The release of these hormones is either stimulated or inhibited by chemical secretions from the hypothalamus of the brain called **regulating factors.** Regulating factors are synthesized for each of the anterior pituitary hormones. These factors constitute an important link between the nervous system and the endocrine system.

The hypothalamic regulating factors are delivered to the adenohypophysis in the following way. The blood supply to the adenohypophysis and infundibulum is derived principally from several *superior hypophyseal* (hī'-po-FIZ-ē-al) *arteries.* These arteries are branches of the internal carotid and posterior communicating arteries (Figure 19-2). The superior hypophyseal arteries form a network or plexus of capillaries, the *primary plexus,* in the infundibulum near the inferior portion of the hypothalamus. Regulating factors from the hypothalamus diffuse into this plexus. This plexus drains into the *hypophyseal portal veins,* that pass down the infundibulum. At

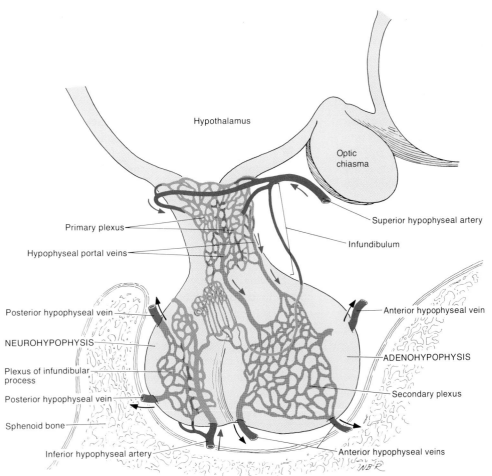

FIGURE 19-2 Blood supply of the pituitary gland.

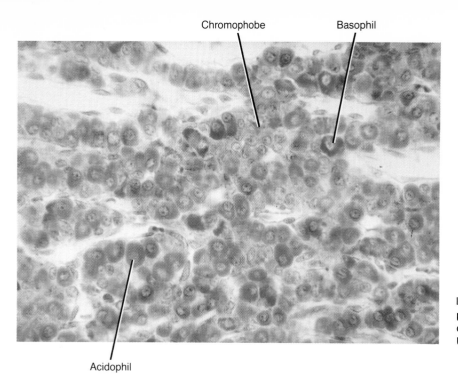

Chromophobe   Basophil

Acidophil

FIGURE 19-3 Histology of the adenohy-pophysis. Photomicrograph at a magnification of 600×. (© 1983 by Michael H. Ross. Used by permission.)

the inferior portion of the infundibulum, the veins form a *secondary plexus* in the adenohypophysis. From this plexus, hormones of the adenohypophysis pass into the anterior hypophyseal veins for distribution to tissue cells.

When the anterior lobe receives the proper chemical stimulation from the hypothalamus, its glandular cells secrete any one of seven hormones. The glandular cells themselves are called **acidophils** (a-SID-ō-fils), **basophils** (BĀ-sō-fils), or **chromophobes** (KRŌ-mō-fōbs), depending on the way their cytoplasm reacts to laboratory stains (Figure 19-3). The acidophils, which stain pink, secrete two hormones: **human growth hormone (HGH)** or **so-matotropin,** which controls general body growth, and **pro-lactin (PR),** which initiates milk production by the breasts. The basophils stain darkly and release the other five hormones. These are **thyroid-stimulating hormone (TSH)** or **thyrotropin,** which controls the thyroid gland; **adrenocorticotropic hormone (ACTH),** which stimulates the adrenal cortex to secrete its hormones; **follicle-stimu-lating hormone (FSH),** which stimulates the production of eggs and sperm in the ovaries and testes, respectively; **luteinizing hormone (LH),** which stimulates other sexual and reproductive activities; and **melanocyte-stimulating hormone (MSH),** which is related to skin pigmentation. The chromophobes may also be involved in the secretion of the adrenocorticotropic hormone.

Except for the human growth hormone, melanocyte-stimulating hormone, and prolactin, all the secretions are referred to as **tropic hormones,** which means that their target organs are other endocrine glands. Follicle-stimu-lating hormone and luteinizing hormone are also called **gonadotropic** (gō'-nad-ō-TRŌ-pik) **hormones** because they regulate the functions of the gonads. The gonads (ovaries and testes) are the endocrine glands that produce sex hormones.

NEUROHYPOPHYSIS

In a strict sense, the posterior lobe, or **neurohypophysis,** is not an endocrine gland, since it does not synthesize hormones. The posterior lobe consists of cells called **pitui-cytes** (pi-TOO-i-sītz), which are similar in appearance to the neuroglia of the nervous system. It also contains axon terminations of secretory nerve cells of the hypo-thalamus (Figure 19-4). The cell bodies of the neurons, called **neurosecretory cells,** originate in nuclei in the hy-pothalamus. The fibers project from the hypothalamus, form the **hypothalamic-hypophyseal tract,** and terminate on blood capillaries in the neurohypophysis. The cell bod-ies of the neurosecretory cells produce two hormones: **oxytocin** (ok'-sē-TŌ-sin) and the **antidiuretic hormone (ADH).** Oxytocin stimulates contraction of the smooth muscle of the pregnant uterus during labor and stimulates contractile cells in the mammary glands to eject milk. ADH prevents excessive urine production by bringing about water reabsorption and secondarily causes blood pressure to rise.

Following their production, the hormones are trans-ported in the neuron fibers by a carrier protein called *neurophysin* into the neurohypophysis and stored in the axon terminals. Later, when the hypothalamus is properly stimulated, it sends impulses over the neurosecretory cells. The impulses cause the release of the hormones from the axon terminals into the blood.

The blood supply to the neurohypophysis is from the *inferior hypophyseal arteries,* derived from the internal carotid arteries. In the neurohypophysis, the inferior hy-pophyseal arteries form a plexus of capillaries called the *plexus of the infundibular process.* From this plexus hor-mones stored in the neurohypophysis pass into the *poste-rior hypophyseal veins* for distribution to tissue cells.

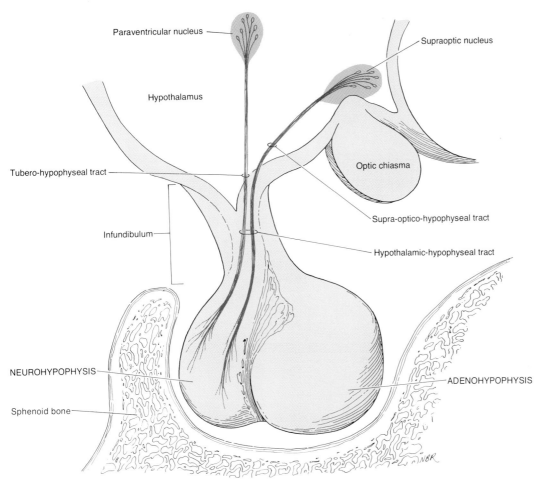

Paraventricular nucleus

Hypothalamus

Supraoptic nucleus

Tubero-hypophyseal tract

Optic chiasma

Infundibulum

Supra-optico-hypophyseal tract

Hypothalamic-hypophyseal tract

NEUROHYPOPHYSIS

ADENOHYPOPHYSIS

Sphenoid bone

FIGURE 19-4 Hypothalamic-hypophyseal tract.

THYROID

The **thyroid gland** is located just below the larynx. The right and left **lateral lobes** lie one on either side of the trachea and are connected by a mass of tissue called an **isthmus** (IS-mus) that lies in front of the trachea just below the cricoid cartilage (Figure 19-5). The **pyramidal lobe,** when present, extends upward from the isthmus. The gland weighs about 25 g (almost 1 oz) and has a rich blood supply, receiving about 80 to 120 ml of blood per minute.

The main blood supply is from the superior thyroid artery, a branch of the external carotid artery, and the inferior thyroid artery, a branch of the subclavian artery. The thyroid is drained by the superior and middle thyroid veins, which pass into the internal jugular veins, and the inferior thyroid veins, which join the brachiocephalic veins or internal jugular veins.

Histologically, the thyroid gland is composed of spherical-shaped sacs called **thyroid follicles** (Figure 19-6). The walls of each follicle consist of cells that reach the surface of the lumen of the follicle (**follicular cells**) and cells that do not reach the lumen (**parafollicular** or **C cells**). The follicular cells manufacture **thyroxin** (thī-ROK-sin)

and **triiodothyronine** (trī-ī'-od-ō-THI-ro-nēn). Together these hormones are referred to as the **thyroid hormones.** Thyroxin is considered to be the major hormone produced by the follicular cells. It controls metabolism, helps regulate growth and development, and increases reactivity of the nervous system. The parafollicular cells produce **thyrocalcitonin** (thī'-rō-kal-si-TŌ-nin) or **TCT.** This hormone decreases blood calcium level by inhibiting bone resorption. The interior of each thyroid follicle is filled with **thyroglobulin,** a stored form of the thyroid hormones.

The nerve supply of the thyroid consists of postganglionic fibers from the superior and middle cervical sympathetic ganglia. Preganglionic fibers from the ganglia are derived from the second through seventh thoracic segments of the spinal cord.

PARATHYROIDS

Typically embedded on the posterior surfaces of the lateral lobes of the thyroid are small, round masses of tissue called the **parathyroid glands.** Usually, two parathyroids, superior and inferior, are attached to each lateral thyroid lobe (Figure 19-7). They measure about 3 to 6 mm (0.1

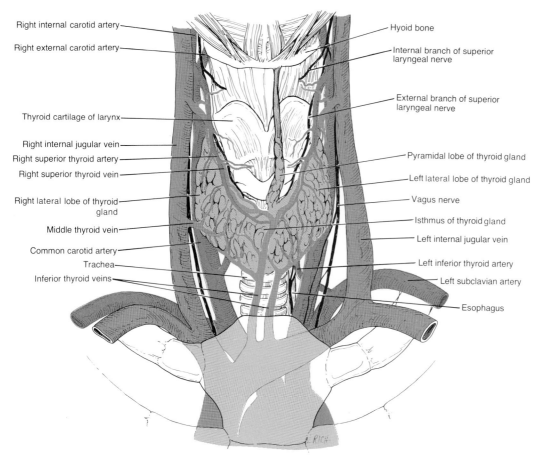

Right internal carotid artery

Right external carotid artery

Thyroid cartilage of larynx

Right internal jugular vein

Right superior thyroid artery

Right superior thyroid vein

Right lateral lobe of thyroid gland

Middle thyroid vein

Common carotid artery

Trachea

Inferior thyroid veins

Hyoid bone

Internal branch of superior laryngeal nerve

External branch of superior laryngeal nerve

Pyramidal lobe of thyroid gland

Left lateral lobe of thyroid gland

Vagus nerve

Isthmus of thyroid gland

Left internal jugular vein

Left inferior thyroid artery

Left subclavian artery

Esophagus

FIGURE 19-5 Location and blood supply of the thyroid gland in anterior view. The pyramidal lobe of the thyroid is not constant and, when present, it may be attached to the hyoid bone by a muscle.

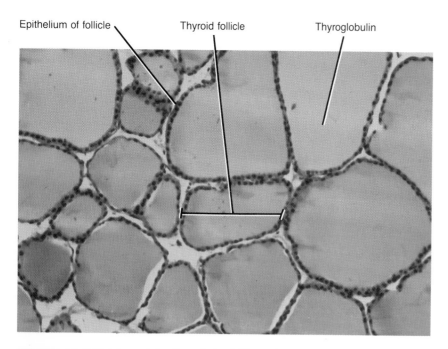

Epithelium of follicle

Thyroid follicle

Thyroglobulin

FIGURE 19-6 Histology of the thyroid gland. Photomicrograph at a magnification of 230×. (© 1983 by Michael H. Ross. Used by permission.)

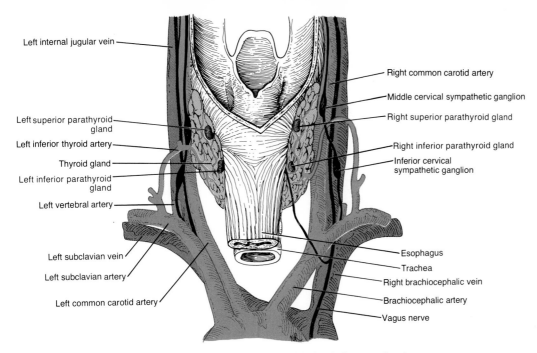

FIGURE 19-7 Location and blood supply of the parathyroid glands in posterior view.

to 0.3 inches) in length, 2 to 5 mm (0.07 to 0.2 inches) in width, and 0.5 to 2 mm (0.02 to 0.07 inches) in thickness.

The parathyroids are abundantly supplied with blood from branches of the superior and inferior thyroid arteries. Blood is drained by the superior, middle, and inferior thyroid veins. The nerve supply of the parathyroids is derived from the cervical sympathetic ganglia and the pharyngeal branch of the vagus.

Histologically, the parathyroids contain two kinds of epithelial cells (Figure 19-8). The more numerous cells, called **principal** or **chief cells,** are believed to be the major synthesizer of the **parathyroid hormone (PTH),** which increases blood calcium level and decreases blood phosphate level. Some researchers believe that the other kind of cell, called an **oxyphil cell,** synthesizes a reserve capacity of hormone.

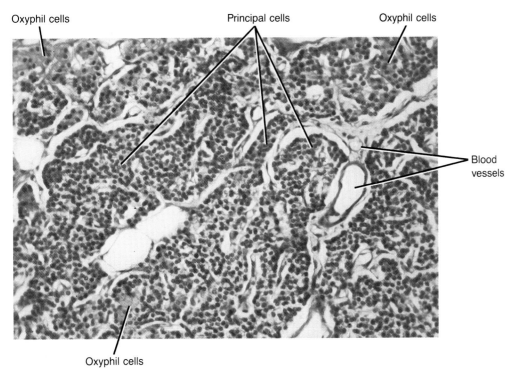

FIGURE 19-8 Histology of the parathyroid glands. Photomicrograph at a magnification of 180×. (© 1983 by Michael H. Ross. Used by permission.)

509

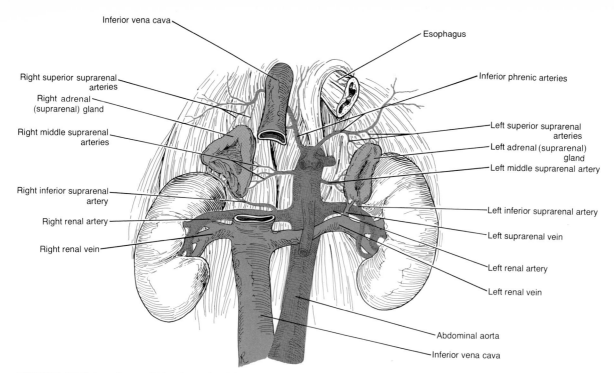

FIGURE 19-9 Location and blood supply of the adrenal (suprarenal) glands.

ADRENALS (SUPRARENALS)

The body has two **adrenal,** or **suprarenal, glands,** one of which is located superior to each kidney (Figure 19-9). Each adrenal gland is structurally and functionally differentiated into two sections: the outer **adrenal cortex,** which makes up the bulk of the gland, and the inner **adrenal medulla** (Figure 19-10a). Whereas the adrenal cortex is derived from mesoderm, the adrenal medulla is derived from the ectoderm. Covering the gland are an inner, thick layer of fatty connective tissue and an outer, thin fibrous capsule.

The average dimensions of the adult adrenal gland are about 50 mm (2 inches) in length, 30 mm (1.1 inch) in width, and 10 mm (0.4 inches) in thickness.

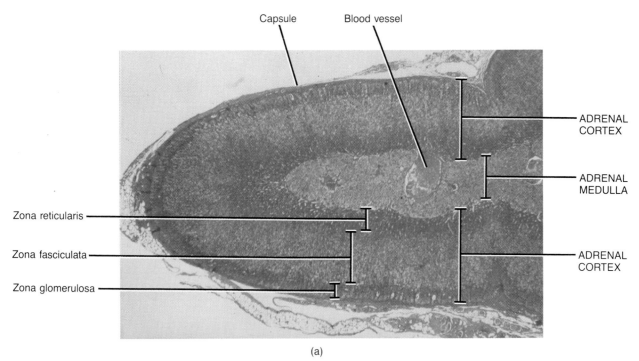

(a)

FIGURE 19-10 Histology of the adrenal (suprarenal) glands. (a) Photomicrograph of a section of the adrenal gland showing the subdivisions and zones at a magnification of 25×.

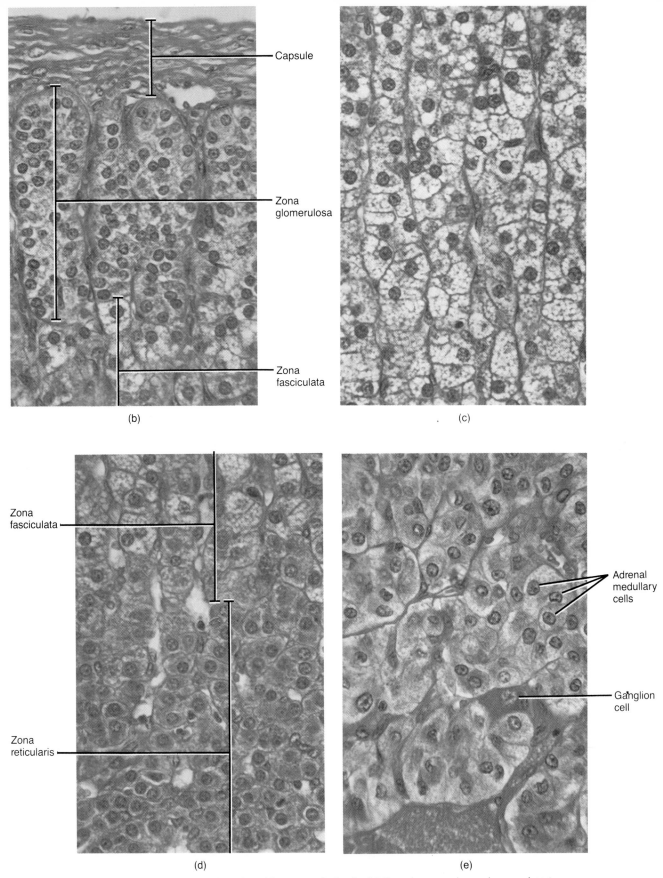

FIGURE 19-10 (*Continued*) Histology of the adrenal (suprarenal) glands. (b) Capsule, zona glomerulosa, and part of the zona fasciculata at a magnification of 400×. (c) Zona fasciculata at a magnification of 400×. (d) Zona fasciculata and zona reticularis at a magnification of 400×. (e) Medulla at a magnification of 400×. (© 1983 by Michael H. Ross. Used by permission.)

The adrenals, like the thyroid, are among the most vascular organs of the body. The main arteries of supply are the several superior suprarenal arteries arising from the inferior phrenic artery, the middle suprarenal artery from the aorta, and the inferior suprarenal arteries from the renal arteries. The suprarenal vein of the right adrenal gland drains into the inferior vena cava, whereas the suprarenal vein of the left adrenal gland empties into the left renal vein.

The principal nerve supply to the adrenal glands is from preganglionic fibers from the splanchnic nerves and from the celiac and associated sympathetic plexuses. These myelinated fibers end on the secretory cells of the gland found in a region of the medulla.

ADRENAL CORTEX

Histologically, the cortex is subdivided into three zones (Figure 19-10a–d). Each zone has a different cellular arrangement and secretes different groups of steroid hormones. The outer zone, directly underneath the connective tissue capsule, is referred to as the **zona glomerulosa,** which comprises about 15% of the total cortical volume. Its cells are arranged in arched loops or round balls. Its primary secretions are a group of hormones called **mineralocorticoids** (min'-er-al-ō-KOR-ti-koyds). The principal mineralocorticoid is **aldosterone** (al-dō-STER-ōn), which causes the kidneys to reabsorb sodium and increase potassium excretion.

The middle zone, or **zona fasciculata,** is the widest of the three zones and consists of cells arranged in long, straight cords. The zona fasciculata secretes mainly **glucocorticoids** (gloo'-kō-KOR-ti-koyds). These include **hydrocortisone, corticosterone,** and **cortisone.** Together the hormones work with other hormones to promote normal metabolism, help resist stress, and decrease edema.

The inner zone, the **zona reticularis,** contains cords of cells that branch freely. This zone synthesizes minute amounts of hormones, mainly the sex hormones called **gonadocorticoids** (gō-na-dō-KOR-ti-koyds) and of these chiefly male hormones called **androgens.** Their functions will be discussed shortly.

ADRENAL MEDULLA

The adrenal medulla consists of hormone-producing cells, called **chromaffin** (krō-MAF-in) **cells,** which surround large blood-containing sinuses (Figure 19-10e). Chromaffin cells develop from the same source as the postganglionic cells of the sympathetic division of the nervous system. They are directly innervated by preganglionic cells of the sympathetic division of the autonomic nervous system and may be regarded as postganglionic cells that are specialized to secrete. In all other visceral effectors, preganglionic sympathetic fibers first synapse with post-

ganglionic neurons before innervating the effector. In the adrenal medulla, however, the preganglionic fibers pass directly into the chromaffin cells of the gland. The secretion of hormones from the chromaffin cells is directly controlled by the autonomic nervous system, and innervation by the preganglionic fibers allows the gland to respond rapidly to a stimulus.

The two principal hormones synthesized by the adrenal medulla are **epinephrine** and **norepinephrine** (also called adrenaline and noradrenaline). Epinephrine constitutes about 80 percent of the total secretion of the gland and is more potent in its action than norepinephrine. Both hormones are **sympathomimetic** (sim'-pa-thō-mi-MET-ik), that is, they produce effects that mimic those brought about by the sympathetic division of the autonomic nervous system during stress.

PANCREAS

Because of its functions, the **pancreas** can be classified as both an endocrine and an exocrine gland. Since the exocrine functions of the gland will be discussed in the chapter on the digestive system, we shall treat only its endocrine functions at this point. The pancreas is a flattened organ located posterior and slightly inferior to the stomach (Figure 19-11). The adult pancreas consists of a head, body, and tail. Its average length is about 12.5 cm (6 inches), and its average weight is about 85 g (3 oz).

The arterial supply of the pancreas is from the superior and inferior pancreaticoduodenal arteries and from the splenic and superior mesenteric arteries. The nerves to the pancreas are branches of the celiac plexus. The glandular portion of the pancreas is innervated by the craniosacral division of the autonomic nervous system, whereas the blood vessels of the pancreas are innervated by the thoracolumbar division of the autonomic nervous system.

The endocrine portion of the pancreas consists of cluster of cells called **islets of Langerhans** (LAHNG-er-hanz) (Figure 19-12). Three kinds of cells are found in these clusters: (1) **alpha cells,** which secrete the hormone glucagon; (2) **beta cells,** which secrete the hormone insulin, and (3) **delta cells,** which secrete human growth hormone inhibiting factor (HGHIF) or somatostatin, a hormone that inhibits secretion of human growth hormone. The islets are surrounded by blood capillaries and by the cells that form the exocrine part of the gland. Glucagon and insulin are concerned with regulation of blood sugar level. **Glucagon** increases blood sugar level stimulating the conversion of glycogen to glucose in the liver, stimulating the release of glucose from the liver, and inhibiting glucose uptake by body cells. **Insulin** decreases blood sugar level by stimulating transportation of glucose into the liver for storage as glycogen and stimulating glucose uptake by body cells.

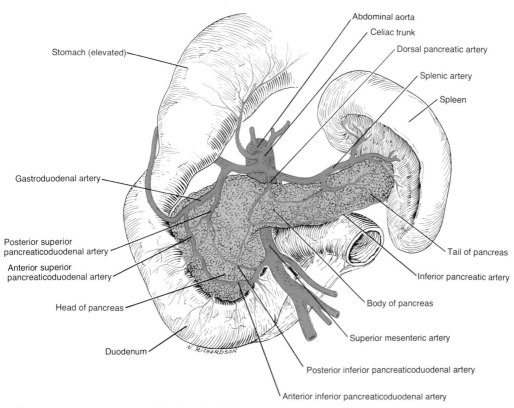

FIGURE 19-11 Location and blood supply of the pancreas.

OVARIES AND TESTES

The female gonads, called the **ovaries,** are paired oval bodies located in the pelvic cavity. The ovaries produce female sex hormones called **estrogens** and **progesterone.** These hormones are responsible for the development and maintenance of the female sexual characteristics. Along with the gonadotropic hormones of the pituitary, the sex hormones also regulate the menstrual cycle, maintain pregnancy, and prepare the mammary glands for lactation.

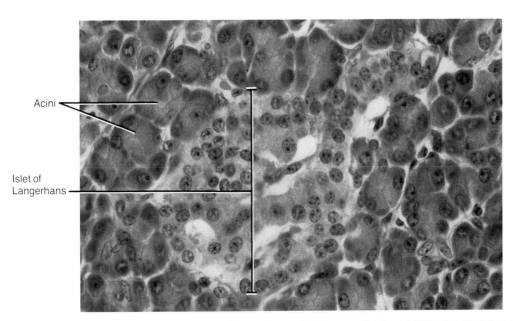

FIGURE 19-12 Histology of the pancreas. Photomicrograph at a magnification of 600×. (© 1983 by Michael H. Ross. Used by permission.)

The male has two oval glands, called **testes,** that lie in the scrotum. The testes produce **testosterone,** the primary male sex hormone, that stimulates the development and maintenance of the male sexual characteristics. The detailed structure of the ovaries and testes will be discussed in Chapter 23.

PINEAL (EPIPHYSIS CEREBRI)

The cone-shaped gland attached to the roof of the third ventricle is known as the **pineal gland,** or **epiphysis cerebri** (see Figure 19-1). The gland is about 5 to 8 mm (0.2 to 0.3 inches) long and 5 mm wide. It weighs about 0.2 g. It is covered by a capsule formed by the pia mater. It consists of masses of parenchymal secretory cells called **pinealocytes** and **neuroglial cells.** Around the cells are scattered preganglionic sympathetic fibers. The pineal gland starts to calcify at about the time of puberty. Such calcium deposits are referred to as **brain sand.** Contrary to a once widely held belief, there is no evidence that the pineal atrophies with age and that the presence of brain sand is an indication of atrophy. In fact, the presence of brain sand may indicate increased secretory activity. The posterior cerebral artery supplies the pineal with blood, and the great cerebral vein drains it.

Although many anatomical facts concerning the pineal gland have been known for years, its physiology is still somewhat obscure. One hormone secreted by the pineal gland is **melatonin,** which appears to inhibit the activities of the ovaries, thus regulating the menstrual cycle. Some evidence also exists that the pineal secretes a second hormone called **adrenoglomerulotropin** (a-drē'-nō-glō-mer'-yoo-lō-TRŌ-pin). This hormone may stimulate the adrenal cortex to secrete aldosterone. Other functions attributed to the pineal gland are the secretion of growth-inhibiting factor and the secretion of a hormone called **serotonin** that is involved in normal brain physiology.

THYMUS

Usually a bilobed lymphatic organ, the **thymus gland** is located in the superior mediastinum posterior to the sternum and between the lungs (Figure 19-13). The two thymic lobes are held in close proximity by an enveloping layer of connective tissue. Each lobe is enclosed by a fibrous connective tissue **capsule.** The capsule gives off extensions into the lobes called **trabeculae** (tra-BEK-yoo-lē), which divide the lobes into **lobules** (Figure 19-14a). Each lobule consists of a deeply staining peripheral **cortex** and a lighter staining central **medulla.** The cortex is composed almost entirely of small, medium, and large tightly packed lymphocytes held in place by reticular tissue fibers (Figure 19-14b). Since the reticular tissue differs in origin and structure from that usually found in other lymphatic organs, it is referred to as **epithelioreticular** supporting tissue. The medulla consists mostly of epithelial cells and more widely scattered lymphocytes, and its reticulum is more cellular than fibrous. The medulla also contains characteristic **thymic (Hassall's) corpuscles,** concentric layers of epithelial cells. Their significance is unknown.

The thymus gland is conspicuous in the infant, and during puberty it reaches its maximum size of about 40 grams. After puberty, most of the thymic tissue is replaced by fat and connective tissue. By the time the person reaches maturity, the gland has atrophied substantially.

The arterial supply of the thymus is derived mainly from the internal thoracic and inferior thyroid vessels. The veins that drain the thymus are the internal thoracic, brachiocephalic, and thyroid veins. Postganglionic sympathetic and parasympathetic fibers supply the gland.

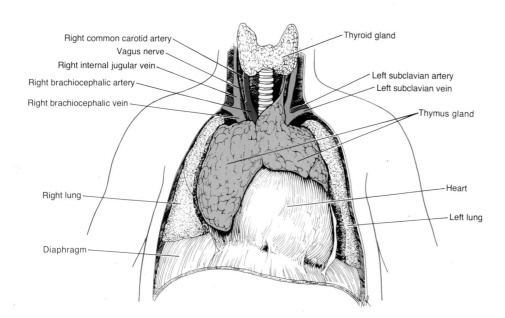

Right common carotid artery
Vagus nerve
Right internal jugular vein
Right brachiocephalic artery
Right brachiocephalic vein
Right lung
Diaphragm

Thyroid gland
Left subclavian artery
Left subclavian vein
Thymus gland
Heart
Left lung

FIGURE 19-13 Location of the thymus gland in a young child.

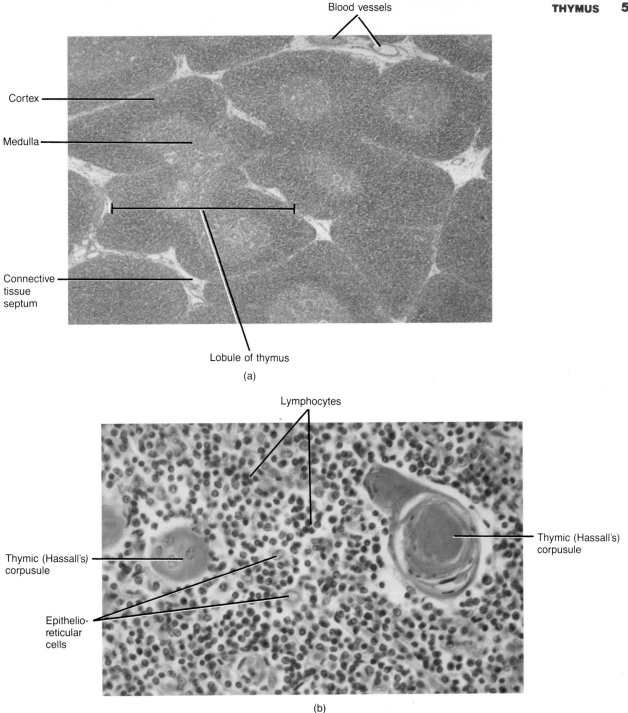

Blood vessels

Cortex

Medulla

Connective
tissue
septum

Lobule of thymus

(a)

Lymphocytes

Thymic (Hassall's)
corpusule

Thymic (Hassall's)
corpusule

Epithelio-
reticular
cells

(b)

FIGURE 19-14 Histology of the thymus gland. (a) Photomicrograph of several lobules at a magnification of 40×. (b) Photomicrograph of an enlarged aspect of thymic (Hassall's) corpuscles at a magnification of 600×. (© 1983 by Michael H. Ross. Used by permission.)

The function of the thymus gland is related to immunity. Lymphoid tissue of the body consists primarily of lymphocytes that may be distinguished into two kinds: B cells and T cells. Both are derived originally in the embryo from lymphocytic stem cells in bone marrow. Before migrating to their positions in lymphoid tissue, the descendants of the stem cells follow two distinct pathways. About half of them migrate to the thymus gland, where they are processed to become thymus-dependent lymphocytes, or **T cells.** The thymus gland confers on them the ability to destroy antigens (foreign microbes and substances) directly. The remaining stem cells are processed in some as yet undetermined area of the body, possibly the fetal liver and spleen, and are known as **B cells.** These cells, perhaps under the influence of a hormone produced by the thymus gland, called **thymosin,** develop into plasma cells. Plasma cells, in turn, produce antibodies against antigens.

OTHER ENDOCRINE TISSUES

Before leaving our discussion of hormones, it should be noted that body tissues other than endocrine glands also secrete hormones. The gastrointestinal tract synthesizes several hormones that regulate digestion in the stomach and small intestine. Among these hormones are **stomach gastrin, intestinal gastrin, pancreatic gastrin, secretin, cholecystokinin (pancreozymin), enterocrinin,** and **enterogastrone.** The placenta produces **human chorionic gonadotropic hormone (HCG), estrogens, progesterone,** and **relaxin,** all of which are related to pregnancy (Chapter 24). Finally, when the kidneys and liver to a lesser extent become hypoxic, it is believed that they release an enzyme called **renal erythropoietic factor.** It is secreted into the blood where it acts on a plasma protein to bring about the production of a hormone called **erythropoietin** (ē-rith'-rō-POY-ē-tin), which stimulates red blood cell production (Chapter 11).

APPLICATIONS TO HEALTH

Disorders of the endocrine system, in general, are based upon under- or overproduction of hormones. The term **hyposecretion** describes an underproduction, whereas the term **hypersecretion** means an oversecretion.

DISORDERS OF THE PITUITARY

The anterior pituitary gland produces many hormones. With the exception of HGH, MSH, and PR, these hormones directly control the activities of other endocrine glands. Thus hyposecretion or hypersecretion of an anterior pituitary hormone produces widespread and complicated abnormalities.

Among the clinically interesting disorders related to the adenohypophysis are those involving human growth hormones. HGH builds up cells, particularly those of bone tissue. If it is hyposecreted during the growth years, bone growth is slow and the epiphyseal plates close before normal height is reached. This condition is called **pituitary dwarfism.** Other organs of the body also fail to grow, and the pituitary dwarf is childlike in many physical respects. Treatment requires administration of HGH during childhood before the epiphyseal plates close.

Hypersecretion of HGH produces completely different disorders. Hypersecretion during childhood results in **giantism,** an abnormal increase in the length of long bones. Hypersecretion during adulthood is called **acromegaly** (ak'-rō-MEG-a-lē). Acromegaly cannot produce further lengthening of the long bones because the epiphyseal plates are already closed. Instead, the bones of the hands, feet, cheeks, and jaws thicken. Other tissues also grow. The eyelids, lips, tongue, and nose enlarge and the skin thickens and furrows, especially on the forehead and soles of the feet.

The principal abnormality associated with dysfunction of the neurohypophysis is **diabetes insipidus** (in-SIP-idus). Diabetes means overflow and insipidus means tasteless. This disorder should not be confused with diabetes mellitus ("sugar"), a disorder of the pancreas characterized by sugar in the urine. Diabetes insipidus is the result of a hyposecretion of ADH, usually caused by damage to the neurohypophysis or the hypothalamus. Symptoms include excretion of large amounts of urine and subsequent thirst. Diabetes insipidus is treated by administering ADH.

DISORDERS OF THE THYROID

Hyposecretion of thyroxin during the growth years results in **cretinism** (KRĒ-tin-izm). Two outstanding clinical symptoms of the cretin are dwarfism and mental retardation. The first is caused by failure of the skeleton to grow and mature. The second is caused by failure of the brain to develop fully. Recall that one function of thyroid hormone is to control tissue growth and development. Cretins also exhibit retarded sexual development and a yellowish skin color. Flat pads of fat develop, giving the cretin a characteristic round face and thick nose; a large, thick, protruding tongue; and protruding abdomen. Because the energy-producing metabolic reactions are slow, the cretin has a low body temperature and general lethargy. Carbohydrates are stored rather than utilized. Heart rate is also slow. If the condition is diagnosed early, the symptoms can be eliminated by administering thyroid hormone.

Hypothyroidism during the adult years produces **myxedema.** A hallmark of this disorder is an edema that causes the facial tissues to swell and look puffy. Like the cretin, the person with myxedema suffers from slow heart rate, low body temperature, muscular weakness, general lethargy, and a tendency to gain weight easily. The long-term effect of a slow heart rate may overwork the heart muscle, causing the heart to enlarge. Because the brain has already reached maturity, the person with myxedema does not experience mental retardation. However, in moderately severe cases, nerve reactivity may be dulled so that the person lacks mental alertness. Myxedema occurs eight times more frequently in females than in males. Its symptoms are abolished by the administration of thyroxin.

Hypersecretion of thyroxin gives rise to **exophthalmic** (ek'-sof-THAL-mik) **goiter** (GOY-ter). This disease, like myxedema, is also more frequent in females. One of its primary symptoms is an enlarged thyroid, called a *goiter,* which may be two to three times its original size. Two other symptoms are an edema behind the eye, which causes the eye to protrude (**exophthalmos),** and an abnormally high metabolic rate. The high metabolic rate produces a range of effects that are generally opposite to those of myxedema—increased pulse, high body temperature, and moist, flushed skin. The person loses weight and is usually full of "nervous" energy. The thyroxin also increases the responsiveness of the nervous system, causing the person to become irritable and exhibit tremors of the extended fingers. Hyperthyroidism is usually treated by administering drugs that suppress thyroxin synthesis or by surgically removing part of the gland.

Goiter is a symptom of many thyroid disorders. It may also occur if the gland does not receive enough iodine to produce sufficient thyroxin for the body's needs. The follicular cells then enlarge in a futile attempt to produce more thyroxin, and they secrete large quantities of colloid. This condition is called **simple goiter.** Simple goiter is most often caused by a lower than average amount of iodine in the diet. It may also develop if iodine intake is not increased during certain conditions that put a high demand on the body for thyroxin—such as frequent exposure to cold and high fat and protein diets.

DISORDERS OF THE PARATHYROIDS

A normal amount of calcium in the extracellular fluid is necessary to maintain the resting state of neurons. A deficiency of calcium caused by *hypoparathyroidism* causes neurons to depolarize without the usual stimulus. As a result, nervous impulses increase and result in muscle twitches, spasms, and convulsions. This condition is called **tetany.** The effects of hypocalcemic tetany are observed in the *Trousseau* (troo-SŌ) and *Chvostek* (VOS-tek) *signs.* Trousseau sign is observed when the binding of a blood pressure cuff around the upper arm produces contraction of the fingers and inability to open the hand. The Chvostek sign is a contracture of the facial muscles elicited by tapping the facial nerves at the angle of the jaw. Hypoparathyroidism results from surgical removal of the parathyroids or from parathyroid damage caused by parathyroid disease, infection, hemorrhage, or mechanical injury.

Hyperparathyroidism causes demineralization of bone. This condition is called **osteitis fibrosa cystica** because the areas of destroyed bone tissue are replaced by cavities that fill with fibrous tissue. The bones thus become deformed and are highly susceptible to fracture. Hyperparathyroidism is usually caused by a tumor in the parathyroids.

DISORDERS OF THE ADRENALS

Hyposecretion of glucocorticoids results in the condition called **Addison's disease.** Clinical symptoms include hypoglycemia, which leads to muscular weakness, mental lethargy, and weight loss. Increased potassium and decreased sodium lead to low blood pressure and dehydration. **Cushing's syndrome** is a hypersecretion of glucocorticoids, especially hydrocortisone and cortisone. The condition is characterized by the redistribution of fat. The result is spindly legs accompanied by a characteristic "moon face," "buffalo hump" on the back, and pendulous abdomen. Facial skin is flushed, and the skin covering the abdomen develops stretch marks. The individual also bruises easily, and wound healing is poor.

Hypersecretion of the mineralocorticoid aldosterone results in **aldosteronism,** characterized by a decrease in the body's potassium concentration. If potassium depletion is great, neurons cannot depolarize and muscular paralysis results. Hypersecretion also brings about excessive retention of sodium and water. The water increases the volume of the blood and causes high blood pressure. It also increases the volume of the interstitial fluid, producing edema.

The **adrenogenital syndrome** usually refers to a group of enzyme deficiencies that block the synthesis of glucocorticoids. In an attempt to compensate, the anterior pituitary secretes more ACTH. As a result, excess androgenic (male) hormones are produced, causing intense masculinizing effects throughout the body. For instance, the female develops extremely virile characteristics such as growth of a beard, development of a much deeper voice, occasionally development of baldness, development of a masculine distribution of hair on the body and on the pubis, growth of the clitoris that resembles a penis, and deposition of proteins in the skin and muscles producing typical masculine characteristics.

In the prepubertal male, the syndrome causes the same characteristics as in the female, plus rapid development of the male sexual organs and creation of male sexual desires. In the adult male, the virilizing characteristics of the adrenogenital syndrome are usually completely obscured by the normal virilizing characteristics of the testosterone secreted by the testes. As a result, it is often difficult to make a diagnosis of adrenogenital syndrome in the male adult. However, an occasional adrenal tumor secretes sufficient quantities of feminizing hormones that the male patient develops **gynecomastia** (*gyneca* = woman, *mast* = breast), which means excessive growth of the male mammary glands.

Tumors of the chromaffin cells of the adrenal medulla, called **pheochromocytomas** (fē-ō-krō'-mō-sī-TŌ-mas), cause hypersecretion of the medullary hormones. The oversecretion causes high blood pressure, high levels of sugar in the blood and urine, an elevated basal metabolic rate, nervousness, and sweating. Since the medullary hormones create the same effects as does sympathetic nervous stimulation, hypersecretion puts the individual into a prolonged version of the fight-or-flight response. This condition ultimately wears out the body, and the individual eventually suffers from general weakness.

ENDOCRINE DISORDERS OF THE PANCREAS

Hyposecretion of insulin results in a number of clinical symptoms referred to as **diabetes mellitus** (MEL-it-us). Research has led many to conclude that diabetes mellitus is not a single hereditary disease but rather a heterogeneous group of hereditary diseases, all of which ultimately lead to an elevation of glucose in the blood (hyperglycemia) and excretion of glucose in the urine as hyperglycemia increases. There is also an inability to reabsorb water, resulting in increased urine production, dehydration, loss of sodium, and thirst. Indeed, two major types of diabetes have been distinguished on the basis of clinical evidence: the maturity-onset type and the juvenile-onset type.

Maturity-onset diabetes is the much more common type, representing more than 90 percent of all the cases. It most often occurs in people who are over 40 and overweight. In this type clinical symptoms are mild, and the high glucose levels in the blood can usually be controlled by diet alone. Instead of having a deficiency of insulin many maturity-onset diabetics have a sufficiency or even a surplus of the hormone in the blood. For these individuals diabetes arises not from a shortage of insulin but probably from defects in the molecular machinery that mediates the action of insulin on its target cells. Maturity-onset diabetes is therefore called *non-insulin-dependent diabetes*.

Juvenile-onset diabetes is much less common than the maturity-onset type, representing well under 10 percent of all the cases. It develops in people younger than age 20 and its onset is more abrupt. The disease is characterized by a marked decline in the number of beta cells in the pancreas leading to a deficiency of insulin and an elevation of glucose in the blood. The deficiency of insulin accelerates the breakdown of the body's reserve of fat resulting in the production of organic acids called ketones. This causes a form of acidosis called *ketosis,* which lowers the pH of the blood and can result in death. The catabolism of stored fats and proteins also causes weight loss. As lipids are transported by the blood from storage depots to hungry cells, lipid particles are deposited on the walls of blood vessels. The deposition leads to atherosclerosis and a multitude of circulatory problems. Because injections of insulin are required to regulate the level of blood glucose, this form of diabetes, which is generally more severe, is called *insulin-dependent diabetes.* This type of diabetes is usually preceded by a virus infection like measles or mumps.

Some cases of juvenile-onset diabetes may be explained primarily on a genetic basis and others on an environmental basis. Still other cases appear to rise from a complex interaction between the genetic background of the individual and his environment.

Hyperinsulinism is much rarer than hyposecretion and is generally the result of a malignant tumor in an islet. The principal symptom is a decreased blood glucose level, which stimulates the secretion of epinephrine, glucagon, and HGH. As a consequence, anxiety, sweating, tremor, increased heart rate, and weakness occur. Moreover, brain cells do not have enough glucose to function efficiently. This condition leads to mental disorientation, convulsions, unconsciousness, shock, and eventual death as the vital centers in the medulla are affected.

KEY MEDICAL TERMS ASSOCIATED WITH THE ENDOCRINE SYSTEM

Antidiuretic (*anti* = against; *diuresis* = urine production) Any chemical substance that prevents excessive urine production.

Feminizing adenoma (*aden* = gland; *oma* = tumor) Malignant tumors of the adrenal gland that secrete abnormally high amounts of female sex hormones and produce female secondary sexual characteristics in the male.

Hyperplasia (*hyper* = over; *plas* = grow, form) Excessive development of tissue.

Hypoplasia (*hypo* = under) Defective development of tissue.

Neuroblastoma (*neuro* = nerve) Malignant tumor arising from the adrenal medulla associated with metastases to bones.

Thyroid storm An aggravation of all symptoms of hyperthyroidism, resulting from trauma, surgery, and unusual emotional stress or labor.

Virilism Masculinization.

Virilizing adenoma Malignant tumors of the adrenal gland that secrete high amounts of male sex hormones and produce male secondary sexual characteristics in the female.

STUDY OUTLINE

Endocrine Glands

1. Exocrine glands (sweat, sebaceous, digestive) secrete their products through ducts into body cavities or onto body surfaces.

2. Endocrine glands are ductless and secrete hormones into the blood.

Pituitary (Hypophysis)

1. The pituitary is located in the sella turcica and is differentiated into the adenohypophysis (the anterior lobe and glandular portion) and the neurohypophysis (the posterior lobe and nervous portion).
2. The adenohypophysis consists of acidophils, basophils, and chromophobes; the neurohypophysis consists of neurosecretory cells and supporting pituicytes.
3. The adenohypophysis secretes tropic hormones and gonadotropic hormones.
4. The blood supply to the adenohypophysis is from the superior hypophyseal arteries.
5. Hormones of the adenohypophysis are: (a) human growth hormone (regulates growth), (b) thyroid-stimulating hormone (regulates activities of thyroid), (c) adrenocorticotropic hormone (regulates adrenal cortex), (d) follicle-stimulating hormone (regulates ovaries and testes), (e) luteinizing hormone (regulates female and male reproductive activities), (f) prolactin (initiates milk secretion), and (g) melanocyte-stimulating hormone (increases skin pigmentation).
6. The neural connection between the hypothalamus and neurohypophysis is via the hypothalamic-hypophyseal tract.
7. Hormones of the neurohypophysis are oxytocin (stimulates contraction of uterus and ejection of milk) and antidiuretic hormone (stimulates arteriole constriction and water reabsorption by the kidneys).

Thyroid

1. The thyroid gland is located below the larynx.
2. Histologically, the thyroid consists of thyroid follicles; follicular cells secrete thyroxin and triiodothyronine (thyroid hormones) and parafollicular cells secrete thyrocalcitonin.
3. Thyroxin controls the rate of metabolism by increasing the catabolism of carbohydrates and proteins.
4. Thyrocalcitonin regulates the homeostasis of calcium by decreasing blood calcium level.

Parathyroids

1. The parathyroids are embedded on the posterior surfaces of the lateral lobes of the thyroid.
2. Histologically, the parathyroids consist of principal and oxyphil cells.
3. Parathyroid hormone regulates the homeostasis of calcium and phosphate by increasing blood calcium level and decreasing blood phosphate level.

Adrenals (Suprarenals)

1. The adrenal glands are located superior to the kidneys. They consist of an outer cortex and inner medulla.
2. Histologically, the cortex is divided into a zona glomerulosa, zona fasciculata, and zona reticularis; the medulla consists of chromaffin cells.
3. Cortical secretions are mineralocorticoids (regulate sodium reabsorption and potassium excretion), glucocorticoids (regulate normal metabolism and resistance to stress), and gonadocorticoids (male and female sex hormones).
4. Medullary secretions are epinephrine and norepinephrine, which produce effects similar to sympathetic responses.

Pancreas

1. The pancreas is posterior and slightly inferior to the stomach.
2. Histologically, it consists of islets of Langerhans (endocrine cells) and acini (enzyme-producing cells). Three types of cells in the endocrine portion are alpha cells, beta cells, and delta cells.
3. Alpha cells secrete glucagon, which increases blood glucose level; beta cells secrete insulin, which decreases blood glucose level; and delta cells secrete human growth hormone inhibiting factor (HGHIF) or somatostatin, which inhibits secretion of human growth hormone.

Ovaries and Testes

1. Ovaries are located in the pelvic cavity and produce sex hormones related to development and maintenance of female sexual characteristics, menstrual cycle, pregnancy, and lactation.
2. Testes lies inside the scrotum and produce sex hormones related to the development and maintenance of male sexual characteristics.

Pineal (Epiphysis Cerebri)

1. The pineal is attached to the roof of the third ventricle.
2. Histologically, it consists of secretory parenchymal cells called pinealocytes, neuroglial cells, and scattered preganglionic sympathetic fibers; calcified deposits are referred to as brain sand.
3. It secretes melatonin (possibly regulates menstrual cycle), adrenoglomerulotropin (may stimulate adrenal cortex), and serotonin (involved in normal brain physiology).

Thymus

1. The thymus is a bilobed lymphatic gland located in the superior mediastinum posterior to the sternum and between the lungs.
2. Histologically, it consists primarily of various sizes of lymphocytes.
3. The gland is necessary for the maturation of the thymus-dependent lymphoctyes (T cells) of the immune system and its hormone, thymosin, may cause B cells to develop into antibody-producing plasma cells.

Applications to Health

1. Disorders related to the pituitary are pituitary dwarfism, giantism, acromegaly, and diabetes insipidus.

2. Disorders related to the thyroid are cretinism, myxedema, exophthalmic goiter, exophthalmos, and simple goiter.
3. Disorders related to the parathyroids are tetany and osteitis fibrosa cystica.
4. Disorders related to the adrenals are Addison's disease, Cushing's syndrome, aldosteronism, adrenogenital syndrome, gynecomastia, and pheochromocytoma.
5. Disorders related to the endocrine portion of the pancreas are diabetes mellitus and hyperinsulinism.

REVIEW QUESTIONS

1. Distinguish between an endocrine gland and an exocrine gland.
2. What is a hormone? Distinguish between tropic and gonadotropic hormones.
3. In what respect is the pituitary gland actually two separate glands? Describe the histology of the adenohypophysis.
4. Why does the anterior lobe of the pituitary gland have such an abundant blood supply?
5. What hormones are produced by the adenohypophysis, and what are their functions?
6. Discuss the histology of the neurohypophysis and the function and regulation of the hormones produced by the neurohypophysis.
7. Describe the structure and importance of the hypothalamic-hypophyseal tract.
8. Describe the location and histology of the thyroid gland.
9. List the functions of the thyroid hormones.
10. Where are the parathyroids located? What is their histology? What are the functions of the parathyroid hormone?
11. Compare the adrenal cortex and adrenal medulla with regard to location and histology.
12. Describe the function of the hormones produced by the adrenal cortex.
13. What relationship does the adrenal medulla have to the autonomic nervous system?
14. Describe the location of the pancreas and the histology of the islets of Langerhans. What are the actions of glucagon and insulin?
15. Where is the pineal gland located? What are its assumed functions?
16. Describe the location and histology of the thymus gland. What is its proposed function?
17. Distinguish between hyposecretion and hypersecretion.
18. What are the principal clinical symptoms of pituitary dwarfism, giantism, and acromegaly?
19. In diabetes insipidus, why does the patient exhibit high urine production and thirst?
20. What clinical symptoms are present in cretinism, myxedema, exophthalmic goiter, and simple goiter? Relate these symptoms to the normal activity of thyroxin.
21. Distinguish between the cause and symptoms of tetany and the cause and symptoms of osteitis fibrosa cystica.
22. What are the effects of hypersecretion of aldosterone?
23. Describe Addison's disease, Cushing's syndrome, the adrenogenital syndrome, and gynecomastia. What is a pheochromocytoma?
24. What are the principal effects of hypoinsulinism and hyperinsulinism?
25. Refer to the glossary of key medical terms associated with the endocrine system. Be sure that you can define each term.

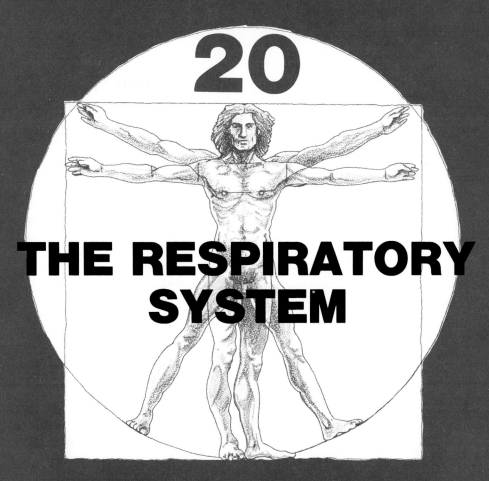

20

THE RESPIRATORY SYSTEM

STUDENT OBJECTIVES

- Identify the organs of the respiratory system.

- Compare the structure of the external and internal nose.

- Differentiate the three anatomical subdivisions of the pharynx and describe their roles in respiration.

- Identify the anatomical features of the larynx related to respiration and voice production.

- Describe the location and structure of the tubes that form the bronchial tree.

- Contrast tracheostomy and intubation as alternative methods for clearing air passageways.

- Identify the coverings of the lungs and the gross anatomical features of the lungs.

- Describe the structure of a bronchopulmonary segment and a lobule of the lung.

- Explain the structure of the alveolar-capillary membrane and its function in the diffusion of respiratory gases.

- Compare the volumes and capacities of air exchanged in respiration.

- Explain how the medullary rhythmicity area, apneustic area, and pneumotaxic area control respiration.

- Define coughing, sneezing, sighing, yawning, sobbing, crying, laughing, and hiccuping as modified respiratory movements.

- List the basic steps involved in cardiopulmonary resuscitation (CPR).

- Explain how the Heimlich maneuver is performed.

- Describe the effects of pollutants on the epithelium of the respiratory system.

- Define nasal polyps, hay fever, bronchial asthma, emphysema, pneumonia, tuberculosis, hyaline membrane disease, sudden infant death syndrome, and carbon monoxide poisoning as disorders of the respiratory system.

- Define key medical terms associated with the respiratory system.

ells need a continuous supply of oxygen to carry out the activities that are vital to their survival. Many of these activities release quantities of carbon dioxide. Since an excessive amount of carbon dioxide produces acid conditions that are poisonous to cells, the gas must be eliminated quickly and efficiently. The two systems that supply oxygen and eliminate carbon dioxide are the cardiovascular system and the respiratory system. The **respiratory system** consists of organs that exchange gases between the atmosphere and blood. These organs are the nose, pharynx, larynx, trachea, bronchi, and lungs (Figure 20-1). The cardiovascular system transports the gases in the blood between the lungs and the cells.

The overall exchange of gases between the atmosphere, the blood, and the cells is **respiration.** Three basic processes are involved. The first process, **ventilation,** or breathing, is the movement of air between the atmosphere and the lungs. The second and third processes involve the exchange of gases within the body. **External respiration** is the exchange of gases between the lungs and blood. **Internal respiration** is the exchange of gases between the blood and the cells.

The respiratory and cardiovascular systems participate equally in respiration. Failure of either system has the same effect on the body: disruption of homeostasis and rapid death of cells from oxygen starvation.

The developmental anatomy of the respiratory system is considered in Exhibit 24-9.

ORGANS

NOSE

The **nose** has an external portion and an internal portion inside the skull (Figure 20-2). Externally, the nose consists of a supporting framework of bone and cartilage covered with skin and lined with mucous membrane. The bridge of the nose is formed by the nasal bones, which hold it in a fixed position. Because it has a framework of pliable cartilage, the rest of the external nose is quite flexible. On the undersurface of the external nose are two openings called the **nostrils** or **external nares** (NA-rēz; singular, **naris**). The surface anatomy of the nose may be examined by referring to Exhibit 25-1.

The internal region of the nose is a large cavity in the skull that lies inferior to the cranium and superior to the mouth. Anteriorly, the internal nose merges with the external nose, and posteriorly it communicates with the throat (pharynx) through two openings called the **internal nares (choanae).** Four paranasal sinuses (frontal, sphenoidal, maxillary, and ethmoidal) and the nasolacrimal ducts also open into the internal nose. The lateral walls of the internal nose are formed by the ethmoid, maxillae, and inferior conchae bones. The ethmoid also forms the roof. The floor is formed by the palatine bones and the palatine process of the maxilla, which together comprise the hard palate.

The inside of both the external and internal nose is divided into right and left **nasal cavities** by a vertical partition called the **nasal septum.** Cartilage is the primary material making up the anterior portion of the septum. The remainder is formed by the vomer and the perpendicular plate of the ethmoid (see Figure 6-7a). The anterior portions of the nasal cavities, which are just inside the nostrils, are called the **vestibules.** The vestibules are surrounded by cartilage as opposed to bone of the upper nasal cavity. The interior structures of the nose are specialized for three functions: incoming air is warmed, moistened, and filtered; olfactory stimuli are received; and large hollow resonating chambers are provided for speech sounds.

When air enters the nostrils, it passes first through the vestibule. The vestibule is lined by skin containing

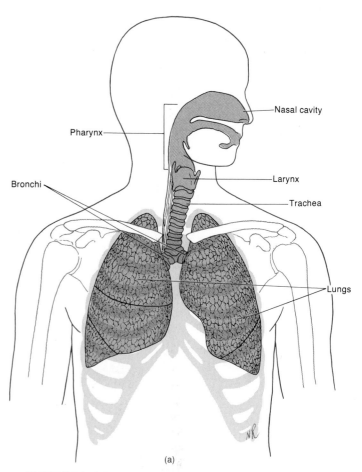

Pharynx

Bronchi

Nasal cavity

Larynx

Trachea

Lungs

(a)

FIGURE 20-1 Organs of the respiratory system in relation to surrounding structures. (a) Diagram.

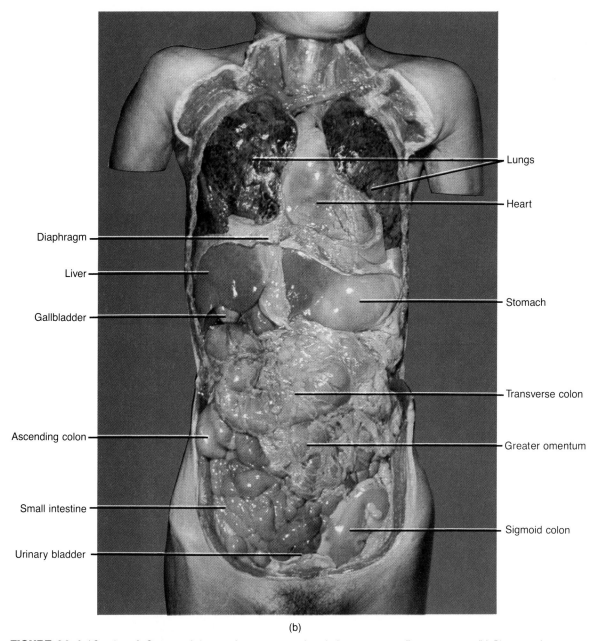

FIGURE 20-1 (*Continued*) Organs of the respiratory system in relation to surrounding structures. (b) Photograph. (Courtesy of C. Yokochi and J. W. Rohen, *Photographic Anatomy of the Human Body,* 1st ed., 1969, IGAKU-SHOIN, Ltd., Tokyo, New York.)

coarse hairs that filter out large dust particles. The air then passes into the rest of the cavity. Three shelves formed by projections of the superior, middle, and inferior conchae, or turbinates, extend out of the lateral wall of the cavity. The conchae, almost reaching the septum, subdivide each nasal cavity into a series of groovelike passageways—the **superior, middle,** and **inferior meatuses.** Mucous membrane lines the cavity and its shelves. The olfactory receptors lie in the membrane lining the area superior to the superior conchae and is also called

the olfactory region. Below the olfactory region, the membrane contains pseudostratified ciliated columnar cells with many goblet cells and capillaries. As the air whirls around the turbinates and meati, it is warmed by the capillaries. Mucus secreted by the goblet cells moistens the air and traps dust particles. Drainage from the lacrimal ducts, and perhaps secretions from the paranasal sinuses, also help moisten the air. The cilia move the mucus-dust packages along to the pharynx so they can be eliminated from the body.

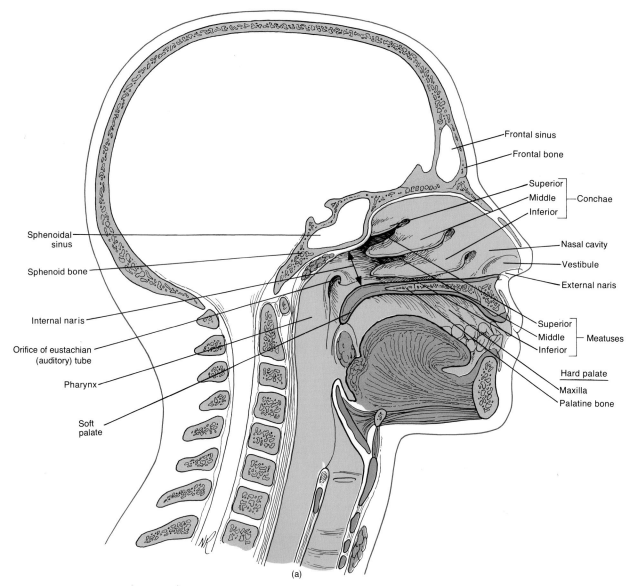

FIGURE 20-2 Nose. (a) Left nasal cavity seen in sagittal section.

CLINICAL APPLICATION

Nosebleed, or **epistaxis** (ep'-i-STAK-sis), is common because of the exposure of the nose to trauma and the extensive blood supply of the nose. Bleeding, either arterial or venous, usually occurs on the anterior part of the septum and can be arrested by firm packing of the external nares. If the point of bleeding is in the posterior region, plugging of both the external and internal nares may be necessary. In extreme emergency, the external carotid artery may have to be ligated (tied) in order to control the hemorrhage.

The arterial supply to the nasal cavity is principally from the sphenopalatine branch of the maxillary artery. The remainder is supplied by the ophthalmic artery. The veins of the nasal cavity drain into the sphenopalatine vein, the facial vein, and the ophthalmic vein.

The nerve supply of the nasal cavity consists of olfactory cells in the olfactory epithelium associated with the olfactory nerve (see Figure 18-7b) and the nerves of general sensation. These nerves are branches of the ophthalmic division of the trigeminal nerve and the maxillary division of the trigeminal nerve.

PHARYNX

The **pharynx** (FAR-inks), or throat, is a somewhat funnel-shaped tube about 13 cm (5 inches) long that starts at the internal nares and extends partway down the neck (Figure 20-3). It lies just posterior to the nasal cavity

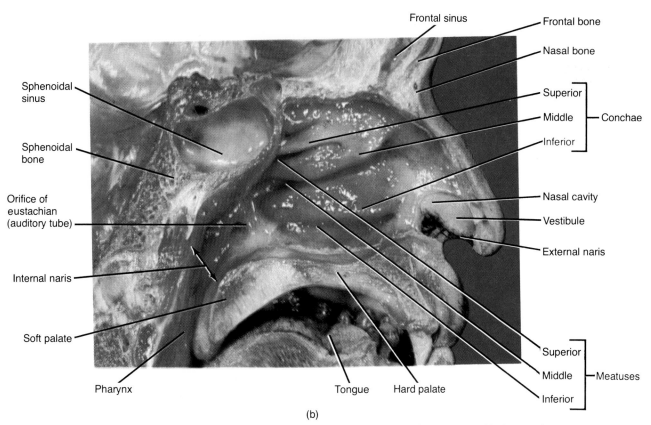

Sphenoidal sinus

Sphenoidal bone

Orifice of eustachian (auditory tube)

Internal naris

Soft palate

Pharynx

Frontal sinus — Frontal bone

Nasal bone

Superior

Middle — Conchae

Inferior

Nasal cavity

Vestibule

External naris

Tongue — Hard palate

Superior

Middle — Meatuses

Inferior

(b)

FIGURE 20-2 (*Continued*) Nose. (b) Photograph of the left nasal cavity seen in sagittal section with the nasal septum removed. (Courtesy of C. Yokochi and J. W. Rohen, *Photographic Anatomy of the Human Body,* 2nd ed., 1979, IGAKU-SHOIN, Ltd., Tokyo, New York.)

and oral cavity and just anterior to the cervical vertebrae. Its walls are composed of skeletal muscles and are lined with mucous membrane. The functions of the pharynx are limited to serving as a passageway for air and food and providing a resonating chamber for speech sounds.

The uppermost portion of the pharynx, called the **nasopharynx,** lies posterior to the internal nasal cavity and extends to the plane of the soft palate. There are four openings in its walls: two internal nares and two openings that lead into the eustachian (auditory) tubes. The posterior wall also contains the pharyngeal tonsil, or adenoid. Through the internal nares the nasopharynx exchanges air with the nasal cavities and receives the packages of dust-laden mucus. It is lined with pseudostratified ciliated epithelium, and the cilia move the mucus down toward the mouth. The nasopharynx also exchanges small amounts of air with the auditory tubes so that the air pressure inside the middle ear equals the pressure of the atmospheric air flowing through the nose and pharynx.

The middle portion of the pharynx, the **oropharynx,** lies posterior to the oral cavity and extends from the soft palate inferior to the level of the hyoid bone. It receives only one opening, the *fauces* (FAW-sēz), or opening from the mouth. It is lined by stratified squamous

epithelium. This portion of the pharynx is both respiratory and digestive in function, since it is a common passageway for both air and food. Two pairs of tonsils, the palatine tonsils and the lingual tonsils, are found in the oropharynx. The lingual tonsils lie at the base of the tongue (also see Figure 21-5a).

The lowest portion of the pharynx, the **laryngopharynx** (la-rin'-gō-FAR-inks), extends downward from the hyoid bone and becomes continuous with the esophagus (food tube) posteriorly and the larynx (voice box) anteriorly. Like the oropharynx, the laryngopharynx is both a respiratory and a digestive pathway and is lined by stratified squamous epithelium.

The arteries of the pharynx are the ascending pharyngeal, the ascending palatine branch of the facial, the descending palatine and pharyngeal branches of the maxillary, and the muscular branches of the superior thyroid artery. The veins of the pharynx drain into the pterygoid plexus and the internal jugular vein.

Most of the muscles of the pharynx (see Figure 10-8) are innervated by the pharyngeal plexus. This plexus is formed by the pharyngeal branches of the glossopharyngeal and vagal nerves and the superior cervical sympathetic ganglion.

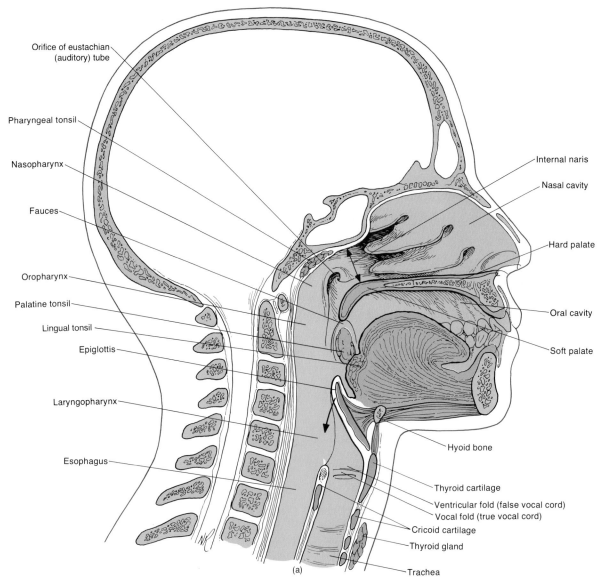

Orifice of eustachian (auditory) tube

Pharyngeal tonsil

Nasopharynx

Fauces

Oropharynx

Palatine tonsil

Lingual tonsil

Epiglottis

Laryngopharynx

Esophagus

Internal naris

Nasal cavity

Hard palate

Oral cavity

Soft palate

Hyoid bone

Thyroid cartilage

Ventricular fold (false vocal cord)

Vocal fold (true vocal cord)

Cricoid cartilage

Thyroid gland

Trachea

(a)

FIGURE 20-3 Head and neck seen in sagittal section. (a) Diagram.

LARYNX

The **larynx,** or voice box, is a short passageway that connects the pharynx with the trachea. It lies in the midline of the neck anterior to the fourth through sixth cervical vertebrae. The walls of the larynx are supported by nine pieces of cartilage (Figure 20-4a-c). Three are single and three are paired. The three single pieces are the large thyroid cartilage and the smaller epiglottic and cricoid cartilage. Of the paired cartilages, the arytenoid cartilages are the most important. The paired corniculate and cuneiform cartilages are of lesser significance.

The **thyroid cartilage,** or Adam's apple, consists of two fused plates that form the anterior wall of the larynx and give it its triangular shape. It is larger in males than in females.

The **epiglottis** is a large, leaf-shaped piece of cartilage lying on top of the larynx. The "stem" of the epiglottis is attached to the thyroid cartilage, but the "leaf" portion is unattached and free to move up and down like a trapdoor. During swallowing, there is elevation of the larynx. This causes the free edge of the epiglottis to form a lid over the glottis and the closure of the glottis. The **glottis** is the space between the vocal folds (true vocal cords) in the larynx. In this way, the larynx is closed off and liquids and foods are routed into the esophagus and kept out of the trachea. If anything but air passes into the larynx, a cough reflex attempts to expel the material.

The **cricoid** (KRĪ-koyd) **cartilage** is a ring of cartilage forming the inferior walls of the larynx. It is attached to the first ring of cartilage of the trachea.

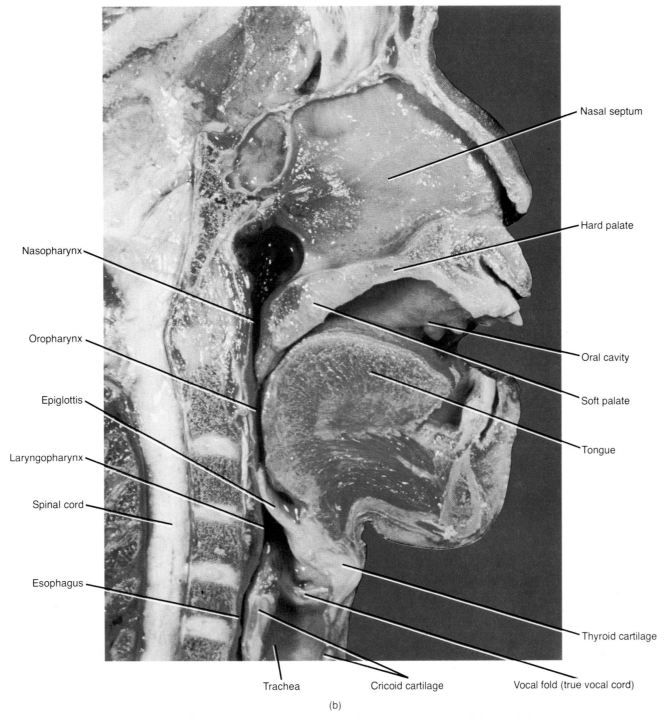

Nasal septum

Hard palate

Nasopharynx

Oropharynx

Oral cavity

Epiglottis

Soft palate

Laryngopharynx

Tongue

Spinal cord

Esophagus

Thyroid cartilage

Trachea   Cricoid cartilage   Vocal fold (true vocal cord)

(b)

FIGURE 20-3 (*Continued*) Head and neck seen in sagittal section. (b) Photograph. (Courtesy of C. Yokochi and J. W. Rohen, *Photographic Anatomy of the Human Body,* 2nd ed., 1979, IGAKU-SHOIN, Ltd., Tokyo, New York.)

The paired **arytenoid** (ar'-i-TĒ-noyd) **cartilages** are pyramidal in shape and located at the superior border of the cricoid cartilage. They attach to the vocal folds and pharyngeal muscles and by their action can move the vocal cords. The **corniculate** (kor-NIK-yoo-lāt) **cartilages** are paired, cone-shaped cartilages located at the apex of each arytenoid cartilage. The paired **cuneiform** (kyoo-NĒ-i-form) **cartilages** are rod-shaped cartilages in the mucous membrane fold that connect the epiglottis to the arytenoid cartilages.

The epithelium lining the larynx below the vocal folds is pseudostratified. It consists of ciliated columnar cells, goblet cells, and basal cells, and it helps to trap dust not removed in the upper passages.

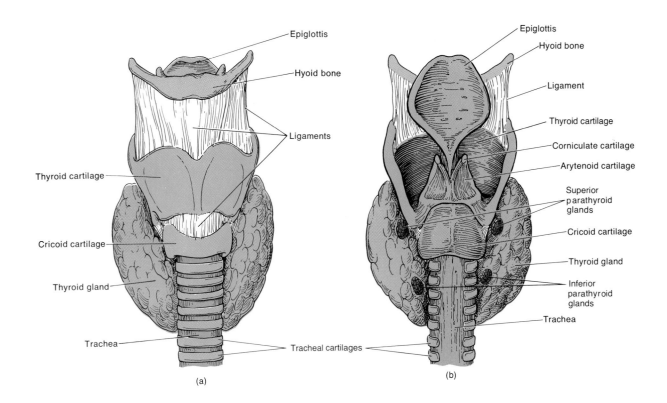

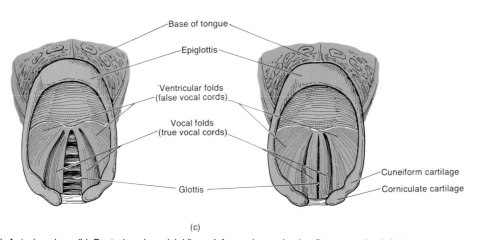

FIGURE 20-4 Larynx. (a) Anterior view. (b) Posterior view. (c) Viewed from above. In the figure on the left the true vocal cords are relaxed and the glottis is open. In the figure on the right the true vocal cords are pulled taut and the glottis is closed.

The mucous membrane of the larynx is arranged into two pairs of folds—an upper pair called the **ventricular folds** (or **false vocal cords**) and a lower pair called simply the **vocal folds** (or **true vocal cords**) (Figure 20-4c, d). Under the mucous membrane of the true vocal cords, which is lined by nonkeratinized stratified squamous epithelium, lie bands of elastic ligaments stretched between pieces of rigid cartilage like the strings on a guitar. Skeletal muscles of the larynx, called intrinsic muscles, are attached internally to the pieces of rigid cartilage and to the vocal folds themselves (see Figure 10-9). When the muscles contract, they pull the strings of elastic ligaments tight and stretch the cords out into the air passageways so that the glottis is narrowed. If air is directed against the vocal folds, they vibrate and set up sound waves in the column of air in the pharynx, nose, and mouth. The greater the pressure of air, the louder the sound.

Pitch is controlled by the tension on the true vocal cords. If the cords are pulled taut by the muscles, they vibrate more rapidly and a higher pitch results. Lower

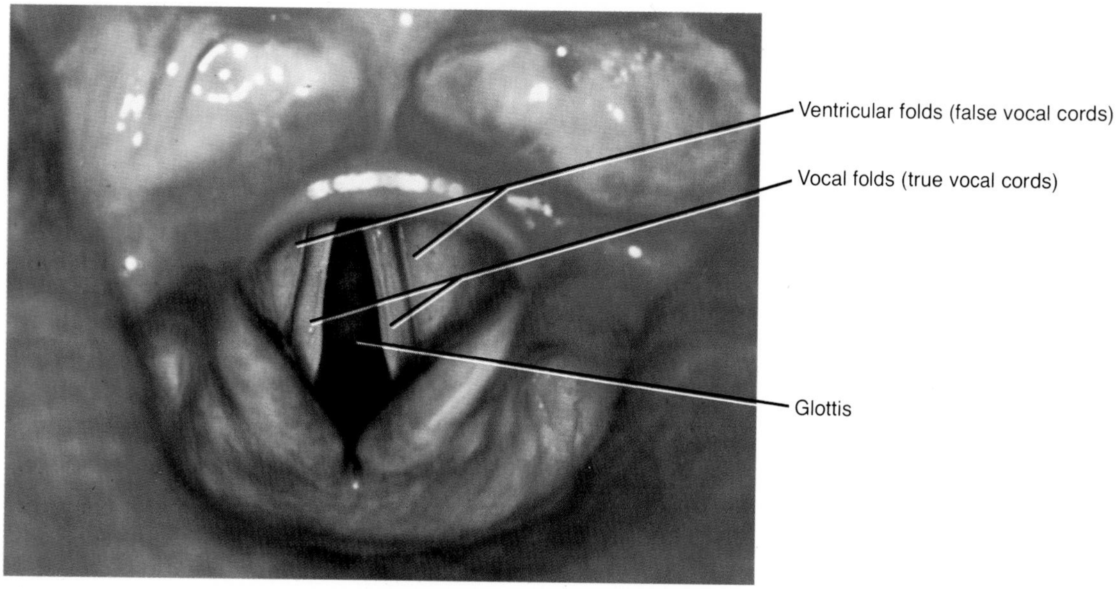

(d)

FIGURE 20-4 (*Continued*) Larynx. (d) Photograph of the true vocal cords, false vocal cords, and open glottis from above. (Courtesy of C. Yokochi and J. W. Rohen, *Photographic Anatomy of the Human Body,* 2nd ed., 1979, IGAKU-SHOIN, Ltd., Tokyo, New York.)

sounds are produced by decreasing the muscular tension on the cords. True vocal cords are usually thicker and longer in males than in females, and therefore they vibrate more slowly. Thus men have a lower range of pitch than women.

Sound originates from the vibration of the true vocal cords, but other structures are necessary for converting the sound into recognizable speech. The pharynx, mouth, nasal cavities, and paranasal sinuses all act as resonating chambers that give the voice its human and individual quality. By constricting and relaxing the muscles in the walls of the pharynx, we produce the vowel sounds. Muscles of the face, tongue, and lips help us to enunciate words.

CLINICAL APPLICATION

Laryngitis is an inflammation of the larynx that is most often caused by a respiratory infection or irritants such as cigarette smoke. Inflammation of the vocal folds themselves causes hoarseness or loss of voice by interfering with the contraction of the cords or by causing them to swell to the point where they cannot vibrate freely. Many long-term smokers acquire a permanent hoarseness from the damage done by chronic inflammation.

The arteries of the larynx are the superior laryngeal, inferior laryngeal, and cricothyroid. The superior and inferior laryngeal and cricothyroid veins accompany the arteries. The superior cricoid vein and cricothyroid vein empty into the superior thyroid vein, and the inferior vein empties into the inferior thyroid vein.

The nerves of the larynx are the superior and recurrent laryngeal branches of the vagus.

TRACHEA

The **trachea** (TRĀ-kē-a), or windpipe, is a tubular passageway for air about 12 cm (4½ inches) in length and 2.5 cm (1 inch) in diameter. It is located anterior to the esophagus and extends from the larynx to the fifth thoracic vertebra, where it divides into right and left primary bronchi (see Figure 20-6a).

The tracheal epithelium is pseudostratified. It consists of ciliated columnar cells, goblet cells, and basal cells. The epithelium provides the same protection against dust as the membrane lining the larynx (Figure 20-5). The walls of the trachea are composed of smooth muscle and elastic connective tissue. They are encircled by a series of 16 to 20 horizontal incomplete rings of hyaline cartilage that look like a series of letter Cs stacked one on top of another. The open parts of the Cs face the esophagus and permit it to expand into the trachea during swallowing. Transverse smooth muscle fibers, called the *trachealis muscle,* attach the open ends of the cartilage rings (Figure 20-5e). The open ends of the rings of cartilage are also attached by elastic connective tissue. The solid parts of the Cs provide a rigid support so the tracheal walls do not collapse inward and obstruct the air passageway.

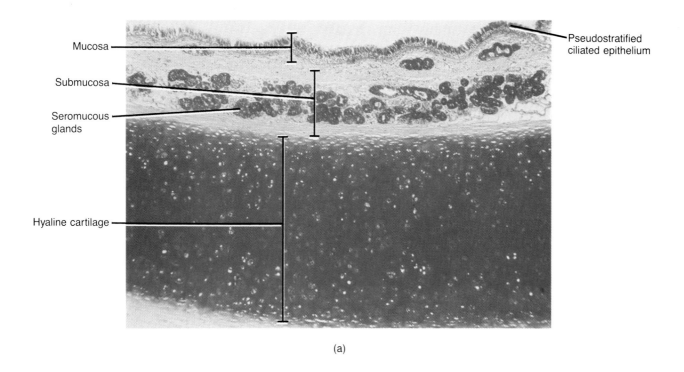

(a)

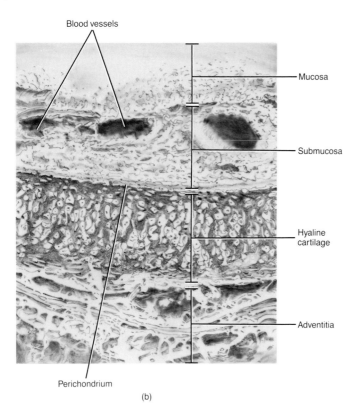

(b)

FIGURE 20-5 Histology of the trachea. (a) Photomicrograph of a portion of the tracheal wall at a magnification of 80×. (© 1983 by Michael H. Ross. Used by permission.) (b) Drawing of the tracheal wall based on a scanning electron micrograph at a magnification of 150×.

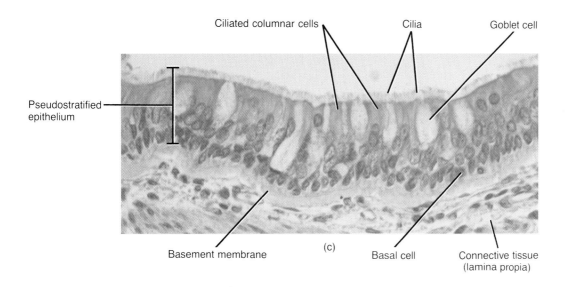

Ciliated columnar cells   Cilia   Goblet cell

Pseudostratified epithelium

Basement membrane   (c)   Basal cell   Connective tissue (lamina propia)

CLINICAL APPLICATION

At the point where the trachea undergoes bifurcation into right and left primary bronchi, there is a ridge on the inside called the **carina** (ka-RĪ-na). It is formed by a posterior and somewhat inferior projection of the last tracheal cartilage. Widening and distortion of the carina, which can be seen in an examination by bronchoscopy, is a serious prognostic sign, since it usually indicates a carcinoma of the lymph nodes around the bifurcation of the trachea.

Occasionally the respiratory passageways are unable to protect themselves from obstruction. The rings of cartilage may be accidentally crushed; the mucous membrane may become inflamed and swell so much that it closes off the air space; inflamed membranes secrete a great deal of mucus that may clog the lower respiratory passageways; or a large object may be breathed in (aspirated) while the glottis is open. In any case, the passageways must be cleared quickly. If the obstruction is above the level of the chest, **tracheostomy** (trā-ke-OS-tō-mē) may be performed. The first step in a tracheostomy is to make an incision in the neck and into the part of the trachea below the obstructed area. The patient breathes through a tube inserted through the incision. Another method is **intubation.** A tube is inserted into the mouth or nose and passed down through the larynx and trachea. The firm walls of the tube push back any flexible obstruction, and the inside of the tube provides a passageway for air. If mucus is clogging the trachea, it can be suctioned through the tube.

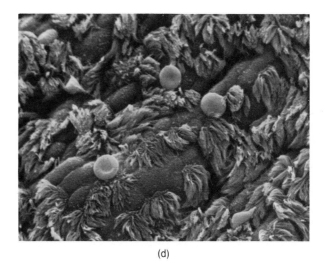

(d)

FIGURE 20-5 (*Continued*) Histology of the trachea. (c) Photomicrograph of an enlarged aspect of the tracheal epithelium at a magnification of 600×. (© 1983 by Michael H. Ross. Used by permission.) (d) Scanning electron micrograph of the tracheal epithelium at a magnification of 200×. (Courtesy of Fisher Scientific Company and S.T.E.M. Laboratories, Inc., Copyright, 1975.)

The arteries of the trachea are branches of the inferior thyroid, internal thoracic, and bronchial arteries. The veins of the trachea terminate in the inferior thyroid veins.

The smooth muscle and glands of the trachea are innervated parasympathetically via the vagus nerve directly and by its recurrent laryngeal branches. Sympathetic innervation is through branches from the sympathetic trunk and its ganglia.

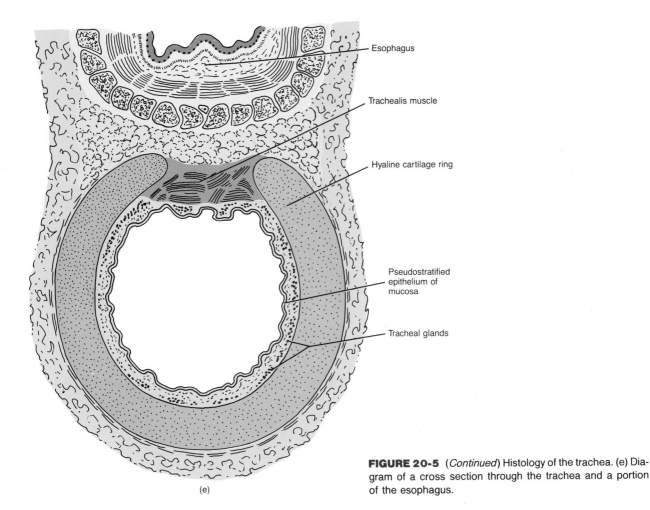

Esophagus

Trachealis muscle

Hyaline cartilage ring

Pseudostratified epithelium of mucosa

Tracheal glands

(e)

FIGURE 20-5 (*Continued*) Histology of the trachea. (e) Diagram of a cross section through the trachea and a portion of the esophagus.

BRONCHI

The trachea terminates in the chest by dividing at the sternal angle into a **right primary bronchus** (BRON-kus), which goes to the right lung, and a **left primary bronchus,** which goes to the left lung (Figure 20-6a, b). The right primary bronchus is more vertical, shorter, and wider than the left. As a result, foreign objects that enter the air passageways frequently lodge in it. Like the trachea, the primary bronchi (BRON-kī) contain incomplete rings of cartilage and are lined by a pseudostratified ciliated epithelium.

The blood supply to the bronchi is via the left bronchial and right bronchial arteries. The veins that drain the bronchi are the right bronchial vein, which enters the azygous vein, and the left bronchial vein, which empties into the hemiazygous vein or the left superior intercostal vein.

Upon entering the lungs, the primary bronchi divide to form smaller bronchi—the **secondary** or **lobar bronchi,** one for each lobe of the lung (the right lung has three lobes; the left lung has two). The secondary bronchi continue to branch, forming still smaller bronchi, called **tertiary** or **segmental bronchi** which divide into **bronchioles.** Bronchioles, in turn, branch into even smaller tubes called **terminal bronchioles.** The continuous branching of the trachea into primary bronchi, secondary bronchi, bron-

chioles, and terminal bronchioles resembles a tree trunk with its branches and is commonly referred to as the **bronchial tree.**

As the branching becomes more extensive in the bron-

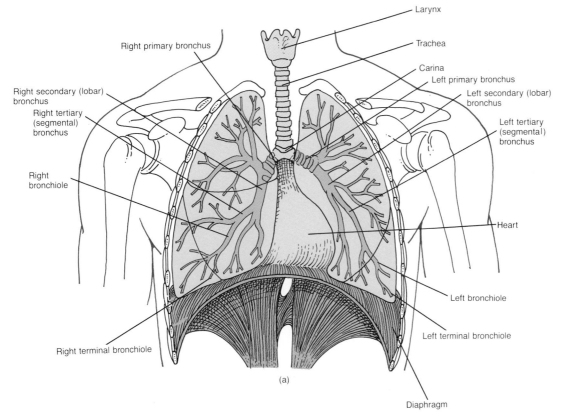

Larynx
Trachea
Carina
Left primary bronchus
Left secondary (lobar) bronchus
Left tertiary (segmental) bronchus
Heart
Left bronchiole
Left terminal bronchiole
Diaphragm

Right primary bronchus
Right secondary (lobar) bronchus
Right tertiary (segmental) bronchus
Right bronchiole
Right terminal bronchiole

(a)

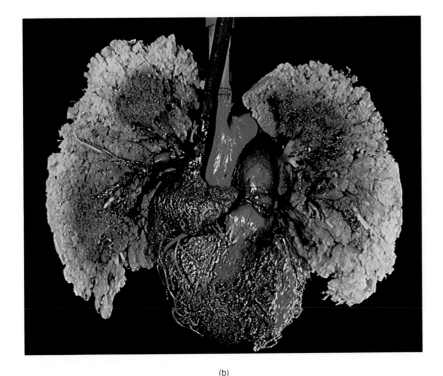

(b)

FIGURE 20-6 Air passageways to the lungs. (a) Diagram of the bronchial tree in relation to the lungs. (b) Photograph of a cast of the bronchial tree. (Courtesy of Lester V. Bergman & Associates.)

chial tree, several structural changes may be noted. First, rings of cartilage are replaced by plates of cartilage that finally disappear in the bronchioles. Second, as the carti-

lage decreases, the amount of smooth muscle increases. In addition, the epithelium changes from ciliated columnar to simple cuboidal in the terminal bronchioles.

Glands

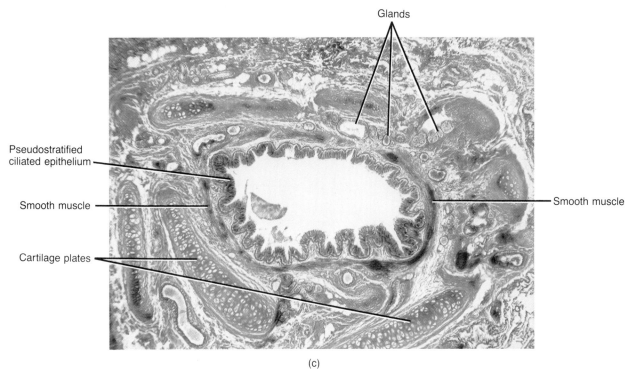

Pseudostratified
ciliated epithelium

Smooth muscle

Smooth muscle

Cartilage plates

(c)

FIGURE 20-6 (*Continued*) Air passageways to the lungs. (c) Photomicrograph of a cross section of a primary bronchus at a magnification of 100×. (© 1983 by Michael H. Ross. Used by permission.) (d) Anteroposterior bronchogram. (Courtesy of Lester W. Paul and John H. Juhl, *The Essentials of Roentgen Interpretation*, 3rd ed., Harper & Row, Publishers, Inc., New York, 1972.)

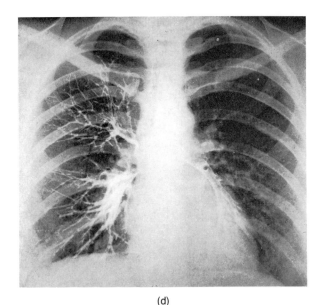

(d)

LUNGS

The **lungs** are paired, cone-shaped organs lying in the thoracic cavity (see Figure 12-1a). They are separated from each other by the heart and other structures in the mediastinum. Two layers of serous membrane, collec-

tively called the **pleural membrane,** enclose and protect each lung. The outer layer is attached to the walls of the thoracic cavity and is called the **parietal pleura.** The inner layer, the **visceral pleura,** covers the lungs themselves. Between the visceral and parietal pleura is a small potential space, the **pleural cavity,** which contains a lubricating fluid secreted by the membranes (see Figure 1-4b). This fluid prevents friction between the membranes and allows them to move easily on one another during breathing.

CLINICAL APPLICATION

In certain conditions, the pleural cavity may fill with air (**pneumothorax**), blood (**hemothorax**), or pus. Inflammation of the pleural membrane, or **pleurisy,** causes friction during breathing that can be quite painful when the swollen membranes rub against each other. Fluid can be drained from the pleural cavity by inserting a needle, usually posteriorly through the seventh intercostal space. The needle is passed along the superior border of the lower rib to avoid damage to the intercostal nerves and blood vessels. Below the seventh intercostal space there is danger of penetrating the diaphragm.

The lungs extend from the diaphragm to a point about 1.5 to 2.5 cm (¾ to 1 inch) superior to the clavicles and lie against the ribs anteriorly and posteriorly. The broad inferior portion of the lung, the **base,** is concave and fits over the convex area of the diaphragm (Figure 20-7). The narrow superior portion of the lung is termed

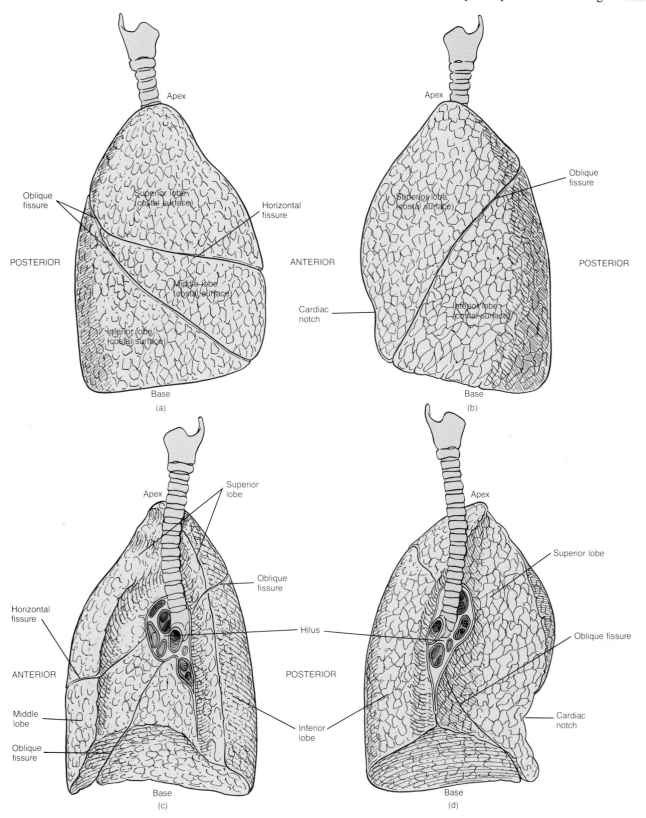

FIGURE 20-7 External features of the lungs. (a) Right lung in lateral view. (b) Left lung in lateral view. (c) Right lung in medial view. (d) Left lung in medial view.

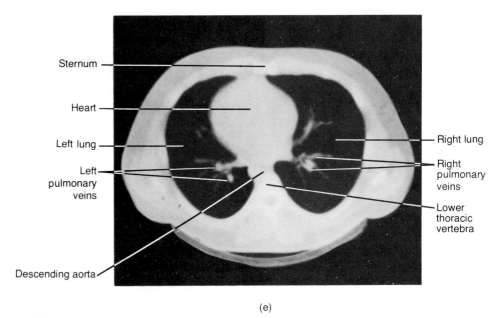

(e)

FIGURE 20-7 (*Continued*) External features of the lungs. (e) CT scan of the lungs. (Courtesy of General Electric Co.)

the **apex** or **cupula.** The surface of the lung lying against the ribs, the **costal surface,** is rounded to match the curvature of the ribs. The **mediastinal (medial) surface** of each lung contains a vertical slit, the **hilus,** through which bronchi, pulmonary vessels, and nerves enter and exit. The blood vessels, bronchi, and nerves are held together by the pleura and connective tissue, and they constitute the **root** of the lung. Medially, the left lung also contains a concavity, the **cardiac notch,** in which the heart lies.

The right lung is thicker and broader than the left. It is also somewhat shorter than the left because the diaphragm is higher on the right side to accommodate the liver that lies below it. The left lung is thinner, narrower, and longer than the right.

Each lung is divided into lobes by one or more fissures. Both lungs have an **oblique fissure,** which extends downward and forward. The right lung also has a **horizontal fissure.** The oblique fissure in the left lung separates the **superior lobe** from the **inferior lobe.** The upper part of the oblique fissure of the right lung separates the superior lobe from the inferior lobe, whereas the lower part of the oblique fissure separates the inferior lobe from the **middle lobe.** The horizontal fissure of the right lung subdivides the superior lobe, thus forming a middle lobe.

Each lobe receives its own secondary or lobar bronchus. Thus the right primary bronchus gives rise to three lobar bronchi called the **superior, middle,** and **inferior lobar** or **secondary bronchi.** The left primary bronchus gives rise to a **superior** and an **inferior lobar** or **secondary bronchus.** Within the substance of the lung, the lobar bronchi give rise to the **tertiary** or **segmental bronchi,** which are constant in both origin and distribution. The segment of lung tissue that each supplies is called a **bronchopulmonary segment** (Figure 20-8).

Each bronchopulmonary segment of the lungs is broken up into many small compartments called **lobules** (Figure 20-9a). Each lobule is wrapped in elastic connective tissue and contains a lymphatic vessel, an arteriole, a venule, and a branch from a terminal bronchiole. Terminal bronchioles subdivide into microscopic branches called **respiratory bronchioles** (Figure 20-9a, b). In the respiratory bronchioles, the epithelial lining changes from cuboidal to squamous as they become more distal. Respiratory bronchioles, in turn, subdivide into several (2 to 11) **alveolar ducts** or **atria.**

Around the circumference of the alveolar ducts are numerous alveoli and alveolar sacs. An **alveolus** (al-VĒ-ō-lus) is a cup-shaped outpouching lined by epithelium and supported by a thin elastic basement membrane. **Alveolar sacs** are two or more alveoli that share a common opening (Figure 20-9b-d). The alveolar walls consist of two principal types of epithelial cells: (1) *squamous pulmonary epithelial cells* (*type I alveolar cells* or *small alveolar cells*) and (2) *septal cells* (*type II alveolar cells* or *great alveolar cells*). The squamous pulmonary epithelial cells are the larger of the two types of cells and form a continuous lining of the alveolar wall, except for occasional septal cells. Septal cells are much smaller, cuboidal in shape, and are dispersed among the squamous pulmonary epithelial cells. Septal cells produce a phospholipid substance called *surfactant* (sur-FAK-tant), which lowers surface tension. Inadequate amounts of surfactant result in hyaline membrane disease (discussed at the end of the chapter). Also found within the alveolar wall are free *alveolar macrophages,* or *dust cells.* They are highly phagocytic and serve to remove dust particles or other debris that gain entrance to alveolar spaces. Deep to the layer of squamous pulmonary epithelial cells is an elastic basement

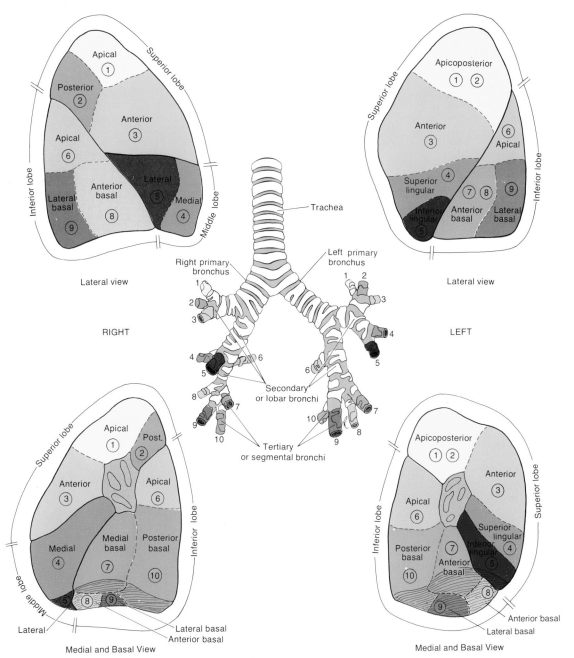

FIGURE 20-8 Bronchopulmonary segments of the lungs. The bronchial branches are shown in the center of the figure. The bronchopulmonary segments are numbered and named for convenience.

membrane. Over the alveoli, the arteriole and venule disperse into a capillary network (Figure 20-9a). The blood capillaries consist of a single layer of endothelial cells and a basement membrane.

The exchange of respiratory gases between the lungs and blood takes place by diffusion across the alveoli and capillary walls. This membrane, through which the respiratory gases move, is collectively known as the **alveolar-capillary (respiratory) membrane** (Figure 20-9e). It consists of: (1) a layer of squamous pulmonary epithelial cells with septal cells and free alveolar macrophages that constitute the alveolar (epithelial) wall, (2) an epithelial basement membrane underneath the alveolar wall, (3) a capillary basement membrane that is often fused to the epithelial basement membrane, and (4) the endothelial cells of the capillary. Despite the large number of layers, the alveolar-capillary membrane averages only 0.5 μm in thickness. This is of considerable importance to the efficient diffusion of respiratory gases. Moreover, it has been estimated that the lungs contain 300 million alveoli, providing an immense surface area (70 square meters) for the exchange of gases.

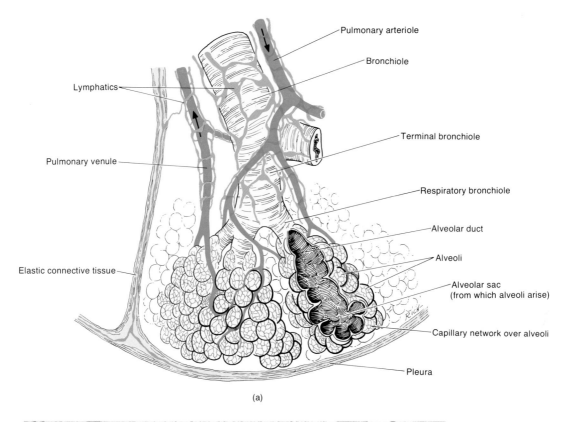

(a)

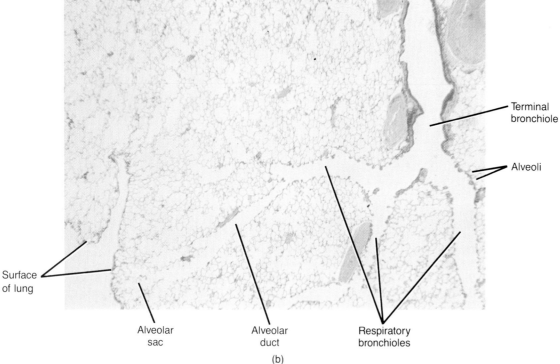

(b)

FIGURE 20-9 Histology of the lungs. (a) Diagram of a lobule of the lung. (b) Photomicrograph of a terminal bronchiole, respiratory bronchioles, alveolar duct, alveolar sac, and alveoli at a magnification of 50×. (© 1983 by Michael H. Ross. Used by permission.)

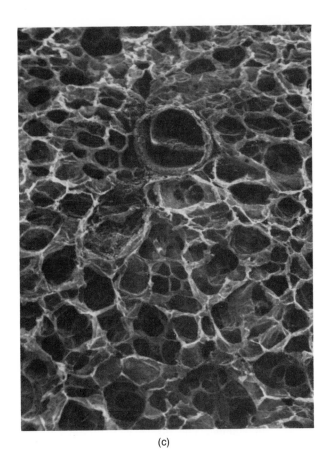

(c)

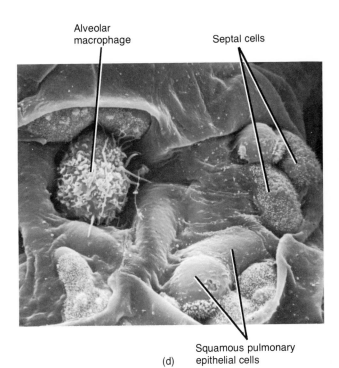

Alveolar
macrophage

Septal cells

Squamous pulmonary
epithelial cells

(d)

FIGURE 20-9 (*Continued*) Histology of the lungs. (c) Scanning electron micrograph of alveolar sacs at a magnification of 100×. (Courtesy of Fisher Scientific Company and S.T.E.M. Laboratories, Inc., Copyright, 1975.) (d) Scanning electron micrograph of an alveolus showing squamous pulmonary alveolar cells and an alveolar macrophage at a magnification of 3430×. (Courtesy of Richard G. Kessel and Randy H. Kardon, *Tissues and Organs: A Text-Atlas of Scanning Electron Microscopy*, W. H. Freeman and Company, San Francisco, 1979.) (e) Diagram of the structure of the alveolar-capillary membrane.

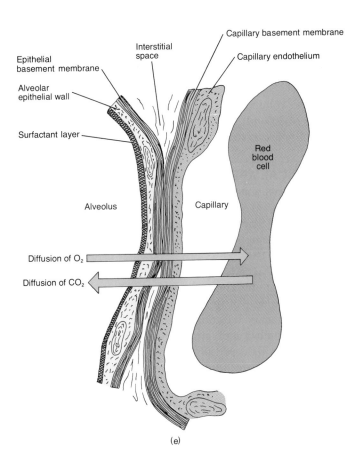

Capillary basement membrane

Interstitial
space

Capillary endothelium

Epithelial
basement membrane

Alveolar
epithelial wall

Red
blood
cell

Surfactant layer

Alveolus

Capillary

Diffusion of O_2

Diffusion of CO_2

(e)

The arterial supply of the lungs is derived from the pulmonary trunk. It divides into a left pulmonary artery which enters the left lung and a right pulmonary artery which enters the right lung. The venous return of the oxygenated blood is by way of the pulmonary veins, typically two in number on each side—the right and left superior and inferior pulmonary veins. All four veins drain into the left atrium.

The nerve supply of the lungs is derived from the autonomic nervous system. Parasympathetic fibers come from the vagus nerve and sympathetic fibers are derived from the second, third, and fourth thoracic ganglia.

AIR VOLUMES EXCHANGED

In clinical practice the word **respiration** means one inspiration plus one expiration. The average healthy adult has 14 to 18 respirations a minute while at rest. During each

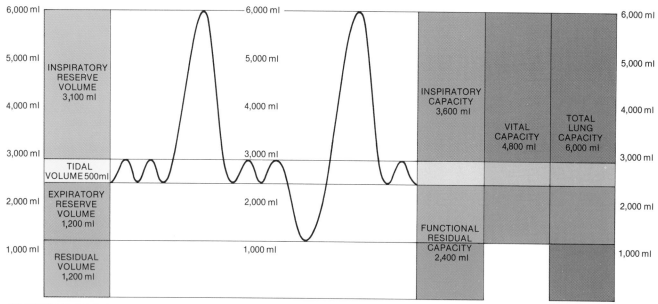

FIGURE 20-10 Spirogram of lung volumes and capacities.

respiration the lungs exchange volumes of air with the atmosphere. A lower than normal exchange volume is usually a sign of pulmonary malfunction.

The apparatus commonly used to measure the amount of air exchanged during breathing is called a **respirometer** or **spirometer.** The record is called a **spirogram** (Figure 20-10). Spirometric studies measure lung capacities and rates and depths of ventilation for diagnostic purposes. Spirometry is usually indicated for individuals who exhibit labored breathing. It is also used in the diagnosis of respiratory disorders such as emphysema and bronchial asthma.

During the process of normal quiet breathing, about 500 ml of air moves into the respiratory passageways with each inspiration. The same amount moves out with each expiration. This volume of air inspired (or expired) is called **tidal volume** (Figure 20-10). Only about 350 ml of the tidal volume reaches the alveoli. The other 150 ml remains in the dead spaces of the nose, pharynx, larynx, trachea, and bronchi and is known as **dead air.**

By taking a very deep breath, we can inspire a good deal more than 500 ml. This excess inhaled air, called the **inspiratory reserve volume,** averages 3,100 ml above the 500 ml of tidal volume. Thus the respiratory system can pull in as much as 3,600 ml of air. If we inhale normally and then exhale as forcibly as possible, we should be able to push out 1,200 ml of air in addition to the 500 ml tidal volume. This extra 1,200 ml is called the **expiratory reserve volume.** Even after the expiratory reserve volume is expelled, a good deal of air still remains in the lungs because the lower intrathoracic pressure keeps the alveoli slightly inflated and some air also remains in the noncollapsible air passageways. This air, the **residual volume,** amounts to about 1,200 ml.

One final measurement is not made with the spirometer. Opening the thoracic cavity allows the intrathoracic pressure to equal the atmospheric pressure, forcing out some of the residual volume. The air remaining is called the **minimal volume.** The presence of minimal volume can be determined by placing a piece of lung in water and watching it float.

Lung capacity can be calculated by combining various lung volumes. **Inspiratory capacity,** the total inspiratory ability of the lungs, is the sum of tidal volume plus inspiratory reserve volume (3,600 ml). **Functional residual capacity** is the sum of residual volume plus expiratory reserve volume (2,400 ml). **Vital capacity** is the sum of inspiratory reserve volume, tidal volume, and expiratory reserve volume (4,800 ml). Finally, **total lung capacity** is the sum of all volumes (6,000 ml).

CLINICAL APPLICATION

The **measurement of respiratory volumes and capacities** is an essential tool for determining how well the lungs are functioning. For instance, during the early stages of emphysema, many of the alveoli lose their elasticity. During expiration they fail to snap inward, and consequently they fail to force out a normal amount of air. Thus the residual volume is increased at the expense of the expiratory reserve volume.

Pulmonary infections can cause inflammation and an accumulation of fluid in the air spaces of the lungs. The fluid reduces the amount of space available for air and consequently decreases the vital capacity.

Minimal volume provides a medical and legal tool for determining whether a baby was born dead or died after birth. Fetal lungs contain no air, and so the lung of a stillborn will not float in water.

As soon as the lungs fill with air, oxygen diffuses from the alveoli into the blood, through the interstitial fluid, and finally into the cells. Carbon dioxide diffuses in the opposite direction—from the cells, through interstitial fluid to the blood, and to the alveoli.

NERVOUS CONTROL OF RESPIRATION

Respiration is controlled by several mechanisms that help the body maintain homeostasis. Inspiration and expiration depend on a pressure differential. If the volume of the lungs is increased, the intrathoracic or intrapleural pressure decreases, and air from the atmosphere enters the lungs. If the volume decreases, intrathoracic pressure increases, and air is expelled. The volume of the lungs is controlled by the action of the respiratory muscles—the diaphragm and external intercostal muscles. Their contraction pulls the diaphragm down and the ribs out, thus increasing the size of the thoracic cavity. Their relaxation decreases the size of the cavity.

The respiratory muscles contract and relax in turn as a result of nerve impulses transmitted to them from centers in the brain. The area from which nerve impulses are sent to respiratory muscles is located in the reticular formation of the brain stem; it is referred to as the **respiratory center.** The respiratory center consists of a widely dispersed group of neurons and is functionally divided into three areas: (1) the **medullary rhythmicity area** in the medulla, (2) the **apneustic** (ap-NOO-stik) **area** in the pons, and (3) the **pneumotaxic** (noo-mō-TAK-sik) **area,** also in the pons (Figure 20-11a).

The function of the medullary rhythmicity area is to control the basic rhythm of respiration. In the normal resting state, inspiration usually lasts for about 2 seconds and expiration for about 3 seconds. Let us consider a proposed mechanism of control of the basic rhythm of respiration. Within the medullary rhythmicity area are found both inspiratory and expiratory neurons arranged in circuits (Figure 20-11b). When one of the neurons in the inspiratory circuit becomes excited, it excites the next one, which, in turn, excites the next, and so on. In this way, the nerve impulses travel around the circuit from one neuron to the next. This continues for about 2 seconds, at the end of which time the neurons become dormant. However, as some nerve impulses travel around the circuit, other impulses travel from the circuit to the muscles of respiration—the diaphragm and intercostal muscles. These impulses reach the diaphragm by the phrenic nerves and the external intercostal muscles by the intercostal nerves. When the impulses reach the respiratory muscles, the muscles contract and inspiration occurs.

In addition to sending impulses to the muscles of respiration, the inspiratory center also sends inhibitory impulses to the expiratory neurons. These impulses inhibit the circuit of expiratory neurons. As long as the inspiratory neurons maintain nerve impulses through their circuit, expiration is inhibited. But when the inspiratory neurons become dormant, the expiratory neurons are no longer inhibited. This causes the expiratory neurons to become excited, and nerve impulses travel around the circuit of expiratory neurons from one neuron to another.

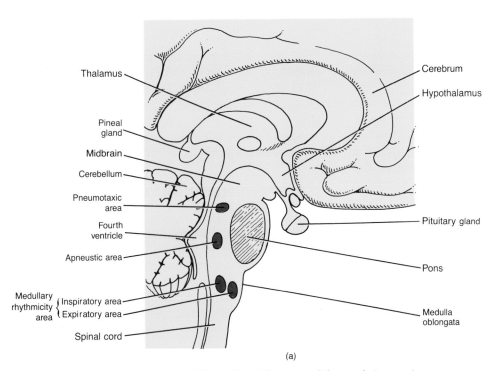

(a)

FIGURE 20-11 Respiratory center. (a) Location of the areas of the respiratory center.

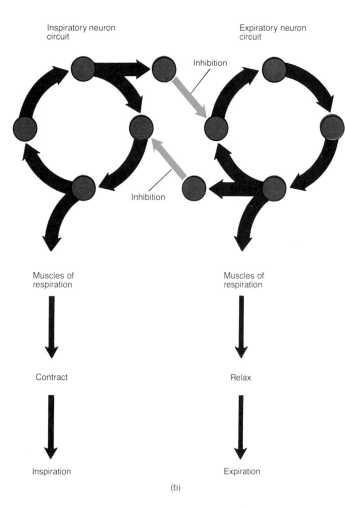

Inspiratory neuron circuit

Expiratory neuron circuit

Inhibition

Inhibition

Muscles of respiration

Muscles of respiration

Contract

Relax

Inspiration

Expiration

(b)

FIGURE 20-11 (*Continued*) Respiratory center. (b) Proposed function of the medullary rhythmicity area in controlling the basic rhythm of respiration.

This continues for about 3 seconds until the neurons become dormant. While the nerve impulses are maintained by the circuit of expiratory neurons, some impulses are transmitted to the respiratory muscles and others inhibit the inspiratory neurons. The impulses to the respiratory muscles, via the phrenic and intercostal nerves, cause the muscles to relax, and expiration occurs. When the expiratory neurons become dormant, the inspiratory neurons are no longer inhibited, and the inspiratory neurons become excited once again. Thus, the basic rhythm of respiration is established by circuits of inspiratory and expiratory neurons that alternately act upon the muscles of respiration and mutually inhibit each other.

Although the medullary rhythmicity area controls the basic rhythm of respiration, other parts of the nervous system can modify the rhythm. For example, in experiments where impulses from the apneustic and pneumotaxic areas are prevented from reaching the medullary rhythmicity area, respirations consist of short inspirations and prolonged expirations. Stimulation of the apneustic area alone sends impulses to the medullary rhythmicity area that result in forceful, prolonged inspirations and weak, short expirations. When the pneumotaxic area is

stimulated, impulses to the medullary rhythmicity area result in the cessation of apneustic breathing. Stimulation of the pneumotaxic area can also alter the rate of respiration.

The term applied to normal quiet breathing is **eupnea** (yoop-NĒ-a; *eu* = normal). Eupnea involves shallow, deep, or combined shallow and deep breathing. Shallow, or chest, breathing is called **costal breathing.** It consists of an upward and outward movement of the chest as a result of contraction of the intercostal muscles. Deep, or abdominal, breathing is called **diaphragmatic breathing.** It consists of the outward movement of the abdomen as a result of the contraction and descent of the diaphragm.

The respiratory center has connections with the cerebral cortex, which means we can voluntarily alter our pattern of breathing. We can even refuse to breathe at all for a short time. Voluntary control is protective because it enables us to prevent water or irritating gases from entering the lungs. The ability to stop breathing is limited by the buildup of CO_2 in the blood, however. When the CO_2 increases to a certain level, the inspiratory center is stimulated, impulses are sent to inspiratory muscles, and breathing resumes whether or not the person wishes. It is impossible for people to kill themselves by holding their breath.

MODIFIED RESPIRATORY MOVEMENTS

Respirations provide human beings with methods for expressing emotions such as laughing, yawning, sighing, and sobbing. Moreover, respiratory air can be used to expel foreign matter from the upper air passages through actions such as sneezing and coughing. Some of the modified respiratory movements that express emotion or clear the air passageways are listed in Exhibit 20-1. All these movements are reflexes, but some of them also can be initiated voluntarily.

EXHIBIT 20-1 MODIFIED RESPIRATORY MOVE-MENTS

MOVEMENT	COMMENT
Coughing	A long-drawn and deep inspiration followed by a complete closure of the glottis, which results in a strong expiration that suddenly pushes the glottis open and sends a blast of air through the upper respiratory passages. Stimulus for this reflex act may be a foreign body lodged in the larynx, trachea, or epiglottis.
Sneezing	Spasmodic contraction of muscles of expiration which forcefully expels air through the nose and mouth. Stimulus may be an irritation of the nasal mucosa.
Sighing	A deep and long-drawn inspiration immediately followed by a shorter but forceful expiration.
Yawning	A deep inspiration through the widely opened mouth producing an exaggerated depression of the lower jaw. It may be stimulated by drowsiness or fatigue, but precise stimulus-receptor cause is unknown.
Sobbing	A series of convulsive inspirations followed by a single prolonged expiration. The glottis closes earlier than normal after each inspiration so only a little air enters the lungs with each inspiration.
Crying	An inspiration followed by many short convulsive expirations, during which the glottis remains open and the vocal cords vibrate; accompanied by characteristic facial expressions.
Laughing	The same basic movements as crying, but the rhythm of the movements and the facial expressions usually differ from those of crying. Laughing and crying are sometimes indistinguishable.
Hiccuping	Spasmodic contraction of the diaphragm followed by a spasmodic closure of the glottis to produce a sharp inspiratory sound. Stimulus is usually irritation of the sensory nerve endings of the digestive tract.

INTERVENTION IN RESPIRATORY CRISES

CARDIOPULMONARY RESUSCITATION (CPR)

A serious decrease in respiration or heart rate presents an urgent crisis because the body's cells cannot survive long if they are starved of oxygenated blood. In fact, if oxygen is withheld from the cells of the brain for 5 to 6 minutes, there is usually severe and permanent brain injury or death. **Cardiopulmonary resuscitation (CPR)** is the artificial reestablishment of normal or near normal respiration and circulation.

The A, B, C's of cardiopulmonary resuscitation are **Airway, Breathing,** and **Circulation.** The rescuer must establish an airway, provide artificial ventilation if breathing has stopped, and reestablish circulation if there is inadequate cardiac action. Cardiopulmonary resuscitation can be administered at the site of emergency and can be used for any heart or respiratory failure, whether the cause be drowning, strangulation, carbon monoxide or insecticide poisoning, overdose of a drug or anesthesia, electrocution, or myocardial infarction. The success of cardiopulmonary resuscitation is directly related to speed and efficiency. Delay may be fatal. The sequential steps must be continued uniformly and without interruption until the patient recovers or is pronounced dead.

Airway

The first step in cardiopulmonary resuscitation is the immediate opening of the airway. One method of doing this is the **head-tilt maneuver.** With the victim on his or her back, the rescuer places one hand under the victim's neck and the other hand on the forehead. The rescuer simultaneously lifts the neck with one hand and tilts the head backward by pressure on the forehead with the other (Figure 20-12a). The tilted position opens the upper air passageways to their maximum size.

Breathing

If the patient does not resume spontaneous breathing after the head has been tilted backward, artificial ventilation must be given mouth-to-mouth or mouth-to-nose. This is called **external air ventilation.** In the more usual mouth-to-mouth method, the rescuer maintains the head tilt on the victim and the victim's nostrils are pinched together with the thumb and index finger (Figure 20-12b). The rescuer then opens his or her mouth widely, takes a deep breath, makes a tight seal with his or her mouth around the patient's mouth, and blows in about twice the amount the patient normally breathes. The rescuer then removes his or her mouth and allows the patient to exhale passively. This cycle is repeated approximately 12 times per minute for adults.

Even though the air supplied to the victim is air exhaled by the rescuer, it can still provide sufficient oxygen. Atmospheric air contains about 21 percent O_2 and a trace of CO_2. Exhaled air still contains about 16 percent O_2 and 5 percent CO_2. This amount is more than adequate to maintain a victim's blood O_2 and CO_2 at normal levels if air is given at the prescribed rate and amount.

If the rescuer observes the following three signs, ventilation is adequate.

1. The chest rises and falls with every breath.
2. The lungs can be felt to resist as they expand.
3. Air can be heard escaping during exhalation.

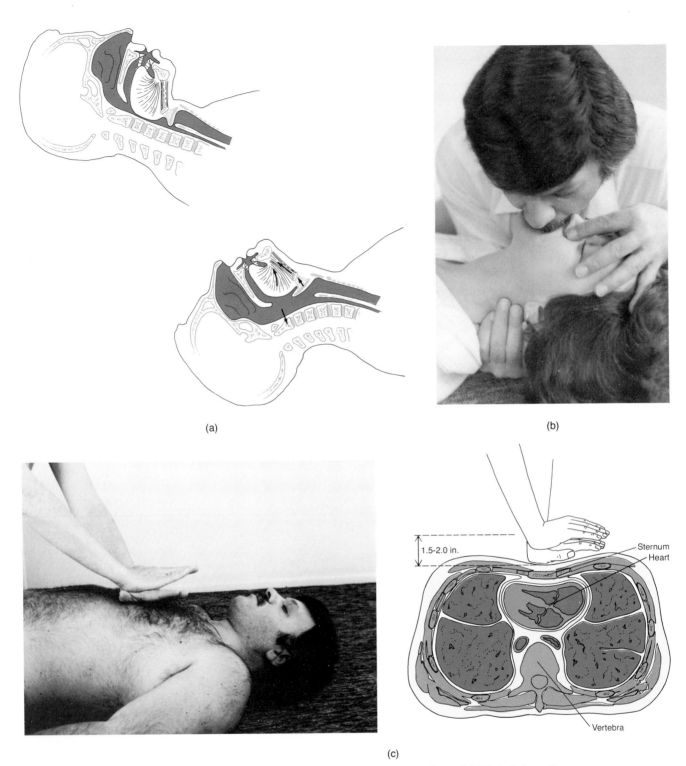

(a)

(b)

1.5-2.0 in.

Sternum

Heart

Vertebra

(c)

FIGURE 20-12 Cardiopulmonary resuscitation (CPR). (a) Head-tilt maneuver to open airway. (b) Exhaled air ventilation. Shown is the procedure for mouth-to-mouth resuscitation. (Courtesy of Matt Iacobino and Geraldine C. Tortora.) (c) External cardiac compression technique. On the left is shown the position of the hands. (Courtesy of Matt Iacobino.) On the right is shown the effect of external cardiac compression on the heart.

CLINICAL APPLICATION

The three most **common errors in external air ventilation** are *inadequate extension* of the victim's head, *inadequate opening* of the rescuer's mouth, and an *inadequate seal* around the patient's mouth or nose. If the rescuer has not made these errors and is still unable to inflate the lungs, a foreign object is probably lodged in the respiratory passages. The rescuer should sweep his or her fingers through the patient's mouth to remove such material. An adult victim should next be rolled quickly onto his or her side, and firm blows should be delivered over the spine between the shoulder blades to dislodge the obstruction. Then exhaled-air ventilation should be resumed quickly. A small child with an obstructive foreign object should be picked up quickly and inverted over the rescuer's forearm while firm blows are delivered over the spine between the shoulder blades. Then ventilation can be resumed.

If the patient is an infant or small child, the rescuer should make the following adjustments in technique. An infant's neck is so pliable that forceful backward tilting of the head may obstruct breathing passages, so the tilted position should not be exaggerated. The lungs of a small child do not have a large capacity. To avoid overinflating the child's lungs, the rescuer should cover both the victim's mouth and nose with his or her mouth and blow gently, using less volume, at a rate of 20 to 30 times a minute.

Circulation

One method of reestablishing circulation is called **external cardiac compression,** or **closed-chest cardiac compression (CCCC).** It consists of the application of rhythmic pressure over the sternum (Figure 20-12c). The rescuer places the heels of the hands on the lower half of the sternum and presses down firmly and smoothly at least 60 times a minute. This action compresses the heart and produces an artificial circulation because the heart lies almost in the middle of the chest between the lower portion of the sternum and the spine. When properly done, external cardiac compression can produce systolic blood pressure peaks of over 100 mm Hg. It can also bring carotid arterial blood flow up to 35 percent of normal.

Complications that can occur from the use of cardiac compression include fracture of the ribs and sternum, laceration of the liver, and the formation of fat emboli. They can be minimized by the following precautions.

1. Never compress over the xiphoid process at the tip of the sternum. This cartilaginous prominence extends down over the abdomen, and pressure on it may cause laceration of the liver, which can be fatal.

2. Never let your fingers touch the patient's ribs when you compress. Keep your fingers off the patient, and place the heel of your hand in the middle of the patient's chest over the lower half of the sternum.

3. Never compress the abdomen and chest simultaneously since this action traps and may rupture the liver.

4. Never use sudden or jerking movements to compress the chest. Compression should be smooth, regular, and uninterrupted, with 50 percent of the cycle compression and 50 percent relaxation.

Compression of the sternum produces some artificial ventilation but not enough for adequate oxygenation of the blood. Therefore exhaled air ventilation must always be used with it. When there are two rescuers, the best technique is to have one rescuer apply at least 60 cardiac compressions per minute. The other rescuer should exhale into the patient's mouth between every fifth and sixth compression.

HEIMLICH MANEUVER

Food choking (café coronary), the sixth leading cause of accidental death, is often mistaken for myocardial infarction. However, it can be recognized easily. The victim cannot speak or breathe and may become panic-stricken and run from the room. He or she becomes pale, then deeply cyanotic, and collapses. Without intervention, severe brain damage or death can occur in 5 to 6 minutes.

The food or other obstructing object causing asphyxiation may lodge in the back of the throat or enter the trachea to occlude the airway. Tracheostomy, even when a physician performs it, can be hazardous in a nonclinical setting. An instrument for removing food from the back of the throat has been developed but is seldom at hand in an emergency. However, there is a first aid procedure that does not require special instruments and can be performed by any informed layman. This procedure is called the **Heimlich maneuver.**

Food choking probably occurs most frequently during inspiration, which causes the bolus of food to be sucked against the opening into the larynx. At the time of the accident, the lungs are therefore expanded. Even during normal expiration, however, some tidal air (500 ml) and the entire expiratory reserve volume (1,200 ml) are in the lungs. The principle of the Heimlich maneuver is to expel the air forcibly in order to dislodge the obstruction.

Essentially, the Heimlich maneuver consists of pressing one's fist upward into the epigastric region of the abdomen to elevate the diaphragm suddenly. This thrust compresses the lungs in the rib cage and increases the air pressure in the bronchopulmonary tree. This forces air out through the trachea and will eject the food (or object) blocking the airway. The action can be simulated by inserting a cork in a compressible plastic bottle and then squeezing it suddenly—the cork flies out because of the increased pressure. Figure 20-13 shows how to administer the Heimlich maneuver.

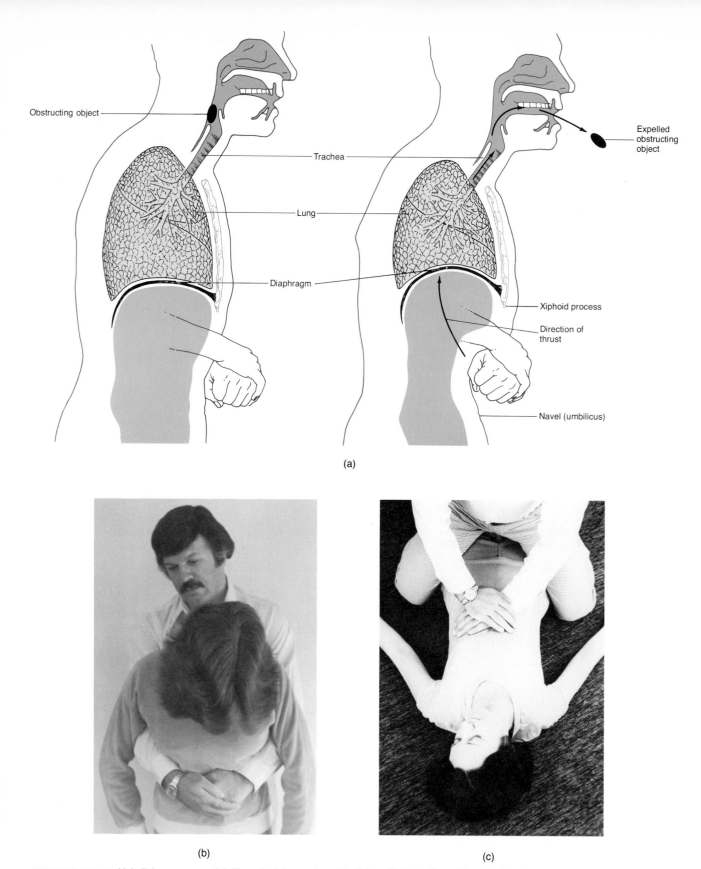

Obstructing object

Trachea

Lung

Diaphragm

Expelled obstructing object

Xiphoid process

Direction of thrust

Navel (umbilicus)

(a)

(b)   (c)

FIGURE 20-13 Heimlich maneuver. (a) The principle employed in this anti-choke first aid method is the application of a quick upward thrust that causes sudden elevation of the diaphragm and forceful, rapid expulsion of air in the lungs. Air forced through the trachea and larynx expels the obstructing object. (b) If the victim is upright, stand behind the victim and place both arms around the waist just above the belt line. Make a fist with one hand and place the thumb side of the fist against the victim's abdomen, slightly above the navel and below the rib cage. Next, grasp your fist with your other hand and press into the victim's abdomen with a quick upward thrust. This should be done by a sharp flexion at the elbows rather than a "bear hug" which compresses the rib cage. Repeat the thrust several times, if necessary. (Courtesy of Matt Iacobino and Lynne Tortora.) (c) If the victim is recumbent, kneel astride the victim's hips, placing the hands one on top of the other with the heel of the bottom hand on the victim's abdomen slightly above the navel and below the rib cage. Then apply the quick upward thrust. (Courtesy of Matt Iacobino and Geraldine C. Tortora.)

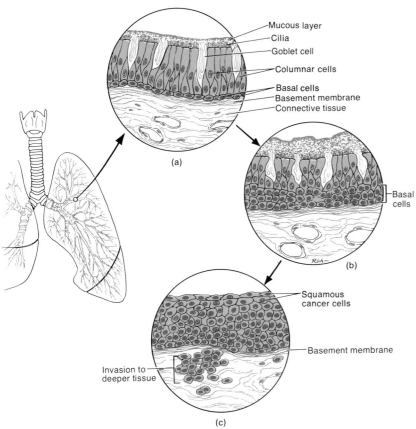

FIGURE 20-14 Effects of smoking on the respiratory epithelium. (a) Microscopic view of the normal epithelium of a bronchial tube. (b) Initial response of the bronchial epithelium to irritation by pollutants. (c) Advanced response of the bronchial epithelium.

SMOKING AND THE RESPIRATORY SYSTEM

As part of ordinary breathing, many irritating substances are inhaled. Almost all pollutants, including inhaled smoke, have an irritating effect on the bronchial tubes and lungs and may be regarded as stresses or irritating stimuli.

Microscopic examination of the epithelium of a bronchial tube reveals three kinds of cells (Figure 20-14). The surface cells are columnar cells that contain cilia. At intervals between the ciliated columnar cells are the mucus-secreting goblet cells. The bottom of the epithelium normally contains a row of basal cells above the basement membrane. The basal cells divide continuously, replacing the ciliated columnar epithelium as they wear down and are sloughed off. The bronchial epithelium is important clinically because a common lung cancer, **bronchogenic carcinoma,** starts in the walls of the bronchi.

The constant irritation by inhaled smoke and pollutants causes an enlargement of the goblet cells of the bronchial epithelium. They respond by secreting excessive mucus. The basal cells also respond to the stress by undergoing cell division so fast that the basal cells push into the area occupied by the goblet and columnar cells. As

many as 20 rows of basal cells may be produced. Many researchers believe that if the stress is removed at this point, the epithelium can return to normal.

If the stress persists, more and more mucus is secreted and the cilia become less effective. As a result, mucus is not carried toward the throat but remains trapped in the bronchial tubes. The individual then develops a "smoker's cough." Moreover, the constant irritation from the pollutant slowly destroys the alveoli, which are replaced with thick, inelastic connective tissue. Mucus that has accumulated becomes trapped in the air sacs. Millions of sacs rupture—reducing the diffusion surface for the exchange of oxygen and carbon dioxide. The individual has now developed emphysema. If the stress is removed at this point, there is little chance for improvement. Alveolar tissue that has been destroyed cannot be repaired. But removal of the stress can stop further destruction of lung tissue.

If the stress continues, the emphysema gets progressively worse, and the basal cells of the bronchial tubes continue to divide and break through the basement membrane. At this point the stage is set for bronchogenic carcinoma. Columnar and goblet cells disappear and may be replaced with squamous cancer cells. If this happens, the malignant growth spreads throughout the lung and may block a bronchial tube. If the obstruction occurs in a large bronchial tube, very little oxygen enters the lung, and disease-producing bacteria thrive on the mucoid secretions. In the end, the patient may develop emphysema, carcinoma, and a host of infectious diseases. Treatment involves surgical removal of the diseased lung. However, metastasis (spreading) of the growth through the lymphatic or blood system may result in new growths in other parts of the body such as the brain and liver.

Other factors may be associated with lung cancer. For instance, breast, stomach, and prostate malignancies can metastasize to the lungs. People who apparently have not been exposed to pollutants do occasionally develop bronchogenic carcinoma. However, the occurrence of bronchogenic carcinoma is probably over 20 times higher in heavy cigarette smokers than it is in nonsmokers.

APPLICATIONS TO HEALTH

NASAL POLYPS

Nasal polyps are protruding growths of the mucous membrane that usually hang down from the posterior wall of the nasal septum. The polyps appear as bluish-white tumors and may fill the nasopharynx as they become larger. Nasal polyps usually undergo atrophy if untreated, but they are easily removed by a physician with a nasal snare and cautery.

HAY FEVER

An allergic reaction to the proteins contained in foreign substances such as plant pollens, dust, and certain foods is called **hay fever**. Allergic reactions are a special antigen-antibody response that initiate either a localized or a systemic inflammatory response. In hay fever the response is localized in the respiratory membranes. The membranes become inflamed, and a watery fluid drains from the eyes and nose.

BRONCHIAL ASTHMA

Bronchial asthma is a reaction, usually allergic, characterized by attacks of wheezing and difficult breathing. Attacks are brought on by spasms of the smooth muscles that lie in the walls of the smaller bronchi and bronchioles, causing the passageways to close partially. The patient has trouble exhaling, and the alveoli may remain inflated during expiration. Usually the mucous membranes that line the respiratory passageways become irritated and secrete excessive amounts of mucus that may clog the bronchi and bronchioles and worsen the attack. About three out of four asthma victims are allergic to edible or airborne substances as common as wheat or dust. Others are sensitive to the proteins of harmless bacteria that inhabit the sinuses, nose, and throat. Asthma might also have a psychosomatic origin.

EMPHYSEMA

In **emphysema** (em'-fi-SĒ-ma), the alveolar walls lose their elasticity and remain filled with air during expiration. The name means "blown up" or "full of air." Reduced forced expiratory volume is the first symptom. Later, alveoli in other areas of the lungs are damaged. Many alveoli may merge to form larger air sacs with a reduced overall volume. The lungs become permanently inflated because they have lost elasticity. To adjust to the increased lung size, the size of the chest cage increases. The patient has to work voluntarily to exhale. Oxygen diffusion does not occur as easily across the damaged alveolar-capillary membrane, blood O_2 is somewhat lowered, and any mild exercise that raises the oxygen requirements of the cells leaves the patient breathless. As the disease progresses, the alveoli are replaced with thick fibrous connective tissue. Even carbon dioxide does not diffuse easily through this fibrous tissue. If the blood cannot buffer all the hydrogen ions that accumulate, the blood pH drops or unusually high amounts of carbon dioxide may dissolve in the plasma. High carbon dioxide levels produce acid conditions that are toxic to brain cells. Consequently, the inspiratory center becomes less active and the respiration rate slows down, further aggravating the problem. The compressed and damaged capillaries around the deteriorating alveoli may no longer be able to receive blood. As a result, resistance to blood flow increases in the pulmonary trunk and the right ventricle overworks as it attempts to force blood through the remaining capillaries.

Emphysema is generally caused by a long-term irritation. Air pollution, occupational exposure to industrial dust, and cigarette smoke are the most common irritants. Chronic bronchial asthma also may produce alveolar damage. Cases of emphysema are becoming more and more frequent in the United States. The irony is that the disease can be prevented and the progressive deterioration can be stopped by eliminating the harmful stimuli.

PNEUMONIA

The term **pneumonia** means an acute infection or inflammation of the alveoli. In this disease, the alveolar sacs fill up with fluid and dead white blood cells, reducing the amount of air space in the lungs. (Remember that one of the cardinal signs of inflammation is edema.) Oxygen has difficulty diffusing through the inflamed alveoli, and the blood O_2 may be drastically reduced. Blood CO_2 usually remains normal because carbon dioxide diffuses through the alveoli more easily than oxygen does. If all the alveoli of a lobe are inflamed, the pneumonia is called *lobar pneumonia*. If only parts of the lobe are involved, it is called *lobular*, or *segmental, pneumonia*. If both the alveoli and the bronchial tubes are included, it is called *bronchopneumonia*.

The most common cause of pneumonia is the pneumococcus bacterium, but other bacteria or a fungus may be the source of trouble. Viral pneumonia is caused by several viruses, including the influenza virus.

TUBERCULOSIS

The bacterium called *Mycobacterium tuberculosis* produces an inflammation called **tuberculosis.** This disease still ranks as the number-one killer in the communicable disease category. Tuberculosis most often affects the lungs and the pleurae. The bacteria destroy parts of the lung tissue, and the tissue is replaced by fibrous connective tissue. Because the connective tissue is inelastic and thick, the affected areas of the lungs do not snap back during expiration and larger amounts of air are retained. Gases no longer diffuse easily through the fibrous tissue.

Tuberculosis bacteria are spread by inhalation. Although they can withstand exposure to many disinfectants, they die quickly in sunlight. Thus, tuberculosis is sometimes associated with crowded, poorly lit housing. Many drugs are successful in treating tuberculosis. Rest, sunlight, and good diet are vital parts of treatment.

HYALINE MEMBRANE DISEASE (HMD)

Sometimes called glassy-lung disease or infant respiratory distress syndrome (RDS), **hyaline membrane disease (HMD)** is responsible for approximately 20,000 newborn infant deaths per year. Before birth, the respiratory passages are filled with fluid. Part of this fluid is amniotic fluid inhaled during respiratory movements in utero. The remainder is produced by the submucosal glands and the goblet cells of the respiratory epithelium.

At birth, this fluid-filled airway must become an air-filled airway, and the collapsed primitive alveoli (terminal sacs) must expand and function in gas exchange. The success of this transition depends largely on surfactant, the phospholipid produced by alveolar cells that lowers surface tension in the fluid layer lining the primitive alveoli once air enters the lungs. Surfactant is present in the fetus's lungs as early as the 23rd week, and by 28 to 32 weeks the amount present is sufficient to prevent alveolar collapse during breathing. Surfactant is produced continuously by septal alveolar cells. The presence of surfactant can be detected by amniocentesis.

Although in a normal, full-term infant the second and subsequent breaths require less respiratory effort than the first, breathing is not completely normal until about 40 minutes after birth. The entire lung is not inflated fully with the first one or two breaths. In fact, for the first 7 to 10 days, small areas of the lungs may remain uninflated.

In the newborn whose lungs are deficient in surfactant, the effort required for the first breath is essentially the same as that required in normal newborns. However, the surface tension of the alveolar fluid is 7 to 14 times higher than the surface tension of alveolar fluid with a monomolecular layer of surfactant. Consequently, during expiration after the first inspiration, the surface tension of the alveoli increases as the alveoli deflate. The alveoli collapse almost to their original uninflated state.

Idiopathic HMD usually appears within a few hours after birth. (An idiopathic condition is one that occurs spontaneously in an individual from an unknown or obscure cause.) Affected infants show difficult and labored breathing with withdrawal of the intercostal and subcostal spaces. Death may occur soon after onset of respiratory difficulty or may be delayed for a few days, although many infants survive. At autopsy, the lungs are underinflated and areas of atelectasis (collapse of lung) are prominent. If the infant survives for at least a few hours after developing respiratory distress, the alveoli are often filled with a fluid of high-protein content that resembles a hyaline (or glassy) membrane. HMD occurs frequently in premature infants and also in infants of diabetic mothers, particularly if the diabetes is untreated or poorly controlled.

A new treatment currently being developed, called PEEP (positive end expiratory pressure) could reverse the mortality rate from 90 percent deaths to 90 percent survival. This treatment consist of passing a tube through the air passage to the top of the lungs to provide needed oxygen-rich air at continuous pressures of up to 14 mm Hg. Continuous pressure keeps the baby's alveoli open and available for gas exchange.

SUDDEN INFANT DEATH SYNDROME (SIDS)

Sudden infant death syndrome (SIDS), also called crib death or cot death, kills approximately 10,000 American babies every year. It kills more infants between the ages of 1 week and 12 months than any other disease. Its sudden nature, not adequately explained by classical autopsy methods, arouses feelings of guilt and suspicion that can be quieted only by "reasons" such as pneumonia, aspiration, or suffocation.

SIDS occurs without warning. Although about half of its victims had an upper respiratory infection within two weeks of death, the babies tended to be remarkably healthy. Two other findings are important: the apparently *silent* nature of death and the disarray frequently found at the scene.

Given this death scene, it is not surprising that for centuries "crib death" was mistakenly thought to be due to suffocation. Most medical people, however, believe that a healthy child *cannot* smother in its bedclothes. The diapers of SIDS victims are usually full of stool and urine, and the urinary bladder is almost always empty. External examination reveals only a blood-tinged froth that often exudes from the nostrils. These signs all point to a sudden lethal episode associated with tonic motor activity, not to a gradually deepening coma leading to death.

The similarity of findings in SIDS cases strengthens the theory that there is one underlying cause. The seasonal distribution of SIDS incidences being lowest during the summer months and highest in the late fall strongly suggests an infectious agent, viruses being the most likely possibility. Most cases occur at a time in life when gamma globulin levels are low and the infant is at a critical period of susceptibility.

At least thirteen different hypotheses have been proposed as explanations for SIDS. Although the precise pathogenic mechanisms have yet to be clarified, several valuable observations have emerged as a result of investigations during the past decade. Most, if not all, such cases appear to share a common mechanism ("final common pathway"). Thus, SIDS is probably a single disease entity. According to a recent study, the most likely cause of death is laryngospasm, possibly triggered by a previous, mild, nonspecific virus infection of the upper respiratory tract. Another possible factor includes periods of apnea as a result of malfunction of the respiratory centers.

CARBON MONOXIDE POISONING

Carbon monoxide (CO) is a colorless and odorless gas found in exhaust fumes from automobiles and in tobacco smoke. It is a by-product of burning carbon-containing materials such as coal and wood. One of its interesting features is that it combines with hemoglobin very much as oxygen does, except that the combination of carbon monoxide and hemoglobin is over 210 times as tenacious as the combination of oxygen and hemoglobin. It combines easily and the combination does not break down readily. In concentrations as small as 0.1 percent, carbon monoxide will combine with half the hemoglobin molecules. Thus, the oxygen-carrying capacity of the blood is reduced by half. Increased levels of carbon monoxide lead to hypoxia, and the result is **carbon monoxide poisoning.**

The condition may be treated by administering pure oxygen, which slowly replaces the carbon monoxide combined with the hemoglobin.

KEY MEDICAL TERMS ASSOCIATED WITH THE RESPIRATORY SYSTEM

Apnea (*a* = without; *pnoia* = breath)   Absence of ventilatory movements.

Asphyxia (*sphyxis* = pulse)   Oxygen starvation due to low atmospheric oxygen or interference with ventilation, external respiration, or internal respiration.

Aspiration (*spirare* = breathe)   The act of breathing or drawing in.

Atelectasis (*ateles* = incomplete; *ektasis* = dilation)   A collapsed lung or portion of a lung.

Bronchiectasis   A chronic dilation of the bronchi or bronchioles.

Bronchitis (*bronch* = bronchus, trachea; *itis* = inflammation of)   Inflammation of the bronchi and bronchioles.

Cheyne-Stokes respiration   Irregular breathing beginning with shallow breaths that increase in depth and rapidity, then decrease and cease altogether for 15 to 20 seconds. The cycle repeats itself again and again. Cheyne-Stokes is normal in infants. It is also often seen just before death from pulmonary, cerebral, cardiac, and kidney disease and is referred to as the "death rattle."

Diphtheria (*diphthera* = membrane)   An acute bacterial infection that causes the mucous membranes of the oropharynx, nasopharynx, and larynx to enlarge and become leathery. Enlarged membranes may obstruct airways and cause death from asphyxiation.

Dyspnea (*dys* = painful, difficult)   Labored breathing (short-windedness).

Hypoxia (*hypo* = below, under)   Reduction in oxygen supply to cells.

Influenza   Viral infection that causes inflammation of respiratory mucous membranes as well as fever.

Orthopnea (*ortho* = straight)   Inability to breathe in a horizontal position.

Pneumonectomy (*pneumo* = lung; *tome* = cutting)   Surgical removal of a lung.

Pneumothorax   Air in pleural space causing collapse of the lung. Most common cause is surgical opening of chest during heart surgery, making intrathoracic pressure equal atmospheric pressure.

Pulmonary (*pulmo* = lung) **edema**   Excess amounts of interstitial fluid in the lungs producing cough and dyspnea. Common in failure of the left side of the heart.

Pulmonary embolism   Presence of a blood clot or other foreign substance in a pulmonary arterial vessel stopping circulation to a part of the lungs.

Rales   Sounds sometimes heard in the lungs that resemble bubbling or rattling. May be caused by air or an abnormal secretion in the lungs.

Respirator   A metal chamber that entombs the chest; also called an "iron lung." Used to produce inspiration and expiration in patient with paralyzed respiratory muscles. Pressure inside the chamber is rhythmically alternated to suck out and push in chest walls.

Rhinitis (*rhino* = nose)   Chronic or acute inflammation of the mucous membrane of the nose.

STUDY OUTLINE

Organs

1. Respiratory organs include the nose, pharynx, larynx, trachea, bronchi, and lungs.
2. They act with the cardiovascular system to supply oxygen and remove carbon dioxide from the blood.

Nose

1. The external portion is made of cartilage and skin and lined with mucous membrane. Openings to the exterior are the external nares.
2. The internal portion communicates with the nasopharynx (through the internal nares) and the paranasal sinuses.
3. The nose is divided into cavities by a septum. Anterior portions of the cavities are called the vestibules.
4. The nose is adapted for warming, moistening, and filtering air, for olfaction, and for speech.

Pharynx

1. The pharynx, or throat, is a muscular tube lined by a mucous membrane.
2. The anatomic regions are nasopharynx, oropharynx, and laryngopharynx.
3. The nasopharynx functions in respiration. The oropharynx and laryngopharynx function both in digestion and in respiration.

Larynx

1. The larynx is a passageway that connects the pharynx with the trachea.
2. Prominent cartilages are the thyroid, or Adam's apple, the epiglottis, which prevents food from entering the larynx, the cricoid, which connects the larynx and trachea, the arytenoid, the corniculate, and the cuneiform cartilages.
3. The larynx contains true vocal cords, which produce sound. Taut cords produce high pitches, and relaxed cords produce low pitches.

Trachea

1. The trachea extends from the larynx to the primary bronchi.
2. It is composed of smooth muscle and C-shaped rings of cartilage and is lined with pseudostratified epithelium.

Bronchi

1. The bronchial tree consists of the trachea, primary bronchi, secondary bronchi, tertiary bronchi, bronchioles, and terminal bronchioles. Walls of bronchi contain rings of cartilage; walls of bronchioles do not.
2. A developed picture of the tree is called a bronchogram.

Lungs

1. Lungs are paired organs in the thoracic cavity. They are enclosed by the pleural membrane. The parietal pleura is the outer layer; the visceral pleura is the inner layer.
2. The right lung has three lobes separated by two fissures; the left lung has two lobes separated by one fissure and a depression, the cardiac notch.
3. The secondary bronchi give rise to branches called segmental bronchi which supply segments of lung tissue called bronchopulmonary segments.
4. Each bronchopulmonary segment consists of lobules, which contain lymphatics, arterioles, venules, terminal bronchioles, respiratory bronchioles, alveolar ducts, alveolar sacs, and alveoli.
5. Gas exchange occurs across the alveolar-capillary membranes.

Air Volumes Exchanged

1. Among the air volumes exchanged in ventilation are tidal volume, inspiratory reserve, expiratory reserve, residual volume, and minimal volumes.
2. Most volumes are measured with a respirometer.

Nervous Control of Respiration

1. Nervous control is regulated by the respiratory center.
2. The center is divided into the medullary rhythmicity area in the medulla, the apneustic area in the pons, and the pneumotaxic area in the pons.

Modified Respiratory Movements

1. Modified respiratory movements may be used to express emotions or expel foreign matter from the respiratory system.
2. Coughing, sneezing, sighing, yawning, sobbing, crying, laughing, and hiccuping are types of modified respiratory movements.

Intervention in Respiratory Crises

1. Cardiopulmonary resuscitation (CPR) is the artificial reestablishment of respiration and circulation. The A, B, C's of CPR are Airway, Breathing, and Circulation.
2. The Heimlich maneuver is a first aid procedure used in case of food choking. It consists of an abdominal thrust that elevates the diaphragm, compresses the lungs, and increases air pressure in the bronchial tree.

Smoking

1. Pollutants, including cigarette smoke, act as stresses on the epithelium of the bronchi and lungs. Constant irritation results in excessive secretion of mucus and rapid division of bronchial basal cells.
2. Additional irritation may cause retention of mucus in bronchioles, loss of elasticity of alveoli, and less surface area for gaseous exchange.
3. In the final stages, bronchial epithelial cells may be

replaced by cancer cells. The growth may block a bronchial tube and spread throughout the lung and other body tissues.

Applications to Health

1. Nasal polyps are growths of mucous membrane in the nasal cavity.
2. Hay fever is an allergic reaction of respiratory membranes.
3. Bronchial asthma occurs when spasms of smooth muscle in bronchial tubes result in partial closure of air passageways, inflammation, inflated alveoli, and excess mucus production.
4. Emphysema is characterized by deterioration of alveoli leading to loss of their elasticity. Symptoms are reduced expiratory volume, inflated lungs, and enlarged chest.
5. Pneumonia is an acute inflammation or infection of alveoli.
6. Tuberculosis is an inflammation of pleura and lungs produced by a specific bacterium.
7. Hyaline membrane disease (HMD) is an infant disorder in which surfactant is lacking and alveolar ducts and alveoli have a glassy appearance.
8. Sudden infant death syndrome (SIDS) has recently been linked to laryngospasm, possibly triggered by a viral infection of the upper respiratory tract.
9. Carbon monoxide poisoning occurs when CO combines with hemoglobin instead of oxygen. The result is hypoxia.

REVIEW QUESTIONS

1. What organs make up the respiratory system? What function do the respiratory and cardiovascular systems have in common?
2. Describe the structures of the external and internal nose and describe their functions in filtering, warming, and moistening air.
3. What is the pharynx? Differentiate the three regions of the pharynx and indicate their roles in respiration.
4. Describe the structures of the larynx and explain how they function in respiration and voice production.
5. Describe the location and structure of the trachea. What is tracheostomy? Intubation?
6. What is the bronchial tree? Describe its structure. What is a bronchogram?
7. Where are the lungs located? Distinguish the parietal pleura from the visceral pleura. What is pleurisy?
8. Define each of the following parts of a lung: base, apex, costal surface, medial surface, hilus, root, cardiac notch, and lobe.
9. What is a lobule of the lung? Describe its composition and function in respiration.
10. What is a bronchopulmonary segment?
11. Describe the histology of the alveolar-capillary membrane.
12. What is a respirometer? Define the various lung volumes and capacities.
13. How does the medullary rhythmicity area function in controlling respiration? How are the apneustic area and pneumotaxic area related to the control of respiration?
14. Define the various kinds of modified respiratory movements.
15. What is the object of cardiopulmonary resuscitation (CPR)?
16. What cautions must be taken in exhaled air ventilation and external cardiac compression? Why?
17. Describe the steps involved in the Heimlich maneuver.
18. Discuss the stages involved in the destruction of the respiratory epithelium as a result of continued irritation by pollutants.
19. For each of the following disorders, list the principal clinical symptoms: nasal polyps, hay fever, bronchial asthma, emphysema, pneumonia, tuberculosis, hyaline membrane disease, sudden infant death syndrome, and carbon monoxide poisoning.
20. Refer to the key medical terms associated with the respiratory system. Be sure that you can define each term.

21

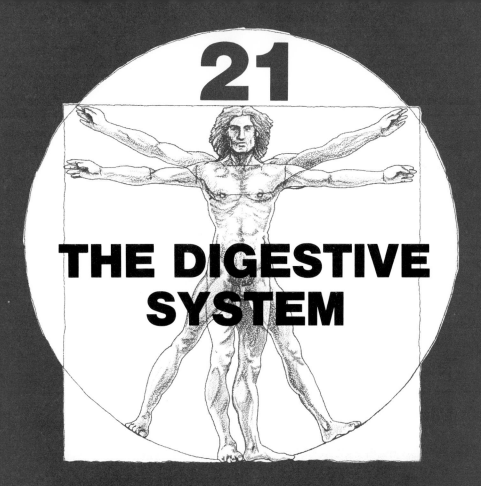

THE DIGESTIVE SYSTEM

STUDENT OBJECTIVES

- Define digestion as a chemical and mechanical process.

- Identify the organs of the gastrointestinal tract and the accessory organs of digestion.

- Describe the structure of the wall of the gastrointestinal tract.

- Define the mesentery, mesocolon, falciform ligament, lesser omentum, and greater omentum.

- Describe the structure of the mouth and its role in mechanical digestion.

- Identify the location and histology of the salivary glands and define the composition and function of saliva.

- Identify the parts of a typical tooth and compare deciduous and permanent dentitions.

- Describe the location, anatomy, and histology of the stomach and compare mechanical and chemical digestion.

- Describe the location, structure, and histology of the pancreas.

- Define the position, structure, and histology of the liver.

- Describe the structure and histology of the gallbladder.

- Describe those structural features of the small intestine that adapt it for digestion and absorption.

- Describe the mechanical movements of the small intestine and define absorption.

- Describe those structural features of the large intestine that adapt it for absorption and feces formation and elimination and describe the mechanical movements of the large intestine.

- Describe the processes involved in feces formation and discuss the mechanisms involved in defecation.

- List the causes and symptoms of dental caries, periodontal disease, and peptic ulcers.

- Explain cirrhosis and gallstones as disorders of accessory organs of digestion.

- Explain the causes and symptoms of appendicitis, diverticulitis, and peritonitis.

- Compare the location of tumors of the gastrointestinal tract.

- Describe the symptoms of anorexia nervosa.

- Define key medical terms associated with the digestive system.

We all know that food is vital to life. Food is required for the chemical reactions that occur in every cell—both those that synthesize new enzymes, cell structures, bone, and all the other components of the body and those that release the energy needed for the building processes. However, the vast majority of foods we eat are simply too large to pass through the plasma membranes of the cells. The breaking down of food molecules for use by body cells is called **digestion** and the organs that collectively perform this function comprise the **digestive system.**

The developmental anatomy of the digestive system is considered in Exhibit 24-10.

DIGESTIVE PROCESSES

The digestive system prepares food for consumption by the cells through five basic activities.

1. Ingestion, or eating. Taking food into the body.

2. Peristalsis. The movement of food along the digestive tract.

3. Mechanical and chemical **digestion.**

4. Absorption. The passage of digested food from the digestive tract into the circulatory and lymphatic systems for distribution to cells.

5. Defecation. The elimination of indigestible substances from the body.

Chemical digestion is a series of catabolic reactions that break down the large carbohydrate, lipid, and protein molecules that we eat into molecules usable by body cells. These products of digestion are small enough to pass through the walls of the digestive organs, into the blood and lymph capillaries, and eventually into the body's cells. **Mechanical digestion** consists of various movements that aid chemical digestion. Food is prepared by the teeth before it can be swallowed. Then the smooth muscles of the stomach and small intestine churn the food so it is thoroughly mixed with the enzymes that catalyze the reactions.

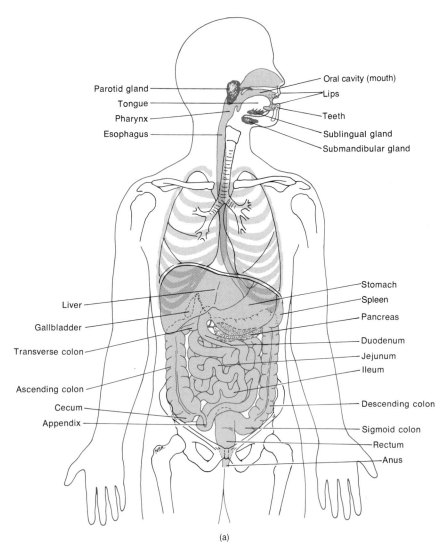

Parotid gland
Tongue
Pharynx
Esophagus

Oral cavity (mouth)
Lips
Teeth
Sublingual gland
Submandibular gland

Liver
Gallbladder
Transverse colon
Ascending colon
Cecum
Appendix

Stomach
Spleen
Pancreas
Duodenum
Jejunum
Ileum
Descending colon
Sigmoid colon
Rectum
Anus

(a)

FIGURE 21-1 Organs of the digestive system and related structures. (a) Diagram.

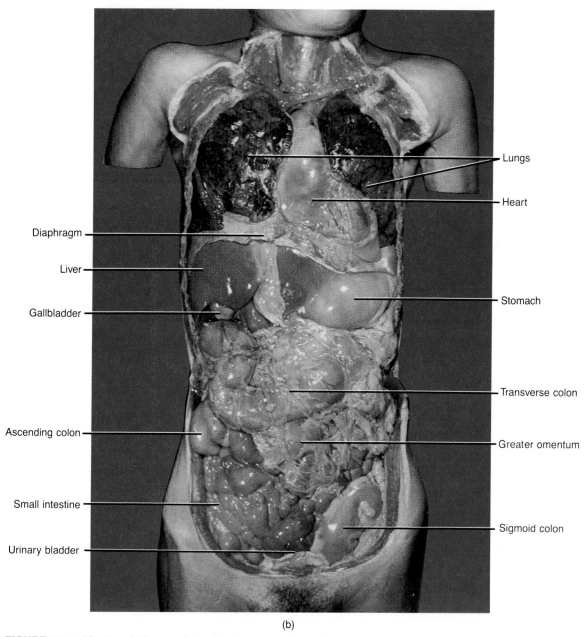

Lungs

Heart

Diaphragm

Liver

Gallbladder

Stomach

Transverse colon

Ascending colon

Greater omentum

Small intestine

Sigmoid colon

Urinary bladder

(b)

FIGURE 21-1 (*Continued*) Organs of the digestive system and related structures. (b) Photograph of thoracic and abdominal viscera. (Courtesy of C. Yokochi and J. W. Rohen, *Photographic Anatomy of the Human Body,* 2nd ed., 1979, IGAKU-SHOIN, Ltd., Tokyo, New York.)

GENERAL ORGANIZATION

The organs of digestion are traditionally divided into two main groups. First is the **gastrointestinal (GI) tract** or **alimentary canal,** a continuous tube running through the ventral body cavity and extending from the mouth to the anus (Figure 21-1). The relationship of the digestive organs to the nine regions of the abdominopelvic cavity may be reviewed in Figure 1-5b. The length of a tract taken from a cadaver is about 9 m (30 ft). In a living person it is somewhat shorter because the muscles in its walls are in a state of tone. Organs composing the gastrointestinal tract include the mouth, pharynx, esophagus, stomach, small intestine, and large intestine. The GI tract contains the food from the time it is eaten until it is digested and prepared for elimination. Muscular contractions in the walls of the GI tract break down the food physically by churning it. Secretions produced by cells along the tract break down the food chemically.

The secondary group of organs composing the digestive system consists of the **accessory organs**—the teeth, tongue, salivary glands, liver, gallbladder, and pancreas. Teeth are cemented to bone, protrude into the GI tract, and aid in the physical breakdown of food. The other accessory organs except for the tongue and gastric and intestinal glands lie totally outside the tract and produce or store secretions that aid in the chemical breakdown of food. These secretions are released into the tract through ducts.

GENERAL HISTOLOGY

The walls of the GI tract, especially from the esophagus to the anal canal, have the same basic arrangement of tissues. The four tunics or coats of the tract from the inside out are the mucosa, submucosa, muscularis, and serosa or adventitia (Figure 21-2).

The **tunica mucosa,** or inner lining of the tract, is a mucous membrane attached to a thin layer of visceral muscle. Two layers compose the membrane: a *lining epithelium,* which is in direct contact with the contents of the GI tract, and an underlying layer of loose connective tissue called the *lamina propria.* Under the lamina propria is visceral muscle called the *muscularis mucosa.*

The epithelial layer is composed of nonkeratinized cells that are stratified in the mouth and esophagus, but are simple throughout the rest of the tract. The functions of the stratified epithelium are protection and secretion. The functions of the simple epithelium are secretion and absorption. However, the lack of keratin allows some absorption to occur in all parts of the tract.

The lamina propria is made of loose connective tissue containing many blood and lymph vessels and scattered lymph nodules. This layer supports the epithelium, binds it to the muscularis mucosa, and provides it with a blood

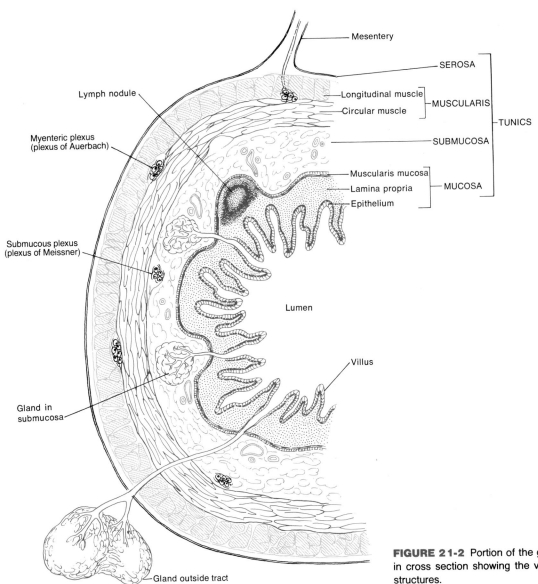

Mesentery

SEROSA

Lymph nodule

Longitudinal muscle

MUSCULARIS

Circular muscle

TUNICS

Myenteric plexus
(plexus of Auerbach)

SUBMUCOSA

Muscularis mucosa

Lamina propria — MUCOSA

Epithelium

Submucous plexus
(plexus of Meissner)

Lumen

Villus

Gland in
submucosa

Gland outside tract

FIGURE 21-2 Portion of the gastrointestinal tract seen in cross section showing the various tunics and related structures.

and lymph supply. The blood and lymph vessels are the avenues by which nutrients in the tract reach the other tissues of the body. The lymph tissue also protects against disease. Remember that the GI tract is in contact with the outside environment and contains food which often carries harmful bacteria. Unlike the skin, the mucous membrane of the tract is not protected from bacterial entry by keratin. The lamina propria also contains glandular epithelium that secretes products necessary for chemical digestion.

The muscularis mucosa contains visceral muscle fibers that throw the mucous membrane of the intestine into small folds which increase the digestive and absorptive area. With one exception, which will be described later, the other three coats of the intestine contain no glandular epithelium.

The **tunica submucosa** consists of loose connective tissue binding the tunica mucosa to the tunica muscularis. It is highly vascular and contains a portion of the *submucous plexus* or *plexus of Meissner* (MIS-ner), which is part of the autonomic nerve supply to the muscularis mucosa.

The **tunica muscularis** of the mouth, pharynx, and esophagus consists in part of skeletal muscle that produces voluntary swallowing. Throughout the rest of the tract, the muscularis consists of smooth muscle that is generally found in two sheets: an inner ring of circular fibers and an outer sheet of longitudinal fibers. Contractions of the smooth muscles help to break down food physically, mix it with digestive secretions, and propel it through the tract. The muscularis also contains the major nerve supply to the alimentary tract—the *myenteric plexus* or *plexus of Auerbach* (OW-er-bak), which consists of fibers from both autonomic divisions.

The **tunica serosa,** the outermost layer of the canal, is a serous membrane composed of connective tissue and epithelium. This covering, also called the **visceral peritoneum,** is worth discussing in detail.

PERITONEUM

The **peritoneum** (per'-i-tō-NĒ-um) is the largest serous membrane of the body. Serous membranes are also associated with the heart (pericardium) and lungs (pleurae). Serous membranes consist of a layer of simple squamous epithelium (called mesothelium) and an underlying supporting layer of connective tissue. The **parietal peritoneum** lines the walls of the abdominal cavity. The **visceral peritoneum** covers some of the organs and constitutes their serosa. The potential space between the parietal and visceral portions of the peritoneum is called the **peritoneal cavity** and contains serous fluid.

Unlike the pericardium and pleurae, the peritoneum contains large folds that weave in between the viscera. The folds bind the organs to each other and to the walls of the cavity and contain the blood and lymph vessels and the nerves that supply the abdominal organs. One extension of the peritoneum is called the **mesentery** (MEZ-en-ter'ē). It is an outward fold of the serous coat of the small intestine (Figure 21-3). The tip of the fold

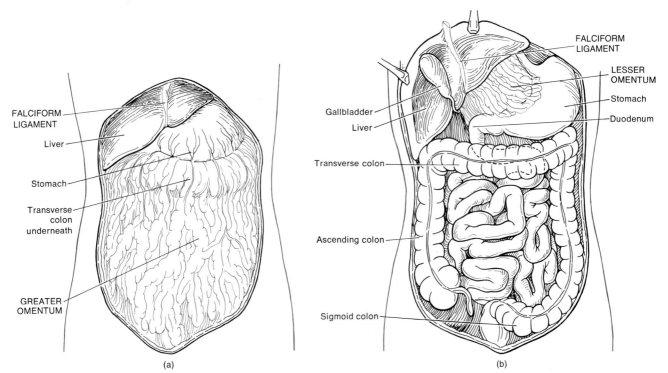

FALCIFORM LIGAMENT
Liver
Stomach
Transverse colon underneath
GREATER OMENTUM

FALCIFORM LIGAMENT
LESSER OMENTUM
Stomach
Duodenum
Gallbladder
Liver
Transverse colon
Ascending colon
Sigmoid colon

(a)

(b)

FIGURE 21-3 Extensions of the peritoneum. (a) Greater omentum. (See also Figure 21-1b.) (b) Lesser omentum. The liver and gallbladder have been lifted.

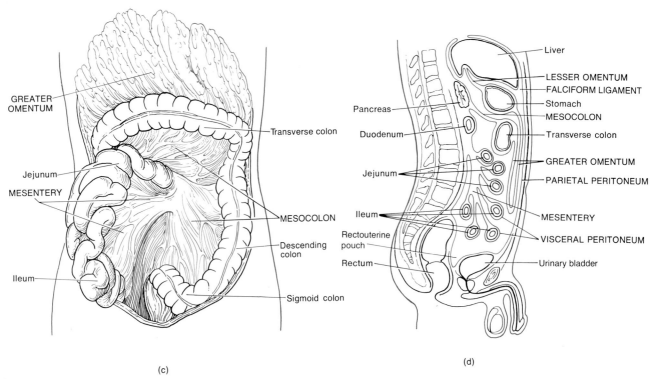

(c)   (d)

FIGURE 21-3 (*Continued*) Extensions of the peritoneum. (c) Mesentery. The greater omentum has been lifted. (d) Sagittal section through the abdomen and pelvis indicating the relationship of the peritoneal extensions to each other.

is attached to the posterior abdominal wall. The mesentery binds the small intestine to the wall. A similar fold of parietal peritoneum, called the **mesocolon** (mez'-ō-KŌ-lon), binds the large intestine to the posterior body wall. It also carries blood vessels and lymphatics to the intestines.

Other important peritoneal folds are the falciform ligament, the lesser omentum, and the greater omentum. The **falciform** (FAL-si-form) **ligament** attaches the liver to the anterior abdominal wall and diaphragm. The **lesser omentum** (ō-MENT-um) arises as two folds in the serosa of the stomach and duodenum suspending the stomach and duodenum from the liver. The **greater omentum** is a four-layered fold in the serosa of the stomach that hangs down like an apron over the front of the intestines. It then passes up to part of the large intestine (the transverse colon), wraps itself around it, and finally attaches to the parietal peritoneum of the posterior wall of the abdominal cavity. Because the greater omentum contains large quantities of adipose tissue, it commonly is called the "fatty apron." The greater omentum contains numerous lymph

nodes. If an infection occurs in the intestine, plasma cells formed in the lymph nodes combat the infection and help prevent it from spreading to the peritoneum.

MOUTH OR ORAL CAVITY

The **mouth,** also referred to as the **oral** or **buccal** (BUK-al) **cavity,** is formed by the cheeks, hard and soft palates, and tongue (Figure 21-4). Forming the lateral walls of the oral cavity are the cheeks—muscular structures covered on the outside by skin and lined by nonkeratinized stratified squamous epithelium. The anterior portions of the cheeks terminate in the superior and inferior lips.

The lips are fleshy folds surrounding the orifice of the mouth. They are covered on the outside by skin and on the inside by a mucous membrane. The transition zone where the two kinds of covering tissue meet is called the **vermilion** (ver-MIL-yon). This portion of the lips is not keratinized and the color of the blood in the underlying blood vessels is visible through the transparent surface layer of the vermilion. The inner surface of each lip is attached to its corresponding gum by a midline fold of mucous membrane called the **labial frenulum** (LĀ-bē-al FREN-yoo-lum).

The orbicularis oris muscle and connective tissue lie between the external integumentary covering and the internal mucosal lining. During chewing, the cheeks and lips help to keep food between the upper and lower teeth. They also assist in speech.

CLINICAL APPLICATION

Inflammation of the peritoneum, called **peritonitis,** is a serious condition because the peritoneal membranes are contiguous with each other, enabling the infection to spread to all the organs in the cavity.

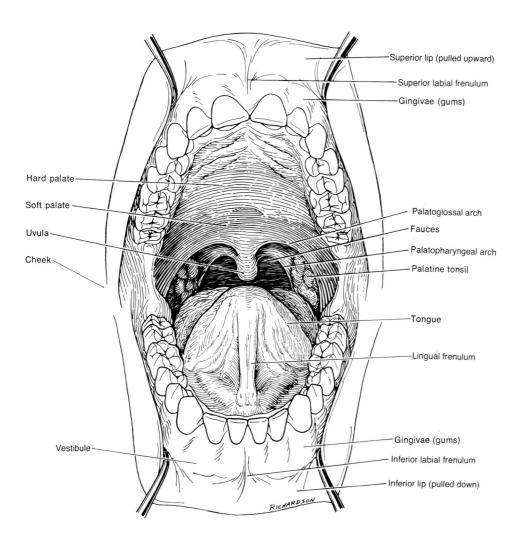

Superior lip (pulled upward)

Superior labial frenulum

Gingivae (gums)

Hard palate

Soft palate

Uvula

Cheek

Palatoglossal arch

Fauces

Palatopharyngeal arch

Palatine tonsil

Tongue

Lingual frenulum

Gingivae (gums)

Inferior labial frenulum

Inferior lip (pulled down)

Vestibule

RICHARDSON

FIGURE 21-4 Mouth or oral cavity.

The **vestibule** of the oral cavity is bounded externally by the cheeks and lips and internally by the gums and teeth. The **oral cavity proper** extends from the vestibule to the **fauces** (FAW-sēs), the opening between the oral cavity and the pharynx or throat.

The **hard palate,** the anterior portion of the roof of the mouth, is formed by the maxillae and palatine bones, covered by mucous membrane, and forms a bony partition between the oral and nasal cavities. The **soft palate** forms the posterior portion of the roof of the mouth. It is an arch-shaped muscular partition between the oropharynx and nasopharynx and is lined by mucous membrane.

Hanging from the free border of the soft palate is a conical muscular process called the **uvula** (OO-vyoo-la). On either side of the base of the uvula are two muscular folds that run down the lateral side of the soft palate. Anteriorly, the **palatoglossal arch (anterior pillar)** extends inferiorly, laterally, and anteriorly to the side of the base of the tongue. Posteriorly, the **palatopharyngeal (PAL-a-tō-fa-rin'-jē-al) arch (posterior pillar)** projects inferiorly, laterally, and posteriorly to the side of the pharynx. The palatine tonsils are situated between the arches, and

the lingual tonsils are situated at the base of the tongue. At the posterior border of the soft palate, the mouth opens into the oropharynx through the fauces.

TONGUE

The **tongue,** together with its associated muscles, forms the floor of the oral cavity. It is composed of skeletal muscle covered with mucous membrane (Figure 21-5). The tongue is divided into symmetrical lateral halves by a medium septum that extends throughout its entire length and is attached inferiorly to the hyoid bone. Each half of the tongue consists of an identical complement of muscles called extrinsic and intrinsic muscles.

The **extrinsic muscles** of the tongue originate outside the tongue and insert into it. They include the hyoglossus, chondroglossus, genioglossus, styloglossus, and palatoglossus (see Figure 10-7). The extrinsic muscles move the tongue from side to side and in and out. These movements maneuver food for chewing, shape the food into a rounded mass, called a **bolus,** and force the food to the back of the mouth for swallowing. They also form

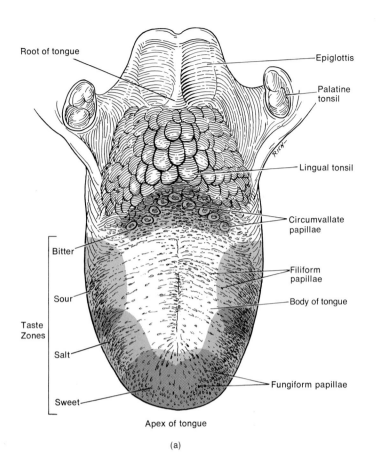

(a)

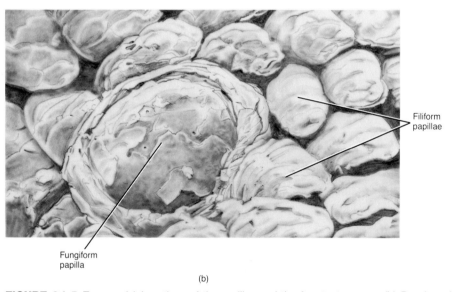

(b)

FIGURE 21-5 Tongue. (a) Locations of the papillae and the four taste zones. (b) Drawing of a fungiform papilla surrounded by filiform papillae based on a scanning electron micrograph at a magnification of 245×.

the floor of the mouth and hold the tongue in position. The **intrinsic muscles** originate and insert within the tongue and alter the shape and size of the tongue for speech and swallowing. The intrinsic muscles include the longitudinalis superior, longitudinalis inferior, transver-

sus linguae, and verticalis linguae (Figure 21-5d). The **lingual frenulum,** a fold of mucous membrane in the midline of the undersurface of the tongue, aids in limiting the movement of the tongue posteriorly.

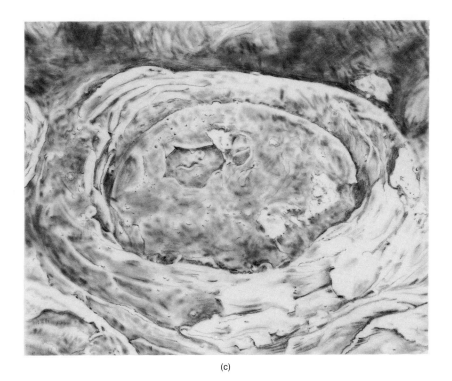

(c)

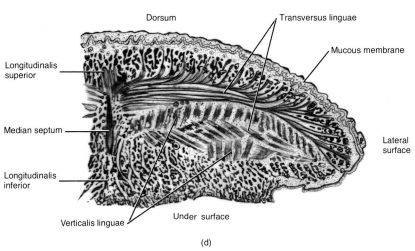

Dorsum

Transversus linguae

Mucous membrane

Longitudinalis superior

Median septum

Longitudinalis inferior

Lateral surface

Verticalis linguae

Under surface

(d)

FIGURE 21-5 (*Continued*) Tongue. (c) Drawing of a circumvallate papilla based on a scanning electron micrograph at a magnification of 910×. (d) Frontal section through the left side of the tongue showing its intrinsic muscles.

CLINICAL APPLICATION

If the lingual frenulum is too short, tongue movements are restricted, speech is faulty, and the person is said to be "tongue-tied." This congenital problem is referred to as **ankyloglossia** (ang-ki-lō-GLOSS-ē-a). It can be corrected by cutting the lingual frenulum.

The upper surface and sides of the tongue are covered with **papillae** (pa-PIL-ē), projections of the lamina propria covered with epithelium (Figure 21-5b, c). **Filiform papillae** are conical projections distributed in parallel rows over the anterior two-thirds of the tongue. They contain no taste buds. **Fungiform papillae** are mushroomlike elevations distributed among the filiform papillae and more numerous near the tip of the tongue. They appear as red dots on the surface of the tongue, and most of them contain taste buds. **Circumvallate papillae** are arranged in the form of an inverted V on the posterior surface of the tongue, and all of them contain taste buds. Note the taste zones of the tongue in Figure 21-5a.

The tongue has a rich blood supply. The lingual artery is the only blood supply, and its veins eventually drain into the internal jugular vein. Most of the muscles of the tongue are supplied by the hypoglossal nerve. Taste sensations from the tongue are conveyed by the facial and glossopharyngeal nerves.

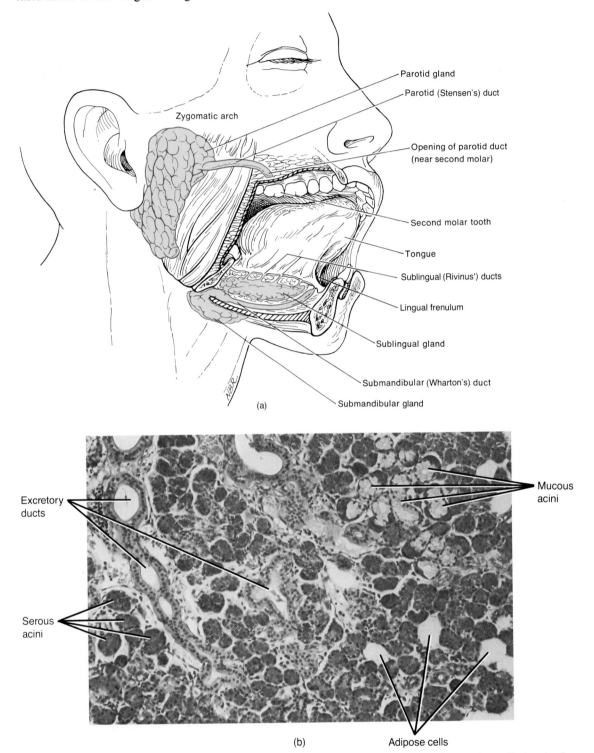

FIGURE 21-6 Salivary glands. (a) Location of the salivary glands. (b) Photomicrograph of the submandibular gland showing mostly serous acini and a few mucous acini at a magnification of 120×.

SALIVARY GLANDS

Saliva is a fluid that is continuously secreted by glands in or near the mouth. Ordinarily, just enough saliva is secreted to keep the mucous membranes of the mouth moist, but when food enters the mouth, secretion increases so the saliva can lubricate, dissolve, and chemically break down some of the food. The mucous membrane lining the mouth contains many small glands, the **buccal glands,** that secrete small amounts of saliva. However, the major portion of saliva is secreted by the **salivary glands,** which lie outside the mouth and pour their contents into ducts

that empty into the oral cavity. There are three pairs of salivary glands: the parotid, submandibular (submaxillary), and sublingual glands (Figure 21-6).

The **parotid glands** are located under and in front of the ears between the skin and the masseter muscle. They are compound tubuloacinar glands. Each secretes into the oral cavity vestibule via a duct, called Stensen's duct, that pierces the buccinator muscle to open into the vestibule opposite the upper second molar tooth. The **submandibular glands,** which are compound acinar glands, are found beneath the base of the tongue in the posterior part of the floor of the mouth. Their ducts (Wharton's

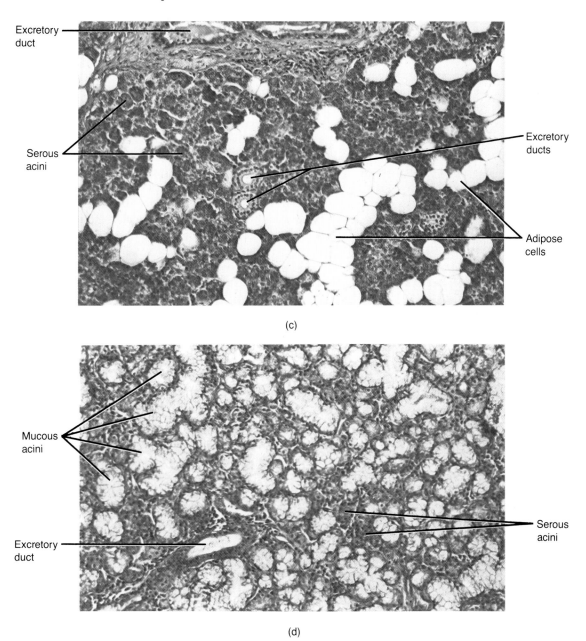

(c)

(d)

FIGURE 21-6 (*Continued*) Salivary glands. (c) Photomicrograph of the parotid gland showing all serous acini at a magnification of 120×. (d) Photomicrograph of the sublingual gland showing mostly mucous acini and a few serous acini at a magnification of 120×. (© 1983 by Michael H. Ross. Used by permission.)

ducts) run superficially under the mucosa on either side of the midline of the floor of the mouth and enter the oral cavity proper just behind the central incisors. The **sublingual glands,** also compound acinar glands, are anterior to the submandibular glands, and their ducts (ducts of Rivinus) open into the floor of the mouth in the oral cavity proper.

The parotid gland receives its blood supply from branches of the external carotid artery and is drained by vessels that are tributaries of the external jugular vein. The submandibular gland is supplied by branches of the facial artery and drained by tributaries of the facial vein. The sublingual gland is supplied by the sublingual branch of the lingual artery and the submental branch of the facial artery and is drained by tributaries of the sublingual and submental veins.

The salivary glands receive both sympathetic and parasympathetic innervation. The sympathetic fibers form plexuses on the blood vessels that supply the glands and serve as vasoconstrictors. The parotid gland receives sympathetic fibers from the plexus on the external carotid artery, whereas the submandibular and sublingual glands receive sympathetic fibers that contribute to the sympathetic plexus and accompany the facial artery to the glands. The parasympathetic fibers of the glands consist of secretomotor fibers to the glands.

The fluids secreted by the buccal glands and the three pairs of salivary glands constitute **saliva.** Amounts of saliva secreted daily vary considerably but range from 1,000 to 1,500 ml. Chemically, saliva is 99.5 percent water and 0.5 percent solutes. Among the solutes are salts—chlorides, bicarbonates, and phosphates of sodium and potassium. Some dissolved gases and various organic substances including urea and uric acid, serum albumin and globulin, mucin, the bacteriolytic enzyme lysozyme, and the digestive enzyme salivary amylase are also present.

Depending on the cells the gland contains, each saliva-producing gland supplies different ingredients to saliva. The parotids contain cells that secrete a thin watery serous liquid containing the enzyme salivary amylase. The submandibular glands contain cells similar to those found in the parotids plus some mucous cells. Therefore, they secrete a fluid that is thickened with mucus but still contains quite a bit of enzyme. The sublingual glands contain mostly mucous cells, so they secrete a much thicker fluid that contributes only a small amount of enzyme to the saliva.

The water in saliva provides a medium for dissolving foods so they can be tasted and digestive reactions can take place. The chlorides in the saliva activate the salivary amylase. The bicarbonates and phosphates buffer chemicals that enter the mouth and keep the saliva at a slightly acidic pH of 6.35 to 6.85. Urea and uric acid are found in saliva because the saliva-producing glands (like the sweat glands of the skin) help the body to get rid of

wastes. Mucin is a protein that forms mucus when dissolved in water. Mucus lubricates the food so it can be easily turned in the mouth, formed into a ball or bolus, and swallowed. The enzyme lysozyme destroys bacteria, thereby protecting the mucous membrane from infection and thus the teeth from decay.

The enzyme **salivary amylase** (AM-i-lās) initiates the breakdown of polysaccharides (carbohydrates). This is the only chemical digestion that occurs in the mouth. Carbohydrates are starches and sugars and are classified as either monosaccharides, disaccharides, or polysaccharides. Monosaccharides are small molecules containing several carbon, hydrogen, and oxygen atoms—an example is glucose. Disaccharides consist of two monosaccharides linked together; polysaccharides are chains of three or more monosaccharides. Most carbohydrates we eat are polysaccharides. Since only monosaccharides can be absorbed into the bloodstream, ingested disaccharides and polysaccharides must be broken down. The function of salivary amylase is to break the chemical bonds between some of the monosaccharides that make up the polysaccharides. In this way, the enzyme breaks the long-chain polysaccharides into shorter polysaccharides called *dextrins*. Given sufficient time, salivary amylase also can break down the dextrins into the disaccharide maltose. Food usually is swallowed too quickly for more than 3 to 5 percent of the carbohydrates to be reduced to disaccharides in the mouth. However, salivary amylase in the swallowed food continues to act on polysaccharides for another 15 to 30 minutes in the stomach before the stomach acids eventually inactivate it.

Saliva continues to be secreted heavily some time after food is swallowed. This flow of saliva washes out the mouth and dilutes and buffers the chemical remnants of irritating substances.

TEETH

The **teeth,** or **dentes,** are located in sockets of the alveolar processes of the mandible and maxillae. The alveolar processes are covered by the **gingivae** (jin-JI-vē) or gums, which extend slightly into each socket forming the gingival sulcus (Figure 21-7). The sockets are lined by the **periodontal ligament,** which consists of dense fibrous connective tissue and is attached to the socket walls and the cemental surface of the roots. Thus it anchors the teeth in position and also acts as a shock absorber to dissipate the forces of chewing.

A typical tooth consists of three principal portions. The **crown** is the portion above the level of the gums. The **root** consists of one to three projections embedded in the socket. The **cervix** is the constricted junction line of the crown and the root.

Teeth are composed primarily of **dentin,** a bonelike substance that gives the tooth its basic shape and rigidity.

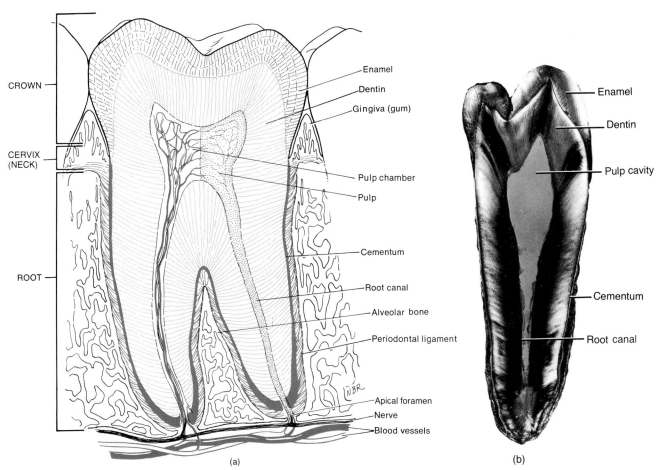

FIGURE 21-7 Parts of a typical tooth. (a) Sagittal section through a molar. (b) Photograph of a sagittal section of a tooth showing the relationship of the pulp cavity to the root canal. (Courtesy of Carolina Biological Supply Company.)

The dentin encloses a cavity. The enlarged part of the cavity, the **pulp cavity,** lies in the crown and is filled with **pulp,** a connective tissue containing blood vessels, nerves, and lymphatics. Narrow extensions of the pulp cavity run through the root of the tooth and are called **root canals.** Each root canal has an opening at its base, the **apical foramen.** Through the foramen enter blood vessels bearing nourishment, lymphatics affording protection, and nerves providing sensation. The dentin of the crown is covered by **enamel** that consists primarily of calcium phosphate and calcium carbonate. Enamel is the

hardest substance in the body and protects the tooth from the wear of chewing. It is also a barrier against acids that easily dissolve the dentin. The dentin of the root is covered by **cementum,** another bonelike substance, which attaches the root to the periodontal ligament.

Dental Terminology

Because of the curvature of the dental arches, it is necessary to use terms other than anterior, posterior, medial, and lateral in describing the surfaces of the teeth. Accordingly, the following directional terms are used. **Labial** refers to the surface of a tooth in contact with or directed toward the lips. **Buccal** refers to the surface in contact with or directed toward the cheeks. **Lingual** is restricted to the teeth of the lower jaw and refers to the surface directed toward the tongue. **Palatal,** on the other hand, is restricted to the teeth of the upper jaw and refers to the surface directed toward the palate. **Mesial** designates the anterior or medial side of the tooth relative to its position in the dental arch. **Distal** refers to the posterior or lateral side of the tooth relative to its position in the dental arch. Essentially, mesial and distal refer to the

CLINICAL APPLICATION

Pyorrhea (pī'-ō-RĒ-a) is inflammation of the periodontal ligament and adjacent gums. A prolonged, severe case of pyorrhea can weaken the periodontal ligament, erode the alveolar bone, and thereby loosen the tooth. Pyorrhea is completely avoidable with proper gum and tooth prophylaxis. This includes daily dental flossing and proper brushing, as well as periodic visits to the dentist.

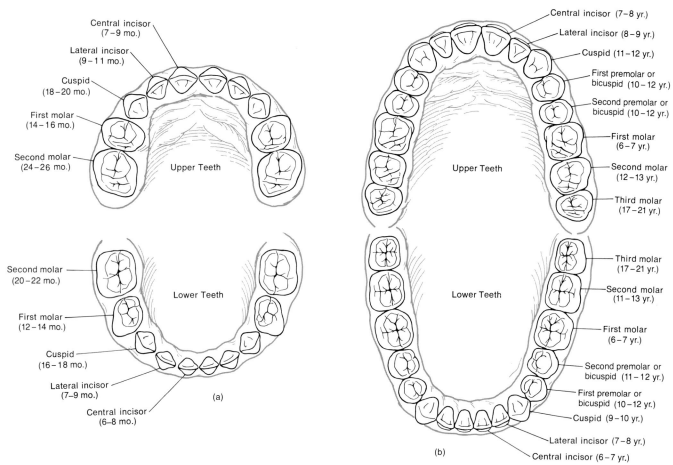

FIGURE 21-8 Dentitions and times of eruptions. The times of eruptions are indicated in parentheses. (a) Deciduous dentition. (b) Permanent dentition.

sides of adjacent teeth that are in contact with each other. Finally, **occlusal** refers to the biting surface of a tooth.

Dentitions

Everyone has two **dentitions,** or sets of teeth. The first of these—the **deciduous teeth, milk teeth,** or **baby teeth—** begin to erupt at about 6 months of age, and one pair appears at about each month thereafter until all 20 are present. Figure 21-8a illustrates the deciduous teeth. The *incisors,* which are closest to the midline, are chisel-shaped and adapted for cutting into food. Next to the incisors, moving posteriorly, are the *cuspids* or *canines,* which have a pointed surface called the cusp. Cuspids are used to tear and shred food. The incisors and cuspids have only one root apiece. Behind them lie the *first* and *second molars,* which have four cusps. Upper molars have three roots; lower molars have two roots. The molars crush and grind food.

All the deciduous teeth are lost—generally between 6 and 12 years of age—and are replaced by the **permanent dentition** (Figure 21-8b). The permanent dentition contains 32 teeth that appear between the age of 6 and adult-

hood. It resembles the deciduous dentition with the following exceptions: the deciduous molars are replaced with *premolars* or *bicuspids* that have two cusps and one root (upper first bicuspids have two roots and are used for crushing and grinding). The permanent molars erupt into the mouth behind the bicuspids. They do not replace any primary teeth and erupt as the jaw grows to accommodate them—the first molars at age 6, the second molars at age 12, the third molars after age 18. The human jaw has become smaller through time and often does not afford enough room behind the second molars for the eruption of the *third molars* or *wisdom teeth.* In this case, the third molars remain embedded in the alveolar bone and are said to be "impacted." Most often they cause pressure and pain and must be surgically removed. In some individuals, however, third molars may be dwarfed in size or may not develop at all.

The arteries that supply blood to the teeth are distributed to the pulp cavity and surrounding periodontal membrane. The upper incisors and cuspids are supplied by anterior superior alveolar branches of the maxillary artery; the upper premolars and molars are supplied by the posterior superior alveolar branches of the maxillary artery; the lower incisors and cuspids are supplied by the incisive branches of the inferior alveolar artery; and the lower premolars and molars are supplied by the dental branches of the inferior alveolar artery.

The teeth receive sensory fibers from branches of the maxillary and mandibular divisions of the trigeminal (V) nerve—the upper teeth from branches of the maxillary division and the lower teeth from branches of the mandibular division.

Through chewing, or **mastication,** the teeth grind food and mix it with saliva. As a result, the food is reduced to a soft, flexible mass that is easily swallowed.

PHARYNX

Swallowing, or **deglutition** (dē-gloo-TISH-un) moves food from the mouth to the stomach. The process is facilitated by saliva. Swallowing starts when the tongue forms a *bolus.* The bolus is forced to the back of the mouth cavity and into the oropharynx by the movement of the tongue upward and backward against the palate. This represents the voluntary stage of swallowing (Figure 21-9).

With the passage of the bolus into the oropharynx, the involuntary pharyngeal stage of swallowing begins. During this stage, the respiratory passageways close, and breathing is temporarily interrupted. The bolus stimulates receptors in the oropharynx, which send impulses to the

medulla and lower pons of the brain stem. The returning impulses cause the soft palate and uvula to move upward to close off the nasopharynx, and the larynx is pulled forward and upward under the tongue. As the larynx rises, it meets the epiglottis, which seals off the glottis. The movement of the larynx also pulls the vocal cords together, further sealing off the respiratory tract, and widens the opening between the laryngopharynx and esophagus. The bolus passes through the laryngopharynx and enters the esophagus in 1 second. The respiratory passageways then reopen and breathing resumes.

ESOPHAGUS

The **esophagus** (e-SOF-a-gus), the third organ involved in deglutition, is a muscular, collapsible tube that lies behind the trachea. It is about 23 to 25 cm (10 inches) long and begins at the end of the laryngopharynx, passes through the mediastinum anterior to the vertebral column, pierces the diaphragm through an opening called the *esophageal hiatus,* and terminates in the superior portion of the stomach.

The *mucosa* of the esophagus consists of nonkeratinized stratified squamous epithelium, lamina propria, and a muscularis mucosae (Figure 21-10a, b, e). The *submucosa* contains connective tissue and blood vessels. The

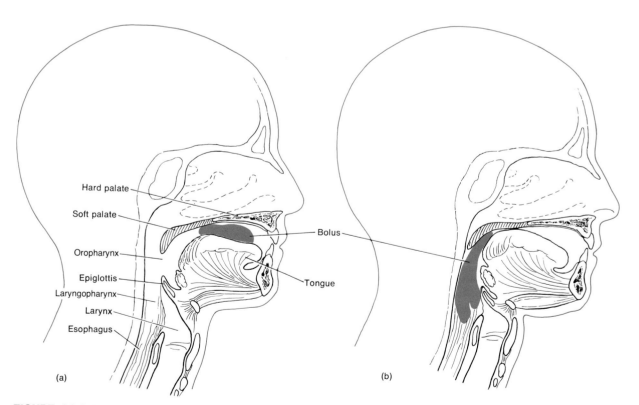

Hard palate
Soft palate
Oropharynx
Epiglottis
Laryngopharynx
Larynx
Esophagus
Bolus
Tongue

(a)   (b)

FIGURE 21-9 Deglutition. (a) Position of structures prior to deglutition. (b) During deglutition, the tongue rises against the palate, the nose is closed off, the larynx rises, the epiglottis seals off the larynx, and the bolus is passed into the esophagus.

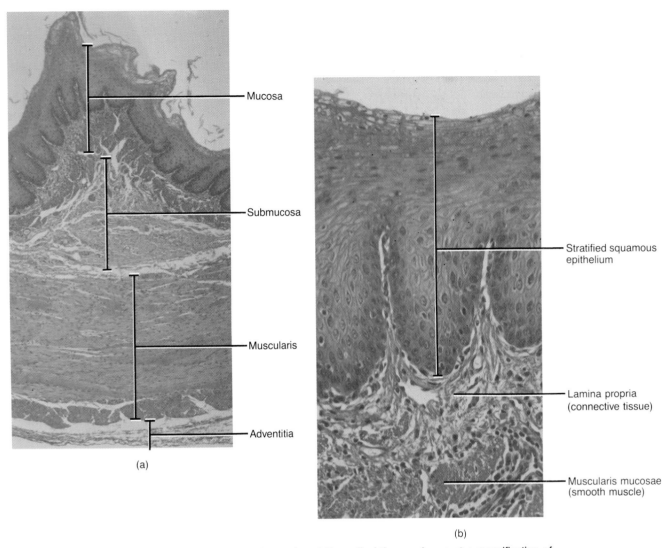

FIGURE 21-10 Esophagus. (a) Photomicrograph of a portion of the wall of the esophagus at a magnification of 60×. (b) Photomicrograph of an enlarged aspect of the mucosa of the esophagus at a magnification of 180×. (© 1983 by Michael H. Ross. Used by permission.)

muscularis of the upper third is striated, the middle third is striated and smooth, and the lower third is smooth. The outer layer is known as the *tunica adventitia* (ad-ven-TISH-a) since it is not covered by peritoneum.

The esophagus does not produce digestive enzymes and does not carry on absorption. It secretes mucus and transports food to the stomach. During the esophageal stage of swallowing, which is also involuntary, food is pushed through the esophagus by muscular movements called **peristalsis** (per'-i-STAL-sis) (Figure 21-10c, d). Peristalsis is a function of the tunica muscularis and is controlled by the medulla. In the section of the esophagus lying just above and around the top of the bolus, the circular muscle fibers contract. The contraction constricts the esophageal (e-sof'-a-JĒ-al) wall and squeezes the bolus downward. Meanwhile, longitudinal fibers lying around the bottom of and just below the bolus also contract. Contraction of the longitudinal fibers shortens this lower

section, pushing its walls outward so it can receive the bolus. The contractions are repeated in a wave that moves down the esophagus, pushing the food toward the stomach. Passage of the bolus is further facilitated by glands secreting mucus. The passage of solid or semisolid food from the mouth to the stomach takes 4 to 8 seconds. Very soft foods and liquids pass through in about 1 second.

Just above the level of the diaphragm, the esophagus is slightly narrowed. This narrowing has been attributed to a physiological sphincter in the inferior part of the esophagus known as the **lower esophageal** or **gastro-esophageal sphincter** (SFINGK-ter). A **sphincter** is a thick circle of muscle around an opening. The lower esophageal sphincter relaxes during swallowing and thus aids the passage of the bolus from the esophagus into the stomach. The movement of the diaphragm against the stomach during breathing presses on the stomach

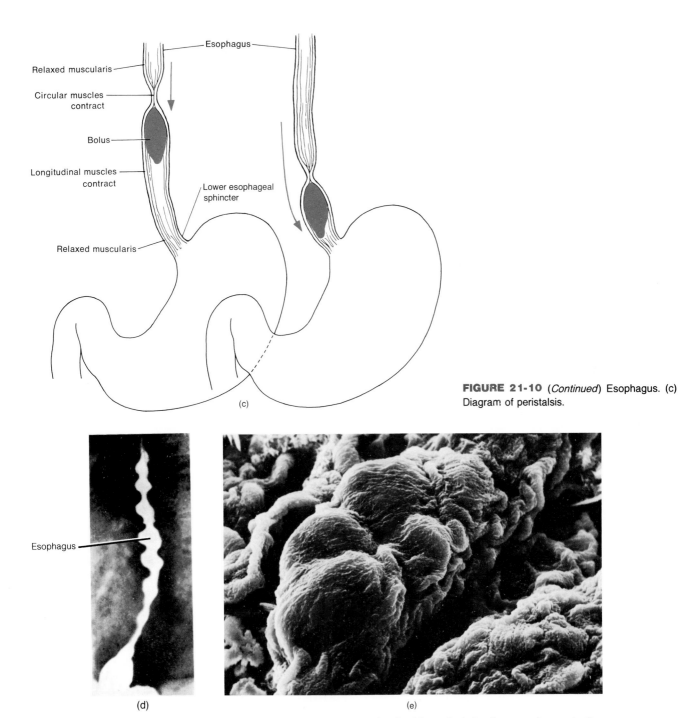

Esophagus

Relaxed muscularis

Circular muscles
contract

Bolus

Longitudinal muscles
contract

Lower esophageal
sphincter

Relaxed muscularis

(c)

FIGURE 21-10 (*Continued*) Esophagus. (c)
Diagram of peristalsis.

Esophagus

(d)

(e)

FIGURE 21-10 (*Continued*) Esophagus. (d) Anteroposterior projection of peristalsis made during fluoroscopic examination while a patient was swallowing barium. (Courtesy of Lester W. Paul and John H. Juhl, *The Essentials of Roentgen Interpretation*, 3rd ed., Harper & Row, Publishers, Inc., New York, 1972.) (e) Scanning electron micrograph of the mucosa of the esophagus at a magnification of 2,000×. (Courtesy of Fisher Scientific Company and S.T.E.M. Laboratories, Inc., Copyright, 1975.)

and helps prevent the regurgitation of gastric contents from the stomach to the esophagus.

The arteries of the esophagus are derived from the arteries along its length: inferior thyroid, thoracic aorta, intercostal arteries, phrenic, and left gastric arteries. It is drained by the adjacent veins. Innervation of the esophagus is by recurrent laryngeal nerves, the cervical sympathetic chain, and vagi.

STOMACH

The **stomach** is a J-shaped enlargement of the GI tract directly under the diaphragm in the epigastric, umbilical, and left hypochondriac regions of the abdomen (see Figure 1-5b). The superior portion of the stomach is a continuation of the esophagus. The inferior portion empties into the duodenum, the first part of the small intestine. Within each individual, the position and size of the stomach vary continually. For instance, the diaphragm pushes the stomach downward with each inspiration and pulls it upward with each expiration. Empty, it is about the size of a large sausage, but it can stretch to accommodate large amounts of food.

ANATOMY

The stomach is divided into four areas: cardia, fundus, body, and pylorus (Figure 21-11). The **cardia** surrounds the lower esophageal sphincter. The rounded portion above and to the left of the cardia is the **fundus.** Below the fundus is the large central portion of the stomach, called the **body.** The narrow, inferior region is the **pylorus.** The concave medial border of the stomach is called the **lesser curvature,** and the convex lateral border is the **greater curvature.** The pylorus communicates with the duodenum of the small intestine via a sphincter called the **pyloric valve.**

The stomach wall is composed of the same four basic layers as the rest of the alimentary canal, with certain modifications. When the stomach is empty, the mucosa lies in large folds, called **rugae** (ROO-gē), that can be seen with the naked eye. As the stomach fills and distends,

> ### CLINICAL APPLICATION
> Two abnormalities of the pyloric valve can occur in infants. **Pylorospasm** is characterized by failure of the muscle fibers encircling the opening to relax normally. It can be caused by hypertrophy or continuous spasm of the sphincter and usually occurs between the second and twelfth weeks of life. Ingested food does not pass easily from the stomach to the small intestine, the stomach becomes overly full, and the infant vomits frequently to relieve the pressure. Pylorospasm is treated by adrenergic drugs that relax the muscle fibers of the valve. **Pyloric stenosis** is a narrowing of the pyloric valve caused by a tumorlike mass that apparently is formed by enlargement of the circular muscle fibers. It must be surgically corrected.

the rugae gradually smooth out and disappear. Microscopic inspection of the mucosa reveals a layer of simple columnar epithelium containing many narrow openings that extend down into the lamina propria (Figure 21-12a-d). These pits—**gastric glands**—are lined with three kinds of secreting cells: zymogenic, parietal, and mucous. The **zymogenic,** or **chief, cells** secrete the principal gastric enzyme pepsinogen. Hydrochloric acid, which activates

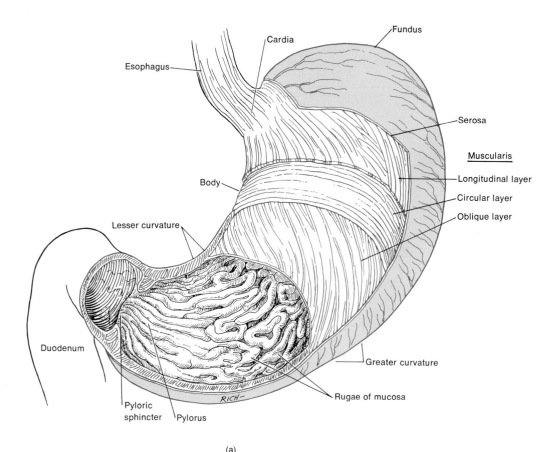

(a)

FIGURE 21-11 External and internal anatomy of the stomach. (a) Diagram in anterior view.

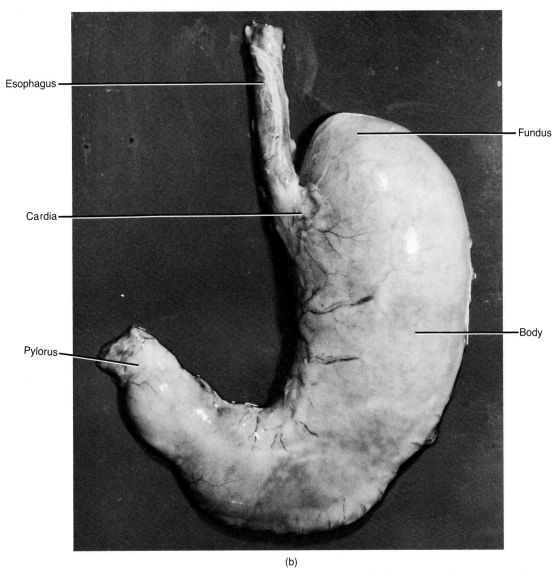

Esophagus

Cardia

Pylorus

Fundus

Body

(b)

FIGURE 21-11 (*Continued*) External and internal anatomy of the stomach. (b) Photograph of the external surface in anterior view.

the pepsinogen, is produced by the **parietal cells.** The **mucous cells** secrete mucus and intrinsic factor, a substance involved in the absorption of vitamin B$_{12}$. Secretions of the gastric glands are together called **gastric juice.**

The submucosa of the stomach is composed of loose areolar connective tissue, which connects the mucosa to the muscularis.

The muscularis, unlike that in other areas of the alimentary canal, has three layers of smooth muscle: an outer longitudinal layer, a middle circular layer, and an inner oblique layer. This arrangement of fibers allows the stomach to contract in a variety of ways to churn food, break it into small particles, mix it with gastric juice, and pass it to the duodenum.

The serosa covering the stomach is part of the visceral

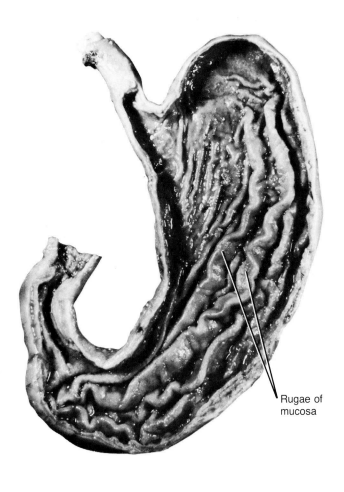

(c)

Rugae of mucosa

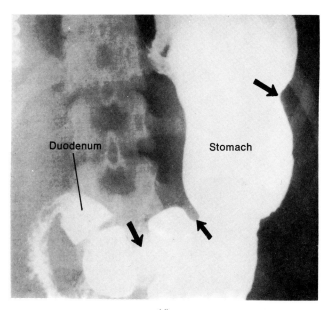

Duodenum

Stomach

(d)

FIGURE 21-11 (*Continued*) External and internal anatomy of the stomach. (c) Photograph of the internal surface in anterior view showing rugae. (Photographs courtesy of C. Yokochi and J. W. Rohen, *Photographic Anatomy of the Human Body,* 2nd ed., 1979, IGAKU-SHOIN, Ltd., Tokyo, New York.) (d) Anteroposterior projection. Arrows indicate peristaltic waves. (Courtesy of Lester W. Paul and John H. Juhl, *The Essentials of Roentgen Interpretation,* 3rd ed., Harper & Row, Publishers, Inc., New York, 1972.)

peritoneum. At the lesser curvature the two layers of the visceral peritoneum come together and extend upward to the liver as the lesser omentum. At the greater curvature, the visceral peritoneum continues downward as the greater omentum hanging over the intestines.

The arterial supply of the stomach is derived from the celiac artery. The right and left gastric arteries form an anastomosing arch along the lesser curvature, and the right and left gastroepiploic arteries form a similar arch on the greater curvature. Short gastric arteries supply the fundus. The veins of the same name accompany the arteries and drain, directly or indirectly, into the hepatic portal vein.

The vagi convey parasympathetic fibers to the stomach. These fibers form synapses within the submucous plexus in the submucosa. The sympathetic nerves arise from the celiac ganglia.

ACTIVITIES

Several minutes after food enters the stomach, gentle, rippling, peristaltic movements called **mixing waves** pass over the stomach every 15 to 25 seconds. These waves macerate food, mix it with the secretions of the gastric

glands, and reduce it to a thin liquid called **chyme** (kīm). Few mixing waves are observed in the fundus, which is primarily a storage area. Foods may remain in the fundus for an hour or more without becoming mixed with gastric juice. During this time, salivary digestion continues.

The principal chemical activity of the stomach is to begin the digestion of proteins. In the adult, digestion is achieved primarily through the enzyme **pepsin.** Pepsin breaks certain peptide bonds between the amino acids making up proteins. Thus a protein chain of many amino acids is broken down into long fragments called *proteoses* and somewhat shorter ones called *peptones.* Pepsin is most effective in the very acidic environment of the stomach (pH of 2). It becomes inactive in an alkaline environment.

What keeps pepsin from digesting the protein in stomach cells along with the food? First, pepsin is secreted in an inactive form called *pepsinogen,* so it cannot digest the proteins in the zymogenic cells that produce it. It is not converted into active pepsin until it comes in contact with the hydrochloric acid secreted by the parietal cells. Second, once pepsin has been activated, the stomach cells are protected by mucus. The mucus coats the mucosa to form a barrier between it and the gastric juices.

CLINICAL APPLICATION

If the stomach mucus fails to protect the stomach cells, the pepsin and hydrochloric acid erode the stomach wall and produce a **gastric ulcer** (see Figure 21-27).

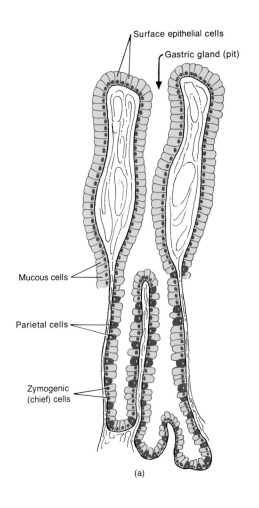

Surface epithelial cells

Gastric gland (pit)

Mucous cells

Parietal cells

Zymogenic (chief) cells

(a)

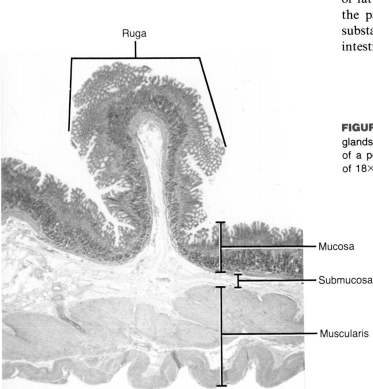

Ruga

Mucosa

Submucosa

Muscularis

(b)

Another enzyme of the stomach is *gastric lipase.* Gastric lipase splits the butterfat molecules found in milk. This enzyme operates best at a pH of 5 to 6 and has a limited role in the adult stomach. Adults rely almost exclusively on an enzyme found in the small intestine to digest fats.

The infant stomach secretes the enzyme *rennin,* which is important in the digestion of milk. Rennin and calcium act on the casein of milk to produce a curd. The coagulation prevents too rapid a passage of milk from the stomach. Rennin is absent in the gastric secretions of adults.

As digestion proceeds in the stomach, more vigorous peristaltic waves begin at about the middle of the stomach, pass downward, reach the pyloric valve, and sometimes go into the duodenum. The movement of chyme from the stomach into the duodenum depends on a pressure gradient between the two organs. When the pressure in the stomach (intragastric pressure) is greater than that in the duodenum (intraduodenal pressure), chyme is forced into the duodenum. Peristaltic waves are largely responsible for increased intragastric pressure. It is estimated that 2 to 5 ml of chyme are passed into the duodenum with each peristaltic wave. When intraduodenal pressure exceeds intragastric pressure, the pyloric valve closes and prevents the regurgitation of chyme from the duodenum to the stomach.

The stomach empties all its contents into the duodenum 2 to 6 hours after ingestion. Foods rich in carbohydrate pass through the stomach first, protein foods are somewhat slower, and foods containing large amounts of fat take longest. The stomach wall is impermeable to the passage of most materials into the blood, so most substances are not absorbed until they reach the small intestine. However, the stomach does participate in the

FIGURE 21-12 Histology of the stomach. (a) Diagram of gastric glands from the fundic wall of the stomach. (b) Photomicrograph of a portion of the fundic wall of the stomach at a magnification of 18×. (© 1983 by Michael H. Ross. Used by permission.)

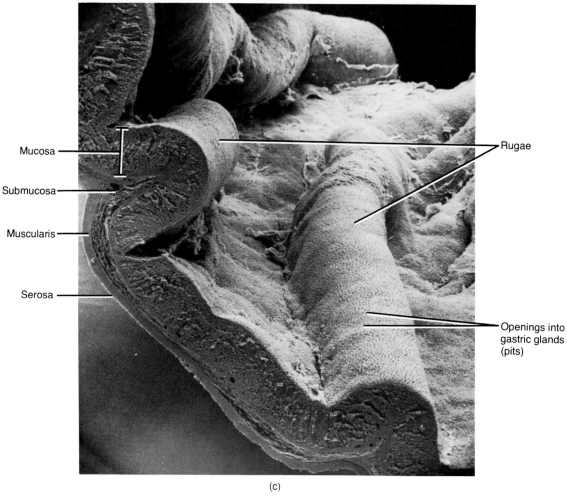

Mucosa

Submucosa

Muscularis

Serosa

Rugae

Openings into
gastric glands
(pits)

(c)

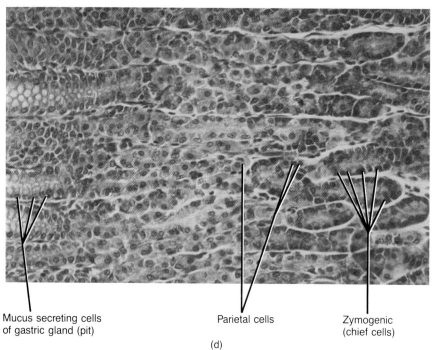

Mucus secreting cells
of gastric gland (pit)

Parietal cells

Zymogenic
(chief cells)

(d)

FIGURE 21-12 (*Continued*) Histology of the stomach. (c) Scanning electron micrograph of the stomach wall at a magnification of 55×. (Courtesy of Richard G. Kessel and Randy H. Kardon, *Tissues and Organs: A Text-Atlas of Scanning Electron Microscopy*, W. H. Freeman and Company, San Francisco, 1979.) (d) Photomicrograph of an enlarged aspect of the mucosa of the fundic wall of the stomach at a magnification of 180×. (© 1983 by Michael H. Ross. Used by permission.)

absorption of some water and salts, certain drugs, and alcohol.

The next step in the breakdown of food is digestion in the small intestine. Chemical digestion in the small intestine depends not only on its own secretions but on activities of three organs outside the alimentary canal: the pancreas, liver, and gallbladder.

PANCREAS

The **pancreas** is a soft, oblong tubuloacinar gland about 12.5 cm (6 inches) long and 2.5 cm (1 inch) thick. It lies posterior to the greater curvature of the stomach and is connected by a duct (sometimes two) to the duodenum (Figure 21-13). The pancreas is divided into a head, body, and tail. The **head** is the expanded portion near the C-shaped curve of the duodenum. Moving superiorly and

to the left of the head are the centrally located **body** and the terminal tapering **tail.**

The pancreas is made up of small clusters of glandular epithelial cells. Some clusters, the **islets of Langerhans,** form the endocrine portions of the pancreas and consist of alpha, beta, and delta cells that secrete hormones. The other masses of cells, called **acini** (AS-i-nē), are the exocrine portions of the organ (see Figure 19-12). Secreting cells of the acini release a mixture of digestive enzymes called **pancreatic juice,** which is dumped into small ducts attached to the acini. Pancreatic juice eventually leaves the pancreas through a large main tube called the **pancreatic duct,** or **duct of Wirsung** (VĒR-sung). In most people the pancreatic duct unites with the common bile duct from the liver and gallbladder and enters the duodenum in a common duct, called the **ampulla of Vater** (am-POOL-la of VA-ter). The ampulla opens on an elevation

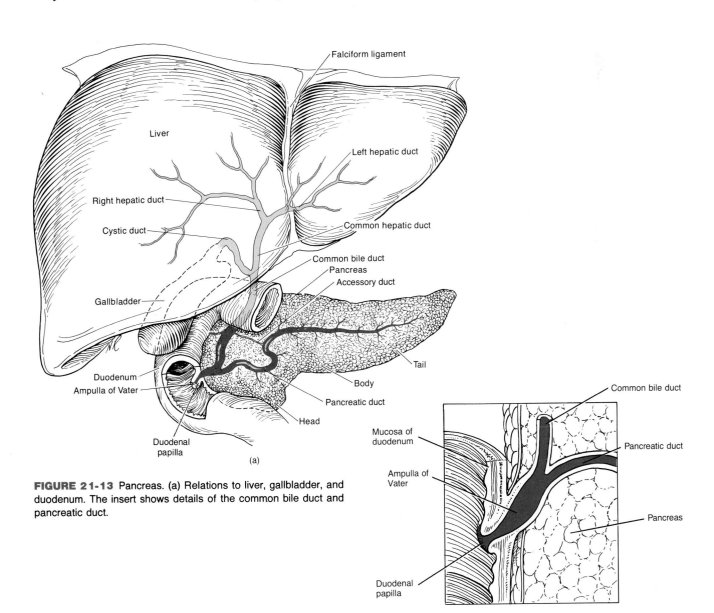

FIGURE 21-13 Pancreas. (a) Relations to liver, gallbladder, and duodenum. The insert shows details of the common bile duct and pancreatic duct.

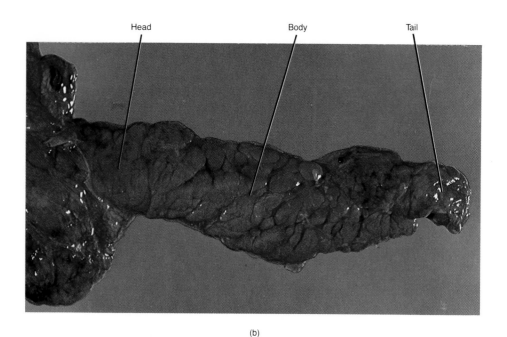

(b)

(c)

FIGURE 21-13 (*Continued*) Pancreas. (b) Photograph of the pancreas. (© 1982, Rotker, Science Photo Library Int.) (c) CT scan. (Courtesy of General Electric Co.)

of the duodenal mucosa known as the **duodenal papilla,** about 10 cm (4 inches) below the pylorus of the stomach. An **accessory duct (duct of Santorini)** may also lead from the pancreas and empty into the duodenum about 2.5 cm (1 inch) above the ampulla of Vater.

The arterial supply of the pancreas is from the splenic artery through its pancreatic branches from the hepatic artery and from the superior mesenteric artery by way of the inferior and superior pancreaticoduodenal arteries. The veins, in general, correspond to the arteries. Venous blood reaches the hepatic portal vein by means of the splenic and superior mesenteric veins.

The nerves to the pancreas are branches of the celiac plexus that accompany the arteries entering the gland. The glandular innervation is from the parasympathetic

division of the autonomic nervous system, and the innervation of the blood vessels is from the sympathetic division of the autonomic nervous system.

The functions of the pancreas are twofold. The acini secrete enzymes that digest food in the small intestine, and the alpha and beta cells secrete the hormones glucagon and insulin, which regulate blood sugar level.

LIVER

The liver weighs about 1.4 kg (4 lb) in the average adult. It is located under the diaphragm and occupies most of the right hypochondrium and part of the epigastrium of the abdomen (see Figure 1-5c).

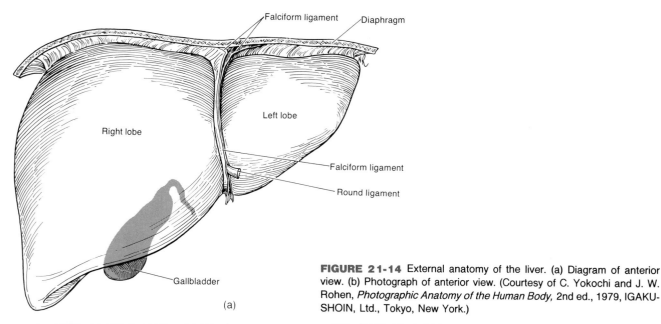

Falciform ligament   Diaphragm

Right lobe

Left lobe

Falciform ligament

Round ligament

Gallbladder

(a)

FIGURE 21-14 External anatomy of the liver. (a) Diagram of anterior view. (b) Photograph of anterior view. (Courtesy of C. Yokochi and J. W. Rohen, *Photographic Anatomy of the Human Body,* 2nd ed., 1979, IGAKU-SHOIN, Ltd., Tokyo, New York.)

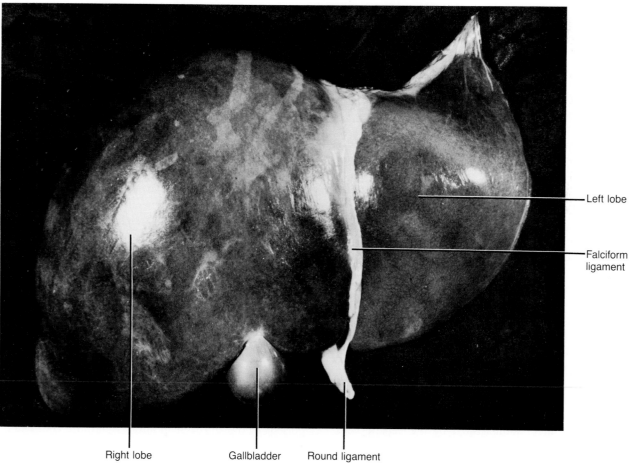

Left lobe

Falciform ligament

Right lobe   Gallbladder   Round ligament

(b)

ANATOMY

The liver is almost completely covered by peritoneum and completely by a dense connective tissue layer that lies beneath the peritoneum. It is divided into two principal lobes—the **right lobe** and the **left lobe**—separated by the **falciform ligament** (Figure 21-14). Associated with

the right lobe are the inferior **quadrate lobe** and posterior **caudate lobe.** The falciform ligament is a reflection of the parietal peritoneum, which extends from the undersurface of the diaphragm to the superior surface of the liver, between the two principal lobes of the liver. In the free border of the falciform ligament is the **ligamentum**

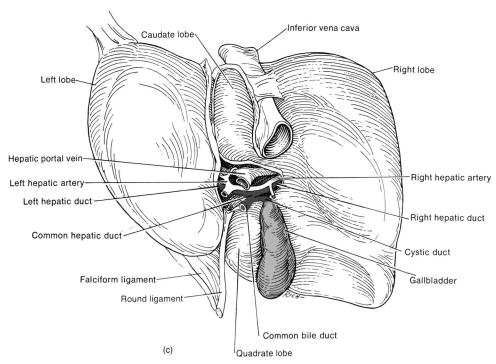

Caudate lobe

Inferior vena cava

Right lobe

Left lobe

Hepatic portal vein

Left hepatic artery

Left hepatic duct

Right hepatic artery

Common hepatic duct

Right hepatic duct

Falciform ligament

Cystic duct

Round ligament

Gallbladder

(c)

Common bile duct

Quadrate lobe

FIGURE 21-14 (*Continued*) External anatomy of the liver. (c) Diagram of posteroinferior view.

teres (**round ligament**). It extends from the liver to the umbilicus. The ligamentum teres is a fibrous cord derived from the umbilical vein of the fetus.

The lobes of the liver are made up of numerous functional units called **lobules,** which may be seen under a microscope (Figure 21-15). A lobule consists of cords of **hepatic** (liver) **cells** arranged in a radial pattern around a **central vein.** Between the cords are endothelial-lined spaces called **sinusoids,** through which blood passes. The sinusoids are also partly lined with phagocytic cells, termed **Kupffer** (KOOP-fer) **cells,** that destroy worn-out white and red blood cells and bacteria.

The liver receives a double supply of blood. From the hepatic artery it obtains oxygenated blood, and from the hepatic portal vein it receives deoxygenated blood containing newly absorbed nutrients (Figure 21-16). Branches of both the hepatic artery and the hepatic portal vein carry the blood into the sinusoids of the lobules, where oxygen, most of the nutrients, and certain poisons are extracted by the hepatic cells. Nutrients are stored or used to make new materials. The poisons are stored or detoxified. Products manufactured by the hepatic cells and nutrients needed by other cells are secreted back into the blood. The blood then drains into the central vein and eventually passes into a hepatic vein. Unlike the other products of the liver, bile normally is not secreted into the bloodstream.

Bile is manufactured by the hepatic cells and secreted into **bile capillaries** or **canaliculi** (kan'-a-LIK-yoo-lī) that empty into small ducts. These small ducts eventually

merge to form the larger **right** and **left hepatic ducts,** which unite to leave the liver as the **common hepatic duct.** Further on, the common hepatic duct joins the **cystic duct** from the gallbladder. The two tubes become the **common bile duct,** which empties into the duodenum at the ampulla of Vater (see Figure 21-13a).

The nerve supply to the liver consists of vagal preganglionic parasympathetic fibers and postganglionic sympathetic fibers from the celiac ganglia.

ACTIVITIES

The **liver** performs so many vital functions that we cannot live without it. Among these are the following:

1. The liver, together with mast cells, manufactures the anticoagulant heparin and most of the other plasma proteins, such as prothrombin, fibrinogen, and albumin.

2. The Kupffer cells of the liver phagocytose worn-out red and white blood cells and some bacteria.

3. Liver cells contain enzymes that either break down poisons or transform them into less harmful compounds. When amino acids are burned for energy, for example, they leave behind toxic nitrogenous wastes (such as ammonia), that are converted to urea by the liver cells. Moderate amounts of urea are harmless to the body and are easily excreted by the kidneys and sweat glands.

4. Newly absorbed nutrients are collected in the liver. It can change any excess monosaccharides into glycogen or fat, both of which can be stored. In addition it can

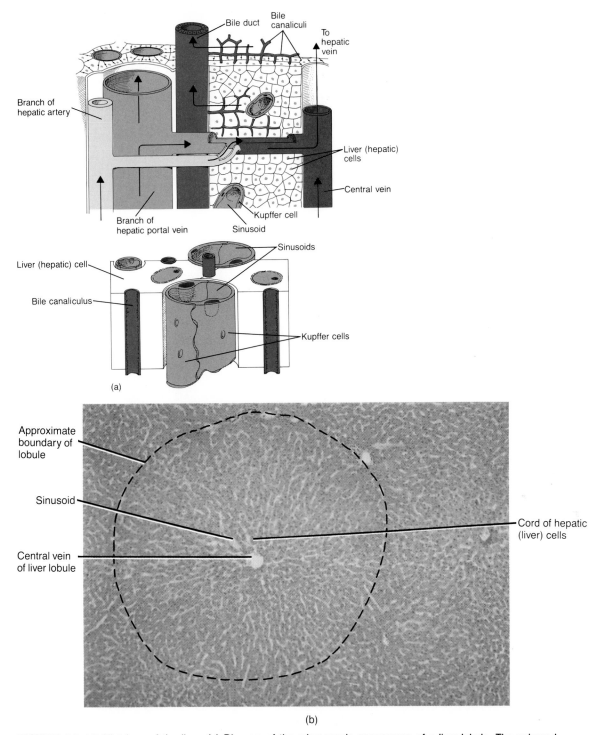

FIGURE 21-15 Histology of the liver. (a) Diagram of the microscopic appearance of a liver lobule. The enlarged insert shows Kupffer cells lining the sinusoids. (b) Photomicrograph of a liver lobule at a magnification of 65×. (© 1983 by Michael H. Ross. Used by permission.)

transform glycogen, fat, and protein into glucose, depending on the body's needs.

5. The liver stores glycogen, copper, iron, and vitamins A, D, E, and K. It also stores some poisons that cannot be broken down and excreted. (High levels of DDT are found in the livers of animals, including humans, who eat sprayed fruits and vegetables.)

6. The liver manufactures bile salts which are used in the small intestine for the emulsification and absorption of fats.

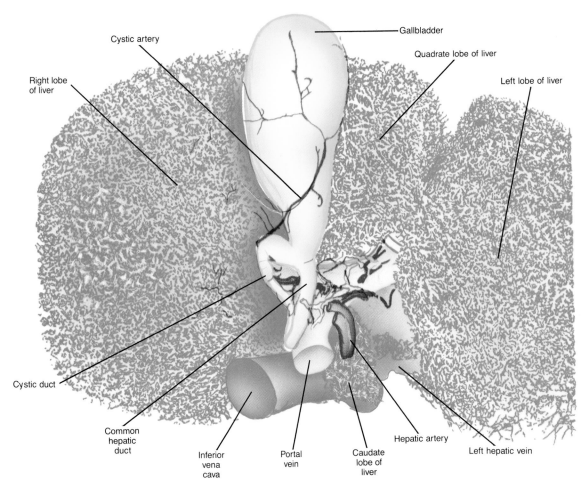

Cystic artery

Gallbladder

Quadrate lobe of liver

Right lobe of liver

Left lobe of liver

Cystic duct

Common hepatic duct

Inferior vena cava

Portal vein

Caudate lobe of liver

Hepatic artery

Left hepatic vein

FIGURE 21-16 Drawing of the cast of the blood vessels of the liver from behind.

GALLBLADDER

The **gallbladder** is a pear-shaped sac about 7 to 10 cm (3 to 4 inches) long located in a fossa of the visceral surface of the liver (see Figure 21-13a). The relationship of the gallbladder to the stomach is shown in the roentgenogram in Figure 21-17a. Figure 21-18 presents a cross section of the abdomen at the level of the pancreas. Note the relationship of the digestive organs to one another.

The inner wall of the gallbladder consists of a mucous membrane arranged in rugae resembling those of the stomach (Figure 21-17b-d) that allow it to expand to the size and shape of a pear. The middle, muscular coat of the wall consists of smooth muscle fibers. Contraction of these fibers by hormonal stimulation ejects the contents

of the gallbladder into the cystic duct. The outer coat is the visceral peritoneum.

The gallbladder is supplied by the cystic artery, which arises from the right hepatic artery. The cystic veins drain the gallbladder. The nerves to the gallbladder include branches from the celiac plexus, the vagus, and the right phrenic nerve.

The function of the gallbladder is to store and concentrate bile (up to 10-fold) until it is needed in the small intestine. Bile from the liver enters the small intestine through the common bile duct. When the small intestine is empty, a valve in the duct, called the **sphincter of Oddi** (OD-dē), closes, and the backed-up bile overflows into the cystic duct to the gallbladder.

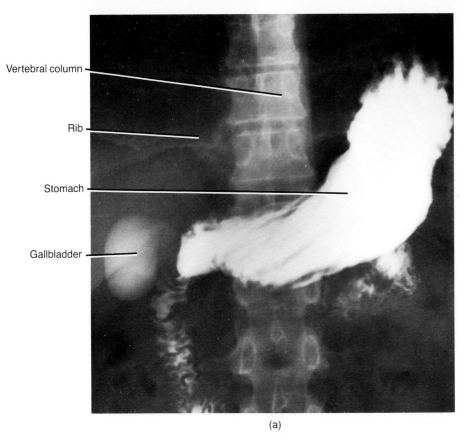

Vertebral column

Rib

Stomach

Gallbladder

(a)

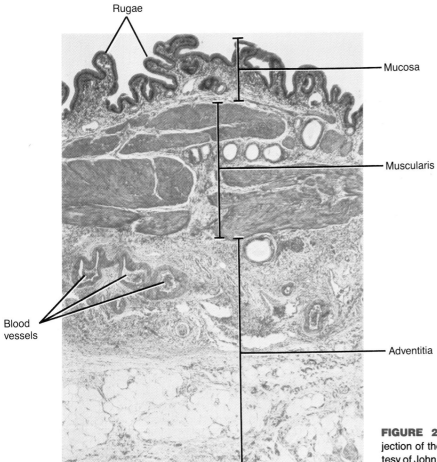

Rugae

Mucosa

Muscularis

Blood
vessels

Adventitia

(b)

FIGURE 21-17 Gallbladder. (a) Anteroposterior projection of the gallbladder in relation to the stomach. (Courtesy of John C. Bennett, St. Mary's Hospital, San Francisco.) (b) Photomicrograph of a portion of the wall of the gallbladder at a magnification of 40×.

Lamina propria   Blood vessel   Simple columnnar
epithelium

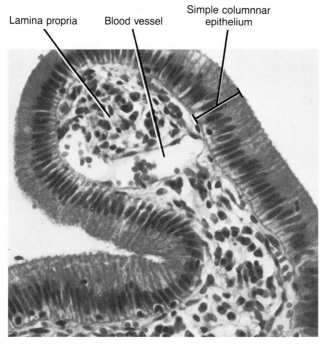

(c)

Rugae

Simple columnar
epithelial cells

Lamina
propria

(d)

FIGURE 21-17 (*Continued*) Gallbladder. (c)
Photomicrograph of an enlarged aspect of the mucosa of the gallbladder at a magnification of 600×.
(© 1983 by Michael H. Ross. Used by permission.)
(d) Drawing of the interior of the gallbladder based
on a scanning electron micrograph at a magnification of 695×.

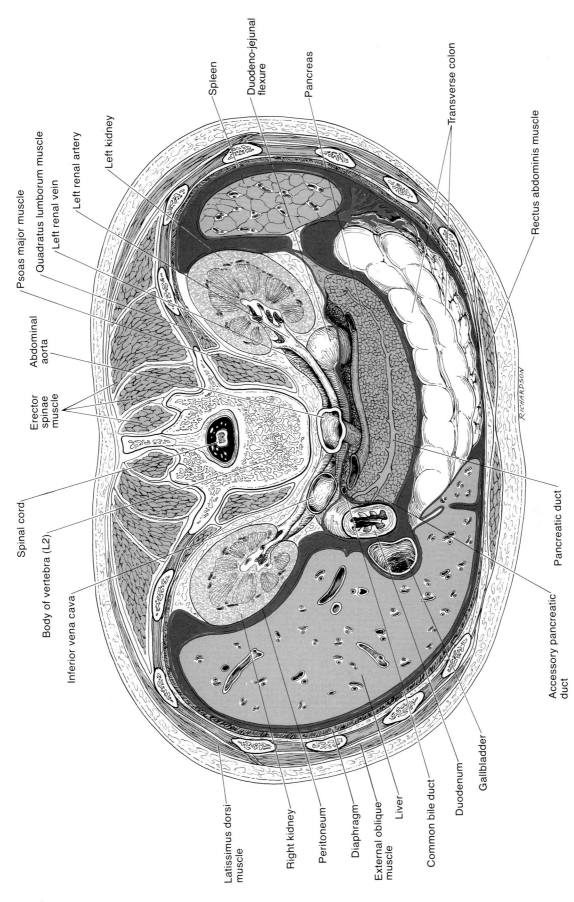

Spleen

Duodeno-jejunal flexure

Pancreas

Transverse colon

Rectus abdominis muscle

Left kidney

Left renal artery

Left renal vein

Quadratus lumborum muscle

Psoas major muscle

Abdominal aorta

Erector spinae muscle

Spinal cord

Body of vertebra (L2)

Inferior vena cava

Latissimus dorsi muscle

Right kidney

Peritoneum

Diaphragm

External oblique muscle

Liver

Common bile duct

Duodenum

Gallbladder

Accessory pancreatic duct

Pancreatic duct

RICHARDSON

FIGURE 21-18 Abdomen seen in cross section at the level of the pancreas.

CLINICAL APPLICATION

As bile is concentrated in the gallbladder, it becomes supersaturated with cholesterol. The precipitation of cholesterol crystals marks the beginning of most **gallstones.**

SMALL INTESTINE

The major portions of digestion and absorption occur in a long tube called the **small intestine.** The small intestine begins at the pyloric valve of the stomach, coils through the central and lower part of the abdominal cavity, and eventually opens into the large intestine. It averages 2.5 cm (1 inch) in diameter and about 6.35 m (21 ft) in length.

ANATOMY

The small intestine is divided into three segments: duodenum (doo'-ō-DĒ-num), jejunum (jē-JOO-num), and ileum (IL-ē-um) (see Figure 21-1). The **duodenum,** the shortest part, originates at the pyloric valve of the stomach and extends about 25 cm (10 inches) until it merges with the jejunum. The **jejunum** is about 2.5 m (8 ft) long and extends to the ileum. The final portion of the small intestine, the **ileum,** measures about 3.6 m (12 ft) and joins the large intestine at the **ileocecal** (il'-ē-ō-SĒ-kal) **valve.** A roentgenogram of the normal small intestine is shown in Figure 21-19.

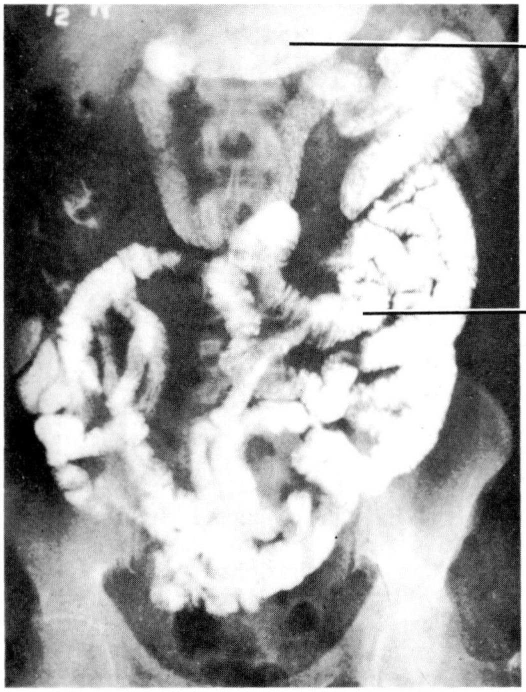

FIGURE 21-19 Small intestine. Anteroposterior projection of the normal small intestine ½ hour after taking a barium "meal." (Courtesy of Lester W. Paul and John H. Juhl, *The Essentials of Roentgen Interpretation,* 3rd ed., Harper & Row, Publishers, Inc., New York, 1972.)

The wall of the small intestine is composed of the same four tunics or coats that make up most of the GI tract. However, both the mucosa and the submucosa are modified to allow the small intestine to complete the processes of digestion and absorption (Figure 21-20).

The mucosa contains many pits lined with glandular epithelium. These pits—the **intestinal glands,** or **crypts**

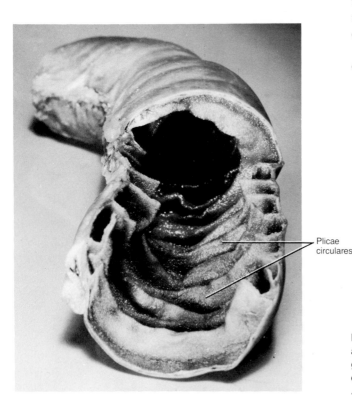

(a)

FIGURE 21-20 Small intestine. Shown are various structures that adapt the small intestine for digestion and absorption. (a) Photograph of a section of the jejunum cut open to expose the plicae circulares. (Courtesy of C. Yokochi and J. W. Rohen, *Photographic Anatomy of the Human Body,* 1st ed., 1969, IGAKU-SHOIN, Ltd., Tokyo, New York.)

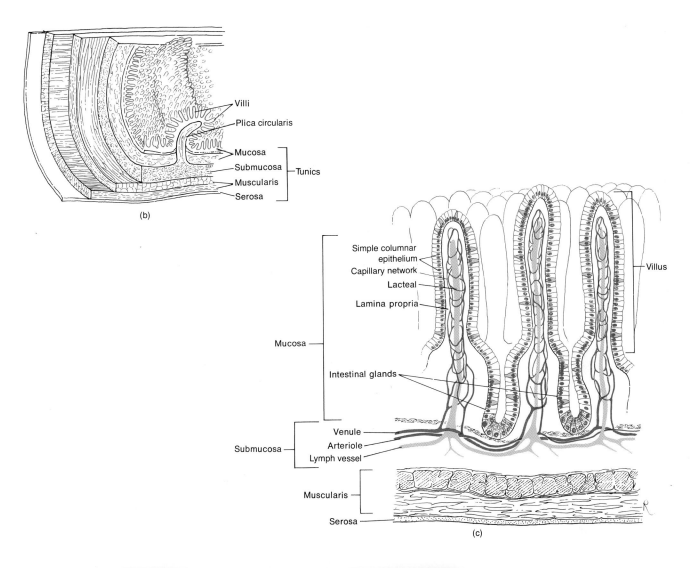

Villi
Plica circularis
Mucosa
Submucosa
Muscularis
Serosa

Tunics

(b)

Mucosa

Simple columnar epithelium
Capillary network
Lacteal
Lamina propria

Villus

Intestinal glands

Submucosa

Venule
Arteriole
Lymph vessel

Muscularis

Serosa

(c)

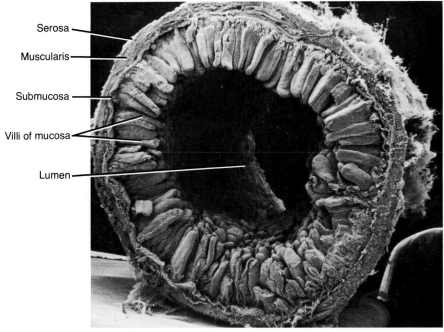

Serosa
Muscularis
Submucosa
Villi of mucosa
Lumen

(d)

FIGURE 21-20 (*Continued*) Small intestine. (b) Villi in relation to the tunics. (c) Enlarged aspect of several villi. (d) Scanning electron micrograph of a cross section of the small intestine at a magnification of 30×. (Courtesy of Richard G. Kessel and Randy H. Kardon, *Tissues and Organs: A Text-Atlas of Scanning Electron Microscopy,* W. H. Freeman and Company, San Francisco, 1979.)

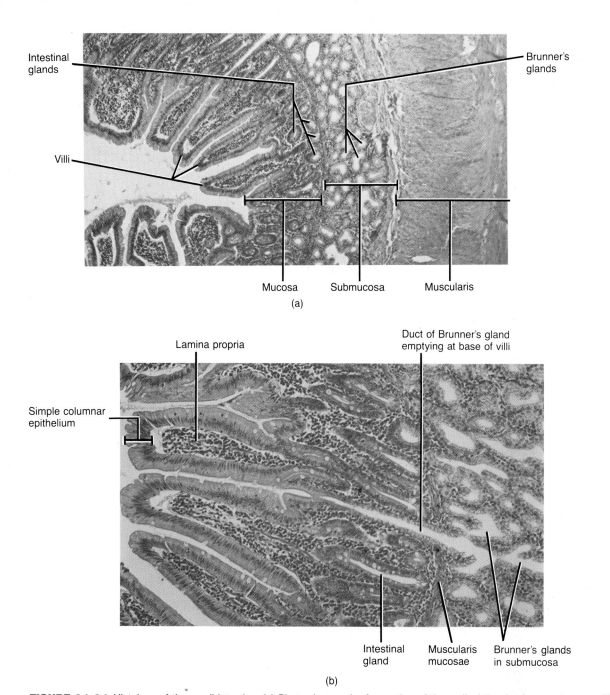

FIGURE 21-21 Histology of the small intestine. (a) Photomicrograph of a portion of the wall of the duodenum at a magnification of 40×. (b) Photomicrograph of an enlarged aspect of the duodenal mucosa at a magnification of 165×.

of Lieberkühn (LĒ-ber-kyoon)—secrete the intestinal digestive enzymes. The submucosa of the duodenum contains **Brunner's glands,** which secrete an alkaline mucus to protect the walls of the small intestine from the action of the enzymes and to aid in neutralizing acid in the chyme. Some of the epithelial cells in the mucosa and submucosa have been transformed to goblet cells, which secrete additional mucus.

Since almost all the absorption of nutrients occurs in the small intestine, its structure is specially adapted for this function. Its length alone provides a large surface area for absorption and that area is further increased by modifications in the structure of its walls. The epithelium covering and lining the mucosa consists of simple columnar epithelium. These epithelial cells, except those transformed into goblet cells, contain **microvilli,** fingerlike projections of the plasma membrane. Larger amounts of digested nutrients diffuse into the intestinal wall because the microvilli increase the surface area of the plasma membrane.

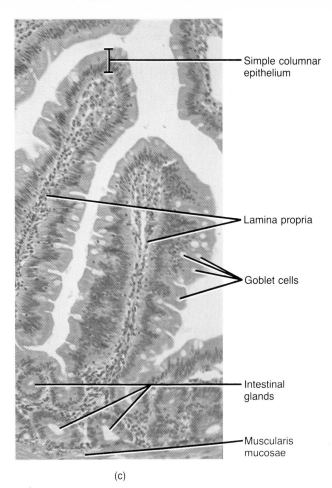

— Simple columnar epithelium

— Lamina propria

— Goblet cells

— Intestinal glands

— Muscularis mucosae

(c)

FIGURE 21-21 (*Continued*) Histology of the small intestine. (c) Photomicrograph of an enlarged aspect of two villi from the ileum at a magnification of 120×. (© 1983 by Michael H. Ross. Used by permission.) (d) Scanning electron micrograph of several villi. (Courtesy of Biophoto Associates.)

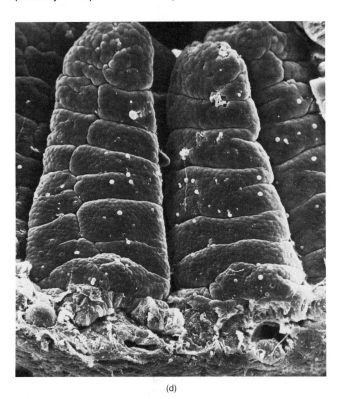

(d)

The mucosa lies in a series of **villi,** projections 0.5 to 1 mm high, giving the intestinal mucosa its velvety appearance. The enormous number of villi (4 to 5 million) vastly increases the surface area of the epithelium available for the epithelial cells specializing in absorption. Each villus has a core of lamina propria, the connective tissue layer of the mucosa. Embedded in this connective tissue are an arteriole, a venule, a capillary network, and a **lacteal** (LAK-tē-al) or lymphatic vessel. Nutrients that dif-

fuse through the epithelial cells which cover the villus are able to pass through the capillary walls and the lacteal and enter the blood and lymphatic system.

In addition to the microvilli and villi, a third set of projections called **plicae circulares** (PLĪ-kē SER-kyoo-lar-es) further increases the surface area for absorption. The plicae are permanent deep folds in the mucosa and submucosa. Some of the folds extend all the way around the intestine, and others extend only part way around. The plicae circulares enhance absorption by causing the chyme to spiral, rather than moving in a straight line, as it passes through the small intestine.

The muscularis of the small intestine consists of two layers of smooth muscle. The outer, thinner layer contains longitudinally arranged fibers. The inner, thicker layer contains circularly arranged fibers. Except for a major portion of the duodenum, the serosa (or visceral peritoneum) completely covers the small intestine. Additional histological aspects of the small intestine are shown in Figure 21-21.

There is an abundance of lymphatic tissue in the wall of the small intestine. Single lymph nodules, called **solitary lymph nodules,** are most numerous in the lower part of the ileum. Aggregated lymph nodules, referred to as **Peyer's** (PĪ-erz) **patches,** are numerous in the ileum.

The arterial blood supply of the small intestine is from the superior mesenteric artery and the gastroduodenal artery, coming from the hepatic artery of the celiac trunk.

Blood is returned by way of the superior mesenteric vein which, with the splenic vein, forms the hepatic portal vein.

The nerves to the small intestine are supplied by the superior mesenteric plexus. The branches of the plexus contain postganglionic sympathetic fibers, preganglionic parasympathetic fibers, and afferent fibers. The afferent fibers are both vagal and of spinal nerves. In the wall of the small intestine are two autonomic plexuses, the myenteric plexus between the muscular layers and the submucous plexus in the submucosa. The nerve fibers are derived chiefly from the sympathetic division of the autonomic nervous system and partly from the vagus.

ACTIVITIES

Chemical Digestion

The digestion of carbohydrates, proteins, and lipids in the small intestine requires secretions from the pancreas, liver, and intestinal glands.

Each day the liver secretes 800 to 1,000 ml (almost 1 qt) of the yellow, brownish, or olive-green liquid called **bile.** It has a pH of 7.6–8.6. Bile consists of water, bile salts, bile acids, a number of lipids, and two pigments called biliverdin and bilirubin. Bile is partially an excretory product and partially a digestive secretion. When red blood cells are broken down, iron, globin, and bilirubin are released. The iron and globin are recycled, but some of the bilirubin is excreted into the bile ducts. Bilirubin eventually is broken down in the intestines, and its breakdown products give feces their color. Other substances found in bile aid in the digestion of fats by emulsifying them and are required for their absorption.

CLINICAL APPLICATION

If the liver is unable to remove bilirubin from the blood or if the bile ducts are obstructed, large amounts of bilirubin circulate through the bloodstream and collect in other tissues, giving the skin and eyes a yellow color. This condition is called **jaundice.**

Each day the pancreas produces 1,200 to 1,500 ml (about 1 to 1.5 qt) of a clear, colorless liquid called **pancreatic juice.** Pancreatic juice consists mostly of water, some salts, sodium bicarbonate, and enzymes. The sodium bicarbonate gives pancreatic juice a slightly alkaline pH (7.1–8.2) that stops the action of pepsin from the stomach and creates the proper environment for the enzymes in the small intestine. The enzymes in pancreatic juice include a carbohydrate-digesting enzyme, several protein-digesting enzymes, the only active fat-digesting enzyme in the adult body, and a nucleic acid-digesting enzyme.

The **intestinal juice,** or **succus entericus,** is a clear

yellow fluid secreted in amounts of about 2 to 3 liters (2 to 3 qt) a day. It has a pH of 7.6, which is slightly alkaline, and contains water, mucus, and enzymes that complete the digestion of carbohydrates, proteins, and nucleic acids.

When chyme reaches the small intestine, the carbohydrates and proteins have been only partly digested and are not ready for absorption. Lipid digestion has not even begun. Digestion of these substances proceeds as follows:

1. Carbohydrates. In the mouth, some polysaccharides are broken down into dextrins containing several monosaccharide units. Sucrose and lactose, two disaccharides, are ingested as such and are not acted upon until they reach the small intestine. Even though the action of salivary amylase may continue in the stomach, few polysaccharides are reduced to disaccharides by the time chyme leaves the stomach. *Pancreatic amylase,* an enzyme in pancreatic juice, breaks dextrins into the disaccharide maltose. Next, enzymes in the intestinal juice digest the disaccharides into monosaccharides. *Maltase* splits maltose into two molecules of glucose. *Sucrase* breaks sucrose into a molecule of glucose and a molecule of fructose. *Lactase* digests lactose into a molecule of glucose and a molecule of galactose. This completes the digestion of carbohydrates.

CLINICAL APPLICATION

A condition called **lactose intolerance** results when the mucosal cells of the small intestine fail to produce sufficient lactase for the digestion of lactose. This deficiency may lead to an electrolyte imbalance. People with this condition complain of diarrhea, gas, and bloating, which may be accompanied by abdominal cramps. It sometimes follows gastric surgery.

---

2. Proteins. Protein digestion starts in the stomach, where most of the proteins are fragmented by the action of pepsin into short chains of amino acids called peptones and proteoses. Enzymes found in pancreatic juice continue the digestion. *Trypsin* (TRIP-sin) digests any intact proteins into peptones and proteoses, breaks the peptones and proteoses into dipeptides (containing only two amino acids), and breaks some of the dipeptides into single amino acids. *Chymotrypsin* (kī-mō-TRIP-sin) duplicates trypsin's activities. *Carboxypeptidase* (kar-bok'-sē-PEP-ti-dās) reduces whole or partly digested proteins into amino acids. To prevent these enzymes from digesting the proteins in the cells of the pancreas, they are secreted in inactive forms—trypsin as *trypsinogen* (trip-SIN-ō-jen), activated by an intestinal enzyme called *enterokinase* (en'-ter-ō-KĪ-nās); chymotrypsin as *chymotrypsinogen,* activated in the small intestine by trypsin; and carboxypeptidase as *procarboxypeptidase,* also activated in the small intestine by trypsin. Protein digestion is completed by several intestinal enzymes grouped together under the name *erepsin* (ē-REP-sin), which converts all the remaining dipeptides into single amino acids. Single amino acids can be absorbed.

3. Lipids. In an adult, almost all lipid digestion occurs in the small intestine. The first step in the process is the **emulsification** of neutral fats. Neutral fats, or just simply fats, are the most abundant lipids in the diet and are triglycerides, consisting of a molecule of glycerol and three molecules of fatty acid. Bile salts break the globules of fat into droplets (emulsification) so the fat-splitting enzyme can get at the lipid molecules. In the second step, *pancreatic lipase,* an enzyme found in pancreatic juice, hydrolyzes each fat molecule into fatty acids, glycerol, and glycerides, end products of fat digestion. Glycerides consist of glycerol with one or two fatty acids still attached and are known as monoglycerides and diglycerides, respectively.

4. Nucleic acids. Both intestinal juice and pancreatic juice contain enzymes, called *nucleases,* that digest nucleic acids into nucleotides.

Mechanical Digestion

In the small intestine, three distinct movements occur as a result of contractions of the longitudinal and circular muscles. These movements are rhythmic segmentation, pendular movements, and propulsive peristalsis.

Rhythmic segmentation and pendular movements are strictly localized contractions in areas containing food. The two movements mix the chyme with the digestive juices and bring the particles of food into contact with the mucosa for absorption. They do not push the intestinal contents along the tract. **Rhythmic segmentation** starts with the contractions of circular muscle fibers in a portion of the intestine, an action that constricts the intestine into segments. Next, muscle fibers that encircle the middle of each segment also contract, dividing each segment into two. Finally, the fibers that contracted first relax, and each small segment unites with an adjoining small segment so that large segments are re-formed. This sequence of events is repeated 12 to 16 times a minute, sloshing the chyme back and forth. **Pendular movements** consist of alternating contractions and relaxations of the longitudinal muscles. The contractions cause a portion of the intestine to shorten and lengthen, spilling the chyme back and forth.

The third movement, **propulsive peristalsis,** propels the chyme onward through the intestinal tract. Peristaltic movement in the intestine is similar to that in the esophagus. In the intestine, these waves may be as slow as 5 cm (2 inches)/minute or as fast as 50 cm (20 inches)/second.

Absorption

All the chemical and mechanical phases of digestion from the mouth down through the small intestine are directed toward changing foods into forms that can diffuse through the epithelial cells lining the mucosa into the underlying blood and lymph vessels. The diffusible forms are monosaccharides (glucose, fructose, and galactose), amino acids, fatty acids, glycerol, and glycerides. Passage of these digested nutrients from the alimentary canal into the blood or lymph is called **absorption.**

About 90 percent of all absorption of nutrients takes place throughout the length of the small intestine. The other 10 percent occurs in the stomach and large intestine. Large amounts of water, electrolytes, mineral salts, and some vitamins are also absorbed in the small intestine. Absorption of materials in the small intestine occurs specifically through the villi and depends on diffusion, facilitated diffusion, osmosis, and active transport.

Monosaccharides and amino acids are absorbed into the blood capillaries of the villi and transported in the bloodstream to the liver via the hepatic portal system (Figure 21-22). Fatty acids, glycerol, and glycerides do not enter the bloodstream immediately. They cluster together and become surrounded by bile salts to form water-soluble particles called **micelles** (mī-SELZ). These are absorbed into the intestinal epithelial cell. Here they enter the smooth endoplasmic reticulum, where triglycerides are resynthesized. The triglycerides, along with small quantities of phospholipids and cholesterol, are then organized into protein-coated lipid droplets called **chylomicrons** (kī'-lō-MĪ-krons). The protein coat keeps the chylomicrons suspended and keeps them from sticking to each other or to the walls of the lymphatics or blood vessels. Small chylomicrons leave the intestinal cells and enter the blood capillaries in the villus. Larger chylomicrons enter the lacteal in the villus, are transported by way of lymphatic vessels to the thoracic duct, and enter the cardiovascular system at the left subclavian vein. Finally, they arrive at the liver through the hepatic artery. Most of the products of carbohydrate, protein, and lipid digestion are processed by the liver before they are delivered to the other cells of the body.

In summary, then, the principal chemical activity of the small intestine is to digest all foods into forms that are usable by body cells. Any undigested materials that are left behind are processed in the large intestine.

LARGE INTESTINE

The overall functions of the large intestine are the completion of absorption, the manufacture of certain vitamins, the formation of feces, and the expulsion of feces from the body.

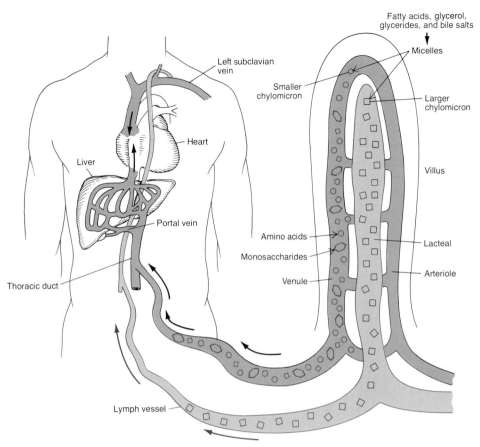

FIGURE 21-22 Absorption.

ANATOMY

The **large intestine** is about 1.5 m (5 ft) in length and averages 6.5 cm (2.5 inches) in diameter. It extends from the ileum to the anus and is attached to the posterior abdominal wall by its **mesocolon** of visceral peritoneum. Structurally, the large intestine is divided into four principal regions: cecum, colon, rectum, and anal canal (Figure 21-23).

The opening from the ileum into the large intestine is guarded by a fold of mucous membrane called the **ileocecal valve.** This structure allows materials from the small intestine to pass into the large intestine. Hanging below the ileocecal valve is the **cecum,** a blind pouch about 6 cm (2 or 3 inches) long. Attached to the cecum is a twisted, coiled tube, measuring about 8 cm (3 inches) in length, called the **vermiform appendix** (*vermis* = worm). The visceral peritoneum of the appendix, called the **mesoappendix,** attaches the appendix to the inferior part of the ileum and adjacent part of the posterior abdominal wall.

The open end of the cecum merges with a long tube called the **colon.** The colon is divided into ascending, transverse, descending, and sigmoid portions. The **ascend-**

ing colon ascends on the right side of the abdomen, reaches the undersurface of the liver, and turns abruptly to the left. Here it forms the **right colic (hepatic) flexure.** The colon continues across the abdomen to the left side as the **transverse colon.** It curves beneath the lower end of the spleen on the left side as the **left colic (splenic) flexure** and passes downward to the level of the iliac crest as the **descending colon.** The **sigmoid colon** begins at the left iliac crest, projects inward to the midline, and terminates as the rectum at about the level of the third sacral vertebra.

The **rectum,** the last 20 cm (7 to 8 inches) of GI tract, lies anterior to the sacrum and coccyx. The terminal 2 to 3 cm (1 inch) of the rectum is called the **anal canal** (Figure 21-24). The mucous membrane of the anal canal is arranged in longitudinal folds called **anal columns** that contain a network of arteries and veins. The opening of the anal canal to the exterior is called the **anus.** It is guarded by an internal sphincter of smooth muscle and an external sphincter of skeletal muscle. Normally the anus is closed except during the elimination of the wastes of digestion.

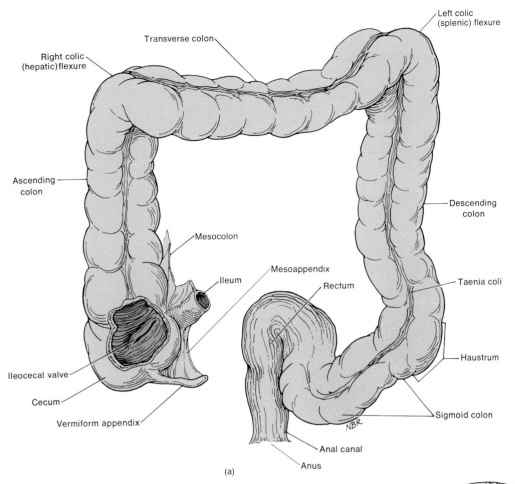

FIGURE 21-23 Large intestine. (a) Anatomy of the large intestine. (b) Anteroposterior projection of the large intestine in which several haustra are clearly visible. (Courtesy of Lester W. Paul and John H. Juhl, *The Essentials of Roentgen Interpretation,* 3rd ed., Harper & Row, Publishers, Inc., New York, 1972.)

(a)

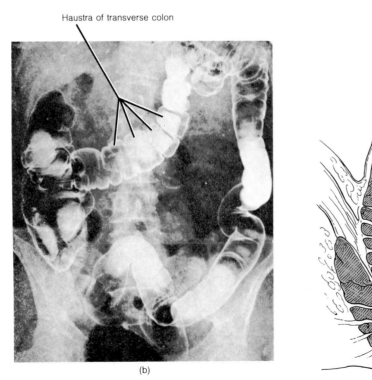

(b)

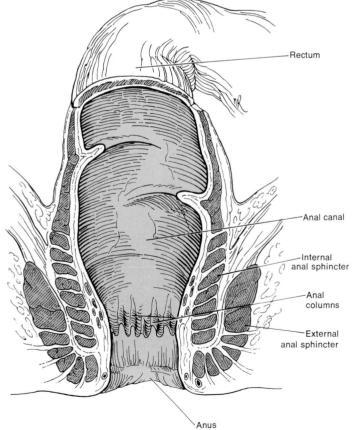

FIGURE 21-24 Anal canal seen in longitudinal section.

The wall of the large intestine differs from that of the small intestine in several respects (Figure 21-25). No villi or permanent circular folds are found in the mucosa, which does, however, contain simple columnar epithelium with numerous goblet cells. These cells secrete mucus that lubricates the colonic contents as they pass through the colon. Solitary lymph nodes also are found in the mucosa. The submucosa of the large intestine is similar to that found in the rest of the alimentary canal. The muscularis consists of an external layer of longitudinal muscles and an internal layer of circular muscles. Unlike other parts of the digestive tract, the longitudinal muscles

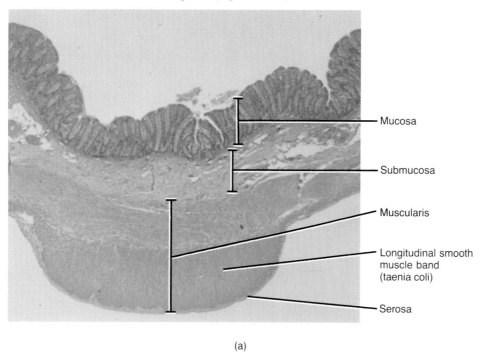

(a)

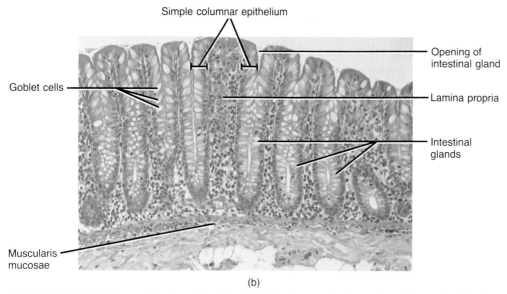

(b)

FIGURE 21-25 Histology of the large intestine. (a) Photomicrograph of a portion of the wall of the large intestine at a magnification of 25×. (b) Photomicrograph of an enlarged aspect of the mucosa of the large intestine at a magnification of 140×. (© 1983 by Michael H. Ross. Used by permission.)

Mucosa   Muscularis   Submucosa   Serosa

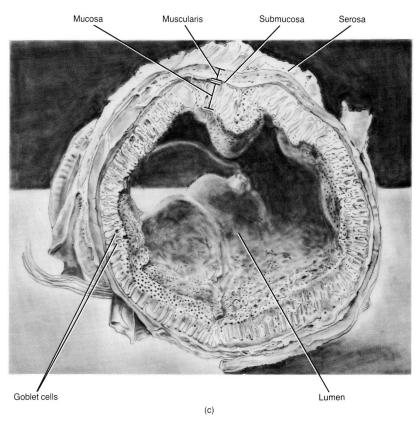

Goblet cells

Lumen

(c)

FIGURE 21-25 (*Continued*) Histology of the large intestine. (c) Drawing of a cross section of the large intestine based on a scanning micrograph at a magnification of 31×.

do not form a continuous sheet around the wall, but are broken up into three flat bands called **taeniae coli** (TĒ-ni-ē KŌ-lī). Each band runs the length of most of the large intestine. Tonic contractions of the bands gather the colon into a series of pouches called **haustra** (HAWS-tra), which give the colon its puckered appearance. The serosa of the large intestine is part of the visceral peritoneum.

The arterial supply of the cecum and colon is derived from branches of the superior mesenteric and inferior mesenteric arteries. The venous return is by way of the superior and inferior mesenteric veins ultimately to the hepatic portal vein and into the liver. The arterial supply of the rectum and anal canal is derived from the superior, middle, and inferior rectal arteries. The rectal veins correspond to the rectal arteries.

The nerves to the large intestine consist of sympathetic, parasympathetic, and afferent components. The sympathetic innervation is derived from the celiac, superior and inferior mesenteric, and internal iliac plexuses. The fibers reach the plexuses by way of the thoracic and lumbar splanchnic nerves. The parasympathetic innervation is derived from the vagus and pelvic splanchnic nerves.

ACTIVITIES

The principal activities of the large intestine are concerned with mechanical movements, absorption, and the formation and elimination of feces.

Mechanical Digestion

Movements of the colon begin when substances enter through the ileocecal valve. Since chyme moves through the small intestine at a fairly constant rate, the time required for a meal to pass into the colon is determined by gastric evacuation time. As food passes through the ileocecal valve, it fills the cecum and accumulates in the ascending colon.

One movement characteristic of the large intestine is **haustral churning.** In this process the haustra remain relaxed and distended while they fill up. When the distension reaches a certain point, the walls contract and squeeze the contents into the next haustrum. **Peristalsis** also occurs, although at a slower rate than in other portions of the tract (3 to 12 contractions per minute). A final type of movement is **mass peristalsis,** a strong peristaltic wave that begins in about the middle of the transverse colon and drives the colonic contents into the rectum. Food in the stomach initiates this reflex action in the colon. Thus mass peristalsis usually takes place three or four times a day, during a meal or immediately after.

Absorption and Feces Formation

By the time the intestinal contents arrive at the large intestine, digestion and absorption are almost complete. By the time the chyme has remained in the large intestine 3 to 10 hours, it has become solid or semisolid as a result of absorption and is now known as **feces.** Chemically, feces consist of water, inorganic salts, epithelial cells from the mucosa of the alimentary canal, bacteria, products of bacterial decomposition, and undigested parts of food not attacked by bacteria. Mucus is secreted by the glands of the large intestine, but no enzymes are secreted.

Chyme is prepared for elimination in the large intestine by the action of bacteria. These bacteria ferment any remaining carbohydrates and release hydrogen, carbon dioxide, and methane gas. They also convert remaining proteins to amino acids and break down the amino acids into simpler substances: indole, skatole, hydrogen sulfide, and fatty acids. Some of the indole and skatole is carried off in the feces and contributes to its odor. The rest is absorbed and transported to the liver, where they are converted to less toxic compounds and excreted in the urine. Bacteria also decompose bilirubin to simpler pigments, which give feces their brown color. Several vitamins needed for normal metabolism, including some B vitamins and vitamin K, are synthesized by bacterial action and absorbed.

Although most water absorption occurs in the small intestine, the large intestine absorbs enough to make it an important organ in maintaining the body's water balance. Intestinal water absorption is greatest in the cecum and ascending colon. The large intestine also absorbs inorganic solutes.

Defecation

Mass peristaltic movements push fecal material into the rectum. The resulting distension of the rectal walls stimulates pressure-sensitive receptors, initiating a reflex for **defecation,** which is emptying of the rectum. Contraction of the longitudinal rectal muscles shortens the rectum, thereby increasing the pressure inside it. The pressure along with voluntary contractions of the diaphragm and abdominal muscles forces the sphincters open, and the feces are expelled through the anus. Voluntary contractions of the diaphragm and abdominal muscles aid defecation by increasing the pressure inside the abdomen, which pushes the walls of the sigmoid colon and rectum inward. If defecation does not occur, the feces remain in the rec-

tum until the next wave of mass peristalsis again stimulates the pressure-sensitive receptors, creating the desire to defecate.

APPLICATIONS TO HEALTH

DENTAL CARIES

Dental caries, or tooth decay, involve a gradual demineralization (softening) of the enamel and dentin (Figure 21-26a). If this condition remains untreated, various microorganisms may invade the pulp, causing inflammation and infection with subsequent death (necrosis) of the dental pulp and abscess of the alveolar bone surrounding the root's apex. Such teeth are treated by root canal therapy.

The process of dental caries is initiated when bacteria act on carbohydrates deposited on the tooth, giving off acids which demineralize the enamel. Microbes that digest carbohydrates include two bacteria, *Lactobacillus acidophilus* and *Streptococcus mutans.* Research suggests that the streptococci break down carbohydrates into *dental plaque,* a polysaccharide that adheres to the tooth surface. When other bacteria digest the plaque, acid is produced. Saliva cannot reach the tooth surface to buffer the acid because the plaque covers the teeth.

Preventive measures include prenatal diet supplements (chiefly vitamin D, calcium, and phosphorus) and fluoride treatments to protect against acids during the period when teeth are being calcified. Naturally occurring excessive fluoride may cause a light brown to brownish black discoloration of the enamel of the permanent teeth called mottling.

Brushing the teeth immediately after eating removes the plaque from flat surfaces before the bacteria have a chance to go to work. Dentists also suggest that the plaque between the teeth be removed every 24 hours with dental floss or by flushing with a water irrigation device.

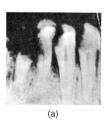 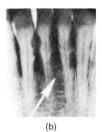

(a)　　　(b)

FIGURE 21-26 Diseases of the teeth. (a) Anteroposterior projection of dental caries. (b) Anteroposterior projection of periodontal disease in which the alveolar bone (arrow) has been destroyed. (Courtesy of Lester W. Paul and John H. Juhl, *The Essentials of Roentgen Interpretation,* 3rd ed., Harper & Row, Publishers, Inc., New York, 1972.)

PERIODONTAL DISEASE

Periodontal disease is a collective term for a variety of conditions characterized by inflammation and degeneration of the gingivae, alveolar bone, periodontal ligament, and cementum. The initial symptoms are enlargement and inflammation of the soft tissue and bleeding gums. Without treatment, the soft tissue may deteriorate and the alveolar bone may be resorbed, causing loosening of the teeth and receding of the gums (Figure 21-26b).

Periodontal diseases are frequently caused by poor oral hygiene, local irritants, such as bacteria, impacted food, and cigarette smoke, or by a poor "bite." The latter may put a strain on the tissues supporting the teeth. Periodontal diseases may also be caused by allergies, vitamin deficiencies, and a number of systemic disorders, especially those that affect bone, connective tissue, or circulation.

PEPTIC ULCERS

An **ulcer** is a craterlike lesion in a membrane. Ulcers that develop in areas of the alimentary canal exposed to acid gastric juice are called *peptic ulcers*. Peptic ulcers occasionally develop in the lower end of the esophagus, but most occur on the lesser curvature of the stomach, where they are called *gastric ulcers*, or in the first part of the duodenum, where they are called *duodenal ulcers* (Figure 21-27).

Hypersecretion of acid gastric juice seems to be the immediate cause of duodenal ulcers. In gastric ulcer patients, because the stomach walls are highly adapted to resist gastric juice through their secretion of mucus, the cause may be hyposecretion of mucus. Hypersecretion of pepsin also may contribute to ulcer formation.

Among the factors believed to stimulate an increase in acid secretion are emotions, certain foods or medications (alcohol, coffee, aspirin), and overstimulation of the vagus nerve. Normally, the mucous membrane lining the stomach and duodenal walls resists the secretions of hydrochloric acid and pepsin. In some people, however, this resistance breaks down and an ulcer develops.

The danger inherent in ulcers is the erosion of the muscular portion of the wall of the stomach or duodenum. This erosion may damage blood vessels and produce fatal hemorrhaging. If an ulcer erodes all the way through the wall, the condition is called *perforation*. Perforation allows bacteria and partially digested food to pass into the peritoneal cavity, producing peritonitis.

CIRRHOSIS

Cirrhosis is a chronic disease of the liver in which the parenchymal (functional) liver cells are replaced by fibrous or adipose connective tissue, a process called *stromal repair*. The liver has a high capacity for parenchymal regeneration, and stromal repair occurs whenever a parenchymal cell is killed or cells are damaged continuously for a long time. These conditions may be caused by hepatitis (inflammation of the liver), certain chemicals that destroy liver cells, parasites that infect the liver, and alcoholism.

GALLSTONES

Gallstones or **biliary calculi** are a major health problem in the United States today, affecting more than 15 million people. *Cholecystectomy* (removal of the gallbladder) is the most frequently performed major operation. The ailment is four times more common in women than in men, and the number of persons suffering with gallstones is highest in the 55 to 65 year old age group. The cholesterol in bile may crystallize at any point between bile canaliculi, where it is first apparent, and where the bile enters the duodenum. The fusion of single crystals is the beginning of 95 percent of all gallstones.

Cholesterol gallstones can cause obstruction to the outflow of bile from the gallbladder and irritate the mucosal surface. These two factors, combined with the presence of bacteria, may lead to precipitation of other substances upon the pure cholesterol core of the gallstone. Following their formation, gallstones gradually grow in size and number and may cause minimal, intermittent, or complete obstruction to the flow of bile from the gallbladder into the duct system. If obstruction of the outlet occurs, and the gallbladder cannot empty as it normally does after eating, the pressure within it increases and the individual may have intense pain or discomfort (*biliary colic*). The pain may radiate to the right shoulder or to the lower back. Bacterial content in the bile is great. With obstruction and stasis, bacteria increase in number. The products of bacterial action may produce a number of additional symptoms, including fever. The longer the calculi are in the gallbladder, the greater the incidence of calculi in the duct system. Small calculi may pass from the gallbladder to the cystic duct and into the common bile duct. Calculi may also ascend into the intrahepatic duct system and obstruct segments of the liver. Complete obstruction of the flow of bile into the duodenum may result in death.

For years, the standard test for gallstones has been oral cholecystography. In this procedure, tablets are taken orally and become concentrated as a radiopaque dye in the gallbladder. If stones are present, the dye makes them visible by outlining them on x-ray pictures.

Intravenous cholangiography is performed when oral cholecystography demonstrates a nonfunctioning gallbladder. This procedure supplies evidence that resolves the questions of whether the bile ducts are dilated, whether stones are present, and whether there is actual obstruction of the cystic duct, in which case the gallbladder is not visualized.

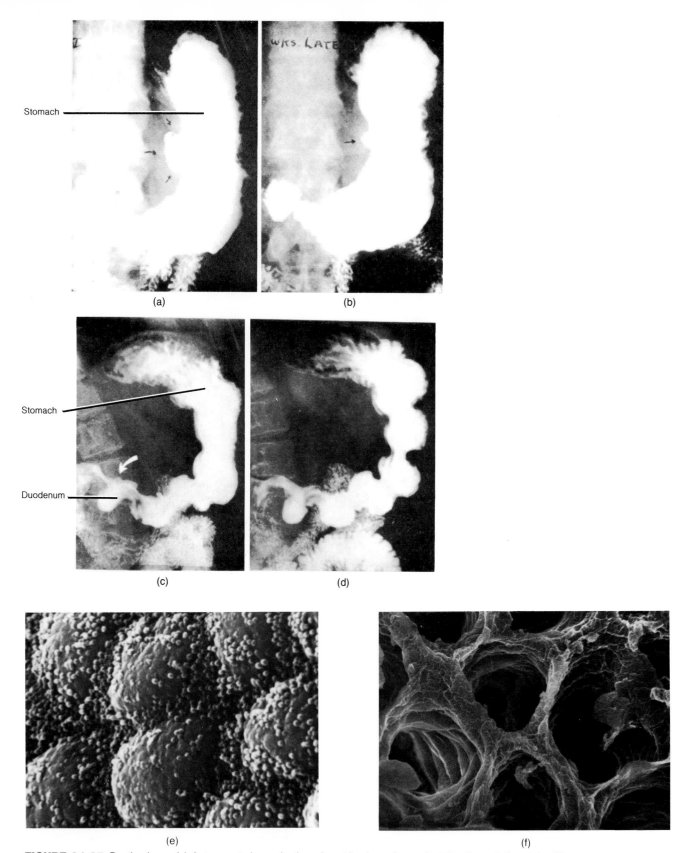

FIGURE 21-27 Peptic ulcers. (a) Anteroposterior projection of gastric ulcers (arrows) at the time of diagnosis. (b) The same ulcers three weeks after treatment. (c) Anteroposterior projection of a duodenal ulcer (arrow) prior to treatment. (d) The same ulcer after treatment. (Courtesy of Lester W. Paul and John H. Juhl, *The Essentials of Roentgen Interpretation,* 3rd ed., Harper & Row, Publishers, Inc., New York, 1972.) (e) Scanning electron micrograph of the normal stomach mucosa at a magnification of 5000×. (f) Scanning electron micrograph of ulceration of the stomach mucosa at a magnification of 1000×. (Courtesy of Fisher Scientific Company and S.T.E.M. Laboratories, Inc., Copyright 1975.)

More recently, the ultrasound technique has been used to diagnose gallstones. Unlike x-rays, ultrasound does not involved ionizing radiation, so it is not dangerous. It uses very high-frequency sound waves which echo off the internal organs. The result is a picture of what is going on inside. Ultrasonography has been applied to the biliary tree. Gallbladder distension, distension of the bile ducts, and calculi within these structures may be demonstrated by this technique. Sonography will demonstrate at least 90 percent of gallbladder stones regardless of the presence or absence of jaundice and is most effective in demonstrating large stones.

Several gallstone-dissolving drugs are being studied. One of them, chenodeoxycholic acid, appears to be effective and capable of desaturating bile and dissolving cholesterol gallstones. Unfortunately, it seems to work only in patients with well-functioning gallbladders and even then only about half the time. The second drug, ursodeoxycholic acid, seems to be more effective in lower doses. Another gallstone-dissolving drug being tested is monooctanoin. However, much more research is needed on these drugs to determine the long-term effects of interfering with cholesterol's metabolic pathways. At present, it seems that surgery offers the only definitive treatment of gallstones.

APPENDICITIS

Appendicitis is an inflammation of the vermiform appendix. It is preceded by obstruction of the lumen of the appendix by fecal material, inflammation, a foreign body, carcinoma of the cecum, stenosis, or kinking of the organ. The infection that follows as a result of obstruction causes inflammation of all layers of the appendix. This may result in edema, ischemia, gangrene, and perforation. Rupture of the appendix results in *peritonitis*. Loops of the bowel, the omentum, and the parietal peritoneum may become adherent and form an abscess, either at the site of the appendix or elsewhere in the abdominal cavity.

Typically, appendicitis begins with pain in the umbilical region of the abdomen followed by anorexia (lack or loss of appetite for food), nausea, and vomiting. After several hours, the pain shifts to the lower right quadrant and is continuous, dull or severe, and intensified by coughing, sneezing, or body movements.

Early appendectomy is recommended in all suspected cases because it is safer to operate than to risk gangrene, rupture, and peritonitis.

DIVERTICULITIS

Diverticulitis is inflammation of diverticula, small blind pouches that form in the lining and wall of the colon when the muscles become weak. When bacteria or other irritating agents collect in these pouches, the ensuing inflammation may cause muscle spasms and cramplike pains in the abdomen, especially in the lower left quadrant. A barium enema will reveal the presence of diverticula.

Research done by placing inflated balloons in the colon and measuring the pressure created by segmentation indicates that diverticula form because of lack of sufficient bulk in the colon during segmentation. The powerful contractions, working against insufficient bulk, create a pressure so high that it causes the colonic walls to blow out.

The increase in diverticular disease, from a rarity at the turn of the century to a disorder of an estimated 25 to 33 percent of middle-aged and older Americans today, has been attributed to a shift to a low residue diet. A recent test result reported that 62 of 70 patients treated for diverticular disease with a high residue diet containing unprocessed bran showed marked relief of symptoms. People who choose a fiber-rich, unrefined diet will greatly reduce their chances of developing the diseases of an underfed large bowel such as diverticular disease.

Treatment consists of bed rest, cleansing enemas, and drugs to reduce infection. In severe cases, portions of the affected colon may require surgical removal and a temporary colostomy.

PERITONITIS

Peritonitis is an acute inflammation of the serous membrane lining the abdominal cavity and covering the abdominal viscera. One possible cause is contamination of the peritoneum by pathogenic bacteria from the external environment. This contamination could result from accidental or surgical wounds in the abdominal wall or from perforation or rupture of organs with consequent exposure to the outside environment. Another possible cause is perforation of the walls of organs that contain bacteria or chemicals beneficial to the organ but toxic to the peritoneum. For example, the large intestine contains colonies of bacteria that live on undigested nutrients and break them down so they can be eliminated. But if the bacteria enter the peritoneal cavity, they attack the cells of the peritoneum for food and produce acute infection. Moreover, the peritoneum has no natural barriers that keep it from being irritated or digested by chemical substances such as bile and digestive enzymes. Although it contains a great deal of lymphatic tissue and can combat infection fairly well, the peritoneum is in contact with most of the abdominal organs. If infection gets out of hand, it may destroy vital organs and bring on death. For these reasons, perforation of the alimentary canal from an ulcer is considered serious. A surgeon planning to do extensive surgery on the colon may give the patient high doses of antibiotics for several days prior to surgery to kill intestinal bacteria and reduce the risk of peritoneal contamination.

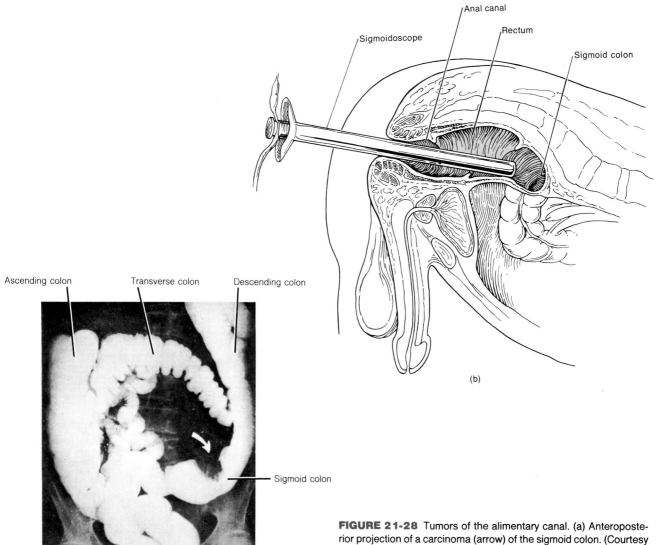

(a)

(b)

FIGURE 21-28 Tumors of the alimentary canal. (a) Anteroposterior projection of a carcinoma (arrow) of the sigmoid colon. (Courtesy of Lester W. Paul and John H. Juhl, *The Essentials of Roentgen Interpretation,* 3rd ed., Harper & Row, Publishers, Inc., New York, 1972.) (b) Detection of carcinomas by use of a sigmoidoscope.

TUMORS

Both benign and malignant **tumors** can occur in all parts of the gastrointestinal tract. The benign growths are much more common, but malignant tumors are responsible for 30 percent of all deaths from cancer in the United States.

For early diagnosis, complete routine examinations are necessary. Cancers of the mouth usually are detected through routine dental checkups. A regular physical checkup should include rectal examination. Fifty percent of all rectal carcinomas are within reach of the finger, and 75 percent of all colonic carcinomas can be seen with the sigmoidoscope (Figure 21-28).

Both the fiberoptic sigmoidoscope and the more recent fiberoptic endoscope are flexible tubular instruments composed of a light and many tiny glass fibers. They allow visualization, magnification, biopsy, electrosurgery, and even photography of the entire length of the gastrointesti-

nal tract. Colonoscopy makes possible the identification and removal of malignant polyps of the colon (gastric polypectomy) before invasion of the bowel wall or lymphatic metastasis. It has proved to be a safe and effective treatment that avoids the significant expense, risk, and discomfort of major surgery. Colonoscopy may be the greatest advance yet toward lowering the death rate from cancer of the colon. Unfortunately, this type of cancer has shown a considerable increase in incidence over the past 20 years.

Another test in a routine examination for intestinal disorders is the filling of the gastrointestinal tract with barium, which is either swallowed or given in an enema. Barium, a mineral, shows up on x-rays the same way that calcium appears in bones. Tumors as well as ulcers can be diagnosed this way. The only definitive treatment of gastrointestinal carcinomas, if they cannot be removed using the endoscope, is surgery.

ANOREXIA NERVOSA

Anorexia nervosa is a disorder characterized by an aversion to food and bizarre patterns of eating. The subconsciously self-imposed starvation appears to be a response to emotional conflicts about self-identification and acceptance of a normal adult sex role. The disorder is found predominantly in young, single females. The physical consequence of the disorder is severe and progressive starvation. Amenorrhea (absence of menstruation) and a lowered basal metabolic rate reflect the depressant effects of the starvation. Individuals may become emaciated and may ultimately die of starvation or one of its complications. Treatment consists of psychotherapy and dietary regulation.

KEY MEDICAL TERMS ASSOCIATED WITH THE DIGESTIVE SYSTEM

Cholecystitis (*chole* = bile; *kystis* = bladder; *itis* = inflammation of) Inflammation of the gallbladder that often leads to infection. Some cases are caused by obstruction of the cystic duct with bile stones. Stagnating bile salts irritate the mucosa. Dead mucosal cells provide a medium for the growth of bacteria.

Cholelithiasis (*lithos* = stone) The presence of gallstones.

Colitis Inflammation of the colon and rectum. Inflammation of the mucosa reduces absorption of water and salts, producing watery, bloody feces, and, in severe cases, dehydration and salt depletion. Irritated muscularis spasms produce cramps.

Colonoscopy (*skopein* = to examine) The endoscopic examination of the entire colon.

Colostomy (*stomoun* = provide an opening) An incision of the colon to create an artificial opening or "stoma" to the exterior. This opening serves as a substitute anus through which feces are eliminated. A temporary colostomy may be done to allow a badly inflamed colon to rest and heal. If the rectum is removed for malignancy, the colostomy provides a permanent outlet for feces.

Concretion (*crescere* = to grow) A hardened mass; calculus.

Constipation (*constipatio* = crowding together) Infrequent or difficult defecation.

Diarrhea (*dia* = through; *rhein* = to flow) Frequent defecation of liquid feces.

Flatus Excessive amounts of air (gas) in the stomach or intestine, usually expelled through the anus. If the gas is expelled through the mouth, it is called *belching* (burping). Flatus may result from gas released during the breakdown of foods in the stomach or from swallowing air or gas-containing substances such as carbonated drinks.

Heartburn A burning sensation in the region of the esophagus and stomach. It may result from regurgitation of gastric contents into the lower end of the esophagus or from distension stemming from retention of regurgitated food and gastric contents in the lower esophagus.

Hepatitis (*hepato* = liver) Inflammation of the liver. It may be caused by organisms, such as viruses, bacteria, and protozoa, or by the absorption of materials toxic to liver cells, such as carbon tetrachloride and certain anesthetics and drugs.

Hernia Protrusion of an organ or part of an organ through a membrane or cavity wall, usually the abdominal cavity. *Diaphragmatic* or *hiatal hernia* is the protrusion of the lower esophagus, stomach, or intestine into the thoracic cavity through the hole in the diaphragm that allows passage of the esophagus. *Umbilical hernia* is the protrusion of abdominal organs through the navel area of the abdominal wall. *Inguinal hernia* is the protrusion of the hernial sac containing the intestine into the inguinal opening. It may extend into the scrotal compartment, causing strangulation of the herniated part.

Mumps Viral disease causing painful inflammation and enlargement of the salivary glands, particularly the parotids. In adults, the sex glands and pancreas may be involved. Inflammation of testes in postpubescent males may cause sterility.

Nausea (*nausia* = seasickness) Discomfort preceding vomiting. Possibly, it is caused by distension or irritation of the gastrointestinal tract, most commonly the stomach.

Pancreatitis Inflammation of the pancreas. The pancreas secretes active trypsin instead of trypsinogen, and the trypsin digests the pancreatic cells and blood vessels.

Vomiting The forcible expulsion of stomach (and sometimes duodenal) contents through the mouth. The abdominal muscles squeeze the stomach, and the lower esophageal sphincter opens so that the gastric contents can be expelled.

STUDY OUTLINE

Digestive Processes

1. Food is prepared for use by cells by five basic activities: ingestion, peristalsis, mechanical and chemical digestion, absorption, and defecation.
2. Chemical digestion is a series of catabolic reactions that break down the large carbohydrate, lipid, and protein molecules of food into molecules that are usable by body cells.
3. Mechanical digestion consists of movements that aid chemical digestion.

General Organization

1. The organs of digestion are usually divided into two main groups: those composing the gastrointestinal (GI) tract, or alimentary canal, a continuous tube running through the ventral body cavity from the mouth to the anus, and the accessory organs (teeth, tongue, salivary glands, liver, gallbladder, pancreas).
2. The basic arrangement of tissues in the alimentary canal from the inside outward is the mucosa, submucosa, muscularis, and serosa (peritoneum).
3. Extensions of the peritoneum include the mesentery, lesser omentum, greater omentum, falciform ligament, and mesocolon.

Mouth, Salivary Glands, and Teeth

1. The mouth is formed by the cheeks, palates, lips, and tongue, which aid mechanical digestion.
2. The tongue, together with its associated muscles, forms the floor of the oral cavity. It is composed of skeletal muscle covered with mucous membrane. The upper surface and sides of the tongue are covered with papillae, some of which contain taste buds.
3. Three pairs of salivary glands (parotid, submandibular [submaxillary], and sublingual) lie outside the mouth and pour their contents into ducts that empty into the oral cavity.
4. The salivary glands produce saliva that lubricates food and starts the chemical digestion of carbohydrates.
5. The teeth, or dentes, project into the mouth and are adapted for mechanical digestion.
6. A typical tooth consists of three principal portions: crown, root, and cervix. Teeth are composed primarily of dentin covered by enamel, the hardest substance in the body.

Pharynx and Esophagus

1. Both pharynx and esophagus assume a role in deglutition, or swallowing.
2. When a bolus is swallowed, the respiratory tract is sealed off and the bolus moves into the esophagus.
3. Peristaltic movements of the esophagus pass the bolus into the stomach.

Stomach

1. The stomach begins at the bottom of the esophagus and ends at the pyloric valve.
2. Adaptations of the stomach for digestion include rugae that permit distension; glands that produce mucus, hydrochloric acid, and enzymes; and a three-layered muscularis for efficient mechanical movement.
3. Nervous and hormonal mechanisms initiate the secretion of gastric juice.
4. Proteins are chemically digested into peptones and proteoses through the action of pepsin in the stomach.
5. The stomach also stores food, produces the intrinsic factor, and carries on some absorption.

Pancreas, Liver, and Gallbladder

1. Pancreatic acini produce enzymes that enter the duodenum via the pancreatic duct. Pancreatic enzymes digest proteins, carbohydrates, and fats.
2. Cells of the liver produce bile, which is needed to emulsify fats.
3. Bile is stored in the gallbladder and passed into the duodenum via the common bile duct.

Small Intestine

1. The small intestine extends from the pyloric valve to the ileocecal valve.
2. It is highly adapted for digestion and absorption. Its glands produce enzymes and mucus, and the microvilli, villi, and plicae circulares of its wall provide a large surface area for absorption.
3. Mechanical digestion in the small intestine involves rhythmic segmentation, pendular movements, and propulsive peristalsis.
4. The entrance of chyme into the small intestine stimulates the secretion of several hormones that coordinate the secretion and release of bile, pancreatic juice, intestinal juice, and inhibit gastric activity.
5. The enzymes in the small intestine digest carbohydrates, proteins, and fats into the end products of digestion: monosaccharides, amino acids, fatty acids, glycerol, and glycerides.
6. Absorption is the passage of the end products of digestion from the alimentary canal into the blood or lymph. Monosaccharides and amino acids pass into the blood capillaries, small aggregations (chylomicrons) of fatty acids and glycerol pass into the blood capillaries, and large chylomicrons enter the lacteal.

Large Intestine

1. The large intestine extends from the ileocecal valve to the anus.
2. Mechanical movements of the large intestine include haustral churning, peristalsis, and mass peristalsis.

3. The last stages of chemical digestion occur in the large intestine through bacterial, rather than enzymatic, action. Substances are further broken down and some vitamins are synthesized. Water absorption leads to feces formation.

4. The elimination of feces from the large intestine is called defecation. Defecation is a reflex action aided by voluntary contractions of the diaphragm and abdominal muscles.

Applications to Health

1. Dental caries are started by acid-producing bacteria.

2. Periodontal diseases are characterized by inflammation and degeneration of gingivae, alveolar bone, periodontal membrane, and cementum.

3. Peptic ulcers are craterlike lesions that develop in the mucous membrane of the alimentary canal in areas exposed to gastric juice.

4. In cirrhosis, parenchymal cells of the liver are replaced by fibrous connective tissue.

5. The fusion of individual crystals of cholesterol is the beginning of 95 percent of all gallstones. Gallstones can cause obstruction to the outflow of bile in any portion of the duct system. Diagnostic tests for gallstones include cholecystography, intravenous cholangiography, and ultrasonography.

6. Appendicitis is an inflammation of the vermiform appendix. Some of its causes include obstruction of the lumen of the appendix from inflammation, a foreign body, carcinoma of the cecum, stenosis, or kinking of the organ.

7. Diverticulitis is inflammation of diverticula in the colon.

8. Peritonitis is inflammation of the peritoneum.

9. Tumors of the gastrointestinal tract may be detected by sigmoidoscopy, colonoscopy, and barium x-ray.

10. Anorexia nervosa is characterized by a self-induced aversion to food.

REVIEW QUESTIONS

1. Define digestion. Distinguish between chemical and mechanical digestion.

2. Identify the organs of the alimentary canal in sequence. How does the alimentary canal differ from the accessory organs of digestion?

3. Describe the structure of each of the four tunics of the alimentary canal.

4. What is the peritoneum? Describe the location and function of the mesentery, lesser omentum, greater omentum, and falciform ligament.

5. What structures form the oral cavity? How does each structure contribute to digestion?

6. Make a simple diagram of the tongue. Indicate the location of the papillae and the four taste zones.

7. Describe the location of the salivary glands and their ducts. What are buccal glands?

8. Describe the composition of saliva and the role of each of its components in digestion. What is the pH of saliva?

9. What are the principal portions of a typical tooth? What are the functions of each part?

10. Compare deciduous and permanent dentitions with regard to numbers of teeth and times of eruption.

11. Contrast the functions of incisors, cuspids, premolars, and molars. What is pyorrhea?

12. What is a bolus? How is it formed?

13. Define deglutition. List the sequence of events involved in passing a bolus from the mouth to the stomach.

14. Describe the location of the stomach. List and briefly explain the anatomical features of the stomach.

15. Distinguish between pyloric stenosis and pylorospasm.

16. What is the importance of rugae, zymogenic cells, parietal cells, and mucous cells in the stomach?

17. Where is the pancreas located? Describe the duct system connecting the pancreas to the duodenum.

18. What are pancreatic acini? Contrast their functions with those of the islets of Langerhans.

19. Where is the liver located? What are its principal functions?

20. Draw a labeled diagram of a liver lobule.

21. How is blood carried to and from the liver?

22. Once bile has been formed by the liver, how is it collected and transported to the gallbladder for storage?

23. Where is the gallbladder located? How is it connected to the duodenum?

24. What are the subdivisions of the small intestine? How are the coats of the small intestine adapted for digestion and absorption?

25. Describe the movements in the small intestine.

26. Define absorption. How are the end products of carbohydrate and protein digestion absorbed? How are the end products of fat digestion absorbed?

27. What routes are taken by absorbed nutrients to reach the liver?

28. What are the principal subdivisions of the large intestine? How does the muscularis of the large intestine differ from that of the rest of the digestive tract?

29. Describe the mechanical movements that occur in the large intestine.

30. Explain the activities of the large intestine that change its contents into feces.

31. Define defecation. How does it occur?

32. Define dental caries. How are they started? What

is dental plaque? What are some preventive measures that can be taken against dental caries?

33. Define periodontal disease and describe how it is caused.

34. What is a peptic ulcer? Distinguish between gastric and duodenal ulcers. Describe some of the suspected causes of ulcers. What is perforation?

35. Define cirrhosis. How is it caused?

36. Explain how gallstones are formed. What are the possible serious consequences of gallstones?

37. Describe some diagnostic tests for the detection of gallstones.

38. Define appendicitis. What are its possible causes?

39. Why is early appendectomy usually indicated?

40. What is diverticulitis? How does it develop?

41. Define peritonitis. Why is it a potentially dangerous condition?

42. How are tumors of the alimentary canal detected?

43. Define anorexia nervosa.

44. Refer to the glossary of key medical terms associated with the digestive system. Be sure that you can define each term.

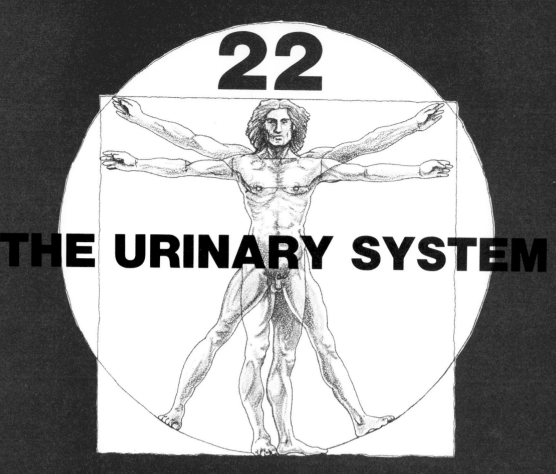

22

THE URINARY SYSTEM

STUDENT OBJECTIVES

- ■ Identify the external and internal gross anatomical features of the kidneys.

- ■ Define the structural features of a nephron.

- ■ Describe the blood and nerve supply to the kidneys.

- ■ Describe the location, structure, and physiology of the ureters.

- ■ Describe the location, structure, histology, and function of the urinary bladder.

- ■ Compare the causes of incontinence, retention, and suppression.

- ■ Describe the location, structure, and physiology of the urethra.

- ■ Discuss the principle of hemodialysis.

- ■ Contrast the abnormal constituents of urine and their clinical significance.

- ■ Discuss the causes of ptosis, kidney stones, gout, glomerulonephritis, pyelitis, cystitis, nephrosis, and polycystic disease.

- ■ Define key medical terms associated with the urinary system.

The metabolism of nutrients results in the production of wastes by body cells, including carbon dioxide and excess water and heat. Protein catabolism produces toxic nitrogenous wastes such as ammonia and urea. In addition, many of the essential ions such as sodium, chloride, sulfate, phosphate, and hydrogen tend to accumulate in excess of the body's needs. All the toxic materials and the excess essential materials must be eliminated.

The primary function of the **urinary system** is to help keep the body in homeostasis by controlling the composition and volume of blood. It does so by removing and restoring selected amounts of water and solutes. It also excretes selected amounts of various wastes. Two kidneys, two ureters, one urinary bladder, and a single urethra make up the system (Figure 22-1). The kidneys regulate the composition and volume of the blood and remove wastes from the blood in the form of urine. Urine is excreted from each kidney through its ureter and is stored in the urinary bladder until it is expelled from the body through the urethra. Other systems that aid in waste elimination are the respiratory, integumentary, and digestive systems (Exhibit 22-1).

The developmental anatomy of the urinary system is considered in Exhibit 24-11.

KIDNEYS

The paired **kidneys** are reddish organs that resemble kidney beans in shape. They are found just above the waist between the parietal peritoneum and the posterior wall of the abdomen. Since they are external to the peritoneal lining of the abdominal cavity, their placement is described as *retroperitoneal* (re'-trō-per-i-tō-NĒ-al) (see Figure 21-18). Other retroperitoneal structures include the ureters and suprarenal glands. Relative to the vertebral column, the kidneys are located between the levels of the last thoracic and third lumbar vertebrae and are partially protected by the eleventh and twelfth pairs of ribs. The right kidney is slightly lower than the left because of the large area occupied by the liver.

EXTERNAL ANATOMY

The average adult kidney measures about 10 to 12 cm (4 to 5 inches) long, 5.0 to 7.5 cm (2 to 3 inches) wide, and 2.5 cm (1 inch) thick. Its concave medial border faces the vertebral column. Near the center of the concave border is a notch called the **hilus,** through which the ureter leaves the kidney. Blood and lymph vessels and nerves also enter and exit the kidney through the hilus (Figure 22-2). The hilus is the entrance to a cavity in the kidney called the **renal sinus.**

Three layers of tissue surround each kidney. The innermost layer, the **renal capsule,** is a smooth, transparent, fibrous membrane that can easily be stripped off the kidney and is continuous with the outer coat of the ureter at the hilus. It serves as a barrier against trauma and the spread of infection to the kidney. The second layer, the **adipose capsule,** is a mass of fatty tissue surrounding the renal capsule. It also protects the kidney from trauma and holds it firmly in place in the abdominal cavity. The outermost layer, the **renal fascia,** is a thin layer of fibrous connective tissue that anchors the kidneys to their surrounding structures and to the abdominal wall.

EXHIBIT 22-1 EXCRETORY ORGANS AND PRODUCTS ELIMINATED

EXCRETORY ORGANS	PRODUCTS ELIMINATED	
	PRIMARY	SECONDARY
Kidneys	Water, soluble salts from protein catabolism, inorganic salts.	Heat and carbon dioxide.
Lungs	Carbon dioxide.	Heat and water.
Skin	Heat.	Water and salts.
Alimentary tract	Solid wastes and secretions.	Carbon dioxide, water, salts, heat.

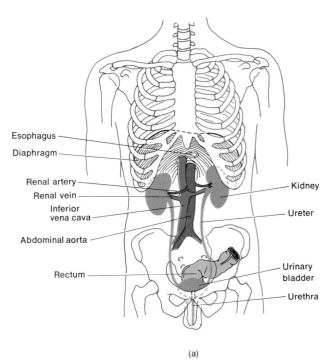

(a)

FIGURE 22-1 Organs of the male urinary system. (a) Diagram.

Kidneys

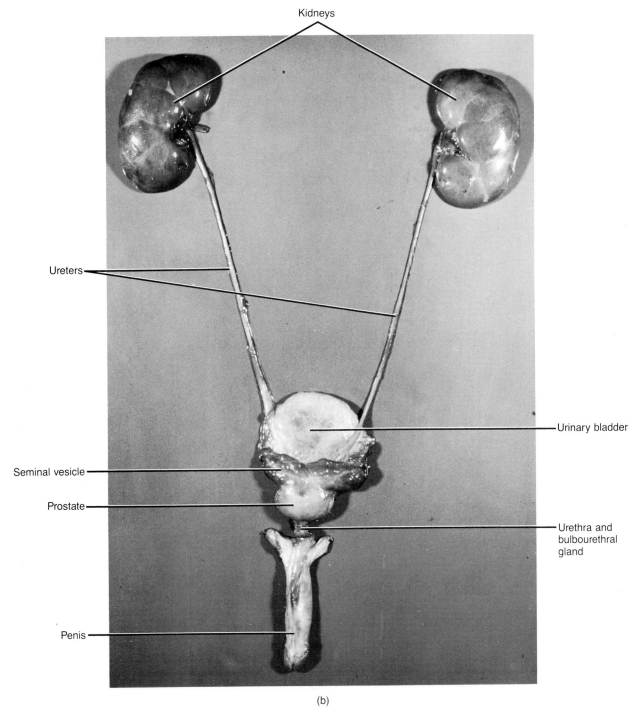

Ureters

Urinary bladder

Seminal vesicle

Prostate

Urethra and
bulbourethral
gland

Penis

(b)

FIGURE 22-1 (*Continued*) Organs of the male urinary system. (b) Photograph. (Courtesy of C. Yokochi and J. W. Rohen, *Photographic Anatomy of the Human Body,* 2nd ed., 1979, IGAKU-SHOIN, Ltd., Tokyo, New York.)

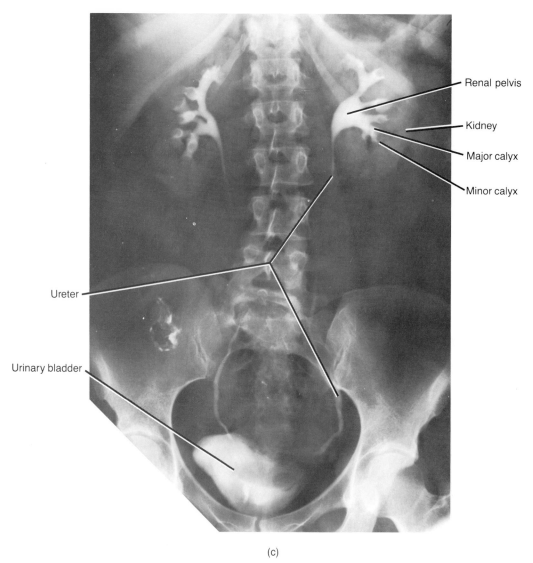

(c)

FIGURE 22-1 (*Continued*) Organs of the male urinary system. (c) Anteroposterior projection. (Courtesy of John C. Bennett, St. Mary's Hospital, San Francisco.)

INTERNAL ANATOMY

A coronal (frontal) section through a kidney reveals an outer, reddish area called the **cortex** and an inner, reddish-brown region called the **medulla** (Figure 22-2b, c). The cortex is arbitrarily divided into an outer cortical zone and an inner juxtamedullary zone. Likewise, the medulla is divided into an outer zone (one-third) and an inner zone (two-thirds). Within the medulla are 8 to 18 striated, triangular structures termed **renal,** or **medullary, pyramids.** The striated appearance is due to the presence of straight tubules and blood vessels. The bases of the pyramids face the cortical area, and their apices, called **renal papillae,** are directed toward the center of the kidney. The cortex is the smooth-textured area extending from the renal capsule to the bases of the pyramids and into the spaces between them. The cortical substance between the renal pyramids forms the **renal columns.** Together the cortex and renal pyramids constitute the parenchyma of the kidney. Structurally, the parenchyma of each kidney consists of approximately 1 million microscopic units called nephrons, collecting ducts, and their associated vascular supply. Nephrons are the functional units of the kidney. They help regulate blood composition and form urine.

In the renal sinus of the kidney is a large cavity called the **renal pelvis.** The edge of the pelvis contains cuplike extensions called the **major** and **minor calyces** (KĀ-li-sēz). There are 2 or 3 major calyces and 8 to 18 minor calyces. Each minor calyx collects urine from collecting ducts of the pyramids. From the major calyces, the urine drains into the pelvis and out through the ureter.

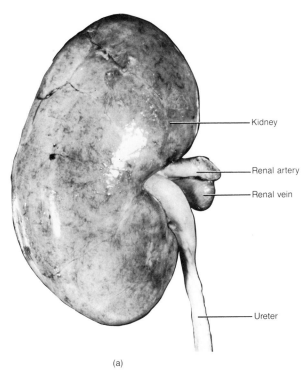

Kidney

Renal artery

Renal vein

Ureter

(a)

FIGURE 22-2 Kidney. (a) Photograph of the external view of the right kidney. (Courtesy of C. Yokochi and J. W. Rohen, *Photographic Anatomy of the Human Body,* 1st ed., 1969, IGAKU-SHOIN, Ltd., Tokyo, New York.) (b) Diagram of a coronal section of the right kidney illustrating the internal anatomy.

NEPHRON

The functional unit of the kidney is the **nephron** (NEF-ron) (Figure 22-3). Essentially, a nephron is a **renal tubule** and its vascular component. It begins as a double-walled cup, called **Bowman's (glomerular) capsule,** lying in the cortex of the kidney. The outer wall, or *parietal layer,* is composed of simple squamous epithelium. It is separated from the inner wall, known as the *visceral layer,* by the *capsular space.* The visceral layer consists of epithelial cells called *podocytes.* It surrounds a capillary network called the **glomerulus** (glō-MER-yoo-lus). Collectively, a Bowman's capsule and its enclosed glomerulus constitute a **renal corpuscle** (KŌR-pus-sul).

The visceral layer of the Bowman's capsule and the endothelium of the glomerulus form an **endothelial-capsular membrane.** This membrane consists of the following parts in the order in which substances filtered by the kidney must pass through (Figure 22-4).

1. Endothelium of the glomerulus. This single layer of endothelial cells has completely open pores averaging 500 to 1,000 Å in diameter.

2. Basement membrane of the glomerulus. This extracellular membrane lies beneath the endothelium and con-

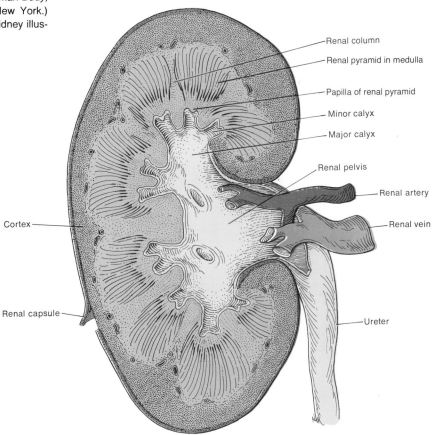

Renal column

Renal pyramid in medulla

Papilla of renal pyramid

Minor calyx

Major calyx

Renal pelvis

Renal artery

Renal vein

Ureter

Cortex

Renal capsule

(b)

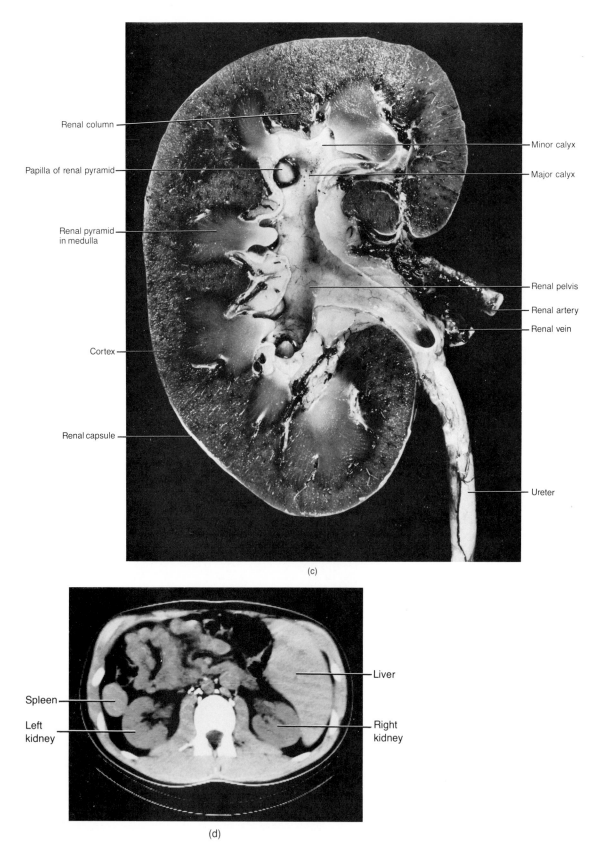

(c)

(d)

FIGURE 22-2 (*Continued*) Kidney. (c) Photograph of a coronal section of the right kidney illustrating the internal anatomy. (Courtesy of C. Yokochi and J. W. Rohen, *Photographic Anatomy of the Human Body,* 2nd ed., 1979, IGAKU-SHOIN, Ltd., Tokyo, New York.) (d) CT scan. (Courtesy of General Electric Co.)

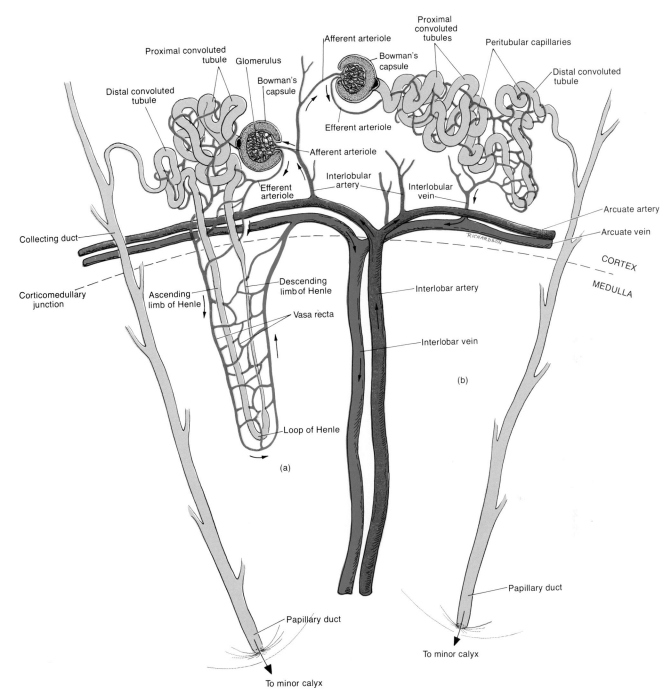

FIGURE 22-3 Nephrons. (a) Juxtamedullary nephron. (b) Cortical nephron.

tains no pores. It consists of fibrils in a glycoprotein matrix. It serves as the dialyzing membrane.

3. Epithelium of the visceral layer of the Bowman's capsule. These epithelial cells, because of their peculiar shape, are called **podocytes.** The podocytes contain foot-like structures called **pedicels** (PED-i-sels). The pedicels are arranged parallel to the circumference of the glomerulus and cover the basement membrane except for spaces between them called **filtration slits** or **slit pores.**

The endothelial-capsular membrane filters water and solutes in the blood. Large molecules, such as proteins, and the formed elements in blood do not normally pass through it. The water and solutes that are filtered out of the blood pass into the capsular space between the visceral and parietal layers of the Bowman's capsule and then into the renal tubule.

The Bowman's capsule opens into the first section of the renal tubule, called the **proximal convoluted tubule,**

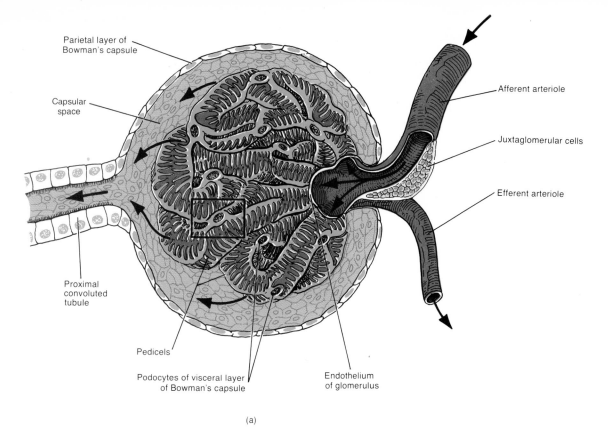

Parietal layer of
Bowman's capsule

Capsular
space

Proximal
convoluted
tubule

Afferent arteriole

Juxtaglomerular cells

Efferent arteriole

Pedicels

Podocytes of visceral layer
of Bowman's capsule

Endothelium
of glomerulus

(a)

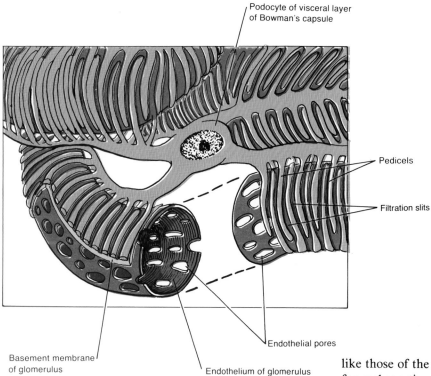

Podocyte of visceral layer
of Bowman's capsule

Pedicels

Filtration slits

Endothelial pores

Basement membrane
of glomerulus

Endothelium of glomerulus

(b)

FIGURE 22-4 Endothelial-capsular membrane. (a) Parts of a renal corpuscle. (b) Enlarged aspect of a portion of the endothelial-capsular membrane.

which also lies in the cortex. Convoluted means the tubule is coiled rather than straight; proximal signifies that the Bowman's capsule is the origin of the tubule. The wall of the proximal convoluted tubule consists of cuboidal epithelium with microvilli. These cytoplasmic extensions,

like those of the small intestine, increase the surface area for reabsorption and secretion.

Nephrons are frequently classified into two kinds. A **cortical nephron** usually has its glomerulus in the outer cortical zone, and the remainder of the nephron rarely penetrates the medulla. A **juxtamedullary nephron** usually has its glomerulus close to the corticomedullary junction, and other parts of the nephron penetrate deeply into the medulla (Figure 22-3). In the juxtamedullary

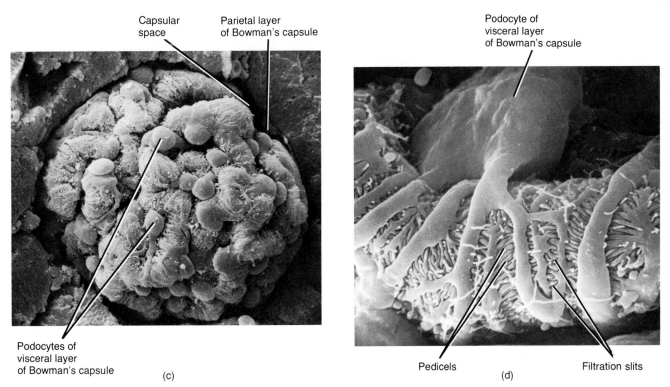

Capsular
space

Parietal layer
of Bowman's capsule

Podocyte of
visceral layer
of Bowman's capsule

Podocytes of
visceral layer
of Bowman's capsule

(c)

Pedicels

(d)

Filtration slits

FIGURE 22-4 (*Continued*) Endothelial-capsular membrane. (c) Scanning electron micrograph of a renal corpuscle at a magnification of 1230×. (Courtesy of Richard G. Kessel and Randy H. Kardon, *Tissues and Organs: A Text-Atlas of Scanning Electron Microscopy,* W. H. Freeman and Company, San Francisco, 1979.) (d) Scanning electron micrograph of a podocyte at a magnification of 7800×. (Courtesy of Richard G. Kessel and Randy H. Kardon, *Tissues and Organs: A Text-Atlas of Scanning Electron Microscopy,* W. H. Freeman and Company, San Francisco, 1979.)

nephron the renal tubule straightens and dips into the medulla, where it is called the **descending limb of Henle** (HEN-lē). This section consists of squamous epithelium. The tubule then bends into a U-shaped structure called the **loop of Henle.** As the tubule straightens, it increases in diameter and ascends toward the cortex as the **ascending limb of Henle,** which consists of cuboidal and low columnar epithelium.

In the cortex the tubule again becomes convoluted. Because of its distance from the point of origin at the Bowman's capsule, this section is referred to as the **distal convoluted tubule.** The cells of the distal tubule, like those of the proximal tubule, are cuboidal. Unlike the cells of the proximal tubule, however, the cells of the distal tubule have few microvilli. In the cortical nephron the proximal section runs into the distal tubule without the intervening limb of Henle. The distal tubule terminates by merging with a straight **collecting duct.**

In the medulla, the collecting ducts receive the distal tubules of several nephrons, pass through the renal pyramids, and open at the renal papillae into the minor calyces through a number of large **papillary ducts.** On the average, there are 30 papillary ducts per renal papilla. Cells of the collecting ducts are cuboidal; those of the papillary ducts are columnar.

The histology of a nephron and glomerulus is shown in Figure 22-5.

BLOOD AND NERVE SUPPLY

The nephrons are largely responsible for removing wastes from the blood and regulating its fluid and electrolyte content. Thus they are abundantly supplied with blood vessels. The right and left **renal arteries** transport about one-fourth the total cardiac output to the kidneys (Figure 22-6). Approximately 1,200 ml passes through the kidneys every minute.

Before or immediately after entering through the hilus, the renal artery divides into several branches that enter the parenchyma and pass as the **interlobar arteries** between the renal pyramids in the renal columns. At the base of the pyramids, the interlobar arteries arch between the medulla and cortex; here they are known as the **arcuate arteries.** Divisions of the arcuate arteries produce a series of **interlobular arteries,** which enter the cortex and divide into **afferent arterioles** (see Figure 22-3).

One afferent arteriole is distributed to each glomerular capsule, where the arteriole divides into the tangled capillary network termed the **glomerulus.** The glomerular capillaries then reunite to form an **efferent arteriole,**

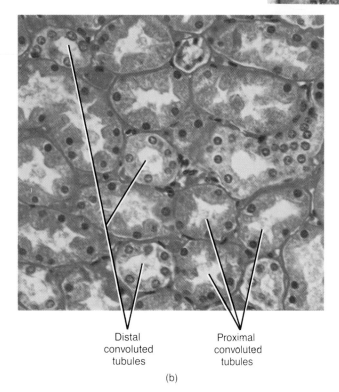

Renal corpuscle

Capsular space

Glomerulus

Bowman's (glomerular) capsule

Macula densa

Distal convoluted tubule

(a)

Distal convoluted tubules

Proximal convoluted tubules

(b)

FIGURE 22-5 Histology of a nephron. (a) Photomicrograph of the cortex of the kidney showing a renal corpuscle and surrounding renal tubules at a magnification of 400×. (b) Photomicrograph of the cortex of the kidney showing renal tubules at a magnification of 400×. (© 1983 by Michael H. Ross. Used by permission.)

which leads away from the capsule and is smaller in diameter than the afferent arteriole. This variation in diameter helps raise the glomerular pressure. The afferent-efferent arteriole situation is unique because blood usually flows out of capillaries into venules and not into other arterioles.

Each efferent arteriole of a cortical nephron divides to form a network of capillaries, called the **peritubular capillaries,** around the convoluted tubules. The efferent arteriole of a juxtamedullary nephron also forms peritubular capillaries. In addition, it forms long loops of thin-walled vessels called **vasa recta** that dip down alongside the loop of Henle into the medullary region of the papilla.

The peritubular capillaries eventually reunite to form **interlobular veins.** The blood then drains through the **arcuate veins** to the **interlobar veins** running between the pyramids and leaves the kidney through a single **renal vein** that exits at the hilus. The vasa recta pass blood into the interlobular veins. From here, it goes to the arcuate veins, the interlobar veins, and then into the renal vein.

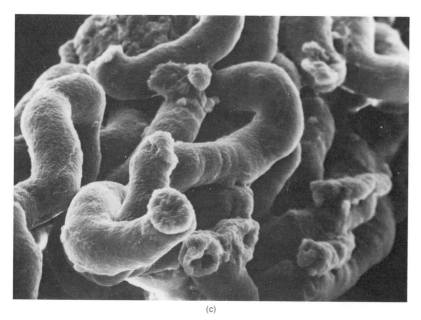

(c)

FIGURE 22-5 (*Continued*) Histology of a nephron. (c) Scanning electron micrograph of several renal tubules at a magnification of 500×. (Courtesy of Fisher Scientific Company and S.T.E.M. Laboratories, Inc., Copyright, 1975.)

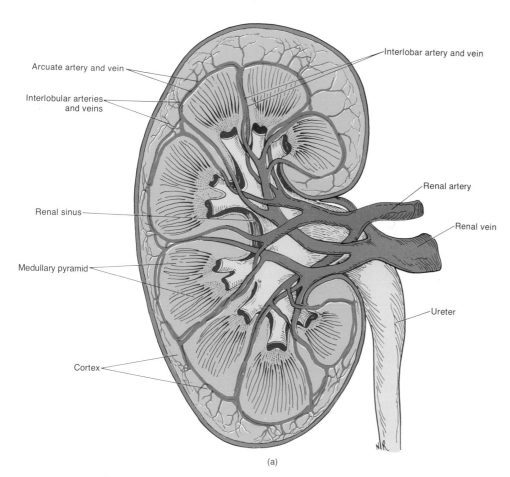

Arcuate artery and vein

Interlobular arteries and veins

Renal sinus

Medullary pyramid

Cortex

Interlobar artery and vein

Renal artery

Renal vein

Ureter

(a)

FIGURE 22-6 Blood supply of the kidney. (a) Right kidney seen in coronal section.

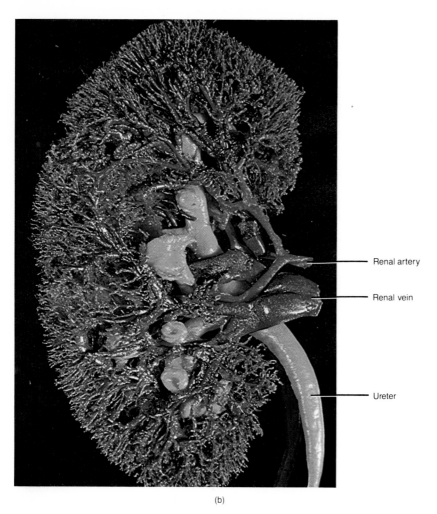

(b)

FIGURE 22-6 (*Continued*) Blood supply of the kidney. (b) Photograph of a cast of the blood supply of the kidneys. (Courtesy of Lester V. Bergman & Associates.)

The nerve supply to the kidneys is derived from the **renal plexus** of the autonomic system. Nerves from the plexus accompany the renal arteries and their branches and are distributed to the vessels. Because the nerves are vasomotor, they regulate the circulation of blood in the kidney by regulating the diameters of the arterioles.

JUXTAGLOMERULAR APPARATUS

As the afferent arteriole approaches the renal corpuscle, the smooth muscle cells of the tunica media become modified. Their nuclei become rounded (instead of elongated), and their cytoplasm contains granules (instead of myofibrils). Such modified cells are called **juxtaglomerular cells.** The cells of the distal convoluted tubule adjacent to the afferent arteriole become considerably narrower. Collectively, these cells are known as the **macula densa.** Together with the modified cells of the afferent arteriole they constitute the **juxtaglomerular apparatus** (Figure 22-7). The juxtaglomerular apparatus helps to regulate renal blood pressure.

PHYSIOLOGY

The major work of the urinary system is done by the nephrons. The other parts of the system are primarily passageways and storage areas. Nephrons carry out three important functions. They control blood concentration and volume by removing selected amounts of water and solutes. They help regulate blood pH. And they remove toxic wastes from the blood. As the nephrons go about these activities, they remove many materials from the blood, return the ones that the body requires, and eliminate the remainder. The eliminated materials are collectively called **urine.** The entire volume of blood in the body is filtered by the kidneys approximately 60 times a day.

URETERS

Once urine is formed by the nephrons and collecting ducts, it drains through papillary ducts into the calyces surrounding the renal papillae. The minor calyces join

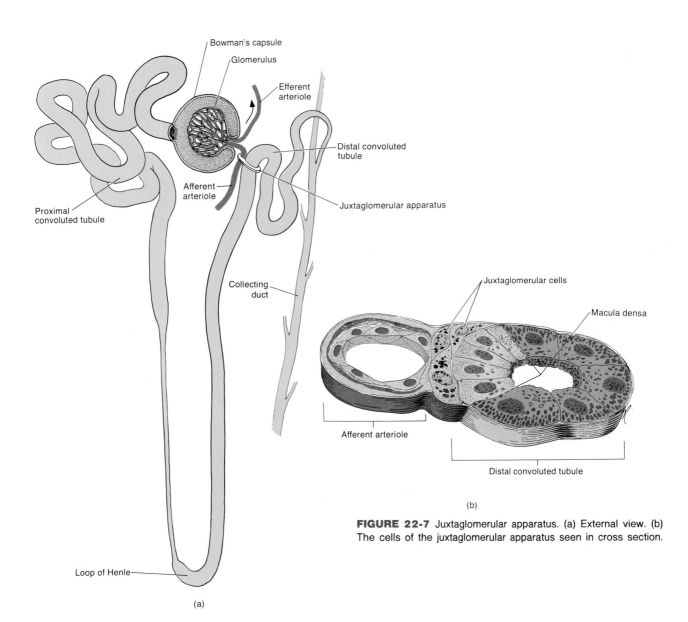

FIGURE 22-7 Juxtaglomerular apparatus. (a) External view. (b) The cells of the juxtaglomerular apparatus seen in cross section.

to become the major calyces, which unite to become the renal pelvis. From the pelvis, the urine drains into the ureters (yoo-RĒ-ters) and is carried by peristalsis to the urinary bladder. From the urinary bladder, the urine is discharged from the body through the single urethra.

STRUCTURE

The body has two **ureters**—one for each kidney. Each ureter is a continuation of the pelvis of the kidney and extends 25 to 30 cm (10 to 12 inches) to the urinary bladder (see Figure 22-1). As the ureters descend, their thick walls increase in diameter, but at their widest point they measure less than 1.7 cm (½ inch) in diameter. Like the kidneys, the ureters are retroperitoneal in placement. The ureters enter the urinary bladder at the superior lateral angle of its base.

Although there are no anatomical valves at the openings of the ureters into the urinary bladder, there are functional ones that are quite effective. Since the ureters pass under the urinary bladder for several centimeters, pressure in the urinary bladder compresses the ureters and prevents backflow of urine when pressure builds up in the urinary bladder during urination. When these physiological valves are not operating, it is possible for cystitis (urinary bladder inflammation) to develop into kidney infection.

Three coats of tissue form the walls of the ureters (Figure 22-8). The inner coat, or mucosa, is mucous membrane with transitional epithelium. The solute concentration and pH of urine differ drastically from the internal environment of cells that form the walls of the ureters. Mucus secreted by the mucosa prevents the cells from coming in contact with urine. Throughout most of the length of the ureters, the second or middle coat, the muscularis, is composed of inner longitudinal and outer circular layers of smooth muscle. The muscularis of the proximal third of the ureters also contains a layer of outer

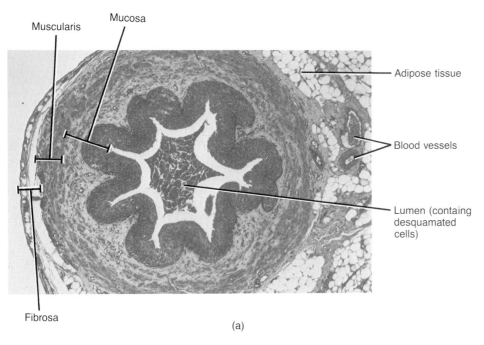

Muscularis

Mucosa

Adipose tissue

Blood vessels

Lumen (containg desquamated cells)

Fibrosa

(a)

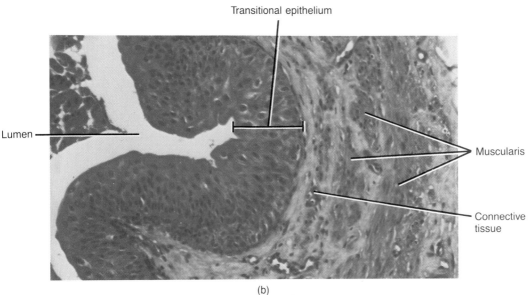

Transitional epithelium

Lumen

Muscularis

Connective tissue

(b)

FIGURE 22-8 Histology of the ureter. (a) Photomicrograph of the ureter seen in cross section at a magnification of 50×. (b) Photomicrograph of an enlarged aspect of the mucosa of the ureter at a magnification of 160×. (© 1983 by Michael H. Ross. Used by permission.)

longitudinal muscle. Peristalsis is the major function of the muscularis. The third, or external, coat of the ureters is a fibrous coat. Extensions of the fibrous coat anchor the ureters in place.

The arterial supply of the ureters is from the renal, testicular or ovarian, common iliac, and inferior vesical arteries. The veins terminate in the corresponding trunks.

The ureters are innervated by the renal and vesical plexuses.

PHYSIOLOGY

The principal function of the ureters is to transport urine from the renal pelvis into the urinary bladder. Urine is carried through the ureters primarily by peristaltic con-

tractions of the muscular walls of the ureters, but hydrostatic pressure and gravity also contribute. Peristaltic waves pass from the kidney to the urinary bladder, varying in rate from 1 to 5/minute depending on the amount of urine formation.

URINARY BLADDER

The **urinary bladder** is a hollow muscular organ situated in the pelvic cavity posterior to the symphysis pubis. In the male, it is directly anterior to the rectum. In the female it is anterior to the vagina and inferior to the uterus. It is a freely movable organ held in position by folds of the peritoneum. The shape of the urinary bladder depends on how much urine it contains. When empty

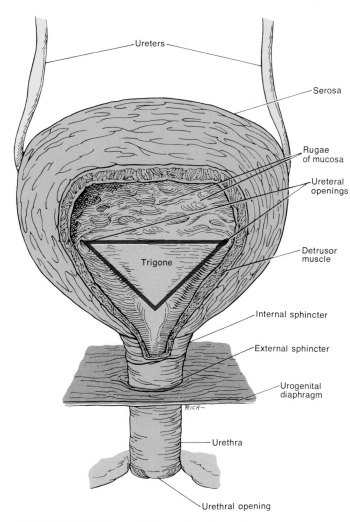

Ureters

Serosa

Rugae
of mucosa

Ureteral
openings

Detrusor
muscle

Trigone

Internal sphincter

External sphincter

Urogenital
diaphragm

Urethra

Urethral opening

FIGURE 22-9 Urinary bladder and female urethra.

it looks like a deflated balloon. It becomes spherical when slightly distended. As urine volume increases, it becomes pear-shaped and rises into the abdominal cavity.

STRUCTURE

At the base of the urinary bladder is a small triangular area, the **trigone** (TRĪ-gōn), that points anteriorly (Figure 22-9). The opening to the urethra is found in the apex of this triangle. At the two points of the base, the ureters drain into the bladder. It is easily identified because the mucosa is firmly bound to the muscularis so that the trigone is typically smooth.

Four coats make up the walls of the urinary bladder (Figure 22-10). The mucosa, the innermost coat, is a mucous membrane containing transitional epithelium. Transitional epithelium is able to stretch—a marked advantage for an organ that must continually inflate and deflate. Stretchability is further enhanced by the rugae (folds in the mucosa) that appear when the urinary bladder is empty. The second coat, the submucosa, is a layer of dense connective tissue that connects the mucosa and

muscular coats. The third coat—a muscular one called the **detrusor** (de-TROO-ser) **muscle**—consists of three layers: inner longitudinal, middle circular, and outer longitudinal muscles. In the area around the opening to the urethra, the circular fibers form an **internal sphincter** muscle. Below the internal sphincter is the **external sphincter,** which is composed of skeletal muscle. The outermost coat, the serous coat, is formed by the peritoneum and covers only the superior surface of the organ.

The arteries of the urinary bladder are the superior vesical, the middle vesical, and the inferior vesical. The veins from the urinary bladder pass to the internal iliac trunk.

The nerves are derived partly from the hypogastric sympathetic plexus and partly from the second and third sacral nerves. The fibers from the sacral nerves constitute the nervi erigentes.

PHYSIOLOGY

Urine is expelled from the urinary bladder by an act called **micturition** (mik'-too-RISH-un), commonly known as urination or voiding. This response is brought about by a combination of involuntary and voluntary nervous impulses. The average capacity of the bladder is 700 to 800 ml. When the amount of urine in the urinary bladder

---

CLINICAL APPLICATION

A lack of voluntary control over micturition is referred to as **incontinence.** In infants about 2 years old and under, incontinence is normal because neurons to the external sphincter muscle are not completely developed. Infants void whenever the urinary bladder is sufficiently distended to arouse a reflex stimulus. Proper training overcomes incontinence if the latter is not caused by emotional stress or irritation of the urinary bladder.

Involuntary micturition in the adult may occur as a result of unconsciousness, injury to the spinal nerves controlling the urinary bladder, irritation due to abnormal constituents in urine, disease of the urinary bladder, and inability of the detrusor muscle to relax due to emotional stress.

Retention, a failure to void urine, may be due to an obstruction in the urethra or neck of the urinary bladder, nervous contraction of the urethra, or lack of sensation to urinate. Far more serious than retention is **suppression,** or **anuria** (a-NOO-rē-a)—failure of the kidneys to secrete urine. It usually occurs when blood plasma is prevented from reaching the glomerulus as a result of inflammation of the glomeruli. Anuria also may be caused by insufficient blood pressure to accomplish filtration. A minimum pressure is required for the kidneys to filter blood as it passes through them.

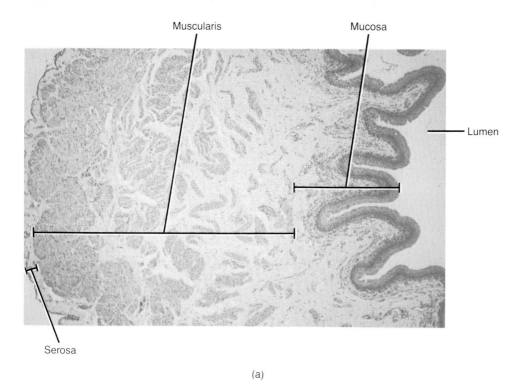

Muscularis · Mucosa

Lumen

Serosa

(a)

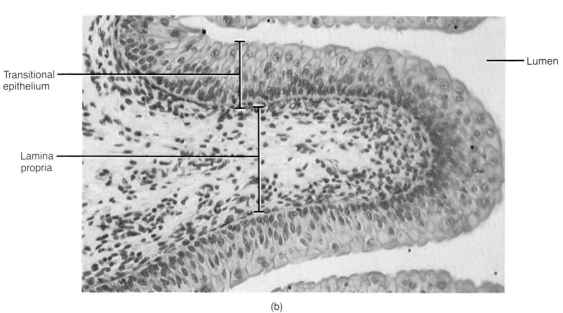

Lumen

Transitional epithelium

Lamina propria

(b)

FIGURE 22-10 Histology of the urinary bladder. (a) Photomicrograph of a portion of the wall of the urinary bladder at a magnification of 50×. (b) Photomicrograph of an enlarged aspect of the mucosa of the urinary bladder at a magnification of 200×. (© 1983 by Michael H. Ross. Used by permission.)

exceeds 200 to 400 ml, stretch receptors in the bladder wall transmit impulses to the lower portion of the spinal cord. These impulses initiate a conscious desire to expel urine and an unconscious reflex referred to as the **micturition reflex.** Parasympathetic impulses transmitted from the sacral area of the spinal cord reach the urinary bladder wall and internal urethral sphincter, bringing about contraction of the detrusor muscle of the urinary bladder and relaxation of the internal sphincter. Then the conscious portion of the brain sends impulses to the external

sphincter, the sphincter relaxes, and urination takes place. Although emptying of the urinary bladder is controlled by reflex, it may be initiated voluntarily and stopped at will because of cerebral control of the external sphincter.

URETHRA

The **urethra** is a small tube leading from the floor of the urinary bladder to the exterior of the body (see Figure

22-9). In females, it lies directly posterior to the symphysis pubis and is embedded in the anterior wall of the vagina. Its undilated diameter is about 6 mm (¼ inch), and its length is approximately 3.8 cm (1½ inch). The female urethra is directed obliquely, inferiorly, and anteriorly. The opening of the urethra to the exterior, the **urethral orifice** is located between the clitoris and vaginal opening.

In males, the urethra is about 20 cm (8 inches) long, and it follows a different course from that of the female urethra. Immediately below the bladder it passes vertically through the prostate gland, then pierces the urogenital diaphragm, and finally traverses the penis and takes a curved course through its body (see Figures 23-1a and 23-8a).

STRUCTURE

The wall of the female urethra consist of three coats: an inner mucous coat that is continuous externally with that of the vulva, an intermediate thin layer of spongy tissue containing a plexus of veins, and an outer muscular coat that is continuous with that of the bladder and consists of circularly arranged fibers of smooth muscle.

The male urethra is composed of two membranes. An inner mucous membrane is continuous with the mucous membrane of the bladder. An outer submucous tissue connects the urethra with the structures through which it passes.

PHYSIOLOGY

Since the urethra is the terminal portion of the urinary system, it serves as the passageway for discharging urine from the body. The male urethra also serves as the duct through which reproductive fluid (semen) is discharged from the body.

APPLICATIONS TO HEALTH

HEMODIALYSIS THERAPY

If the kidneys are so impaired by disease or injury that they are unable to excrete nitrogenous wastes and regulate pH and electolyte concentration of the plasma, the blood must be filtered by an artificial device. Such filtering of the blood is called **hemodialysis.** *Dialysis* means using a semipermeable membrane to separate large nondiffusible particles from smaller diffusible ones. One of the best-known devices for accomplishing dialysis is the kidney machine (Figure 22-11). Blood from the patient's radial artery is pumped through a tube to one side of a semipermeable cellophane membrane. The other side of the membrane is continually washed with an artificial solution called the dialyzing solution. The blood that passes

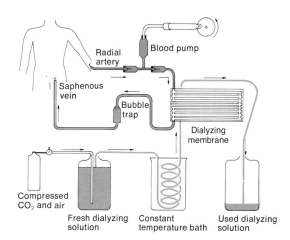

FIGURE 22-11 Operation of an artificial kidney. The blood route is indicated in color. The route of the dialyzing solution is indicated in gray.

through the artificial kidney is treated with an anticoagulant. Only about 500 ml of the patient's blood is in the machine at a time. This volume is easily compensated for by vasoconstriction and increased cardiac output.

All substances (including wastes) in the blood except protein molecules and blood cells can diffuse back and forth across the semipermeable membrane. The electrolyte level of the plasma is controlled by keeping the electrolytes in the dialyzing solution at the concentration found in normal plasma. Any excess plasma electrolytes move down the concentration gradient and into the dialyzing solution. If the plasma electrolyte level is normal, it is in equilibrium with the dialyzing solution and no electrolytes are gained or lost. Since the dialyzing solution contains no wastes, substances such as urea move down the concentration gradient and into the dialyzing solution. Thus wastes are removed and normal electrolyte balance is maintained.

A great advantage of the kidney machine is that nutrition can be bolstered by placing large quantities of glucose in the dialyzing solution. While the blood gives up its wastes, the glucose diffuses into the blood.

There are obvious drawbacks to the artificial kidney, however. Anticoagulants must be added to the blood during dialysis, and a large amount of the patient's blood must flow through the apparatus to make it work. To date, no artificial kidney has been capable of becoming a permanent implant.

ABNORMAL URINE CONSTITUENTS

If the body's chemical processes are not operating efficiently, traces of substances not normally present may appear in the urine or normal constituents may appear in abnormal amounts. Analyzing the physical and chemical properties of urine often provides information that aids diagnosis. Such an analysis is called a **urinalysis.**

Albumin

Protein albumin is a normal constituent of plasma, but it usually does not appear in significant quantities in urine because the particles are too large to pass through the pores in the capillary walls. The presence of albumin in the urine, called **albuminuria,** indicates an increase in the permeability of the glomerular membrane. Conditions that lead to albuminuria include injury to the glomerular membrane as a result of disease, increased blood pressure, and irritation of kidney cells by substances such as bacterial toxins, ether, or heavy metals. Other proteins, such as globulin and fibrinogen, may also appear in the urine under certain conditions.

Glucose

The presence of sugar in the urine is termed **glycosuria.** Normal urine contains such small amounts of glucose that clinically it may be considered absent. The most common cause of glycosuria is a high blood sugar level. Glucose is filtered into the Bowman's capsule. In the proximal convoluted tubules, the tubule cells actively transport the glucose back into the blood. However, the number of glucose carrier molecules is limited. If more carbohydrates are ingested than can be converted into glycogen or fat, more sugar is filtered into the Bowman's capsule than can be removed by the carriers. This condition, called **temporary** or **alimentary glycosuria,** is not considered pathological. Another nonpathological cause is emotional stress. Stress can cause excessive amounts of epinephrine to be secreted. Epinephrine stimulates the breakdown of glycogen and the liberation of glucose from the liver. A **pathological glycosuria** results from diabetes mellitus. In this case, there is a frequent or continuous elimination of glucose because the pancreas fails to produce sufficient insulin. When glycosuria occurs with a normal blood sugar level, the problem lies in failure of the kidney tubular cells to reabsorb glucose.

Erythrocytes

The appearance of red blood cells in the urine is called **hematuria.** Hematuria generally indicates a pathological condition. One cause is acute inflammation of the urinary organs as a result of disease or irritation from kidney stones. Whenever blood is found in the urine, additional tests are performed to ascertain the part of the urinary tract that is bleeding. One should also make sure the sample was not contaminated with menstrual blood from the vagina.

Leucocytes

The presence of leucocytes and other components of pus in the urine, referred to as **pyuria,** indicates infection in the kidney or other urinary organs. Again the source of the pus must be located, and care should be taken that the urine was not contaminated.

Ketone Bodies

Ketone, or acetone, bodies appear in normal urine in small amounts. Their appearance in high quantities, a condition called **ketosis,** or **acetonuria,** may indicate abnormalities. It may be caused by diabetes mellitus, starvation, or simply too little carbohydrate in the diet. Whatever the cause, excessive quantities of fatty acids are oxidized in the liver and the ketone bodies are filtered from the plasma into Bowman's capsule.

Casts

Microscopic examination of urine may reveal **casts**—tiny masses of material that have hardened and assumed the shape of the lumens of the tubules. They are flushed out of the tubules by a buildup of filtrate behind them. Casts are named after the substances that compose them or for their appearance. There are white-blood-cell casts, red-blood-cell casts, epithelial casts that contain cells from the walls of the tubes, granular casts that contain decomposed cells which form granules, and fatty casts from cells that have become fatty.

Renal Calculi

Occasionally, the salts found in urine may solidify into insoluble stones called **renal calculi** or **kidney stones** (Figure 22-12). They may be formed in any portion of the urinary tract from the kidney tubules to the external opening. Conditions leading to calculi formation include the ingestion of excessive mineral salts, a decrease in the

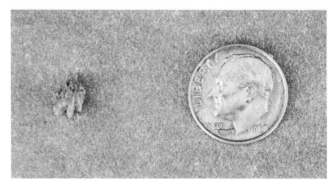

FIGURE 22-12 Photograph of a renal calculus adjacent to a dime for comparative size. (Courtesy of Matt Iacobino.)

amount of water, abnormally alkaline or acid urine, and overactivity of the parathyroid glands. Common constituents of stones are uric acid, calcium oxalate, and calcium phosphate. The stones usually form in the pelvis of the kidney, where they cause pain, hematuria, and pyuria. A *staghorn calculus* may fill the entire collecting system; a *dendritic stone* involves some but not all of the calyces. Severe pain occurs when a stone passes through a ureter and stretches its walls. *Ureteral stones* are seldom completely obstructive because they are usually needle-shaped and urine can flow around them.

FLOATING KIDNEY (PTOSIS)

Floating kidney (ptosis) occurs when the kidney is no longer held in place securely by the adjacent organs or its covering of fat and slips from its normal position. Some individuals, especially thin people in whom either the adipose capsule or renal fascia is deficient, may develop ptosis. It is dangerous because it may cause kinking of the ureter with reflux of urine and retrograde pressure. Pain occurs if the ureter is twisted. Also, if the kidneys drop below the rib cage, they become susceptible to blows and penetrating injuries.

GOUT

Gout is a hereditary condition associated with an excessively high level of uric acid in the blood. When the purine-type nucleic acids are catabolized, a certain amount of uric acid is produced as a waste. Some people seem to produce excessive amounts of uric acid, and others seem to have trouble excreting normal amounts. In either case, uric acid accumulates in the body and tends to solidify into crystals that are deposited in the joints and kidney tissue. Gout is aggravated by excessive use of diuretics, dehydration, and starvation.

GLOMERULONEPHRITIS (BRIGHT'S DISEASE)

Glomerulonephritis, or **Bright's disease,** is an inflammation of the kidney that involves the glomeruli. One of the most common causes of glomerulonephritis is an allergic reaction to the toxins given off by streptococci bacteria that have recently infected another part of the body, especially the throat. The glomeruli become so inflamed, swollen, and engorged with blood that the glomerular membranes become highly permeable and allow blood cells and proteins to enter the filtrate. Thus the urine contains many erythrocytes and much protein. The glomeruli may be permanently changed, leading to chronic renal disease and renal failure.

PYELITIS AND PYELONEPHRITIS

Pyelitis is an inflammation of the kidney pelvis and its calyces. **Pyelonephritis** is the interstitial inflammation of one or both kidneys. It usually involves both the parenchyma and the renal pelvis and is due to bacterial invasion from the middle and lower urinary tracts or the bloodstream.

CYSTITIS

Cystitis is an inflammation of the urinary bladder involving principally the mucosa and submucosa. It may be caused by bacterial infection, chemicals, or mechanical injury.

NEPHROSIS

Nephrosis results from leaking of the glomerular membrane, which permits large amounts of protein to escape from the blood into the urine. Water and sodium then accumulate in the body, especially around the ankles and feet, abdomen, and eyes—a condition called edema. Nephrosis is more common in children than adults, but occurs in all ages. Although it cannot always be cured, certain synthetic steroid hormones such as cortisone and prednisone, which are similar to the natural hormones secreted by the adrenal glands, can suppress some forms of it.

POLYCYSTIC DISEASE

Polycystic disease may be caused by a defect in the renal tubular system that deforms nephrons and results in cyst-like dilations along their course. It is the most common inherited disorder of the kidneys. The kidney tissue is riddled with cysts, small holes, and fluid-filled bubbles ranging in size from a pinhead to the diameter of an egg. These cysts gradually increase until they squeeze out the normal tissue, interfering with kidney function and causing uremia. The chief symptom is weight gain. The kidneys themselves may enlarge from their normal ½ lb to as much as 30 lb. Many people lead normal lives without ever knowing they have the disease, however, and the condition is sometimes discovered only after death. Although the disease is progressive, its advance can be slowed by diet, drugs, and fluid intake. Kidney failure as a result of polycystic disease seldom occurs before the mid-40s and frequently can be postponed until the 60s.

KEY MEDICAL TERMS ASSOCIATED WITH THE URINARY SYSTEM

Azotemia (*azo* = nitrogen-containing; *emia* = condition of blood) Presence of urea or other nitrogenous elements in the blood.

Cystocele (*cyst* = bladder; *cele* = cyst) Hernia of the urinary bladder.

Cystoscope (*skopein* = to examine) Instrument used to examine the urinary bladder.

Dysuria (*dys* = painful; *uria* = urine) Painful urination.

Enuresis (*enourein* = to void urine) Bed-wetting; may be due to faulty toilet training, to some psychological or emotional disturbance, or rarely to some physical disorder.

Intravenous pyelogram (*intra* = within; *veno* = vein; *pyelo* = pelvis of kidney; *gram* = written or recorded) X-ray film of the kidneys after injection of a dye.

Nephroblastoma (*neph* = kidney; *blastos* = germ or forming; *oma* = tumor) Embryonal carcinosarcoma or Wilms's tumor.

Nephrocele Hernia of the kidney.

Oliguria (*olig* = scanty) Scanty urine.

Polyuria (*poly* = much) Excessive urine.

Stricture Narrowing of the lumen of a canal or hollow organ, as the ureter or urethra.

Uremia (*emia* = condition of blood) Toxic levels of urea in the blood resulting from severe malfunction of the kidneys.

Urethritis Inflammation of the urethra, caused by highly acid urine, the presence of bacteria, or constriction of the urethral passage.

STUDY OUTLINE

1. The primary function of the urinary system is to regulate the concentration and volume of blood by removing and restoring selected amounts of water and solutes. It also excretes wastes.
2. The organs of the urinary system are the kidneys, ureters, urinary bladder, and urethra.

Kidneys

1. The kidneys are retroperitoneal organs attached to the posterior abdominal wall.
2. Three layers of tissue surround the kidneys: renal capsule, adipose capsule, and renal fascia.
3. Internally, the kidneys consist of a cortex, medulla, pyramids, papillae, columns, calyces, and a pelvis.
4. The nephron is the functional unit of the kidneys.
5. Each juxtamedullary nephron consists of a Bowman's capsule, glomerulus, proximal convoluted tubule, descending limb of Henle, loop of Henle, ascending limb of Henle, distal convoluted tubule, and collecting duct. The limb of Henle is absent in a cortical nephron.
6. The filtering unit of a nephron is the endothelial-capsular membrane.
7. The extensive flow of blood through the kidney begins in the renal artery and terminates in the renal vein.
8. The nerve supply to the kidney is derived from the renal plexus.
9. The juxtaglomerular apparatus consists of the juxtaglomerular cells of the afferent arteriole and the macula densa of the distal convoluted tubule.

Ureters

1. The ureters are retroperitoneal and consist of a mucosa, muscularis, and fibrous coat.
2. The ureters transport urine from the renal pelvis to the urinary bladder, primarily by peristalsis.

Urinary Bladder

1. The urinary bladder is posterior to the symphysis pubis. Its function is to store urine prior to micturition.
2. Histologically, the urinary bladder consists of a mucosa (with rugae), a muscularis (detrusor muscle), and a serous coat.
3. A lack of control over micturition is called incontinence, failure to void urine is referred to as retention, and inability of the kidneys to produce urine is called suppression (anuria).

Urethra

1. The urethra is a tube leading from the floor of the urinary bladder to the exterior.
2. It is considerably longer in the male. Its function is to discharge urine from the body.

Applications to Health

1. Filtering blood through an artificial device is called hemodialysis. The kidney machine filters the blood of wastes and adds nutrients.
2. A chemical and physical analysis of urine is called urinalysis. Abnormal conditions diagnosed through

urinalysis include albuminuria, glycosuria, hematuria, pyuria, ketosis, casts, and calculi.
3. Ptosis, or floating kidney, occurs when the kidney slips from its normal position.
4. Renal calculi are kidney stones.
5. Gout is a high level of uric acid in the blood.
6. Glomerulonephritis is an inflammation of the glomeruli of the kidney.
7. Pyelitis is an inflammation of the kidney pelvis and calyces. Pyelonephritis is the interstitial inflammation of one or both kidneys involving the parenchyma and renal pelvis.
8. Cystitis is an inflammation of the urinary bladder.
9. Nephrosis leads to protein in the urine due to glomerular membrane permeability.
10. Polycystic disease is an inherited kidney disease in which nephrons are deformed.

REVIEW QUESTIONS

1. What are the functions of the urinary system? What organs compose the system?
2. Contrast the functions of the lungs, integument, and alimentary tract as excretory organs.
3. Describe the location of the kidneys. Why are they said to be retroperitoneal?
4. Prepare a labeled diagram that illustrates the principal external and internal features of the kidney.
5. What is a nephron? List and describe the parts of a nephron from the Bowman's capsule to the collecting duct.
6. How are nephrons supplied with blood?
7. Describe the structure, histology, and function of the ureters.
8. How is the urinary bladder adapted to its storage function?
9. What is micturition?
10. Contrast the causes of incontinence, retention, and suppression.
11. Compare the urethra in the male and female.
12. What is hemodialysis? Briefly describe the operation of an artificial kidney.
13. Define each of the following: albuminuria, glycosuria, hematuria, pyuria, ketosis, casts, and calculi.
14. Define each of the following: ptosis, renal calculi, gout, glomerulonephritis, pyelitis, pyelonephritis, cystitis, nephrosis, and polycystic disease.
15. Refer to the glossary of key medical terms associated with the urinary system. Be sure that you can define each term.

23

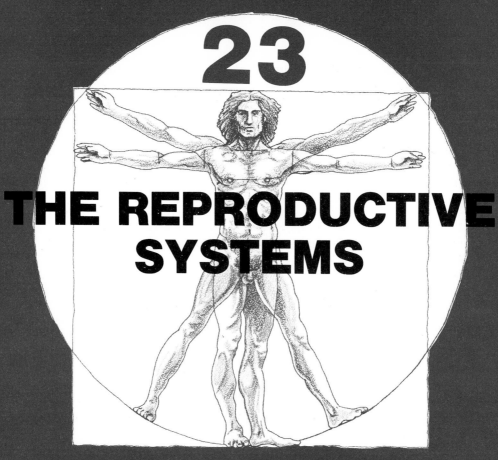

THE REPRODUCTIVE SYSTEMS

STUDENT OBJECTIVES

- Define reproduction and classify the organs of reproduction by function.
- Describe the structure and function of the scrotum.
- Explain the structure, histology, and functions of the testes.
- Describe the effects of testosterone.
- Describe the straight tubules and rete testis as components of the duct system of the testes.
- Describe the location, structure, histology, and functions of the ductus epididymis.
- Explain the structure and functions of the ductus deferens.
- Describe the structure and function of the ejaculatory duct.
- List and describe the three anatomical subdivisions of the male urethra.
- Explain the location and functions of the seminal vesicles, prostate gland, and bulbourethral glands, the accessory reproductive glands.
- Explain the structure and functions of the penis.
- Describe the location, histology, and functions of the ovaries.
- Explain the location, structure, histology, and functions of the uterine tubes.

- Describe the location and ligamentous attachments of the uterus.
- Explain the histology and blood supply of the uterus.
- Describe the effects of estrogens and progesterone.
- Describe the location, structure, and functions of the vagina.
- List and explain the components of the vulva and note the function of each component.
- Describe the anatomical landmarks of the perineum.
- Explain the structure and histology of the mammary glands.
- Explain the symptoms and causes of disorders of the reproductive systems.
- Contrast ovarian cysts, endometriosis, and leukorrhea.
- Discuss breast and cervical cancer. Explain special detection procedures such as mammography, thermography, ultrasonic scanning, and CT/M.
- Describe pelvic inflammatory disease.
- Define key medical terms associated with the reproductive systems.

eproduction is the mechanism by which the thread of life is sustained. It is the process by which a single cell duplicates its genetic material, allowing an organism to grow and repair itself. In this sense, reproduction maintains the life of the individual. But reproduction is also the process by which genetic material is passed from generation to generation. In this regard, reproduction maintains the continuation of the species.

The organs of the male and female reproductive systems may be grouped by function. The testes and ovaries, also called **gonads,** function in the production of gametes—sperm cells and ova, respectively. The gonads also secrete hormones. The **ducts** transport, receive, and store gametes. Still other reproductive organs, called **accessory glands,** produce materials that support gametes.

The developmental anatomy of the reproductive system is considered in Exhibit 24-12.

MALE REPRODUCTIVE SYSTEM

The organs of the male reproductive system (Figure 23-1) are the testes, or male gonads, which produce sperm, a number of ducts that either store or transport sperm

to the exterior, accessory glands that add secretions constituting the semen, and several supporting structures, including the penis.

SCROTUM

The **scrotum** is a cutaneous outpouching of the abdomen consisting of loose skin and superficial fascia (Figure 23-1). It is the supporting structure for the testes. Externally, it looks like a single pouch of skin separated into lateral portions by a median ridge called the **raphe** (RĀ-fē). Internally, it is divided by a septum into two sacs, each containing a single testis. The septum consists of superficial fascia and contractile tissue called the **dartos** (DAR-tōs), which consists of bundles of smooth muscle fibers. The dartos is also found in the subcutaneous tissue of the scrotum and is directly continuous with the subcutaneous tissue of the abdominal wall. The dartos causes wrinkling of the skin of the scrotum.

The location of the scrotum and the contraction of its muscle fibers regulate the temperature of the testes, organs that produce sperm and the male hormone testosterone. The production and survival of sperm and the

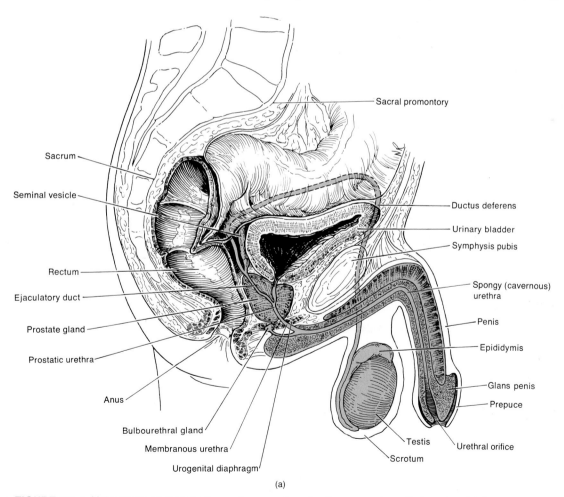

(a)

FIGURE 23-1 Male organs of reproduction and surrounding structures seen in sagittal section. (a) Diagram.

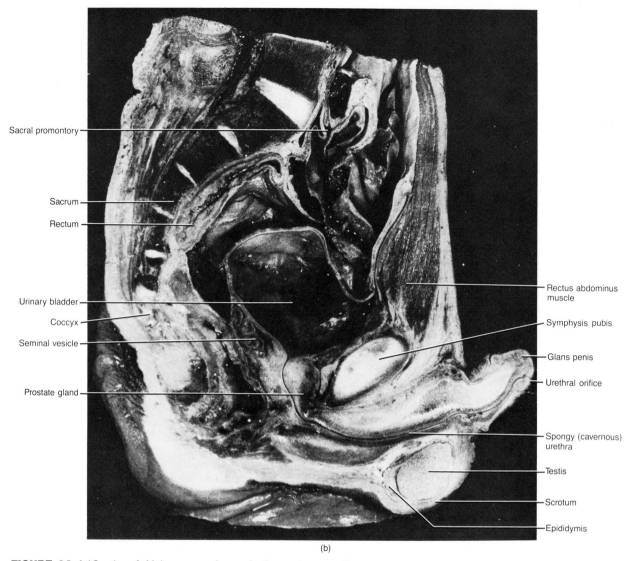

(b)

FIGURE 23-1 (*Continued*) Male organs of reproduction and surrounding structures seen in sagittal section. (b) Photograph. (Courtesy of C. Yokochi and J. W. Rohen, *Photographic Anatomy of the Human Body*, 2nd ed., 1979, IGAKU-SHOIN, Ltd., Tokyo, New York.)

hormone require a temperature that is lower than body temperature. Because the scrotum is outside the body cavities, it supplies an environment about 3° F below body temperature. A skeletal muscle surrounding the testes and in the spermatic cord, the **cremaster muscle,** elevates the testes upon exposure to cold, moving the testes closer to the pelvic cavity where they can absorb body heat. Exposure to warmth reverses the process.

The blood supply of the scrotum is derived from the internal pudendal branch of the internal iliac, the cremasteric branch of the inferior epigastric artery, and the external pudendal artery from the femoral artery. The scrotal veins follow the arteries.

The scrotal nerves are derived from the pudendal, posterior cutaneous of the thigh, and ilioinguinal nerves.

TESTES

The **testes** are paired oval glands measuring about 5 cm (2 inches) in length and 2.5 cm (1 inch) in diameter. They weigh between 10 and 15 g (Figure 23-2). The testes develop high on the embryo's posterior abdominal wall, and usually enter the scrotum by 32 weeks. Full descent is not complete until just prior to birth.

The testes are covered by a dense layer of white fibrous tissue, the **tunica albuginea** (al'-byoo-JIN-ē-a), that extends inward and divides each testis into a series of internal compartments called **lobules.** Each of the 200 to 300 lobules contains one to three tightly coiled tubules, the convoluted **seminiferous tubules,** that produce sperm by a process called **spermatogenesis.**

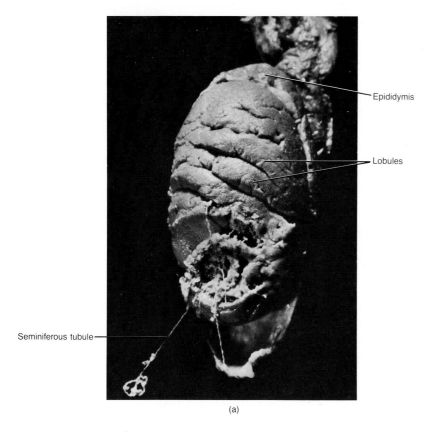

Epididymis

Lobules

Seminiferous tubule

(a)

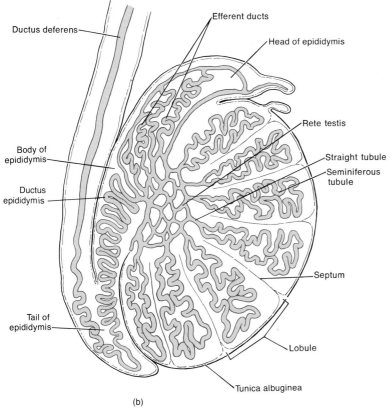

Ductus deferens

Efferent ducts

Head of epididymis

Body of epididymis

Rete testis

Straight tubule

Seminiferous tubule

Ductus epididymis

Septum

Tail of epididymis

Lobule

Tunica albuginea

(b)

FIGURE 23-2 External and internal anatomy of testes. (a) Photograph of external view. (Courtesy of C. Yokochi and J. W. Rohen, *Photographic Anatomy of the Human Body,* 1st ed., 1969, IGAKU-SHOIN, Ltd. Tokyo, New York.) (b) Diagram of a sagittal section illustrating the internal anatomy.

CLINICAL APPLICATION

When the testes do not descend, the condition is referred to as **cryptorchidism** (krip-TOR-ki-dizm). Cryptorchidism results in sterility because the cells involved in the initial development of sperm cells are destroyed by the higher body temperature of the pelvic cavity. Undescended testes can be placed in the scrotum by administering hormones or by surgical means prior to puberty without ill effects.

A cross section through a seminiferous tubule reveals that it is packed with sperm cells in various stages of development (Figure 23-3). The most immature cells, the **spermatogonia,** are located against the basement membrane. Toward the lumen in the center of the tube, one can see layers of progressively more mature cells. In order of advancing maturity, these are primary spermatocytes, secondary spermatocytes, and spermatids. By the time a **sperm cell,** or **spermatozoon** (sper'-ma-tō-ZŌ-on), has reached full maturity, it is in the lumen of the tubule and begins to be moved through a series of ducts. Embedded between the developing sperm cells in the tubules are **Sertoli** (ser-TŌ-lē) **cells** that produce secretions for supplying nutrients to the spermatozoa.

Between the seminiferous tubules are clusters of **interstitial cells of Leydig** (LĪ-dig). These cells secrete the male hormone testosterone, the most important androgen. Since the testes produce sperm and testosterone, they are both exocrine and endocrine glands.

The testes are supplied by the testicular arteries, which arise from the aorta immediately below the origin of the renal arteries. Blood from the testes is drained on the right side by the testicular vein which enters the inferior vena cava and on the left side by the testicular vein which enters the left renal vein.

The nerve supply to the testes is from the testicular plexus which contains vagal parasympathetic fibers and sympathetic fibers from the tenth and eleventh thoracic segment of the spinal cord.

Spermatozoa

Spermatozoa are produced or matured at the rate of about 300 million per day and, once ejaculated, have a life expectancy of about 48 hours within the female reproductive tract. A spermatozoon is highly adapted for reaching and penetrating a female ovum. It is composed of a head, a middle piece, and a tail (Figure 23-4). Within the *head* are the nuclear material and the *acrosome,* which contains enzymes that effect penetration of the sperm cell into

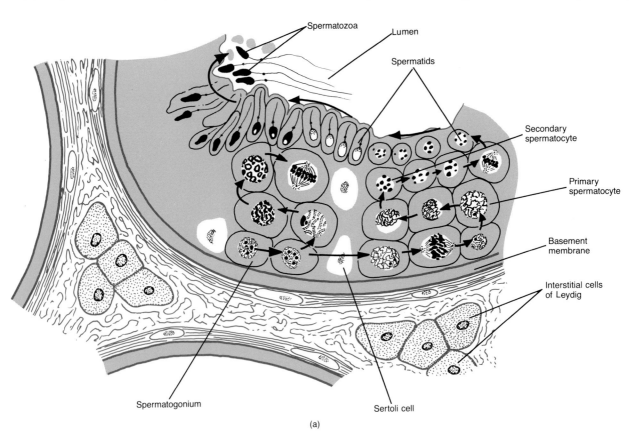

(a)

FIGURE 23-3 Histology of the testes. (a) Diagram of a cross section of a portion of a seminiferous tubule showing the stages of spermatogenesis.

Lumina of seminiferous tubules

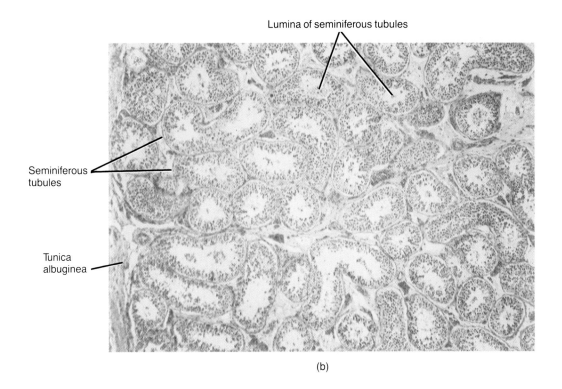

Seminiferous
tubules

Tunica
albuginea

(b)

Spermatids (immature)   Spermatozoa

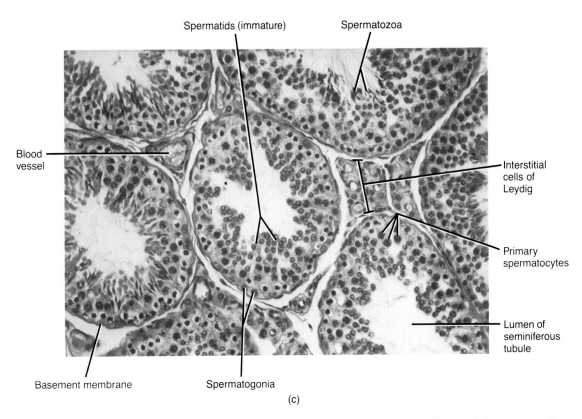

Blood
vessel

Interstitial
cells of
Leydig

Primary
spermatocytes

Lumen of
seminiferous
tubule

Basement membrane   Spermatogonia

(c)

FIGURE 23-3 (*Continued*) Histology of the testes. (b) Photomicrograph of several seminiferous tubules at a magnification of 60×. (c) Photomicrograph of an enlarged aspect of several seminiferous tubules at a magnification of 350X. (© 1983 by Michael H. Ross. Used by permission.)

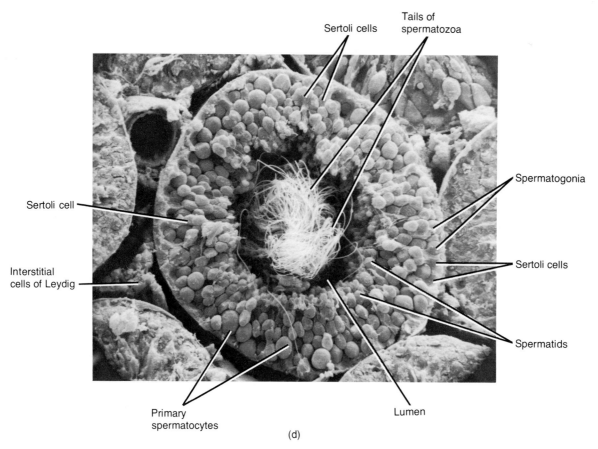

Sertoli cells

Tails of spermatozoa

Spermatogonia

Sertoli cells

Spermatids

Sertoli cell

Interstitial cells of Leydig

Primary spermatocytes

Lumen

(d)

FIGURE 23-3 (*Continued*) Histology of the testes. (d) Scanning electron micrograph of a seminiferous tubule at a magnification of 663×. (Courtesy of Richard G. Kessel and Randy H. Kardon, *Tissues and Organs: A Text-Atlas of Scanning Electron Microscopy,* W. H. Freeman and Company, San Francisco, 1979.)

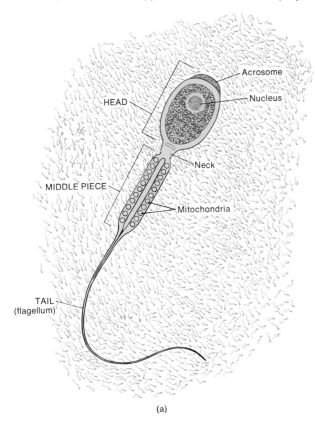

Acrosome

Nucleus

HEAD

Neck

MIDDLE PIECE

Mitochondria

TAIL (flagellum)

(a)

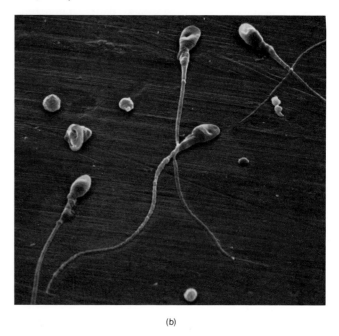

(b)

FIGURE 23-4 Spermatozoa. (a) Parts of a spermatozoon. (b) Scanning electron micrograph of several spermatozoa at a magnification of 2000×. (Courtesy of Fisher Scientific Company and S.T.E.M. Laboratories, Inc., Copyright 1975.)

the ovum. Numerous mitochondria in the *middle piece* carry on the metabolism that provides energy for locomotion. The *tail,* a typical flagellum, propels the sperm along its way.

Testosterone

Secretions of the anterior pituitary gland assume a major role in the developmental changes associated with puberty. At the onset of puberty the anterior pituitary starts to secrete gonadotropic hormones called follicle-stimulating hormone (FSH) and interstitial cell-stimulating hormone (ICSH). Their release is controlled from the hypothalamus by follicle-stimulating hormone releasing factor (FSHRF) and interstitial cell-stimulating hormone releasing factor (ICSHRF). Once secreted, the gonadotropic hormones have profound effects on male reproductive organs. FSH acts on the seminiferous tubules to initiate spermatogenesis. ICSH acts on the seminiferous tubules and further assists the tubules to develop mature sperm, but its chief function is to stimulate the interstitial cells of Leydig to secrete testosterone (tes-TOS-te-rōn).

Testosterone has a number of effects on the body. It controls the development, growth, and maintenance of the male sex organs. It also stimulates bone growth, protein anabolism, sexual behavior, final maturation of sperm, and the development of male secondary sex characteristics. These characteristics, which appear at puberty, include muscular and skeletal development resulting in wide shoulders and narrow hips; body hair patterns that include pubic hair, axillary and chest hair (within hereditary limits), facial hair, and temporal hairline recession; and enlargement of the thyroid cartilage producing deepening of the voice. Testosterone also stimulates descent of the testes just prior to birth.

DUCTS

Ducts of the Testis

When the sperm mature, they are moved through the convoluted seminiferous tubules to the **straight tubules** (see Figure 23-2b). The straight tubules lead to a network of ducts in the testis called the **rete (RĒ-tē) testis.** Some of the cells lining the rete testis possess cilia that probably help move the sperm along. The sperm are next transported out of the testis through the epididymis.

Epididymis

The sperm are transported out of the testis through a series of coiled **efferent ducts** in the epididymis that empty into a single tube called the ductus epididymis (ep'-i-DID-i-mis). At this point, the sperm are morphologically mature.

The two **epididymides** (ep'i-DID-i-mi-dēz') are comma-shaped organs. Each lies along the posterior border of the testis (see Figures 23-1 and 23-2) and consists mostly of a tightly coiled tube, the **ductus epididymis.** The larger, superior portion of the epididymis is known as the *head.* It consists of the efferent ducts that empty into the ductus epididymis. The *body* of the epididymis contains the ductus epididymis. The *tail* is the smaller, inferior portion. Within the tail, the ductus epididymis continues as the ductus deferens.

The ductus epididymis measures about 6 m (20 ft) in length and 1 mm in diameter. It is tightly packed within the epididymis, which measures only about 3.8 cm (1.5 inches). The ductus epididymis is lined with pseudostratified columnar epithelium, and its wall contains smooth muscle. The free surfaces of the columnar cells contain long, branching microvilli called *stereocilia* (Figure 23-5).

Functionally, the ductus epididymis is the site of sperm maturation. Here they require between 18 hours and 10 days to complete their maturation, that is, they become capable of fertilizing an ovum. The ductus epididymis also stores spermatozoa and propels them toward the urethra during ejaculation by peristaltic contraction of its smooth muscle. Spermatozoa may remain in storage in the ductus epididymis for up to four weeks. After that, they are reabsorbed.

Ductus Deferens

Within the tail of the epididymis, the ductus epididymis becomes less convoluted, its diameter increases, and at this point it is referred to as the **ductus (vas) deferens** or **seminal duct** (see Figure 23-2b). The ductus deferens, about 45 cm long (18 inches), ascends along the posterior border of the testis, penetrates the inguinal canal, and enters the pelvic cavity, where it loops over the side and down the posterior surface of the urinary bladder. The dilated terminal portion of the ductus deferens is known as the **ampulla** (am-POOL-la). Functionally, the ductus deferens stores sperm for up to several months and propels them toward the urethra during ejaculation.

Histologically, the ductus deferens is lined with pseudostratified epithelium and contains a heavy coat of three layers of muscle (Figure 23-6). Peristaltic contractions of the muscular coat propel the spermatozoa toward the urethra during ejaculation.

CLINICAL APPLICATION

One method of sterilization of males is called **vasectomy,** in which a portion of each ductus deferens is removed. In the procedure, an incision is made in the scrotum, the ducts are located, and each is tied in two places. Then the portion between the ties is excised (see Figure 24-29a). Although sperm production continues in the testes, the sperm cannot reach the exterior because the ducts are cut.

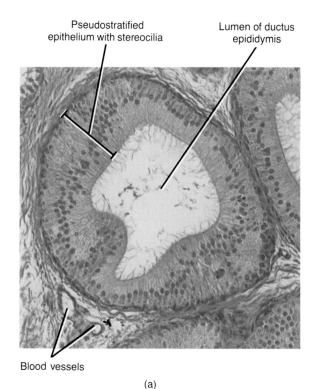

Pseudostratified epithelium with stereocilia

Lumen of ductus epididymis

Blood vessels

(a)

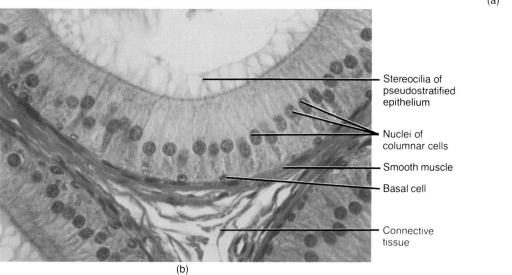

Stereocilia of pseudostratified epithelium

Nuclei of columnar cells

Smooth muscle

Basal cell

Connective tissue

(b)

FIGURE 23-5 Histology of the ductus epididymis. (a) Photomicrograph of the ductus epididymis seen in cross section at a magnification of 160×. (b) Photomicrograph of an enlarged aspect of the mucosa of the ductus epididymis at a magnification of 350×. (© 1983 by Michael H. Ross. Used by permission.)

Traveling with the ductus deferens as it ascends in the scrotum are the testicular artery, autonomic nerves, veins that drain the testes, lymphatics, and a small circular band of skeletal muscle, the cremaster (krē-MAS-ter) muscle. These structures constitute the **spermatic cord,** a supporting structure of the male reproductive system. The cremaster muscle elevates the testes during sexual stimulation and exposure to cold. The spermatic cord passes through the **inguinal** (IN-gwin-al) **canal,** a slitlike passageway in the anterior abdominal wall just superior to the medial half of the inguinal ligament.

CLINICAL APPLICATION
The area of the inguinal canal and spermatic cord represents a weak spot in the abdominal wall. It is frequently the site of an **inguinal hernia**—a rupture or separation of a portion of the abdominal wall resulting in the protrusion of a part of an organ.

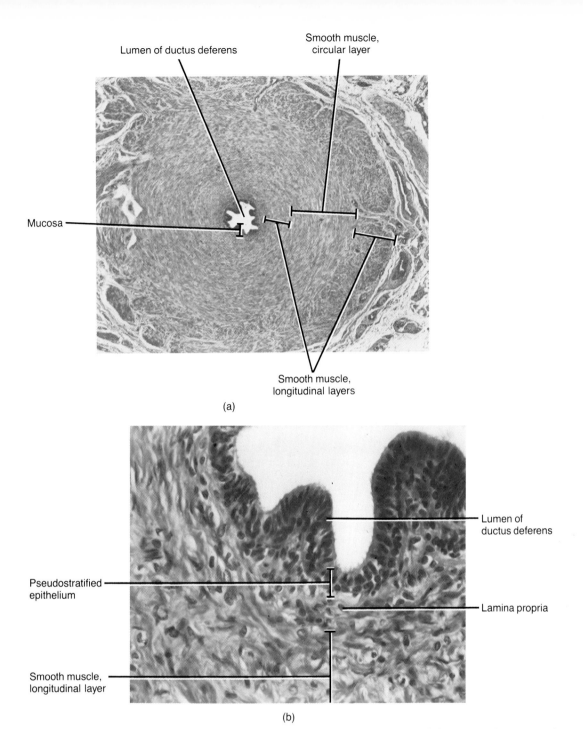

Lumen of ductus deferens

Smooth muscle, circular layer

Mucosa

Smooth muscle, longitudinal layers

(a)

Lumen of ductus deferens

Pseudostratified epithelium

Lamina propria

Smooth muscle, longitudinal layer

(b)

FIGURE 23-6 Histology of the ductus deferens. (a) Photomicrograph of the ductus deferens seen in cross section at a magnification of 40×. (b) Photomicrograph of an enlarged aspect of the mucosa of the ductus deferens at a magnification of 160×. (© 1983 by Michael H. Ross. Used by permission.)

Ejaculatory Duct

Posterior to the urinary bladder are the **ejaculatory** (ē-JAK-yoo-la-tō'-rē) **ducts** (Figure 23-7). Each duct is about 2 cm (1 inch) long and is formed by the union of the duct from the seminal vesicle and ductus deferens. The ejaculatory ducts eject spermatozoa into the prostatic urethra.

Urethra

The **urethra** is the terminal duct of the system, serving as a common passageway for spermatozoa or urine. In the male, the urethra passes through the prostate (PROS-tāt) gland, the urogenital diaphragm, and the penis. It measures about 20 cm (8 inches) in length and is subdivided into three parts (see Figures 23-1 and 23-8). The **prostatic urethra** is 2 to 3 cm (1 inch) long and passes through the prostate gland. It continues inferiorly and as it passes through the urogenital diaphragm, a muscular partition between the two ischiopubic rami, it is known

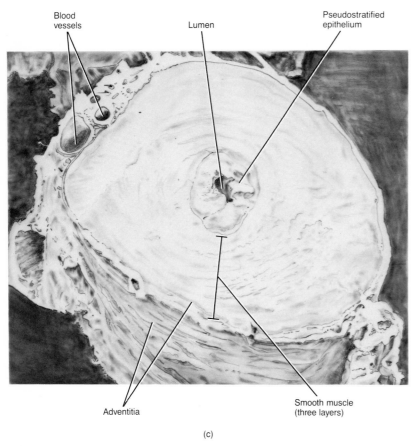

Blood vessels · Lumen · Pseudostratified epithelium · Adventitia · Smooth muscle (three layers)

(c)

FIGURE 23-6 (*Continued*) Histology of the ductus deferens. (c) Drawing of the ductus deferens based on a scanning electron micrograph at a magnification of 78×.

as the **membranous** (MEM-bra-nus) **urethra.** The membranous portion is about 1 cm (½ inch) in length. As it passes through the corpus spongiosum of the penis, it is known as the **spongy (cavernous) urethra.** This portion is about 15 cm (6 inches) long. The spongy urethra enters the bulb of the penis and terminates at the external **urethral orifice.**

ACCESSORY GLANDS

Whereas the ducts of the male reproductive system store and transport sperm cells, the **accessory glands** secrete the liquid portion of semen. The paired **seminal vesicles** (VES-i-kuls) are convoluted pouchlike structures, about 5 cm (2 inches) in length, lying posterior to and at the base of the urinary bladder in front of the rectum (see Figure 23-7). They secrete the alkaline viscous component of semen rich in the sugar fructose and pass it into the ejaculatory duct. The seminal vesicles contribute about 60 percent of the volume of semen.

The **prostate gland** is a single, doughnut-shaped gland about the size of a chestnut (see Figure 23-7). It is inferior to the urinary bladder and surrounds the superior portion of the urethra. The prostate secretes an alkaline fluid that constitutes 13 to 33 percent of the semen into the prostatic urethra.

CLINICAL APPLICATION

In older men, the prostate sometimes enlarges to the point where it compresses the urethra and obstructs urine flow. At this stage, surgical removal of part of or the entire gland, called **prostatectomy** (pros'-ta-TEK-tō-mē), usually is indicated. The prostate gland is also a common tumor site in older males.

The paired **bulbourethral** (bul'-bō-yoo-RĒ-thral), or **Cowper's, glands** are about the size of peas. They are located beneath the prostate on either side of the membranous urethra (see Figure 23-7). The bulbourethral glands secrete mucus for lubrication; their ducts open into the spongy urethra.

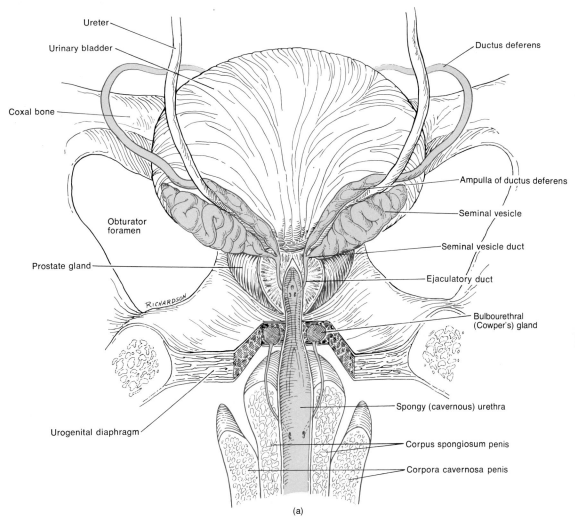

Ureter

Urinary bladder

Coxal bone

Obturator foramen

Prostate gland

RICHARDSON

Urogenital diaphragm

Ductus deferens

Ampulla of ductus deferens

Seminal vesicle

Seminal vesicle duct

Ejaculatory duct

Bulbourethral (Cowper's) gland

Spongy (cavernous) urethra

Corpus spongiosum penis

Corpora cavernosa penis

(a)

FIGURE 23-7 Male reproductive organs in relation to surrounding structures seen in posterior view. (a) Diagram.

SEMEN

Semen, or **seminal fluid,** is a mixture of sperm and the secretions of the seminal vesicles, the prostate gland, and the bulbourethral glands. The average volume of semen for each ejaculation is 2.5 to 6 ml, and the average range of spermatozoa ejaculated is 50 to 100 million per milliliter. When the number of spermatozoa falls below approximately 20 million per milliliter, the male is likely to be sterile. Though only a single spermatozoon fertilizes an ovum, fertilization seems to require the combined action of a large number of spermatozoa. The ovum is enclosed by cells that present a barrier to the sperm. An enzyme called *hyaluronidase* is secreted by the acrosomes of sperm. Hyaluronidase is believed to digest intercellular materials of the cells covering the ovum, giving the sperm a passageway into the ovum.

Semen has a pH range of 7.35 to 7.50—slightly alkaline. The prostatic secretion gives semen a milky appearance, and fluids from the seminal vesicles and bulboure-

thral glands give it a mucoid consistency. Semen provides spermatozoa with a transportation medium and nutrients. It also neutralizes the acid environment of the male urethra and the female vagina. Semen contains enzymes which activate sperm after ejaculation.

Recently, semen has been shown to contain an antibiotic, called *seminalplasmin,* which has the ability to destroy a number of bacteria. Its antimicrobial activity has been described as similar to that currently exerted by penicillin, streptomycin, and tetracyclines. Since both semen and the lower female reproductive tract contain bacteria, seminalplasmin may keep these bacteria under control to help ensure fertilization.

Once ejaculated into the vagina, liquid semen coagulates rapidly because of a clotting enzyme produced by the prostate that acts on a substance produced by the seminal vesicle. This clot liquefies in a few minutes because of another enzyme produced by the prostate gland. It is not clear why semen coagulates and liquefies in this manner.

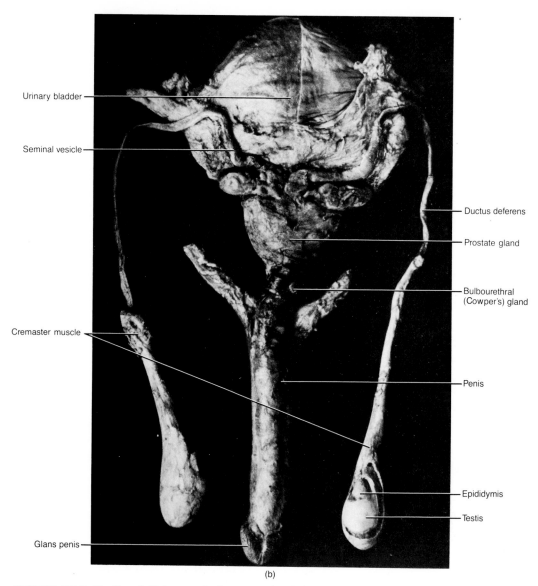

Urinary bladder

Seminal vesicle

Ductus deferens

Prostate gland

Bulbourethral (Cowper's) gland

Cremaster muscle

Penis

Epididymis

Testis

Glans penis

(b)

FIGURE 23-7 (*Continued*) Male reproductive organs in relation to surrounding structures seen in posterior view. (b) Photograph. (Courtesy of C. Yokochi and J. W. Rohen, *Photographic Anatomy of the Human Body,* 2nd ed., 1979, IGAKU-SHOIN, Ltd., Tokyo, New York.)

PENIS

The **penis** is used to introduce spermatozoa into the vagina (Figure 23-8). The distal end of the penis is a slightly enlarged region called the **glans,** which means shaped like an acorn. Covering the glans is the loosely fitting **prepuce** (PRĒ-pyoos), or **foreskin.** Internally, the penis is composed of three cylindrical masses of tissue bound together by fibrous tissue. The two dorsolateral masses are called the **corpora cavernosa penis.** The smaller midventral mass, the **corpus spongiosum penis,** contains the spongy urethra. All three masses of tissue are erectile and contain blood sinuses. Under the influence of sexual stimulation, the arteries supplying the penis dilate, and large quantities of blood enter the blood sinuses. Expansion of these spaces compresses the veins draining the penis so most entering blood is retained. These vascular changes result in an **erection,** a parasympathetic reflex. The penis returns to its flaccid state when the arteries constrict and pressure on the veins is relieved. During ejaculation, a sympathetic reflex, the smooth muscle sphincter at the base of the urinary bladder is closed because of the higher pressure in the urethra caused by expansion of the corpus spongiosum penis. Thus urine is not expelled during ejaculation and semen does not enter the urinary bladder.

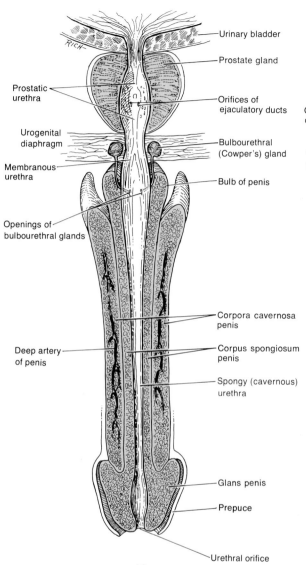

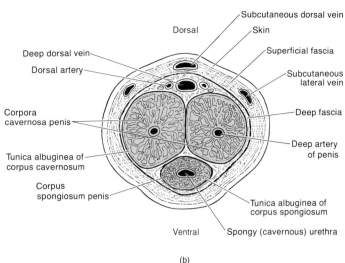

FIGURE 23-8 Internal structure of the penis. (a) Coronal section. (b) Cross section.

The penis has a very rich blood supply from the internal pudendal artery and the femoral artery. The veins drain into corresponding vessels.

The sensory nerves to the penis are branches from the pudendal and ilioinguinal nerves. The corpora have a sympathetic and parasympathetic supply. As a result of parasympathetic stimulation, the blood vessels dilate and the muscles contract. The result is that blood is trapped within the penis and erection is maintained. The musculature of the penis is supplied by the pudendal nerve. The muscles include the bulbocavernosus muscle, which overlies the bulb of the penis, the ischiocavernosus muscles on either side of the penis, and the superficial transverse perineus muscles on either side of the bulb of the penis (see Figures 10-13 and 23-18a).

FEMALE REPRODUCTIVE SYSTEM

The female organs of reproduction (Figure 23-9) include the ovaries, which produce ova (eggs); the uterine or fallopian tubes, which transport the ova to the uterus (womb); the vagina; and external organs that constitute the vulva, or pudendum. The mammary glands, or breasts, also are considered part of the female reproductive system.

OVARIES

The **ovaries,** or female gonads, are paired glands resembling unshelled almonds in size and shape. They are positioned in the upper pelvic cavity, one on each side of the uterus. The ovaries are maintained in position by a series of ligaments (Figures 23-10 and 23-11). They are attached to the broad ligament of the uterus, which is itself part of the parietal peritoneum, by a double-layered fold of peritoneum called the **mesovarium** which surrounds the ovary and ovarian ligament. The ovaries are anchored to the uterus by the **ovarian ligament** and are attached to the pelvic wall by the **suspensory ligament.** Each ovary also contains a **hilus,** the point of entrance for blood vessels and nerves.

The microscope reveals that each ovary consists of the following parts (Figure 23-12).

1. Germinal epithelium. A layer of simple cuboidal epithelium that covers the free surface of the ovary.

2. Tunica albuginea. A capsule of collagenous connective tissue immediately deep to the germinal epithelium.

3. Stroma. A region of connective tissue deep to the tunica albuginea and composed of an outer, dense layer called the *cortex* and an inner, loose layer known as the *medulla.* The cortex contains ovarian follicles.

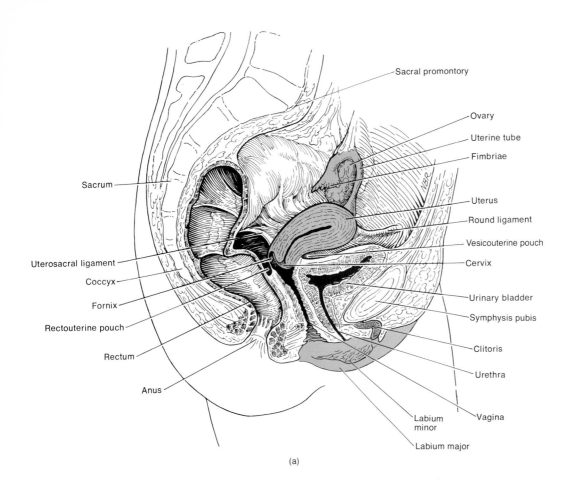

(a)

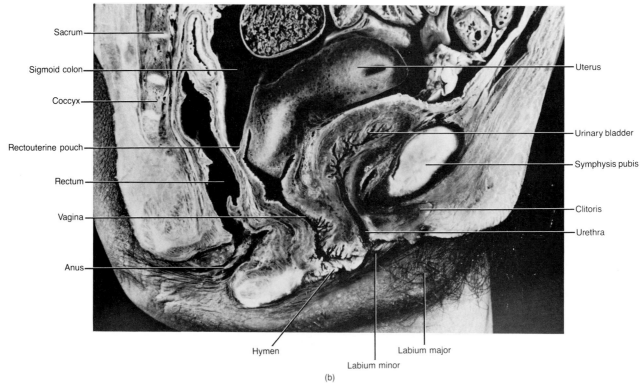

(b)

FIGURE 23-9 Female organs of reproduction and surrounding structures seen in sagittal section. (a) Diagram. (b) Photograph. (Courtesy of C. Yokochi and J. W. Rohen, *Photographic Anatomy of the Human Body,* 2nd ed., 1979, IGAKU-SHOIN, Ltd., Tokyo, New York.)

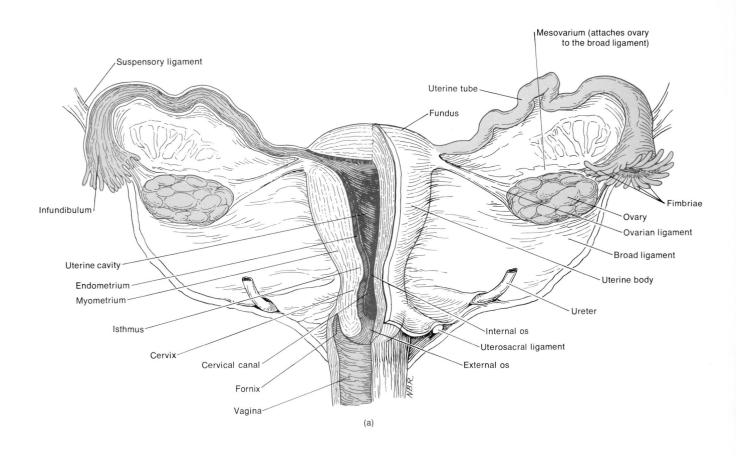

(a)

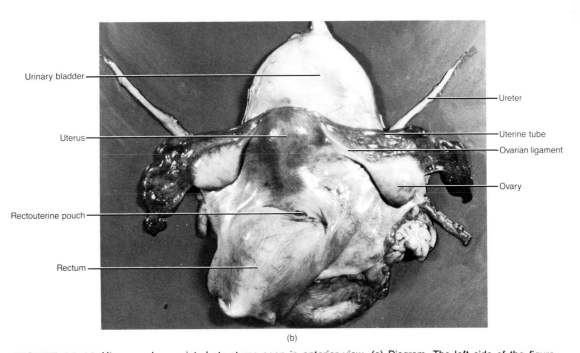

(b)

FIGURE 23-10 Uterus and associated structures seen in anterior view. (a) Diagram. The left side of the figure has been sectioned to show internal structures. (b) Photograph. (Courtesy of C. Yokochi and J. W. Rohen, *Photographic Anatomy of the Human Body,* 2nd ed., 1979, IGAKU-SHOIN, Ltd., Tokyo, New York.)

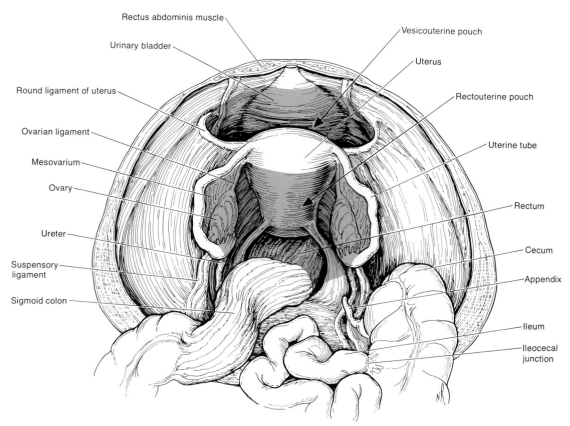

FIGURE 23-11 Female pelvis viewed from above.

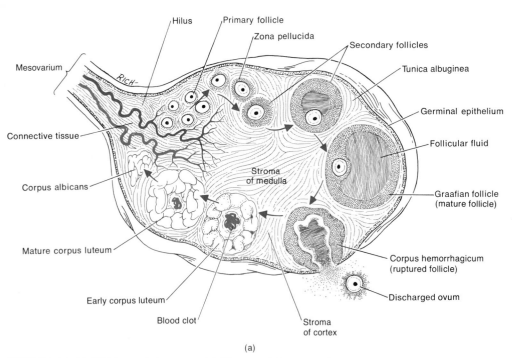

(a)

FIGURE 23-12 Histology of the ovary. (a) Diagram of the parts of an ovary seen in sectional view. The arrows indicate the sequence of developmental stages that occur as part of the ovarian cycle.

Corpus albicans

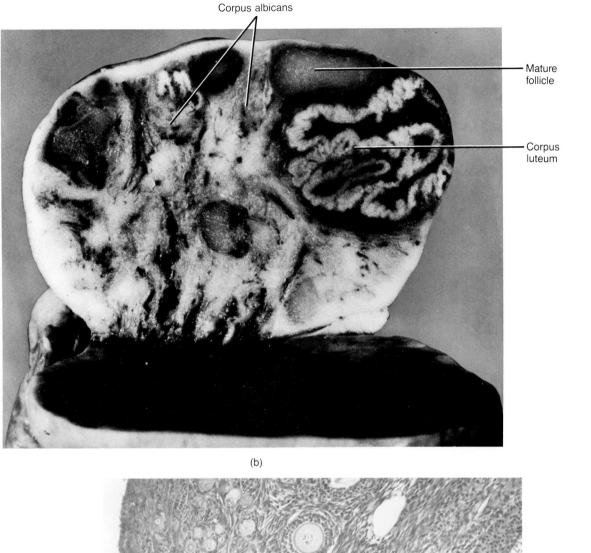

Mature
follicle

Corpus
luteum

(b)

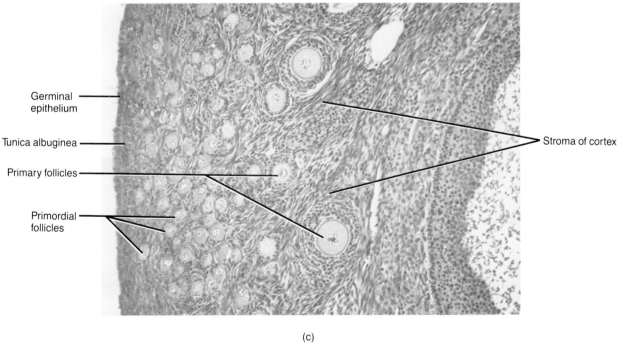

Germinal
epithelium

Tunica albuginea

Primary follicles

Primordial
follicles

Stroma of cortex

(c)

FIGURE 23-12 (*Continued*) Histology of the ovary. (b) Photograph of the ovary seen in sectional view. (Courtesy of C. Yokochi and J. W. Rohen, *Photographic Anatomy of the Human Body,* 1st ed., 1969, IGAKU-SHOIN, Ltd., Tokyo, New York.) (c) Photomicrograph of the cortex of an ovary at a magnification of 60×. (© 1983 by Michael H. Ross. Used by permission.)

4. Ovarian follicles. Ova and their surrounding tissues in various stages of development.

5. Graafian (GRAF-ē-an) follicle. An endocrine gland made up of a mature ovum and its surrounding tissues. The Graafian follicle secretes hormones called estrogens.

6. Corpus luteum. Glandular body that develops from a Graafian follicle after extrusion of an ovum (ovulation).

The corpus luteum produces the hormones progesterone and estrogens.

The ovarian blood supply is furnished by the ovarian arteries which anastomose with branches of the uterine arteries. The ovaries are drained by the ovarian veins. On the right side, they drain into the inferior vena cava, and on the left side, they drain into the renal vein.

Sympathetic and parasympathetic nerve fibers to the ovaries are said to terminate on the blood vessels and not enter the substance of the ovaries.

The ovaries produce ova, discharge ova (ovulation), and secrete the female sexual hormones. The ovaries are analogous to the testes of the male reproductive system.

UTERINE TUBES

The female body contains two **uterine,** or **fallopian, tubes** that extend laterally from the uterus and transport the ova from the ovaries to the uterus (see Figure 23-10). Measuring about 10 cm (4 inches) long, the tubes are positioned between the folds of the broad ligaments of the uterus. The funnel-shaped open distal end of each tube, called the **infundibulum,** lies close to the ovary but is not attached to it and is surrounded by a fringe of fingerlike projections called **fimbriae (FIM-brē-ē).** From the infundibulum the uterine tube extends medially and inferiorly and attaches to the superior lateral angle of the uterus. The **ampulla** of the uterine tube is the widest,

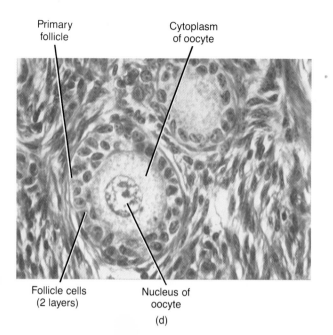

Primary follicle

Cytoplasm of oocyte

Follicle cells (2 layers)

Nucleus of oocyte

(d)

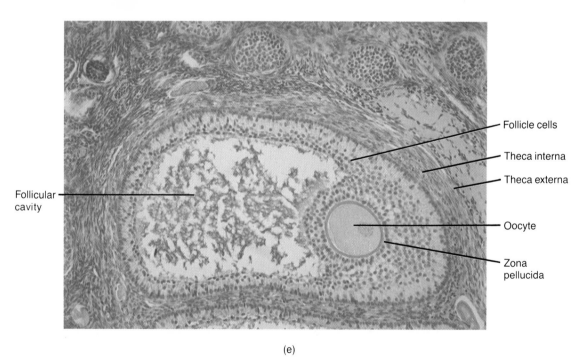

Follicular cavity

Follicle cells

Theca interna

Theca externa

Oocyte

Zona pellucida

(e)

FIGURE 23-12 (*Continued*) Histology of the ovary. (d) Photomicrograph of an enlarged aspect of a primary follicle at a magnification of 300×. (e) Photomicrograph of an enlarged aspect of a secondary follicle at a magnification of 160×. The unfamiliar terms that appear in the photomicrograph will be explained in Chapter 24. (© 1983 by Michael H. Ross. Used by permission.)

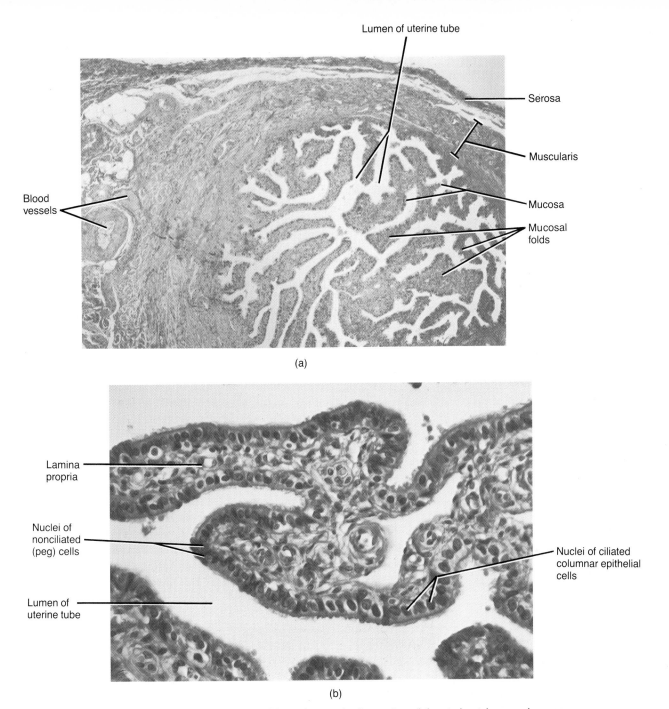

Lumen of uterine tube

Serosa

Muscularis

Mucosa

Mucosal folds

Blood vessels

(a)

Lamina propria

Nuclei of nonciliated (peg) cells

Nuclei of ciliated columnar epithelial cells

Lumen of uterine tube

(b)

FIGURE 23-13 Histology of the uterine tube. (a) Photomicrograph of a portion of the uterine tube seen in cross section at a magnification of 25×. (b) Photomicrograph of an enlarged aspect of the mucosa of the uterine tube at a magnification of 220X. (© 1983 by Michael H. Ross. Used by permission.)

longest portion, making up about two-thirds of its length. The **isthmus** of the uterine tube is the short, narrow, thick-walled portion that joins the uterus.

Histologically, the uterine tubes are composed of three layers (Figure 23-13). The internal *mucosa* contains ciliated columnar cells and secretory cells, which are believed to aid the movement and nutrition of the ovum. The middle layer, the *muscularis,* is composed of a thick, circular region of smooth muscle and an outer, thin, longitudinal region of smooth muscle. Wavelike contractions of the muscularis and the ciliated action of the mucosa

help move the ovum down into the uterus. The outer layer of the uterine tubes is a *serous membrane.*

About once a month an immature ovum ruptures from the surface of the ovary near the infundibulum of the uterine tube, a process called **ovulation.** The ovum is swept into the tube by the ciliary action of the epithelium of the infundibulum. It is then moved along the tube by ciliary action supplemented by the wavelike contractions of the muscularis. If the ovum is fertilized by a sperm cell, it usually occurs in the ampulla of the uterine tube. Fertilization may occur at any time up to about 24 hours following ovulation. The fertilized ovum, now referred to as a blastocyst, descends into the uterus within 7 days. An unfertilized ovum disintegrates.

CLINICAL APPLICATION

Fertilization may occur while the ovum is free in the pelvic cavity, and implantation may take place on one of the pelvic viscera. *Pelvic implantations* usually fail because the developing fertilized ovum does not make vascular connection with the maternal blood supply. On occasion, a fertilized ovum fails to descend to the uterus and implants in the uterine tube. In the case of a *tubular implantation,* the pregnancy must be terminated surgically before the tube ruptures. Both pelvic and tubular implantations are referred to as **ectopic** (ek-TOP-ik) **pregnancies.**

The uterine tubes are supplied by branches of the uterine and ovarian arteries. Venous return is via the uterine veins.

The uterine tubes are supplied with sympathetic and parasympathetic nerve fibers from the hypogastric plexus and the pelvic splanchnic nerves. The fibers are distributed to the muscular coat of the tubes and their blood vessels.

UTERUS

The site of menstruation, implantation of a fertilized ovum, development of the fetus during pregnancy, and labor is the **uterus.** Situated between the urinary bladder and the rectum, the uterus is shaped like an inverted pear (see Figures 23-9, 23-10, and 23-11). Before the first pregnancy, the adult uterus measures approximately 7.5 cm (3 inches) long, 5 cm (2 inches) wide, and 2.5 cm (1 inch) thick.

Anatomical subdivisions of the uterus include the dome-shaped portion above the uterine tubes called the **fundus,** the major tapering central portion called the **body,** and the inferior narrow portion opening into the vagina called the **cervix.** Between the body and the cervix is the **isthmus** (IS-mus), a constricted region about 1 cm (½ inch) long. The interior of the body of the uterus is called the **uterine cavity,** and the interior of the narrow cervix is called the **cervical canal.** The junction of the isthmus with the cervical canal is the **internal os.** The **external os** is the place where the cervix opens into the vagina.

Normally the uterus is flexed between the uterine body and the cervix. In this position, the body of the uterus projects anteriorly and slightly superiorly over the urinary bladder, and the cervix projects inferiorly and posteriorly and enters the anterior wall of the vagina at nearly a right angle. Several structures that are either extensions of the parietal peritoneum or fibromuscular cords, re-

ferred to as ligaments, maintain the position of the uterus. The paired **broad ligaments** are double folds of parietal peritoneum attaching the uterus to either side of the pelvic cavity. Uterine blood vessels and nerves pass through the broad ligaments. The paired **uterosacral ligaments,** also peritoneal extensions, lie on either side of the rectum and connect the uterus to the sacrum. The **cardinal (lateral cervical) ligaments** extend below the bases of the broad ligaments between the pelvic wall and the cervix and vagina. These ligaments contain smooth muscle, uterine blood vessels, and nerves and are the chief ligaments that maintain the position of the uterus and help to keep it from dropping down into the vagina. The **round ligaments** are bands of fibrous connective tissue between the layers of the broad ligament. They extend from a point on the uterus just below the uterine tubes to a portion of the external genitalia.

CLINICAL APPLICATION

Although the ligaments normally maintain the position of the uterus, they also afford the uterine body some movement. As a result, the uterus may become malpositioned. A posterior tilting of the uterus, called **retroflexion,** or an anterior tilting, called **anteflexion,** may occur.

Histologically, the uterus consists of three layers of tissue. The outer layer, the **serous layer** or **serosa,** is part of the visceral peritoneum. Laterally, the serosa becomes the broad ligament. Anteriorly, the serosa is reflected over the urinary bladder and forms a shallow pouch, the **vesicouterine** (ves'-i-kō-YOO-ter-in) **pouch** (see Figure 23-9). Posteriorly, it is reflected onto the rectum and forms a deep pouch, the **rectouterine** (rek-tō-YOO-ter-in) **pouch,** or **pouch of Douglas**—the lowest point in the pelvic cavity.

The middle layer of the uterus, the **myometrium,** forms the bulk of the uterine wall (Figure 23-14). This layer consists of three layers of smooth muscle fibers and is thickest in the fundus and thinnest in the cervix. During childbirth, coordinated contractions of the muscles help to expel the fetus from the body of the uterus.

The inner layer of the uterus, the **endometrium,** is a mucous membrane composed of two principal layers. The *stratum functionalis,* the layer closer to the uterine cavity, is shed during menstruation. The second layer, the *stratum basalis* (bā-SAL-is), is permanent and produces a new functionalis following menstruation. The endometrium contains numerous glands.

Blood is supplied to the uterus by branches of the internal iliac artery called *uterine arteries.* Branches called *arcuate arteries* are arranged in a circular fashion in the myometrium and give off *radial arteries* that penetrate deeply into the myometrium (Figure 23-15). Just before the branches enter the endometrium, they divide into two

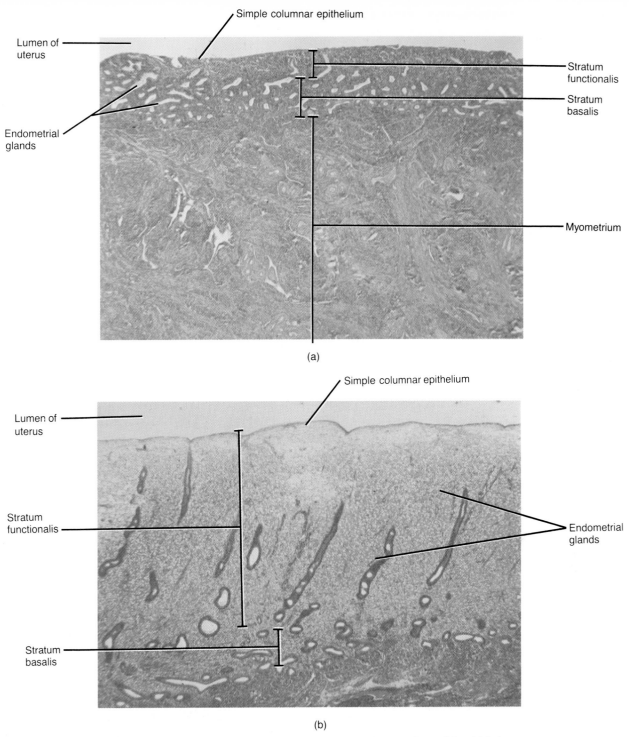

FIGURE 23-14 Histology of the uterus. (a) Photomicrograph of a portion of the uterine wall in which the stratum functionalis is in an early stage of proliferation at a magnification of 25×. (b) Photomicrograph of a portion of the uterine wall in which the stratum functionalis is in an advanced stage of proliferation at a magnification of 25×. Note the increased thickness of the stratum functionalis and the long, straight nature of the glands. (© 1983 by Michael H. Ross. Used by permission.)

kinds of arterioles. The *straight arteriole* terminates in the basalis and supplies it with the materials necessary to regenerate the functionalis. The *spiral arteriole* penetrates the functionalis and changes markedly during the menstrual cycle. The uterus is drained by the *uterine veins.*

Sympathetic and parasympathetic fibers are supplied to the uterus via the hypogastric and pelvic plexuses. Both sets of nerves terminate on the uterine vessels. The myometrium is believed to be innervated by the sympathetic fibers alone.

645

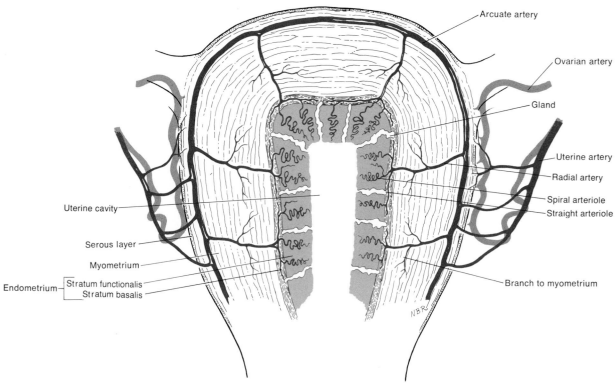

FIGURE 23-15 Blood supply of the uterus.

ENDOCRINE RELATIONS: MENSTRUAL AND OVARIAN CYCLES

The principal events of the menstrual cycle can be correlated with those of the ovarian cycle and changes in the endometrium. All are hormonally controlled events.

The **menstrual cycle** is a series of changes in the endometrium of a nonpregnant female. Each month the endometrium is prepared to receive a fertilized ovum. An implanted fertilized ovum eventually develops into an embryo and then a fetus, which normally remains in the uterus until delivery. If no fertilization occurs, the stratum functionalis portion of the endometrium is shed. The **ovarian cycle** is a monthly series of events associated with the maturation of an ovum.

The menstrual cycle, ovarian cycle, and other changes associated with puberty in the female are controlled by regulating factors from the hypothalamus called the follicle-stimulating hormone releasing factor (FSHRF) and the luteinizing hormone releasing factor (LHRF). FSHRF stimulates the release of follicle-stimulating hormone (FSH) from the anterior pituitary. FSH stimulates the initial development of the ovarian follicles and the secretion of estrogens by the follicles. LHRF stimulates the release of another anterior pituitary hormone—the luteinizing hormone (LH), which stimulates the development of ovarian follicles, brings about ovulation, and stimulates the production of both estrogens and progesterone by ovarian cells.

The female sex hormones—estrogens and progesterone—affect the body in different ways. **Estrogens,** the hormones of growth, have three main functions. First is the development and maintenance of female reproductive structures, especially the endometrial lining of the uterus, secondary sex characteristics, and the breasts. The secondary sex characteristics include fat distribution to the breasts, abdomen, mons pubis, and hips; voice pitch; broad pelvis; and hair pattern. Second, they control fluid and electrolyte balance. Third, they increase protein anabolism. In this regard, estrogens are synergistic with human growth hormone (HGH). High levels of estrogens in the blood inhibit the release of FSHRF by the hypothalamus, which in turn inhibits the secretion of FSH by the anterior pituitary gland. This inhibition provides the basis for the

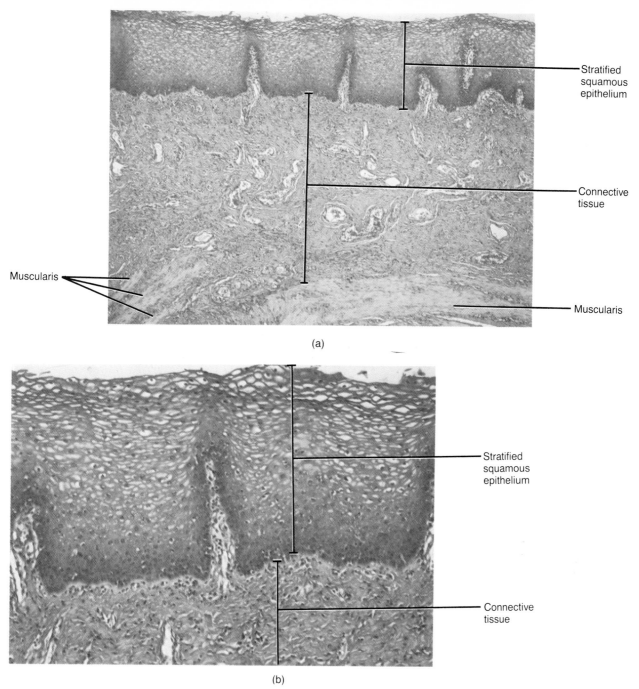

FIGURE 23-16 Histology of the vagina. (a) Photomicrograph of a portion of the wall of the vagina at a magnification of 50×. (b) Photomicrograph of an enlarged aspect of the mucosa of the vagina at a magnification of 125×. (© 1983 by Michael H. Ross. Used by permission.)

action of one kind of contraceptive pill. **Progesterone,** the hormone of maturation, works with estrogens to prepare the endometrium for implantation and the breasts for milk secretion.

The menstrual cycle normally occurs once each month from **menarche** (me-NAR-kē), the first menses, to **menopause,** the last menses. Menopause is preceded by the *climacteric* (klī-mak-TER-ik), the gradual cessation of reproductive function during which menstrual cycles become irregular and less frequent until they stop. The cli-

macteric, which typically begins between ages 40 and 50, results from the failure of the ovaries to respond to the stimulation of gonadotropic hormones from the anterior pituitary. Some women experience hot flashes, copious sweating, headache, muscular pains, and emotional instability. In the postmenopausal woman there will be some atrophy of the ovaries, uterine tubes, uterus, vagina, external genitalia, and breasts.

The cause of menopause is related to a decreasing ability of the aging ovaries to respond to FSH and LH. As a result, there is a decrease in estrogen production by the ovaries. Throughout a woman's period of reproductive function, primary ovarian follicles grow into Graafian follicles with each ovarian cycle, and eventually most of them degenerate. Only about 400 are ovulated. At the age of approximately 45, therefore, few primary follicles remain to be stimulated by FSH and LH. As the number of primary follicles diminishes, the production of estrogens by the ovary decreases.

VAGINA

The **vagina** serves as a passageway for the menstrual flow. It is also the receptacle for the penis during coitus, or sexual intercourse, and the lower portion of the birth canal. It is a muscular, tubular organ lined with mucous membrane and measures about 10 cm (4 inches) in length, extending from the cervix to the vestibule (Figures 23-9 and 23-10). Situated between the bladder and the rectum, it is directed superiorly and posteriorly, where it attaches to the uterus. A recess, called the **fornix,** surrounds the vaginal attachment to the cervix. The dorsal recess, called the posterior fornix, is deeper than the ventral and two lateral fornices. The fornices make possible the use of contraceptive diaphragms.

Histologically, the mucosa of the vagina is continuous with that of the uterus and consists of stratified squamous epithelium and connective tissue that lies in a series of transverse folds, the **rugae,** and is capable of a good deal of distension (Figure 23-16). The muscularis is composed of longitudinal smooth muscle that can stretch considerably. This distension is important because the vagina receives the penis during sexual intercourse and serves as the lower portion of the birth canal. At the lower end of the vaginal opening **(vaginal orifice)** is a thin fold of vascularized mucous membrane called the **hymen,** which forms a border around the orifice, partially closing it (see Figure 23-17).

CLINICAL APPLICATION

Sometimes the hymen completely covers the orifice, a condition called **imperforate** (im-PER-fō-rāt) **hymen.** Surgery is required to open the orifice to permit the discharge of the menstrual flow.

The mucosa of the vagina contains large amounts of glycogen, which upon decomposition produces organic acids. These acids create a low pH environment that retards microbial growth. However, the acidity is also injurious to sperm cells. Semen neutralizes the acidity of the vagina to ensure survival of the sperm.

The blood supply of the vagina is derived from branches of the internal iliac arteries: vaginal, vaginal branch of the uterine, internal pudendal, and vaginal branches of the middle rectal arteries. The veins form vaginal venous plexuses along the sides of the vagina and are drained through their vaginal veins into the internal iliac veins.

The nerves of the vagina are derived from the uterovaginal plexus at the base of the broad ligament. Lower nerve fibers from the plexus supply the cervix and superior vagina; vaginal nerves supply the wall of the vagina.

VULVA

The term **vulva** (VUL-va), or **pudendum** (pyoo-DEN-dum), is a collective designation for the external genitalia of the female (Figure 23-17).

The **mons pubis (veneris),** an elevation of adipose tissue covered by coarse pubic hair, is situated over the symphysis pubis. It lies anterior to the vaginal and urethral openings. From the mons pubis, two longitudinal folds of skin, the **labia majora** (LĀ-bē-a ma-JŌ-ra), extend inferiorly and posteriorly. The labia majora, the female homologue of the scrotum, contain an abundance of adipose tissue and sebaceous and sweat glands; they are covered by hair on their upper outer surfaces. Medial to the labia majora are two folds of skin called the **labia minora** (MĪ-nō-ra). Unlike the labia majora, the labia minora are devoid of hair and fat and have few sweat glands. They do, however, contain numerous sebaceous glands.

The **clitoris** (KLI-to-ris) is a small, cylindrical mass of erectile tissue and nerves. It is located at the anterior junction of the labia minora. A layer of skin called the **prepuce,** or **foreskin,** is formed at the point where the labia minora unite and covers the body of the clitoris. The exposed portion of the clitoris is the **glans.** The clitoris is homologous to the penis of the male in that it is capable of enlargement upon tactile stimulation and assumes a role in sexual excitement of the female.

The cleft between the labia minora is called the **vestibule.** Within the vestibule are the hymen, vaginal orifice, urethral orifice, and the openings of several ducts. The **vaginal orifice** occupies the greater portion of the vestibule and is bordered by the hymen. Anterior to the vaginal orifice and posterior to the clitoris is the **urethral orifice.** Posterior and to either side of the urethral orifice are the openings of the ducts of the **lesser vestibular** (ves-TIB-yoo-lar) or **Skene's glands.** These glands secrete mucus. On either side of the vaginal orifice itself are two small glands: the **greater vestibular** or **Bartholin's** (BAR-tō-linz) **glands.** These glands open by ducts into a groove between the hymen and labia minora and produce a mucoid secretion that supplements lubrication during sexual intercourse. The lesser vestibular glands are homologous to the male prostate. The greater vestibular glands are homologous to the male bulbourethral, or Cowper's, glands.

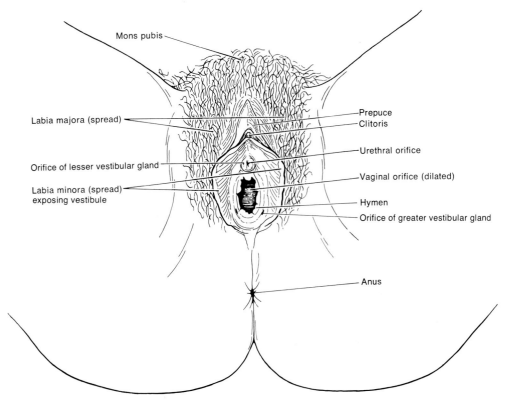

Mons pubis

Labia majora (spread)

Orifice of lesser vestibular gland

Labia minora (spread)
exposing vestibule

Prepuce
Clitoris

Urethral orifice

Vaginal orifice (dilated)

Hymen

Orifice of greater vestibular gland

Anus

FIGURE 23-17 Vulva.

PERINEUM

The **perineum** (per'-i-NĒ-um) is the diamond-shaped area at the inferior end of the trunk between the thighs and buttocks of both males and females (Figure 23-18). It is bounded anteriorly by the symphysis pubis, laterally by the ischial tuberosities, and posteriorly by the coccyx. A transverse line drawn between the ischial tuberosities divides the perineum into an anterior **urogenital** (yoo'-rō-JEN-i-tal) **triangle** that contains the external genitalia and a posterior **anal triangle** that contains the anus.

CLINICAL APPLICATION

In the female, the region between the vagina and anus is known as the *clinical perineum.* If the vagina is too small to accommodate the head of an emerging fetus, the skin, vaginal epithelium, subcutaneous fat, and superficial transverse perineal muscle of the clinical perineum may tear. Moreover, the tissues of the rectum may be damaged. To avoid this, a small incision called an **episiotomy** (ē-piz'-ē-OT-ō-mē) is made in the perineal skin and underlying tissues just prior to delivery. After delivery the episiotomy is carefully sutured in layers.

MAMMARY GLANDS

The **mammary glands** are branched tubuloalveolar glands that lie over the pectoralis major muscles and are attached to them by a layer of connective tissue (Figure 23-19). Internally, each mammary gland consists of 15 to 20 **lobes,** or compartments, separated by adipose tissue. The amount of adipose tissue determines the size of the breasts. However, breast size has nothing to do with the amount of milk produced. In each lobe are several smaller compartments called **lobules,** composed of connective tissue in which milk-secreting cells referred to as **alveoli** are embedded (Figure 23-20). Between the lobules are strands of connective tissue called the **suspensory ligaments of Cooper.** These ligaments run between the skin and deep fascia and support the breast. Alveoli are arranged in grapelike clusters. They convey the milk into a series of **secondary tubules.** From here the milk passes into the **mammary ducts.** As the mammary ducts approach the nipple, expanded sinuses called **ampullae** (am-POOL-ē), where milk may be stored, are present. The ampullae continue as **lactiferous ducts** that terminate in the **nipple.** Each lactiferous duct conveys milk from one of the lobes to the exterior, although some may join before reaching the surface. The circular pigmented area of skin surrounding the nipple is called the **areola** (a-RĒ-ō-la). It appears rough because it contains modified sebaceous glands.

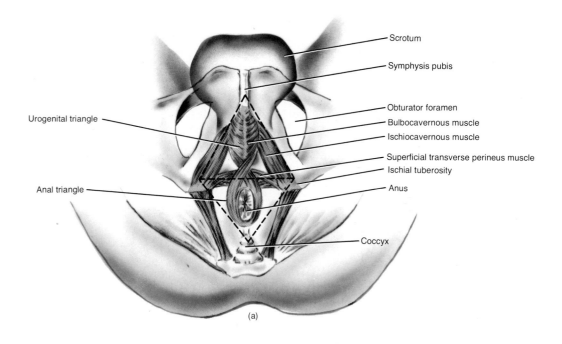

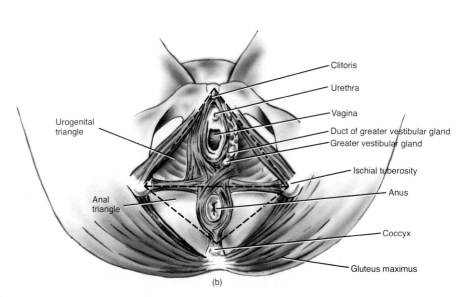

FIGURE 23-18 Perineum. (a) Borders seen in the male. (b) Dissected view of same regions in the female.

At birth, both male and female mammary glands are undeveloped and appear as slight elevations on the chest. With the onset of puberty, the female breasts begin to develop—the mammary ducts elongate, extensive fat deposition occurs, and the areola and nipple grow and become pigmented. These changes are correlated with an increased output of estrogen by the ovary. Further mammary development occurs at sexual maturity with the onset of ovulation and the formation of the corpus luteum. During adolescence, lobules are formed and fat deposition continues, increasing the size of the glands. Although these changes are associated with estrogen and progester-

one secretion by the ovaries, ovarian secretion is ultimately controlled by FSH, which is secreted in response to FSHRF by the hypothalamus.

The essential function of the mammary glands is milk secretion and ejection, together called lactation.

APPLICATIONS TO HEALTH

SEXUALLY TRANSMISSIBLE DISEASES (STD)

The general term **sexually transmissible disease** (STD) is applied to any of the large group of diseases that can be spread by sexual contact. The group includes condi-

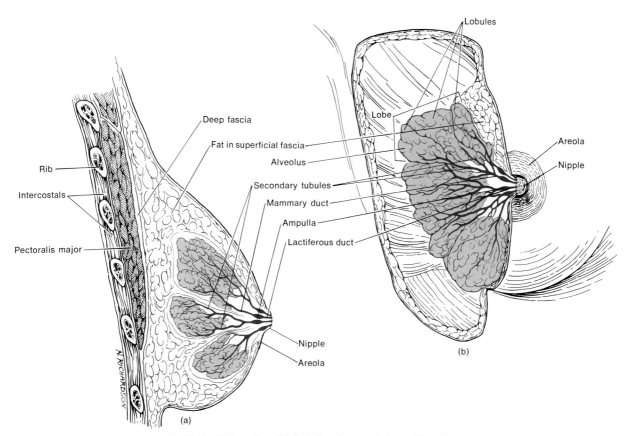

FIGURE 23-19 Mammary glands. (a) Sagittal section. (b) Anterior view, partially sectioned.

tions traditionally specified as **venereal diseases** (from Venus, goddess of love), such as gonorrhea, syphilis, and genital herpes, and several other conditions which are contracted sexually or may be contracted otherwise, but are then transmitted to a sexual partner.

Gonorrhea

Gonorrhea is an infectious disease that affects primarily the mucous membrane of the urogenital tract, the rectum, and occasionally the eyes. The disease is caused by the bacterium *Neisseria gonorrhoeae.* Discharges from the involved mucous membranes are the source of infection, and the bacteria are transmitted by direct contact, usually sexual or during passage of a newborn through the birth canal.

Males usually suffer inflammation of the urethra with pus and painful urination. Fibrosis sometimes occurs in an advanced stage, causing stricture of the urethra. There also may be involvement of the epididymis and prostate gland. In females, infection may occur in the urethra, vagina, and cervix, and there may be a discharge of pus. However, infected females often harbor the disease without any symptoms until it has progressed to a more advanced stage. If the uterine tubes become involved, pelvic inflammation may follow. Peritonitis, or inflammation of

the peritoneum, is a very serious disorder. The infection should be treated and controlled immediately because, if neglected, sterility or death may result. Although antibiotics have greatly reduced the mortality rate of acute peritonitis, it is estimated that between 50,000 and 80,000 women are made sterile by gonorrhea every year. If the bacteria are transmitted to the eyes of a newborn in the birth canal, blindness can result.

Administration of a 1 percent silver nitrate solution or penicillin in the infant's eyes prevents infection. Penicillin is the drug of choice for the treatment of gonorrhea in adults.

Syphilis

Syphilis is an infectious disease caused by the bacterium *Treponema pallidum.* It is acquired through sexual contact or transmitted through the placenta to a fetus. The early stages of the disease affect primarily the organs most likely to have made sexual contact—the genital organs, the mouth, and the rectum. The point where the bacteria enter the body is marked by a lesion called a *chancre.* In males, it usually occurs on the penis; in females, in the vagina or cervix. The chancre heals without scarring and without treatment. Following the initial infection, the bacteria enter the bloodstream and spread throughout

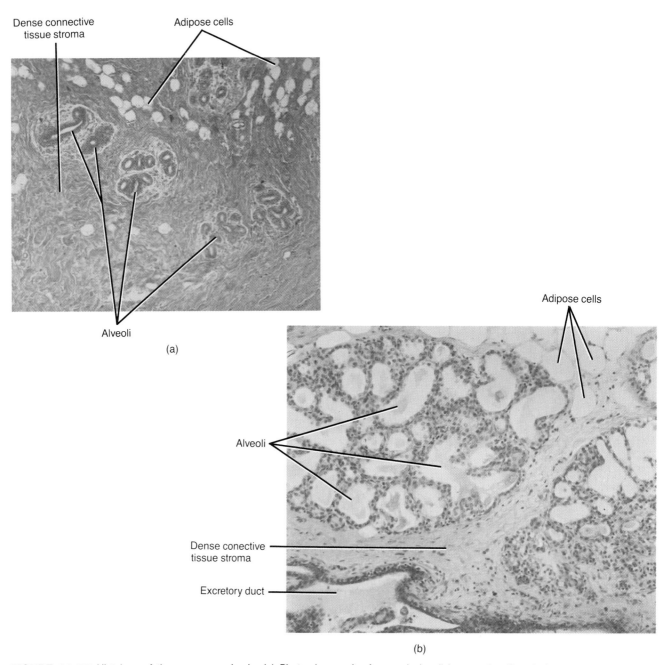

FIGURE 23-20 Histology of the mammary glands. (a) Photomicrograph of several alveoli in a resting (inactive) mammary gland at a magnification of 60×. (b) Photomicrograph of several alveoli during late pregnancy at a magnification of 160×. (© 1983 by Michael H. Ross. Used by permission.)

the body. In some individuals, an active secondary stage of the disease occurs, characterized by lesions of the skin and mucous membranes and often fever. The signs of the secondary stage also go away without medical treatment. During the next several years, the disease progresses without symptoms and is said to be in a latent phase. If symptoms again appear, anywhere from 5 to 40 years after the initial infection, the person is said to be in the tertiary stage of the disease. Tertiary syphilis may involve the circulatory system, skin, bones, viscera, and the nervous system.

Infection of the fetus with syphilis can occur after the fifth month. Infection of the mother is not necessarily followed by fetal infection, provided that the placenta remains intact. But once the bacteria gain access to fetal circulation, there is nothing to impede their growth and multiplication. As many as 80 percent of children born to untreated syphilitic mothers will be infected in the uterus if the fetus is exposed at the onset or in the early stages of the disease. About 25 percent of the fetuses will die within the uterus. Most of the survivors will arrive prematurely, but 30 percent will die shortly after birth.

Of the infected and untreated children surviving infancy, about 40 percent will develop symptomatic syphilis during their lifetimes.

Genital Herpes

Another venereal disease, **genital herpes,** is now common in the United States. The sexual transmission of the herpes simplex virus is well established. Unlike syphilis and gonorrhea, genital herpes is incurable. Type I herpes simplex virus is the virus that causes typical cold sores. Type II herpes simplex virus causes painful genital blisters on the prepuce, glans penis, and penile shaft in males and on the vulva or sometimes high up in the vagina in females. The blisters disappear and reappear, but the disease itself is still present in the body.

Genital herpes virus infection causes considerable discomfort, and there is an extraordinarily high rate of recurrence of the disease. Because genital herpes is statistically associated with later cervical carcinoma, all women with this infection should be reminded of the importance of annual Papanicolaou smears. For pregnant women with genital herpes, a Cesarean section will usually prevent complications in the child. Complications range from a mild asymptomatic infection to CNS damage to death. Treatment of the symptoms involves pain medication, saline compresses, and sexual abstinence for the duration of the eruption.

Trichomoniasis

The microorganism *Trichomonas vaginalis,* a flagellated protozoan, causes **trichomoniasis,** an inflammation of the mucous membrane of the vagina in females and the urethra in males. Symptoms include vaginal discharge and severe vaginal itch in women. Men can have it without symptoms, but can transmit it to women nonetheless. Sexual partners must be treated simultaneously.

Nongonococcal Urethritis (NGU)

The term **nongonococcal urethritis (NGU),** also known as **nonspecific urethritis (NSU),** refers to an inflammation of the urethra not caused by the bacterium *Neisseria gonorrhoeae.* The bacterium *Chlamydia trachomatis* is known to cause some cases of NGU. NGU affects both males and females. The urethra swells and narrows, impeding the flow of urine. Both urination and the urgency to urinate increase. Urination is accompanied by burning pain, and there may be a purulent discharge.

Although nonmicrobial factors such as trauma (passage of a catheter) or chemical agents (alcohol and certain chemotherapeutic agents) can cause this condition, it is estimated that at least 40 percent of the cases of NGU are acquired sexually. In fact, NGU may be the most common sexually transmitted disease in the United States today. Although it is not a disease which must be reported to state departments of health and exact data are lacking, the Center for Disease Control estimates that 4 to 9 million Americans have NGU. Since the symptoms are often mild in males and females are usually asymptomatic, many cases of infection go untreated. Complications are not common, but can be serious. Males may develop inflammation of the epididymis. In females, inflammation may cause sterility by blocking the cervix or uterine tubes. As in gonorrhea, the bacteria can be passed from mother to infant during birth, infecting the eyes. NGU of chlamydial origin responds to treatment with tetracycline.

MALE DISORDERS

Prostate

The prostate gland is susceptible to infection, enlargement, and benign and malignant tumors. Because the prostate surrounds the urethra, any of these disorders can obstruct the flow of urine. Prolonged obstruction may result in serious changes in the urinary bladder, ureters, and kidneys.

Acute and chronic infections of the prostate gland are common in postpubescent males, often in association with inflammation of the urethra. In **acute prostatitis,** the prostate gland becomes swollen and tender. Appropriate antibiotic therapy, bed rest, and above-normal fluid intake are effective treatment.

Chronic prostatitis is one of the most common chronic infections in men of the middle and later years. On examination, the prostate gland feels enlarged, soft, and extremely tender. The surface outline is irregular and may be hard. This disease frequently produces no symptoms, but the prostate is believed to harbor infectious microorganisms responsible for some allergic conditions, arthritis, and inflammation of nerves (neuritis), muscles (myositis), and the iris (iritis).

An **enlarged prostate** gland, increasing to two to four times larger than normal, occurs in approximately one-third of all males over age 60. The cause is unknown, and the enlarged condition usually can be detected by rectal examination.

Tumors of the male reproductive system usually involve the prostate gland. Carcinoma of the prostate is the second leading cause of death from cancer in men in the United States and it is responsible for approximately 19,000 deaths annually. Its incidence is related to age, race, occupation, geography, and ethnic origin. Both benign and malignant growths are common in elderly men. Both types of tumors put pressure on the urethra, making urination painful and difficult. At times, the excessive back pressure destroys kidney tissue and gives rise to an increased susceptibility to infection. Therefore, even when the tumor is benign, surgery is indicated to remove the prostate or parts of it if the growth is obstructive and perpetuates urinary tract infections.

Sexual Functional Abnormalities

Impotence is the inability of an adult male to attain or hold an erection long enough for normal intercourse. Impotence could be the result of physical abnormalities of the penis, systemic disorders such as syphilis, vascular disturbances, neurological disorders, or psychic factors such as fear of causing pregnancy, fear of venereal disease, religious inhibitions, and emotional immaturity.

Infertility, or **sterility,** is an inability to fertilize the ovum. It does not imply impotence. Male fertility requires production of adequate amounts of viable, normal spermatozoa by the testes, unobstructed transportation of sperm through the seminal tract, and satisfactory deposition in the vagina. The tubules of the testes are sensitive to many factors—x-rays, infections, toxins, malnutrition—that may cause degenerative changes and produce male sterility.

If inadequate spermatozoa production is suspected, a sperm analysis should be performed. Analysis includes measuring the volume of semen, counting the number of sperm per milliliter, evaluating sperm motility 4 hours after ejaculation, and determining the percentage of abnormal sperm forms (not to exceed 20 percent).

FEMALE DISORDERS

Menstrual Abnormalities

Because menstruation reflects not only the health of the uterus, but also the health of the glands that control it, the ovaries and the pituitary gland, disorders of the female reproductive system frequently involve menstrual disorders.

Amenorrhea is the absence of menstruation. If a woman has never menstruated, the condition is called *primary amenorrhea.* Primary amenorrhea can be caused by endocrine disorders, most often in the pituitary gland and hypothalamus, or by genetically caused abnormal development of the ovaries or uterus. *Secondary amenorrhea,* the skipping of one or more periods, is commonly experienced by many women sometime during their lives. Changes in body weight, either gains or losses, often cause amenorrhea. Obesity may disturb ovarian function, and similarly, the extreme weight loss which characterizes anorexia nervosa often leads to a suspension of menstrual flow. When amenorrhea is unrelated to weight, analysis of estrogen levels often reveals deficiencies of pituitary and ovarian hormones.

Dysmenorrhea is painful menstruation caused by forceful contraction of the uterus. It is often accompanied by nausea, vomiting, diarrhea, headache, fatigue, and nervousness. Some cases are caused by pathological conditions such as uterine tumors, ovarian cysts, endometriosis, and pelvic inflammatory disease. However, other cases of dysmenorrhea are not related to any pathologies. Although the cause of these cases is unknown, it appears that they may be triggered by an overproduction of prostaglandins by the uterus. Prostaglandins are known to stimulate uterine contractions, but they cannot do so in the presence of high levels of progesterone. As we have noted earlier, progesterone levels are high during the last half of the menstrual cycle. During this time, prostaglandins are apparently inhibited by progesterone from producing uterine contractions. However, if pregnancy does not occur, progesterone levels drop rapidly and prostaglandin production increases. This causes the uterus to contract and slough off its lining and may result in dysmenorrhea. The other symptoms of dysmenorrhea—nausea, vomiting, diarrhea, and headache—may be due to prostaglandins stimulating contractions of the smooth muscle of the stomach, intestines, and blood vessels in the brain. Research is now underway to develop drugs to inhibit prostaglandin synthesis in order to treat dysmenorrhea.

Abnormal uterine bleeding includes menstruation of excessive duration or excessive amount, too frequent menstruation, intermenstrual bleeding, and postmenopausal bleeding. These abnormalities may be caused by disordered hormonal regulation, emotional factors, fibroid tumors of the uterus, and systemic diseases.

Ovarian Cysts and Endometriosis

Ovarian cysts are fluid-containing tumors of the ovary. Follicular cysts may occur in the ovaries of elderly women, in ovaries that have inflammatory diseases, and in menstruating females. They have thin walls and contain a serous albuminous material. Cysts may also arise from the corpus luteum or the endometrium. The endometrium is the inner lining of the uterus that is sloughed off in menstruation.

Endometriosis occurs when the endometrial tissue grows outside the uterus. The tissue enters the pelvic cavity via the open uterine tubes and may be found in any of a dozen sites—on the ovaries, cervix, abdominal wall, and urinary bladder. Causes are unknown. Endometriosis is common in women 30 to 40 years of age. Symptoms include premenstrual pain or unusual menstrual pain. The unusual pain is caused by the displaced tissue sloughing off at the same time the normal uterine endometrium is being shed during menstruation. Infertility can be a consequence. Treatment usually consists of hormone therapy or surgery. Endometriosis disappears at menopause or when the ovaries are removed.

Infertility

Female infertility, or the inability to conceive, occurs in about 10 percent of married females in the United States. Once it is established that ovulation occurs regularly, the reproductive tract is examined for functional and anatomical disorders to determine the possibility of union of the sperm and the ovum in the oviduct.

Disorders Involving the Breasts

The breasts of females are highly susceptible to cysts and tumors. Men are also susceptible to breast tumor, but certain breast cancers are 100 times more common in women.

In the female, the benign *fibroadenoma* is a common tumor of the breast. It occurs most frequently in young women. Fibroadenomas have a firm rubbery consistency and are easily moved about within the mammary tissue. The usual treatment is excision of the growth. The breast itself is not removed.

Breast cancer has the highest fatality rate of all cancers affecting women, but it is rare in men. In the female, breast cancer is rarely seen before age 30, and its occurrence rises rapidly after menopause. Breast cancer is generally not painful until it becomes quite advanced, so often it is not discovered early or, if noted, is ignored. Any lump, no matter how small, should be reported to a doctor at once. Treatment for breast cancer may involve hormone therapy, chemotherapy, a *modified* or *radical mastectomy,* or a combination of these. A radical mastectomy involves removal of the affected breast along with the underlying pectoral muscles and the axillary lymph nodes. Metastasis of cancerous cells is usually through the lymphatics or blood. Radiation treatment and chemotherapy may follow the surgery to ensure the destruction of any stray cancer cells.

The mortality from breast cancer has not improved significantly in the past 50 years. Early detection—especially by breast self-examination and mammography—is still the most promising method to increase the survival rate.

It is estimated that 95 percent of breast cancer is first detected by the women themselves. Each month after the menstrual period the breasts should be thoroughly examined for lumps, puckering of the skin, or discharge. *Mammography* is a sophisticated breast x-ray technique used to detect breast masses and determine whether they are malignant. The examination consists of two x-ray, right-angle views of each breast. Mammographic diagnosis of breast masses is 80 percent reliable. Mammographic x-ray prints are also used to guide surgeons performing mastectomies. As an aid in analyzing mammographic findings, a new x-ray technique, *xeroradiogra-*

phy, is being used. In this photoelectric (rather than photochemical) method, the x-ray image is reproduced on paper instead of film. Xeroradiography provides excellent soft-tissue detail and requires less radiation than film mammography.

Modern x-ray films, xeroradiography, and special x-ray techniques have reduced the problem of mammographic radiation. Ovaries are not exposed to radiation during mammography, and the technique can be used safely on pregnant women. Most cancer experts agree that mammography should be used only after a careful clinical examination and under limited conditions. *Thermography,* a method of measuring and graphically recording heat radiation emitted by the breast, is also frequently used in conjunction with mammography. Tumors, both benign and malignant, emit more heat than nonaffected areas.

One of the most recent breast cancer detecting procedures is an automated *ultrasonic scanner* specifically designed for breast examination. An ultrasonic scan is performed as the patient lies on her stomach in a specially designed hospital bed with her breasts immersed in a tank of water heated to body temperature. Between 150 to 200 three-dimensional views of each breast are recorded on a television monitor. At present, however, this method does not detect lesions less than 1 cm in diameter and will hopefully be improved with future technology.

Another recent technique, called *CT/M,* combines computed tomography with mammography and appears to overcome some of the limitations of mammography. The procedure is based upon the fact that breast carcinoma has an abnormal affinity for iodide. CT scans are made before and after the rapid intravenous infusion of an iodide contrast material. Comparison of the initial density of a suspected lesion with the density following infusion of the iodide gives an indication of the status of the tumor. CT/M affords definitive diagnostic help in instances where the mammographic and physical examinations are inconclusive and appears to be a significantly improved new method of breast cancer diagnosis.

Cervical Cancer

Another common disorder of the female reproductive system is cancer of the uterine cervix. It ranks third in frequency after breast and skin cancers. **Cervical cancer** starts with a change in the shape, growth, and number of the cervical cells called *cervical dysplasia* (dis-PLĀ-sē-a). Cervical dysplasia is not a cancer in itself, but the abnormal cells tend to become malignant.

Cervical cancer may be a venereal disease with a long incubation period. Inciting factors are not known, but herpes virus type II has recently become suspect. Smegma and the DNA of spermatozoa have also been implicated.

Cancer of the cervix (except for adenocarcinoma) rarely occurs in celibate women, and for unknown reasons it is rare in Jewish women.

Early diagnosis of cancer of the uterus is accomplished by the *Papanicolaou* (pap'-a-NIK-ō-la-oo) *test,* or "Pap" smear. In this generally painless procedure, a few cells from the vaginal fornix (that part of the vagina surrounding the cervix) and the cervix are removed with a swab and examined microscopically. Malignant cells have a characteristic appearance and indicate an early stage of cancer, even before symptoms occur. Estimates indicate that the Pap smear is more than 90 percent reliable in detecting cancer of the cervix.

To rule out invasive carcinoma a *cone biopsy* of the cervix is performed. A cone biopsy is a hospital procedure in which an inverted cone of tissue is excised. It requires an anesthetic and is usually done only when the field of dysplasia extends beyond the field of vision afforded by the colposcope or when abnormal cells have been detected. In another procedure, *punch biopsy* is combined with an *endocervical curettage* (ku'-re-TAZH) or *ECC;* this combination has a high degree of diagnostic accuracy.

In a punch biopsy, a disc or segment of tissue is excised. Curettage is the scraping of a diseased surface with a spoon shaped instrument called a curette. If the carcinoma has spread beyond the mucous membrane, treatment may involve complete or partial removal of the uterus, called a *hysterectomy,* or radiation treatment.

Pelvic Inflammatory Disease (PID)

Pelvic inflammatory disease (PID) is a collective term for any extensive bacterial infection of the pelvic organs, especially the uterus, uterine tubes, or ovaries. A vaginal or uterine infection may spread into the uterine tube (*salpingitis*) or even farther into the abdominal cavity, where it infects the peritoneum (*peritonitis*). PID is most commonly caused by gonorrhea, but any bacteria can trigger infection. Often the early symptoms of PID, which include increased vaginal discharge and pelvic pain, occur just after menstruation. As infection spreads, fever may develop in advanced cases along with painful abscesses of the reproductive organs. Early treatment with antibiotics can stop the spread of PID.

KEY MEDICAL TERMS ASSOCIATED WITH THE REPRODUCTIVE SYSTEMS

Leukorrhea (*leuco* = white; *rrhea* = discharge) A nonbloody vaginal discharge that may occur at any age and affects most women at some time.

Neoplasia (*neo* = new; *plas* = form, grow) A condition characterized by the presence of new growths (tumors).

Oophorectomy (*oophoro* = ovary; *ectomy* = removal of) Excision of an ovary. Bilateral oophorectomy refers to the removal of both ovaries.

Pruritus Itching.

Purulent (*puris* = pus) Containing or consisting of pus.

Salpingectomy (*salpingo* = tube) Excision of a uterine tube.

Salpingitis (*itis* = inflammation of) Inflammation or infection of the uterine tube.

Smegma (*smegma* = soap) The secretion, consisting principally of desquamated epithelial cells, found chiefly about the external genitalia and especially under the foreskin of the male.

Vaginitis Inflammation of the vagina.

STUDY OUTLINE

Male Reproductive System

1. The scrotum is a cutaneous outpouching of the abdomen that supports the testes and regulates the temperature of the testes by contraction of the dartos to elevate them closer to the pelvic cavity.

2. The testes are oval-shaped glands in the scrotum containing seminiferous tubules, in which sperm cells are made; Sertoli cells, which nourish sperm cells; and interstitial cells of Leydig, which produce the male sex hormone testosterone.

3. Failure of the testes to descend is called cryptorchidism.

4. Mature spermatozoa consist of a head, middle piece, and tail. Their function is to fertilize an ovum.

5. Spermatozoa are moved through the testes through the seminiferous tubules, straight tubules, rete testis, and efferent ducts.

6. The ductus epididymis is the site of maturation of spermatozoa.

7. The ductus epididymis and the ductus deferens store the spermatozoa and propel them toward the urethra during ejaculation.

8. Alteration of the ductus deferens to prevent fertilization is called vasectomy.

9. The ejaculatory ducts eject spermatozoa into the urethra.
10. The male urethra is subdivided into three portions: prostatic, membranous, and spongy (cavernous).
11. The accessory glands are the seminal vesicles, which secrete an alkaline, viscous fluid that constitutes about 60 percent of the volume of semen; the prostate gland, which secretes an alkaline fluid that constitutes about 13 to 33 percent of the volume of semen; and the bulbourethral (Cowper's) glands, which secrete mucus for lubrication.
12. Semen (seminal fluid) is a mixture of spermatozoa and accessory gland secretions that provide the fluid in which spermatozoa are transported, provide nutrients, and neutralize the acidity of the female vagina.
13. The penis is the male organ of copulation. Expansion of its blood sinuses under the influence of sexual excitation is called erection.

Female Reproductive System

1. The ovaries are female gonads located in the upper pelvic cavity. They produce mature ova, discharge ova (ovulation), and secrete female sex hormones (estrogens and progesterone).
2. The uterine tubes transport ova from the ovaries to the uterus and are the normal site of fertilization.
3. The uterus is an inverted, pear-shaped organ that functions in menstruation, implantation of a fertilized ovum, development of a fetus during pregnancy, and labor.
4. The uterus is normally held in position by a series of ligaments.
5. The menstrual and ovarian cycles are controlled by hypothalamic regulating factors and pituitary hormones (FSH and LH).
6. The vagina is a passageway for the menstrual flow, the receptacle for the penis during sexual intercourse, and the lower portion of the birth canal.

7. The vulva is a collective term for the external genitals of the female, consisting of the mons veneris, labia majora, labia minora, clitoris, vestibule, vaginal and urethral orifices, and greater and lesser vestibular glands.
8. The perineum is a diamond-shaped area at the inferior end of the trunk between the thighs and buttocks.
9. An incision in the perineal skin prior to delivery is called an episiotomy.
10. The mammary glands are branched tubuloalveolar glands over the pectoralis major muscle. Their function is to secrete and eject milk (lactation).

Applications to Health

1. Sexually transmissible diseases (STD) are a group of infectious diseases spread primarily through sexual contact. Among those termed venereal diseases are gonorrhea, syphilis, and genital herpes.
2. Other STDs are trichomoniasis, an inflammation of the mucous membrane of the vagina in females and the urethra in males, and nongonococcal urethritis (NGU), a urethral inflammation.
3. Conditions that affect the prostate are prostatitis, enlarged prostate, and tumors.
4. Impotence is the inability of the male to attain or hold an erection long enough for intercourse.
5. Infertility is the inability of a male's sperm to fertilize an ovum.
6. Menstrual disorders include amenorrhea, dysmenorrhea, and abnormal bleeding.
7. Ovarian cysts are tumors that contain fluid.
8. The mammary glands are susceptible to benign fibroadenomas and malignant tumors. The removal of a malignant breast, pectoral muscles, and lymph nodes is called a radical mastectomy.
9. Cervical cancer can be diagnosed by a "Pap" test.
10. Pelvic inflammatory disease refers to bacterial infection of pelvic organs.

REVIEW QUESTIONS

1. Define reproduction. List the male and female organs of reproduction.
2. Describe the function of the scrotum in protecting the testes from temperature fluctuations. What is cryptorchidism?
3. Describe the internal structure of a testis. Where are the sperm cells made?
4. Identify the principal parts of a spermatozoon. List the function of each.
5. Explain the effects of FSH and ICSH on the male reproductive system.
6. Describe the physiological effects of testosterone on the male reproductive system.
7. Trace the course of a sperm cell through the male

system of ducts from the seminiferous tubules through the urethra.
8. Explain the location of the three subdivisions of the male urethra.
9. What is the spermatic cord?
10. Briefly explain the functions and locations of the seminal vesicles, prostate gland, and bulbourethral glands.
11. What is seminal fluid? What is its function?
12. How is the penis structurally adapted as an organ of copulation?
13. How are the ovaries held in position in the pelvic cavity? What is ovulation?
14. What is the function of the uterine tubes? Briefly

describe their histology. Define an ectopic pregnancy.

15. Diagram the principal parts of the uterus.

16. Describe the arrangement of ligaments that hold the uterus in its normal position. What are the two major malpositions of the uterus?

17. Discuss the blood supply to the uterus. Why is an abundant blood supply important?

18. Define menstrual cycle and ovarian cycle. What is the function of each?

19. Explain the physiological cause of the climacteric and menopause.

20. What is the function of the vagina? Describe its histology.

21. List the parts of the vulva and the functions of each part.

22. What is the perineum? Define episiotomy.

23. Describe the structure of the mammary glands. How are they supported?

24. Describe the passage of milk from the areolar cells of the mammary gland to the nipple.

25. Define a sexually transmissible disease (STD). Distin-guish between gonorrhea, syphilis, and genital herpes.

26. What is trichomoniasis?

27. What are the symptoms of nongonococcal urethritis (NGU)?

28. Describe several disorders that affect the prostate gland.

29. Distinguish between impotence and infertility.

30. What are some of the causes of amenorrhea, dys-menorrhea, and abnormal uterine bleeding?

31. What are ovarian cysts? Define endometriosis.

32. Distinguish between a fibroadenoma and a malignant tumor of the breast.

33. What is a radical mastectomy? Describe the impor-tance of mammography, thermography, ultrasonic scanning, and CT/M.

34. What is a "Pap" smear? What is a hysterectomy and colposcopy?

35. Define pelvic inflammatory disease (PID).

36. Refer to the glossary of key medical terms at the end of the chapter. Be sure that you can define each term.

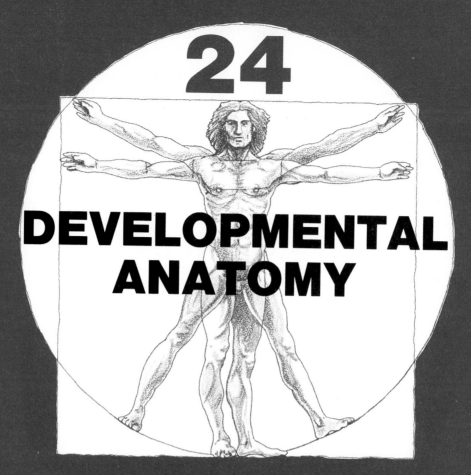

24

DEVELOPMENTAL ANATOMY

STUDENT OBJECTIVES

■ Define meiosis.

■ Contrast the events of spermatogenesis and oogenesis.

■ Compare the role of the male and female in sexual intercourse.

■ Explain the activities associated with fertilization and implantation.

■ Discuss the formation of the three primary germ layers, embryonic membranes, and placenta and umbilical cord as the principal events of the embryonic period.

■ Outline the principal developmental events associated with the major systems of the body.

■ List several body structures produced by the three primary germ layers.

■ Compare the sources and functions of the hormones secreted during pregnancy.

■ Explain the events associated with the three stages of labor.

■ Explain the respiratory and cardiovascular adjustments that occur in an infant at birth.

■ Discuss potential hazards to the embryo and fetus associated with chemicals and drugs, irradiation, alcohol, and cigarette smoking.

■ Discuss the mechanism of lactation.

■ Contrast the various kinds of birth control and their effectiveness.

■ Explain the procedure of amniocentesis and its value in diagnosing diseases in the newborn.

■ Define key medical terms associated with developmental anatomy.

Developmental anatomy is the study of the sequence of events starting with the fertilization of an egg and ending with the formation of a complete organism. As we look at this sequence, we will consider how reproductive cells are produced, the role of the male and female in sexual intercourse, and events associated with pregnancy, birth, and lactation. Finally, we will say a few words about birth control.

GAMETE FORMATION

DIPLOID AND HAPLOID CELLS

In sexual reproduction, each human is produced by the union and fusion of two different sex cells, each produced in the gonads of the respective parents. Such sex cells are called **gametes** and are referred to as ova produced in the female gonads (ovaries) and sperm produced in the male gonads (testes). The union and fusion of gametes is called fertilization and the cell thus produced is known as a zygote. The zygote contains a mixture of chromosomes (DNA) from the two parents and, through its repeated mitotic division, develops into a new organism.

Gametes differ from all other body cells (somatic cells) with respect to the number of chromosomes in their nuclei. Somatic cells, such as brain cells, stomach cells, and kidney cells, contain 46 chromosomes in their nuclei. Of the 46 chromosomes, 23 are a complete set that contain all the genes necessary for carrying out the activities of the cell. In a sense, the other 23 chromosomes are a duplicate set. The symbol n is used to designate the number of different chromosomes within the nucleus. Since somatic cells contain two sets of chromosomes, they are referred to as **diploid** (DIP-loyd) **cells** (di = two), symbolized as $2n$. In a diploid cell, two chromosomes that belong to a pair are called **homologous** (ho-MOL-o-gus) **chromosomes.**

If gametes had the same number of chromosomes as somatic cells, the zygote formed from their fusion would have double the number. The somatic cells of the resulting individual would have twice the number of chromosomes ($4n$) as the somatic cells of the parents. Further, at every succeeding generation, the number of chromosomes would double. In time, there would be so many chromosomes in the nucleus that there would literally be no room for cell division or other cellular activities, including gamete production.

Fortunately, the chromosome number does not double with each generation because of a special nuclear division called **meiosis.** Meiosis occurs only in the production of gametes. It causes a developing sperm or ovum to relinquish its duplicate set of chromosomes so that the mature gamete has only 23. The number of chromosomes in a sex cell is called the **haploid** (HAP-loyd) **number,** meaning "one-half" and symbolized n.

The formation in the testes of haploid spermatozoa by meiosis is called **spermatogenesis** (sper'-ma-tō-JEN-e-sis). The formation in the ovary of haploid ova by meiosis is referred to as **oogenesis** (ō'-ō-JEN-e-sis).

SPERMATOGENESIS

In humans, spermatogenesis takes about 2–3 weeks. The seminiferous tubules are lined with immature cells called **spermatogonia** (sper'-ma-tō-GŌ-nē-a) or sperm mother cells (Figure 24-1). Spermatogonia contain the diploid chromosome number and are the precursor cells for all the spermatozoa the male will produce. At puberty, when the anterior pituitary secretes FSH in response to FSHRF from the hypothalamus, the spermatogonia embark on a lifetime of active mitotic division. Some of the daughter cells are pushed inward toward the lumen of the seminiferous tubule. These cells lose contact with the basement membrane of the seminiferous tubule, and as a result of certain developmental changes, become known as **primary spermatocytes** (SPER-ma-tō-sītz'). Primary spermatocytes, like spermatogonia, are diploid, that is, they have 46 chromosomes. The other daughter cells formed by mitosis of the spermatogonia remain near the basement membrane and form a reservoir of precursor cells for future meiotic divisions.

Each primary spermatocyte enlarges before dividing. Then two nuclear divisions take place as part of meiosis. In the first, DNA is replicated and 46 chromosomes form and move toward the equatorial plane of the nucleus. There they line up by homologous pairs so that there are 23 pairs of duplicated chromosomes (each of two chromatids) in the center of the nucleus. The four chromatids of each homologous pair then twist around each other to form a **tetrad.** In a tetrad, portions of one chromatid may be exchanged with portions of another. This process, called **crossing-over,** permits an exchange of genes among chromatids (Figure 24-2) that results in the recombination of genes. Thus the spermatozoa eventually produced may be genetically unlike each other and unlike the cell that produced them—hence the great variation among humans. Next the spindle forms and the threads attach to the centromeres of the paired chromosomes. As the pairs separate, one member of each pair migrates to opposite poles of the dividing nucleus. The cells thus formed by the first nuclear division (reduction division) are called **secondary spermatocytes.** Each cell has 23 chromosomes—the haploid number. Each chromosome of the secondary spermatocytes, however, is made up of two chromatids. Moreover, the genes of the chromosomes of secondary spermatocytes may be rearranged as a result of crossing-over.

The second nuclear division of meiosis is equatorial division. There is no replication of DNA. The chromo-

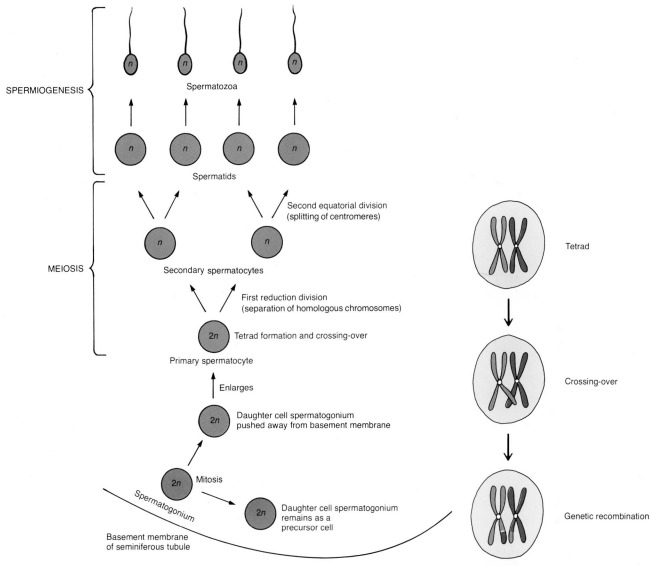

FIGURE 24-1 Spermatogenesis.

FIGURE 24-2 Crossing-over within a tetrad resulting in genetic recombination.

somes (each of two chromatids) line up in single file around the equatorial plane, and the chromatids of each chromosome separate from each other. The cells thus formed from the equatorial division are called **spermatids.** Each contains half the original chromosome number, or 23 chromosomes, and is haploid. Each primary spermatocyte therefore produces four spermatids by meiosis (reduction division and equatorial division). Spermatids lie close to the lumen of the seminiferous tubule.

The final stage of spermatogenesis, called **spermiogenesis** (sper'-mē-ō-JEN-e-sis), involves the maturation of spermatids into spermatozoa. Each spermatid embeds in a Sertoli cell and develops a head with an acrosome and a flagellum (tail). Sertoli cells extend from the basement membrane to the interior of the seminiferous tubule where they nourish the developing spermatids. Since there is

no cell division in spermiogenesis, each spermatid develops into a single spermatozoon, or sperm cell.

In summary, then, a single primary spermatocyte develops into four spermatozoa by spermatogenesis, a process that involves meiosis and spermiogenesis. Spermatozoa enter the lumen of the seminiferous tubule and migrate to the ductus epididymis, where in 18 hours to 10 days they complete their maturation and become capable of fertilizing an ovum. Spermatozoa are also stored in the ductus deferens. Here, they can retain their fertility for up to several months.

OOGENESIS

The formation of a haploid ovum by meiosis in the ovary is referred to as oogenesis. With some exceptions, oogen-

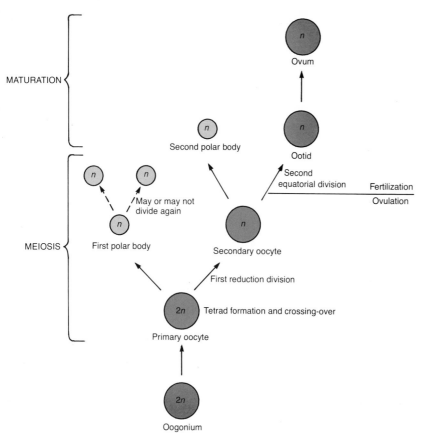

FIGURE 24-3 Oogenesis.

esis occurs in essentially the same manner as spermatogenesis. It involves meiosis and maturation.

The precursor cell in oogenesis is a diploid cell called the **oogonium** (ō'-o-GŌ-nē-um) or egg mother cell (Figure 24-3). Even before the female fetus is ready for birth, the oogonia in the primary follicles have lost their ability to carry on mitosis. Such immature cells containing the diploid number of chromosomes are called **primary oocytes** (Ō-o-sītz). They remain in this stage until their follicular cells respond to FSH from the anterior pituitary, which in turn has responded to FSHRF from the hypothalamus. Starting with puberty, several follicles respond each month to the rising level of FSH. As the cycle proceeds and LH is secreted from the anterior pituitary, one of the follicles reaches a stage in which the now mature, diploid primary oocyte undergoes reduction division. Tetrad formation and crossing-over occur, and two cells of unequal size, both with 23 chromosomes of two chromatids each, are produced. The smaller cell, called the **first polar body,** is essentially a packet of discarded nuclear material. The larger cell, known as the **secondary oocyte,** receives most of the cytoplasm.

At this stage in the ovarian cycle, ovulation takes place. Since the secondary oocyte with its polar body and surrounding supporting cells is discharged at the time of ovulation, the "ovum" discharged is not yet mature. The discharged secondary oocyte enters the uterine tube and, if spermatozoa are present and fertilization occurs, the second division takes place: the equatorial division. The secondary oocyte produces two cells of unequal size, both of them haploid. The larger cell is called the **ootid;** the smaller is the **second polar body.** In time, the ootid develops into an **ovum,** or mature egg.

The first polar body may undergo another division. If it does, meiosis of the primary oocyte results in a single haploid ovum and three polar bodies. In any event, the polar bodies disintegrate. Thus each oogonium produces a single ovum, whereas each spermatocyte produces four spermatozoa.

FERTILIZATION AND IMPLANTATION

Once spermatozoa and ova are developed through meiosis and the spermatozoa are deposited in the vagina, pregnancy can occur. **Pregnancy** is a sequence of events including fertilization, implantation, embryonic growth, and, normally, fetal growth that terminates in birth.

SEXUAL INTERCOURSE

Fertilization of an ovum is accomplished through **sexual intercourse,** or **copulation** (in human beings, called **coitus**), in which spermatozoa are deposited in the vagina. The male role in the sexual act starts with **erection,** the enlargement and stiffening of the penis. An erection may be initiated in the cerebrum by stimuli such as anticipation, memory, and visual sensation, or it may be a reflex brought on by stimulation of the touch receptors in the penis, especially in the glans. In any case, parasympathetic impulses that pass from the sacral portion of the spinal cord to the penis cause dilation of the arteries of the penis, allowing blood to fill the cavernous spaces of the spongy bodies. These impulses also cause the bulbourethral (Cowper's) glands to secrete mucus, which affords some **lubrication** for intercourse. The major portion of lubricating fluid is produced by the female.

Tactile stimulation of the penis brings about emission and ejaculation. When sexual stimulation becomes intense, rhythmic sympathetic impulses leave the spinal cord at the levels of the first and second lumbar vertebrae and pass to the genital organs. These impulses cause peristaltic contractions of the ducts in the testes, the epididymides, the ductus deferens, and the seminal ducts that propel spermatozoa into the urethra—a process called **emission.** Simultaneously, peristaltic contractions of the seminal vesicles and prostate expel seminal and prostatic fluid along with the spermatozoa. All these mix with the mucus of the bulbourethral glands, resulting in the fluid called semen. Other rhythmic impulses sent from the spinal cord at the levels of the first and second sacral vertebrae reach the skeletal muscles at the base of the penis, and the penis expels the semen from the urethra to the exterior. The propulsion of semen from the urethra to the exterior constitutes an **ejaculation.** A number of sensory and motor activities accompany ejaculation, including a rapid heart rate, an increase in blood pressure, an increase in respiration, and pleasurable sensations. These activities, together with the muscular events involved in ejaculation, are referred to as an **orgasm.**

The female role in the sex act also involves erection, lubrication, and orgasm. Stimulation of the female, as in the male, depends on both psychic and tactile responses. Under appropriate conditions, stimulation of the female genitalia, especially the clitoris, results in **erection** and widespread sexual arousal. This response is controlled by parasympathetic impulses sent from the spinal cord to the external genitalia. These impulses also pass to the greater vestibular (Bartholin's) glands which secrete some of the **lubrication** during sexual intercourse. When tactile stimulation of the genitalia reaches maximum intensity, reflexes are initiated that cause the female **orgasm** or **climax.** Female orgasm is analogous to male ejaculation. The perineal muscles contract rhythmically from spinal reflexes similar to those that occur in the male ejaculation.

FERTILIZATION OF THE OVUM

The term **fertilization** is applied to the penetration of an ovum by a spermatozoon and the subsequent union of the sperm nucleus and the nucleus of the ovum. Of the hundreds of millions of sperm cells introduced into the vagina, only about 2,000 arrive in the vicinity of the ovum. Fertilization normally occurs in the uterine tube when the ovum is about one-third of the way down the tube, usually within 24 hours after ovulation. Peristaltic contractions and the action of cilia transport the ovum through the uterine tube. The mechanism by which sperm reach the uterine tube is still unclear. Some believe that sperm swim up the female tract by means of whiplike movements of their flagella; others believe sperm are transported by muscular contractions of the uterus. Their motility is probably a combination of both.

Sperm must remain in the female genital tract 4 to 6 hours before they are capable of fertilizing an ovum. During this time, the enzyme hyaluronidase is activated and secreted by the acrosomes of the spermatozoa. Hyaluronidase apparently dissolves parts of the membrane covering the ovum. The mature ovum is surrounded by a gelatinous covering called the **zona pellucida** (pe-LOO-si-da) and several layers of cells, the innermost of which are follicle cells, known as the **corona radiata** (Figure 24-4a). Normally, only one spermatozoon fertilizes an ovum, because once union is achieved, the ovum develops a fertilization membrane that is impermeable to the entrance of other spermatozoa. When the spermatozoon has entered the ovum, the tail is shed and the nucleus in the head develops into a structure called the **male pronucleus.** The nucleus of the ovum develops into a **female pronucleus** (Figure 24-4b). After the pronuclei are formed, they fuse to produce a **segmentation nucleus.** The segmentation nucleus contains 23 chromosomes from the male pronucleus and 23 chromosomes from the female pronucleus. Thus the fusion of the haploid pronuclei restores the diploid number. The fertilized ovum, consisting of a segmentation nucleus, cytoplasm, and enveloping membrane, is called a **zygote.**

FORMATION OF THE MORULA

Immediately after fertilization, rapid cell division of the zygote takes place. This early division of the zygote is called **cleavage.** During this time the dividing cells are contained by the zona pellucida. The first cleavage is completed after about 36 hours, and each succeeding division takes slightly less time (Figure 24-5). By the second day after conception, the second cleavage is completed. By the end of the third day there are 16 cells. The progressively smaller cells produced are called **blastomeres** (BLAS-tō-mērz).

Successive cleavages produce a solid mass of cells, the **morula** (MOR-yoo-la) or mulberry, which is only slightly

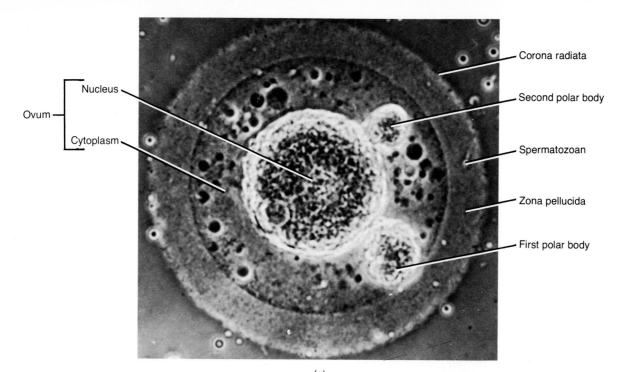

(a)

FIGURE 24-4 Fertilization. (a) Photomicrograph of a spermatozoon moving through zona pellucida on its way to reach the nucleus of the ovum. (Courtesy of *The Rand McNally Atlas of the Body and Mind,* Rand McNally and Company, New York, Chicago, San Francisco, in association with Mitchell Beazley Publishers Limited, London, 1976.) (b) Photomicrograph showing male and female pronuclei. (Courtesy of Carolina Biological Supply Company.)

FIGURE 24-5 Formation of the morula. (a) Diagram of successive cleavages of the zygote.

Pronuclei

(b)

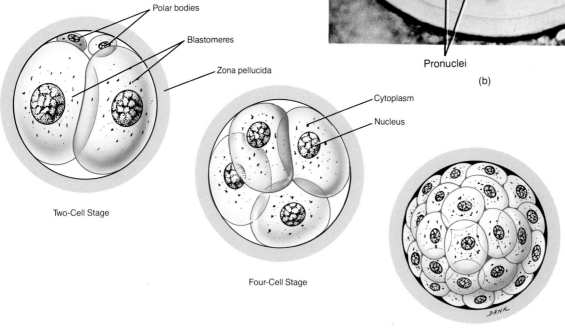

Two-Cell Stage

Four-Cell Stage

Morula

(a)

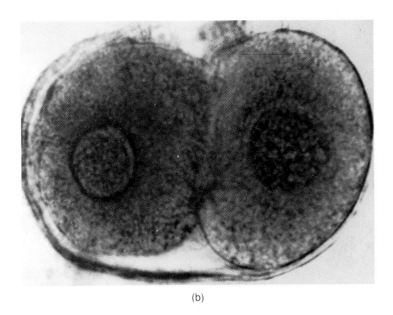

(b)

(c)

FIGURE 24-5 (*Continued*) Formation of the morula. (b) Photomicrograph of an ovum at two-cell stage. (Courtesy of Roberts Rugh and Landrum B. Shettles, M.D., with Richard N. Einhorn, from *From Conception to Birth: The Drama of Life's Beginnings,* Harper & Row, Publishers, Inc., New York, 1971.) (c) Photomicrograph of a morula at 3½ days. (Courtesy of Roberts Rugh and Landrum B. Shettles, M.D., with Richard N. Einhorn, from *From Conception to Birth: The Drama of Life's Beginnings,* Harper & Row, Publishers, Inc., New York, 1971.)

larger than the original zygote. The morula is formed a few days after fertilization.

DEVELOPMENT OF THE BLASTOCYST

As the number of cells in the morula increases, it moves from the original site of fertilization down through the ciliated uterine tube toward the uterus and enters the uterine cavity. By this time, the dense cluster of cells is altered to form a hollow ball of cells. The mass is now referred to as a **blastocyst** (Figure 24-6).

The blastocyst is differentiated into an outer covering of cells called the **trophectoderm** (trō-FEK-tō-derm), an **inner cell mass,** and the internal fluid-filled cavity referred to as the **blastocoel** (BLAS-tō-sēl). The trophectoderm ultimately forms part of the membranes composing the fetal portion of the placenta; the inner cell mass develops into the embryo.

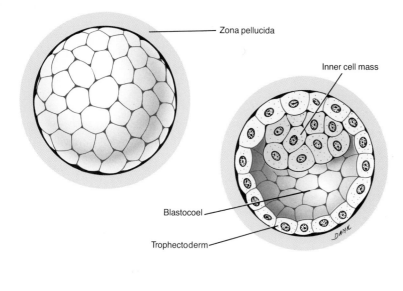

Zona pellucida

Inner cell mass

Blastocoel

Trophectoderm

(a)

(b)

FIGURE 24-6 Blastocyst. (a) Diagram of external view. (b) Diagram of internal view. (c) Photomicrograph of a blastocyst at 4 days. (Courtesy of Carnegie Institution of Washington, Department of Embryology, Davis Division.)

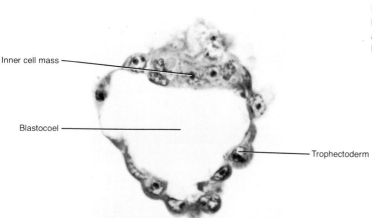

Inner cell mass

Blastocoel

Trophectoderm

(c)

IMPLANTATION

The attachment of the blastocyst to the endometrium occurs 7 to 8 days after fertilization and is called **implantation** (Figure 24-7). At this time, the endometrium is in its postovulatory phase. During implantation, the cells of the trophectoderm secrete an enzyme that enables the blastocyst to penetrate the uterine lining and become buried in the endometrium, usually on the posterior wall of the fundus or body of the uterus. The blastocyst, now only about $\frac{1}{100}$ of an inch in diameter, becomes oriented so that the inner cell mass is toward the endometrium. Implantation enables the blastocyst to absorb nutrients from the glands and blood vessels of the endometrium for its subsequent growth and development.

A summary of the principal events associated with fertilization and implantation is shown in Figure 24-8.

EXTERNAL HUMAN FERTILIZATION

On July 12, 1978, Louise Joy Brown was born near Manchester, England. Her birth was the first recorded case of **external human fertilization,** fertilization in a glass dish. The procedure developed for external human fertilization is believed to have been carried out as follows. The female is given FSH soon after menstruation, so that several ova, rather than the typical single one, will be produced (superovulation). Administration of LH may also ensure the maturation of the ova. Next, a small incision is made near the umbilicus, and the ova are aspirated from the follicles and placed in a medium that simulates the fluids in the female reproductive tract. The ova are then transferred to a solution of the male's sperm. Once fertilization has taken place, the fertilized ovum is put in another medium and is observed for cleavage. When

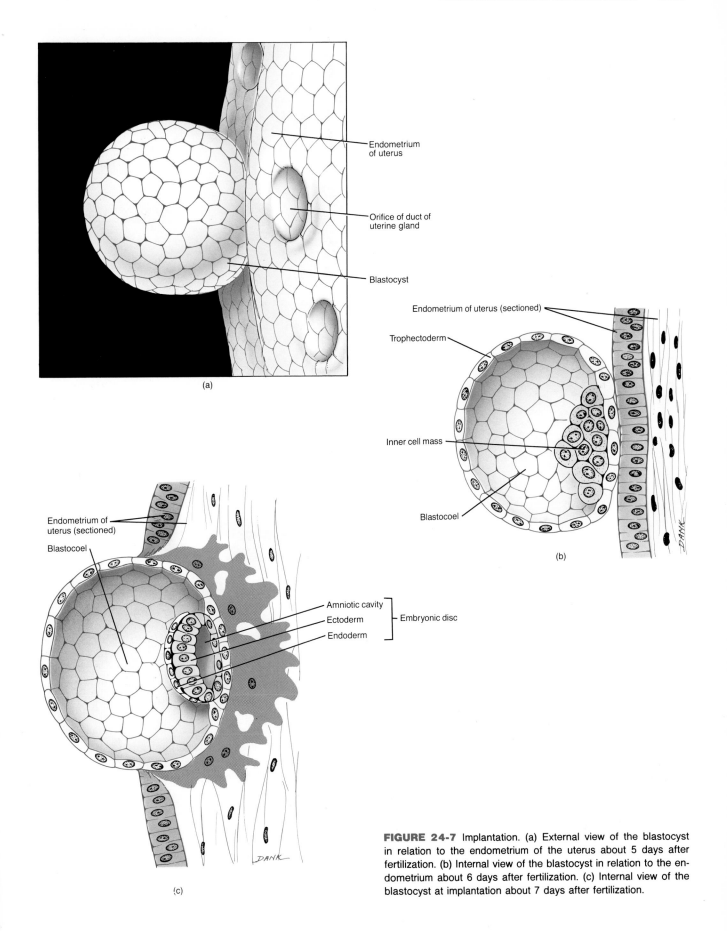

Endometrium
of uterus

Orifice of duct of
uterine gland

Blastocyst

(a)

Endometrium of uterus (sectioned)

Trophectoderm

Inner cell mass

Blastocoel

(b)

Endometrium of
uterus (sectioned)

Blastocoel

Amniotic cavity
Ectoderm   Embryonic disc
Endoderm

(c)

FIGURE 24-7 Implantation. (a) External view of the blastocyst
in relation to the endometrium of the uterus about 5 days after
fertilization. (b) Internal view of the blastocyst in relation to the endometrium about 6 days after fertilization. (c) Internal view of the
blastocyst at implantation about 7 days after fertilization.

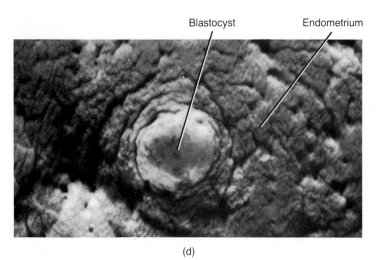

(d)

FIGURE 24-7 (*Continued*) Implantation. (d) Photomicrograph of implantation. (Courtesy of Roberts Rugh and Landrum B. Shettles, M.D., with Richard N. Einhorn, from *From Conception to Birth: The Drama of Life's Beginnings,* Harper & Row, Publishers, Inc., New York, 1971.)

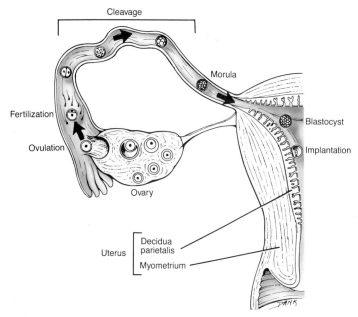

FIGURE 24-8 Summary of events associated with fertilization and implantation.

the fertilized ovum reaches the 8-cell or 16-cell stage, it is introduced into the uterus for implantation and subsequent growth. The growth and developmental sequences that occur are similar to those in internal fertilization.

EMBRYONIC DEVELOPMENT

The first two months of development are considered the **embryonic period.** During this period the developing human is called an **embryo.** The months of development after the second month are considered the **fetal period,** and during this time the developing human is called a **fetus.** By the end of the embryonic period the rudiments of all the principal adult organs are present, the embryonic membranes are developed, and the placenta is functioning.

BEGINNINGS OF ORGAN SYSTEMS

Following implantation, the inner cell mass of the blastocyst begins to differentiate into the three **primary germ layers:** ectoderm, endoderm, and mesoderm. They are the embryonic tissues from which all tissues and organs of the body will develop.

In the human being, the germ layers form so quickly that it is difficult to determine the exact sequence of events. Before implantation, a layer of **ectoderm** (the trophectoderm) already has formed around the blastocoel (see Figure 24-7). The trophectoderm will become part of the chorion, one of the fetal membranes. Within eight days after fertilization the inner cell mass moves downward so a space called the **amniotic cavity** lies between the inner cell mass and the trophectoderm. The bottom layer of the inner cell mass develops into an **endodermal** germ layer.

About the twelfth day after fertilization striking changes appear (Figure 24-9a). A layer of cells from the inner cell mass has grown around the top of the amniotic cavity. These cells will become the amnion, another fetal membrane. The cells below the cavity are called the **embryonic disc.** They will form the embryo. At this stage the embryonic disc contains ectodermal and endodermal cells; mesodermal cells are scattered external to the disc. The cells of the endodermal layer have been dividing

rapidly, so that groups of them now extend downward in a circle, forming the yolk sac, another fetal membrane. The **mesodermal** cells also have been dividing, and many have left the area of the embryonic disc and can be seen around the structures that are becoming fetal membranes.

About the fourteenth day the cells of the embryonic disc differentiate into three distinct layers: the upper ectoderm, the middle mesoderm, and the lower endoderm (Figure 24-9b). At this time the two ends of the embryonic disc draw together, squeezing off the yolk sac. The resulting cavity inside the disc is the endoderm-lined **primitive gut.** The mesoderm in the disc soon splits into two layers, and the space between the layers becomes the **extraembryonic coelom.**

As the embryo develops, (Figure 24-9c), the endoderm becomes the epithelium lining the digestive tract and respiratory tract and a number of other organs. The mesoderm forms the peritoneum, muscle, bone, and other connective tissue. The ectoderm develops into the skin and nervous system. More details about the fates of these primary germ layers are presented later in the chapter.

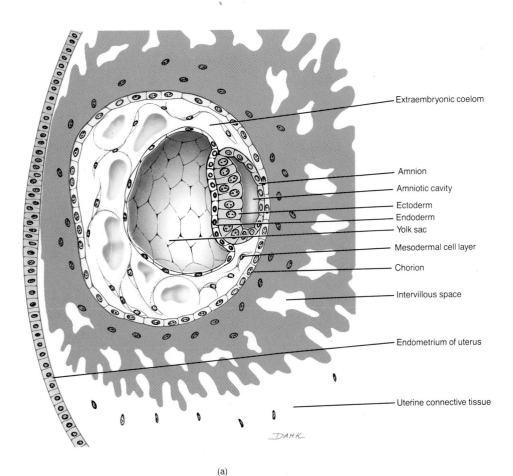

- Extraembryonic coelom

- Amnion
- Amniotic cavity

- Ectoderm
- Endoderm
- Yolk sac

- Mesodermal cell layer

- Chorion

- Intervillous space

- Endometrium of uterus

- Uterine connective tissue

DANK

(a)

FIGURE 24-9 Formation of the primary germ layers and associated structures in a developing embryo. (a) Internal view about 12 days after fertilization.

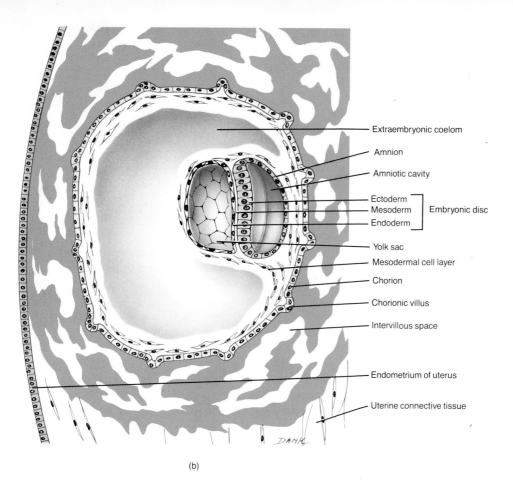

Extraembryonic coelom

Amnion

Amniotic cavity

Ectoderm ⎤
Mesoderm ⎬ Embryonic disc
Endoderm ⎦

Yolk sac

Mesodermal cell layer

Chorion

Chorionic villus

Intervillous space

Endometrium of uterus

Uterine connective tissue

(b)

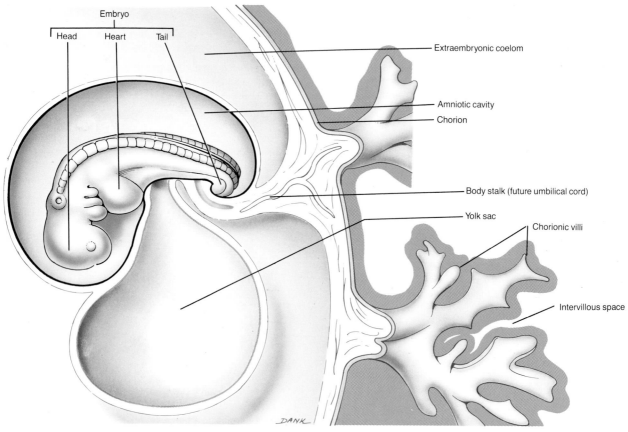

Embryo

Head   Heart   Tail

Extraembryonic coelom

Amniotic cavity

Chorion

Body stalk (future umbilical cord)

Yolk sac

Chorionic villi

Intervillous space

(c)

FIGURE 24-9 (*Continued*) Formation of the primary germ layers and associated structures in a developing embryo. (b) Internal view about 14 days after fertilization. (c) External view about 25 days after fertilization.

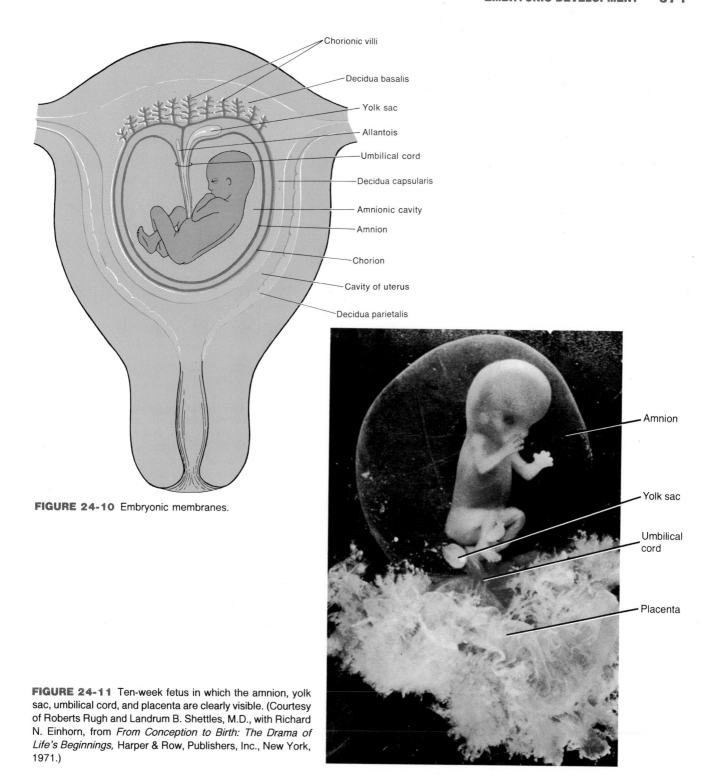

FIGURE 24-10 Embryonic membranes.

FIGURE 24-11 Ten-week fetus in which the amnion, yolk sac, umbilical cord, and placenta are clearly visible. (Courtesy of Roberts Rugh and Landrum B. Shettles, M.D., with Richard N. Einhorn, from *From Conception to Birth: The Drama of Life's Beginnings,* Harper & Row, Publishers, Inc., New York, 1971.)

EMBRYONIC MEMBRANES

During the embryonic period, the **embryonic membranes** form. These membranes lie outside the embryo and protect and nourish the embryo and later the fetus. The membranes are the yolk sac, the amnion, the chorion (KŌ-rē-on), and the allantois (a-LAN-tō-is) (Figure 24-10).

The **yolk sac** is an endoderm-lined membrane that, in many species, provides the primary or exclusive nutrient for the embryo (Figure 24-11). However, the human embryo receives its nourishment from the endometrium and the yolk sac remains small (Figure 24-11). During an early stage of development it becomes a nonfunctional part of the umbilical cord.

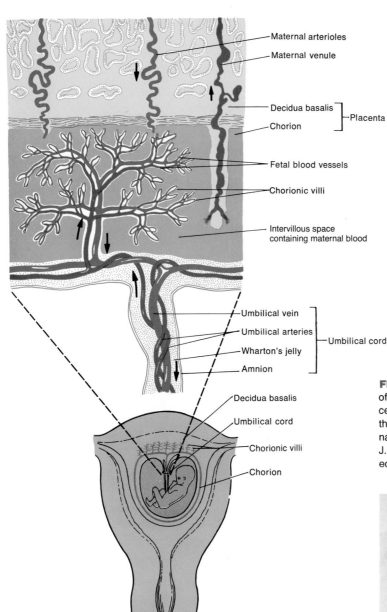

- Maternal arterioles
- Maternal venule
- Decidua basalis] Placenta
- Chorion
- Fetal blood vessels
- Chorionic villi
- Intervillous space containing maternal blood
- Umbilical vein
- Umbilical arteries] Umbilical cord
- Wharton's jelly
- Amnion
- Decidua basalis
- Umbilical cord
- Chorionic villi
- Chorion

(a)

FIGURE 24-12 Placenta and umbilical cord. (a) Diagram of the structure of the placenta and umbilical cord. The placenta can also be seen, along with the umbilical cord, in the photograph in Figure 24-14. (b) Photograph of the maternal aspect of the placenta. (Courtesy of C. Yokochi and J. W. Rohen, *Photographic Anatomy of the Human Body,* 2nd ed., 1979, IGAKU-SHOIN, Ltd., Tokyo, New York.)

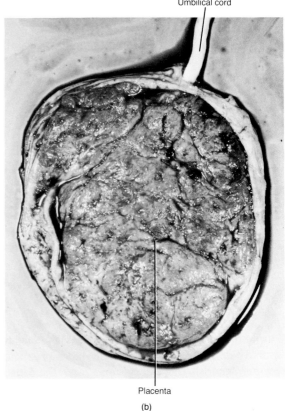

Umbilical cord

Placenta

(b)

The **amnion** is a thin, protective membrane that initially overlies the embryonic disc and is formed by the embryo by the eighth day following fertilization. As the embryo grows, the amnion entirely surrounds the embryo and becomes filled with **amniotic fluid** (Figure 24-11). Amniotic fluid serves as a shock absorber for the fetus. The amnion usually ruptures just before birth and with its fluid constitutes the "bag of waters."

The **chorion** is derived from the trophectoderm of the blastocyst and its associated mesoderm. It surrounds the embryo and, later, the fetus. Eventually the chorion becomes the principal part of the placenta, the structure through which materials are exchanged between mother

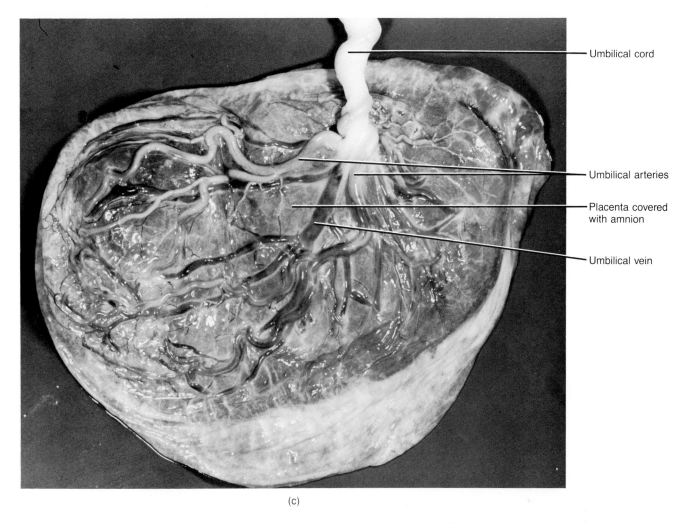

Umbilical cord

Umbilical arteries

Placenta covered
with amnion

Umbilical vein

(c)

FIGURE 24-12 (*Continued*) Placenta and umbilical cord. (c) Photograph of the fetal aspect of the placenta. (Courtesy of C. Yokochi and J. W. Rohen, *Photographic Anatomy of the Human Body,* 2nd ed., 1979, IGAKU-SHOIN, Ltd., Tokyo, New York.)

and fetus. The amnion also surrounds the fetus and eventually fuses to the inner layer of the chorion.

The **allantois** is a small vascularized membrane. Later its blood vessels serve as connections in the placenta between mother and fetus. This connection is the umbilical cord.

PLACENTA AND UMBILICAL CORD

Development of the placenta, the next major event of the embryonic period, is accomplished by the third month of pregnancy. The **placenta** has the shape of a flat cake when fully developed and is formed by the chorion of the embryo and a portion of the endometrium of the mother (Figures 24-12 and 24-14). It provides an exchange of nutrients and wastes between fetus and mother and secretes the hormones necessary to maintain pregnancy.

If implantation occurs, a portion of the endometrium becomes modified and is known as the **decidua** (dē-SID-yoo-a). The decidua includes all but the deepest layer of the endometrium and is shed when the fetus is delivered. Different regions of the decidua are named on the basis of their positions relative to the site of the implanted ovum (Figures 24-10 and 24-13). The *decidua parietalis* (pa-rī-e-TAL-is) is the portion of the modified endometrium that lines the entire pregnant uterus, except for the area where the placenta is forming. The *decidua capsularis* is the portion of the endometrium between the embryo and the uterine cavity. The *decidua basalis* is the portion of the endometrium between the chorion and the muscularis of the uterus. The decidua basalis becomes the maternal part of the placenta.

During embryonic life, fingerlike projections of the chorion, called **chorionic villi** (kō'-rē-ON-ik VIL-ī), grow

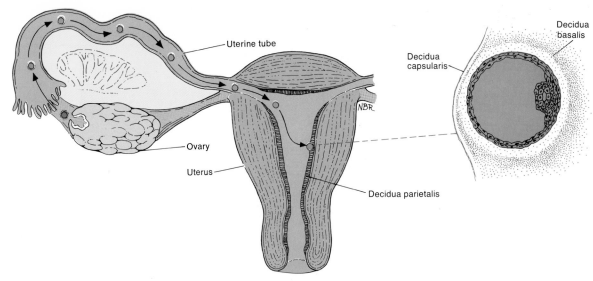

FIGURE 24-13 Portions of the decidua.

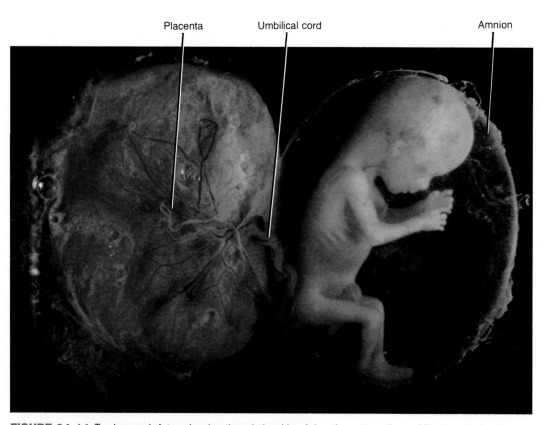

FIGURE 24-14 Twelve-week fetus showing the relationship of the placenta to the umbilical cord. (Courtesy of Roberts Rugh and Landrum B. Shettles, M.D., with Richard N. Einhorn, from *From Conception to Birth: The Drama of Life's Beginnings,* Harper & Row, Publishers, Inc., New York, 1971.)

into the decidua basalis of the endometrium (see Figure 24-12). These will contain fetal blood vessels of the allantois. They continue growing until they are bathed in maternal blood sinuses called **intervillous** (in-ter-VIL-us) **spaces.** Thus, maternal and fetal blood vessels are brought into proximity. It should be noted, however, that maternal and fetal blood do not normally mix. Oxygen and nutrients from the mother's blood diffuse across the walls and into the capillaries of the villi. From the capillaries the nutrients circulate into the umbilical vein. Wastes leave the fetus through the umbilical arteries, pass into the capillaries of the villi, and diffuse into the maternal blood. The **umbilical cord** consists of an outer layer of amnion containing the umbilical arteries and umbilical vein, supported internally by mucous connective tissue from the allantois called Wharton's jelly (Figure 24-14).

CLINICAL APPLICATION

At delivery, the placenta detaches from the uterus and is referred to as the **afterbirth.** At this time, the umbilical cord is severed, leaving the baby on its own. The scar that marks the site of the entry of the fetal umbilical cord into the abdomen is the **umbilicus (navel).**

DEVELOPMENTAL ANATOMY OF BODY SYSTEMS

Now that you have a basic understanding of the principal events associated with embryonic development, we will turn our attention to the development of the body as it continues through the fetal period. The salient features of the developmental anatomy of the organ systems are presented in the following exhibits.

EXHIBIT 24-1 THE INTEGUMENTARY SYSTEM

The *epidermis* is derived from the **ectoderm.** At the beginning of the second month, the ectoderm consists of simple cuboidal epithelium. These cells become flattened and are known as the **periderm.** By the fourth month, all layers of the epidermis are formed and each layer assumes its characteristic structure.

Nails are developed during the third month. Initially, they consist of a thick layer of epithelium called the **primary nail field.** The nail itself is keratinized epithelium and grows forward from its base. It is not until the ninth month that the nails actually reach the tips of the digits.

Hair follicles develop as downgrowths of the stratum basale of the epidermis into the dermis below, between the third and fourth months. The downgrowths soon differentiate into the bulb, papilla of the hair, beginnings of the epithelial portions of sebaceous glands, and other structures associated with hair follicles. By the fifth or sixth month, the follicles produce **lanugo** (delicate fetal hair) first on the head and then on other parts of the body. The lanugo is usually shed prior to birth.

The epithelial (secretory) portions of *sebaceous* (*oil*) *glands* develop from the sides of the hair follicles and remain connected to the follicles.

The epithelial portions of *sudoriferous* (*sweat*) *glands* are also derived from downgrowths of the stratum basale of the epidermis into the dermis. They appear during the fourth month on the palms and soles and a little later in other regions. The connective tissue and blood vessels associated with the glands develop from **mesoderm.**

The *dermis* is derived from wandering **mesenchymal (mesodermal) cells.** The mesenchyme becomes arranged in a zone beneath the ectoderm and there undergoes changes into the connective tissues that form the dermis.

EXHIBIT 24-2 THE SKELETAL SYSTEM

Bone forms about the sixth or seventh week by either of two processes, **intramembranous ossification** or **endochondral ossification.** Both processes begin when **mesenchymal (mesodermal) cells** migrate into the area where bone formation will occur. In some skeletal structures, the mesenchymal cells develop into **chondroblasts** that form *cartilage.* In other skeletal structures, the mesenchymal cells develop into **osteoblasts** that form *bone tissue* by intramembranous or endochondral ossification. (Refer to Chapter 5 for the details of ossification, and also note the salient features of the developing skeleton in Figure 5-4.)

Discussion of the development of the skeletal system provides us with an excellent opportunity to trace the development of the extremities. The *extremities* make their appearance about the fifth week as small elevations at the sides of the trunk called **limb buds** (Figure 24-15). They consist of masses of general **mesoderm** surrounded by **ectoderm.** At this point, a mesenchymal skeleton exists in the limbs; some of the masses of mesoderm surrounding the developing bones will develop into the skeletal muscles of the extremities. By the sixth week, the limb buds develop a constriction around the middle portion. The constriction produces distal segments of the upper buds called **hand plates** and distal segments of the lower buds called **foot plates.** These plates represent the beginnings of the *hands* and *feet,* respectively. At this stage of limb development, a cartilaginous skeleton is present. By the seventh week, the *arm, forearm,* and *hand* are evident in the upper limb bud and the *thigh, leg,* and *foot* appear in the lower limb bud. Endochondral ossification has begun. By the eighth week, the upper limb bud is appropriately called the *upper extremity* as the *shoulder, elbow,* and *wrist* areas become apparent and the lower limb bud is referred to as the *lower extremity* with the appearance of the *knee* and *ankle* areas.

The *notochord* is a flexible rod of tissue that lies in a position where the future vertebral column will develop (see Figures 24-16b and 24-19c). As the vertebrae develop, the notochord becomes surrounded by the developing vertebral bodies and the notochord eventually disappears, except for remnants that persist as the *nucleus pulposus* of the intervertebral discs.

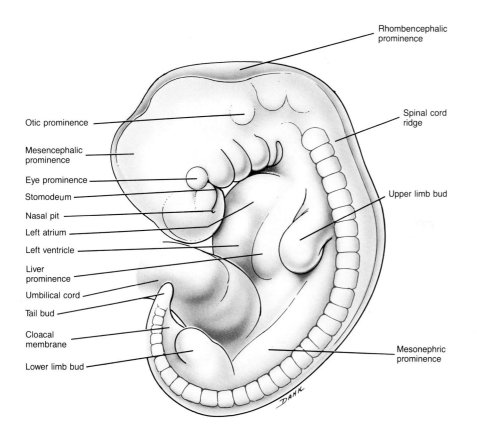

Rhombencephalic prominence

Otic prominence

Mesencephalic prominence

Eye prominence

Stomodeum

Nasal pit

Left atrium

Left ventricle

Liver prominence

Umbilical cord

Tail bud

Cloacal membrane

Lower limb bud

Spinal cord ridge

Upper limb bud

Mesonephric prominence

(a)

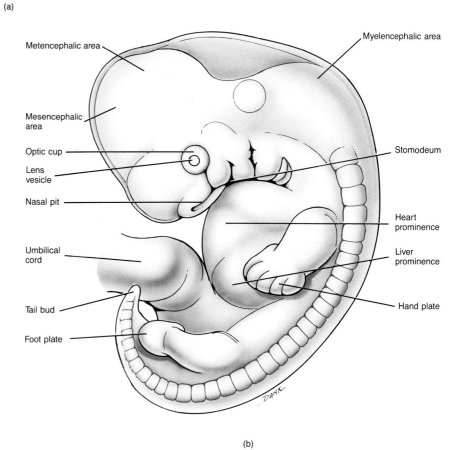

Metencephalic area

Mesencephalic area

Optic cup

Lens vesicle

Nasal pit

Umbilical cord

Tail bud

Foot plate

Myelencephalic area

Stomodeum

Heart prominence

Liver prominence

Hand plate

(b)

FIGURE 24-15 External features of a developing embryo at various stages of development. (a) Fifth week. (b) Sixth week.

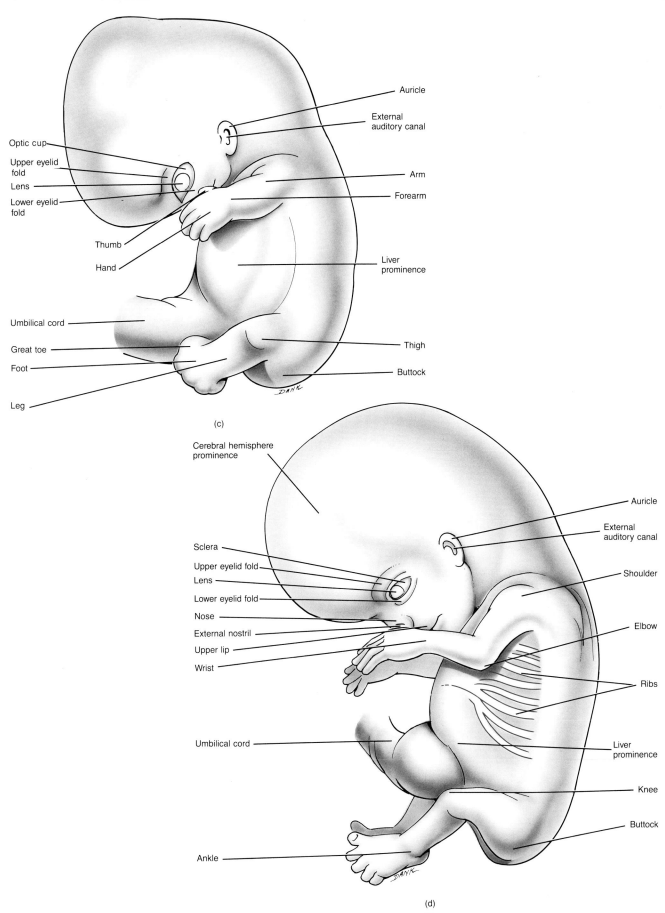

FIGURE 24-15 (*Continued*) External features of a developing embryo at various stages of development. (c) Seventh week. (d) Eighth week.

EXHIBIT 24-3 **THE MUSCULAR SYSTEM**

In this brief discussion of the development of the muscular system, we will concentrate mostly on skeletal muscles. Except for the muscles of the iris of the eyes and the arrector pili muscles attached to hairs, all muscles of the body are derived from **mesoderm.** As the mesoderm develops, a portion of it becomes arranged in dense columns on either side of the developing nervous system. These columns of mesoderm undergo segmentation into a series of blocks of cells called **somites** (Figure 24-16a). The first pair of somites appears on the twentieth day. Eventually, 44 pairs of somites are formed by the thirtieth day.

With the exception of the skeletal muscles of the head and extremities, *skeletal muscles* develop from the **mesoderm of somites.** Since there are very few somites in the head region of the embryo, most of the skeletal muscles there develop from the **general mesoderm** in the head region. The skeletal muscles of the limbs develop from masses of general mesoderm around developing bones in embryonic limb buds (origins of future extremities).

The cells of a somite are differentiated into three regions: (1) **myotome,** which forms most of the skeletal muscles; (2) **dermatome,** which forms the connective tissues, including the dermis, under the epidermis; and (3) **sclerotome,** which gives rise to the vertebrae (Figure 24-16b).

In the development of a skeletal muscle from a myotome of a somite, certain patterns emerge. For example, a myotome may split longitudinally into two or more portions. This represents the manner in which the trapezius muscle forms. Other myotomes split into two or more layers. Such a pattern represents how the external oblique, internal oblique, and transversus abdominis muscles develop. In other instances, several myotomes fuse to form a single muscle. Such an example is the rectus abdominis muscle. Muscles may also migrate, wholly or in part, from their sites of origin. The latissimus dorsi, for example, originates from the cervical myotomes, but extends all the way down to the thoracic and lumbar vertebrae and hip bone. Another trend that may be observed is a change in the direction of muscle fibers. Initially, muscle fibers in a myotome are parallel to the long axis of the embryo. However, nearly all developed skeletal muscles do not have fibers that are parallel to the long axis. One example is the external oblique. Finally, *fasciae, ligaments,* and *aponeuroses* may form as a result of degeneration of all or parts of myotomes.

Visceral or *smooth muscle* develops from **mesodermal cells** that migrate to and envelop the developing gastrointestinal tract and viscera.

Cardiac muscle develops from **mesodermal cells** that migrate to and envelop the developing heart while it is still in the form of primitive heart tubes (see Figure 24-17b).

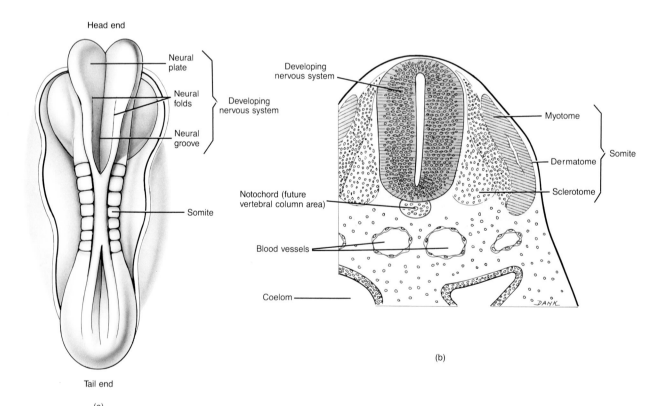

FIGURE 24-16 Development of the muscular system. (a) Dorsal aspect of an embryo indicating the location of somites. (b) Cross section of a somite.

EXHIBIT 24-4 THE CARDIOVASCULAR SYSTEM

Since the human egg and yolk sac have little yolk to nourish the developing embryo, blood and blood vessel formation starts as early as 15 to 16 days. The development begins in the **mesoderm** of the yolk sac, chorion, and body stalk.

Blood vessels develop from isolated masses and cords of mesenchyme in the mesoderm called **blood islands** (Figure 24-17a; see also Figure 24-24). Spaces soon appear in the islands and become the lumens of the blood vessels. Some of the mesenchymal cells immediately around the spaces give rise to the *endothelial lining of the blood vessels*. Mesenchyme around the endothelium forms the *tunics* (intima, media, externa) of the larger blood vessels. Growth and fusion of blood islands form an extensive network of blood vessels throughout the embryo.

Blood plasma and *blood cells* are produced by the endothelial cells and appear in the blood vessels of the yolk sac and allantois quite early. Blood formation in the embryo itself begins at about the second month in the liver, spleen, bone marrow, and lymph nodes.

The *heart,* also a derivative of **mesoderm,** begins to develop before the end of the third week. It begins its development in the ventral region of the embryo beneath the foregut (see Figure 24-24). The first step is the formation of a pair of tubes, the **endothelial**

(endocardial) tubes, from mesodermal cells (Figure 24-17b). These tubes then unite to form a common tube, the **primitive heart tube.** Next, the primitive heart tube undergoes a series of constrictions so that five regions develop: **truncus arteriosus, bulbus cordis, ventricle, atrium,** and **sinus venosus.** Since the bulbus cordis and ventricle subsequently grow more rapidly than the others, the heart assumes a U-shape and later an S-shape. The flexures of the heart reorient the regions so that the atrium and sinus venosus eventually come to lie superior to the bulbus cordis, ventricle, and truncus arteriosus.

At about the seventh week a partition, the **atrial septum,** forms in the atrial region dividing it into a *right* and *left atrium.* The opening in the partition is the **foramen ovale,** which normally closes at birth and later forms a depression called the *fossa ovalis.* A **ventricular septum** also develops and partitions the ventricular region into a *right* and *left ventricle.* The bulbus cordis and truncus arteriosus divide into two vessels, the *aorta* (arising from the left ventricle) and the *pulmonary trunk* (arising from the right ventricle). Recall that the ductus arteriosus is a temporary vessel between the aorta and pulmonary trunk until birth. The great veins of the heart, *superior vena cava* and *inferior vena cava,* develop from the venous end of the primitive heart tube.

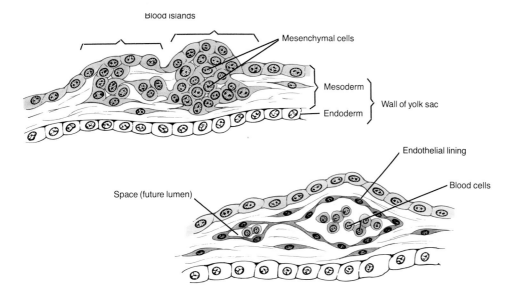

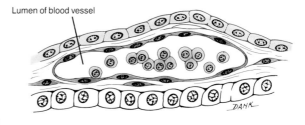

(a)

FIGURE 24-17 Development of the cardiovascular system. (a) Development of blood vessels and blood cells from blood islands.

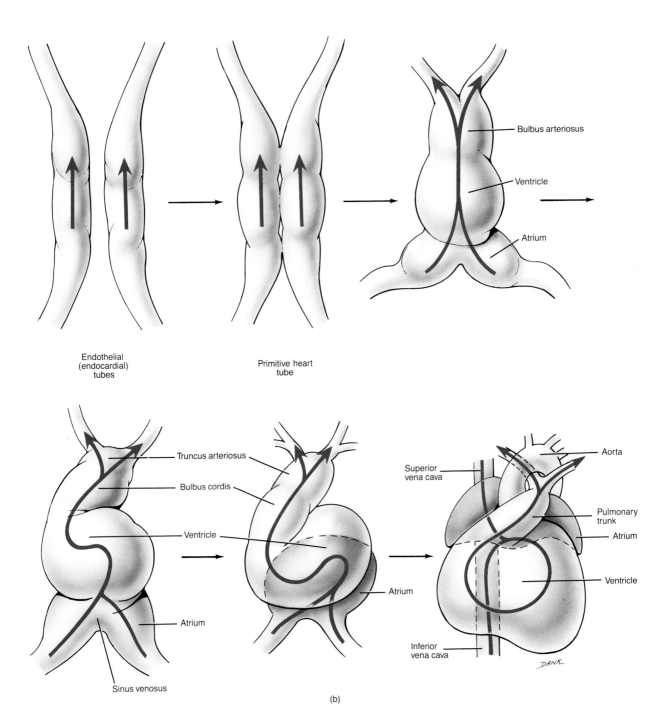

FIGURE 24-17 (*Continued*) Development of the cardiovascular system. (b) Development of the heart. The arrows show the direction of the flow of blood from the venous to the arterial end.

EXHIBIT 24-5 THE LYMPHATIC SYSTEM

The lymphatic system begins its development by the end of the fifth week. *Lymphatic vessels* develop from **lymph sacs** which arise from developing veins. Thus, the lymphatic system is also derived from **mesoderm.**

The first lymph sacs to appear are the paired **jugular lymph sacs** at the junction of the internal jugular and subclavian veins (Figure 24-18). From the jugular lymph sacs, capillary plexuses spread to the *thorax, upper extremities, neck,* and *head.* Some of the plexuses enlarge and form lymphatics in their respective regions. Each jugular lymph sac retains at least one connection with its jugular vein, the left one developing into the superior portion of the thoracic duct (left lymphatic duct). The next lymph sac to appear is the unpaired **retroperitoneal lymph sac** at the root of the mesentery of the intestine. It develops from the primitive vena cava and mesonephric (primitive kidney) veins. Capillary plexuses and lymphatics spread from the retroperitoneal lymph sac to the *abdominal viscera* and *diaphragm.* The sac establishes connections with the cisterna chyli but loses its connections with neighboring veins.

At about the time the retroperitoneal lymph sac is developing, another lymph sac, the **cisterna chyli,** develops below the diaphragm on the posterior abdominal wall. It gives rise to the inferior portion of the *thoracic duct* and the *cisterna chyli* of the thoracic duct. Like the retroperitoneal lymph sac, the cisterna chyli also loses its connections with surrounding veins.

The last of the lymph sacs, the paired **posterior lymph sacs,** develop from the iliac veins at their union with the posterior cardinal veins. The posterior lymph sacs produce capillary plexuses and lymphatics of the *abdominal wall, pelvic region,* and *lower extremity.* The posterior lymph sacs join the cisterna chyli and lose their connections with adjacent veins.

With the exception of the anterior part of the sac from which the cisterna chyli develops, all lymph sacs become invaded by **mesenchymal cells** and are converted into groups of *lymph nodes.*

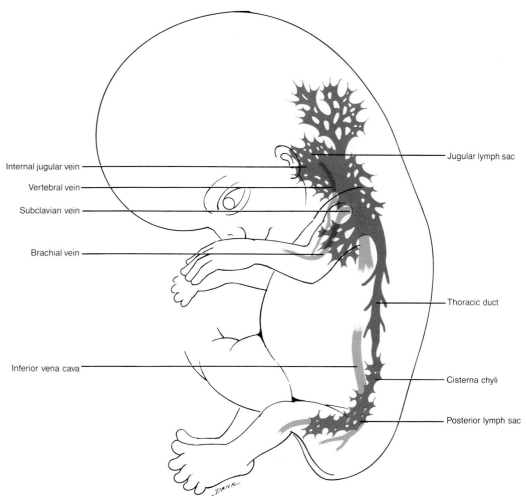

Internal jugular vein

Vertebral vein

Subclavian vein

Brachial vein

Inferior vena cava

Jugular lymph sac

Thoracic duct

Cisterna chyli

Posterior lymph sac

FIGURE 24-18 Development of the lymphatic system.

EXHIBIT 24-6 THE NERVOUS SYSTEM

The development of the nervous system occurs early in the third week of development and begins with a thickening of the **ectoderm** called the **neural plate** (Figure 24-19a–c). The plate folds inward and forms a longitudinal groove, the **neural groove.** The raised edges of the neural plate are called **neural folds.** As development continues, the neural folds increase in height, meet, and form a tube, the **neural tube.**

The cells of the wall that encloses the neural tube differentiate into three kinds. The outer or **marginal layer** develops into the *white matter* of the nervous system; the middle or **mantle layer** develops into the *gray matter* of the system; and the inner or **epen-**

dymal layer eventually forms the *lining of the ventricles* of the central nervous system.

The **neural crest** is a mass of tissue between the neural tube and the ectoderm (Figure 24-19c). It becomes differentiated and eventually forms the *posterior (dorsal) root ganglia of spinal nerves, spinal nerves, ganglia of cranial nerves, cranial nerves, ganglia of the autonomic nervous system,* and the *adrenal medulla.*

When the neural tube is formed from the neural plate, the anterior portion of the neural tube develops into three enlarged areas called vesicles: (1) **forebrain vesicle (prosencephalon),** (2) **midbrain vesicle (mesencephalon),** and (3) **hindbrain vesicle (rhomben-**

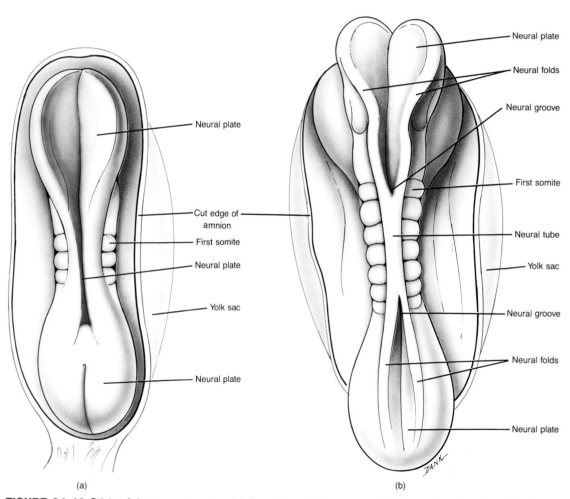

(a)   (b)

FIGURE 24-19 Origin of the nervous system. (a) Dorsal view of an embryo with three pairs of somites showing the neural plate. (b) Dorsal view of an embryo with seven pairs of somites in which the neural folds have united just medial to the somites forming the early neural tube.

cephalon) (Figure 24-20). The vesicles are fluid-filled enlargements that develop by the fourth week of gestation. Since they are the first vesicles to form, they are called **primary vesicles.** As development progresses, the vesicular region undergoes several flexures (bends). This results in subdivision of the three primary vesicles so that by the fifth week of development the embryonic brain consists of five **secondary vesicles.** The forebrain vesicle (prosencephalon) divides into an anterior **telencephalon** and a posterior **diencephalon;** the midbrain vesicle (mesencephalon) remains unchanged; the hindbrain vesicle (rhombencephalon) divides into an anterior **metencephalon** and a posterior **myelencephalon.**

Ultimately, the telencephalon develops into the *cerebral hemispheres* and *basal ganglia;* the diencephalon develops into the *thalamus, hypothalamus,* and *pineal gland;* the midbrain vesicle (mesencephalon) develops into the *midbrain;* the metencephalon develops into the *pons varolii* and *cerebellum;* and the myelencephalon develops into the *medulla oblongata.* The cavities within the vesicles develop into the *ventricles* of the brain, whereas the fluid within them is *cerebrospinal fluid.* The area of the neural tube posterior to the myelencephalon gives rise to the *spinal cord.*

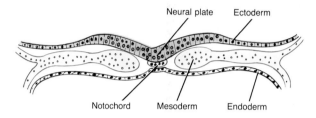

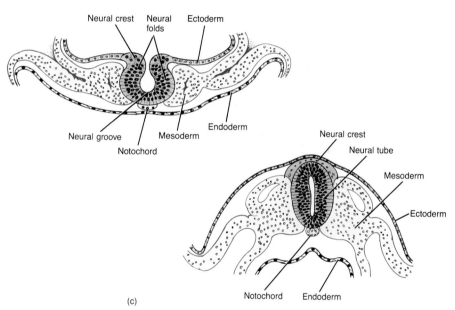

(c)

FIGURE 24-19 (*Continued*) Origin of the nervous system. (c) Cross section through the embryo showing the formation of the neural tube.

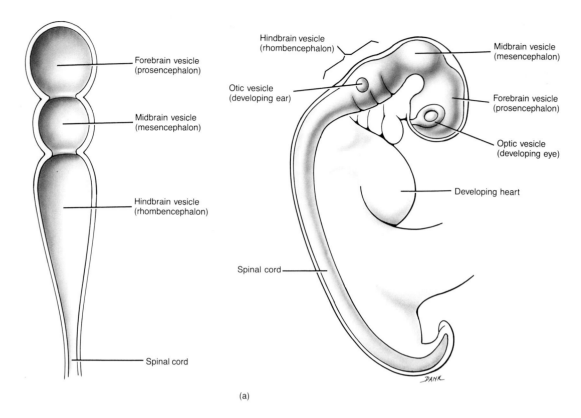

Forebrain vesicle (prosencephalon)

Midbrain vesicle (mesencephalon)

Hindbrain vesicle (rhombencephalon)

Spinal cord

Hindbrain vesicle (rhombencephalon)

Otic vesicle (developing ear)

Midbrain vesicle (mesencephalon)

Forebrain vesicle (prosencephalon)

Optic vesicle (developing eye)

Developing heart

Spinal cord

(a)

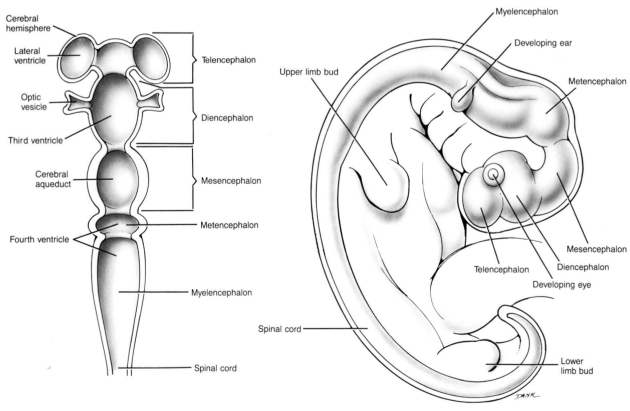

Cerebral hemisphere

Lateral ventricle

Optic vesicle

Third ventricle

Cerebral aqueduct

Fourth ventricle

Telencephalon

Diencephalon

Mesencephalon

Metencephalon

Myelencephalon

Spinal cord

Myelencephalon

Developing ear

Metencephalon

Upper limb bud

Mesencephalon

Diencephalon

Telencephalon

Developing eye

Spinal cord

Lower limb bud

(b)

FIGURE 24-20 Development of the brain and spinal cord. (a) Primary vesicles of the neural tube seen in frontal section (left) and right lateral view (right) at about 3 to 4 weeks. (b) Secondary vesicles seen in frontal section (left) and right lateral view (right) at about 5 weeks.

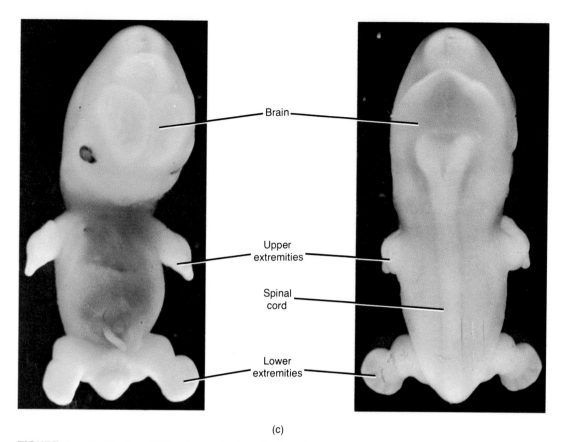

(c)

FIGURE 24-20 (*Continued*) Development of the brain and spinal cord. (c) Photograph of a 40-day embryo seen in anterior view (left) and posterior view (right). Note the brain, spinal cord, and upper and lower extremities. (Courtesy of Roberts Rugh and Landrum B. Shettles, M.D., with Richard N. Einhorn, from *From Conception to Birth: The Drama of Life's Beginnings,* Harper & Row, Publishers, Inc., New York, 1971.)

EXHIBIT 24-7 THE SPECIAL SENSE ORGANS: EYE AND EAR

Recall from Chapter 18 that smell, taste, sight, hearing, and equilibrium are the special senses. We will consider the development of only two special sense organs: the eye and the ear.

The first appearance of the eyes, at about the fourth week of development, is a broad, shallow depression on the medial surface of the diencephalon called the **optic groove** (Figure 24-21a). As development continues, lateral outgrowths from either side of the diencephalon arise. Each outgrowth is called an **optic vesicle** (see Figure 24-20b). The distal part of each vesicle expands to form a hollow bulb, while the proximal part remains narrow and is referred to as the **optic stalk** (Figure 24-21b). Surface **ectoderm** adjacent to the optic vesicle thickens and is known as the **lens placode.**

As each optic vesicle enlarges, it folds in on itself and becomes a double-layered **optic cup.** The outer, thinner layer becomes the *pigmented layer of the retina;* the inner, thicker layer becomes the *nervous layer of the retina.* Along with the invagination of each optic cup, the lens placode invaginates to form the hollow **lens vesicle.** Each lens vesicle separates from the surface ectoderm and becomes surrounded by its optic cup. The lens vesicle will eventually develop into the *lens* of the eye and the overlying **ectoderm** will develop into the *cornea* of the eye.

Fibers from the nervous layer of the retina pass through the optic stalk on their way to the brain and retinal blood vessels infiltrate the stalk. As a result, the *optic nerve* is formed. The *fibrous tunic* and *vascular tunic* are formed from local **mesenchymal (mesodermal) cells.**

The first rudiment of the *internal ear* is a thickened region of surface **ectoderm** on either side of the head near the hindbrain vesicle. This region, which develops shortly after the first appearance of the optic groove, is called the **otic placode** (Figure 24-21 a, c). As the placode develops, it invaginates to form the **otic pit.** As invagination continues, the otic pit separates from the surface ectoderm, forming a closed sac, referred to as the **otic vesicle.** The otic vesicle forms the various parts of the *membranous labyrinth.*

Lateral pouches, called **pharyngeal pouches,** develop from the sides of the pharynx. Over these pouches, the surface **ectoderm** indents forming what are called **branchial grooves.** The *middle ear cavity* and *eustachian tube* develop from the first pharyngeal pouch (see Figure 24-22a). The *auditory ossicles* (malleus, incus, and stapes) develop in the middle ear cavity from **mesenchymal (mesodermal) cells.** The *external ear* and *external auditory canal* develop from the first branchial groove.

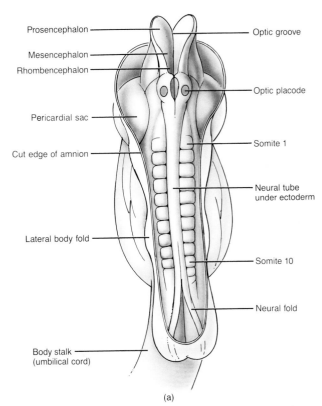

(a)

FIGURE 24-21 Development of the eye and ear. (a) Dorsal aspect of an embryo with 10 pairs of somites showing the position of the optic groove and otic placode.

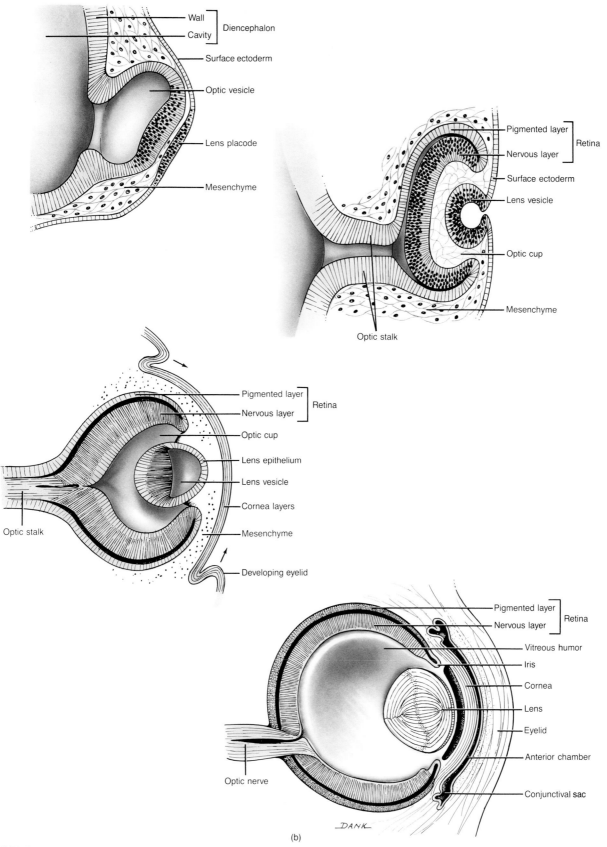

(b)

FIGURE 24-21 (*Continued*) Development of the eye and ear. (b) Development of the eye.

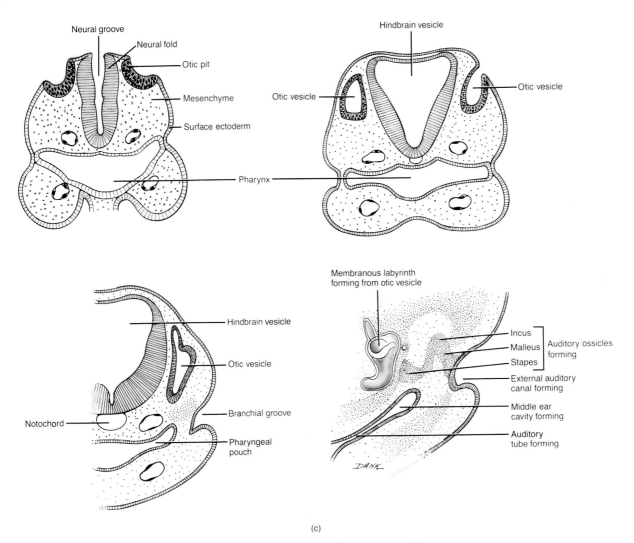

(c)

FIGURE 24-21 (*Continued*) Development of the eye and ear. (c) Development of the ear.

EXHIBIT 24-8 THE ENDOCRINE SYSTEM

The development of the endocrine system is not as localized as the development of other systems. The endocrine organs develop in widely separated parts of the embryo.

The *pituitary gland* (hypophysis) has a double origin from different regions of the **ectoderm.** The *neurohypophysis* (posterior lobe) is derived from an outgrowth of ectoderm called the **neurohypophyseal bud,** located on the floor of the hypothalamus (Figure 24-22a, b). The *infundibulum,* also an outgrowth of the neurohypophyseal bud, connects the neurohypophysis to the hypothalamus. The *adenohypophysis* (anterior lobe) is derived from an outgrowth of **ectoderm** from the roof of the stomodeum (mouth) called **Rathke's pouch.** The pouch grows toward the neurohypophyseal bud and the pouch loses its connection with the roof of the mouth.

The *thyroid gland* develops as a midventral outgrowth of **endoderm,** called **thyroid diverticulum,** from the floor of the pharynx at the level of the second pair of pharyngeal pouches (Figure 24-22a). The outgrowth projects inferiorly and differentiates into the right and left lateral lobes and the isthmus of the gland.

The *parathyroid glands* develop as outgrowths from the third and fourth **pharyngeal pouches** (Figure 24-22a).

The adrenal cortex and adrenal medulla have completely different embryological origins. The *adrenal cortex* is derived from intermediate **mesoderm** from the same region that produces the gonads (see Figure 24-25). The *adrenal medulla* is **ectodermal** in origin and is derived from the **neural crest** which also produces sympathetic ganglia and other structures of the nervous system (see Figure 24-19c).

The *pancreas* develops from **endoderm** from dorsal and ventral outgrowths of the part of the **foregut** that later becomes the duodenum (see Figure 24-24). The two outgrowths eventually fuse to form the pancreas. The origin of the ovaries and testes is discussed in the section on the reproductive system.

The *pineal gland* arises from **ectoderm** from the **diencephalon** (see Figure 24-20), as an outgrowth between the thalamus and colliculi.

The *thymus gland* arises from **endoderm** from the third **pharyngeal pouches** (see Figure 24-22a).

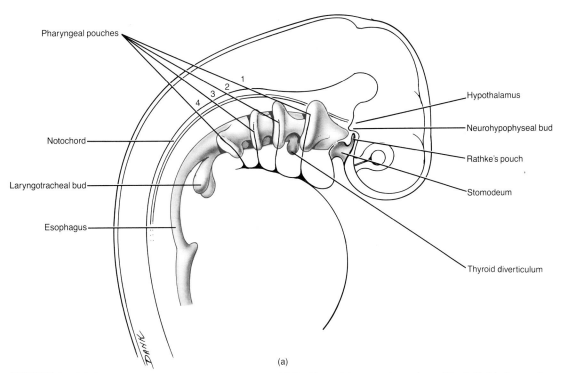

FIGURE 24-22 Development of the endocrine system. (a) Position of the neurohypophyseal bud, Rathke's pouch, and thyroid diverticulum.

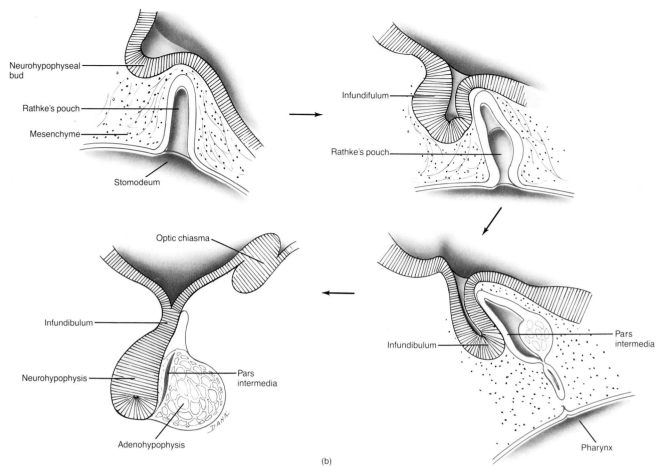

(b)

FIGURE 24-22 (*Continued*) Development of the endocrine system. (b) Development of the pituitary gland.

EXHIBIT 24-9 THE RESPIRATORY SYSTEM

The development of the mouth and pharynx is considered in the exhibit on the digestive system. Here, we will consider the remainder of the respiratory system.

At about four weeks, the respiratory system begins as an outgrowth of the **endoderm** of the foregut (precursor of some digestive organs) just behind the pharynx. This outgrowth is called the **laryngotracheal bud** (see Figure 24-22a). As the bud grows, it elongates and differentiates into the future *larynx*. Its proximal end maintains a slitlike opening into the pharynx called the *glottis*. The middle portion of the bud gives rise to the *trachea*. The distal portion divides into two **lung buds,** which grow into the *bronchi* and *lungs* (Figure 24-23).

As the lung buds develop, they branch and rebranch and give rise to all the *bronchial tubes.* After the sixth month, the closed terminal portions of the tubes dilate and become the *alveoli* of the lungs. The smooth muscle, cartilage, and connective tissues of the bronchial tubes and the pleural sacs of the lungs are contributed by **mesenchymal (mesodermal) cells.**

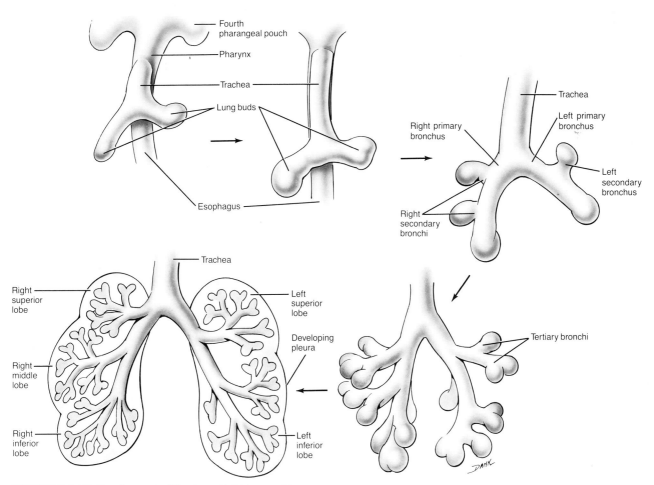

FIGURE 24-23 Development of the bronchial tubes and lungs.

EXHIBIT 24-10 **THE DIGESTIVE SYSTEM**

About the fourteenth day after fertilization the cells of the endoderm form a cavity referred to as the **primitive gut** (Figure 24-24). Soon after the mesoderm forms and splits into two layers (somatic and splanchnic), the splanchnic mesoderm associates with the endoderm of the primitive gut. Thus, the primitive gut has a double-layered wall. The **endodermal layer** gives rise to the *epithelial lining* and *glands* of most of the digestive tract, and the **mesodermal layer** produces its *smooth muscle* and *connective tissue*.

The primitive gut elongates, and about the latter part of the third week, it differentiates into an anterior **foregut,** a central **midgut,** and a posterior **hindgut.** Until the fifth week of development, the midgut opens into the yolk sac. After that time, the yolk sac constricts, detaches from the midgut, and the midgut seals. In the region of the foregut, a depression consisting of **ectoderm,** the **stomodeum,** appears. This develops into the *oral cavity*. The **oral membrane** that separates the foregut from the stomodeum ruptures during the fourth week of development, so that the foregut is continuous with the outside of the embryo through the oral cavity. Another depression consisting of **ectoderm,** the **proctodeum,** forms in the hindgut and goes on to develop into the *anus*. The **cloacal membrane,** which separates the hindgut from the proctodeum, ruptures, so that the hindgut is continuous with the outside of the embryo through the anus. Thus, the digestive tract forms a continuous tube from the mouth to the anus.

The foregut develops into the *pharynx, esophagus, stomach,* and a *portion of the duodenum*. The midgut is transformed into the *remainder of the duodenum,* the *jejunum,* the *ileum,* and a *portion of the large intestine* (cecum, appendix, ascending colon, and most of the transverse colon). The hindgut develops into the *remainder of the large intestine,* except for a portion of the anal canal which is derived from the proctodeum.

As development progresses, the endoderm at various places along the foregut develops into hollow buds that grow into the mesoderm. These buds will develop into the *salivary glands, liver, gallbladder,* and *pancreas*. Each of the glands retains a connection with the gastrointestinal tract via a duct.

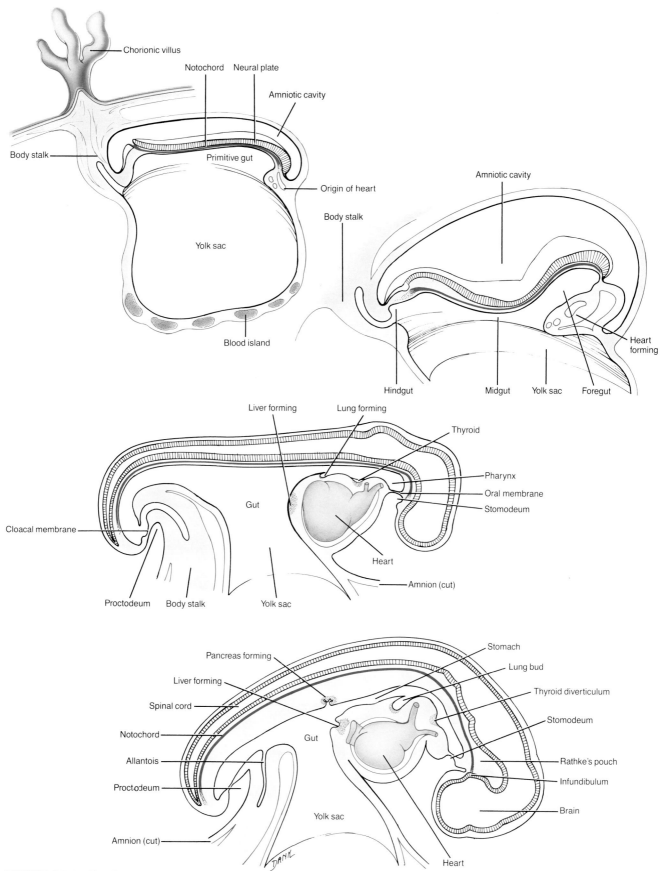

FIGURE 24-24 Development of the digestive system.

EXHIBIT 24-11 THE URINARY SYSTEM

As early as the third week of development, a portion of the mesoderm along the posterior half of the dorsal side of the embryo, the **intermediate mesoderm,** differentiates into the kidneys. Three pairs of kidneys form within the intermediate mesoderm in successive time periods: pronephros, mesonephros, and metanephros (Figure 24-25). Only the last remains as the functional kidneys of the adult.

The first kidney to form, the **pronephros,** is the superior of the three. Associated with its formation is a tube, the **pronephric duct.** This duct empties into the **cloaca,** which is the dilated caudal end of the gut derived from **endoderm.** The pronephros begins to degenerate during the fourth week and is completely gone by the sixth week. The pronephric ducts, however, remain.

The pronephros is replaced by the second kidney, the **mesonephros.** With its appearance, the retained portion of the pronephric duct, which connects to the mesonephros, becomes known as the **mesonephric duct.** The mesonephros begins to degenerate by the sixth week and is just about completely gone by the eighth week.

At about the fifth week, an outgrowth, called a **ureteric bud,** develops from the distal end of the mesonephric duct near the cloaca. This bud is the developing **metanephros.** As it grows toward the head of the embryo, its end widens to form the *pelvis* of the kidney with its *calyces* and associated *collecting tubules.* The unexpanded portion of the bud, the **metanephric duct,** becomes the *ureter.* The *nephrons,* the functional units of the kidney, arise from the intermediate mesoderm around each ureteric bud.

During development, the cloaca divides into a **urogenital sinus,** into which urinary and genital ducts empty, and a *rectum* that discharges into the anal canal. The *urinary bladder* develops from the urogenital sinus. In the female, the *urethra* develops from lengthening of the short duct that extends from the urinary bladder to the urogenital sinus. The *vestibule,* into which the urinary and genital ducts empty, is also derived from the urogenital sinus. In the male, the urethra is considerably longer and more complicated, but is also derived from the urogenital sinus.

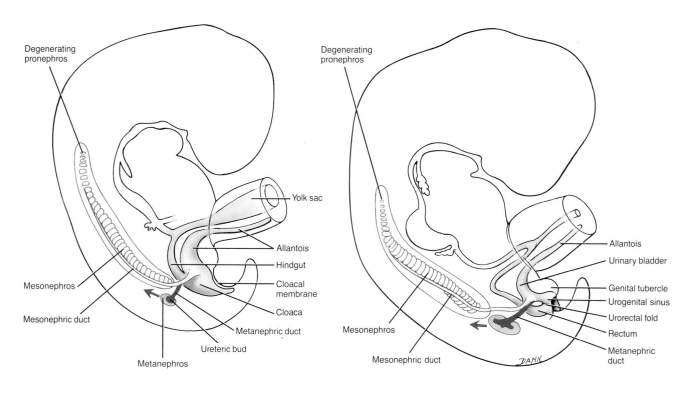

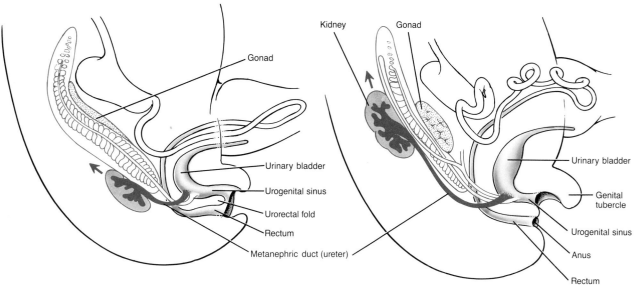

FIGURE 24-25 Development of the urinary system.

EXHIBIT 24-12 THE REPRODUCTIVE SYSTEMS

The *gonads* also develop from the **intermediate mesoderm.** By the sixth week, they appear as bulges that protrude into the coelom (Figure 24-26). The gonads develop near the mesonephric ducts. A second pair of ducts, the **Mullerian (paramesonephric) ducts,** develop lateral to the mesonephric ducts. Both sets of ducts empty into the urogenital sinus. At about the eighth week, the gonads are clearly differentiated into ovaries or testes.

In the male embryo, the *testes* connect to the mesonephric duct through a series of tubules. These tubules become the *seminiferous tubules.* Continued development of the mesonephric duct produces the *efferent ducts, ductus epididymis, ductus deferens, ejaculatory ducts,* and *seminal vesicle.* The *prostate* and *bulbourethral glands* are **endodermal** outgrowths of the urethra. Shortly after the gonads differentiate into testes, the Mullerian ducts degenerate without contributing any functional structures to the male reproductive system.

In the female embryo, the gonads develop into *ovaries.* At about the same time, the distal ends of the Mullerian ducts fuse to form the *uterus* and *vagina.* The unfused portions become the *uterine tubes.* The *greater* and *lesser vestibular glands* develop from **endo-**dermal outgrowths of the vestibule. The mesonephric ducts in the female degenerate without contributing any functional structures to the female reproductive system.

The *external genitals* of both male and female embryos also remain undifferentiated until about the eighth week. Before differentiation, all embryos have an elevated region, the **genital tubercle,** a point between the tail (future coccyx) and the umbilical cord where the mesonephric and Mullerian ducts open to the exterior (Figure 24-27). The tubercle consists of a **urethral groove** (opening into the urogenital sinus), paired **urethral folds,** and paired **labioscrotal swellings.**

In the male embryo, the genital tubercle elongates and develops into a *penis.* Fusion of the urethral folds forms the *spongy (cavernous) urethra* and leaves an opening only at the distal end of the penis, the *urethral orifice.* The labioscrotal swellings develop into the *testes.* In the female, the genital tubercle gives rise to the *clitoris.* The urethral folds remain open as the *labia minora* and the labioscrotal swellings become the *labia majora.* The urethral groove becomes the *vestibule.*

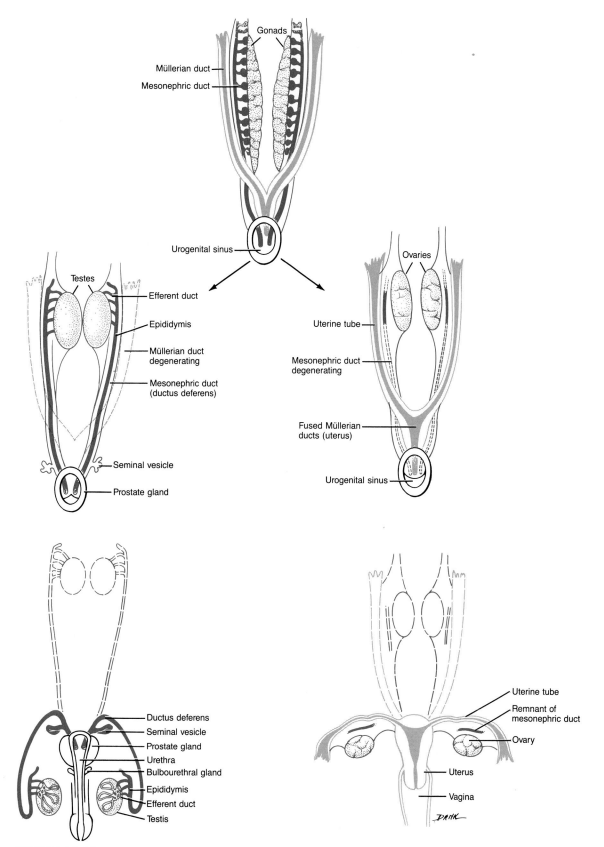

FIGURE 24-26 Development of the internal reproductive system.

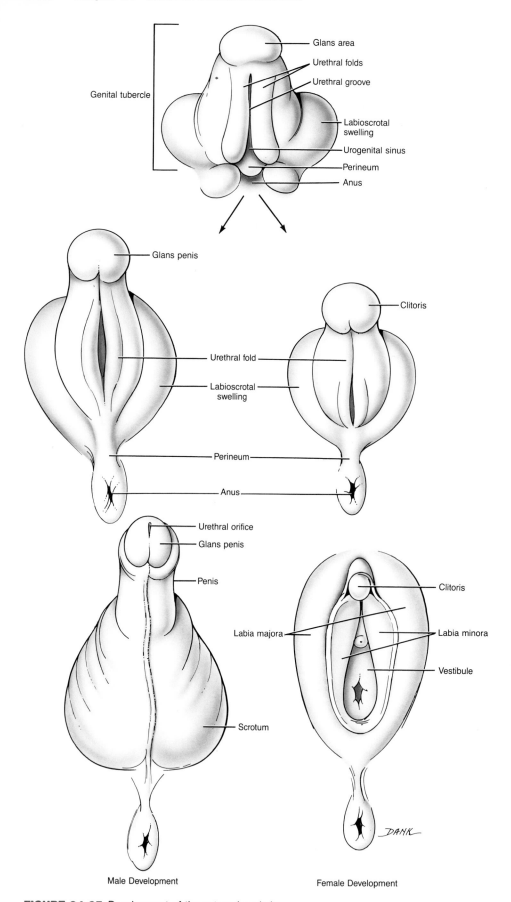

FIGURE 24-27 Development of the external genitals.

EXHIBIT 24-13 STRUCTURES PRODUCED BY THE THREE PRIMARY GERM LAYERS

ENDODERM	MESODERM	ECTODERM
Epithelium of digestive tract (except the oral cavity and anal canal) and the epithelium of its glands.	All skeletal, most smooth, and all cardiac muscle.	Entire nervous tissue.
	Cartilage, bone, and other connective tissues.	Epidermis of skin.
Epithelium of urinary bladder, gallbladder, and liver.	Blood, bone marrow, and lymphoid tissue.	Hair follicles, arrector pili muscles, nails, and epithelium of skin glands (sebaceous and sudoriferous).
Epithelium of pharynx, eustachian tube, tonsils, larynx, trachea, bronchi, and lungs.	Endothelium of blood vessels and lymphatics.	Lens, cornea, and optic nerve of eye and internal eye muscles.
Epithelium of thyroid, parathyroid, pancreas, and thymus glands.	Dermis of skin.	Internal and external ear.
Epithelium of prostate and bulbourethral glands, vagina, vestibule, urethra, and associated glands such as the greater vestibular and lesser vestibular glands.	Fibrous tunic and vascular tunic of eye.	Neuroepithelium of sense organs.
	Middle ear.	Epithelium of oral cavity, nasal cavity, paranasal sinuses, salivary glands, and anal canal.
	Mesothelium of coelomic and joint cavities.	Epithelium of pineal gland, hypophysis, and adrenal medulla.
	Epithelium of kidneys and ureters.	
	Epithelium of adrenal cortex.	
	Epithelium of gonads and genital ducts.	

A summary of structures that develop from each of the three primary germ layers is presented in Exhibit 24-13, and Exhibit 24-14 presents length and weight changes associated with embryonic and fetal growth.

HORMONES OF PREGNANCY

Following fertilization, the corpus luteum is maintained until about the fourth month of pregnancy. For most of this time it continues to secrete **estrogens** and **progesterone.** Both these hormones maintain the lining of the uterus during pregnancy and prepare the mammary glands to secrete milk. The amounts of estrogens and progesterone secreted by the corpus luteum, however, are only slightly higher than those produced after ovulation in a normal menstrual cycle. The high levels of estrogens and progesterone needed to maintain pregnancy and develop the breasts for lactation are provided by the placenta.

During pregnancy, the chorion of the placenta secretes a hormone called **human chorionic gonadotropin (HCG).** This hormone is excreted in the urine of pregnant women from about the middle of the first month of pregnancy, reaching its peak of excretion during the third month. The HCG level decreases sharply during the fourth and fifth months and then levels off until childbirth. Excretion of HCG in the urine serves as the basis for most **pregnancy tests.** The primary role of HCG seems to be to maintain the activity of the corpus luteum, especially with regard to continuous progesterone secretion—an activity neces-

EXHIBIT 24-14 EMBRYONIC AND FETAL GROWTH

END OF MONTH	APPROXIMATE LENGTH AND WEIGHT
1	0.6 cm (3/16 inch)
2	3 cm (1¼ inches) 1 g (1/30 oz)
3	7.5 cm (3 inches) 28 g (1 oz)
4	18 cm (6½–7 inches) 113 g (4 oz)
5	25–30 cm (10–12 inches) 227–454 g (½–1 lb)
6	27–35 cm (11–14 inches) 567–681 g (1¼–1½ lb)
7	32–42 cm (13–17 inches) 1,135–1,362 g (2½–3 lb)
8	41–45 cm (16½–18 inches) 2,043–2,270 g (4½–5 lb)
9	50 cm (20 inches) 3,178–3,405 g (7–7½ lb)

sary for the continued attachment of the fetus to the lining of the uterus.

The placenta begins to secrete estrogens and progesterone no later than the sixtieth day of pregnancy. They are secreted in increasing quantities until the time of birth.

Once the placenta is established, the secretion of HCG is cut back drastically at about the fourth month. The corpus luteum then disintegrates; it is no longer needed because the placenta supplies the levels of estrogens and progesterone needed to maintain the pregnancy. The fetal hormones thus take over the management of the mother's body in preparation for parturition (birth) and lactation. Following delivery, estrogens and progesterone in the blood decrease to normal levels.

Recent evidence indicates that the placenta produces a regulating factor called *placental luteotropic releasing factor (pLRF)*. It is chemically similar to the luteotropic releasing factor (LRF) produced by the hypothalamus, which causes the synthesis and release of luteinizing hormone (LH) by the anterior pituitary. The proposed function of pLRF is to stimulate the secretion of HCG by the placenta.

PARTURITION AND LABOR

The time the embryo or fetus is carried in the uterus is called **gestation** (jes-TĀ-shun). The total human gestation period is about 280 days from the beginning of the last menstrual period. The term **parturition** (par'-too-RISH-un) refers to birth. Parturition is preceded by a sequence of events commonly called **labor.** The onset of labor is apparently related to a complex interaction of many factors. Just prior to birth, the muscles of the uterus contract rhythmically and forcefully. Both placental and ovarian hormones seem to play a role in these contractions. Since progesterone inhibits uterine contractions, labor cannot take place until its effects are diminished. At the end of gestation, there is just enough estrogen in the mother's blood to overcome the inhibiting effects of progesterone and labor commences. It has been suggested that some factor released by the placenta, fetus, or mother rather suddenly overcomes the inhibiting effects of progesterone so that estrogen can exert its effect. Prostaglandins may also play a role in labor. Oxytocin from the posterior pituitary gland also stimulates uterine contractions.

Uterine contractions occur in waves, quite similar to peristaltic waves, that start at the top of the uterus and move downward. These waves expel the fetus. **True labor** begins when pains occur at regular intervals. The pains correspond to uterine contractions. As the interval between contractions shortens, the contractions intensify. Another sign of true labor in some females is localization of pain in the back, which is intensified by walking. A reliable indication of true labor is the "show" and dilation of the cervix. The "show" is a discharge of a blood-containing mucus that accumulates in the cervical canal during pregnancy. In **false labor,** pain is felt in the abdomen at irregular intervals. The pain does not intensify and is not altered significantly by walking. There is no "show" and no cervical dilation.

The first stage of labor, the **stage of dilation,** is the time from the onset of labor to the complete dilation of the cervix (Figure 24-28). During this stage there are regular contractions of the uterus, usually a rupturing of the amnionic sac, and complete dilation (10 cm) of the cervix. If the amnionic sac does not rupture spontaneously, it is done artificially. The next stage of labor, the **stage of expulsion,** is the time from complete cervical dilation to delivery. In the final stage, the **placental stage,** the placenta or "afterbirth" is expelled a few minutes after delivery by powerful uterine contractions. These contractions also constrict blood vessels that were torn during delivery. In this way, the possibility of hemorrhage is reduced.

CLINICAL APPLICATION

Pudendal (pyoo-DEN-dal) **nerve block** is used for procedures such as episiotomy. The primary innervation to the skin and muscles of the perineum is the pudendal nerve. In the transvaginal approach, the needle is passed through the lateral vaginal wall to a point just medial to the ischial spine. The anesthesia results in loss of the anal reflex, relaxation of the muscles of the floor of the pelvis, and loss of sensation to the vulva and lower one-third of the vagina.

ADJUSTMENTS OF THE INFANT AT BIRTH

During pregnancy, the embryo and later the fetus is totally dependent on the mother for its existence. The mother supplies the fetus with O_2 and nutrients, eliminates its CO_2 and other wastes, and protects it against shocks, temperature changes, and certain harmful microbes. At birth the baby becomes self-supporting, and the newborn's body systems must make various adjustments. Following are some changes that occur in the respiratory and cardiovascular systems.

RESPIRATORY SYSTEM

The respiratory system is fairly well developed at least 2 months before birth as evidenced by the fact that premature babies delivered at 7 months are able to breathe and cry. In the uterus, the fetus depends entirely on the mother for obtaining O_2 and eliminating CO_2. The fetal lungs are either collapsed or partially filled with amniotic fluid, which is absorbed at birth. After delivery the baby's supply of O_2 from the mother is stopped. Circulation in the baby continues and as the blood level of CO_2 increases, the respiratory center in the medulla is stimulated. This causes the respiratory muscles to contract and the baby draws its first breath. Since the first inspiration

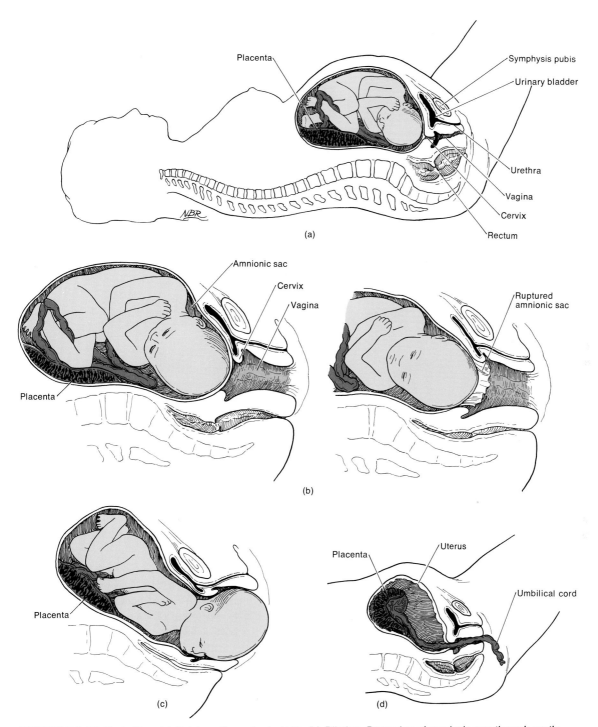

Placenta
Symphysis pubis
Urinary bladder
Urethra
Vagina
Cervix
Rectum

(a)

Amnionic sac
Cervix
Vagina
Placenta

Ruptured amnionic sac

(b)

Placenta

Placenta
Uterus
Umbilical cord

(c)   (d)

FIGURE 24-28 Parturition. (a) Fetal position prior to birth. (b) Dilation. Protrusion of amnionic sac through partly dilated cervix. Amnionic sac ruptured and complete dilation of cervix. (c) Stage of expulsion. (d) Placental stage.

is unusually deep because the lungs contain no air, the baby exhales vigorously and naturally cries. A full-term baby may breathe 45 times a minute for the first two weeks after exposure to air. The rate is gradually reduced until it approaches a normal rate.

CARDIOVASCULAR SYSTEM

Following the first inspiration by the baby, the cardiovascular system must make several adjustments. The foramen ovale of the fetal heart between atria closes at the moment of birth. This diverts deoxygenated blood to the

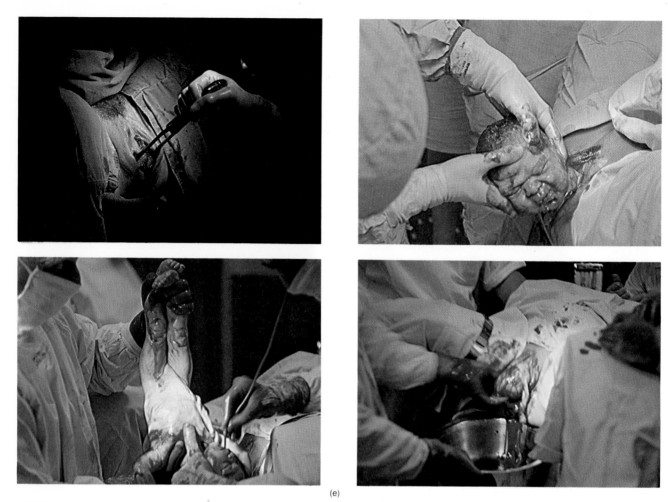

(e)

FIGURE 24-28 (*Continued*) Parturition. (e) Sequential photographs of parturition. (Top left and right Kinne, Photo Researchers; bottom left Thomas, Photo Researchers; bottom right McCartney, Photo Researchers.)

lungs for the first time. The foramen ovale is closed by two flaps of heart tissue that fold together and permanently fuse. The remnant of the foramen ovale is the fossa ovalis. Once the lungs begin to function, the ductus arteriosus is shut off by contractions of the muscles in its wall. The ductus arteriosus generally does not completely and irreversibly close for about 3 months following birth. Incomplete closing, as you already know, results in patent ductus arteriosus.

The ductus venosus of fetal circulation connects the umbilical vein directly with the inferior vena cava. It forces any remaining umbilical blood directly into the fetal liver and from there to the heart. When the umbilical cord is severed, all visceral blood of the fetus goes directly to the fetal heart via the inferior vena cava. This shunting of blood usually occurs within minutes after birth, but may take a week or two to complete. The ligamentum venosum, the remnant of the ductus venosus, is well established by the eighth postnatal week.

At birth, the infant's pulse may be from 120 to 160 per minute and may go as high as 180 following excitation.

Several days after birth, there is a greater independent need for O_2 which stimulates an increase in the rate of erythrocyte and hemoglobin production. This increase usually lasts for only a few days. Moreover, the white blood cell count at birth is very high, sometimes as much as 45,000 cells per cubic millimeter, but this decreases rapidly by the seventh day.

Finally, the infant's liver may not be adjusted at birth to control the production of bile pigment. As a result of this and other complicating factors, a temporary jaundice may result in as many as 50 percent of normal newborns by the third or fourth day after birth.

POTENTIAL HAZARDS TO THE DEVELOPING EMBRYO AND FETUS

The developing fetus is susceptible to a number of potential hazards that can be transmitted from the mother. Such hazards include infectious microbes, chemicals and drugs, irradiation, alcohol, and cigarette smoking. In addition, certain environmental conditions and pollutants can damage the fetus or even cause fetal death.

CHEMICALS AND DRUGS

Since the placenta is known to be an ineffective barrier between the maternal and fetal circulations, actually any drug or chemical dangerous to the infant may be considered potentially dangerous to the fetus when given to the mother. Many chemicals and drugs have been proven to be toxic and teratogenic to the developing embryo and fetus. A **teratogen** (*terato* = monster) is any agent or influence that causes physical defects in the developing embryo. Examples are some hormones, antibiotics, oral anticoagulants, anticonvulsants, antitumor agents, thyroid drugs, thalidomide, LSD, and marijuana. In addition, accumulating evidence suggests that many agents assumed to be nonteratogenic can produce long-term adverse effects on reproductive ability and neurobehavioral development.

IRRADIATION

Ionizing radiations are potent teratogens. Treatment of pregnant mothers with large doses of x-rays and radium during the embryo's susceptible period of development may cause microcephaly (small size of head in relation to the rest of the body), mental retardation, and skeletal malformations. Caution is advised for diagnostic x-rays during the first trimester of pregnancy.

ALCOHOL

Alcohol has been a suspected teratogen for centuries, but only recently has a relationship been recognized between maternal alcohol intake and the characteristic pattern of malformations in the fetus. The term applied to the effects of intrauterine exposure to alcohol is **fetal alcohol syndrome (FAS).** Studies to date have indicated that in the general population the incidence of FAS may exceed 1 per 1,000 live births, and may be by far the number one fetal teratogen. The symptoms shown by children may include slow growth before and after birth, small head, facial irregularities such as narrow eye slits and sunken nasal bridge, defective heart and other organs, malformed arms and legs, genital abnormalities, and mental retardation. There are also behavioral problems, such as hyperactivity, extreme nervousness, and a poor attention span.

CIGARETTE SMOKING

The latest evidence not only indicates a causal relationship between cigarette smoking during pregnancy and low infant birth weight, but also points to a strong probable association between smoking and a higher fetal and infant mortality. Infants nursing from smoking mothers have also been found to have an increased incidence of gastrointestinal disturbances. Other pathologies of infants of smoking mothers include an increased incidence of respiratory problems during the first year of life, including bronchitis and pneumonia. Cigarette smoking may be teratogenic and cause cardiac abnormalities and anencephaly (a developmental anomaly with absence of neural tissue in the cranium). Maternal smoking also appears to be a significant etiologic factor in the development of cleft lip and palate and has been tentatively linked with sudden infant death syndrome (SIDS).

LACTATION

The term **lactation** refers to the secretion and ejection of milk by the mammary glands. The major hormone in promoting lactation is prolactin from the anterior pituitary. It is released in response to the hypothalamic prolactin releasing factor (PRF). Even though prolactin levels increase as the pregnancy progresses, there is no milk secretion because estrogens and progesterone cause the hypothalamus to release prolactin-inhibiting factor (PIF). Following delivery, the levels of estrogens and progesterone in the mother's blood decrease and the inhibition is removed.

The principal stimulus in maintaining prolactin secretion during lactation is the sucking action of the infant. Sucking initiates impulses from receptors in the nipples to the hypothalamus. The impulses inhibit PIF production, and prolactin is released by the anterior pituitary. The sucking action also initiates impulses to the posterior pituitary via the hypothalamus. These impulses stimulate the release of the hormone oxytocin by the posterior pituitary. This hormone induces certain cells surrounding the outer walls of the alveoli to contract, thereby compressing the alveoli. The compression moves milk from the alveoli of the mammary gland into the ducts, where it can be sucked. This process is referred to as **milk letdown.** Lactation often prevents the occurrence of female ovarian cycles for the first few months following delivery by inhibiting FSH and LH release by the anterior pituitary.

During late pregnancy and the first few days after birth, the mammary glands secrete a cloudy fluid called *colostrum.* Although it is not as nutritious as true milk, since it contains less lactose and virtually no fat, it serves adequately until the appearance of true milk on about the fourth day. Colostrum is thought to contain antibodies that protect the infant during the first few months of life.

AMNIOCENTESIS

Amniocentesis is a technique of withdrawing some of the amniotic fluid that bathes the developing fetus to diagnose genetic disorders or to determine fetal maturity or well-being. The fluid is removed by hypodermic needle puncture of the uterus, usually 16 to 20 weeks after conception. Cells in the fluid are examined for biochemical defects and abnormalities in chromosome number or structure. There are over 50 biochemical inheritable disorders and close to 300 chromosomal disorders that can be detected through amniocentesis—hemophilia, certain muscular dystrophies, Tay-Sachs disease, myelocytic leukemia, Klinefelter's and Turner's syndromes, sickle cell anemia, thalassemia, and cystic fibrosis. When both parents are known or suspected to be genetic carriers of any one of these disorders, amniocentesis is advised.

One chromosome disorder that may be diagnosed through amniocentesis is **Down's syndrome.** This disorder is characterized by mental retardation, retarded physical development (short stature and stubby fingers), distinctive facial structures (large tongue, broad skull, slanting eyes, and round head), and malformation of the heart, ears, hands, and feet. Sexual maturity is rarely attained. Individuals with the disorder usually have 47 chromosomes instead of the normal 46. The extra chromosome is responsible for the syndrome. All the chromosomes of a person with Down's syndrome are in pairs except the twenty-first pair. Here the chromosomes are present in triplicate.

CLINICAL APPLICATION

A new technology has allowed physicians to see into the uteri of pregnant women without exposing them to the known dangers of x-rays and without pain or intrusion. This technique, called **ultrasound** (also *sonography* or *ultrasonography*), uses high frequency, inaudible sound waves, which are directed into the abdomen of the mother-to-be and then reflected back to a receiver. The reflected waves give a visual "echo" of what is inside the uterus. This echo is transformed electronically into an image on a screen.

The diagnostic ultrasound unit is actually one of two ultrasound devices commonly used on pregnant women. The other is the **Doppler Detector,** a fetal monitor that registers the baby's heart rate. It is used routinely in many hospitals during labor.

In the United States today, most obstetricians prescribe ultrasound examinations only when there is some clinical question about the normal progress of the pregnancy. By far the most common use of diagnostic ultrasound is to determine true fetal age when the date of conception is unknown or mistaken by the mother.

Ultrasound has gained a wide application beyond obstetrics as well, in detecting the presence of tumors, gallstones, and other abnormal internal masses. Research is going on to determine and ensure that ultrasound is absolutely safe for the developing embryo and fetus.

BIRTH CONTROL

Methods of **birth control** include removal of the gonads and uterus, sterilization, contraception, and abstinence. **Castration** (removal of the testes), **hysterectomy** (removal of the uterus), and **oophorectomy** (ō-of-ō-REK-tō-mē; removal of the ovaries) are all absolute preventive methods. Once performed, these operations cannot be reversed and it is impossible to produce offspring. However, removal of the testes or ovaries has adverse effects because of the importance of these organs in the endocrine system. Generally these operations are performed only if the organs are diseased. Castration before puberty prevents the development of secondary sex characteristics.

One means of **sterilization** of males is **vasectomy**—a simple operation in which a portion of each ductus deferens is removed (Figure 24-29a). An incision is made in the scrotum, the tubes are located, and each is tied in two places. Then the portion between the ties is cut out. Sperm production can continue in the testes, but the sperm cannot reach the exterior.

Sterilization in females generally is achieved by performing a **tubal ligation** (lī-GA-shun). An incision is made into the abdominal cavity, the uterine tubes are squeezed, and a small loop called a knuckle is made (Figure 24-29b). A suture is tied tightly at the base of the knuckle and the knuckle is then cut. After four or five days the suture is digested by body fluids and the two severed ends of the tubes separate. The ovum thus is prevented from passing to the uterus, and the sperm cannot reach the ovum. Sterilization normally does not affect sexual performance or enjoyment.

Another method of sterilizing women is the **laparoscopic** (lap'-a-rō-SKŌ-pik) **technique.** After a woman receives local or general anesthesia, a harmless gas is introduced into her abdomen to create a gas bubble. The bubble expands the abdominal cavity and pushes the intestines away from the pelvic organs, permitting safe, easy access to the uterine tubes. The doctor makes a small incision at the lower rim of the umbilicus and inserts a laparoscope to view the inside of the abdominal cavity and the uterine tubes. The tubes can be closed with this instrument or a second incision can be made at the pubic hairline to insert a cautery forceps. Once the uterine tubes are sealed, the instrument is removed, the gas is released, and the incision is covered with a bandage. After a few hours the patient can usually go home.

Contraceptives include all methods—natural, mechanical, or chemical—of preventing fertilization without destroying fertility. The natural methods include complete or periodic abstinence. An example of periodic abstinence is the rhythm method, which takes advantage of the fact that a fertilizable ovum is available only during a period of three to five days in each menstrual cycle. During this time the couple refrains from intercourse. Its effectiveness is limited by the fact that few women have abso-

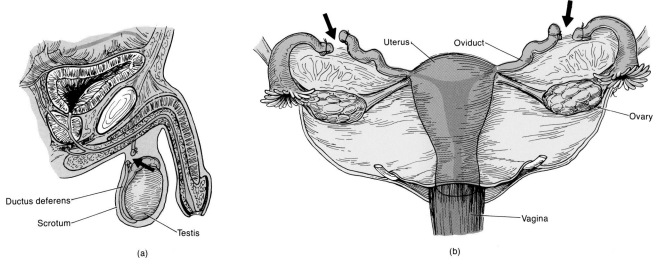

FIGURE 24-29 Sterilization. (a) Vasectomy. The ductus deferens of each testis is cut and tied after an incision is made into the scrotum. (b) Tubal ligation. Each uterine tube is cut and tied after an incision is made into the abdomen.

lutely regular cycles. Moreover, some women occasionally ovulate during the "safe" times of the month, such as during menstruation.

Mechanical means of contraception include the condom used by the male and the diaphragm used by the female. The **condom** is a nonporous, elastic (rubber or similar material) covering placed over the penis that prevents deposition of sperm in the female reproductive tract.

CLINICAL APPLICATION

Microsurgery is any operation done with the aid of microscopes. Surgeons work with spectacle-mounted magnifiers or with microscopes. Microsurgery's impact on curing fertility problems has been nothing short of revolutionary. Using this technique, surgeons can remove a prolactin-secreting adenoma, a hormone-producing tumor of the pituitary gland that causes infertility in women, without damaging the pituitary gland itself. Microsurgery is also used to open blocked uterine tubes and spermatic ducts. Natural obstructions are a major cause of sterility in both sexes, and in the United States alone, more than 10 million people have been sterilized by artificial closures. For men who have had vasectomies, reversal through microsurgery has meant about a 90 percent chance of regaining a normal sperm count. Limited success has been attained in reversing women who have undergone sterilization surgery through microsurgery.

Neurosurgeons and ear and eye surgeons all rely on microsurgery, and thousands of people today already owe their health, children, limbs, and lives to microsurgery.

The **diaphragm** (Figure 24-30) is a dome-shaped structure that fits over the cervix and is generally used in conjunction with a sperm-killing chemical. The diaphragm stops the sperm from passing into the cervix. The chemical kills the sperm cells.

Another mechanical method of contraception is an **intrauterine device (IUD).** The device is a small object made of plastic, copper, or stainless steel and shaped like a loop, coil, T, or 7. It is inserted into the cavity of the uterus (Figure 24-31). It is not clear how IUDs operate. Some investigators believe they cause changes in the uterus lining that, in turn, produce a substance which destroys either the sperm or the fertilized ovum. Some IUDs release contraceptive agents in minute amounts; one new device secretes progesterone.

Chemical means of contraception include spermicidal and hormonal methods. Various foams, creams, jellies, suppositories, and douches make the vagina and cervix unfavorable for sperm survival. The hormonal method, **oral contraception** (the pill), has found rapid and widespread use. Although several pills are available, the one most commonly used contains a high concentration of progesterone and a low concentration of estrogens. These two hormones act on the anterior pituitary to decrease the secretion of FSH and LH by inhibiting FSHRF and LHRF production, respectively, by the hypothalamus. The low levels of FSH and LH are not adequate to initiate follicle maturation or ovulation. In the absence of a mature ovum, pregnancy cannot occur. A summary of contraceptive methods is presented in Exhibit 24-15.

Women for whom all oral contraceptives are contraindicated include those with a history of thromboembolic disorders (predisposition to blood clotting), cerebral blood vessel damage, hypertension, liver malfunction, heart disease, or cancer of the breast or reproductive system.

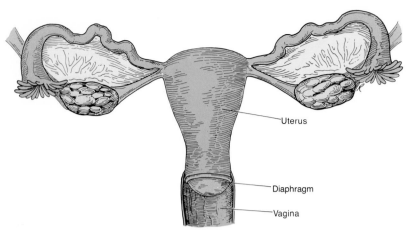

FIGURE 24-30 Diaphragm. The thin spring around the margin of the diaphragm opens outward, presses against the wall of the vagina, and stretches across the cervix.

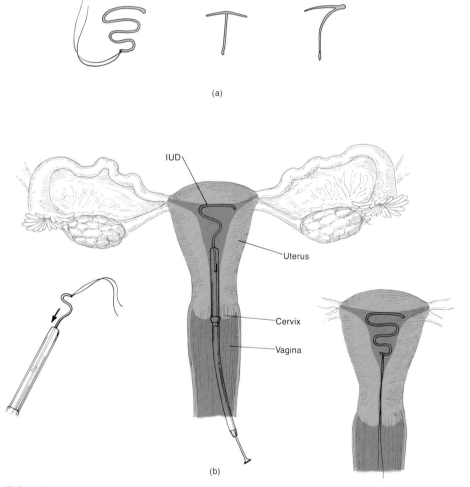

FIGURE 24-31 Intrauterine devices. (a) Representative designs of three intrauterine devices. (b) Procedure for insertion. The device is compressed to fit into a long, narrow-bore tube that is passed through the slightly dilated cervix. Once in position in the uterus, it spreads out to its former shape. IUD's have a thread or chain projecting into the vagina that is used as an indication that the device is still in place and for removal by the physician.

About 40 percent of all pill users experience side effects—generally minor problems such as nausea, weight gain, headaches, irregular menses, spotting between periods, and amenorrhea. The statistics on the life-threatening conditions associated with the pill such as blood clots, heart attacks, liver tumors, and gallbladder disease are somewhat more reassuring. For all the problems combined, fewer than 2 deaths occurred per 100,000 users under age 30, 4 among those 30 to 35, 10 among those 35 to 39, and 18 among women over 40.

The major exception is that women who take the pill and smoke face far higher odds of developing heart attack and stroke than do nonsmoking pill users.

> **CLINICAL APPLICATION**
>
> The quest for an efficient male oral contraceptive has been disappointing until recently. An oral contraceptive, **gossypol,** which is a phenolic compound derived from the cotton plant, has achieved an efficacy (power to produce effects) rate of 99.89 percent in clinical tests in more than 4,000 healthy males in China. Levels of blood luteinizing hormone (LH) and testosterone remained unchanged and potency was not impaired. Preliminary data from the study suggest that the number and morphology of spermatozoa gradually recover and that fertility is restored to normal within three months after termination of therapy.

EXHIBIT 24-15 CONTRACEPTIVE METHODS

METHOD	COMMENTS
Removal of gonads and uterus	Irreversible sterility. Generally performed if organs are diseased rather than as contraceptive method because of importance of hormones produced by gonads.
Sterilization	Procedure involving severing ductus deferens in males and uterine tubes in females.
Natural contraception	Abstinence from intercourse during time of month woman is fertile. Under ideal circumstances, effectiveness in women with regular menstrual cycles may approach that of mechanical and chemical contraceptives. Extremely difficult to determine fertile period. Effectiveness can be increased by recording body temperature each morning before getting up; a small rise in temperature indicates that ovulation has occurred one or two days earlier and ovum can no longer be fertilized.
Mechanical contraception **Condom**	Thin, strong sheath of rubber or similar material worn by male to prevent sperm from entering vagina. Failures caused by sheath tearing or slipping off after climax or not putting the sheath on soon enough. If used correctly and consistently, effectiveness similar to that of diaphragm.
Diaphragm	Flexible rubber dome inserted into vagina to cover cervix, providing barrier to sperm. Usually used with spermicidal cream or jelly. Must be left in place at least 6 hours after intercourse and may be left in place as long as 24 hours. Must be fitted by physician or other trained personnel and refitted every two years and after each pregnancy. Offers high level of protection if used with spermicide; rate of 2 to 3 pregnancies per 100 women per year is estimated for consistent users. If not used consistently, much higher pregnancy rates must be expected. Occasional failures caused by improper insertion or displacement during sexual intercourse.
Intrauterine device	Small object (loop, coil, T, or 7) made of plastic, copper, or stainless steel and inserted into uterus by physician. May be left in place for long periods of time (some must be changed every two to three years). Does not require continued attention by user. Some women cannot use them because of expulsion, bleeding, or discomfort. Not recommended for women who have not had child because uterus is too small and cervical canal too narrow. Infrequently, insertion may lead to inflammation of pelvic organs.
Chemical contraception **Foams, creams, jellies, suppositories, vaginal douches**	Sperm-killing chemicals inserted into vagina to coat vaginal surfaces and cervical opening. Provide protection for about 1 hour. Effective when used alone, but significantly more effective when used with diaphragm or condom.
Oral contraceptive	Except for total abstinence or surgical sterilization, most effective contraceptive known. Side effects include nausea, occasional light bleeding between periods, breast tenderness or enlargement, fluid retention, and weight gain. Should not be used by women who have cardiovascular conditions (thromboembolic disorders, cerebrovascular disease, heart disease, hypertension), liver malfunction, cancer or neoplasia of breast or reproductive organs, or by women who smoke.

KEY MEDICAL TERMS ASSOCIATED WITH DEVELOPMENTAL ANATOMY

Abortion Premature expulsion from the uterus of the products of conception—embryo or nonviable fetus.

Autosome (*auto* = self, *some* = body) Any chromosome that is not a sex chromosome (not an X or Y chromosome). Humans have 22 pairs of autosomes.

Cautery Application of a caustic (burning) substance or instrument for the destruction of tissue.

Cesarean section (*caedere* = to cut) Removal of the baby and placenta through an abdominal incision in the uterine wall.

Colpotomy (*colp* = vagina; *tome* = cutting) Incision of the vagina.

Culdoscopy (*skopein* = to examine) A procedure in which a culdoscope (endoscope) is used to view the pelvic cavity. The approach is through the vagina.

Endometrectomy (*metro* = uterus) Removal of the uterine mucous lining; also called curettage. A variation of this procedure is dilation and curettage (D and C).

Hermaphroditism Presence of both male and female sex organs in one individual.

Karyotype (*karyon* = nucleus) The chromosomal elements typical of a cell, drawn in their true proportions, based on the average of measurements determined in a number of cells. Useful in judging whether or not chromosomes are normal in number and structure.

Laparoscope (*lapar* = abdomen) A long, thin instrument containing a light-reflecting material as well as a lens for examination of the uterine tubes and abdomen.

Lethal gene (*lethum* = death) A gene that, when expressed, results in death either in the embryonic state or shortly after birth.

Lochia The discharge from the uterus through the birth canal occurring after childbirth.

Mutation (*mutare* = change) A permanent, heritable change in a gene that causes it to have a different effect than it had previously.

Puerperal (*puer* = child, *parere* = to bring forth) **fever** Infectious disease of childbirth, also called puerperal sepsis and childbed fever. The disease results from an infection originating in the birth canal and affects the endometrium. It may spread to other pelvic structures and lead to septicemia.

STUDY OUTLINE

Gamete Formation
1. Ova and sperm are collectively called gametes.
2. Whereas gametes are haploid (*n*), somatic cells are diploid (2*n*).
3. Meiosis occurs in the process of producing sex cells. It causes a developing sperm or ovum to cut in half its duplicate set of chromosomes so that the mature gamete has only 23 chromosomes (the haploid number) instead of 46 (the diploid number).
4. Spermatogenesis occurs in the testes. It is the formation of haploid spermatozoa by meiosis.
5. Oogenesis occurs in the ovaries. It is the formation of haploid ova by meiosis.

Fertilization and Implantation
1. Pregnancy is a sequence of events that includes fertilization, implantation, embryonic growth, fetal growth, and birth.
2. Fertilization of an ovum is accomplished by sexual intercourse.
3. The role of the male in the sex act involves erection, emission, and ejaculation; the female role involves erection, lubrication, and climax.
4. Fertilization, the penetration of an ovum by a spermatozoon and the subsequent union of the ovum and sperm nuclei, normally occurs in the uterine tubes.

5. The embedding of a blastocyst in the endometrium of the uterus is called implantation.

Embryonic Development
1. During embryonic growth, the primary germ layers and embryonic membranes are formed and the placenta is functioning.
2. The primary germ layers—ectoderm, mesoderm, and endoderm—form all tissues of the developing organism.
3. Embryonic membranes include the yolk sac, amnion, chorion, and allantois.
4. Fetal and maternal materials are exchanged through the placenta.

Developmental Anatomy of Body Systems
Review Exhibits 24-1 through 24-12 for development of body organs from embryonic primary germ layers.

Hormones of Pregnancy
1. The corpus luteum secretes small amounts of estrogens and progesterone until about the fourth month of pregnancy.
2. Pregnancy is maintained primarily by the human chorionic gonadotropic hormone and large amounts of estrogens and progesterone secreted by the placenta.

Parturition and Labor
1. The time an embryo or fetus is carried in the uterus is called gestation.
2. Parturition refers to birth and is preceded by a sequence of events called labor.
3. The birth of a baby involves dilation of the cervix, expulsion of the fetus, and delivery of the placenta.

Adjustments of the Infant at Birth
1. The fetus depends on the mother for oxygen and nutrients, removal of wastes, and protection.
2. Following birth the respiratory and cardiovascular systems undergo changes in adjusting to self-supporting postnatal life.

Potential Hazards to the Developing Embryo and Fetus
1. The developing embryo and fetus is susceptible to many potential hazards that can be transmitted from the mother.
2. Examples are infections, microbes, chemicals and drugs, irradiation, alcohol, and smoking.

Lactation
1. Lactation refers to the secretion and ejection of milk by the mammary glands.
2. It is influenced by estrogens and progesterone, prolactin, and oxytocin.

Amniocentesis
1. Amniocentesis is the withdrawal of amniotic fluid. It can be used to diagnose hemophilia, Tay-Sachs disease, sickle cell anemia, and Down's syndrome.
2. Down's syndrome is a chromosomal abnormality characterized by mental retardation and retarded physical development.

Birth Control
1. Methods include removal of gonads; sterilization; rhythm; use of condom, diaphragm, and intrauterine devices; chemicals that kill sperm; and the contraceptive pill.
2. Contraceptive pills of the combination type contain estrogens and progesterone in concentrations that decrease the secretion of FSH and LH and thereby inhibit ovulation.

REVIEW QUESTIONS

1. Explain the terms haploid number and diploid number. What is the importance of meiosis to chromosome number?
2. Compare the events associated with spermatogenesis and oogenesis.
3. Explain the role of the male's erection, emission, and ejaculation in the sex act. How do the female's erection, lubrication, and orgasm contribute to the sex act?
4. Define fertilization. Where does it normally occur? How is a morula formed?
5. What is implantation? How does the fertilized ovum implant itself?
6. Define the embryonic period and the fetal period.
7. What is an embryonic membrane? Describe the functions of the four embryonic membranes.
8. Explain the importance of the placenta and umbilical cord to fetal growth.
9. Using Exhibits 24-1 through 24-12, briefly outline the events associated with the development of various body systems.
10. Describe several body structures formed by the ectoderm, mesoderm, and endoderm.
11. Compare the sources and functions of estrogens, progesterone, and the human chorionic gonadotropic hormone. How does the placenta serve as an endocrine organ during pregnancy?
12. Define gestation and parturition.
13. Distinguish between false and true labor. Describe what happens during the stage of dilation, the stage of expulsion, and the placental stage of delivery.
14. Discuss the principal respiratory and cardiovascular adjustments made by an infant at birth.
15. Explain in detail some of the potential hazards for the developing embryo and fetus.
16. What is lactation? How is the female prepared for it?
17. Briefly describe the following methods of birth control: removal of the gonads, sterilization, rhythm.
18. Distinguish between a condom, a diaphragm, and an IUD as methods of mechanical contraception.
19. List several examples and functions of chemical contraceptives.
20. Explain the operation of the combination contraceptive pill.
21. What is amniocentesis? What is its value?
22. Define Down's syndrome. What are its principal clinical symptoms?
23. Refer to the glossary of key medical terms associated with developmental anatomy. Be sure that you can define each term.

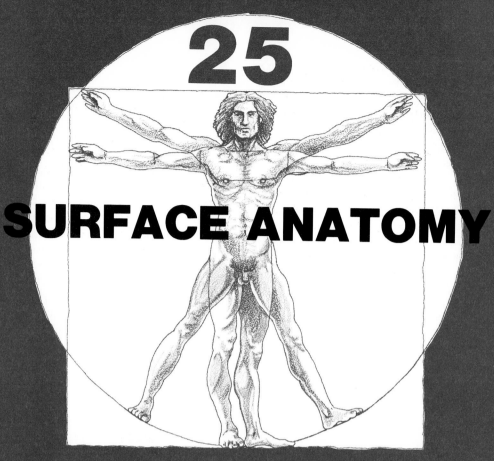

25

SURFACE ANATOMY

STUDENT OBJECTIVES

- Define surface anatomy.
- Identify the principal regions of the human body.
- Describe the surface anatomy features of the head.
- Describe the surface anatomy features of the neck.
- Describe the surface anatomy features of the trunk.
- Describe the surface anatomy features of the upper extremities.
- Describe the surface anatomy features of the lower extremities.

n Chapter 1, several branches of anatomy were defined and their importance to an understanding of the structure of the body was noted. Now that you have studied all of the body systems and have a fairly good idea how the body is organized, we can take a closer look at surface anatomy.

Very simply, **surface anatomy** is the study of the form and markings of the surface of the body. A knowledge of surface anatomy will help you to identify certain superficial structures through visual inspection or palpation through the skin. **Palpation** means to feel with the hand.

To introduce you to surface anatomy, we will consider some key features of each of the principal regions of the body. These regions are outlined as follows and may be reviewed in Figure 1-7.

Head (cephalic region or caput)
    Cranium (skull or brain case)
    Face (facies)

Neck (collum)

Trunk
    Back (dorsum)
    Chest (thorax)
    Abdomen (venter)
    Pelvis

Upper extremity
    Armpit (axilla)
    Shoulder (acromial or omos)
    Arm (brachium)
    Elbow (cubitus)
    Forearm (antebrachium)
    Wrist (carpus)
    Hand (manus)
        1. Metacarpus (dorsum and palm)
        2. Fingers (phalanges or digits)

Lower extremity
    Buttocks (gluteal region)
    Thigh (femoral region)
    Knee (genu)
    Leg (crus)
    Ankle (tarsus)
    Foot (pes)
        1. Metatarsus (dorsum and sole)
        2. Toes (phalanges or digits)

HEAD

The **head (cephalic region** or **caput)** is divided into the cranium and face. The **cranium (skull** or **brain case)** consists of a *frontal region* (front of skull or sinciput that includes the forehead), *parietal region* (crown of skull or vertex), *temporal region* (side of skull or tempora), and *occipital region* (base of skull or occiput). The **face (facies)** is subdivided into an *orbital* or *ocular region* which includes the eyeballs (bulbi oculorum), eyebrows (supercilia), and eyelids (palpebrae); *nasal region* (nose or nasus); *infraorbital region* (inferior to orbit); *oral region* (mouth); *mental region* (anterior portion of mandible); *buccal region* (cheek); *parotid-masseteric region* (external to the parotid gland and masseter muscle); *zygomatic region* (inferolateral to orbit); and *auricular region* (ear).

Examine Exhibit 25-1. It contains illustrations of various surface features of the head with an accompanying description of each of the features.

NECK

The **neck (collum)** is divided into an *anterior cervical region*, two *lateral cervical regions*, and a *posterior region* or *nucha*.

The most prominent structure in the midline of the anterior cervical region is the *thyroid cartilage* or *Adam's apple* of the larynx. Just superior to it, the *hyoid bone* can be palpated. Inferior to the Adam's apple, the *cricoid cartilage* of the larynx can be felt.

A major portion of the lateral cervical regions is formed by the *sternocleidomastoid muscles*. Each muscle extends from the mastoid process of the temporal bone, felt as a bump behind the auricle of the ear, to the sternum and clavicle. Each sternocleidomastoid muscle divides its portion of the neck into an anterior triangle and a posterior (lateral) triangle. The *anterior triangle* is bordered superiorly by the mandible, inferiorly by the sternum, medially by the cervical midline, and laterally by the anterior border of the sternocleidomastoid muscle. The *posterior (lateral) triangle* is bordered inferiorly by the clavicle, anteriorly by the posterior border of the sternocleidomastoid muscle, and posteriorly by the anterior border of the trapezius muscle.

A muscle that extends downward and outward from the base of the skull and occupies a portion of the lateral cervical region is the *trapezius muscle*.

EXHIBIT 25-1 SURFACE ANATOMY OF THE HEAD

HEAD

1. **Zygomatic region.** Inferolateral to orbit.

2. **Orbital (ocular) region.** Includes eyeball, eyelids, and eyebrows.

3. **Infraorbital region.** Inferior to orbit.

4. **Nasal region.** Nose.

5. **Buccal region.** Cheek.

6. **Oral region.** Mouth.

7. **Mental region.** Anterior portion of mandible.

8. **Parotid-masseteric region.** External to parotid gland and masseter muscle.

9. **Occipital region.** Base of skull (occiput).

10. **Auricular region.** Ear.

11. **Parietal region.** Crown of skull (vertex).

12. **Temporal region.** Side of skull (tempora).

13. **Frontal region.** Front of skull (sinciput).

EYE

1. **Pupil.** Opening of center of iris of eyeball for light transmission.

2. **Iris.** Circular pigmented muscular membrane behind cornea.

3. **Sclera.** "White" of eye, a coat of fibrous tissue that covers entire eyeball except for cornea.

4. **Conjunctiva.** Membrane that covers exposed surface of eyeball and lines eyelids.

5. **Palpebrae** or **eyelids.** Folds of skin and muscle lined by conjunctiva.

6. **Palpebral fissure.** Space between eyelids when they are open.

7. **Medial canthus.** Site of union of upper and lower eyelids near nose.

8. **Lateral canthus.** Site of union of upper and lower eyelids away from nose.

9. **Lacrimal caruncle.** Fleshy, yellowish projection of medial commissure that contains modified sweat and sebaceous glands.

10. **Eyelashes.** Hairs on margins of eyelids, usually arranged in two or three rows.

11. **Supercilia** or **eyebrows.** Several rows of hairs superior to upper eyelids.

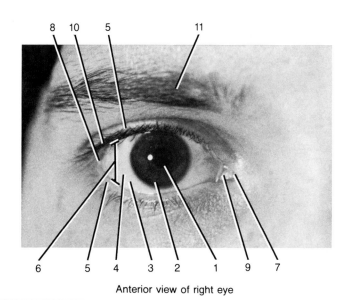

Right lateral view of head

Anterior view of right eye

Photographs courtesy of Victor B. Eichler © 1980.

EXHIBIT 25-1 *(Continued)*

EAR

1. **Auricle.** Portion of external ear not contained in head; also called pinna or trumpet.

2. **Tragus.** Cartilaginous projection anterior to external opening to ear.

3. **Antitragus.** Cartilaginous projection opposite tragus.

4. **Concha.** Hollow of auricle.

5. **Helix.** Superior and posterior free margin of auricle.

6. **Antihelix.** Semicircular ridge posterior and superior to concha.

7. **Triangular fossa.** Depression in superior portion of antihelix.

8. **Lobule.** Interior portion of auricle devoid of cartilage.

9. **External auditory meatus.** Canal extending from external ear to eardrum.

NOSE AND LIPS

1. **Root.** Superior attachment of nose at forehead located between eyes.

2. **Apex.** Tip of nose.

3. **Dorsum nasi.** Rounded anterior border connecting root and apex; in profile, may be straight, convex, concave, or wavy.

4. **Nasofacial angle.** Point at which side of nose blends with tissues of face.

5. **Ala.** Convex flared portion of inferior lateral surface; unites with upper lip.

6. **External nares.** External openings into nose.

7. **Bridge.** Superior portion of dorsum nasi, superficial to nasal bones.

8. **Philtrum.** Vertical groove in medial portion of upper lip.

9. **Lips.** Upper and lower fleshy border of the oral cavity.

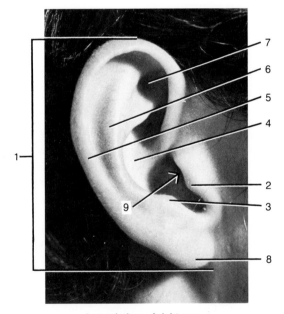

Lateral view of right ear

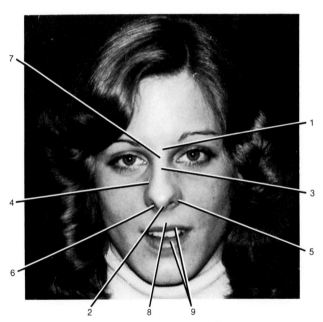

Anterior view of nose and lips

Photographs courtesy of Victor B. Eichler © 1980 (left); Donald Castellaro and Deborah Massimi (right).

EXHIBIT 25-2 SURFACE ANATOMY OF THE NECK

1. **Anterior triangle of neck.** Bordered superiorly by mandible, inferiorly by sternum, medially by cervical midline, and laterally by anterior border of sternocleidomastoid muscle.

2. **Posterior triangle of neck.** Bordered inferiorly by clavicle, anteriorly by posterior border of sternocleidomastoid muscle, and posteriorly by anterior border of trapezius muscle.

3. **Trapezius muscle.** Occupies a portion of lateral surface of neck; helps raise, lower, and draw shoulders backward and extends head.

4. **Sternocleidomastoid muscle.** Forms major portion of lateral surface of neck; flexes head and rotates it to opposite side.

5. **Cricoid cartilage.** Inferior laryngeal cartilage that attaches larynx to trachea; can be palpated by running your fingertip down from your chin over the thyroid cartilage (after you pass the cricoid cartilage your fingertip sinks in).

6. **Thyroid cartilage (Adam's apple).** Triangular laryngeal cartilage in the midline of the anterior cervical region.

7. **Hyoid bone.** Lies just superior to thyroid cartilage; it is the first resistant structure palpated in the midline below the chin.

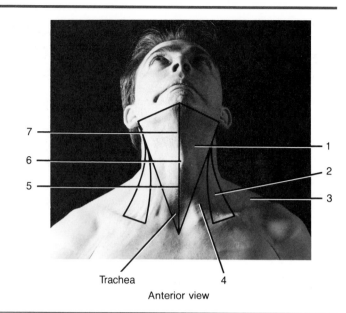

Anterior view

Photograph courtesy of Carroll H. Weiss, Camera M.D. Studios, Inc.

A "stiff neck" is frequently associated with an inflammation of this muscle. A very prominent vein that runs along the lateral surface of the neck is the *external jugular vein.* It is readily seen if you are angry or if your collar is too tight. The locations of a few surface features of the neck are shown in Exhibit 25-2.

TRUNK

The **trunk** is divided into the back, chest, abdomen, and pelvis.

One of the most striking surface features of the **back (dorsum)** is the *vertebral spines,* the dorsally pointed projections of the vertebrae. A very prominent vertebral spine is the *vertebra prominens* of the seventh cervical vertebra. It is easily seen when the head is flexed. Another easily identifiable surface landmark of the back is the *scapula.* In fact, several parts of the scapula (axillary border, vertebral border, inferior angle, spine, and acromion) may also be seen or palpated. In lean individuals the *ribs* may also be seen. Among the superficial muscles of the back that can be seen are the *latissimus dorsi, erector spinae, infraspinatus, trapezius,* and *teres major.* These, as well as the other surface features of the back, are described in Exhibit 25-3.

The **chest (thorax)** presents a number of anatomical landmarks. At its superior region are the *clavicles.* The *sternum* lies in the midline of the chest and is divisible into a superior manubrium, a middle body, and an inferior xiphoid process. Its superior border attaches to the clavicles. Between the medial ends of the clavicles, there is a depression on the superior surface of the sternum called the *jugular notch.* The *sternal angle* is formed by a junction line between the manubrium and body of the sternum and is palpable under the skin. It locates the sternal end of the second rib and is the most reliable surface landmark of the chest. The inferior portion of the sternum, the *xiphoid process,* may be palpated. Also visible or palpable are the *ribs.* The *costal margins,* the inferior edges of the costal cartilages of ribs 7 through 10, can also be seen or palpated. Among the more prominent superficial chest muscles are the *pectoralis major* and *serratus anterior.* These and other surface features of the chest are described in Exhibit 25-3.

The **abdomen (venter)** and **pelvis** have already been discussed in terms of their nine regions or four quadrants (see Figures 1-5 and 1-6). External features of the abdomen and pelvis include the *umbilicus, linea alba, external oblique muscle, rectus abdominis muscle, tendinous intersections,* and *symphysis pubis.* The surface features of the abdomen and pelvis are described in Exhibit 25-3.

EXHIBIT 25-3 SURFACE ANATOMY OF THE TRUNK

BACK

1. **Vertebral spines.** Posteriorly pointed projections of vertebrae.

2. **Scapula.** Shoulder blade; leader is on vertebral border.

3. **Latissimus dorsi muscle.** Broad muscle of back that helps draw shoulders backward and downward.

4. **Erector spinae muscle.** Parallel to vertebral column; moves vertebral column in various directions.

5. **Infraspinatus muscle.** Located inferior to spine of scapula; helps laterally rotate humerus.

6. **Trapezius muscle.** Extends from cervical and thoracic vertebrae to spine of scapula and lateral end of clavicle; occupies portion of lateral surface of neck (see Exhibit 25-2); helps raise, lower, and draw shoulders backward and extend head.

7. **Teres major muscle.** Located inferior to infraspinatus; helps extend, adduct, and medially rotate humerus.

CHEST

1. **Jugular notch of sternum.** Depression on superior border of manubrium of sternum between medial ends of clavicles; trachea can be palpated in the notch.

2. **Clavicle.** Collarbone.

3. **Sternal angle of sternum.** Formed by junction between manubrium and body of sternum.

4. **Pectoralis major muscle.** Principal upper chest muscle; flexes, adducts, and medially rotates humerus.

5. **Body of sternum.** Midportion of sternum.

6. **Xiphoid process of sternum.** Inferior portion of sternum.

7. **Costal margin.** Inferior edges of costal cartilages of ribs 7 through 10.

8. **Serratus anterior muscle.** Inferior and lateral to pectoralis major muscle; helps laterally rotate scapula and elevate ribs; also illustrated in Exhibit 25-3, Abdomen.

Ribs. Form bony cage of thoracic cavity (not illustrated).

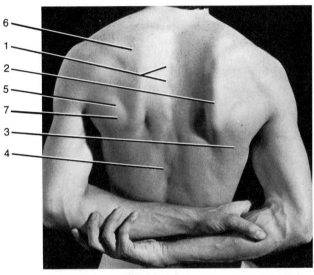

Posterior view

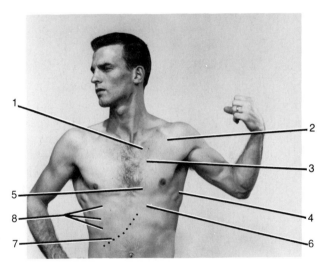

Anterior view

Photographs courtesy of Victor B. Eichler © 1980 (left); Vincent P. Destro, Mayo Foundation (right).

EXHIBIT 25-3

ABDOMEN

1. **Umbilicus.** Also called navel; previous site of attachment of umbilical cord in fetus.

2. **External oblique muscle.** Located inferior to serratus anterior; helps compress abdomen and bend vertebral column laterally.

3. **Rectus abdominis muscle.** Located just lateral to midline of abdomen; helps compress abdomen and flex vertebral column.

4. **Linea alba.** Flat, tendinous raphe along midline between rectus abdominis muscles.

5. **Tendinous Intersections.** Fibrous bands that run transversely or obliquely across the rectus abdominis muscles.

   Symphysis pubis. Anterior joint of hipbones; palpated as a firm resistance in the midline at the inferior portion of the anterior abdominal wall (not illustrated).

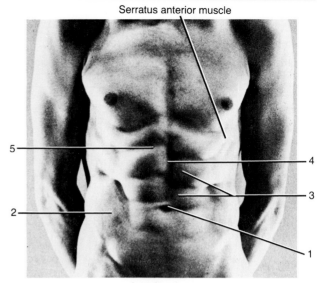

Serratus anterior muscle

Anterior view

Photograph courtesy of R. D. Lockhart, *Living Anatomy*, Faber & Faber, 1974.

UPPER EXTREMITY

The **upper extremity** consists of the armpit, shoulder, arm, elbow, forearm, wrist, and hand.

At the **shoulder (acromial** or **omos)** moving laterally along the top of the clavicle, it is possible to palpate a slight elevation at the lateral end of the clavicle, the *acromioclavicular joint.* Less than 2.5 cm (1 inch) distal to this joint, one can also feel the *acromion* of the scapula which forms the tip of the shoulder. The rounded prominence of the shoulder is formed by the *deltoid muscle,* a frequent site for intramuscular injections (Exhibit 25-4).

Most of the anterior surface of the **arm (brachium)** is occupied by the *biceps brachii muscle,* whereas most of the posterior surface is occupied by the *triceps brachii muscle* (Exhibit 25-4).

At the **elbow (cubitus)** it is possible to locate three bony protuberances. The *medial* and *lateral epicondyles* of the humerus form visible eminences on the dorsum of the elbow. The *olecranon* of the ulna forms the large eminence in the middle of the dorsum of the elbow and lies between and slightly superior to the epicondyles when the forearm is extended. The *ulnar nerve* can be palpated in a groove behind the medial epicondyle. The triangular space of the anterior region of the elbow is the *cubital fossa.* The *median cubital vein* usually crosses the cubital fossa obliquely. This vein is the one frequently selected for removal of blood for diagnosis, transfusions, and intravenous therapy (Exhibit 25-4).

One of the most prominent landmarks of the **forearm (antebrachium)** is the *styloid process* of the ulna. It may be seen as a protuberance on the medial side of the wrist. The ulna is the medial bone of the forearm, and the radius is the lateral bone of the forearm. On the lateral side of the upper forearm is the *brachioradialis muscle.* Next to it is the *flexor carpi radialis muscle.* On the medial side is the *flexor carpi ulnaris muscle* (Exhibit 25-4).

At the **wrist (carpus),** several structures may be palpated or seen. On the anterior surface are *bracelet flexure lines* where the skin is firmly attached to the underlying deep fascia. On the anterior surface of the wrist, it is possible to see the *tendon of the palmaris longus muscle* by making a fist. Next to this tendon, as you move toward the thumb, you can feel the *tendon of the flexor carpi radialis muscle.* If you continue toward the thumb, you can palpate the *radial artery* just medial to the styloid process of the radius. This artery is frequently used to take the pulse. The *pisiform bone,* the medial bone of the proximal carpals, can be palpated as a projection distal to the styloid process of the ulna. By bending the thumb backward, two prominent tendons may be located along the posterior surface of the wrist. The one closer to the styloid process of the radius is the *tendon of the extensor pollicus brevis muscle.* The one closer to the styloid process of the ulna is the *tendon of the extensor pollicus longus muscle.* The depression between these two tendons is known as the *"anatomical snuffbox."* By palpating the depression, you can feel the radial artery (Exhibit 25-4).

EXHIBIT 25-4 SURFACE ANATOMY OF THE UPPER EXTREMITY

SHOULDER

1. **Acromion.** Expanded end of spine of scapula; forms tip of shoulder; clearly visible in some individuals and can be palpated about 2.5 cm (1 inch) distal to acromioclavicular joint.

2. **Deltoid muscle.** Triangular muscle that forms rounded prominence of shoulder; abducts arm.

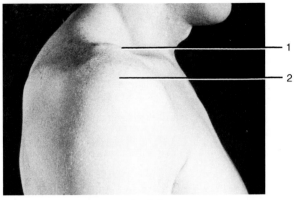

Lateral view

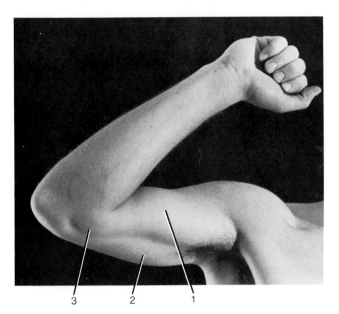

Medial view of upper extremity

ARM AND ELBOW

1. **Biceps brachii muscle.** Forms bulk of anterior surface of arm; helps flex forearm.

2. **Triceps brachii muscle.** Forms bulk of posterior surface of arm; helps extend forearm.

3. **Medial epicondyle.** Medial projection at distal end of humerus.

4. **Lateral epicondyle.** Lateral projection at distal end of humerus.

5. **Olecranon.** Projection of proximal end of ulna; forms elbow.

6. **Cubital fossa.** Triangular space in anterior region of elbow; contains tendon of biceps brachii muscle, brachial artery and its terminal branches (radial and ulnar arteries), and parts of median and radial nerves.

7. **Median cubital vein.** Crosses cubital fossa obliquely.

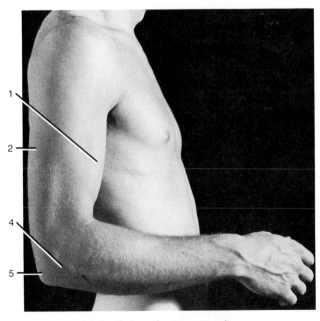

Lateral view of upper extremity

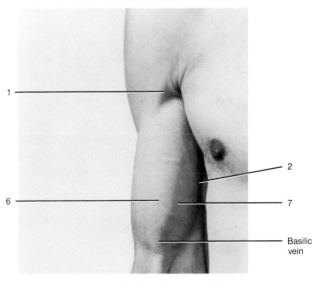

Anteromedial view of arm and forearm

Photographs courtesy of Victor B. Eichler © 1980; Vincent P. Destro, Mayo Foundation (bottom right).

EXHIBIT 25-4 *(Continued)*

FOREARM

1. **Styloid process of ulna.** Projection of distal end of ulna at medial side of wrist.

2. **Brachioradialis muscle.** Located at superior and lateral aspect of forearm; helps flex forearm.

3. **Flexor carpi radialis muscle.** Located along midportion of forearm; helps flex wrist.

4. **Flexor carpi ulnaris muscle.** Located at medial aspect of forearm; helps flex wrist.

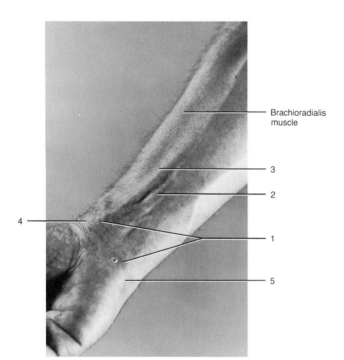

Anterior view

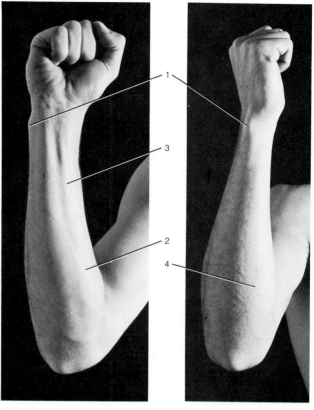

Medial view   Anterior view

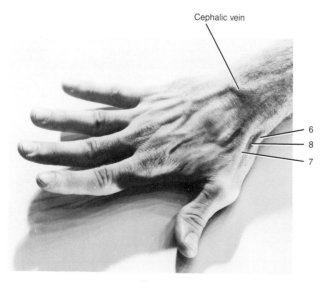

Dorsum

WRIST

1. **Bracelet flexure lines.** Several more or less constant lines on anterior aspect of wrist where skin is firmly attached to underlying deep fascia.

2. **Tendon of palmaris longus muscle.** By making a fist, the tendon can be seen on anterior surface of wrist nearer ulna; muscle helps to flex wrist.

3. **Tendon of flexor carpi radialis muscle.** Tendon on anterior surface of wrist lateral to tendon of palmaris longus.

4. **Radial artery.** Can be palpated just medial to styloid process of ulna; frequently used to take pulse.

5. **Pisiform bone.** Medial bone of proximal carpals; easily palpated as a projection distal to styloid process of ulna.

6. **Tendon of extensor pollicus brevis muscle.** Tendon closer to styloid process of radius along posterior surface of wrist, best seen when thumb is bent backward; muscle extends thumb.

7. **Tendon of extensor pollicus longus muscle.** Tendon closer to styloid process of ulna along posterior surface of wrist, best seen when thumb is bent backward; muscle extends thumb.

8. **"Anatomical snuffbox."** Depression between tendons of extensor pollicus brevis and extensor pollicus longus muscles; radial artery can be palpated in the depression.

EXHIBIT 25-4 *(Continued)*

HAND

1. **"Knuckles."** Commonly refers to dorsal aspects of distal ends of metacarpals II, III, IV, and V; also includes dorsal aspects of metacarpophalangeal and interphalangeal joints.

2. **Thenar eminence.** Lateral rounded contour on palm of hand formed by muscles of thumb.

3. **Hypothenar eminence.** Medial rounded contour on palm of hand formed by muscles of little finger.

4. **Digital flexion creases.** Skin ceases on anterior surface of fingers.

5. **Palmar flexion creases.** Skin creases on palm of hand.

6. **Tendon of extensor digiti minimi muscle.** Extensor tendon in line with phalanx V (little finger); muscle extends little finger.

7. **Tendons of extensor digitorum muscle.** Extensor tendons in line with phalanges II, III, and IV; muscle extends fingers and wrist.

8. **Dorsal venous arch.** Superficial veins on dorsum of hand that form cephalic vein.

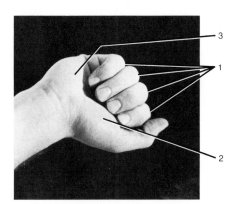

Palmar and dorsal surfaces

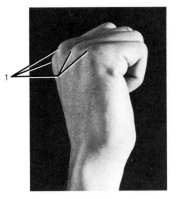

Medial view

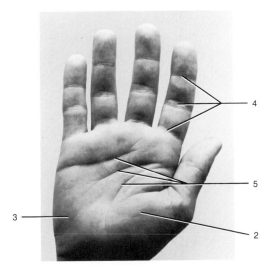

Palmar surface

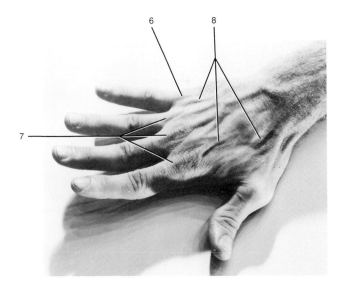

Dorsum

Photographs courtesy of Victor B. Eichler © 1980 (top); Carroll H. Weiss, Camera M.D. Studios, Inc. (bottom left); Mayo Foundation (bottom right).

On the dorsum of the **hand (manus)** the distal ends of the second through fifth metacarpal bones are commonly referred to as the *"knuckles."* The term also includes the joints between metacarpals and phalanges and the joints between the phalanges of the fingers. The *dorsal venous arch,* the superficial veins on the dorsum of the hand, can be displayed by compressing the blood vessels at the wrist for a few moments as the hand is opened and closed. The *tendon of the extensor digiti minimi muscle* can be seen in line with phalanx V (little finger) and the *tendons of the extensor digitorum muscle* can be seen in line with phalanges II, III, and IV. By examining the palm of the hand, it is possible to see a number of *skin creases,* as well as the location of the joints of the fingers. The skin creases of the palm are collectively called *palmar flexion creases.* The skin creases on the anterior surface of the fingers are called *digital flexion creases.* Also on the palm are the *thenar eminence,* a rounded contour formed by thumb muscles, and the *hypothenar eminence,* a rounded contour formed by little finger muscles (Exhibit 25-4).

LOWER EXTREMITY

The **lower extremity** consists of the buttocks, thigh, knee, leg, ankle, and foot.

The outline of the superior border of the **buttock (gluteal region)** is formed by the *iliac crest.* The *posterior superior iliac spine,* the posterior termination of the iliac crest, lies deep to a dimple (skin depression) about 4 cm (1.5 inches) lateral to the midline. Most of the prominence of the buttocks is formed by the *gluteus maximus* and *gluteus medius muscles.* The depression that separates the buttocks is the *gluteal cleft* whereas the inferior limit of the buttock formed by the inferior margin of the gluteus maximus muscle is called the *gluteal fold.* The bony prominence in each buttock is the *ischial tuberosity* of the hipbone. This structure bears the weight of your body when you are seated. About 20 cm (8 inches) below the highest portion of the iliac crest, the *greater trochanter* of the femur can be felt on the lateral side of the thigh. As you have already seen, the iliac crest and greater trochanter are useful landmarks when giving an intramuscular injection in the gluteal muscle (Exhibit 25-5).

Among the prominent superficial muscles on the anterior surface of the **thigh (femoral region)** are the *sartorius* and three of the four components of the *quadriceps femoris* (*vastus lateralis, vastus medialis,* and *rectus femoris*). The vastus lateralis is frequently used as an injection site for diabetics when administering insulin. The superficial medial thigh muscles include the *adductor magnus, adductor brevis, adductor longus, gracilis, obturator externus,* and *pectineus.* The superficial posterior thigh muscles are the hamstrings (*semitendinosus, semimembranosus,* and *biceps femoris*). See Exhibit 25-5.

On the anterior surface of the **knee (genu)**, the *patella,* or kneecap, is observable. Below it is the *patellar ligament.* On the posterior surface of the knee is a diamond-shaped space, the *popliteal fossa.* Just below the patella, on either side of the patellar ligament, the *medial* and *lateral condyles* of the femur and tibia can be felt (Exhibit 25-5).

The bony prominence inferior to the patella in the middle of the **leg (crus)** is the *tibial tuberosity.* The tibia is the medial bone of the leg, and the fibula is the lateral bone of the leg. Prominent superficial muscles of the leg include the *tibialis anterior, gastrocnemius,* and *soleus* (Exhibit 25-5).

At the **ankle (tarsus)** the *medial malleolus* of the tibia and the *lateral malleolus* of the fibula can be noted as two prominent eminences (Exhibit 25-5).

On the dorsum of the **foot (pes)** you can see the *tendon of the extensor hallucis longus muscle* in line with phalanx V (great toe), the *tendons of the extensor digitorum longus muscle* in line with phalanges II through V, and the *dorsal venous arch,* superficial veins that unite to form the small and great saphenous veins. Arising from the *heel bone (calcaneus)* is the *calcaneal* (Achilles) *tendon* (Exhibit 25-5).

EXHIBIT 25-5 **SURFACE ANATOMY OF THE LOWER EXTREMITY**

BUTTOCKS AND THIGH

1. **Iliac crest.** Superior margin of ilium of hipbone; forms outline of superior border of buttock; when you rest your hands on your hips, they rest on the iliac crests.

2. **Posterior superior iliac spine.** Posterior termination of iliac crest; lies deep to a dimple (skin depression) about 4 cm (1.5 inches) lateral to midline; dimple forms because skin and underlying fascia are attached to bone.

3. **Gluteus maximus muscle.** Forms major portion of prominence of buttock; extends and laterally rotates thigh.

4. **Gluteus medius muscle.** Superolateral to gluteus maximus; abducts and medially rotates thigh; frequent site for intramuscular injections.

5. **Gluteal cleft.** Depression along midline that separates the buttocks.

6. **Gluteal fold.** Inferior limit of buttock formed by inferior margin of gluteus maximus muscle.

7. **Ischial tuberosity.** Bony prominence of ischium of hipbone; bears weight of body when seated.

8. **Greater trochanter.** Projection of proximal end of femur on lateral surface of thigh; can be palpated about 20 cm (8 inches) inferior to iliac crest.

9. **Hamstrings.** Superficial posterior thigh muscles that flex leg and extend thigh.

10. **Sartorius muscle.** Superficial anterior thigh muscle that flexes leg and flexes and laterally rotates thigh.

11. **Rectus femoris muscle.** Component of quadriceps femoris group located at midportion of anterior aspect of thigh; extends leg (in conjunction with other components of quadriceps femoris) and flexes thigh (acting alone).

12. **Vastus lateralis muscle.** Component of quadriceps femoris group located at anterolateral aspect of thigh; extends leg.

13. **Vastus medialis muscle.** Component of quadriceps femoris group located at anteromedial aspect of thigh; extends leg.

14. **Adductor magnus muscle.** Located on medial aspect of thigh; adducts, flexes, and extends thigh.

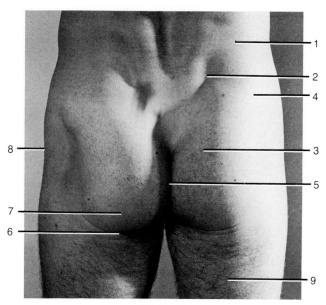

Posterior view

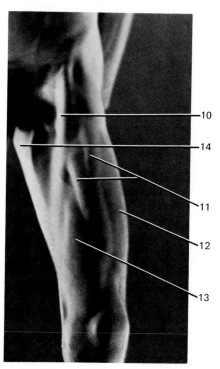

Anterior view

Photographs courtesy of Lester Bergman & Associates (left); R. D. Lockhart, *Living Anatomy*, Faber & Faber, 1974 (right).

EXHIBIT 25-5 (Continued)

KNEE

1. **Patella.** Kneecap; located within quadriceps femoris tendon on anterior surface of knee along midline; margins of condyles (described shortly) can be felt on either side of it.

2. **Medial condyle of femur.** Medial projection of distal end of femur.

3. **Medial condyle of tibia.** Medial projection of proximal end of tibia.

4. **Patellar ligament.** Continuation of quadriceps femoris tendon inferior to patella.

5. **Popliteal fossa.** Diamond-shaped space on posterior aspect of knee visible when knee is flexed; fossa is bordered superolaterally by the biceps femoris muscle, superomedially by the semimembranosus and semitendinosus muscles, and inferolaterally and inferomedially by the lateral and medial heads of the gastrocnemius muscle, respectively.

6. **Lateral condyle of tibia.** Lateral projection of proximal end of tibia.

7. **Lateral condyle of femur.** Lateral projection of distal end of femur.

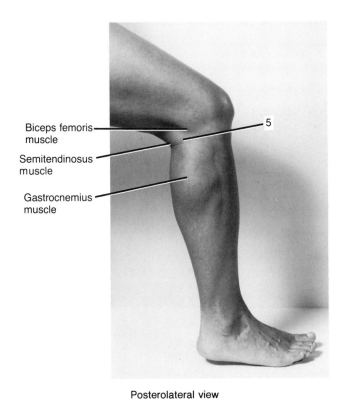

Anterior view

Posterolateral view

Photographs courtesy of Carroll H. Weiss, Camera M.D. Studios, Inc. (left); J. J. Martin, Mayo Clinic (right).

EXHIBIT 25-5 (Continued)

LEG AND ANKLE

Tibial tuberosity. Bony prominence of tibia into which patellar ligament inserts (see anterior view of knee).

Tibialis anterior muscle. Located at anterior surface of leg along midportion; dorsiflexes and inverts foot (see anterior view of knee).

1. **Gastrocnemius muscle.** Forms bulk of mid and upper portion of posterior aspect of leg; plantar flexes foot.

2. **Soleus muscle.** Located deep to gastrocnemius muscle; plantar flexes foot.

3. **Calcaneal (Achilles) tendon.** Conspicuous tendon of gastrocnemius and soleus muscles that inserts into calcaneus (heel) bone of foot.

4. **Lateral malleolus of fibula.** Projection of distal end of fibula that forms lateral prominence of ankle.

5. **Medial malleolus of tibia.** Projection of distal end of tibia that forms medial prominence of ankle.

FOOT

1. **Calcaneus.** Heel bone.

2. **Dorsal venous arch.** Superficial veins on dorsum of foot that unite to form small and great saphenous veins.

3. **Tendons of extensor digitorum longus muscles.** Visible in line with phalanges II through V; muscle extends toes and dorsiflexes and everts foot.

4. **Tendon of extensor hallucis longus muscle.** Visible in line with phalanx I (great toe); muscle extends great toe and dorsiflexes ankle.

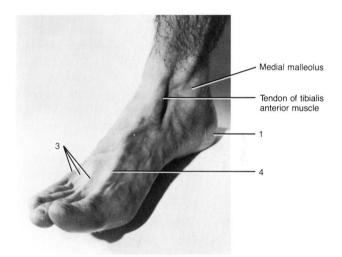

Dorsomedial view

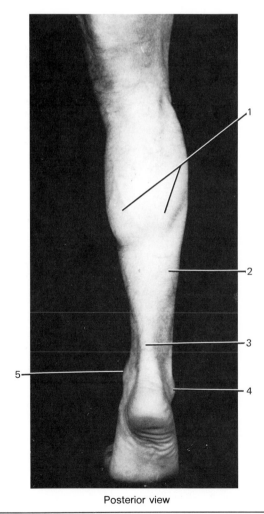

Posterior view

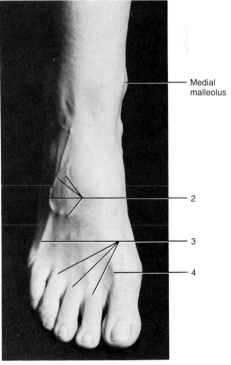

Dorsum

Photographs courtesy of Donald Castellaro and Richard Sollazzo (left); Mayo Foundation (top right); Carroll H. Weiss, Camera M.D. Studios, Inc. (bottom right).

STUDY OUTLINE

1. Surface anatomy is the study of the form and markings of the surface of the body.
2. Surface anatomy features may be noted by visual inspection or palpation.
3. The principal regions of the body used to study surface anatomy are the head, neck, trunk, upper extremity, and lower extremity.
4. A review of surface anatomy is presented in Exhibits 25-1 through 25-5.

REVIEW QUESTIONS

1. Define surface anatomy. What is meant by palpation?
2. List the principal regions of the body and their subdivisions.
3. Using Exhibit 25-1 as an outline, locate as many of the surface features of the head as you can on your partner's body, wall charts, models, photographs, and skeletons.
4. Using Exhibit 25-2 as an outline, locate as many of the surface features of the neck as you can on your partner's body, wall charts, models, photographs, and skeletons.
5. Using Exhibit 25-3 as an outline, locate as many of the surface features of the trunk as you can on your partner's body, wall charts, models, photographs, and skeletons.
6. Using Exhibit 25-4 as an outline, locate as many of the surface features of the upper extremity as you can on your partner's body, wall charts, models, photographs, and skeletons.
7. Using Exhibit 25-5 as an outline, locate as many of the surface features of the lower extremity as you can on your partner's body, wall charts, models, photographs, and skeletons.

SELECTED READINGS

Barr, M. L., *The Human Nervous System,* 3d ed. New York: Harper & Row, 1979.

Basmajian, J. V., *Grant's Method of Anatomy,* 9th ed. Baltimore: Williams & Wilkins, 1975.

Basmajian, J. V., *Surface Anatomy,* Baltimore: Williams & Wilkins, 1977.

Bloom, W., and D. W. Fawcett, *A Textbook of Histology,* 10th ed. Philadelphia: Saunders, 1975.

Carpenter, M. B., *Human Neuroanatomy,* 7th ed. Baltimore: Williams & Wilkins, 1976.

Christensen, J. B., and I. R. Telford, *Synopsis of Gross Anatomy,* 3d ed. New York: Harper & Row, 1978.

Clemente, C. D., *Anatomy: A Regional Atlas of the Human Body,* 2d ed. Baltimore: Urban and Schwarzenberg, 1981.

Cunningham's Textbook of Anatomy, 12th ed. Edited by G. J. Romanes. London: Oxford University Press, 1981.

DiFiore, M. S. H., *An Atlas of Human Histology,* 5th ed. Philadelphia: Lea & Febiger, 1981.

Ellis, H., *Clinical Anatomy,* 5th ed. Philadelphia: Lippincott, 1971.

Gardner, E. D., D. J. Gray, and R. O'Rahilly, *Anatomy,* 4th ed. Philadelphia: Saunders, 1975.

Gluhbegovic, N., and T. H. Williams, *The Human Brain: A Photographic Guide.* New York: Harper & Row, 1980.

Grant, J. C. B., *Grant's Atlas of Anatomy,* 6th ed. Baltimore: Williams & Wilkins, 1972.

Gray, H., *Anatomy of the Human Body,* 29th ed. Edited by C. M. Goss. Philadelphia: Lea & Febiger, 1973.

Ham, A. W. and D. H. Cormack, *Histology,* 8th ed., Philadelphia: J. B. Lippincott Co., 1979.

Hamilton, W. J., and H. W. Mossman, *Hamilton, Boyd, and Mossman's Human Embryology,* 4th ed. Cambridge: Heffer & Sons, 1972.

Hamilton, W. J., G. Simon, and S. G. I. Hamilton, *Surface and Radiological Anatomy,* 5th ed. Cambridge: Heffer & Sons, 1971.

Hollinshead, W. H., *Textbook of Anatomy,* 3d ed. New York: Harper & Row, 1974.

Kessel, R. G., and R. H. Kardon, *Tissues and Organs: A Text-Atlas of Scanning Electron Microscopy.* San Francisco: Freeman, 1979.

Kieffer, S. A., and E. R. Heitzman, *An Atlas of Cross-Sectional Anatomy.* New York: Harper & Row, 1979.

Lachman, E., *Case Studies in Anatomy,* 2d ed. New York: Oxford University Press, 1971.

Langman, J., *Medical Embryology,* 4th ed. Baltimore: Williams & Wilkins, 1981.

Langman, J., and M. W. Woerdeman, *Atlas of Medical Anatomy.* Philadelphia: Saunders, 1978.

Lockhart, R. D., *Living Anatomy,* 7th ed. London: Faber & Faber, 1974.

Lockhart, R. D., G. F. Hamilton, and F. W. Fyfe, *Anatomy of the Human Body,* 2d ed. Philadelphia: Lippincott, 1969.

McMinn, R. M. H., and R. T. Hutchings, *Color Atlas of Human Anatomy.* Chicago: Year Book Medical Publishers, 1977.

Montgomery, R. L., *Basic Anatomy for the Allied Health Professions,* Baltimore: Urban & Schwarzenberg, 1981.

Moore, K. L., *Clinically Oriented Anatomy,* Baltimore: Williams & Wilkins, 1980.

Moyer, K. E., *Neuroanatomy.* New York: Harper & Row, 1980.

Netter, F. H., *CIBA Collection of Medical Illustrations,* Vols. 1–7. Summit, N.J.: CIBA, 1962–1979.

Pernkopf, E., *Atlas of Topographical and Applied Human Anatomy.* Edited by H. Ferner; translated by H. Monsen. Philadelphia: Saunders, Vol. 1, 1963; Vol. 2, 1964.

Ranson, S. W., and S. L. Clark, *Anatomy of the Nervous System, Its Development and Function,* 10th ed. Philadelphia: Saunders, 1959.

Reith, E. J., and M. N. Ross, *Atlas of Descriptive Histology,* 3d ed. New York: Harper & Row, 1977.

Royce, J., *Surface Anatomy.* Philadelphia: Davis, 1965.

Snell, R. S., *Clinical Anatomy for Medical Students,* Boston: Little, Brown & Co., 1981.

Sobotta/Hammersen, *Histology.* Philadelphia: Lea & Febiger, 1976.

Woodburne, R. T., *Essentials of Human Anatomy,* 6th ed. New York: Oxford University Press, 1978.

Yokochi, C., and J. W. Rohen, *Photographic Anatomy of the Human Body,* 2d ed. Tokyo, New York: Igaku-Shoin, Ltd., 1979.

GLOSSARY

1. The strongest accented syllable appears in capital letters, for example, bilateral (bī-LAT-er-al) and diagnosis (dī-ag-NŌ-sis).
2. If there is a secondary accent, it is noted by a single quote mark ('), for example, constitution (kon'-sti-TOO-shun) and physiology (fiz'-ē-OL-ō-jē). Any additional secondary accents are also noted by a single quote mark, for example, decarboxylation (dē'-kar-bok'-si-LĀ-shun).
3. Vowels marked with a line above the letter are pronounced with the long sound as in the following common words.

    ā as in *māke*
    ē as in *bē*
    ī as in *īvy*
    ō as in *pōle*

4. Vowels not so marked are pronounced with the short sound, as in the following words.

    e as in *bet*
    i as in *sip*
    o as in *not*
    u as in *bud*

5. Other phonetic symbols are used to indicate the following sounds:

    a as in *above*
    oo as in *sue*
    yoo as in *cute*
    oy as in *oil*

Abatement (a-BĀT-ment)   A decrease in the seriousness of a disorder or in the severity of pain or other symptoms.

Abdomen (ab-DŌ-men)   The area between the diaphragm and pelvis.

Abduction (ab-DUK-shun)   Movement away from the axis or midline of the body or one of its parts.

Abortion (a-BOR-shun)   Premature loss or removal of an embryo or nonviable fetus; any failure in the normal process of developing or maturing.

Abscess (AB-ses)   A localized collection of pus and liquefied tissue in a cavity.

Absorption (ab-SORP-shun)   The taking up of liquids by solids or of gases by solids or liquids.

Accommodation (a-kom-mō-DĀ-shun)   A change in the shape of the eye lens so that vision is more acute; an adjustment of the eye lens for various distances; focusing.

Accretion (a-KRĒ-shun)   A mass of material that has accumulated in a space or cavity; the adhesion of parts.

Acetabulum (as'-e-TAB-yoo-lum)   The rounded cavity on the external surface of the innominate bone that receives the head of the femur.

Acetone bodies (as-e-tōne)   Organic compounds that may be found in excessive amounts in the urine and blood of diabetics and whenever too much fat in proportion to carbohydrate is being oxidized.

Also called ketone bodies (KĒ-tōne).

Acetylcholine (ACh) (as'-ē-til-KŌ-lēn) A chemical transmitter substance, liberated at synapses in the central nervous system, that stimulates skeletal muscle contraction.

Achilles reflex (a-KIL-ēz) Extension of the foot and contraction of calf muscles following a tap upon the Achilles tendon.

Achilles tendon The tendon of the soleus and gastrocnemius muscles at the back of the heel.

Achlorhydria (ā-klōr-HĪ-dre-a) Absence of hydrochloric acid in the gastric juice.

Acid (AS-id) A proton donor; excess hydrogen ions producing a pH less than 7.

Acidophil (a-SID-o-fil) Glandular cell of the adenohypophysis.

Acidosis (as-i-DŌ-sis) A serious disorder in which the normal alkaline substances of the blood are reduced in amount.

Acinar (AS-i-nar) Flasklike.

Acini (AS-i-nē) Masses of cells in the pancreas.

Acinus (AS-i-nus) A small sac; a grapelike dilation.

Acoustic (a-KOOS-tik) Pertaining to sound or the sense of hearing.

Acromegaly (ak'-rō-MEG-a-lē) A pathological condition caused by hypersecretion of human growth hormone during adulthood.

Acromion (a-KRŌ-me-on) The lateral triangular projection of the spine of the scapula, forming the point of the shoulder and articulating with the clavicle.

Actin (AK-tin) One of the contractile proteins in muscle fiber (the other is myosin).

Action potential An impulse or wave of negativity along a conducting neuron.

Active transport The movement of substances across cell membranes, against a concentration gradient, requiring the expenditure of energy.

Actomyosin (ak-tō-MĪ-ō-sin) The combination of actin and myosin in a muscle cell.

Acuity (a-KYOO-i-tē) Clearness, or sharpness, usually of vision.

Acute Having rapid onset, severe symptoms, and a short course; not chronic.

Adam's apple The laryngeal prominence of the thyroid cartilage.

Adaptation The adjustment of the pupil of the eye to light variations.

Addison's disease (AD-i-sonz) Disorder due to deficiency in the secretion of adrenocortical hormones.

Adduction (ad-DUK-shun) Movement toward the axis or midline of the body or one of its parts.

Adenohypophysis (ad'i-nō-hī-POF-i-sis) The anterior portion of the pituitary gland.

Adenoids (AD-i-noyds) The pharyngeal tonsils.

Adenosine triphosphate (ATP) (a-DEN-ō-sēn trī-FOS-fāt) A compound containing sugar, adenine, nitrogen, and three phosphoric acids; the breadown of ATP provides the energy for cellular work.

Adenyl cyclase (AD-e-nil SĪ-klās) An enzyme that converts ATP into cyclic AMP.

Adhesion (ad-HĒ-zhun) Abnormal joining of parts to each other.

Adrenal cortex (a-DRĒ-nal KOR-teks) The outer portion of an adrenal gland; the cortex is divided into three zones, each of which has a different cellular arrangement and secretes different hormones.

Adrenal glands Two glands, one superior to each kidney; also called the suprarenal glands.

Adrenalin (a-DREN-a-lin) Proprietary name for epinephrine (adrenaline); one of the active secretions of the medulla of the adrenal gland.

Adrenal medulla (me-DUL-a) The inner portion of an adrenal gland, consisting of cells which secrete epinephrine and norepinephrine in response to the stimulation of preganglionic sympathetic neurons.

Adrenergic (ad'-ren-ER-jik) Applied to nerve fibers that, when stimulated, release norepinephrine (noradrenaline) at their terminations.

Adrenocorticotropic hormone (ACTH) (ad-rē'-nō-kor-ti-kō-TROP-ik) Hormone produced by the anterior pituitary gland, which influences the cortex of the adrenal glands.

Adrenoglomerulotropin (a-drē'-nō-glō-mer'-yoo-lō-TRŌ-pin) A hormone secreted by the pineal gland.

Adsorption (ad-SORP-shun) Process whereby a gas or a dissolved substance becomes concentrated at the surface of a solid or at the interfaces of a colloid system.

Adventitia (ad-ven-TISH-ah) The outermost covering of a structure or organ.

Afferent arteriole (AF-er-ent ar-TĒ-rē-ōl) A blood vessel of a kidney that breaks up into the capillary network called a glomerulus; there is one afferent arteriole for each glomerulus.

Afferent neuron (NOO-ron) A nerve cell that carries an impulse toward the central nervous system.

Agglutination (a-gloo'-ti-NĀ-shun) Clumping of microorganisms, blood corpuscles, or particles; an immunity response; an antigen-antibody reaction.

Agglutinin (a-GLOO-ti-nin) A specific principle or antibody in blood serum of an animal affected with a microbial disease that is capable of causing the clumping of bacteria, blood corpuscles, or particles.

Agglutinogen (a-GLOOT-in-o-gen) A genetically determined antigen located on the surface of erythrocytes. These proteins are responsible for the two major blood group classifications: the ABO group and the Rh system.

Agnosia (ag-NŌ-zē-a) A loss of the ability to recognize the meaning of stimuli from the various senses (visual, auditory, touch).

Agonist (AG-ō-nist) The prime mover—the muscle directly engaged in contraction as distinguished from muscles that are relaxing at the same time.

Agraphia (a-GRAF-ē-a) An inability to write.

Albinism (AL-bi-nizm) Abnormal, nonpathological, partial or total absence of pigment in skin, hair, and eyes.

Albumin (al-BYOO-min) A protein substance found in nearly every animal or plant tissue and fluid.

Albuminuria (al-byoo'-min-UR-ēa) Presence of albumin in the urine.

Aldosterone (al-dō-STĒR-ōn) Powerful salt-retaining hormone of the adrenal cortex.

Alimentary (al-i-MEN-ta-rē) Pertaining to nutrition.

Alkaline (AL-ka-līn) Containing more hydroxyl than hydrogen ions and producing a pH of more than 7.

Alkalosis (al-ka-LŌ-sis) Increased bicarbonate content of the blood due to excess of alkalies or withdrawal of acid or chlorides from the blood.

Allantois (a-LAN-tō-is) A kind of elongated bladder between the chorion and amnion of the fetus.

Allergic (a-LER-jik) Pertaining to or sensitive to an allergen.

All-or-none principle In muscle physiology: muscle fibers of a motor unit contract to their fullest extent or not at all. In neuron physiology: if a stimulus is strong enough to initiate an action potential, an impulse is transmitted along the entire neuron at a constant and maximum strength.

Alveolar sacs (al-VĒ-ō-lar) A collection or cluster of alveoli opening into a central atrium or chamber.

Alveolus (al-VĒ-ō-lus) A small hollow or cavity; an air cell in the lungs.

Amenorrhea (ā-men-ō-RĒ-a) Absence or suppression of menstruation.

Amino acids (a-MĒ-nō) Any one of a class of organic compounds occurring naturally in plant and animal tissues and forming the chief constituents of protein. About 22 different ones are known.

Amniocentesis (am'-nē-ō-sen-TĒ-sis) Removal of amniotic fluid by inserting a needle transabdominally into the amniotic cavity.

Amnion (AM-nē-on) The inner of the fetal membranes, a thin transparent sac that holds the fetus suspended in amniotic fluid. Also called the "bag of waters."

Amorphous (a-MOR-fus) Without definite shape or differentiation in structure; pertains to solids without crystalline structure.

Amphiarthrosis (am'fē-ar-THRŌ-sis) An articulation midway between a diarthrosis and synarthrosis, in which the articulating bony surfaces are separated by an elastic substance to which both are attached, so that the mobility is slight, but may be exerted in all directions.

Ampulla (am-POOL-la) A saclike dilation of a canal.

Ampulla of Vater (VA-ter) A small, raised area in the duodenum where the combined common bile duct and main pancreatic duct empty into the duodenum.

Anabolism (a-NAB-ō-lizm) Synthetic reactions whereby small molecules are built up into larger ones.

Anal canal (Ā-nal) The terminal 2 or 3 cm of the rectum; opens to the exterior at the anus.

Anal column A longitudinal fold in the mucous membrane of the anal canal that contains a network of arteries and veins.

Analgesia (an-al-JĒ-zē-a) Absence of normal sense of pain.

Anal triangle The subdivision of the male or female perineum that contains the anus.

Anaphase (AN-a-fāz) The third stage of mitosis when the chromatids that have separated at the centromeres move to opposite poles.

Anamnestic (an'-am-NES-tik) The second more intense response of lymphocytes to an antigen.

Anaphylactic (an'-a-fi-LAK-tik) Pertaining to increasing susceptibility to any foreign protein introduced into the body; decreasing immunity.

Anastomosis (a-nas-tō-MŌ-sis) An end-to-end union or joining together of blood vessels, lymphatics, or nerves.

Anatomical position (an'-a-TOM-i-kal) The body is erect, facing the observer, with the arms at the sides and the palms of the hands facing forward.

Anatomy (a-NAT-o-mē) The structure or study of structure of the body and the relationship of its parts to each other.

Androgen (AN-drō-jen) Substance producing or stimulating male characteristics, such as the male hormone.

Anemia (a-NĒ-mē-a) Condition of the blood in which the number of functional red blood cells or their hemoglobin content is below normal.

Aneurysm (AN-yoo-rizm) A saclike enlargement of a blood vessel caused by a weakening of the wall.

Angina pectoris (an-JĪ-na *or* AN-ji-na PEK-tō-ris) An agonizing pain in the chest resulting from deficient coronary circulation. The cause of the reduced circulation may or may not involve heart or artery disease.

Angiography (an-jē-OG-ra-fē) X-ray examination of blood vessels after injection of a radiopaque substance into the common carotid or vertebral artery; used to demonstrate cerebral blood vessels, and may detect brain tumors with specific vascular patterns.

Angstrom (Å) (ANG-strum) One-tenth of a millimicron or about 1/250,000,000 inch.

Ankyloglossia (ang'-ki-lō-GLOSS-ē-a) "Tongue-tied"; restriction of tongue movements by a short lingual frenulum.

Ankylose (ANG-ke-lōs) Immobilization of a joint by pathological or surgical process.

Anomaly (a-NOM-a-lē) An abnormality that may be a developmental (congenital) defect; a variant from the usual standard.

Anorexia (an-ō-REK-sē-a) Loss of appetite.

Anorexia nervosa A condition of psychological origin in which a person's eating habits are disorganized and total food intake is severely restricted, which may lead to serious physical conditions resulting from nutritional deficiencies.

Anoxia (an-OK-sē-a) Deficiency of oxygen.

Antepartum (an-tē-PAR-tum) Before delivery of the child; occurring (to the mother) before childbirth.

Anterior (an-TĒR-ē-or) In front of or the ventral surface.

Anterior root The structure composed of axons of motor or efferent fibers that emerges from the anterior aspect of the spinal cord and extends laterally to join a posterior root, forming a spinal nerve. There are 31 pairs of anterior or motor or ventral roots.

Antibiotic (an'-ti-bī-OT-ik) Destructive of living things, especially chemicals produced by living organisms that act against disease-producing organisms.

Antibody (AN-ti-bod'-ē) A substance produced by certain cells in the presence of a specific antigen that combines with that antigen to neutralize, inhibit, or destroy it.

Antidiuretic (an'-ti-dī'-yoo-RET-ik) Substance that inhibits urine formation.

Antigen (AN-ti-jen) Any substance that when introduced into the tissues or blood induces the formation of antibodies or reacts with them.

Antrum (AN-trum) Any nearly closed cavity or chamber, especially one within a bone, such as a sinus.

Anulus fibrosus (AN-yoo-lus fī-BRŌ-sus) A ring of fibrous tissue and fibrocartilage that encircles the pulpy substance (nucleus pulposus) of an intervertebral disc.

Anuria (a-NOO-rē-a) Absence of urine formation.

Anus (Ā-nus) The distal end and outlet of the rectum.

Aorta (ā-OR-ta) The main systemic trunk of the arterial system of the body; emerges from the left ventricle.

Aperture (AP-er-chur) An opening or orifice.

Apex (Ā-peks) The pointed end of a conical structure.

Aphasia (a-FĀ-zē-a) Loss of ability to express oneself properly through speech or loss of verbal comprehension.

Apnea (ap-NĒ-a) Temporary cessation of breathing.

Apneustic (ap-NOO-stik) Pertaining to sustained inspiratory effort, even to the point of asphyxia.

Apocrine (AP-ō-krin) Pertaining to cells that lose part of their cytoplasm while secreting.

Apocrine gland A type of gland in which the secretory products gather at the free end of the secreting cell and are pinched off, along with some of the cytoplasm, to become the secretion, as in mammary glands.

Aponeurosis (ap'-ō-nyoo-RŌ-sis) A sheetlike layer of connective tissue joining a muscle to the part that it moves or functioning as a sheath enclosing a muscle.

Appendage (a-PEN-dij) A part or thing attached to the body.

Appositional (exogenous) growth Growth due to surface deposition of material as in the growth in diameter of cartilage and bone.

Aqueduct (AK-we-duct) A canal or passage, especially for the conduction of a liquid.

Aqueous humor (AK-wē-us HYOO-mor) The watery fluid that fills the anterior cavity of the eye.

Arachnoid (a-RAK-noyd) The middle of the three coverings (meninges) of the brain.

Arachnoid villi (VIL-ē) Berrylike tufts of arachnoid that protrude into the superior sagittal sinus and through which the cerebrospinal fluid enters the bloodstream; arachnoid granulations.

Arbor vitae (AR-bōr VĒ-tē) The treelike appearance of the white matter tracts of the cerebellum when seen in midsagittal section. Also a series of branching ridges within the cervix of the uterus.

Arch of aorta (ā-OR-ta) The most superior portion of the aorta, lying between the ascending and descending segments of the aorta. The brachiocephalic, left common carotid, and left subclavian arteries are the three branches of the arch of the aorta.

Areflexia (a'-rē-FLEK-sē-a) Absence of reflexes.

Areola (a-RĒ-ō-la) Any tiny space in a tissue; the pigmented ring around the nipple of the breast.

Areolar (a-RĒ-ō-lar) A type of connective tissue.

Arm The portion of the upper extremity from the shoulder to the elbow.

Arrector pili (a-REK-tor PI-lē) Smooth muscles attached to hairs. Contraction of the muscles pulls the hairs into a more vertical position, resulting in "goose bumps."

Arrhythmia (a-RITH-mē-a) Irregular heart action causing absence of rhythm.

Arteriogram (ar-TĒR-ē-ō-gram) A roentgenogram of

an artery obtained by injecting radiopaque substances into the blood.

Arteriole (ar-TĒ-rē-ōl) A very small arterial branch.

Artery (AR-ter-ē) A blood vessel carrying blood away from the heart.

Arthritis (ar-THRĪ-tis) Inflammation of a joint.

Arthrology (ar-THROL-ō-jē) The scientific study or description of joints.

Arthrosis (ar-THRŌ-sis) A joint or articulation.

Articular cartilage (ar-TIK-yoo-lar KAR-ti-lij) The gristle or white elastic substance attached to articular bone surfaces.

Articulate (ar-TIK-yoo-lāt) To join together as a joint to permit motion between parts.

Articulation (ar-tik'-yoo-LĀ-shun) A joint.

Arytenoid (ar'-i-TĒ-noyd) Ladle-shaped.

Arytenoid cartilages A pair of small cartilages of the larynx that articulates with the cricoid cartilage.

Ascending colon (KŌ-lon) The portion of the large intestine that passes upward from the cecum to the lower edge of the liver where it bends at the hepatic flexure to become the transverse colon.

Ascites (as-SĪ-tēz) Serous fluid in the peritoneal cavity.

Aseptic (ā-SEP-tik) Free from any infectious or septic material.

Asphyxia (as-FIX-ē-a) Unconsciousness due to interference with the oxygen supply of the blood.

Aspirate (AS-pir-āt) To remove by suction.

Association area A portion of the cerebral cortex connected by many motor and sensory fibers to other parts of the cortex. The association areas are concerned with motor patterns, memory, concepts of word-hearing and word-seeing, reasoning, will, judgment, and personality traits.

Association neuron (NOO-ron) A nerve cell lying completely within the central nervous system that carries impulses from sensory neurons to motor neurons; internuncial or connecting or central neuron.

Astereognosis (as-ter'-ē-ōg-NŌ-sis) Inability to recognize objects or forms by touch.

Asthenia (as-THĒ-nē-a) Lack or loss of strength; debility.

Astigmatism (a-STIG-ma-tizm) An irregularity of the lens or cornea of the eye causing the image to be out of focus and producing faulty vision.

Astrocyte (AS-trō-sīt) A neuroglial cell having a star shape.

Ataxia (a-TAK-sē-a) Lack of muscular coordination, lack of precision.

Atelectasis (at'-ē-LEK-ta-sis) A collapsed or airless state of the lung, which may be acute or chronic, and may involve all or part of the lung.

Atherosclerosis (ath'-er-ō-skle-RŌ-sis) A disease involving mainly the lining of large arteries, in which yellow patches of fat are deposited, forming plaques that decrease the size of the lumen. Also called arteriosclerosis.

Atresia (a-TRĒ-zē-a) Abnormal closure of a passage, or absence of a normal body opening.

Atrioventricular (AV) bundle (ā'-trē-ō-ven-TRIK-yoo-lar) The portion of the conduction system of the heart that begins at the atrioventricular node and passes through the cardiac skeleton separating the atria and the ventricles, then runs a short distance down the interventricular septum before splitting into right and left bundle branches. Also called the bundle of His.

Atrioventricular (AV) node The portion of the conduction system of the heart made up of a compact mass of conducting cells located near the orifice of the coronary sinus in the right atrial wall. Also called node of Tawara.

Atrioventricular (AV) valve A structure made up of membranous flaps or cusps that allows blood to flow in one direction only, from an atrium into a ventricle.

Atrium (Ā-trē-um) A cavity or sinus.

Atrophy (AT-rō-fē) Wasting away or decrease in size of a part, due to a failure, abnormality of nutrition, or lack of use.

Aura (OR-a) A feeling or sensation that precedes an epileptic seizure or any paroxysmal attack (like those of bronchial asthma).

Auricle (OR-i-kul) The flap or pinna of the ear; an appendage of an atrium of the heart.

Auscultation (aws-kul-TĀ-shun) Examination by listening to sounds in the body.

Autonomic ganglion (aw'-tō-NOM-ik GANG-lē-on) A cluster of sympathetic or parasympathetic cell bodies located outside the central nervous system.

Autonomic nervous system Visceral efferent neurons, both sympathetic and parasympathetic, that transmit impulses from the central nervous system to smooth muscle, cardiac muscle, and glands; so named because this portion of the nervous system was thought to be self-governing or spontaneous.

Autonomic plexus (PLEK-sus) An extensive network of sympathetic and parasympathetic fibers. The cardiac, celiac, and pelvic plexuses are located in the thorax, abdomen, and pelvis, respectively.

Autopsy (AW-top-sē) The examination of the body after death; a postmortem study of the corpse. Also called necropsy.

Axilla (ak-SIL-a) The small hollow beneath the arm where it joins the body at the shoulder; the armpit.

Axon (AK-son) The process of a nerve cell that carries an impulse away from the cell body.

Azygos (AZ-i-gos)   An anatomical structure that is not paired; occurring singly.

Back   The posterior part of the body; the dorsum.

Bacteriophage (bac-TĒ-rē-ō-fāj)   A nonspecific agent capable of destroying bacteria.

Ball-and-socket joint   A synovial joint in which the rounded surface of one bone moves within a cup-shaped depression or fossa of another bone; the shoulder or hip joint.

Baroreceptor (bar'-ō-rē-SEP-tor)   Receptor stimulated by pressure change.

Bartholin's glands (BAR-tō-linz)   Two small mucous glands located one on each side of the vaginal opening.

Basal ganglia (BĀ-sal GANG-glē-a)   Paired clusters of cell bodies that make up the central gray matter in each cerebral hemisphere. They include the caudate nucleus, the lentiform nucleus, the claustrum, and the amygdaloid body.

Basal metabolic rate (BMR) (BĀ-sel met'-a-BOL-ik)   The rate of metabolism measured under standard or basal conditions.

Base   The broadest part of a pyramidal structure; a nonacid, or a proton acceptor, characterized by excess of OH⁻ ions and a pH greater than 7; a ring-shaped, nitrogen-containing organic molecule which is one of the components of a nucleotide, for example, adenine, guanine, cytosine, thymine, and uracil.

Basement membrane   A thin layer of amorphous substance underlying the epithelium of mucous surfaces.

Basilar membrane (BAS-i-lar)   A membrane in the cochlea of the inner ear that separates the cochlear duct from the scala tympani. The organ of Corti rests on the basilar membrane.

Basophil (BĀ-sō-fil)   A type of white blood cell characterized by a pale nucleus and large granules which stain readily with basic dyes.

Belly   The abdomen; the prominent, fleshy part of a skeletal muscle; gaster.

Benign (bē-NĪN)   Not malignant.

Bicipital (bī-CIP-i-tal)   Having two heads, as in muscle.

Bifurcate (bī-FUR-kāt)   Having two branches or divisions; forked.

Bilateral (bī-LAT-er-al)   Pertaining to two sides of the body.

Bile (bīl)   A secretion of the liver.

Biliary (BIL-ē-a-rē)   Relating to bile, the gallbladder, or the bile ducts.

Bilirubin (bil'-ē-ROO-bin)   An orange or yellowish pigment in bile.

Biliverdin (bil'-ē-VER-din)   A greenish pigment in bile formed in the oxidation of bilirubin.

Biopsy (BĪ-op-sē)   Removal of tissue or other material from the living body for examination, usually microscopic.

Blastocoel (BLAS-tō-sēl)   The internal fluid-filled cavity that develops from the blastocyst.

Blastocyst (BLAS-tō-sist)   A stage in the development of a mammalian embryo.

Blastomere (BLAS-tō-mēr)   One of the cells resulting from the cleavage or segmentation of a fertilized ovum.

Blastula (BLAS-tyoo-la)   An early stage in the development of a zygote.

Blepharism (BLEF-a-rizm)   Spasm of the eyelids; continuous blinking.

Blind spot   Area in the retina at the end of the optic nerve in which there are no light receptor cells.

Blood   The fluid that circulates through the heart, arteries, capillaries, and veins and that constitutes the chief means of transport within the body.

Blood–brain barrier   A mechanism that prevents the passage of materials from the blood to the cerebrospinal fluid and brain.

Body cavity   A space within the body that contains various internal organs.

Bolus (BŌ-lus)   A soft, rounded mass, usually food, that is swallowed.

Bony labyrinth (LAB-ī-rinth)   A series of cavities within the petrous portion of the temporal bone, or the vestibule, cochlea, and semicircular canals of the inner ear.

Bowman's capsule (KAP-sul)   A double-walled globe at the proximal end of a nephron that encloses the glomerulus.

Brachial plexus (BRĀ-kē-al PLEK-sus)   A network of nerve fibers of the anterior rami of spinal nerves C5, C6, C7, C8, and T1. The nerves that emerge from the brachial plexus supply the upper extremity.

Bradycardia (brad'-ē-KAR-dē-a)   Slow heart rate.

Brain   A mass of nerve tissue located in the cranial cavity.

Brain stem   The portion of the brain immediately superior to the spinal cord, made up of the medulla oblongata, pons varolii, and midbrain.

Broad ligament   A double fold of parietal peritoneum attaching the uterus to the side of the pelvic cavity.

Bronchi (BRONG-kē)   The two divisions of the trachea.

Bronchial tree (BRONG-kē-al)   The bronchi and their branching structures.

Bronchiectasis (brong'-kē-EK-tā-sis)   A chronic disorder in which there is a loss of the normal tissue and expansion of lung air passages; characterized by difficult breathing, coughing, expectoration of pus, and foul breath.

Bronchioles (BRONG-kē-ōlz)   The smaller divisions of the bronchi.

Bronchogenic carcinoma (brong'-kō-JEN-ik kar'-si-NŌ-ma)   Cancer originating in the bronchi.

Bronchogram (BRONG-kō-gram)   A roentgenogram of the lungs and bronchi.

Bronchopulmonary segment (brong'-kō-PUL-mō-ner'-ē)   One of the smaller divisions of a lobe of a lung supplied by its own branch of a bronchus.

Bronchoscope (BRONG-kō-skōp)   An instrument used to examine the interior of the large tubes (bronchi) of the lungs.

Bronchus (BRONG-kus)   One of the two large branches of the trachea. Plural form is bronchi (BRONG-kē).

Brunner's glands   Glands in the submucosa of the duodenum that secrete intestinal juice.

Buccal (BUK-al)   Pertaining to the cheek or mouth.

Bulbourethral glands (BUL-bō-yoo-RĒ-thral)   A pair of glands, also known as Cowper's glands, located inferior to the prostate on either side of the urethra that secrete an alkaline fluid into the cavernous urethra.

Bullae (BYOOL-ē)   Blisters beneath or within the epidermis.

Bundle branch   One of the two branches of the bundle of His or the atrioventricular bundle, made up of specialized muscle fibers that transmit electrical impulses to the ventricles.

Bunion (BUN-yun)   Inflammation and thickening of the bursa of a toe joint.

Burn   An injury caused by fire, steam, chemicals, electricity, lightning, or the ultraviolet rays of the sun.

Bursa (BUR-sa)   A sac or pouch of synovial fluid located at friction points, especially about joints.

Bursitis (bur-SĪ-tis)   Inflammation of a bursa.

Buttocks (BUT-oks)   The two fleshy masses on the posterior aspect of the lower trunk, formed by the gluteal muscles.

Cachexia (kah-KEK-sē-ah)   A state of ill health, malnutrition, and wasting.

Calcaneus (kal-KĀ-nē-us)   The heel bone.

Calcify (KAL-si-fī)   To harden by deposits of calcium salts.

Calcitonin (kal-si-TŌ-nin)   Hormone secreted by the parafollicular cells of the thyroid gland.

Calculus (KAL-kyoo-lus)   A stone formed within the body, as in the gallbladder, kidney, or urinary bladder.

Callus (KAL-lus)   A growth of new bone tissue in and around a fractured area, ultimately replaced by mature bone; an acquired, localized thickening

of the stratum corneum, as a result of continued physical trauma.

Calorie (KAL-o-rē)   A unit of heat. The small calorie is the standard unit and is the amount of heat necessary to raise 1 g of water 1° C. The large Calorie (kilocalorie) is used in metabolic and nutrition studies; it is the amount of heat necessary to raise 1 kg of water 1° C.

Calyx (KĀL-iks)   Any cuplike division of the kidney pelvis. Plural form is calyces (KĀ-li-sēz).

Canal (ka-NAL)   A narrow tube, channel, or passageway.

Canaliculus (kan'-a-LIK-yoo-lus)   A small channel or canal; a canal connecting lacunae in bone tissue. Plural form is canaliculi (kan'-a-LIK-yoo-lī).

Canal of Schlemm (shlem)   A circular venous sinus located at the junction of the sclera and the cornea through which aqueous humor drains from the anterior chamber of the eyeball into the blood.

Cancellous (KAN-sel-us)   Having a reticular or latticework structure, as in spongy tissue of bone.

Cancer (KAN-ser)   A malignant tumor of epithelial origin tending to infiltrate and give rise to new growths or metastases. Also called carcinoma (kar'-si-NŌ-ma).

Canthus (KAN-thus)   The angular junction of the eyelids at either corner of the eyes.

Capillary (KAP-i-lar'-ē)   A microscopic blood vessel located between an arteriole and venule.

Capitulum (ka-PIT-yoo-lum)   A rounded knob on the humerus.

Carbohydrate (kar'-bō-HĪ-drāt)   An organic compound containing carbon, hydrogen, and oxygen in a particular amount and arrangement and comprised of sugar subunits.

Carbuncle (KAR-bung-kal)   A hard, pus-filled, and painful inflammation involving the skin and underlying connective tissues; similar to boils but larger and more deeply rooted with several openings.

Carcinogen (kar-SIN-ō-jen)   Any substance that causes cancer.

Cardiac muscle   An organ specialized for contraction, composed of striated muscle cells, forming the walls of the heart, and stimulated by an intrinsic conduction system and visceral efferent neurons.

Cardiac notch (KAR-dē-ak)   An angular notch in the anterior border of the left lung.

Cardinal ligament   A ligament of the uterus, extending laterally from the cervix and vagina as a continuation of the broad ligament.

Caries (KĀR-ēz)   Decay of a tooth or bone.

Carina (ka-RĪ-na)   A ridge on the inside of the division of the right and left primary bronchi.

Carotene (KAR-o-tēn)   A yellow pigment in carrots,

tomatoes, egg yolk, and other substances that can be converted by the body into vitamin A.

Carotid (ka-ROT-id) The main artery in the neck extending to the head.

Carpal (KAR-pul) Pertaining to the carpus or wrist.

Cartilage (KAR-ti-lij) A specialized connective tissue forming part of the skeleton; gristle.

Cartilaginous joint (kar'-ti-LAJ-i-nus) A joint without a joint cavity where the articulating bones are held tightly together by cartilage, allowing little or no movement.

Caruncle (KAR-ung-kul) A small fleshy eminence, often abnormal.

Cast A solid mold of a part; a small mass of hardened material formed within a cavity in the body and then discharged from the body; can originate in different areas and be composed of various materials.

Castration (kas-TRĀ-shun) The removal of the testes or ovaries.

Cataract (KAT-a-rakt) A loss of transparency of the crystalline lens of the eye or its capsule or both.

Catheter (KATH-i-ter) A tube that can be inserted into a body cavity through a canal to remove fluids such as urine or blood; also, a tube inserted into a blood vessel through which opaque materials are injected into blood vessels or the heart for x-ray visualization.

Cauda equina (KAW-da ē-KWĪ-na) A tail-like collection of roots of spinal nerves at the inferior end of the spinal canal.

Caudal (KAW-dal) Pertaining to any tail-like structure; inferior in position.

Cecum (SĒ-kum) A blind pouch at the proximal end of the large intestine to which the ileum is attached.

Celiac (SĒ-lē-ak) Pertaining to the abdomen.

Celiac plexus (PLEK-sus) A large mass of ganglia and nerve fibers located at the level of the upper part of the first lumbar vertebra; the solar plexus.

Cell The basic, living, structural and functional unit of all organisms; the smallest structure capable of performing all the activities vital to life.

Cell inclusion A lifeless, often temporary, constituent in the cytoplasm of a cell as opposed to an organelle.

Cellulitis (sel-yoo-LĪ-tis) Inflammation of cellular or connective tissue, especially the subcutaneous tissue.

Cementum (se-MEN-tum) Calcified tissue covering the root of a tooth.

Center An area in the brain where a particular function is localized.

Center of ossification (os'-i-fi-KĀ-shun) An area in the cartilage model of a future bone where the cartilage cells hypertrophy, secrete enzymes that result in the calcification of their matrix resulting in the death of the cartilage cells, followed by the invasion of the area by osteoblasts that then lay down bone.

Central canal A microscopic tube running the length of the spinal cord in the gray commissure.

Central fovea (FŌ-vē-a) A cuplike depression in the center of the macula lutea of the retina, containing cones only; the area of clearest vision.

Central nervous system The brain and spinal cord.

Centrosome (SEN-tro-sōm) A rather dense area of cytoplasm, near the nucleus of a cell, containing a pair of centrioles.

Cephalic (se-FAL-ik) Pertaining to the head; superior in position.

Cerebellar peduncle (ser-e-BEL-ar pe-DUNG-kul) A bundle of nerve fibers connecting the cerebellum with the brain stem.

Cerebellum (ser-e-BEL-um) The portion of the hindbrain lying posterior to the medulla and pons, concerned with coordination of movements.

Cerebral aqueduct (SER-ē-bral AK-we-dukt) A channel through the midbrain connecting the third and fourth ventricles and containing cerebrospinal fluid; the aqueduct of Sylvius.

Cerebral peduncle (pe-DUNG-kul) One of a pair of nerve fiber bundles located on the ventral surface of the midbrain, conducting impulses between the pons and the cerebral hemispheres.

Cerebrospinal fluid (SER-ē-bro-SPĪ-nal) A fluid produced in the choroid plexuses of the ventricles of the brain.

Cerebrum (SER-ē-brum) The two hemispheres of the forebrain, making up the largest part of the brain.

Cerumen (se-ROO-min) Ear wax.

Cervical ganglion (SER-vi-kul GANG-glē-on) A cluster of nerve cell bodies of postganglionic sympathetic neurons, located in the neck, near the vertebral column.

Cervical plexus (PLEK-sus) A network of neuron fibers formed by the anterior rami of the first four cervical nerves.

Cervix (SER-viks) Neck; any constricted portion of an organ, especially the lower cylindrical part of the uterus.

Chalazion (ka-LĀ-zē-on) A small tumor of the eyelid.

Chemotherapy (kēm'-ō-THER-a-pē) The treatment of illness or disease by chemicals.

Chiasma (kī-AZ-ma) A crossing; especially the crossing of the optic nerve fibers.

Chiropractic (kī'-rō-PRAK-tik) A system of treating disease by manipulation of body parts, mostly the vertebral column.

Choana (KŌ-a-na) A funnel-shaped structure; the posterior opening of the nasal fossa, or internal naris.

Cholecystectomy (kō'-lē-sis-TEK-tō-mē) Surgical removal of the gallbladder.

Cholesterol (kō-LES-te-rol) An organic sterol found in many parts of the body.

Cholinergic (kō'-lin-ER-jik) Applied to nerve endings of the nervous system that liberate acetylcholine at a synapse.

Cholinergic fiber An axon of a neuron that liberates acetylcholine at a synapse.

Cholinesterase (kō'-lin-ES-ter-ās) An enzyme that hydrolyzes acetylcholine.

Chondrocytes (KON-drō-sīts) Cells of mature cartilage.

Chondroitin sulfate (kon-DROY-tin) An amorphous matrix material found outside the cell.

Chordae tendinae (KOR-dē TEN-di-nē) Cords that connect the heart valves with the papillary muscles.

Chorion (KŌ-rē-on) The outermost fetal membrane; serves a protective and nutritive function.

Choroid (KŌ-royd) One of the vascular coats of the eyes.

Choroid plexus (PLEK-sus) A vascular structure located in the roof of each of the four ventricles of the brain; produces cerebrospinal fluid.

Chromaffin cells (kro-MAF-in) Cells that have an affinity for chrome salts, due in part to the presence of the precursors of the chemical transmitter epinephrine; found, among other places, in the adrenal medulla. Also called pheochrome cells.

Chromatin (KRŌ-ma-tin) The threadlike mass of the genetic material consisting principally of DNA, which is present in the nucleus of a nondividing or interphase cell.

Chromatolysis (krō'-ma-TOL-i-sis) The disintegration and disappearance of the chromophil granules of a cell.

Chromophobe (KRŌ-mō-fōb) Glandular cell of the adenohypophysis.

Chromosome (KRŌ-mō-sōm) One of the 46 small, dark-staining bodies that appear in the nucleus of a human diploid cell during cell division.

Chronic (KRON-ik) Long-term; applied to a disease that is not acute.

Chyle (kīl) The milky fluid found in the lacteals of the small intestine after digestion.

Chylomicron (kī'-lō-MĪ-kron) A protein-coated lipid droplet.

Chyme (kīm) The semifluid mixture of partly digested food and digestive secretions found in the stomach and small intestine.

Cicatrix (SIK-a-triks) A scar left by a healed wound.

Ciliary (SIL-ē-ar'-ē) Pertaining to any hairlike processes; an eyelid, an eyelash.

Ciliary body One of the three portions of the vascular tunic of the eyeball, the others being the choroid

and the iris. The ciliary body includes the ciliary muscle and the ciliary processes.

Ciliary ganglion (GANG-glē-on) A very small parasympathetic ganglion whose preganglionic fibers come from the oculomotor nerve and whose postganglionic fibers carry impulses to the ciliary muscle and the sphincter muscle of the iris.

Cilium (SIL-ē-um) A hair or hairlike process. Cilia projecting from cells can be used to move the entire cell or can move substances along the surface of the cell.

Circadian (ser-KĀ-dē-an) Pertaining to a period of 24 hours.

Circle of Willis A ring of arteries forming an anastomosis at the base of the brain between the internal carotid and vertebral arteries and arteries supplying the brain.

Circumcision (ser'-kum-SIZH-un) Removal of the foreskin, the fold over the glans penis.

Circumduction (ser'-kum-DUK-shun) A movement at a synovial joint in which the distal end of a bone moves in a circle while the proximal end remains relatively stable.

Circumvallate papilla (ser'-kum-VAL-āt pa-PIL-a) One of the circular projections which are arranged in an inverted V-shaped row at the posterior portion of the tongue; the largest of the elevations on the upper surface of the tongue.

Cirrhosis (si-RŌ-sis) A liver disorder in which the parenchymal cells are destroyed and replaced by connective tissue.

Cisterna chyli (sis-TER-na KĪ-lē) The origin of the thoracic duct.

Climacteric (klī-mak-TER-ik) Cessation of the reproductive function in the female or diminution of testicular activity in the male.

Clitoris (KLI-to-ris) An erectile organ of the female that is homologous to the male penis.

Coccygeal (kok-SIJ-ē-al) Pertaining to the coccyx bone.

Coccygeal plexus (PLEK-sus) A network of nerves formed by the anterior rami of the coccygeal and the fourth and fifth sacral nerves. Fibers from the plexus supply the skin in the region of the coccyx.

Coccyx (KOK-six) The fused bones at the end of the vertebral column.

Cochlea (KŌK-lē-a) A winding, cone-shaped tube forming a portion of the inner ear and containing the organ of hearing.

Cochlear duct The membranous cochlea consisting of a spirally arranged tube enclosed in the bony cochlea and lying along its outer wall.

Coitus (KŌ-i-tus) Sexual intercourse; also coition or copulation.

Collagen (KOL-a-jen) A protein that is the main organic constituent of connective tissue.

Colliculus (ko-LIK-yoo-lus) A small elevation.

Colostomy (kō-LOS-tō-mē) The surgical creation of a new opening from the colon to the body surface.

Colostrum (kō-LOS-trum) The first milk secreted at the end of pregnancy.

Colposcope (KOL-pō-skōp) An instrument used in the examination of the vagina; a vaginal speculum.

Coma (KŌ-ma) Profound unconsciousness from which one cannot be roused.

Commissure (KOM-mi-shyoor) A joining together.

Common bile duct A tube formed by the union of the common hepatic duct and the cystic duct that empties bile into the duodenum at the ampulla of Vater.

Compact (dense) bone Bone tissue with no apparent spaces in which the layers or lamellae are fitted tightly together. Compact bone is found immediately deep to the periosteum and external to spongy bone.

Concha (KONG-ka) A scroll-like bone found in the skull. Plural form is conchae (KONG-kē).

Cone The light-sensitive receptor in the retina concerned with color vision.

Congenital (kon-JEN-i-tal) Present at the time of birth.

Conjunctiva (kon'-junk-TĪ-va) The delicate membrane covering the eyeball and lining the eyelids.

Connective tissue The most abundant of the four tissue types in the body, performing the functions of binding and supporting. There are relatively few cells and a great deal of intercellular substance.

Contractility (kon'-trak-TIL-i-tē) The ability to shorten.

Contralateral (kon'-tra-LAT-er-al) Situated on or pertaining to the opposite side.

Conus medullaris (KŌ-nus med-yoo-LAR-is) The tapering of the spinal cord below the lumbar enlargement.

Convergence (kon-VER-jens) The coordinated movement of the eyes so that both focus on a single set of objects.

Convoluted (CON-vō-loo-ted) Rolled together or coiled.

Coracoid (KOR-a-koyd) A bony process for muscle attachment.

Cornea (KOR-nē-a) The transparent fibrous coat that covers the iris of the eye.

Coronal plane (kō-RŌ-nal) A plane that runs vertical to the ground and divides the body into anterior and posterior portions. Also called frontal plane.

Coronary circulation (KOR-o-na-rē) The pathway followed by the blood from the ascending aorta through the blood vessels supplying the heart and returning to the right atrium.

Coronary sinus (SI-nus) A wide venous channel on the posterior surface of the heart that collects the blood from the coronary circulation and returns it to the right atrium.

Corpora quadrigemina (KOR-por-a kwad-ri-JEM-in-a) Four small elevations on the dorsal region of the midbrain concerned with visual and auditory functions.

Corpus (KOR-pus) The principal part of any organ; any mass or body.

Corpus callosum (KOR-pus ka-LŌ-sum) The great commissure of the brain between the cerebral hemispheres.

Corpus striatum (strī-Ā-tum) An area in the interior of each cerebral hemisphere composed of the caudate and lentiform nuclei of the basal ganglia and white matter of the internal capsule, arranged in a striated manner.

Cortex (KOR-teks) An outer layer of an organ; the convoluted layer of gray matter covering each cerebral hemisphere.

Costal (KOS-tal) Pertaining to a rib.

Cramp A spasmodic, especially a tonic, contraction of one or many muscles, usually painful.

Cranial nerve (KRĀ-nē-al) One of 12 pairs of nerves that leave the brain, pass through formina in the skull, and supply the head, neck, and part of the trunk. Each is designated by a Roman numeral and a name.

Craniosacral outflow (krā'-nē-ō-SĀ-kral) The fibers of parasympathetic preganglionic neurons, which have their cell bodies located in nuclei in the brain stem and in the lateral gray matter of the sacral portion of the spinal cord.

Craniotomy (krā'-nē-OT-ō-mē) Any operation on the skull, as for surgery on the brain or decompression of the fetal head in difficult labor.

Cranium (KRĀ-nē-um) The bones of the skull that enclose and protect the brain and the organs of sight, hearing, and balance.

Cremaster muscle (krē-MAS-ter) A small band of skeletal muscle that elevates the testes during sexual stimulation or exposure to cold.

Crenation (krē-NĀ-shun) The conversion of normally shaped red blood corpuscles into shrunken, knobbed, starry forms, as when blood is mixed with salt solution of 5 percent strength.

Cretinism (KRĒ-tin-izm) Severe congenital thyroid deficiency during childhood leading to physical and mental retardation.

Crista (KRIS-ta) A crestlike or ridged structure; for example, the crista galli, a triangular process projecting superiorly from the cribriform plate of the ethmoid bone.

Cryosurgery (KRĪ-ō-ser-jer-ē)   The destruction of tissue by application of extreme cold.

Cryptorchidism (krip-TOR-ki-dizm)   The condition of undescended testes.

Crypts of Lieberkühn (LĒ-ber-kyoon)   Simple tubular glands that open onto the surface of the intestinal mucosa and secrete digestive enzymes.

Cupula (KUP-yoo-la)   A mass of gelatinous material covering the hair cells of a crista, a receptor in the ampulla of a semicircular canal stimulated when the head moves.

Curvature (KUR-va-tūr)   A nonangular deviation of a straight line, as in the greater and lesser curvatures of the stomach. Abnormal curvatures of the vertebral column include kyphosis, lordosis, and scoliosis.

Cutaneous (kyoo-TĀ-nē-us)   Pertaining to the skin.

Cyanosis (sī'-a-NŌ-sis)   Slightly bluish or dark purple discoloration of the skin and the mucous membrane due to an oxygen deficiency.

Cystic duct (SIS-tik)   The duct that transports bile from the gallbladder to the common bile duct.

Cystitis (sis-TĪ-tis)   Inflammation of the urinary bladder.

Cystoscope (SIS-to-skōp)   An instrument used to examine the inside of the urinary bladder.

Cytokinesis (sī'-tō-ki-NĒ-sis)   Division of the cytoplasm.

Cytology (sī-TOL-o-jē)   The study of cells.

Cytopenia (sī-tō-PĒ-nē-a)   Reduction or lack of cellular elements in the circulating blood.

Cytoplasm (SĪ-tō-plazm)   The protoplasm of a cell, excluding the nucleus.

Dartos (DAR-tōs)   The contractile tissue under the skin of the scrotum.

Debility (dē-BIL-i-tē)   Weakness of tonicity in functions or organs of the body.

Decidua (dē-SID-yoo-a)   The membranous lining of the uterus that is shed after childbirth or at menstruation.

Deciduous (dē-SID-yoo-us)   Anything that is cast off at maturity; in the body, the first set of teeth.

Decubitus (dē-KYOO-be-tus)   The lying-down position; a decubitus ulcer is caused by pressure when a patient is confined to bed for a long period of time.

Decussation (dē'-kus-SĀ-shun)   Crossing.

Deep   Away from the surface of the body.

Deep fascia (FASH-ē-a)   A sheet of connective tissue wrapped around a muscle to hold it in place.

Defecation (def-e-KĀ-shun)   The discharge of excreta (feces) from the rectum.

Deglutition (dē-gloo-TISH-un)   The act of swallowing.

Dehydration (dē-hī-DRĀ-shun)   A condition due to excessive water loss from the body or its parts.

Demineralization (de-min'-er-al-i-ZĀ-shun)   The loss of calcium and phosphorus from bones.

Dendrite (DEN-drīt)   A nerve cell process carrying an impulse toward the cell body.

Dens (denz)   Tooth.

Denticulate (den-TIK-yoo-lāt)   Finely toothed or serrated; characterized by a series of small, pointed projections.

Dentine (DEN-tēn)   The osseous tissues of a tooth enclosing the pulp cavity.

Dentition (den-TI-shun)   The eruption of teeth; the number, shape, and arrangement of teeth. Also called teething.

Deoxyribonucleic acid (DNA) (dē-ok'-sē-rī'-bō-nyoo-KLĒ-ik)   One of the main nucleic acids found in all living systems.

Dermatology (der-ma-TOL-ō-jē)   The medical specialty dealing with diseases of the skin.

Dermatome (DER-ma-tōm)   An instrument for incising the skin or cutting thin transplants of skin; the cutaneous area developed from one embryonic spinal cord segment and receiving most of its innervation from one spinal nerve.

Dermis (DER-mis)   A layer of dense connective tissue lying deep to the epidermis; the true skin or corium.

Descending colon (KŌ-lon)   The part of the large intestine descending from the level of the lower end of the spleen on the left side to the level of the left iliac crest.

Detritus (de-TRI-tus)   Any broken-down or degenerative tissue or carious matter.

Detrusor muscle (de-TROO-ser)   Muscle in the wall of the urinary bladder.

Developmental anatomy   The study of development from the fertilized egg to the adult form. The branch of anatomy called embryology is generally restricted to the study of development from the fertilized egg through the eighth week in utero, or sometimes to the period just before birth.

Diabetogenic (dī'-a-bet'-ō-JEN-ik)   Producing diabetes.

Diagnosis (dī-ag-NŌ-sis)   Recognition of disease states from symptoms, inspection, palpation, posture, reflexes, general appearance, abnormalities, and other means.

Dialysis (dī-AL-i-sis)   The process of separating crystalloids (smaller particles) from colloids (larger particles) by the difference in their rates of diffusion through a semipermeable membrane.

Diapedesis (dī'-a-pe-DĒ-sis)   The passage of white blood cells through intact blood vessel walls.

Diaphragm (DĪ-a-fram)   Any partition that separates one area from another, especially the dome-shaped skeletal muscle between the thoracic and abdominal cavities.

Diaphysis (dī-AF-i-sis)   The shaft of a long bone.

Diarthrosis (dī'-ar-THRŌ-sis)   An articulation in which opposing bones move freely, as in a hinge joint.

Diastole (dī-AS-tō-lē)   The relaxing dilation period of the heart muscle, especially of the ventricles.

Diencephalon (dī'-en-SEF-a-lon)   A part of the brain consisting of the thalamus and the hypothalamus.

Differentiation (dif'-e-ren'-shē-Ā-shun)   Acquisition of specific functions different from those of the original general type.

Diffusion (dif-yoo-zhun)   A passive process in which there is a net or greater movement of molecules or ions from a region of high concentration to a region of low concentration until equilibrium is reached.

Digestion (di-JES-chun)   The mechanical and chemical breakdown of food to simple molecules that can be absorbed by the body.

Digitalis (dij-i-TAL-is)   The dried leaf of foxglove used in the treatment of heart disease.

Dilate (DĪ-lāte)   To expand or swell.

Diploë (DIP-lō-ē)   The spongy bone of the cranial bones.

Diploid (DIP-loyd)   Having the number of chromosomes characteristically found in the somatic cells of an organism.

Diplopia (di-PLŌ-pē-a)   Double vision.

Dissect (dī-SEKT)   To separate tissues and parts of a cadaver (corpse) or an organ for anatomical study.

Distal (DIS-tal)   Farthest from the center, from the medial line, or from the trunk.

Diuretic (dī'-yoo-RET-ik)   Any agent that increases the secretion or lack of absorption of urine.

Diurnal (dī-UR-nal)   Daily.

Diverticulum (dī'-ver-TIK-yoo-lum)   A sac or pouch in the walls of a canal or organ, especially in the colon.

Dorsal (DOR-sal)   Nearer to, located at, or referring to the back of the body; posterior.

Dorsal ramus (RĀ-mus)   A branch of a spinal nerve containing motor and sensory fibers supplying the muscles, skin, and bones of the posterior part of the head, neck, and trunk.

Dorsiflexion (dor'-si-FLEK-shun)   Flexion of the foot at the ankle.

Dropsy (DROP-sē)   A condition in which there is an abnormal accumulation of water in the tissues and cavities.

Duct of Santorini (san'-tor-Ē-nē)   An accessory duct of the pancreas that empties into the duodenum about 2.5 cm superior to the ampulla of Vater.

Ductus arteriosus (DUK-tus ar-tē-rē-Ō-sus)   A small vessel connecting the pulmonary trunk with the aorta; found only in the fetus.

Ductus deferens (DEF-er-ens)   The vas deferens or seminal duct that conducts spermatozoa from the epididymis to the ejaculatory duct.

Ductus venosus (DUK-tus ve-NO-sus)   A small vessel in the fetus that helps the circulation bypass the liver.

Duodenum (doo'-ō-DĒ-num)   The first portion of the small intestine.

Dura mater (DYOO-ra MĀ-ter)   The outer membrane covering the brain and spinal cord.

Dysentery (DIS-en-ter'-ē)   A painful disorder due to intestinal inflammation and accompanied by frequent, loose, bloody stools.

Dysfunction (dis-FUNK-shun)   Absence of complete, normal function.

Dyslexia (dis-LEK-sē-a)   Impairment of ability to comprehend written language.

Dysmenorrhea (dis'-men-o-RĒ-a)   Painful or difficult menstruation.

Dystrophia (dis-TRŌ-fē-a)   Progressive weakening of a muscle.

Ectopic (ek-TOP-ik)   Out of the normal location.

Eczema (EK-ze-ma)   A skin rash characterized by itching, swelling, blistering, oozing, and scaling of the skin.

Edema (e-DĒ-ma)   An abnormal accumulation of fluid in the body tissues.

Effector (e-FEK-tor)   The organ of the body, either a muscle or a gland, that responds to a motor neuron impulse.

Efferent arteriole (EF-er-ent ar-TĒ-rē-ōl)   A vessel of the renal vascular system that transports blood from the glomerulus to the peritubular capillary.

Efferent ducts   A series of coiled tubes that transport spermatozoa from the rete testis to the epididymis.

Efferent neuron (NOO-ron)   A motor neuron that conveys impulses from the brain and spinal cord to effectors which may be either muscles or glands.

Effusion (e-FYOO-zhun)   The escape of fluid from the lymphatics or blood vessels into a cavity or into tissues.

Ejaculatory duct (e-JAK-ū-la-tor'-ē)   A tube that transports spermatozoa from the ductus deferens (seminal duct) to the prostatic urethra.

Elasticity (e-las-TIS-i-tē)   The ability of muscle to return to its original shape after contraction or extension.

Elastin (e-LAS-tin)   A substance found in elastic fibers of connective tissue that gives elasticity to the skin and to the tissues that form the walls of blood vessels.

Electrocardiogram   (e-lek'-trō-KAR-dē-ō-gram)   A recording of the electrical changes that accompany the cardiac cycle.

Electroencephalogram   (e-lek'-trō-en-SEF-a-lō-gram)   A recording of the electrical impulses of the brain.

Eleidin (el-Ē-i-din)   A translucent substance found in the skin.

Ellipsoidal joint (e-lip-SOY-dal)   A synovial or diarthrotic joint structured so that an oval-shaped condyle of one bone fits into an elliptical cavity of another bone, permitting side-to-side and back-and-forth movements, as at the joint at the wrist between the radius and carpals.

Embolism (EM-bō-lizm)   Obstruction or closure of a vessel by a transported blood clot, a mass of bacteria, or other foreign material.

Embryo (EM-brē-ō)   The young of any organism in an early stage of development; in humans between the second and eighth weeks inclusively.

Embryology (em'-brē-OL-ō-jē)   The study of development from the fertilized egg through the eighth week in utero.

Emesis (EM-e-sis)   Vomiting.

Emmetropia (em'-e-TRŌ-pē-a)   The ideal optical condition of the eyes.

Emphysema (em'-fi-SĒ-ma)   A swelling or inflation of air passages with resulting stagnation of air in parts of the lungs; loss of elasticity in the alveoli.

Emulsification (ē-mul'-si-fi-KĀ-shun)   The dispersion of large fat globules in the intestine to smaller uniformly distributed particles.

Enamel (e-NAM-el)   The very hard, white substance covering the crown of a tooth.

Endocardium (en-dō-KAR-dē-um)   A thin layer of endothelium and smooth muscle that lines the inside of the heart and covers the valves and tendons that hold the valves open.

Endochondral ossification (en'-dō-KON-dral os'-i-fi-KĀ-shun)   The replacement of cartilage by bone.

Endocrine gland (EN-dō-krin)   A gland that secretes a hormone into the blood.

Endogenous (en-DOJ-en-us)   Growing from or beginning within the organism.

Endometrium (en'-dō-MĒ-trē-um)   The mucous membrane lining the uterus.

Endomysium (en'-dō-MIZ-ē-um)   Invaginations of the perimysium separating each individual muscle cell.

Endoneurium (en'-dō-NYOO-rē-um)   Connective tissue wrapping around individual nerve fibers.

Endoplasmic reticulum (en'-dō-PLAZ-mik re-TIK-yoo-lum)   A network of canals running through the cytoplasm of a cell. In the areas where ribosomes are attached to the outer surface of the ER, it is referred to as granular or rough reticulum; portions of the ER that have no ribosomes are called agranular or smooth reticulum.

Endorphin (en-DOR-fin)   A type of peptide occurring naturally in the central nervous system that acts as a pain killer.

Endoscope (EN-dō-skōp')   An instrument used to look inside hollow organs such as the stomach (gastroscope) or urinary bladder (cystoscope).

Endosteum (en-DOS-tē-um)   The membrane that lines the medullary cavity of bones.

Endothelium (en'-dō-THĒ-lē-um)   The layer of simple squamous epithelium that lines the cavities of the heart and blood and lymphatic vessels.

Enkephalin (en-KEF-a-lin)   A type of peptide occurring naturally in the central nervous system that acts as a pain killer.

Enuresis (en'-yoo-RĒ-sis)   Involuntary discharge of urine, complete or partial, after age 3.

Enzyme (EN-zīm)   A substance that affects the occurrence and speed of chemical changes; an organic catalyst, usually a protein.

Eosinophil (ē'-ō-SIN-ō-fil)   A type of white blood cell characterized by granular cytoplasm readily stained by eosin.

Epicardium (ep'-i-KAR-dē-um)   The thin outer layer of the heart; also called the visceral pericardium.

Epidemiology (ep'-i-DĒ-mē-ol'-ō-jē)   Medical science concerned with the occurrence and distribution of disease in human populations.

Epidermis (ep'-i-DERM-is)   The outermost layer of skin, composed of stratified squamous epithelium.

Epididymis (ep'-i-DID-i-mis)   An elongated tube along the posterior bladder of the testis in which sperm are stored. Plural form is epididymides (ep'-i-DID-i-mi-dēz').

Epidural space (ep'-i-DOO-ral)   A space between the spinal dura mater and the vertebral canal, containing loose connective tissue and a plexus of veins.

Epiglottis (ep'-i-GLOT-is)   A large, leaf-shaped piece of cartilage lying on top of the larynx, with its "stem" attached to the thyroid cartilage and its "leaf" portion unattached and free to move up and down to cover the glottis.

Epimysium (ep'-i-MIZ-ē-um)   Fibrous connective tissue around muscles.

Epineurium (ep'-i-NOO-rē-um)   The outermost covering around the entire nerve.

Epiphyseal line (ep'-i-FIZ-ē-al)   The remnant of the epiphyseal plate in a long bone.

Epiphyseal plate   The plate between the epiphysis and diaphysis.

Epiphysis (ē-PIF-i-sis)   The end of a long bone usually larger in diameter than the shaft (the diaphysis). Plural form is epiphyses (ē-PIF-i-sēz).

Epiphysis cerebri (se-RĒ-brī)   Pineal gland.

Episiotomy (ē-piz'-ē-OT-ō-mē)   Incision of the clinical perineum at the end of second stage of labor to avoid tearing.

Epistaxis (ep'-i-STAK-sis)   Hemorrhage from the nose; nosebleed.

Epithelial tissue (ep'-i-THĒ-lē-al)   The tissue that forms glands or the outer part of the skin and lines blood vessels, hollow organs, and passages that lead externally from the body.

Eponychium (ep'-ō-NIK-ē-um)   The horny embryonic structure from which the nail develops.

Eructation (e-ruk'-TĀ-shun)   The forceful expulsion of gas from the stomach; belching.

Erythema (er'-e-THĒ-ma)   Skin redness usually caused by engorgement of the capillaries in the lower layers of the skin.

Erythematosus (er-i'-them-a-TŌ-sis)   Pertaining to redness.

Erythrocyte (e-RITH-rō-sīt)   Red blood cell.

Erythropoiesis (e-rith'-rō-poy-Ē-sis)   The process by which erythrocytes are formed.

Erythropoietin (e-rith'-rō-POY-ē-tin)   A hormone secreted by the kidneys that stimulates red blood cell production.

Esophagus (e-SOF-a-gus)   A hollow muscular tube connecting the pharynx and the stomach.

Estrogen (ES-trō-jen)   Any substance that induces estrogenic activity or stimulates the development of secondary female characteristics; female hormones.

Etiology (ē'-tē-OL-ō-jē)   The study of the causes of disease, including theories of origin and the organisms, if any, involved.

Euphoria (yoo-FŌR-ē-a)   A subjectively pleasant feeling of well-being marked by confidence and assurance.

Eupnea (yoop-NĒ-a)   Normal quiet breathing.

Eustachian tube (yoo-STĀ-kē-an)   The tube that connects the middle ear with the nose and nasopharynx of the throat. Also called the auditory tube.

Euthanasia (yoo'-tha-NĀ-zhē-ā)   The proposed practice of ending a life in case of incurable disease.

Eversion (ē-VER-zhun)   Turning outward.

Exacerbation (eg-zas'-er-BĀ-shun)   An increase in the severity of symptoms of disease.

Excrement (EKS-kre-ment)   Material cast out from the body as waste, especially fecal matter.

Excretion (eks-KRĒ-shun)   The elimination of waste products from the body. It can refer to the expulsion of any matter—whether from a single cell or from the entire body—or to the matter excreted.

Exocrine gland (EK-sō-krin)   A gland that secretes its product into a duct or tube that empties out at the surface of the covering and lining epithelium.

Exogenous (ex-SOJ-e-nus)   Originating outside an organ or part.

Exophthalmos (ek'-sof-THAL-mus)   An abnormal protrusion or bulging of the eyeball.

Expiration (ek'-spi-RĀ-shun)   Breathing out or exhalation.

Extensibility (ek-sten'-si-BIL-i-tē)   The ability to be stretched when pulled.

Extension (ek-STEN-shun)   An increase in the anterior angle between two bones, except in extension of the knee and toes, in which the posterior angle is involved; restoring a body part to its anatomical position after flexion.

External   Located on or near the surface.

External auditory meatus (AW-di-tōr-ē mē-Ā-tus)   A canal in the temporal bone that leads to the middle ear.

External ear   The outer ear, consisting of the pinna, the external auditory canal, and the tympanic membrane or eardrum.

External nares (NA-rēs)   The external nostrils, or the openings into the nasal cavity on the exterior of the body.

Exteroceptor (eks'-ter-ō-SEP-tor)   A sense organ adapted for the reception of stimuli from outside the body.

Extracorporeal (eks'-tra-kor-PŌ-rē-al)   The circulation of blood outside of the body.

Extravasation (eks-trav-a-SĀ-shun)   The escape of fluid from a vessel into the tissues, especially blood, lymph, or serum.

Extrinsic (ek-STRIN-sik)   Of external origin.

Eyebrow   The hairy ridge above the eye.

Exudate (EKS-yoo-dāt)   Escaping fluid or semifluid material that oozes from a space that may contain serum, pus, and cellular debris.

Face   The anterior aspect of the head.

Facilitated diffusion (fah-SIL-i-tā-ted dif-YOO-zhun)   Diffusion in which a substance not soluble by itself in lipids is transported across a semipermeable membrane by combining with a carrier substance.

Falciform ligament (FAL-si-form LIG-a-ment)   A sheet of parietal peritoneum between the two principal lobes of the liver. The ligamentum teres, or remnant of the umbilical vein, lies within its fold.

False vocal folds   A pair of folds of the mucous membrane of the larynx superior to the true vocal cords.

Falx cerebelli (falks ser'-e-BEL-lē)   A small triangular process of the dura mater attached to the occipital bone in the posterior cranial fossa and projecting inward between the two cerebellar hemispheres.

Falx cerebri (SER-e-brē)   A fold of the dura mater

extending down into the longitudinal fissure between the two cerebral hemispheres.

Fascia (FASH-ē-a) A fibrous membrane covering, supporting, and separating muscles.

Fascicle (FAS-i-kul) A small bundle or cluster, especially of nerve or muscle fibers. Plural form is fasciculi (fa-SIK-yoo-lī).

Fauces (FAW-sēs) The opening from the mouth and the pharynx.

Febrile (FĒ-bril) Feverish; pertaining to a fever.

Feces (FĒ-sēz) Material discharged from the bowel and made up of bacteria, secretions, and food residue; stool.

Fenestra cochlea (fe-NES-tra KŌK-lē-a) The round window of the middle ear.

Fenestration (fen-e-STRĀ-shun) Surgical opening made into the labyrinth of the ear for some conditions of deafness.

Fenestra vestibuli (fe-NES-tra ves-TIB-yoo-lī) The oval window of the middle ear.

Fetal circulation (FĒ-tal) The circulatory system of the fetus, including the placenta and special blood vessels involved in the exchange of materials between fetus and mother.

Fetus (FĒ-tus) The latter stages of the developing young of an animal; in human beings, the organism in utero from the third month to birth.

Fibrillation (fi-bre-LĀ-shun) Irregular twitching of individual muscle cells (fibers) or small groups of muscle fibers preventing effective action by an organ or muscle.

Fibroblast (FĪ-brō-blast) A flat, long connective tissue cell that forms the fibrous tissues of the body.

Fibrocyte (FĪ-brō-sīt) A mature fibroblast, found in the fibroelastic connective tissues of the body.

Fibrosis (fi-BRŌ-sis) Abnormal formation of fibrous tissue.

Fibrous joint (FĪ-brus) A joint that allows little or no movement, such as a suture and syndesmosis.

Fibrous tunic (TOO-nik) The outer coat of the eyeball, made up of the posterior sclera and the anterior cornea.

Fight-or-flight response The effect of the stimulation of the sympathetic division of the autonomic nervous system.

Filiform papilla (FIL-i-form pa-PIL-a) One of the conical projections which are distributed in parallel rows over the anterior two-thirds of the tongue and contain no taste buds.

Filtration (fil-TRĀ-shun) The passage of a liquid through a filter or membrane that acts like a filter.

Filum terminale (FĪ-lum ter-mi-NAL-ē) Nonnervous fibrous tissue of the spinal cord that extends inferiorly from the conus medullaris to the coccyx.

Fimbriae (FIM-brē-ē) Fringelike structures, especially the lateral ends of the uterine tubes (oviducts).

Fissure (FISH-ur) A groove, fold, or slit that may be normal or abnormal.

Fistula (FIS-choo-la) An abnormal passage between two organs or between an organ cavity and the outside.

Flaccid (FLAK-sid) Relaxed, flabby, or soft; lacking muscle tone.

Flagellum (fla-JEL-um) A hairlike, motile process on the extremity of a bacterium or protozoon. Plural form is flagella.

Flatfoot A condition in which the ligaments and tendons of the arches of the foot are weakened and the height of the longitudinal arch decreases.

Flatus (FLĀ-tus) Gas or air in the digestive tract; commonly used to denote passage of gas rectally.

Flexion (FLEK-shun) A folding movement in which there is a decrease in the angle between two bones anteriorly, except in flexion of the knee and toes, in which the bones are approximated posteriorly.

Fluoroscope (FLOOR-o-skōp) An instrument for visual observation of the body (heart, bowels) by means of x-ray.

Follicle (FOL-i-kul) A small secretory sac or cavity.

Fontanel (fon'-ta-NEL) A soft area in a baby's skull; a membrane-covered spot where bone formation has not yet occurred.

Foot The terminal part of the lower extremity.

Foramen (fō-RĀ-men) A passage or opening; a communication between two cavities of an organ or a hole in a bone for passage of vessels or nerves.

Foramen ovale (ō-VAL-ē) An opening in the fetal heart in the septum between the right and left atria; a hole in the greater wing of the sphenoid bone that transmits the mandibular branch of the trigeminal nerve (V).

Forearm (FOR-arm) The part of the upper extremity between the elbow and the wrist.

Fornix (FOR-niks) An arch or fold; a tract in the brain made up of association fibers, connecting the hippocampus with the mammillary bodies; a recess around the cervix of the uterus where it protrudes into the vagina.

Fourth ventricle (VEN-tri-kul) A cavity within the brain lying between the cerebellum and the medulla and pons.

Fovea (FŌ-vē-a) A pit or cuplike depression.

Fracture (FRAK-chur) Any break in a bone.

Frenulum (FREN-yoo-lum) A small fold of mucous membrane that connects two parts and limits movement.

Frontal plane A plane that runs vertical to the ground and divides the body into anterior and posterior portions. Also called coronal plane.

Fulminate (FUL-mi-nāt') To occur suddenly with great intensity.

Fundus (FUN-dus) The part of a hollow organ farthest from the opening.

Fungiform papilla (FUN-ji-form pa-PIL-a) A mushroomlike elevation on the upper surface of the tongue appearing as a red dot. Most fungiform papillae contain taste buds.

Furuncle (FYOOR-ung-kul) A boil; painful nodule caused by bacteria infection and inflammation of a hair follicle or oil gland.

Gamete (GAM-ēt) A male or female reproductive cell; the spermatozoon or ovum.

Ganglion (GANG-glē-on) A group of nerve cell bodies that lie outside the central nervous system. Plural form is ganglia (GANG-glē-a)

Gangrene (GANG-grēn) Death of tissue accompanied by bacterial invasion and putrefaction; usually due to blood vessel obstruction.

Gastrointestinal tract (gas'-tro-in-TES-ti-nal) The portion of the digestive tract from the cardia of the stomach to the anus.

Gavage (ga-VAZH) Feeding through a tube passed through the esophagus and into the stomach.

Gene (jēn) One of the biological units of heredity; an ultramicroscopic, self-reproducing DNA particle located in a definite position on a particular chromosome.

Genitalia (jen'-i-TĀL-ya) Reproductive organs.

Genotype (JĒ-nō-tīp) The total hereditary information carried by an individual; the genetic makeup of an organism.

Germinativum (jer'-mi-na-TĒ-vum) Skin layers where new cells are germinated.

Gerontology (jer'-on-TOL-o-jē) The study of old age.

Gestation (jes-TĀ-shun) The period of intrauterine fetal development.

Gingivitis (jin'-je-VĪ-tis) Inflammation of the gums.

Gland A secretory structure.

Glans penis (glanz PĒ-nis) The slightly enlarged region at the distal end of the penis.

Glaucoma (glow-KŌ-ma) An eye disorder in which there is increased pressure due to an excess of fluid within the eye.

Gliding joint A synovial or diarthrotic joint having articulating surfaces that are usually flat, permitting only side-to-side and back-and-forth movements, as between carpal bones, tarsal bones, and the scapula and clavicle.

Glomerulus (glō-MER-yoo-lus) A rounded mass of nerves or blood vessels, especially the microscopic tuft of capillaries that is surrounded by the expanded part of each kidney tubule.

Glottis (GLOT-is) The air passageway between the vocal folds in the larynx.

Glucagon (GLOO-ka-gon) A hormone that increases the blood glucose level.

Glucocorticoids (gloo-kō-KOR-ti-koyds) A group of hormones of the adrenal cortex.

Glycosuria (glī'-kō-SOO-rē-a) Presence of glucose in the urine.

Gnostic area (NOS-tik) Sensory area of the cerebral cortex that receives and integrates sensory input from various parts of the brain so that a common thought can be formed.

Goblet cell A goblet-shaped unicellular gland that secretes mucus.

Goiter (GOY-ter) An enlargement of the thyroid gland.

Golgi complex (GOL-jē) An organelle in the cytoplasm of cells consisting of four to eight flattened channels, stacked upon one another, with expanded areas at their ends.

Golgi tendon organ A receptor found chiefly near the junction of tendons and muscles.

Gomphosis (gom-FŌ-sis) A fibrous joint in which a cone-shaped peg fits into a socket.

Gonad (GŌ-nad) A gland that produces gametes and sex hormones; ovary and testis.

Gonadocorticoids (gō-na-dō-KOR-ti-koyds) Sex hormones secreted by the adrenal cortex.

Gonadotropic (gō'-nad-ō-TRŌ-pik) Hormones that regulate the functions of the gonads.

Gradient (GRĀ-dē-ent) A slope or gradation in the body; difference in concentration or electrical charges across a semipermeable membrane.

Gray commissure (KOM-i-shur) A narrow strip of gray matter connecting the two lateral gray masses within the spinal cord.

Gray matter Areas in the central nervous system consisting of nonmyelinated nerve tissue.

Gray ramus communicans (RĀ-mus ko-MYOO-ni-kans) A short nerve containing postganglionic sympathetic fibers; the cell bodies of the fibers are in a sympathetic chain ganglion, and the nonmyelinated axons run by way of the gray ramus to a spinal nerve and then to the periphery to supply smooth muscle in blood vessels, arrector pili muscles, and sweat glands. Plural form is rami communicantes (RĀ-mē kō-myoo-ni-KAN-tēz).

Greater omentum (ō-MEN-tum) A large fold in the serosa of the stomach that hangs down like an apron over the front of the intestines.

Greater vestibular glands (ves-TIB-yoo-lar) A pair of glands on either side of the vaginal orifice that open by a duct into the space between the hymen and the labia minora. Also called Bartholin's glands.

These glands are homologous to the male Cowper's or bulbourethral glands.

Groin (groyn) The depression between the thigh and the trunk; the inguinal region.

Gross anatomy The branch of anatomy that deals with structures that can be studied without using a microscope.

Gustatory (GUS-ta-tō'-rē) Pertaining to taste.

Gynecology (gī'-ne-KOL-o-jē) The branch of medicine dealing with the study and treatment of disorders of the female reproductive system.

Gyrus (JĪ-rus) One of the convolutions of the cerebral cortex region of the brain. Plural form is gyri (JĪ-rī).

Hair A threadlike structure produced by the specialized epidermal structure developing from a papilla sunk in the dermis.

Hand The terminal portion of an upper extremity, including the carpus, metacarpus, and phalanges.

Haploid (HAP-loyd) Having half the number of chromosomes characteristically found in the somatic cells of an organism; characteristic of mature gametes.

Hard palate (PAL-at) The anterior portion of the roof of the mouth, formed by the maxillae and palatine bones and lined by mucous membrane.

Haustra (HAWS-tra) The sacculated elevations of the colon.

Haversian canal (ha-VER-shun) A circular channel running longitudinally in the center of a Haversian system of mature compact bone, containing blood and lymph vessels and nerves.

Haversian system The basic unit of structure in adult compact bone, consisting of a Haversian canal with its concentrically arranged lamellae, lacunae, osteocytes, and canaliculi. Also called an osteon.

Head The superior part of a human being, cephalic to the neck; the superior or proximal part of a structure.

Heart A hollow muscular organ lying slightly to the left of the midline of the chest that pumps the blood through the cardiovascular system.

Heart block An arrhythmia of the heart in which the atria and ventricles contract independently because of a blocking of electrical impulses through the heart at a critical point in the conduction system.

Hematology (hē'-ma-TOL-o-jē) The study of the blood.

Hematoma (hēm'-a-TŌ-ma) A tumor or swelling filled with blood.

Hematopoiesis (hem'-a-tō-poy-Ē-sis) Blood cell production occurring in the red marrow of bones; also called hemopoiesis (hē-mō-poy-Ē-sis).

Hematuria (hē'-ma-TOOR-ē-a) Presence of blood in the urine.

Hemiballismus (hem'-i-ba-LIZ-mus) Violent muscular restlessness of half of the body, especially of the upper extremity.

Hemocytometer (hē-mō-sī-TOM-e-ter) An instrument used in counting blood cells.

Hemodialysis (hē'-mō-dī-AL-i-sis) Filtering of the blood by means of an artificial device so that certain substances are removed from the blood as a result of the difference in rates of their diffusion through a semipermeable membrane while the blood is being circulated outside the body.

Hemoglobin (hē'-mō-GLŌ-bin) A substance in erythrocytes consisting of the protein globin and the iron-containing, red pigment heme and constituting about 33 percent of the cell volume. It is involved in the transport of oxygen and carbon dioxide.

Hemolysis (hē-MOL-i-sis) The rupture of red blood cells with release of hemoglobin into the plasma.

Hemophilia (hē'-mō-FĒL-ē-a) A hereditary blood disorder where there is a deficient production of certain clotting factors, resulting in excessive bleeding into joints, deep tissues, and elsewhere.

Hemorrhoids (HEM-ō-royds) Dilated or varicosed blood vessels (usually veins) in the anal region. Also called piles.

Hemostasis (hē-MŌ-stā-sis) The stoppage of bleeding.

Hemostat (HĒ-mō-stat) An agent or instrument used to prevent the flow or escape of blood.

Hepatic duct (he-PAT-ik) A duct that receives bile from the bile capillaries. Small hepatic ducts merge to form the larger right and left hepatic ducts that unite to leave the liver as the common hepatic duct.

Hepatic portal circulation (POR-tal) The flow of blood from the digestive organs to the liver before returning to the heart.

Hernia (HER-nē-a) The protrusion or projection of an organ or part of an organ through the wall of the cavity containing it.

Hiatus (hī-Ā-tus) An opening; a foramen.

Hilum (HĪ-lum) An area, depression, or pit where blood vessels and nerves enter or leave the organ.

Hinge joint A synovial or diarthrotic joint characterized by a convex surface of one bone that fits into a concave surface of another bone, such as the elbow, knee, ankle, and interphalangeal joints.

Histiocyte (HISS-tē-ō-sīt) A large phagocytic cell (macrophage) found in loose connective tissue and derived from the reticuloendothelial system.

Histology (his-TOL-o-jē) Microscopic study of the structure of tissues.

Holocrine gland (HŌL-ō-krin) A type of gland in which the entire secreting cell, along with its accumulated secretions, makes up the secretory product of the gland, as in the sebaceous glands.

Homeostasis (hō'-mē-ō-STĀ-sis) The state in which the body's internal environment remains relatively constant within a limited range around optimum conditions.

Homogeneous (hō'-mō-JĒ-nē-us) Having similar or the same consistency and composition throughout.

Homologous (hō-MOL-ō-gus) Similar in structure and origin, but not necessarily in function.

Horizontal plane A plane that runs parallel to the ground and divides the body into superior and inferior portions. Also called transverse plane.

Hormone (HŌR-mōn) A secretion of an endocrine or ductless gland secreted into the blood for transport.

Horn Principal area of gray matter in the spinal cord.

Hyaluronic acid (hīl'-yoo-RON-ik) An amorphous matrix material found outside the cell.

Hyaluronidase (hīl'-yoo-RON-i-dās) An enzyme that breaks down hyaluronic acid, increasing the permeability of connective tissues by dissolving the substances that hold body cells together.

Hydrocele (HĪ-drō-sēl) A fluid-containing sac or tumor; specifically, a collection of fluid formed in the space along the spermatic cord and in the scrotum.

Hydrocephalus (hī-drō-SEF-a-lus) Abnormal accumulation of water on the brain.

Hydrophobia (hī'-drō-FŌ-bē-a) Rabies; a condition characterized by severe muscle spasms when attempting to drink water. Also, an abnormal fear of water.

Hydrostatic (hī'-drō-STAT-ik) Pertaining to the pressure of liquids in equilibrium and that exerted on liquids.

Hymen (HĪ-men) A thin fold of vascularized mucous membrane at the inferior end of the vaginal opening.

Hypercapnia (hī'-per-KAP-nē-a) An abnormal amount of carbon dioxide in the blood.

Hyperemia (hī'-per-Ē-mē-a) An excess of blood in an area or part of the body.

Hyperextension (hī'-per-ek-STEN-shun) Continuation of extension beyond the anatomic position, as in bending the head backward.

Hyperglycemia (hī'-per-glī-SĒ-mē-a) An elevated blood sugar level.

Hypermetropia (hī'-per-mē-TRŌ-pē-a) A condition in which visual images are focused behind the retina with resulting defective vision of near objects; farsightedness.

Hyperplasia (hī'-per-PLĀ-zē-a) An abnormal increase in the number of normal cells in a tissue or organ, increasing its size.

Hypersecretion (hī'-per-se-KRĒ-shun) Overactivity of glands resulting in excessive secretion.

Hypertension (hī'-per-TEN-shun) High blood pressure.

Hypertonic (hī'-per-TON-ik) Having an osmotic pressure greater than that of a solution with which it is compared.

Hypertrophy (hī-PER-trō-fē) An excessive enlargement or overgrowth of an organ or part.

Hyponychium (hī'-pō-NIK-ē-um) Free edge of the fingernail.

Hypophysis (hī-POF-i-sis) Pituitary gland.

Hyposecretion (hī'-pō-se-KRĒ-shun) Underactivity of glands resulting in diminished secretion.

Hypothalamic-hypophyseal tract (hī'-pō-thal-AM-ik–hī'-po-FIZ-ē-al) A bundle of nerve processes made up of fibers that have their cell bodies in the hypothalamus but release their neurosecretions in the posterior pituitary gland or neurohypophysis.

Hypothalamus (hī'-pō-THAL-a-mus) A portion of the forebrain, lying beneath the thalamus and forming the floor and part of the walls of the third ventricle.

Hypothermia (hī-pō-THER-mē-a) Body cooling.

Hypotonic (hī'-pō-TON-ik) Having an osmotic pressure lower than that of a solution with which it is compared.

Hypoxia (hī-POKS-ē-a) Lack of adequate oxygen. Also called anoxia.

Ileocecal valve (il'-ē-o-SĒ-kal) A fold of mucous membrane that guards the opening from the ileum into the large intestine.

Ileum (IL-ē-um) The final portion of the small intestine.

Immunity (i-MYOON-i-tē) The state of being resistant to injury, particularly by poisons, foreign proteins, and invading parasites.

Imperforate (im-PER-fō-rāt) Abnormally closed.

Impetigo (im'-pe-TĪ-gō) A contagious skin disorder characterized by pustular eruptions.

Implantation (im-plan-TĀ-shun) The insertion of a tissue or a part into a part of the body; the attachment of the preembryonic ball of cells (blastula) into the lining of the uterus.

Impotence (IM-pō-tens) Weakness; inability to copulate; failure to maintain an erection.

Inclusion A secretion or storage area in the cytoplasm of a cell.

Incontinence (in-KON-ti-nens) Inability to retain urine, semen, or feces, through loss of sphincter control.

Infarction (in-FARK-shun) The presence of a localized area of necrotic tissue, produced by inadequate oxygenation of the tissue.

Inferior Away from the head or toward the lower part of a structure.

Infraspinatous (in'-fra-SPĪ-na-tus) Bony process found below the spine of the scapula used for muscle attachment.

Infundibulum (in'-fun-DIB-yoo-lum) The stalklike structure that attaches the pituitary gland or hypophysis to the hypothalamus of the brain.

Inguinal (IN-gwi-nal) Pertaining to the groin.

Insertion (in-SER-shun) The manner or place of attachment of a muscle to the bone that it moves.

In situ (in SĪ-too) In position.

Inspiration (in'-spi-RĀ-shun) The drawing of air into the lungs.

Insula (IN-su-la) A triangular area of cerebral cortex that lies deep within the lateral cerebral fissure, under the parietal, frontal, and temporal lobes, and cannot be seen in an external view of the brain. Also called the island or isle of Reil.

Insulin (IN-su-lin) A hormone that decreases the blood glucose level.

Integument (in-TEG-yoo-ment) A covering, especially the skin.

Intercalated disc (in-TER-ka-lāt-ed) An irregular transverse thickening in cardiac muscle.

Intercellular (in'-ter-SEL-yoo-lar) Between the cells of a structure.

Intercostal nerve (in'-ter-KOS-tal) A nerve supplying a muscle located between the ribs.

Internal Away from the surface of the body.

Internal capsule A thick sheet of white matter made up of myelinated fibers connecting various parts of the cerebral cortex and lying between the thalamus and the caudate and lentiform nuclei of the basal ganglia.

Internal ear The inner ear or labyrinth, lying inside the temporal bone, containing the organs of hearing and balance.

Internal nares (NA-rēs) The two openings posterior to the nasal cavities opening into the nasopharynx; the choanae.

Interphase (IN-ter-fāz) The period during its life cycle when a cell is carrying on every life process except division; the stage between two mitotic divisions.

Interstitial cell of Leydig (in'-ter-STISH-al . . . LĪ-dig) A secretory cell that secretes testosterone, located in the connective tissue between seminiferous tubules in a mature testis.

Interstitial fluid The fluid filling the microscopic spaces between the cells of tissues.

Interstitial (endogenous) growth Growth from within, as in the growth of cartilage.

Interventricular foramen (in'-ter-ven-TRIK-yoo-lar) A narrow, oval opening through which the lateral ventricles of the brain communicate with the third ventricle. Also called the foramen of Monro.

Intervertebral disc (in'-ter-VER-te-bral) A pad of fibrocartilage located between the bodies of two vertebrae.

Intracellular (in'-tra-SEL-yoo-lar) Within cells.

Intramembranous ossification (in'-tra-MEM-bra-nus os'-i-fī-KĀ-shun) The method of bone formation in which the bone is formed directly in membranous tissue.

Intrinsic (in-TRIN-sik) Of internal origin; for example, the intrinsic factor, a mucoprotein formed by the gastric mucosa that is necessary for the absorption of vitamin B_{12}.

Intubation (in'-too-BĀ-shun) Insertion of a tube into the larynx through the glottis for entrance of air or to dilate a stricture.

Intussusception (in'-ta-sa-SEP-shun) The infolding (invagination) of one part of the intestine within another segment.

In utero (in YOO-ter-ō) Within the uterus.

Invagination (in-vaj'-i-NĀ-shun) The pushing of the wall of a cavity into the cavity itself.

Inversion (in-VER-zhun) The movement of the sole inward at the ankle joint.

In vitro (in VĒ-trō) In a glass, as in a test tube.

In vivo (in VĒ-vō) In the living body.

Ipsilateral (ip'-si-LAT-er-al) On the same side, affecting the same side of the body.

Irritability (ir'-i-ta-BIL-i-tē) The ability to receive and respond to stimuli.

Ischemia (is-KĒ-mē-a) A lack of sufficient blood to a part due to obstruction of circulation.

Islet of Langerhans (Ī-let of LAHNG-er-hanz) A cluster of endocrine gland cells in the pancreas that secretes insulin and glucagon.

Isotonic (ī'-sō-TON-ik) Having equal tension or tone; having the same osmotic pressure as a solution with which it is compared.

Isthmus (IS-mus) A narrow strip of tissue or narrow passage connecting two larger parts.

Jaundice (JAWN-dis) A condition characterized by yellowness of skin, white of the eyes, mucous membranes, and body fluids.

Jejunum (je-JOO-num) The middle portion of the small intestine.

Joint capsule A sleevelike, fibrous cuff lined with a synovial membrane that encases the articulating bones of a diarthrotic or synovial joint.

Joint kinesthetic receptor (kin'-es-THET-ik) A proprioceptive receptor located in a joint, stimulated by joint movement.

Karyotype (KAR-ē-ō-tīp) Chromosome characteristics of an individual or a group of cells.

Keratin (KER-a-tin) A special insoluble protein

found in the hair, nails, and other horny tissues of the epidermis.

Keratohyalin (ker'-a-tō-HĪ-a-lin) A compound involved in the formation of keratin.

Kidney (KID-nē) One of the paired reddish organs located in the lumbar region that secrete urine.

Kilogram (KIL-o-gram) Equivalent to 1,000 g; about 2.2 lb avoirdupois.

Kinesiology (ki-nē'-sē-OL-ō-jē) The study of the movement of body parts.

Kinesthesia (kin-is-THĒ-szē-a) Ability to perceive extent, direction, or weight of movement; muscle sense.

Krebs cycle The citric acid cycle; a series of energy-yielding steps in the catabolism of carbohydrates.

Kupffer cell (KOOP-fer) Special phagocytic cell of the liver.

Kyphosis (kī-FŌ-sis) An increased curvature of the chest giving a hunchback appearance.

Labia majora (LĀ-bē-a ma-JOR-a) Two longitudinal folds of skin extending downward and backward from the mons pubis of the female.

Labia minora (min-OR-a) Two small folds of skin lying between the labia majora of the female.

Labial frenulum (LĀ-bē-al FREN-yoo-lum) A medial fold of mucous membrane between the inner surface of the lip and the gums.

Labium (LĀ-bē-um) A lip; a liplike structure. Plural form is labia (LĀ-bē-a).

Labyrinth (LAB-i-rinth) Intricate communicating passageways, especially the internal ear.

Lacrimal (LAK-ri-mal) Pertaining to tears.

Lacrimal canal A duct, one on each eyelid, commencing at the punctum at the medial margin of an eyelid and conveying the tears medially into the nasolacrimal sac.

Lacrimal gland Secretory cells located at the superior lateral portion of each orbit that secrete tears into excretory ducts that open onto the surface of the conjunctiva.

Lacrimal sac The superior expanded portion of the nasolacrimal duct that receives the tears from a lacrimal canal.

Lactation (lak-TĀ-shun) The period of milk secretion and release; suckling in mammals.

Lacteal (LAK-tē-al) Related to milk; one of many intestinal lymph vessels that absorb fat from digested food.

Lacuna (la-KOO-na) A small, hollow space, such as that found in bones, in which the osteoblasts lie. Plural form is lacunae (la-KOO-nē).

Lambdoidal suture (lam-DOY-dal) The line of union in the skull between the parietal bones and the occipital bone.

Lamellae (la-MEL-ē) Concentric rings found in compact bone.

Lamina (LAM-i-na) A thin, flat layer or membrane, as the flattened part of either side of the arch of a vertebra. Plural form is laminae (LAM-i-nē).

Lamina propria (PRO-prē-a) The connective tissue layer of a mucous membrane.

Langerhans (LANG-er-hanz), **Islet of** A cluster of cells that forms the endocrine portion of the pancreas.

Lanugo (lan-YOO-gō) Fine downy hairs that cover the fetus.

Large intestine The portion of the digestive tract extending from the ileum of the small intestine to the anus, divided structurally into the cecum, colon, rectum, and anal canal.

Laryngopharynx (la-rin'-gō-FAR-inks) The inferior portion of the pharynx, extending downward from the level of the hyoid bone to empty posteriorly into the esophagus and anteriorly into the larynx.

Larynx (LAR-inks) The voice box, a short passageway that connects the pharynx with the trachea.

Laryngoscope (la-RINJ-ō-skōp) An instrument for examining the larynx.

Lateral (LAT-er-al) To the side.

Lateral apertures (AP-er-choorz) Two of the three openings in the roof of the fourth ventricle through which cerebrospinal fluid enters the subarachnoid space of the brain and spinal cord. Also known as the foramina of Luschka.

Lateral ventricle (VEN-tri-kul) A cavity within a cerebral hemisphere that communicates with the lateral ventricle in the other cerebral hemisphere and with the third ventricle by way of the interventricular foramen. Cerebrospinal fluid flows through the ventricular system.

Leg The part of the lower extremity between the knee and the ankle.

Lens A transparent organ lying posterior to the pupil and iris of the eyeball and anterior to the vitreous humor.

Lesion (LĒ-zhun) Any local, diseased change in tissue formation.

Lesser omentum (ō-MEN-tum) A fold of the peritoneum that extends from the liver to the lesser curvature of the stomach and the commencement of the duodenum.

Lesser vestibular gland (ves-TIB-yoo-lar) One of the paired mucus-secreting glands that have ducts that open on either side of the urethral orifice in the vestibule of the female. This gland is the female homologue of the male prostate gland. Also known as Skene's gland.

Leucocyte, leukocyte (LOO-kō-sīt) A white blood cell.

Leukemia (loo-KĒ-mē-a)   A cancerlike disease of the blood-forming organs characterized by a rapid and abnormal increase in the number of white blood cells and the presence of many immature cells in the circulating blood.

Leukocytosis (loo-kō-sī-TŌ-sis)   An increase in the number of white blood cells, characteristic of many infections and other disorders.

Leukopenia (loo-kō-PĒ-nē-a)   A decrease of the number of white blood cells below 5,000 per cubic millimeter.

Leukoplakia (loo-kō-PLĀ-kē-a)   A disorder in which there are white patches in the mucous membranes of the tongue, gums, and cheeks.

Libido (li-BĒ-dō)   The sexual drive, conscious or unconscious.

Ligament (LIG-a-ment)   Collagenous connective tissue, with numerous collagen molecules arranged parallel to one another, that attaches muscles to bones.

Limbic system   A portion of the forebrain, sometimes termed the visceral brain, concerned with various aspects of emotion and behavior, that includes the limbic lobe (hippocampus and associated areas of gray matter plus the cingulate gyrus), certain parts of the temporal and frontal cortex, some thalamic and hypothalamic nuclei, and parts of the basal ganglia.

Lingual frenulum (LIN-gwal FREN-yoo-lum)   A fold of mucous membrane that connects the tongue to the floor of the mouth.

Lipoma (li-PŌ-ma)   A fatty tissue tumor, usually benign.

Liter (LĒT-er)   Volume occupied by 1 kg of water at standard atmosphere pressure; equivalent to 1.057 qt.

Lobe (lōb)   A curve or rounded projection.

Local   Pertaining to or restricted to one spot or part.

Locus coeruleus (LŌ-kus sē-ROO-lē-us)   A group of neurons in the brain stem where norepinephrine is concentrated.

Lordosis (lor-DŌ-sis)   Abnormal anterior convexity of the spine.

Lower extremity   An inferior limb, including the thigh, knee, leg, ankle, and foot.

Lumbar (LUM-bar)   Region of the back and side between the ribs and pelvis; loin.

Lumbar plexus (PLEK-sus)   A network formed by the anterior branches of spinal nerves L1 through L4.

Lumen (LOO-men)   The space within an artery, vein, intestine, or tube.

Lung   One of the two main organs of respiration, lying on either side of the heart in the thoracic cavity.

Lunula (LOO-nyoo-la)   The moon-shaped white area at the base of a nail.

Lymph (limf)   Tissue fluid confined in lymphatic vessels and flowing through the lymphatic system to be returned to the blood.

Lymph node   An oval- or bean-shaped structure located along the lymphatic vessels.

Lymphangiogram (lim-FAN-jē-ō-gram)   An x-ray of lymphatic vessels and lymphatic organs.

Lymphangiography (lim-fan'-jē-OG-ra-fē)   A procedure by which lymphatic vessels and lymph organs are filled with a radiopaque substance in order to be x-rayed.

Lymphatic (lim-FAT-ik)   Pertaining to lymph or referring to one of the vessels that collects lymph from the tissues and carries it to the blood.

Lymphocyte (LIM-fō-sīt)   A type of white blood cell, found in lymph nodes, associated with the immune system.

Lysosome (LĪ-sō-sōm)   An organelle in the cytoplasm of a cell, enclosed in a single membrane and containing powerful digestive enzymes.

Macrophage (MAK-rō-fāj)   A large phagocytic cell of the reticuloendothelial system common in loose connective tissue. Also called a histiocyte.

Macula (MAK-yoo-la)   A discolored spot or a colored area.

Macula lutea (LOO-tē-a)   The yellow spot in the center of the retina.

Malaise (ma-LĀZ)   Discomfort, uneasiness, indisposition, often indicative of infection.

Malignant (ma-LIG-nant)   Referring to diseases that tend to become worse and cause death; especially the invasion and spreading of cancer.

Mammary gland (MAM-ar-ē)   Gland of the female that secretes milk for the nourishment of the young.

Manubrium (ma-NOO-brē-um)   The upper bone of the sternum.

Marrow (MAR-ō)   Soft, spongelike material in the cavities of bone. Red marrow produces blood cells; yellow marrow, formed mainly of fatty tissue, has no blood-producing function.

Mast cell   A cell found in loose connective tissue along blood vessels that produces heparin, an anticoagulant; the name given to a basophil after it has left the bloodstream and entered the tissues.

Mastication (mas'-ti-KĀ-shun)   Chewing.

Maxilla (mak-SIL-a)   One of the two bones that unite to form the upper jawbone.

Meatus (mē-Ā-tus)   A passage or opening, especially the external portion of a canal.

Medial (MĒ-dē-al)   Relating to the midline.

Medial lemniscus (lem-NIS-kus)   A flat band of myelinated nerve fibers extending through the medulla, pons, and midbrain and terminating in the thalamus on the opposite side. Second-order sensory neurons in this tract transmit impulses for discriminating touch and pressure sensations.

Median aperture (AP-er-choor) One of the three openings in the roof of the fourth ventricle through which cerebrospinal fluid enters the subarachnoid space of the brain and cord. Also called the foramen of Magendie.

Median plane A vertical plane dividing the body into right and left halves. Situated in the middle.

Mediastinum (mē'-dē-as-TĪ-num) A mass of tissue found between the plurae of the lungs.

Medulla oblongata (me-DUL-la ob'-long-GA-ta) The most inferior part of the brain stem.

Medullary cavity (MED-yoo-lar'-ē) The space within the diaphysis of a bone.

Meibomian glands (mī-BŌ-mē-an) Sebaceous glands that open on the edge of each eyelid.

Meissner's corpuscles (MĪS-nerz) Cutaneous receptors for fine touch.

Melanin (MEL-a-nin) The dark pigment found in some parts of the body such as the skin.

Melanocyte (MEL-a-no-sīt') A pigmented cell located in the deepest layer of the epidermis that synthesizes melanin, a black pigment.

Melanoma (mel'-a-NŌ-ma) A tumor containing melanin, usually dark.

Melatonin (mel-a-TŌN-in) A hormone secreted by the pineal gland.

Membrane The combination of an epithelial layer and an underlying connective tissue layer constitute an epithelial membrane, such as a mucous, serous, or cutaneous membrane; a nonepithelial membrane is a synovial membrane.

Membranous labyrinth (mem-BRA-nus LAB-i-rinth) The portion of the labyrinth of the inner ear that is located inside the bony labyrinth and separated from it by the perilymph. It is made up of the membranous semicircular canals, the saccule and utricle, and the cochlear duct.

Menarche (me-NAR-kē) Beginning of the menstrual function.

Meninges (me-NIN-jēz) Three membranes covering the brain and spinal cord, called the dura mater, arachnoid, and pia mater. Singular form is meninx (MEN-inks).

Menisci (me-NIS-kē) Crescent-shaped fibrocartilages in the knee joint. Also called the semilunar cartilages. Singular form is meniscus (me-NIS-kus).

Menopause (MEN-o-pawz) The termination of the menstrual cycles.

Merkel's disc (MER-kuls) An encapsulated, cutaneous receptor for touch attached to deeper layers of epidermal cells.

Merocrine gland (MER-ō-krin) A secretory cell that remains intact throughout the process of formation and discharge of the secretory product, as in the salivary and pancreatic glands.

Mesenchyme (MEZ-en-kīm) An embryonic connective tissue.

Mesentery (MEZ-en-ter'-ē) A fold of peritoneum attaching the small intestine to the posterior abdominal wall.

Mesocolon (mez'-ō-KŌ-lon) A fold of peritoneum attaching the colon to the posterior abdominal wall.

Mesothelium (mez'-ō-THĒ-lē-um) The simple squamous epithelium that lines the serous cavities.

Mesovarium (mez'-ō-VAR-ēum) A short fold of peritoneum that attaches an ovary to the broad ligament of the uterus.

Metabolism (me-TAB-ō-lizm) The physical and chemical changes or processes by which living substance is maintained, producing energy for the use of the organism.

Metacarpus (met'-a-KAR-pus) A collective term for the five bones that make up the palm of the hand.

Metaphase (MET-a-phāz) The second stage in mitosis in which chromatid pairs line up on the equatorial plane of the spindle fibers.

Metaphysis (me-TAF-i-sis) Growing portion of a bone.

Metastasis (me-TAS-ta-sis) The transfer of disease from one organ or part of the body to another.

Metatarsus (met'-a-TAR-sus) A collective term for the five bones located in the foot between the tarsals and the phalanges.

Micelle (mī-SEL) A water-soluble particle involved in fat absorption.

Microcephalus (mī-krō-SEF-a-lus) An abnormally small head.

Micrometer (mī-KRO-me-ter) One one-millionth of a meter or 1/1000 of a millimeter; 1/25,000 of an inch. Formerly called a micron.

Microtubule (mī-kro-TOOB-yool') A small protein tubule that contributes to cell shape and stiffness and forms cilia, flagella, and spindle fibers.

Microvilli (mī'-krō-VIL-ē) Microscopic fingerlike projections of the cell membranes of certain cells; the brush border of light microscopy.

Micturition (mik'-too-RISH-un) The act of expelling urine from the bladder; urination.

Midbrain The part of the brain between the pons and the forebrain.

Middle ear A small, epithelial-lined cavity hollowed out of the temporal bone, separated from the external ear by the eardrum and from the internal ear by a thin bony partition containing the oval and round windows. Extending across the middle ear are the three auditory ossicles. Also called the tympanic cavity.

Midsagittal (mid-SAJ-i-tal)   Median or relating to the midline.

Midsagittal plane   A plane through the midline of the body, running vertical to the ground and dividing the body into equal right and left sides.

Milliliter (MIL-i-lēt-er)   One one-thousandth of a liter; equivalent to 1 cm³.

Millimeter (MIL-i-mēt-er)   One one-thousandth of a meter; about ¹⁄₂₅ inch.

Mineralocorticoids (min'-er-al-ō-KOR-ti-koyds)   A group of hormones of the adrenal cortex.

Mitochondrion (mi'-tō-KON-drē-on)   The powerhouse of the cell; a double-membraned organelle that plays a central role in the production of ATP.

Mitosis (mī-TŌ-sis)   The orderly division of the nucleus of a cell that ensures that each new daughter nucleus has the same number and kind of chromosomes as the original parent nucleus. The process of mitosis includes the replication of chromosomes and the distribution of the two sets of chromosomes into two separate and equal nuclei.

Mittelschmerz (MIT-el-shmerz)   Abdominopelvic pain that supposedly indicates the release of an egg from the ovary.

Modiolus (mō-DĪ-ō'-lus)   The central pillar or column of the cochlea.

Monocyte (MON-ō-sīt')   A type of white blood cell characterized by agranular cytoplasm, the largest of the leucocytes.

Mons pubis (monz PYOO-bis)   The rounded, fatty prominence over the symphysis pubis, covered by coarse pubic hair.

Morbid (MOR-bid)   Diseased; pertaining to disease.

Morphology   See Anatomy.

Motor area   The region of the cerebral cortex that governs muscular movement, particularly the precentral gyrus of the frontal lobe.

Motor unit   A motor neuron, together with the muscle cells it stimulates.

Mucin (MYOO-sin)   A protein found in mucus.

Mucous cell (MYOO-kus)   A unicellular gland that secretes mucus. Also called a goblet cell.

Mucous membrane   A membrane that lines a body cavity that opens to the exterior. Also called the mucosa.

Mucus (MYOO-kus)   The thick fluid secretion of the mucous glands and mucous membranes.

Muscle   An organ composed of one of three types of muscle tissue (skeletal, cardiac, or visceral), specialized for contraction to produce voluntary or involuntary movement of parts of the body.

Muscle tissue   A tissue specialized to produce motion in response to nerve impulses by its qualities of contractibility, extensibility, elasticity, and excitability.

Muscularis (MUS-kyoo-la'-ris)   A muscular layer or tunic of an organ.

Muscularis mucosa (myoo-KŌ-sa)   A thin layer of smooth muscle cells located in the outermost layer of the mucosa of the alimentary canal, underlying the lamina propria of the mucosa.

Musculi pectinati (MUS-kyoo-lī pek-ti-NA-tē)   Projecting muscle bundles of the anterior atrial walls and the lining of the auricles.

Myelin sheath (MĪ-e-lin)   A white, lipid, segmented covering, formed by Schwann cells, around the axons and dendrites of many peripheral neurons.

Myocardium (mī'-ō-KAR-dē-um)   A layer of the heart, made up of cardiac muscle, comprising the bulk of the heart, and lying between the epicardium and the endocardium.

Myofibril (mī'-ō-FĪ-bril)   A threadlike structure, running longitudinally through a muscle cell, consisting mainly of molecules of the proteins actin and myosin.

Myology (mī-OL-ō-jē)   The science or study of the muscles and their parts.

Myometrium (mī'-ō-MĒ-trē-um)   The smooth muscle coat or tunic of the uterus.

Myopia (mī-Ō-pē-a)   A condition in which visual images are focused in front of the retina with resulting defective vision of far objects; near-sightedness.

Myosin (MĪ-ō-sin)   The protein that makes up the thick filaments of myofibrils of muscle cells.

Nail   A horny plate, composed largely of keratin, that develops from the epidermis of the skin to form a protective covering on the dorsal surface of the distal phalanges of the fingers and toes.

Nail matrix (MĀ-triks)   The part of the nail beneath the body and root from which the nail is produced.

Narcosis (nar-KŌ-sis)   Unconscious state due to narcotics.

Nasal cavity (NĀ-zal)   A mucosa-lined cavity on either side of the nasal septum that opens onto the face at an external nostril or naris and into the nasopharynx at an internal nostril or naris.

Nasal septum (SEP-tum)   A vertical partition composed of bone and cartilage, covered with a mucous membrane, separating the right and left nasal cavities.

Nasolacrimal duct (nā'-zō-LAK-ri-mal)   A canal that transports the lacrimal secretion from the nasolacrimal sac into the nose.

Nasopharynx (nā'-zō-FAR-inks)   The uppermost portion of the pharynx, lying posterior to the nose and extending down to the soft palate.

Nebulization (neb'-yoo-li-ZĀ-shun)   Treatment with medication by spray method.

Neck   The part of the body connecting the head and the trunk; a constricted portion of an organ such as the neck of a femur or the neck of the uterus.

Necrosis (ne-KRŌ-sis)   Death of a cell or group of cells as a result of disease or injury.

Neonatal (nē'-ō-NĀ-tal)   Pertaining to the first four weeks after birth.

Neoplasm (NĒ-ō-plazm)   A mass of new, abnormal tissue; a tumor.

Nephritis (ne-FRĪT-is)   Kidney inflammation.

Nephron (NEF-ron)   The functional unit of the kidney.

Nerve   A cordlike bundle of nerve fibers and their associated connective tissue coursing together outside the central nervous system.

Nerve impulse   An action potential or a wave of negativity that sweeps along the outside of the membrane of a neuron.

Nervous tissue   Tissue that initiates and transmits nerve impulses to coordinate homeostasis.

Neurilemma (noo-ri-LEM-a)   A delicate sheath around a nerve fiber composed of the cytoplasm, nucleus, and enclosing cell membrane of a Schwann cell. There may or may not be a layer of myelin between the neurilemma and the axis cylinder (axon or dendrite) of the neuron.

Neuroeffector junction (noo-rō-e-FEK-tor)   A synapse found between an autonomic fiber and a visceral effector.

Neurofibril (noo-rō-FĪ-bril)   One of the delicate threads that forms a complicated network in the cytoplasm of the cell body and processes of a neuron.

Neuroglia (noo-ROG-lē-a)   Cells of the nervous system that are specialized to perform the functions of connective tissue. The neuroglia of the central nervous system are the astrocytes, oligodendrocytes, and microglia; neuroglia of the peripheral nervous system include the Schwann cells and the ganglion satellite cells.

Neurohypophysis (noo-rō-hī-POF-i-sis)   The posterior lobe of the pituitary gland.

Neuromuscular junction (noo-rō-MUS-kyoo-lar)   The area of contact or synapse between a neuron and a muscle fiber.

Neuromuscular spindle   An encapsulated receptor in a skeletal muscle, consisting of specialized muscle cells and nerve endings, stimulated by change in length or tension of muscle cells; a proprioceptor.

Neuron (NOO-ron)   A nerve cell, consisting of a cell body, dendrites, and an axon.

Neurosyphilis (noo-rō-SIF-i-lis)   A form of the tertiary stage of syphilis in which various types of nervous tissue are attacked by the bacteria and degenerate.

Neutrophil (NOO-trō-fil)   A type of white blood cell characterized by granular cytoplasm which stains as readily with acid or basic dyes.

Nipple   A round or cone-shaped projection at the tip of the breast, or any similarly shaped structure.

Nissl bodies (NIS-l)   Rough endoplasmic reticulum in the cell bodies of neurons.

Node of Ranvier (ron-vē-A)   A space, along a myelinated nerve fiber, between the individual Schwann cells that form the myelin sheath and the neurilemma.

Norepinephrine (nor'-ep-i-NEF-rin)   An excitatory chemical transmitter.

Nucleus (NOO-klē-us)   A spherical or oval organelle of a cell that contains the hereditary factors of the cell, called genes; a cluster of nerve cell bodies in the central nervous system.

Nucleus cuneatus (kyoo-nē-Ā-tus)   A group of nerve cells in the inferior portion of the medulla oblongata in which fibers of the fasciculus cuneatus terminate.

Nucleus gracilis (gras-I-lis)   A group of nerve cells in the medulla oblongata in which fibers of the fasciculus gracilis terminate.

Nucleus pulposus (pul-PŌ-sus)   A soft, pulpy, highly elastic substance in the center of an intervertebral disc, a remnant of the notochord.

Nystagmus (nis-TAG-mus)   Constant, involuntary, rhythmic movement of the eyeballs, horizontal, rotary, or vertical.

Obturator (OB-tyoo-rā'-ter)   Anything that obstructs or closes a cavity or opening.

Occlusion (o-KLOO-zhun)   The act of closure or state of being closed.

Odontoid (ō-DON-toyd)   Toothlike.

Olfactory (ol-FAK-tō-rē)   Pertaining to smell.

Olfactory bulb   A mass of gray matter at the termination of an olfactory nerve, lying beneath the frontal lobe of the cerebrum on either side of the crista galli of the ethmoid bone.

Olfactory cell   A bipolar neuron with its cell body lying between supporting cells located in the mucous membrane lining the upper portion of each nasal cavity.

Olfactory tract   A bundle of axons that extends from the olfactory bulb posteriorly to the olfactory portion of the cortex.

Oligospermia (ol'-i-gō-SPER-mē-a)   A deficiency of spermatozoa in the semen.

Olive   A prominent oval mass on each lateral surface of the superior part of the medulla.

Oogenesis (ō'-ō-JEN-e-sis)   Formation and development of the ovum.

Oophorectomy (ō'-of-ō-REK-tō-mē) The surgical removal of the ovaries.

Ophthalmic (of-THAL-mik) Pertaining to the eye.

Optic chiasma (OP-tik kī-AZ-ma) A crossing point of the optic nerves, anterior to the pituitary gland.

Optic disc A small area of the retina containing openings through which the fibers of the ganglion neurons emerge as the optic nerve; the blind spot.

Optic tract A bundle of axons that transmits impulses from the retina of the eye between the optic chiasma and the thalamus.

Ora serrata of the retina (Ō-ra ser-RĀ-ta) The notched anterior edge of sensory portion of the retina.

Orbit (OR-bit) The bony pyramid-shaped cavity of the skull that holds the eyeball.

Organ A group of two or more different kinds of tissue that performs a particular function.

Organ of Corti (KOR-tē) The spiral organ of hearing consisting of supporting cells and hair cells that rest on the basilar membrane and extend into the endolymph of the cochlear duct; the receptor for hearing.

Organelle (or-gan-EL) A component or structure within a cell specialized to serve a specific function in cellular activities.

Organism (OR-ga-nizm) A total living form; one individual.

Orgasm (OR-gazm) A state of highly emotional excitement; muscular, sensory, and motor events involved in ejaculation for the male and involuntary contraction of the perineal muscles in the female at the climax of sexual intercourse.

Orifice (OR-i-fis) Any aperture or opening.

Origin (OR-i-jin) The place of attachment of a muscle to the more stationary bone, or the end opposite the insertion.

Oropharynx (or'-ō-FAR-inks) The second portion of the pharynx, lying posterior to the mouth and extending from the soft palate down to the hyoid bone.

Osmosis (os-MŌ-sis) The net movement of water molecules through a semipermeable membrane from an area of high water concentration to an area of lower water concentration.

Osseous (OS-ē-us) Bony.

Ossicle (OS-si-kul) Any small bone, as the three tiny bones of the ear.

Ossification (os'-i-fi-KĀ-shun) Formation of bone substance.

Osteoblast (OS-tē-ō-blast') A bone-forming cell.

Osteoclast (OS-tē-ō-clast') A large, multinuclear cell that destroys or resorbs bone tissue.

Osteocyte (OS-tē-ō-sīt') A mature osteoblast that has lost its ability to produce new bone tissue.

Osteogenic layer (os'-tē-ō-JEN-ik) The inner layer of the periosteum.

Osteology (os'-tē-OL-ō-jē) The study of bones.

Osteomalacia (os'-tē-ō-ma-LĀ-shē-a) A deficiency of vitamin D causing demineralization and softening of bone.

Osteomyelitis (os'-tē-ō-mī-i-LĪ-tis) Inflammation of bone marrow or of the bone and marrow.

Osteon (OS-tē-on) See Haversian system.

Osteoporosis (os'-tē-ō-pō-RŌ-sis) Increased porosity of bone.

Ostium (OS-tē-um) Any small opening, especially an entrance into a hollow organ or canal.

Otic (Ō-tik) Pertaining to the ear.

Otic ganglion (GANG-glē-on) A ganglion of the parasympathetic subdivision of the autonomic nervous system, made up of cell bodies of postganglionic parasympathetics to the parotid gland.

Otolith (Ō-tō-lith) A particle of calcium carbonate embedded in the gelatinous layer that coats the hair cells of the sensory receptor (the macula) in the saccule and utricle of the inner ear.

Oval window A small opening between the middle and inner ear into which the footplate of the stapes fits. Also called the fenestra vestibuli.

Ovarian follicle (ō-VAR-ē-an FOL-i-kul) A general name for an ovum in any stage of development, along with its surrounding group of epithelial cells.

Ovarian ligament (LIG-a-ment) A rounded cord of connective tissue that attaches the ovary to the uterus.

Ovary (O-var-ē) The female gonad in which the ova are formed.

Ovulation (ov-yoo-LĀ-shun) The discharge of an egg cell (ovum) from the follicle of the ovary.

Ovum (Ō-vum) The female reproductive or germ cell; an egg cell.

Oxidation (ok-si-DĀ-shun) The combining of oxygen with substances such as organic molecules in tissues; the loss of electrons.

Oxytocin (ok'-sē-TŌ-sin) A hormone released by the posterior pituitary gland.

Pacinian corpuscles (pa-SIN-ē-an) Deep pressure receptors in the skin.

Paget's disease (PAJ-ets) A disorder characterized by an irregular thickening and softening of the bones. The cause is unknown, but the bone-producing osteoblasts and the bone-destroying osteoclasts apparently become uncoordinated.

Palate (PAL-at) The horizontal structure separating the mouth and the nasal cavity; the roof of the mouth.

Palliative (PAL-ē-a-tiv) Serving to relieve or alleviate without curing.

Palpate (PAL-pāt) To examine by touch; to feel.

Palpebra (PAL-pe-bra) Eyelid.

Pancreas (PAN-krē-as) A soft, oblong-shaped organ lying along the greater curvature of the stomach and connected by a duct to the duodenum. It is both exocrine (secreting pancreatic juice) and endocrine (secreting insulin and glucagon).

Pancreatic duct (pan'-krē-AT-ik) A single, large tube that drains pancreatic juice into the duodenum at the ampulla of Vater. Also called the duct of Wirsung. Sometimes an accessory duct, the duct of Santorini, is also present and empties into the duodenum above the ampulla of Vater.

Papilla (pa-PIL-a) Any small projection or elevation. Plural form is papillae (pa-PIL-ē).

Paranasal sinus (par'-a-NĀ-zal SĪ-nus) A mucus-lined air cavity in a skull bone that communicates with the nasal cavity. Paranasal sinuses are located in the frontal, maxillary, ethmoid, and sphenoid bones.

Parasympathetic (par'-a-sim-pa-THET-ik) One of the two subdivisions of the autonomic nervous system, having cell bodies of preganglionic neurons in nuclei in the brain stem and in the lateral gray matter of the sacral portion of the spinal cord; primarily concerned with activities that restore and conserve body energy. Also called the craniosacral division.

Parathyroid gland (par'-a-THĪ-royd) One of four small endocrine glands embedded on the posterior surfaces of the lateral lobes of the thyroid. Their secretory product is parathormone, or parathyroid hormone.

Parenchyma (par-EN-ki-ma) The essential parts of any organ concerned with its function.

Parenteral (par-EN-ter-al) Situated or occurring outside the intestines; referring to introduction of substances into the body other than by way of the intestines.

Paries (PĀ-ri-ēz) The enveloping wall of any structure, especially hollow organs.

Parietal (pa-RĪ-e-tal) Pertaining to or forming the wall of a cavity.

Parietal cell The secreting cell of the gastric glands that produces hydrochloric acid.

Parietal pleura (PLOO-ra) The outer layer of the serous pleural membrane that encloses and protects the lung, the layer that is attached to the walls of the pleural cavity.

Parotid gland (pa-ROT-id) One of the paired salivary glands located inferior and anterior to the ears connected to the oral cavity via a duct that opens into the inside of the cheek opposite the upper second molar tooth.

Paroxysm (PAR-ok-sizm) A sudden periodic attack or recurrence of symptoms of a disease.

Parturition (par'-too-RISH-un) Act of giving birth to young; childbirth, delivery.

Pathogenesis (path'-ō-JEN-e-sis) The development of a disease or a morbid or pathological state.

Pathological (path'-ō-LOJ-i-kal) Pertaining to or caused by disease.

Pathological anatomy The study of structural changes caused by disease.

Pathogen (PATH-ō-jen) A disease-producing organism.

Pectoral (PEK-tō-ral) Pertaining to the chest or breast.

Pediatrician (pē'dē-a-TRISH-un) A physician who specializes in the development, diseases, and care of children.

Pedicel (PED-i-sel) A footlike structure; a footlike projection on podocytes in Bowman's capsule.

Pedicle (PED-i-kul) A footlike structure; a short, thick process found on vertebrae.

Pelvic splanchnic nerves (PEL-vik SPLANGK-nik) Preganglionic parasympathetic fibers from the levels of S2, S3, and S4 that supply the urinary bladder, reproductive organs, and the descending and sigmoid colon and rectum.

Pelvis The basinlike structure formed by the two pelvic bones, the sacrum, and the coccyx; the expanded, proximal portion of the ureter, lying within the kidney and into which the major calyces open.

Penis (PĒ-nis) The male copulary organ, used to introduce spermatozoa into the female vagina.

Pericardial cavity (per'-i-KAR-dē-al) The space between the epicardium and the parietal layer of the serous pericardium.

Pericardium (per'-i-KAR-dē-um) A loose-fitting serous membrane that encloses the heart, consisting of an outer fibrous layer and an internal serous layer containing a parietal layer, that lines the inside of the fibrous pericardium, and a visceral layer (epicardium).

Perichondrium (per'-i-KON-drē-um) The membrane that covers cartilage.

Perikaryon (per'-i-KAR-ē-on) The nerve cell exclusive of the nucleus.

Perimysium (per'-i-MĪZ-ē-um) Invaginations of the epimysium that divides muscles into bundles.

Perineum (per'-i-NĒ-um) The pelvic floor; the space between the anus and the scrotum in the male and between the anus and the vulva in the female.

Perineurium (per'-i-NYOO-rē-um) Connective tissue wrapping each muscle bundle.

Periodontal membrane (per'ē-ō-DON-tal) The periosteum lining the sockets for the teeth in the alveolar processes of the mandible and maxillae.

Periosteum (per'-ē-OS-tē-um) The membrane that covers bone.

Peripheral (pe-RIF-er-al) Located on the outer part or a surface of the body.

Peripheral nervous system The part of the nervous system that lies outside the central nervous system—nerves and ganglia.

Periphery (pe-RIF-er-ē) Outer part or a surface of the body; part away from the center.

Peristalsis (per'-i-STAL-sis) Successive muscular movements or contractions along the wall of a hollow muscular structure.

Peritoneum (per'-i-tō-NĒ-um) The second largest serous membrane of the body which lines the abdominal cavity and covers the viscera.

Peritonitis (per'-i-tō-NĪ-tis) Inflammation of the peritoneum.

Pernicious (per-NISH-us) Fatal.

Peyer's patches (PĪ-erz) Aggregated lymph nodules that are most numerous in the ileum.

pH The symbol commonly used to express hydrogen ion concentration in a solution, that is, to describe the degree of acidity or alkalinity.

Phagocytosis (fag'-ō-sī-TŌ-sis) The process by which cells ingest particulate matter; especially the ingestion and destruction of microbes, cell debris, and other foreign matter.

Phalanx (FĀ-lanks) The bone of a finger or toe. The plural form is phalanges (fa-LAN-jēz).

Pharynx (FAR-inks) The throat, a tube that starts at the internal nares and runs partway down the neck where it opens into the esophagus posteriorly and into the larynx anteriorly.

Phenotype (FĒ-nō-tīp) The observable expression of genotype; physical characteristics of an organism determined by genetic makeup and influenced by interaction between genes and internal and external environmental factors.

Phenylketonuria (fen'-il-kēt'-on-YOOR-ē-a) A disorder characterized by an elevation of the amino acid phenylalanine in the blood.

Pheochromocytoma (fē-ō-krō'-mō-sī-TŌ-ma) Tumor of the chromaffin cells of the adrenal medulla.

Phlebotomy (fle-BOT-ō-me) The cutting of a vein to allow the escape of blood.

Physiology (fiz'-ē-OL-ō-jē) The study of the functions of an organism or any of its parts and the physical and chemical processes involved in the activities.

Pia mater (PĒ-a MĀ-ter) The inner membrane covering the brain and spinal cord.

Pilonidal (pī'-lō-NĪ-dal) Containing hairs resembling a tuft inside a cyst or sinus.

Pineal gland (PIN-ē-al) The cone-shaped gland located in the roof of the third ventricle. Also called the epiphysis cerebri.

Pinna (PIN-na) The projecting part of the external ear; the auricle.

Pinocytosis (pi'nō-sī-TŌ-sis) The process by which cells ingest liquid matter.

Pituicytes (pi-TOO-i-sītz) Supporting cells of the posterior lobe of the pituitary gland.

Pituitary gland (pi-TOO-i-tar-ē) The hypophysis, nicknamed the "master gland," a small endocrine gland lying in the sella turcica of the sphenoid bone and attached to the hypothalamus by the infundibulum.

Pivot joint (PIV-ot) A synovial or diarthrotic joint in which a rounded, pointed, or conical surface of one bone articulates with a shallow depression of another bone, as in the joint between the atlas and axis and between the proximal ends of the radius and ulna.

Placenta (pla-SEN-ta) The special structure through which the exchange of materials between fetal and maternal circulations occurs.

Plantar flexion (PLAN-tar FLEK-shun) Extension of the foot at the ankle joint.

Plasma (PLAZ-mah) The extracellular fluid found in blood vessels.

Plasma cell A cell of loose connective tissue that gives rise to antibodies; the name given to a lymphocyte after it leaves the circulatory system and becomes a connective tissue cell.

Pleura (PLOOR-a) The serous membrane that enfolds the lungs and lines the walls of the chest and diaphragm.

Plexus (PLEK-sus) A network of nerves, veins, or lymphatic vessels.

Plexus of Auerbach (OW-ur-bok) A network of nerve fibers from both autonomic divisions located in the muscularis coat or tunic of the small intestine. Also called the myenteric plexus.

Plexus of Meissner (MĪS-ner) A network of autonomic nerve fibers located in the outer portion of the submucous layer or tunic of the small intestine.

Plicae circulares (PLĪ-kē SER-kyoo-lar-es) Permanent, deep, transverse folds in the mucosa and submucosa of the small intestine that increase the surface area for absorption.

Pneumotaxic area (noo-mō-TAK-sik) An area in the pons of the brain that is part of the respiratory center.

Podiatry (pō-DĪ-a-trē) The diagnosis and treatment of foot disorders.

Polycythemia (pol'-ē-sī-THĒ-mē-a) An abnormal increase in red blood cells.

Polysaccharides (pol'-ē-SAK-a-rīds) Three or more monosaccharides joined chemically.

Pons varolii (ponz var-Ō-lē-i)   The portion of the brain stem that forms a "bridge" between the medulla and the midbrain, anterior to the cerebellum.

Posterior (pos-TĒR-ē-or)   Nearer to or at the back of the body; dorsal.

Posterior root   The structure composed of afferent (sensory) fibers lying between a spinal nerve and the dorsolateral aspect of the spinal cord. Also called the dorsal root or sensory root.

Posterior root ganglion   A group of cell bodies of sensory (afferent) neurons and their supporting cells located along the posterior root of a spinal nerve; also called a dorsal or sensory root ganglion.

Postganglionic neuron (pōst'-gang-lē-ON-ik NOO-ron)   The second visceral efferent neuron in an autonomic pathway, having its cell body and dendrites located in an autonomic ganglion and its unmyelinated axon ending at cardiac muscle, smooth muscle, or a gland.

Postpartum (pōst-PAR-tum)   After parturition; occurring after the delivery of a baby.

Preganglionic neuron (prē'-gang-lē-ON-ik)   The first visceral efferent neuron in an autonomic pathway, with its cell body and dendrites in the brain or spinal cord and its myelinated axon ending at an autonomic ganglion where it synapses with a postganglionic neuron.

Premonitory (prē-MON-i-tō-rē)   Giving previous warning; as premonitory symptoms.

Prepuce (PRĒ-pyoos)   Foreskin, the loosely fitting skin covering the glans of the penis.

Presbyopia (prez-bē-Ō-pē-a)   A loss of elasticity of the lens of the eye due to advancing age with resulting inability to focus clearly on near objects.

Pressor (PRES-or)   Stimulating the activity of a function, especially vasomotor, usually accompanied by an increase in blood pressure.

Primigravida (prī-mi-GRAV-i-da)   A woman pregnant for the first time.

Primordial (prī-MŌR-dē-al)   Existing first; especially primordial egg cells in the ovary.

Proctology (prok-TOL-ō-jē)   The branch of medicine that treats the rectum and its disorders.

Progeny (PROJ-e-nē)   Offspring or descendants.

Prognosis (prog-NŌ-sis)   A forecast of the probable results of a disorder; the outlook for recovery.

Prolapse (PRŌ-laps)   A dropping or falling down of an organ, especially the uterus or rectum.

Proliferation (pro-lif'-er-Ā-shun)   Rapid and repeated reproduction of new parts, especially cells.

Promontory (PROM-on-tor'-ē)   A projecting process or part.

Pronation (prō-NĀ-shun)   A movement of the flexed forearm in which the palm of the hand is turned backward (posteriorly).

Prophase (PRŌ-fāz)   The first stage in mitosis during which chromatid pairs are formed and aggregate around the equatorial plane region of the cell.

Proprioceptor (prō'-prē-ō-SEP-tor)   A receptor located in muscles, tendons, or joints that provides information about body position and movements.

Prostaglandin (pros'-ta-GLAN-din)   A lipid produced in cell membranes.

Prostatectomy (pros'-ta-TEK-tō-mē)   The surgical removal of the prostate gland.

Prostate gland (PROS-tāt)   A muscular and glandular doughnut-shaped organ inferior to the urinary bladder that surrounds the upper portion of the male urethra.

Prosthesis (pros-THĒ-sis)   An artificial device to replace a missing body part.

Protraction (prō-TRAK-shun)   The movement of the mandible or clavicle forward on a plane parallel with the ground.

Protuberance (pro-TOO-ber-ans)   A part that is prominent beyond a surface, like a knob.

Proximal (PROK-si-mal)   Near the point of origin; referring to the nearest part.

Pseudopodia (soo'-dō-PŌ-dē-a)   Temporary, protruding projections of cytoplasm.

Psoriasis (sō-RĪ-a-sis)   A chronic skin disease.

Psychosomatic (sī'-kō-sō-MAT-ik)   Pertaining to the relationship between mind and body.

Pterygopalatine ganglion (ter'-i-gō-PAL-a-tin GANG-glē-on)   A cluster of cell bodies of parasympathetic postganglionic neurons ending at the lacrimal and nasal glands.

Ptosis (TŌ-sis)   Drooping, as of the eyelid or the kidney.

Puberty (PYOO-ber-tē)   Period of life at which the reproductive organs become functional.

Pubococcygeus (pyoo'-bō-kok-SIJ-ē-us)   Muscle in the lower pelvic area.

Pudendum (pyoo-DEN-dum)   A collective designation for the external genitalia of the female.

Puerperium (pyoo'-er-PER-ē-um)   The state immediately after childbirth, usually 4 to 6 weeks.

Pulmonary (PUL-mo-ner'-ē)   Concerning or affected by the lungs.

Pulmonary circulation   The flow of deoxygenated blood from the right ventricle to the lungs and the return of oxygenated blood from the lungs to the left atrium.

Pulp cavity   A cavity within the crown and neck of a tooth, filled with pulp, a connective tissue containing blood vessels, nerves, and lymphatics.

Pulse   Throbbing caused by the regular recoil and alternate expansion of an artery; the periodic thrust felt over arteries in time with the heartbeat.

Pupil   The hole in the center of the iris, the area

through which light enters the posterior cavity of the eyeball.

Purkinje fibers (pur-KIN-jē) Muscle fibers in the subendocardial tissue of the heart specialized for conducting an impulse to the myocardium; part of the conduction system of the heart.

Pus The liquid product of inflammation containing leucocytes, or their remains, and debris of dead cells.

Pyemia (pī-Ē-mē-a) Infection of the blood, with multiple abscesses, caused by pus-forming microorganisms.

Pyloric sphincter (pī-LOR-ik SFINGK-ter) A thickened ring of smooth muscle through which the pylorus of the stomach communicates with the duodenum. Also called the pyloric valve.

Pyorrhea (pī-ō-RĒ-a) A discharge or flow of pus, especially in the tooth sockets and the tissues of the gums.

Pyramidal pathways (pi-RAM-i-dal) Collections of motor nerve fibers arising in the brain and passing down through the spinal cord to motor cells in the anterior horns.

Pyramid (PIR-a-mid) A pointed or cone-shaped structure; one of two roughly triangular structures on the ventral side of the medulla composed of the largest motor tracts that run from the cerebral cortex to the spinal cord; a triangular-shaped structure in the renal medulla composed of the straight segments of renal tubules.

Pyrexia (pī-REK-sē-a) A condition in which the temperature is above normal.

Quadrant (KWOD-rant) One of four parts.

Radiographic anatomy (rā'-dē-ō-GRAF-ic) Diagnostic branch of anatomy that includes the use of x-rays.

Radiography (rā'-dē-OG-ra-fē) The making of x-ray pictures.

Rami communicantes (RĀ-mē kō-myoo-ni-KAN-tēz) Branches of a spinal nerve. Singular form is ramus communicans (RĀ-mus ko-MYOO-ni-kans).

Ramus (RĀ-mus) A branch, especially a nerve or blood vessel.

Raphe (RĀ-fē) A crease, ridge, or seam noting union of the halves of a part, as the division of the two lateral halves of the scrotum.

Receptor (rē-SEP-tor) A nerve ending that receives a stimulus.

Rectouterine (rek-tō-YOO-ter-in) Pertaining to the rectum and uterus.

Rectouterine pouch A pocket formed by the parietal peritoneum as it moves posteriorly from the surface of the uterus and is reflected onto the rectum; the lowest point in the pelvic cavity. Also called the pouch or cul de sac of Douglas.

Rectum (REK-tum) The last 20 cm of the gastrointestinal tract, from the sigmoid colon to the anus.

Recumbent (re-KUM-bent) Lying down.

Red nucleus A cluster of cell bodies in the midbrain, occupying a large portion of the tegmentum, and sending fibers into the rubroreticular and rubrospinal tracts.

Reflex arc The most basic conduction pathway through the nervous system, connecting a receptor and an effector and consisting of a receptor, a sensory neuron, a center in the central nervous system for a synapse, a motor neuron, and an effector.

Regimen (REJ-i-men) A strictly regulated scheme of diet, exercise, or activity designed to achieve certain ends.

Regional anatomy The division of anatomy dealing with a specific region of the body, such as the head, neck, chest, or abdomen.

Regurgitation (rē-gur'-ji-TA-shun) Return of solids or fluids to the mouth from the stomach; flowing backward of blood through incompletely closed heart valves.

Relapse (rē-LAPS) The return of a disease weeks or months after its apparent cessation.

Renal (RĒ-nal) Pertaining to the kidney.

Renal pelvis A cavity in the center of the kidney formed by the expanded, proximal portion of the ureter, lying within the kidney, and into which the major calyces open.

Renal pyramid A triangular-shaped structure in the renal medulla composed of the straight segments of renal tubules.

Respiration (res-pi-RĀ-shun) The exchange of gases between the atmosphere and the cells.

Resuscitation (rē-sus'-i-TĀ-shun) Act of bringing a person back to full consciousness.

Rete testis (RĒ-tē) The network of ducts in the testes.

Reticular formation (re-TIK-yoo-lar) A network of small groups of nerve cells scattered among bundles of fibers beginning in the medulla as a continuation of the spinal cord and extending upward through the central part of the brain stem.

Reticulin (re-TIK-yoo-lin) A fiber of small diameter in the matrix of connective tissue that branches to form a netlike supporting framework around fat cells, nerve fibers, muscle cells, and blood vessels. Also called a reticular fiber.

Reticulocyte (re-TIK-yoo-lō-sīt) An immature red blood cell.

Reticulum (re-TIK-yoo-lum) A network.

Retina (RET-i-na) The inner coat of the eyeball, lying only in the posterior portion of the eye and consisting of nervous tissue and a pigmented layer comprised of epithelial cells lying in contact with the choroid.

Retraction (rē-TRAK-shun)   The movement of a protracted part of the body backward on a plane parallel to the ground, as in pulling the lower jaw back in line with the upper jaw.

Retroperitoneal (re'-trō-per-i-tō-NĒ-al)   External to the peritoneal lining of the abdominal cavity.

Retroversion (re-trō-VER-zhun)   A turning backward of an entire organ, especially the uterus.

Rhinology (rī-NOL-ō-jē)   The study of the nose and its disorders.

Rhodopsin (rō-DOP-sin)   A photosensitive, reddish-purple pigment in rods of the retina, consisting of the protein scotopsin plus retinene; visual purple.

Ribosome (RĪ-bo-sōm)   An organelle in the cytoplasm of cells, composed of ribosomal RNA, that synthesizes proteins; nicknamed the "protein factory."

Rickets (RIK-ets)   A condition affecting children, caused by vitamin D deficiency and resulting in soft or deformed bones.

Right lymphatic duct (lim-FAT-ik)   A vessel of the lymphatic system that drains lymph from the upper right side of the body and empties it into the right subclavian vein.

Rod   A visual receptor in the retina of the eye that is specialized for vision in dim light.

Roentgen (RENT-gen)   The international unit of radiation; a standard quantity of x or gamma radiation.

Roentgenogram (RENT-gen-ō-gram)   A photographic image made with x-rays.

Root canal   A narrow extension of the pulp cavity lying within the root of a tooth.

Root hair plexus (PLEK-sus)   A network of dendrites arranged around the root of a hair as free or naked nerve endings that are stimulated when a hair shaft is moved.

Rotation (rō-TĀ-shun)   Moving a bone around its own axis, with no other movement.

Round ligament (LIG-a-ment)   A band of fibrous connective tissue enclosed between the folds of the broad ligament of the uterus, emerging from a point on the uterus just below the uterine tube, extending laterally along the pelvic wall, and penetrating the abdominal wall through the deep inguinal ring to end in the labia majora.

Round window   A small opening between the middle and inner ear, directly below the oval window, covered by the second tympanic membrane. Also called the fenestra cochleae.

Rugae (ROO-jē)   Temporary, large folds in the mucosa of an empty hollow organ, such as the stomach and vagina, that gradually smooth out and disappear as the organ is distended.

Saccule (SAK-yool)   The lower and smaller of the two chambers in the membranous labyrinth inside the vestibule of the inner ear containing a receptor organ for equilibrium.

Sacral (SĀ-kral)   Pertaining to the sacrum.

Sacral plexus (PLEK-sus)   A network formed by the anterior branches of spinal nerves L4 through S3.

Sacral promontory (PROM-on-tor'-ē)   The superior surface of the body of the first sacral vertebra that projects anteriorly into the pelvic cavity; a line from the sacral promontory to the superior border of the symphysis pubis divides the abdominal and pelvic cavities.

Saddle joint   A synovial or diarthrotic joint in which the articular surfaces of both of the bones are saddle-shaped or concave in one direction and convex in the other direction, as in the joint between the trapezium and the metacarpal of the thumb.

Sagittal (SAJ-i-tal)   In an anteroposterior direction.

Salivary gland (SAL-i-ver-ē)   One of three pairs of glands that lie outside the mouth and pour their secretory product (called saliva) into ducts that empty into the oral cavity; the parotid, submandibular, and sublingual glands.

Salpingitis (sal'-pin-JĪ-tis)   Inflammation of the uterine (fallopian) tube or of the auditory (eustachian) tube.

Sarcolemma (sar'-kō-LEM-ma)   The cell membrane of a muscle cell, especially of a skeletal muscle cell.

Sarcoma (sar-KŌ-ma)   A connective tissue tumor, often highly malignant.

Sarcomere (SAR-kō-meer)   A contractile unit in a striated muscle cell extending from one Z-line to the next Z-line.

Sarcoplasm (SAR-kō-plazm)   The cytoplasm of a striated muscle cell.

Sarcoplasmic reticulum (sar'-kō-PLAZ-mik rē-TIK-yoo-lum)   A network of membrane-enclosed tubules within a muscle fiber.

Satiety (sa-TĪ-e-tē)   Fullness or gratification, as of appetite or thirst.

Scala tympani (SKA-la TIM-pan-ē)   The lower spiral-shaped channel of the bony cochlea, filled with perilymph.

Scala vestibuli (ves-TIB-u-lē)   The upper spiral-shaped channel of the bony cochlea, filled with perilymph.

Scapula (SKAP-yoo-la)   Shoulder blade.

Schwann cell (shwon)   A glial cell of the peripheral nervous system that forms the myelin sheath and neurilemma of a nerve fiber by spiraling around a nerve fiber in a jelly-roll fashion.

Sciatic (sī-AT-ik) Pertaining to the ischium.

Sciatica (sī-AT-i-ka) Inflammation and pain along the sciatic nerve, felt at the back of the thigh running down the inside of the leg.

Sclera (SKLĒ-ra) The white coat of fibrous tissue that forms the outer protective covering over the eyeball except in the area of the anterior cornea; the posterior portion of the fibrous tunic.

Sclerosis (skle-RŌ-sis) A hardening with loss of elasticity of the tissues.

Scoliosis (skō'-lē-Ō-sis) An abnormal lateral curvature from the normal vertical line of the spine.

Scrotum (SKRŌ-tum) A skin-covered pouch that contains the testes and their accessory organs.

Sebaceous (se-BĀ-shus) Secreting oil.

Sebaceous gland An exocrine gland in the dermis of the skin, almost always associated with a hair follicle, that secretes sebum.

Sebum (SĒ-bum) Secretion of the sebaceous glands.

Segmental bronchi (seg-MEN-tal BRONG-kē) Branches of the secondary bronchi that lead into the segmental subdivisions of the lobes of the lungs.

Sella turcica (SEL-a TUR-si-ka) A saddlelike depression on the middle upper surface of the sphenoid bone enclosing the pituitary gland.

Semen (SĒ-men) A fluid discharge at ejaculation by a male that consists of a mixture of spermatozoa and the secretions of the seminal vesicles, the prostate gland, and the bulbourethral glands.

Semicircular canals Three bony channels projecting upward and posteriorly from the vestibule of the inner ear, filled with perilymph, in which lie the membranous semicircular canals filled with endolymph. They contain receptors for equilibrium.

Semicircular ducts The membranous semicircular canals filled with endolymph and floating in the perilymph of the bony semicircular canals.

Semilunar valve (sem'-ē-LOO-nar) A valve guarding the entrance into the aorta or the pulmonary trunk from a ventricle of the heart.

Seminal vesicle (SEM-i-nal VES-i-kul) One of a pair of convoluted, pouchlike structures, lying posterior and inferior to the urinary bladder and anterior to the rectum, that secrete their component of semen into the ejaculatory ducts.

Seminiferous tubule (sem'-i-NI-fer-us TOO-byool) A tightly coiled duct, located in a lobule of the testis, where spermatozoa are produced.

Semipermeable Having pores of a size that permits diffusion of some, but not all substances; differentially permeable.

Senescence (se-NES-ens) The process of growing old; the period of old age.

Senility (se-NIL-i-tē) A loss of mental or physical ability due to old age.

Sensation A state of awareness of external or internal conditions of the body.

Sensory area A region of the cerebral cortex concerned with the interpretation of sensory impulses.

Sepsis (SEP-sis) A morbid condition resulting from the presence of pathogenic bacteria and their products.

Septum (SEP-tum) A wall dividing two cavities.

Serosa (ser-Ō-sa) Any serous membrane; the outermost layer or tunic of an organ formed by a serous membrane; the membrane that lines the pleural, pericardial, and peritoneal cavities.

Serous membrane (SIR-us) An epithelial membrane (a combination of an epithelial and a connective tissue layer) that lines body cavities that do not open to the exterior.

Sertoli cell (ser-TŌ-lē) A supporting cell of seminiferous tubules that produces secretions for supplying nutrients to the spermatozoa.

Sesamoid bones (SES-a-moyd) Small bones usually found in tendons.

Shoulder A synovial or diarthrotic joint where the humerus joins the scapula.

Sigmoid (SIG-moyd) S-shaped, from the Greek letter sigma.

Sigmoid colon (KŌ-lon) The S-shaped portion of the large intestine that begins at the level of the left iliac crest, projects inward to the midline, and terminates at the rectum at about the level of the third sacral vertebra.

Simmond's disease (SIM-mundz) A condition in which atrophy of the pituitary gland causes extreme weight loss.

Sinuatrial node (si-noo-Ā-trē-al) A compact mass of cardiac muscle cells specialized for conduction, located in the right atrium beneath the opening of the superior vena cava. Also called the sinoatrial node, SA node, or pacemaker.

Sinus (SĪ-nus) A hollow in a bone or other tissue; a channel for blood; any cavity having a narrow opening.

Sinusoid (SĪ-nus-oyd) A blood space in certain organs as the liver or spleen.

Skeletal muscle An organ specialized for contraction, composed of striated muscle cells, supported by connective tissue, attached to a bone by a tendon or an aponeurosis, and stimulated by somatic efferent neurons.

Sliding-filament hypothesis The most commonly accepted explanation for muscle contraction in which actin and myosin move into interdigitation with each other, decreasing the length of the sarcomeres.

Slipped disc The common name for a rupture of an intervertebral disc so that the nucleus pulposus protrudes into the vertebral cavity. Also called a ruptured or herniated disc.

Small intestine A long tube of the gastrointestinal tract that begins at the pyloric valve of the stomach, coils through the central and lower part of the abdominal cavity, and ends at the large intestine. The small intestine is divided into three segments: duodenum, jejunum, and ileum.

Soft palate (PAL-at) The posterior portion of the roof of the mouth, extending posteriorly from the palatine bones and ending at the uvula. It is a muscular partition lined with mucous membrane.

Somatic nervous system (sō-MAT-ik) The portion of the peripheral nervous system made up of the somatic efferent fibers that run between the central nervous system and the skeletal muscles and skin.

Somatic reflex (RĒ-fleks) A reflex arc in which the effector is a skeletal muscle.

Somatotropin (sō'-ma-tō-TRŌ-pin) Human growth hormone.

Somesthetic (sō'-mes-THET-ik) Pertaining to sensations and sensory structures of the body.

Spasm (spazm) An involuntary, convulsive, muscular contraction.

Spermatic cord (sper-MAT-ik) A supporting structure of the male reproductive system, extending from a testis to the deep inguinal ring, that includes the ductus deferens, arteries, veins, lymphatics, nerves, cremaster muscle, and connective tissue.

Spermatogenesis (sper'-ma-tō-JEN-e-sis) The formation and development of the spermatozoa.

Spermatozoon (sper'-ma-tō-ZŌ-on) A mature male germ cell.

Spermicidal (sper'-mi-SĪ-dal) An agent that kills spermatozoa.

Sphincter (SFINGK-ter) A circular muscle constricting an orifice.

Sphincter of Oddi (Ō-dē) A circular muscle at the opening of the common bile and main pancreatic ducts in the duodenum that protrudes as a mass of tissue called the ampulla of Vater.

Sphygmomanometer (sfig'-mō-ma-NOM-e-ter) An instrument for measuring arterial blood pressure.

Spina bifida (SPĪ-na BIF-i-da) A congenital defect of the vertebral column in which the two halves of the neural arch of a vertebra fail to fuse in midline.

Spinal nerve (SPĪ-nal) One of the 31 pairs of nerves that originates on the spinal cord.

Spinal puncture Withdrawal of some of the cerebrospinal fluid from the subarachnoid space in the lumbar region.

Spinous process or **spine** (SPĪ-nus) A sharp or thornlike process or projection; a sharp ridge running

diagonally across the posterior surface of the scapula.

Spirometer (spī-ROM-e-ter) An apparatus used to measure air capacity of the lungs.

Splanchnic (SPLANK-nik) Pertaining to the viscera.

Spongy bone (SPUN-jē) Bone tissue containing many large spaces filled with marrow. Also called cancellous bone.

Sputum (SPYOO-tum) Substance ejected from the mouth containing saliva and mucus.

Squamosal (skwa-MŌ-sal) Suture found in the skull between the parietal bones and the temporal bones.

Squamous (SKWĀ-mus) Scalelike.

Stasis (STĀ-sis) Stagnation or halt of normal flow of fluids, as blood, urine, or of the intestinal mechanism.

Stereocilia (ste'-rē-ō-SIL-ē-a) Groups of extremely long, slender, nonmotile microvilli projecting from epithelial cells lining the epididymis.

Stereognosis (ste'-rē-og-NŌ-sis) Recognizing the size, shape, and texture of an object.

Sterility (ste-RIL-i-tē) Infertility; absence of reproductive power.

Stomach The J-shaped enlargement of the gastrointestinal tract directly under the diaphragm in the epigastric, umbilical, and left hypochondriac regions of the abdomen, between the esophagus and the small intestine.

Straight tubule (TU-byool) A duct in a testis leading from a convoluted seminiferous tubule to the rete testis.

Stratum (STRĀ-tum) A layer.

Stratum basalis (ba-SAL-is) The outer layer of the endometrium, next to the myometrium, that is maintained during menstruation and gestation and produces a new functionalis following menstruation or parturition.

Stratum functionalis (funk'-shun-AL-is) The inner layer of the endometrium, the layer next to the uterine cavity, that is shed during menstruation and that forms the maternal portion of the placenta during gestation.

Stricture (STRIK-cher) A local contraction of a tubular structure.

Stroma (STRŌ-ma) The tissue that forms the ground substance, foundation, or framework of an organ, as opposed to its functional parts.

Subarachnoid space (sub'-a-RAK-noyd) A wide space between the arachnoid and the pia mater that surrounds the brain and spinal cord and through which cerebrospinal fluid circulates.

Subcutaneous (sub'-kyoo-TĀ-nē-us) Beneath the skin.

Subcutaneous layer A continuous sheet of fibrous connective tissue between the dermis of the skin and the deep fascia of the muscles. Also called superficial fascia.

Subdural space (sub-DOO-ral) A space between the dura mater and the arachnoid of the brain and spinal cord that contains a small amount of fluid.

Sublingual gland (sub-LING-gwal) One of a pair of salivary glands situated in the floor of the mouth under the mucous membrane and to the side of the lingual frenulum, with a duct that opens into the floor of the mouth; the smallest of the three pairs of salivary glands.

Submandibular ganglion (sub-man-DIB-yoo-lar GANG-lē-on) A cluster of cell bodies of postganglionic parasympathetic neurons, located above the submandibular gland with fibers ending at the submandibular and sublingual salivary glands and other small salivary glands in the floor of the mouth.

Submandibular gland One of a pair of salivary glands found beneath the base of the tongue under the mucous membrane in the posterior part of the floor of the mouth, posterior to the sublingual glands, with a duct situated to the side of the lingual frenulum. Also called the submaxillary gland.

Submucosa (sub-myoo-KŌ-sa) A layer of connective tissue located beneath a mucous membrane, as in the alimentary canal or the urinary bladder where a submucosa connects the mucosa to the muscularis tunic.

Subserous fascia (sub-SE-rus FASH-ē-a) A layer of connective tissue internal to the deep fascia, lying between the deep fascia and the serous membrane that lines the body cavities.

Sudoriferous (soo'-dor-IF-er-us) Conveying or producing sweat.

Sulcus (SUL-kus) A groove or depression between parts, especially a fissure between the convolutions of the brain. Plural form is sulci (sul-kē).

Superficial (soo'-per-FISH-al) Located on or near the surface.

Superficial fascia (FASH-ē-a) A continuous sheet of fibrous connective tissue between the dermis of the skin and the deep fascia of the muscles. Also called subcutaneous layer.

Superior (su-PEER-e-or) Toward the head; toward the upper part of a structure.

Supination (soo-pī-NĀ-shun) A movement of the forearm in which the palm of the hand is turned anteriorly.

Suppuration (sup'-yoo-RĀ-shun) The process of pus formation.

Surface anatomy The study of the structures that can be identified from the outside of the body.

Surfactant (sur-FAK-tant) A substance that is active on the surface of a liquid; a substance in the lungs that decreases surface tension.

Susceptibility (sus-sep'-ti-BIL-i-tē) Lack of resistance of a body to the deleterious or other effects of an agent such as pathogenic microorganisms.

Suspensory ligament (sus-PEN-so-rē LIG-a-ment) A fold of peritoneum extending laterally from the surface of the ovary to the pelvic wall.

Sutural bone (SOO-cher-al) A small bone located within a suture of certain cranial bones. Also called Wormian bone.

Suture (SOO-cher) A type of joint, especially in the skull, where bone surfaces are closely united.

Sweat gland A gland widely distributed through the skin, particularly in the skin of the palms, soles, armpits, and forehead, with the secretory portion lying in the subcutaneous tissue and the excretory duct projecting upward through the dermis and epidermis to open in a pore at the surface. Also called sudoriferous gland.

Sympathetic (sim'-pa-THET-ik) One of the two subdivisions of the autonomic nervous system, having cell bodies of preganglionic neurons in the lateral gray columns of the thoracic segment and first two or three lumbar segments of the spinal cord; primarily concerned with processes involving the expenditure of energy. Also called the thoracolumbar division.

Sympathetic trunk ganglion (GANG-glē-on) A cluster of cell bodies of postganglionic sympathetic neurons lateral to the vertebral column, close to the body of a vertebra. These ganglia extend downward through the neck, thorax, and abdomen to the coccyx, on both sides of the vertebral column and are connected to one another to form a chain on each side of the vertebral column. Also called lateral or sympathetic chain or vertebral chain ganglia.

Sympathomimetic (sim'-pa-thō-mi-MET-ik) Producing effects that mimic those brought about by the sympathetic division of the autonomic nervous system.

Symphysis (SIM-fi-sis) A line of union; a cartilaginous joint such as that between the bodies of vertebrae.

Symphysis pubis (PYOO-bis) A slightly movable cartilaginous joint between the anterior surfaces of the pelvic bones.

Symptom (SIMP-tum) An observable abnormality that indicates the presence of a disease or disorder of the body.

Synapse (SIN-aps) A small gap that serves as the functional junction between two neurons, where a nerve impulse is conducted from one neuron to another by a neurohumor.

Synarthrosis (sin'-ar-THRŌ-sēz) An immovable joint.

Synchondrosis (sin'-kon-DRŌ-sis) A cartilaginous

joint in which the connecting material is hyaline cartilage.

Syncytium (sin-SISH-ē-um) A multinucleated mass of protoplasm produced by the merging of cells.

Syndesmosis (sin'-dez-MŌ-sis) A joint in which articulating bones are united by dense fibrous tissue.

Syndrome (SIN-drōm) A group of abnormalities that occur together in a characteristic pattern; the complete picture of a disease.

Synergist (SIN-er-jist) A muscle that assists the prime mover or agonist by reducing undesired action or unnecessary movement. Also called a fixator.

Synostosis (sin'-os-TO-sēz) A joint in which the dense fibrous connective tissue that unites bones at a suture has been replaced by bone, resulting in a complete fusion across the suture line.

Synovial (si-NŌ-vē-al) An articulation where there is a joint cavity.

Synovial cavity The space between the articulating bones of a synovial or diarthrotic joint, filled with synovial fluid.

Synovial joint A fully movable or diarthrotic joint in which a joint or synovial cavity is present between the two articulating bones.

Synovial membrane The inner of the two layers of the articular capsule of a synovial joint, composed of loose connective tissue covered with epithelium that secretes synovial fluid into the joint cavity.

System An association of organs that have a common function.

Systemic (sis-TEM-ik) Affecting the whole body.

Systemic anatomy The study of particular systems of the body, such as the system of nerves, spinal cord, and brain or the system of heart, blood vessels, and blood.

Systemic circulation All of the circulatory routes taken by oxygenated blood that leaves the left ventricle through the aorta and returns to the right atrium, carrying oxygenated blood to all of the organs of the body.

Systole (SIS-tō-lē) Heart muscle contraction, especially that of the ventricles.

T tubule (TOOB-yool) In a muscle cell, an invagination of the sarcolemma that runs transversely through the fiber perpendicular to the sarcoplasmic reticulum.

Tachycardia (tak'-ē-KAR-dē-a) A rapid heart or pulse rate.

Tactile (TAK-tĭl) Pertaining to the sense of touch.

Taenia coli (TĒ-nē-a KŌ-lĭ) One of three flat bands of muscles running the length of the large intestine.

Target organ The organ or group of organs affected by a particular hormone.

Tarsal plate (TAR-sal) A thin, elongated sheet of connective tissue, one in each eyelid, giving the eyelid form and support. The aponeurosis of the levator palpebrae superioris is attached to the tarsal plate of the superior eyelid.

Tarsus (TAR-sus) A collective designation for the seven bones of the ankle.

Tectorial membrane (tek-TŌ-rē-al) A gelatinous membrane projecting over and in contact with the hair cells of the organ of Corti in the cochlear duct.

Telophase (TEL-ō-fāz) The final stage of mitosis in which the nuclei become established.

Tendinitis (ten'-din-Ī-tis) A disorder involving the inflammation of a tendon and synovial membrane at a joint.

Tendon (TEN-don) A white fibrous cord of dense, regular connective tissue that attaches muscle to bone.

Tentorium cerebelli (ten-TŌ-rē-um ser'-e-BEL-ē) A transverse shelf of dura mater that forms a partition between the occipital part of the cerebral hemispheres and the cerebellum and that covers the cerebellum.

Teratology (ter'-a-TOL-ō-jē) The branch of science dealing with the study of monsters.

Teratoma (ter'-a-TŌ-ma) A congenital tumor containing embryonic elements of all three primary germ layers, as hair or teeth.

Terminal ganglion (TER-min-al GANG-lē-on) A cluster of cell bodies of postganglionic parasympathetic neurons either lying very close to the visceral effectors or located within the walls of the visceral effectors supplied by the postganglionic fibers.

Testosterone (tes-TOS-te-rōn) A male sex hormone secreted by the interstitial cells (of Leydig) of a mature testis.

Tetany (TET-a-nē) A nervous condition characterized by intermittent or continuous tonic muscular contractions of the extremities.

Thalamus (THAL-a-mus) A large, oval structure, located above the midbrain, consisting of two masses of gray matter covered by a thin layer of white matter.

Thalassemia (thal'-a-SĒ-mē-a) A group of hereditary hemolytic anemias.

Therapy (THER-a-pē) The treatment of disease or disorder.

Thigh The portion of the lower extremity between the hip and the knee.

Third ventricle (VEN-tri-kul) A slitlike cavity between the right and left halves of the thalamus and between the lateral ventricles.

Thoracic (thō-RAS-ik) Pertaining to the chest.

Thoracic duct A lymphatic vessel that begins as a dilation called the cisterna chyli and receives lymph from the left side of the head, neck, and chest, and the left arm, and the entire body below the ribs, finally emptying the lymph into the left subclavian vein. Also called the left lymphatic duct.

Thoracolumbar (thō'-ra-kō-LUM-bar) Sympathetic division of the autonomic nervous system.

Thoracolumbar outflow The fibers of the sympathetic preganglionic neurons, which have their cell bodies in the lateral gray columns of the thoracic segment and first two or three lumbar segments of the spinal cord.

Thorax (THŌ-raks) The chest.

Thrombocyte (THROM-bō-sīt) A tiny particle found in the circulating blood, a blood platelet that plays a part in the mechanism of blood clotting.

Thrombophlebitis (throm'-bō-fle-BĪ-tis) A disorder in which inflammation of a vein wall is followed by the formation of a blood clot (thrombus).

Thymectomy (thī-MEK-tō-mē) Surgical removal of the thymus.

Thymus gland (THĪ-mus) A bilobed organ, located in the upper mediastinum posterior to the sternum and between the lungs, that plays a role in the immunity mechanism of the body.

Thyrocalcitonin (thī'-rō-kal-si-TŌ-nin) A thyroid gland hormone.

Thyroid cartilage (THĪ-royd KAR-tĭ-lij) The largest single cartilage of the larynx, consisting of two fused plates that form the anterior wall of the larynx. Also called the Adam's apple.

Thyroid gland An endocrine gland with right and left lateral lobes on either side of the trachea connected by an isthmus located in front of the trachea just below the cricoid cartilage.

Thyrotropin (thī-rō-TRŌ-pin) A hormone released by the adenohypophysis.

Thyroxin (thī-ROK-sin) The main hormone of the thyroid gland.

Tinnitus (ti-NĪ-tus) A ringing or tinkling sound in the ears.

Tissue A group of similar cells, and their intercellular substance, joined together to perform a specific function.

Tongue A large skeletal muscle on the floor of the oral cavity.

Tonsil (TON-sil) A mass of lymphoid tissue embedded in mucous membrane.

Toxic (TOK-sik) Pertaining to poison; poisonous.

Trabecula (tra-BEK-yoo-la) A fibrous cord of connective tissue serving as support by forming a septum extending into an organ from its wall or capsule. The plural form is trabeculae (tra-BEK-yoo-lē).

Trabeculae carnae (tra-BEK-yoo-lē KAR-nē) Ridges and folds of the myocardium in the ventricles.

Trachea (TRĀ-kē-a) The windpipe.

Tracheostomy (trā-kē-OS-tō-mē) Creation of an opening into the trachea through the neck, with insertion of a tube to facilitate the passage of air or evacuation of secretions.

Trachoma (tra-KŌ-ma) A chronic, contagious conjunctivitis.

Tract A bundle of nerve fibers in the central nervous system.

Transection (tran-SEK-shun) A cross cut.

Transplantation (trans'-plan-TĀ-shun) The transfer or implantation of body tissue from one part of the body to another or from one person to another.

Transverse (trans-VERS) Lying across; crosswise.

Transverse colon (trans-VERS KO-lon) The portion of the large intestine extending across the abdomen from the hepatic flexure to the splenic flexure.

Transverse fissure (FISH-er) The deep cleft that separates the cerebrum from the cerebellum.

Transverse plane A plane that runs parallel to the ground and divides the body into superior and inferior portions. Also called horizontal plane.

Trauma (TRAW-ma) An injury or wound that may be produced by external force or by shock, as in psychic trauma.

Triad (TRĪ-ad) A complex of three units in a muscle cell composed of a T tubule and the segments of sarcoplasmic reticulum on both sides of it.

Trigone (TRĪ-gōn) A triangular area at the base of the urinary bladder.

Trochanter (trō-KAN-ter) A large, blunt projection on the femur.

Trochlea (TROK-lē-a) A pulleylike surface on the humerus.

Trophectoderm (trō-FEK-tō-derm) The outer covering of cells that develops from the blastocysts.

True vocal cords A pair of folds of the mucous membrane of the larynx enclosing two strong bands of fibrous tissue, the vocal ligaments, that vibrate to make sounds during speaking. The true vocal cords are inferior to the false vocal folds.

Trunk The part of the body to which the upper and lower extremities are attached.

Tubercle (TOO-ber-kul) A small, rounded process.

Tumor (TOO-mor) A swelling or enlargement.

Tunica albuginea (TOO-ni-ka al'-byoo-JIN-ē-a) A dense layer of white fibrous tissue covering the testes.

Tunica externa (eks-TER-na) The outer coat of an artery or vein, composed of loose connective tissue containing elastic, collagenous, and a few smooth muscle fibers.

Tunica interna (in-TER-na) The inner coat of an artery or vein, consisting of a lining of endothelium and its supporting layer of connective tissue. Also called tunica intima.

Tunica media (ME-de-a) The middle coat of an artery or vein, composed of smooth muscle and elastic fibers.

Tympanic antrum (tim-PAN-ik AN-trum) An air space in the posterior wall of the middle ear that leads into the mastoid air cells or sinus.

Tympanic membrane A thin semitransparent partition of fibrous connective tissue between the external auditory meatus and the middle ear; eardrum.

Ulcer (UL-ser) An open lesion of the skin or a mucous membrane of the body with loss of substance and necrosis of the tissue.

Umbilical (um-BIL-i-kal) Pertaining to the umbilicus or navel.

Umbilicus (um-BIL-i-kus) A small scar on the abdomen that marks the former attachment of the umbilical cord to the fetus; the navel.

Upper extremity The appendage attached at the shoulder, consisting of an arm, a forearm, and a hand.

Uremia (yoo-RE-me-a) Accumulation of toxic levels of urea and other waste products in the blood, usually resulting from severe kidney malfunction.

Ureters (yoo-RE-ters) Tubes that connect the kidneys with the urinary bladder.

Urethra (yoo-RE-thra) The duct that conveys urine from the urinary bladder to the exterior of the body.

Urinary bladder (YOO-ri-ner-e) A hollow, muscular organ situated in the pelvic cavity posterior to the symphysis pubis.

Urogenital triangle (YOO'-ro-JEN-i-tal) The region of the pelvic floor below the symphysis pubis, bounded by the symphysis pubis and the ischial tuberosities and containing the external genitalia.

Urticaria (ur'-ti-KA-re-a) A skin reaction to certain foods, drugs, or other substances to which a person may be allergic; hives.

Uterine tubes (YOO-ter-in) Ducts that transport ova from the ovary to the uterus. Also called the fallopian tubes or the oviducts.

Uterosacral ligament (yoo'-ter-o-SA-kral LIG-a-ment) A fibrous band of tissue extending from the cervix of the uterus laterally to attach to the sacrum.

Uterovesical pouch (yoo'-ter-o-VES-i-kal) A shallow pouch formed by the reflection of the peritoneum from the anterior surface of the uterus, at the junction of the cervix and the body, to the posterior surface of the urinary bladder.

Uterus (YOO-ter-us) An inverted, pear-shaped, hollow organ situated between the urinary bladder and the rectum in the female pelvis; the womb.

Utricle (YOO-tri-kul) The larger of the two divisions of the membranous labyrinth located inside the vestibule of the inner ear.

Uvea (YOO-ve-a) The three structures that make up the vascular tunic of the eye.

Uvula (OO-vyoo-la) A soft, fleshy mass, especially the V-shaped pendant part, descending from the soft palate.

Vagina (va-JI-na) A muscular, tubular organ that leads from the uterus to the vestibule, situated between the urinary bladder and the rectum of the female.

Varicocele (VAR-i-ko-sel) A twisted vein; especially, the accumulation of blood in the veins of the spermatic cord.

Varicose (VAR-i-kos) Pertaining to an unnatural swelling, as in the case of a varicose vein.

Vas A vessel or duct.

Vasa vasorum (VA-sa va-SO-rum) Blood vessels supplying nutrients to the larger arteries and veins.

Vascular (VAS-kyoo-lar) Pertaining to or containing many vessels.

Vascular tunic (TOO-nik) The middle layer of the eyeball, composed of the choroid, the ciliary body, and the iris.

Vasectomy (va-SEK-to-me) A means of sterilization of males in which a portion of each ductus deferens is removed.

Vein A blood vessel that conveys blood from the tissues back to the heart.

Vena cava (VE-na KA-va) One of two large veins that open into the right atrium, returning to the heart all of the deoxygenated blood from the systemic circulation except from the coronary circulation.

Ventilation (ven'-ti-LA-shun) Breathing, or the process by which atmospheric gases are drawn down into the lungs and waste gases that have diffused into the lungs are expelled back up through the respiratory passageways.

Ventral (VEN-tral) Pertaining to the anterior or front side of the body; opposite of dorsal.

Ventral ramus (RA-mus) The anterior branch of a spinal nerve, containing sensory and motor fibers to the muscles and skin of the anterior surface of the head, neck, trunk, and the extremities.

Ventricles (VEN-tri-kuls) Cavities in the heart or brain.

Venule (VEN-yool) A small vein.

Vermiform appendix (VER-mi-form a-PEN-diks) A twisted, coiled tube attached to the cecum.

Vermilion (ver-MIL-yon) The area of the mouth where the skin on the outside meets the mucous membrane on the inside.

Vermis (VER-mis) The central constricted area of the cerebellum that separates the two cerebellar hemispheres.

Vertebral canal (VER-te-bral) The cavity formed by the vertebral foramina of all vertebrae together; the inferior portion of the dorsal cavity, containing the spinal cord. Also called the spinal canal.

Vertigo (VER-ti-go) Sensation of dizziness.

Vesicle (VES-i-kul) A small bladder or sac containing liquid.

Vesicouterine (ves'-i-kō-YOO-ter-in) Pertaining to the bladder and uterus.

Vestibular membrane (ves-TIB-yoo-lar) The membrane that separates the cochlear duct from the scala vestibuli. Also called Reissner's membrane.

Vestibule (VES-ti-byool) A small space or cavity at the beginning of a canal, especially the inner ear, larynx, mouth, nose, vagina.

Villus (VIL-lus) One of the short, vascular, hairlike processes found on certain membranous surfaces. Plural form is villi (VIL-ē).

Viscera (VIS-er-a) The organs inside the ventral body cavity.

Visceral (VIS-er-al) Pertaining to the covering of an organ.

Visceral autonomic reflex (aw'-tō-NOM-ik RĒ-fleks) A quick, involuntary response in which the impulse travels over visceral efferent neurons to smooth muscle, cardiac muscle, or a gland.

Visceral effector (e-FEK-tor) Cardiac muscle, smooth muscle, and glandular epithelium.

Visceral muscle An organ specialized for contraction, composed of smooth muscle cells, located in the walls of hollow internal structures, and stimulated by visceral efferent neurons.

Visceral pleura (PLOO-ra) The inner layer of the serous membrane that covers the lungs.

Visceroceptors (vis'-er-ō-SEP-tors) Receptors that provide information about the body's internal environment.

Viscosity (vis-KOS-i-tē) The state of being sticky or thick.

Vitreous humor (VIT-rē-us HYOO-mor) A soft, jellylike substance that fills the posterior cavity of the eyeball, lying between the lens and the retina.

Volkmann's canal (FOLK-mans) A minute passageway by means of which blood vessels and nerves from the periosteum penetrate into compact bone.

Vulva (VUL-va) The external genitalia of the female. Also called the pudendum.

Wallerian degeneration (wal-LE-rē-an) Degeneration of the distal portion of the axon and myelin sheath.

Wheal (hwēl) Elevated lesion of the skin.

White matter Aggregations or bundles of myelinated axons located in the brain and spinal cord.

White pulp The portion of the spleen composed of ovoid masses of lymphoid tissue called lymph follicles that contain germinal centers and where lymphocytes are produced.

White ramus communicans (RĀ-mus co-MYOO-ni-kans) The portion of a preganglionic sympathetic nerve fiber that branches away from the anterior ramus of a spinal nerve to enter the nearest sympathetic trunk ganglion.

Wormian bone A small bone located within a suture of certain cranial bones. Also called sutural bone.

Xiphoid (ZĪ-foyd) Sword-shaped; the lowest portion of the sternum.

Zona fasciculata (ZŌ-na fa-sik'-yoo-LA-ta) The middle zone of the adrenal cortex that consists of cells arranged in long, straight cords and that secretes glucocorticoid hormones.

Zona glomerulosa (glo-mer'-yoo-LŌ-sa) The outer zone of the adrenal cortex, directly under the connective tissue covering, that consists of cells arranged in arched loops or round balls and that secretes mineralocorticoids.

Zona reticularis (ret-ik'-yoo-LAR-is) The inner zone of the adrenal cortex, consisting of cords of branching cells that secrete sex hormones, chiefly androgens.

Zymogenic cell (zī'-mō-JEN-ik) A cell that secretes enzymes; for example, the chief cells of the gastric glands that secrete pepsinogen.

INDEX